Scientific Computation

AF183760

Pierre Sagaut

Large Eddy Simulation for Incompressible Flows

An Introduction

Third Edition
With a Foreword by Massimo Germano

With 99 Figures and 15 Tables

Springer

Prof. Dr. Pierre Sagaut
LMM-UPMC/CNRS
Boite 162, 4 place Jussieu
75252 Paris Cedex 05, France
sagaut@lmm.jussieu.fr

Title of the original French edition:
Introduction à la simulation des grandes échelles pour les écoulements de fluide incompressible, Mathématique & Applications.

ISBN 978-3-642-06580-4 e-ISBN 978-3-540-26403-3

ISSN 1434-8322

Springer is a part of Springer Science+Business Media
springeronline.com

Softcover reprint of the hardcover 3rd edition 2006

Cover design: *design & production* GmbH, Heidelberg

Foreword to the Third Edition

It is with a sense of great satisfaction that I write these lines introducing the third edition of Pierre Sagaut's account of the field of Large Eddy Simulation for Incompressible Flows. Large Eddy Simulation has evolved into a powerful tool of central importance in the study of turbulence, and this meticulously assembled and significantly enlarged description of the many aspects of LES will be a most welcome addition to the bookshelves of scientists and engineers in fluid mechanics, LES practitioners, and students of turbulence in general.

Hydrodynamic turbulence continues to be a fundamental challenge for scientists striving to understand fluid motions in fields as diverse as oceanography, acoustics, meteorology and astrophysics. The challenge also has socio-economic attributes as engineers aim at predicting flows to control their features, and to improve thermo-fluid equipment design. Drag reduction in external aerodynamics or convective heat transfer augmentation are well-known examples. The fundamental challenges posed by turbulence to scientists and engineers have not, in essence, changed since the appearance of the second edition of this book, a mere two years ago. What has evolved significantly is the field of Large Eddy Simulation (LES), including methods developed to address the closure problem associated with LES (also called the problem of subgrid-scale modeling), numerical techniques for particular applications, and more explicit accounts of the interplay between numerical techniques and subgrid modeling.

The original hope for LES was that simple closures would be appropriate, such as mixing length models with a single, universally applicable model parameter. Kolmogorov's phenomenological theory of turbulence in fact supports this hope but only if the length-scale associated with the numerical resolution of LES falls well within the ideal inertial range of turbulence, in flows at very high Reynolds numbers. Typical applications of LES most often violate this requirement and the resolution length-scale is often close to some externally imposed scale of physical relevance, leading to loss of universality and the need for more advanced, and often much more complex, closure models. Fortunately, the LES modeler disposes of large amount of raw materials from which to assemble improved models. During LES, the resolved motions present rich multi-scale fields and dynamics including highly non-trivial nonlinear interactions which can be interrogated to learn about the local state of turbulence. This availability of dynamical information has led to the formulation of a continuously growing number of different closure models and methodologies and associated numerical approaches, including many variations on several basic themes. In consequence, the literature on LES has increased significantly in recent years. Just to mention a quantitative measure of this trend, in 2000 the ISI science citation index listed 164 papers published including the keywords "large-eddy-simulation" during that year. By 2004 this number had doubled to over 320 per year. It is clear, then, that a significantly enlarged version of Sagaut's book, encompassing much of what has been added to the literature since the book's second edition, is a most welcome contribution to the field.

What are the main aspects in which this third edition has been enlarged compared to the first two? Sagaut has added significantly new material in a number of areas. To begin, the introductory chapter is enriched with an overview of the structure of the book, including an illuminating description of three fundamental errors one incurs when attempting to solve fluid mechanics' infinite-dimensional, non-linear differential equations, namely projection error, discretization error, and in the case of turbulence and LES, the physically very important resolution error. Following the chapters describing in significant detail the relevant foundational aspects of filtering in LES, Sagaut has added a new section dealing with alternative mathematical formulations of LES. These include statistical approaches that replace spatial filtering with conditionally averaging the unresolved motions, and alternative model equations in which the Navier-Stokes equations are replaced with mathematically better behaved equations such as the Leray model in which the advection velocity is regularized (i.e. filtered).

In the chapter dealing with functional modeling approaches, in which the subgrid-scale stresses are expressed in terms of local functionals of the resolved velocity gradients, a more complete account of the various versions of the dynamic model is given, as well as extended discussions of new structure-function and multiscale models. The chapter on structural modeling, in which the stress tensor is reconstructed based on its definition and various direct hypotheses about the small-scale velocity field is significantly enhanced: Closures in which full prognostic transport equations are solved for the subgrid-scale stress tensor are reviewed in detail, and entire new subsections have been added dealing with filtered density function models, with one-dimensional turbulence mapping models, and variational multi-scale models, among others. The chapter focussing on numerical techniques contains an interesting new description of the effects of pre-filtering and of the various methods to perform grid refinement. In the chapter on analysis and validation of LES, a new detailed account is given about methods to evaluate the subgrid-scale kinetic energy. The description of boundary and inflow conditions for LES is enhanced with new material dealing with one-dimensional-turbulence models near walls as well as stochastic tools to generate and modulate random fields for inlet turbulence specification. Chapters dealing with coupling of multiresolution, multidomain, and adaptive grid refinement techniques, as well as LES - RANS coupling, have been extended to include recent additions to the literature. Among others, these are areas to which Sagaut and his co-workers have made significant research contributions.

The most notable additions are two entirely new chapters at the end of the book, on the prediction of scalars using LES. Both passive scalars, for which subgrid-scale mixing is an important issue, and active scalars, of great importance to geophysical flows, are treated. The geophysics literature on LES of stably and unstably stratified flows is voluminous - the field of LES in fact traces its origins to simulating atmospheric boundary layer flows in the early 1970s. Sagaut summarizes this vast field using his classifications of subgrid closures introduced earlier, and the result is a conceptually elegant and concise treatment, which will be of significant interest to both engineering and geophysics practitioners of LES.

The connection to geophysical flow prediction reminds us of the importance of LES and subgrid modeling from a broader viewpoint. For the field of large-scale numerical simulation of complex multiscale nonlinear systems is, today, at the center of scientific discussions with important societal and political dimensions. This is most visible in the discussions surrounding the trustworthiness of global change models. Among others, these include boundary-layer parameterizations that can be studied by means of LES done at smaller scales. And LES of turbulence is itself a prime example of large-scale computing applied to prediction of a multi-scale complex system, including issues surrounding the verification of its predictive capabilities, the testing of the cumulative accuracy of individual building blocks, and interesting issues on the interplay of stochastic and deterministic aspects of the problem. Thus the book - as well as its subject - Large Eddy Simulation of Incompressible Flow, has much to offer to one of the most pressing issues of our times.

With this latest edition, Pierre Sagaut has fully solidified his position as the preeminent cartographer of the complex and multifaceted world of LES. By mapping out the field in meticulous fashion, Sagaut's work can indeed be regarded as a detailed and evolving atlas of the world of LES. And yet, it is not a tourist guide: as with any relatively young terrain in which the main routes have not yet been firmly established, what is called for is unbiased, objective, and sophisticated cartography. The cartographer describes the topography, scenery, and landmarks as they appear, without attempting to preach to the traveler which route is best. In return, the traveler is expected to bring along a certain sophistication to interpret the maps and to discern which among the many paths will most likely lead towards particular destinations of interest. The reader of this latest edition will thus be rewarded with a most solid, insightful, and up-to-date account of an important and exciting field of research.

Baltimore, January 2005 *Charles Meneveau*

Foreword to the Second Edition

It is a particular pleasure to present the second edition of the book on Large Eddy Simulation for Incompressible Flows written by Pierre Sagaut: two editions in two years means that the interest in the topic is strong and that a book on it was indeed required. Compared to the first one, this second edition is a greatly enriched version, motivated both by the increasing theoretical interest in Large Eddy Simulation (LES) and the increasing numbers of applications and practical issues. A typical one is the need to decrease the computational cost, and this has motivated two entirely new chapters devoted to the coupling of LES with multiresolution multidomain techniques and to the new hybrid approaches that relate the LES procedures to the classical statistical methods based on the Reynolds Averaged Navier–Stokes equations.

Not that literature on LES is scarce. There are many article reviews and conference proceedings on it, but the book by Sagaut is the first that organizes a topic that by its peculiar nature is at the crossroads of various interests and techniques: first of all the physics of turbulence and its different levels of description, then the computational aspects, and finally the applications that involve a lot of different technical fields. All that has produced, particularly during the last decade, an enormous number of publications scattered over scientific journals, technical notes, and symposium acta, and to select and classify with a systematic approach all this material is a real challenge. Also, by assuming, as the writer does, that the reader has a basic knowledge of fluid mechanics and applied mathematics, it is clear that to introduce the procedures presently adopted in the large eddy simulation of turbulent flows is a difficult task in itself. First of all, there is no accepted universal definition of what LES really is. It seems that LES covers everything that lies between RANS, the classical statistical picture of turbulence based on the Reynolds Averaged Navier–Stokes equations, and DNS, the Direct Numerical Simulations resolved in all details, but till now there has not been a general unified theory that gradually goes from one description to the other. Moreover we should note the different importance that the practitioners of LES attribute to the numerical and the modeling aspects. At one end the supporters of the *no model* way of thinking argue that the numerical scheme should and could capture by itself the resolved scales. At the other end the theoretical modelers try to develop new universal equations for the filtered quantities. In some cases LES is regarded as a technique imposed by the present provisional inability of the computers to solve all the details. Others think that LES modeling is a contribution to the understanding of turbulence and the interactions among different ideas are often poor.

Pierre Sagaut has elaborated on this immense material with an open mind and in an exceptionally clear way. After three chapters devoted to the basic problem of the scale separation and its application to the Navier–Stokes equations, he classifies the various subgrid models presently in use as functional and structural ones. The chapters devoted to this general review are of the utmost interest: obviously some selection has been done, but both the student and the professional engineer will find there a clear unbiased exposition. After this first part devoted to the fundamentals a second part covers many of the interdisciplinary problems created by the practical use of LES and its coupling with the numerical techniques. These subjects, very important obviously from the practical point of view, are also very rich in theoretical aspects, and one great merit of Sagaut is that he presents them always in an attractive way without reducing the exposition to a mere set of instructions. The interpretation of the numerical solutions, the validation and the comparison of LES databases, the general problem of the boundary conditions are mathematically, physically and numerically analyzed in great detail, with a principal interest in the general aspects. Two entirely new chapters are devoted to the coupling of LES with multidomain techniques, a topic in which Pierre Sagaut and his group have made important contributions, and to the new hybrid approaches RANS/LES, and finally in the last expanded chapter, enriched by new examples and beautiful figures, we have a review of the different applications of LES in the nuclear, aeronautical, chemical and automotive fields.

Both for graduate students and for scientists this book is a very important reference. People involved in the large eddy simulation of turbulent flows will find a useful introduction to the topic and a complete and systematic overview of the many different modeling procedures. At present their number is very high and in the last chapter the author tries to draw some conclusions concerning their efficiency, but probably the person who is only interested in the basic question "*What is the best model for LES?*" will remain a little disappointed. As remarked by the author, both the structural and the functional models have their advantages and disadvantages that make them seem complementary, and probably a mixed modeling procedure will be in the future a good compromise. But for a textbook this is not the main point. The fortunes and the misfortunes of a model are not so simple to predict, and its success is in many cases due to many particular reasons. The results are obviously the most important test, but they also have to be considered in a textbook with a certain reserve, in the higher interest of a presentation that tries as much as possible to be not only systematic but also rational.

To write a textbook obliges one in some way or another to make judgements, and to transmit ideas, sometimes hidden in procedures that for some reason or another have not till now received interest from the various groups involved in LES and have not been explored in full detail.

Pierre Sagaut has succeeded exceptionally well in doing that. One reason for the success is that the author is curious about every detail. The final task is obviously to provide a good and systematic introduction to the beginner, as rational as a book devoted to turbulence can be, and to provide useful information for the specialist. The research has, however, its peculiarities, and this book is unambiguously written by a passionate researcher, disposed to explore every problem, to search in all models and in all proposals the germs of new potentially useful ideas. The LES procedures that mix theoretical modeling and numerical computation are often, in an inextricable way, exceptionally rich in complex problems. What about *the problem of the mesh adaptation on unstructured grids for large eddy simulations?* Or *the problem of the comparison of the LES results with reference data? Practice shows that nearly all authors make comparisons with reference data or analyze large eddy simulation data with no processing of the data ....* Pierre Sagaut has the courage to dive deep into procedures that are sometimes very difficult to explore, with the enthusiasm of a genuine researcher interested in all aspects and confident about every contribution. This book now in its second edition seems really destined for a solid and durable success. Not that every aspect of LES is covered: the rapid progress of LES in compressible and reacting flows will shortly, we hope, motivate further additions. Other developments will probably justify new sections. What seems, however, more important is that the basic style of this book is exceptionally valid and open to the future of a young, rapidly evolving discipline. This book is not an encyclopedia and it is not simply a monograph, it provides a framework that can be used as a text of lectures or can be used as a detailed and accurate review of modeling procedures. The references, now increased in number to nearly 500, are given not only to extend but largely to support the material presented, and in some cases the dialogue goes beyond the original paper. As such, the book is recommended as a fundamental work for people interested in LES: the graduate and postgraduate students will find an immense number of stimulating issues, and the specialists, researchers and engineers involved in the more and more numerous fields of application of LES will find a reasoned and systematic handbook of different procedures. Last, but not least, the applied mathematician can finally enjoy considering the richness of challenging and attractive problems proposed as a result of the interaction among different topics.

Torino, April 2002 *Massimo Germano*

Foreword to the First Edition

Still today, turbulence in fluids is considered as one of the most difficult problems of modern physics. Yet we are quite far from the complexity of microscopic molecular physics, since we only deal with Newtonian mechanics laws applied to a continuum, in which the effect of molecular fluctuations has been smoothed out and is represented by molecular-viscosity coefficients. Such a system has a dual behaviour of determinism in the Laplacian sense, and extreme sensitivity to initial conditions because of its very strong non-linear character. One does not know, for instance, how to predict the critical Reynolds number of transition to turbulence in a pipe, nor how to compute precisely the drag of a car or an aircraft, even with today's largest computers.

We know, since the meteorologist Richardson,[1] numerical schemes allowing us to solve in a deterministic manner the equations of motion, starting with a given initial state and with prescribed boundary conditions. They are based on momentum and energy balances. However, such a resolution requires formidable computing power, and is only possible for low Reynolds numbers. These Direct-Numerical Simulations may involve calculating the interaction of several million interacting sites. Generally, industrial, natural, or experimental configurations involve Reynolds numbers that are far too large to allow direct simulations,[2] and the only possibility then is Large Eddy Simulations, where the small-scale turbulent fluctuations are themselves smoothed out and modelled via eddy-viscosity and diffusivity assumptions. The history of large eddy simulations began in the 1960s with the famous Smagorinsky model. Smagorinsky, also a meteorologist, wanted to represent the effects upon large synoptic quasi-two-dimensional atmospheric or oceanic motions[3] of a three-dimensional subgrid turbulence cascading toward small scales according to mechanisms described by Richardson in 1926 and formalized by the famous mathematician Kolmogorov in 1941.[4] It is interesting to note that Smagorinsky's model was a total failure as far as the

[1] L.F. Richardson, *Weather Prediction by Numerical Process*, Cambridge University Press (1922).

[2] More than 10^{15} modes should be necessary for a supersonic-plane wing!

[3] Subject to vigorous inverse-energy cascades.

[4] L.F. Richardson, Proc. Roy. Soc. London, Ser A, **110**, pp. 709–737 (1926); A. Kolmogorov, Dokl. Akad. Nauk SSSR, **30**, pp. 301–305 (1941).

atmosphere and oceans are concerned, because it dissipates the large-scale motions too much. It was an immense success, though, with users interested in industrial-flow applications, which shows that the outcomes of research are as unpredictable as turbulence itself! A little later, in the 1970s, the theoretical physicist Kraichnan[5] developed the important concept of spectral eddy viscosity, which allows us to go beyond the separation-scale assumption inherent in the typical eddy-viscosity concept of Smagorinsky. From then on, the history of large eddy simulations developed, first in the wake of two schools: Stanford–Torino, where a dynamic version of Smagorinsky's model was developed; and Grenoble, which followed Kraichnan's footsteps. Then researchers, including industrial researchers, all around the world became infatuated with these techniques, being aware of the limits of classical modeling methods based on the averaged equations of motion (Reynolds equations).

It is a complete account of this young but very rich discipline, the large eddy simulation of turbulence, which is proposed to us by the young ONERA researcher Pierre Sagaut, in a book whose reading brings pleasure and interest. *Large-Eddy Simulation for Incompressible Flows - An Introduction* very wisely limits itself to the case of incompressible fluids, which is a suitable starting point if one wants to avoid multiplying difficulties. Let us point out, however, that compressible flows quite often exhibit near-incompressible properties in boundary layers, once the variation of the molecular viscosity with the temperature has been taken into account, as predicted by Morkovin in his famous hypothesis.[6] Pierre Sagaut shows an impressive culture, describing exhaustively all the subgrid-modeling methods for simulating the large scales of turbulence, without hesitating to give the mathematical details needed for a proper understanding of the subject.

After a general introduction, he presents and discusses the various filters used, in cases of statistically homogeneous and inhomogeneous turbulence, and their applications to Navier–Stokes equations. He very aptly describes the representation of various tensors in Fourier space, Germano-type relations obtained by double filtering, and the consequences of Galilean invariance of the equations. He then goes into the various ways of modeling isotropic turbulence. This is done first in Fourier space, with the essential wave-vector triad idea, and a discussion of the transfer-localness concept. An excellent review of spectral-viscosity models is provided, with developments going beyond the original papers. Then he goes to physical space, with a discussion of the structure-function models and the dynamic procedures (Eulerian and Lagrangian, with energy equations and so forth). The study is then generalized to the anisotropic case. Finally, functional approaches based on Taylor series expansions are discussed, along with non-linear models, homogenization techniques, and simple and dynamic mixed models.

---

[5] He worked as a postdoctoral student with Einstein at Princeton.

[6] M.V. Morkovin, in *Mécanique de la Turbulence*, A. Favre et al. (eds.), CNRS, pp. 367–380 (1962).

Pierre Sagaut also discusses the importance of numerical errors, and proposes a very interesting review of the different wall models in the boundary layers. The last chapter gives a few examples of applications carried out at ONERA and a few other French laboratories. These examples are well chosen in order of increasing complexity: isotropic turbulence, with the non-linear condensation of vorticity into the "worms" vortices discovered by Siggia;[7] planar Poiseuille flow with ejection of "hairpin" vortices above low-speed streaks; the round jet and its alternate pairing of vortex rings; and, finally, the backward-facing step, the unavoidable test case of computational fluid dynamics. Also on the menu: beautiful visualizations of separation behind a wing at high incidence, with the shedding of superb longitudinal vortices. Completing the work are two appendices on the statistical and spectral analysis of turbulence, as well as isotropic and anisotropic EDQNM modeling.

A bold explorer, Pierre Sagaut had the daring to plunge into the jungle of multiple modern techniques of large-scale simulation of turbulence. He came back from his trek with an extremely complete synthesis of all the models, giving us a very complete handbook that novices can use to start off on this enthralling adventure, while specialists can discover models different from those they use every day. *Large-Eddy Simulation for Incompressible Flows - An Introduction* is a thrilling work in a somewhat austere wrapping. I very warmly recommend it to the broad public of postgraduate students, researchers, and engineers interested in fluid mechanics and its applications in numerous fields such as aerodynamics, combustion, energetics, and the environment.

Grenoble, March 2000 *Marcel Lesieur*

[7] E.D. Siggia, J. Fluid Mech., **107**, pp. 375–406 (1981).

Preface to the Third Edition

Working on the manuscript of the third edition of this book was a very exciting task, since a lot of new developments have been published since the second edition was printed.

The large-eddy simulation (LES) technique is now recognized as a powerful tool and real applications in several engineering fields are more and more frequently found. This increasing demand for efficient LES tools also sustains growing theoretical research on many aspects of LES, some of which are included in this book. Among them, it is worth noting the mathematical models of LES (the convolution filter being only one possiblity), the definition of boundary conditions, the coupling with numerical errors, and, of course, the problem of defining adequate subgrid models. All these issues are discussed in more detail in this new edition. Some good news is that other monographs, which are good complements to the present book, are now available, showing that LES is a topic with a fastly growing audience. The reader interested in mathematics-oriented discussions will find many details in the monoghaphs by Volker John (*Large-Eddy Simulation of Turbulent Incompressible Flows*, Springer) and Berselli, Illiescu and Layton (*Mathematics of Large-Eddy Simulation of Turbulent Flows*, Springer), while people looking for a subsequent description of numerical methods for LES and direct numerical simulation will enjoy the book by Bernard Geurts (*Elements of Direct and Large-Eddy Simulation*, Edwards). More monographs devoted to particular features of LES (implicit LES appraoches, mathematical backgrounds, etc.) are to come in the near future.

My purpose while writing this third edition was still to provide the reader with an up-to-date review of existing methods, approaches and models for LES of incompressible flows. All chapters of the previous edition have been updated, with the hope that this nearly exhaustive review will help interested readers avoid rediscovering old things. I would like to apologize in advance for certainly forgetting some developments. Two entirely new chapters have been added. The first one deals with mathematical models for LES. Here, I believe that the interesting point is that the filtering approach is nothing but a model for the true LES problem, and other models have been developed that seem to be at least as promising as this very popular one. The second new chapter is dedicated to the scalar equation, with both passive scalar and active scalar (stable/unstable stratification effects) cases being discussed. This extension illustrates the way the usual LES can be extended and how new physical mechanisms can be dealt with, but also inspires new problems.

Paris, November 2004 *Pierre Sagaut*

Preface to the Second Edition

The astonishingly rapid development of the Large-Eddy Simulation technique during the last two or three years, both from the theoretical and applied points of view, have rendered the first edition of this book lacunary in some ways. Three to four years ago, when I was working on the manuscript of the first edition, coupling between LES and multiresolution/multilevel techniques was just an emerging idea. Nowadays, several applications of this approach have been succesfully developed and applied to several flow configurations. Another example of interest from this exponentially growing field is the development of hybrid RANS/LES approaches, which have been derived under many different forms. Because these topics are promising and seem to be possible ways of enhancing the applicability of LES, I felt that they should be incorporated in a general presentation of LES.

Recent developments in LES theory also deal with older topics which have been intensely revisited by reseachers: a unified theory for deconvolution and scale similarity ways of modeling have now been established; the "no model" approach, popularized as the MILES approach, is now based on a deeper theoretical analysis; a lot of attention has been paid to the problem of the definition of boundary conditions for LES; filtering has been extended to Navier–Stokes equations in general coordinates and to Eulerian time–domain filtering.

Another important fact is that LES is now used as an engineering tool for several types of applications, mainly dealing with massively separated flows in complex configurations. The growing need for unsteady, accurate simulations, more and more associated with multidisciplinary applications such as aeroacoustics, is a very powerful driver for LES, and it is certain that this technique is of great promise.

For all these reasons, I accepted the opportunity to revise and to augment this book when Springer offered it me. I would also like to emphasize the fruitful interactions between "traditional" LES researchers and mathematicians that have very recently been developed, yielding, for example, a better understanding of the problem of boundary conditions. Mathematical foundations for LES are under development, and will not be presented in this book, because I did not want to include specialized functional analysis discussions in the present framework.

I am indebted to an increasing number of people, but I would like to express special thanks to all my colleagues at ONERA who worked with me on LES: Drs. E. Garnier, E. Labourasse, I. Mary, P. Quéméré and M. Terracol. All the people who provided me with material dealing with their research are also warmly acknowledged. I also would like to thank all the readers of the first edition of this book who very kindly provided me with their remarks, comments and suggestions. Mrs. J. Ryan is once again gratefully acknowledged for her help in writing the English version.

Paris, April 2002 *Pierre Sagaut*

Preface to the First Edition

While giving lectures dealing with Large-Eddy Simulation (LES) to students or senior scientists, I have found difficulties indicating published references which can serve as general and complete introductions to this technique.

I have tried therefore to write a textbook which can be used by students or researchers showing theoretical and practical aspects of the Large Eddy Simulation technique, with the purpose of presenting the main theoretical problems and ways of modeling. It assumes that the reader possesses a basic knowledge of fluid mechanics and applied mathematics.

Introducing Large Eddy Simulation is not an easy task, since no unified and universally accepted theoretical framework exists for it. It should be remembered that the first LES computations were carried out in the early 1960s, but the first rigorous derivation of the LES governing equations in general coordinates was published in 1995! Many reasons can be invoked to explain this lack of a unified framework. Among them, the fact that LES stands at the crossroads of physical modeling and numerical analysis is a major point, and only a few really successful interactions between physicists, mathematicians and practitioners have been registered over the past thirty years, each community sticking to its own language and center of interest. Each of these three communities, though producing very interesting work, has not yet provided a complete theoretical framework for LES by its own means. I have tried to gather these different contributions in this book, in an understandable form for readers having a basic background in applied mathematics.

Another difficulty is the very large number of existing physical models, referred to as subgrid models. Most of them are only used by their creators, and appear in a very small number of publications. I made the choice to present a very large number of models, in order to give the reader a good overview of the ways explored. The distinction between functional and structural models is made in this book, in order to provide a general classification; this was necessary to produce an integrated presentation.

In order to provide a useful synthesis of forty years of LES development, I had to make several choices. Firstly, the subject is restricted to incompressible flows, as the theoretical background for compressible flow is less evolved. Secondly, it was necessary to make a unified presentation of a large number of works issued from many research groups, and very often I have had to change the original proof and to reduce it. I hope that the authors will not feel betrayed by the present work. Thirdly, several thousand journal articles and communications dealing with LES can be found, and I had to make a selection. I have deliberately chosen to present a large number of theoretical approaches and physical models to give the reader the most general view of what has been done in each field. I think that the most important contributions are presented in this book, but I am sure that many new physical models and results dealing with theoretical aspects will appear in the near future.

A typical question of people who are discovering LES is "what is the best model for LES?". I have to say that I am convinced that this question cannot be answered nowadays, because no extensive comparisons have been carried out, and I am not even sure that the answer exists, because people do not agree on the criterion to use to define the "best" model. As a consequence, I did not try to rank the model, but gave very generally agreed conclusions on the model efficiency.

A very important point when dealing with LES is the numerical algorithm used to solve the governing equations. It has always been recognized that numerical errors could affect the quality of the solution, but new emphasis has been put on this subject during the last decade, and it seems that things are just beginning. This point appeared as a real problem to me when writing this book, because many conclusions are still controversial (e.g. the possibility of using a second-order accurate numerical scheme or an artificial diffusion). So I chose to mention the problems and the different existing points of view, but avoided writing a part dealing entirely with numerical discretization and time integration, discretization errors, etc. This would have required writing a companion book on numerical methods, and that was beyond the scope of the present work. Many good textbooks on that subject already exist, and the reader should refer to them.

Another point is that the analysis of the coupling of LES with typical numerical techniques, which should greatly increase the range of applications, such as Arbitrary Lagrangian–Eulerian methods, Adaptive Mesh-Refinement or embedded grid techniques, is still to be developed.

I am indebted to a large number of people, but I would like to express special thanks to Dr. P. Le Quére, O. Daube, who gave me the opportunity to write my first manuscript on LES, and to Prof. J.M. Ghidaglia who offered me the possibility of publishing the first version of this book (in French). I would also like to thank ONERA for helping me to write this new, augmented and translated version of the book. Mrs. J. Ryan is gratefully acknowledged for her help in writing the English version.

Paris, September 2000 *Pierre Sagaut*

Contents

1. Introduction

1.1 Computational Fluid Dynamics

Computational Fluid Dynamics (CFD) is the study of fluids in flow by numerical simulation, and is a field advancing by leaps and bounds. The basic idea is to use appropriate algorithms to find solutions to the equations describing the fluid motion.

Numerical simulations are used for two types of purposes.

The first is to accompany research of a fundamental kind. By describing the basic physical mechanisms governing fluid dynamics better, numerical simulation helps us understand, model, and later control these mechanisms. This kind of study requires that the numerical simulation produce data of very high accuracy, which implies that the physical model chosen to represent the behavior of the fluid must be pertinent and that the algorithms used, and the way they are used by the computer system, must introduce no more than a low level of error. The quality of the data generated by the numerical simulation also depends on the level of resolution chosen. For the best possible precision, the simulation has to take into account all the space-time scales affecting the flow dynamics. When the range of scales is very large, as it is in turbulent flows, for example, the problem becomes a stiff one, in the sense that the ratio between the largest and smallest scales becomes very large.

Numerical simulation is also used for another purpose: engineering analyses, where flow characteristics need to be predicted in equipment design phase. Here, the goal is no longer to produce data for analyzing the flow dynamics itself, but rather to predict certain of the flow characteristics or, more precisely, the values of physical parameters that depend on the flow, such as the stresses exerted on an immersed body, the production and propagation of acoustic waves, or the mixing of chemical species. The purpose is to reduce the cost and time needed to develop a prototype. The desired predictions may be either of the mean values of these parameters or their extremes. If the former, the characteristics of the system's normal operating regime are determined, such as the fuel an aircraft will consume per unit of time in cruising flight. The question of study here is mainly the system's performance. When extreme parameter values are desired, the question is rather the system's characteristics in situations that have a little probability of ever existing, i.e. in the presence of rare or critical phenomena, such as rotating stall in aeronautical engines. Studies like this concern system safety at operating points far from the cruising regime for which they were designed.

The constraints on the quality of representation of the physical phenomena differ here from what is required in fundamental studies, because what is wanted now is evidence that certain phenomena exist, rather than all the physical mechanisms at play. In theory, then, the description does not have to be as detailed as it does for fundamental studies. However, it goes without saying that the quality of the prediction improves with the richness of the physical model.

The various levels of approximation going into the physical model are discussed in the following.

1.2 Levels of Approximation: General

A mathematical model for describing a physical system cannot be defined before we have determined the *level of approximation* that will be needed for obtaining the required precision on a fixed set of parameters (see [307] for a fuller discussion). This set of parameters, associated with the other variables characterizing the evolution of the model, contain the necessary information for describing the system completely.

The first decision that is made concerns the scale of reality considered. That is, physical reality can be described at several levels: in terms of particle physics, atomic physics, or micro- and macroscopic descriptions of phenomena. This latter level is the one used by classical mechanics, especially continuum mechanics, which will serve as the framework for the explanations given here.

A system description at a given scale can be seen as a statistical averaging of the detailed descriptions obtained at the previous (lower) level of description. In fluid mechanics, which is essentially the study of systems consisting of a large number of interacting elements, the choice of a level of description, and thus a level of averaging, is fundamental. A description at the molecular level would call for a definition of a discrete system governed by Boltzmann equations, whereas the continuum paradigm would be called for in a macroscopic description corresponding to a scale of representation larger than the mean free path of the molecules. The system will then be governed by the Navier–Stokes equations, if the fluid is Newtonian.

After deciding on a level of reality, several other levels of approximation have to be considered in order to obtain the desired information concerning the evolution of the system:

- *Level of space-time resolution.* This is a matter of determining the time and space scales characteristic of the system evolution. The smallest pertinent scale is taken as the resolution reference so as to capture all the dynamic mechanisms. The system spatial dimension (zero to three dimensions) has to be determined in addition to this.

– *Level of dynamic description.* Here we determine the various forces exerted on the system components, and their relative importance. In the continuum mechanics framework, the most complete model is that of the Navier–Stokes equations, complemented by empirical laws for describing the dependency of the diffusion coefficients as a function of the other variables, and the state law. This can first be simplified by considering that the elliptic character of the flow is due only to the pressure, while the other variables are considered to be parabolic, and we then refer to the parabolic Navier–Stokes equations. Other possible simplifications are, for example, Stokes equations, which account only for the pressure and diffusion effects, and the Euler equations, which neglect the viscous mechanisms.

The different choices made at each of these levels make it possible to develop a mathematical model for describing the physical system. In all of the following, we restrict ourselves to the case of a Newtonian fluid of a single species, of constant volume, isothermal, and isochoric in the absence of any external forces. The mathematical model consists of the unsteady Navier–Stokes equations. The numerical simulation then consists in finding solutions of these equations using algorithms for Partial Differential Equations. Because of the way computers are structured, the numerical data thus generated is a discrete set of degrees of freedom, and of finite dimensions. We therefore assume that the behavior of the discrete dynamical system represented by the numerical result will approximate that of the exact, continuous solution of the Navier–Stokes equations with adequate accuracy.

1.3 Statement of the Scale Separation Problem

Solving the unsteady Navier–Stokes equations implies that we must take into account all the space-time scales of the solution if we want to have a result of maximum quality. The discretization has to be fine enough to represent all these scales numerically. That is, the simulation is discretized in steps Δx in space and Δt in time that must be smaller, respectively, than the characteristic length and the characteristic time associated with the smallest dynamically active scale of the exact solution. This is equivalent to saying that the space-time resolution scale of the numerical result must be at least as fine as that of the continuous problem. This solution criterion may turn out to be extremely constrictive when the solution to the exact problem contains scales of very different sizes, which is the case for turbulent flows.

This is illustrated by taking the case of the simplest turbulent flow, i.e. one that is statistically homogeneous and isotropic (see Appendix A for a more precise definition). For this flow, the ratio between the characteristic length of the most energetic scale, L, and that of the smallest dynamically active scale, η, is evaluated by the relation:

$$\frac{L}{\eta} = O\left(Re^{3/4}\right) \quad , \tag{1.1}$$

in which Re is the Reynolds number, which is a measure of the ratio of the forces of inertia and the molecular viscosity effect, ν. We therefore need $O\left(Re^{9/4}\right)$ degrees of freedom in order to be able to represent all the scales in a cubic volume of edge L. The ratio of characteristic times varies as $O\left(Re^{1/2}\right)$, but the use of explicit time-integration algorithm leads to a linear dependency of the time step with respect to the mesh size. So in order to calculate the evolution of the solution in a volume L^3 for a duration equal to the characteristic time of the most energetic scale, we have to solve the Navier–Stokes equations numerically $O\left(Re^3\right)$ times!

This type of computation for large Reynolds numbers (applications in the aeronautical field deal with Reynolds numbers of as much as 10^8) requires computer resources very much greater than currently available supercomputer capacities, and is therefore not practicable.

In order to be able to compute the solution, we need to reduce the number of operations, so we no longer solve the dynamics of all the scales of the exact solution directly. To do this, we have to introduce a new, coarser level of description of the fluid system. This comes down to picking out certain scales that will be represented directly in the simulation while others will not be. The non-linearity of the Navier–Stokes equations reflects the dynamic coupling that exists among all the scales of the solution, which implies that these scales cannot be calculated independently of each other. So if we want a quality representation of the scales that are resolved, their interactions with the scales that are not have to be considered in the simulation. This is done by introducing an additional term in the equations governing the evolution of the resolved scales, to model these interactions. Since these terms represent the action of a large number of other scales with those that are resolved (without which there would be no effective gain), they reflect only the global or average action of these scales. They are therefore only statistical models: an individual deterministic representation of the inter-scale interactions would be equivalent to a direct numerical simulation.

Such modeling offers a gain only to the extent that it is universal, i.e. if it can be used in cases other than the one for which it is established. This means there exists a certain universality in the dynamic interactions the models reflect. This universality of the assumptions and models will be discussed all through the text.

1.4 Usual Levels of Approximation

There are several common ways of reducing the number of degrees of freedom in the numerical solution:

- By calculating the statistical average of the solution directly. This is called the Reynolds Averaged Numerical Simulation (RANS)[424], which is used mostly for engineering calculations. The exact solution $\boldsymbol{u}$ splits into the sum of its statistical average $\langle \boldsymbol{u} \rangle$ and a fluctuation $\boldsymbol{u}'$ (see Appendix A):

$$\boldsymbol{u}(\boldsymbol{x},t) = \langle \boldsymbol{u}(\boldsymbol{x},t) \rangle + \boldsymbol{u}'(\boldsymbol{x},t) \quad .$$

  This splitting, or "decomposition", is illustrated by Fig. 1.1. The fluctuation $\boldsymbol{u}'$ is not represented directly by the numerical simulation, and is included only by way of a turbulence model. The statistical averaging operation is in practice often associated with a time averaging:

$$\langle \boldsymbol{u}(\boldsymbol{x},t) \rangle \approx \overline{\boldsymbol{u}}(\boldsymbol{x}) = \lim_{T \to \infty} \frac{1}{T} \int_0^T \boldsymbol{u}(\boldsymbol{x},t) dt \quad .$$

  The mathematical model is then that of the steady Navier–Stokes equations. This averaging operation makes it possible to reduce the number of scales in the solution considerably, and therefore the number of degrees of freedom of the discrete system. The statistical character of the solution prevents a fine description of the physical mechanisms, so that this approach is not usable for studies of a fundamental character, especially so when the statistical average is combined with a time average. Nor is it possible to isolate rare events. On the other hand, it is an appropriate approach for analyzing performance as long as the turbulence models are able to reflect the existence of the turbulent fluctuation $\boldsymbol{u}'$ effectively.

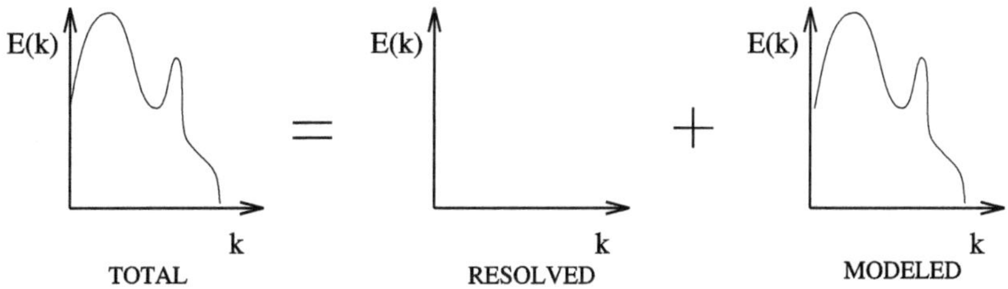

Fig. 1.1. Decomposition of the energy spectrum of the solution associated with the Reynolds Averaged Numerical Simulation (symbolic representation).

- By calculating directly only certain low-frequency modes in time (of the order of a few hundred hertz) and the average field. This approach goes by a number of names: *Unsteady Reynolds Averaged Numerical Simula-*

tion (URANS), *Semi-Deterministic Simulation* (SDS), *Very Large-Eddy Simulation* (VLES), and sometimes *Coherent Structure Capturing* (CSC) [726, 44]. The field $\boldsymbol{u}$ appears here as the sum of three contributing terms [456, 451, 240, 726]:

$$\boldsymbol{u}(\boldsymbol{x},t) = \overline{\boldsymbol{u}}(\boldsymbol{x}) + \langle \boldsymbol{u}(\boldsymbol{x},t) \rangle_{\mathrm{c}} + \boldsymbol{u}'(\boldsymbol{x},t) \quad .$$

The first term is the time average of the exact solution, the second its conditional statistical average, and the third the turbulent fluctuation. This decomposition is illustrated in Fig. 1.2. The conditional average is associated with a predefined class of events. When these events occur at a set time period, this is a phase average. The $\langle \boldsymbol{u}(\boldsymbol{x},t) \rangle_{\mathrm{c}}$ term is interpreted as the contribution of the coherent modes to the flow dynamics, while the $\boldsymbol{u}'$ term, on the other hand, is supposed to represent the random part of the turbulence. The variable described by the mathematical model is now the sum $\overline{\boldsymbol{u}}(\boldsymbol{x}) + \langle \boldsymbol{u}(\boldsymbol{x},t) \rangle_{\mathrm{c}}$, with the random part being represented by a turbulence model. It should be noted that, in the case where there exists a deterministic low-frequency forcing of the solution, the conditional average is conventionally interpreted as a phase average of the solution, for a frequency equal to that of the forcing term; but if this does not exist, the interpretation of the results is still open to debate. Since this is an unsteady approach, it contains more information than the previous one; but it still precludes a deterministic description of a particular event. It is of use for analyzing the performance characteristics of systems in which the unsteady character is forced by some external action (such as periodically pulsed flows).

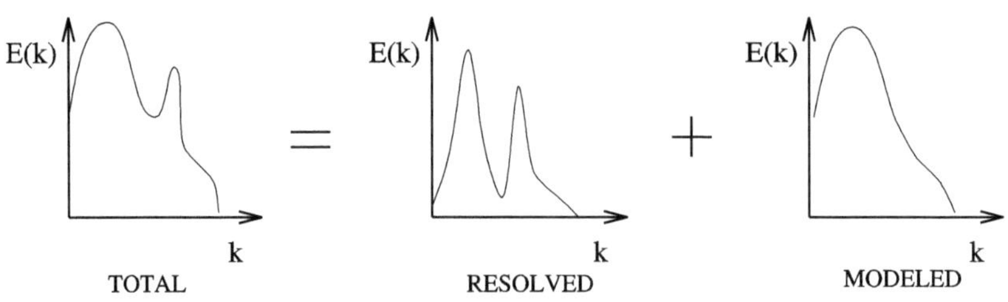

Fig. 1.2. Decomposition of the energy spectrum of the solution associated with the Unsteady Reynolds Averaged Numerical Simulation approach, when a predominant frequency exists (symbolic representation).

– By projecting the solution on the ad hoc function basis and retaining only a minimum number of modes, to get a dynamical system with fewer degrees of freedom. The idea here is to find an optimum decomposition basis for representing the phenomenon, in order to minimize the number of degrees of freedom in the discrete dynamical system. There is no averaging done here, so the space-time and dynamics resolution of the numerical model is

still as fine as that of the continuum model, but is now optimized. Several approaches are encountered in practice.
The first is to use standard basis function (Fourier modes in the spectral space or polynomials in the physical space, for example) and distribute the degrees of freedom as best possible in space and time to minimize the number of them, i.e. adapt the space-time resolution of the simulation to the nature of the solution. We thus adapt the topology of the discrete dynamical system to that of the exact solution. This approach results in the use of self-adapting grids and time steps in the physical space. It is not associated with an operation to reduce the complexity by switching to a higher level of statistical description of the system. It leads to a much less important reduction of the discrete system than those techniques based on statistical averaging, and is limited by the complexity of the continuous solution.
Another approach is to use optimal basis functions, a small number of which will suffice for representing the flow dynamics. The problem is then to determine what these basis functions are. One example is the Proper Orthogonal Decomposition (POD) mode basis, which is optimum for representing kinetic energy (see [55] for a survey). This technique allows very high data compression, and generates a dynamical system of very small dimensions (a few dozen degrees of freedom at most, in practice). The approach is very seldom used because it requires very complete information concerning the solution in order to be able to determine the base functions.
The various approaches above all return complete information concerning the solutions of the exact problem, so they are perfectly suited to studies of a fundamental nature. They may not, on the other hand, be optimal in terms of reducing the complexity for certain engineering analyses that do not require such complete data.
- By calculating only the low-frequency modes in space directly. This is what is done in *Large-Eddy Simulation* (LES). It is this approach that is discussed in the following. It is illustrated in Fig. 1.3.

Typical results obtained by these three approaches are illustrated in Fig. 1.4.

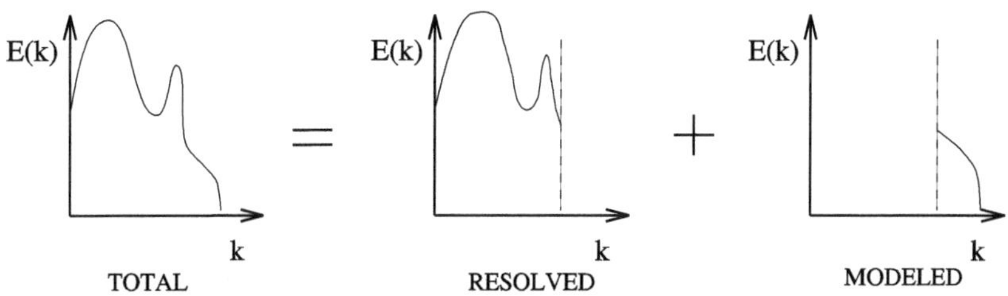

Fig. 1.3. Decomposition of the energy spectrum in the solution associated with large-eddy simulation (symbolic representation).

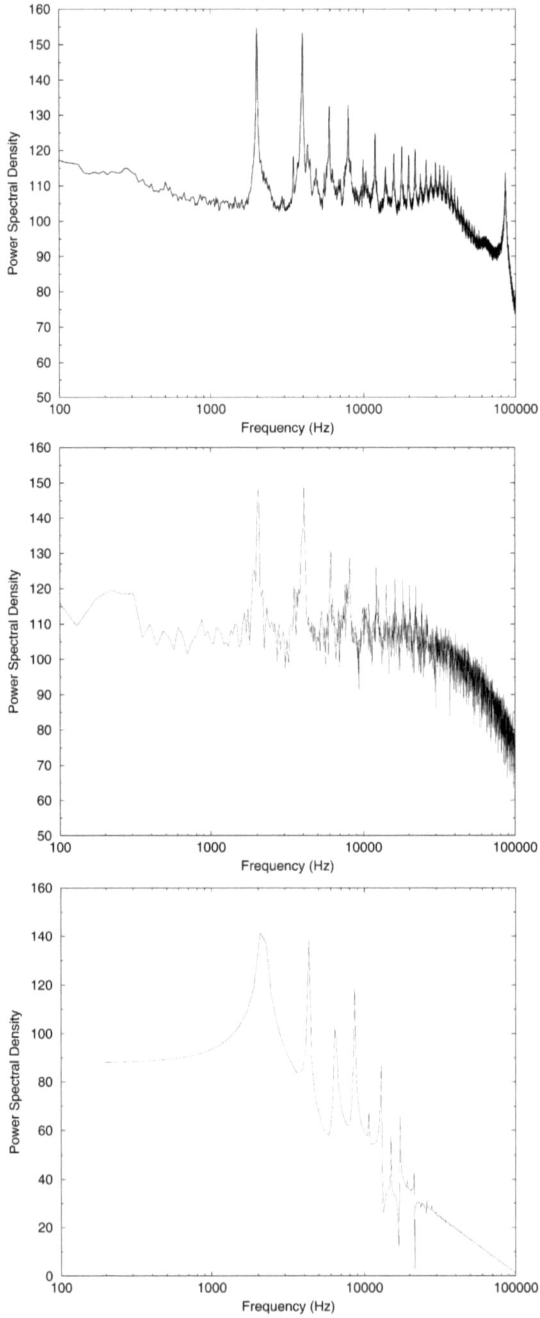

Fig. 1.4. Pressure spectrum inside a cavity. *Top*: experimental data (ideal direct-numerical simulation) (courtesy of L. Jacquin, ONERA); *Middle*: large-eddy simulation (Courtesy of L. Larchevêque, ONERA); *Bottom*: unsteady RANS simulation (Courtesy of V. Gleize, ONERA).

1.5 Large-Eddy Simulation: from Practice to Theory. Structure of the Book

As mentioned above, the Large-Eddy Simulation approach relies on the definition of large and small scales. This fuzzy and empirical concept requires further discussion to become a tractable tool from both the theoretical and practical points of view. Bases for the *theoretical understanding* and *modeling* of this approach are now introduced. In practice, the Large-Eddy Simulation technique consists in solving the set of *ad hoc* governing equations on a computational grid which is too coarse to represent the smallest physical scales. Let Δx and η be the computional mesh size (assumed to be uniform for the sake of simplicity) and the characteristic size of the smallest physical scales. Let u be the exact solution of the following *continuous* generic conservation law (the case of the Navier–Stokes equations will be extensively discussed in the core of the book)

$$\frac{\partial u}{\partial t} + F(u, u) = 0 \tag{1.2}$$

where $F(\cdot, \cdot)$ is a non-linear flux function. The Large-Eddy Simulation problem consists in finding the best approximation of u on the computational grid by solving the following *discrete problem*

$$\frac{\delta u_{\mathrm{d}}}{\delta t} + F_{\mathrm{d}}(u_{\mathrm{d}}, u_{\mathrm{d}}) = 0 \tag{1.3}$$

where u_{d}, $\delta/\delta t$ and $F_{\mathrm{d}}(\cdot, \cdot)$ are the discrete approximations of u, $\partial/\partial t$ and $F(\cdot, \cdot)$ on the computational grid, respectively. Thus, the question arise of defining what is the *best possible approximation* of u, u_{Π}, among all discrete solutions u_{d} associated with Δx.

Let $e(u, u_{\mathrm{d}})$ be a measure of the difference between u and u_{d}, which does not need to be explicitely defined for the present purpose. It is just emphasized here that since Large-Eddy Simulation is used to compute turbulent flows, u exhibits a chaotic behavior and therefore $e(u, u_{\mathrm{d}})$ should rely on statistical moments of the solutions. A consistency constraint on the definition of the error functional is that it must vanish in the limit case of the Direct Numerical Simulation

$$\lim_{\Delta x \longrightarrow \eta} e(u, u_{\mathrm{d}}) = 0 \tag{1.4}$$

A careful look at the problem reveals that the error can be decomposed as

$$e(u, u_{\mathrm{d}}) = e_{\Pi}(u, u_{\mathrm{d}}) + e_{\mathrm{d}}(u, u_{\mathrm{d}}) + e_{\mathrm{r}}(u, u_{\mathrm{d}}) \tag{1.5}$$

where

1. $e_\Pi(u, u_d)$ is the *projection error* which accounts for the fact that the exact solution u is approximated using a finite number of degrees of freedom. The Nyquist theorem tells us that no scale smaller than $2\Delta x$ can be captured in the simulation. As a consequence, u_d can never be strictly equal to u :
$$|u - u_d| \neq 0 \tag{1.6}$$
2. $e_d(u, u_d)$ is the *discretization error* which accounts for the fact that partial derivatives which appear in the continuous problem are approximated on the computational grid using Finite Diffrence, Finite Volume, Finite Element (or other similar) schemes. Putting the emphasis on spatial derivatives, this is expressed as
$$F_d(u, u) \neq F(u, u) \tag{1.7}$$
3. $e_r(u, u_d)$ is the *resolution error*, which accounts for the fact that, some scales of the exact solution being missing, the evaluation of the non-linear flux function cannot be exact, even if the discretization error is driven to zero:
$$F(u_d, u_d) \neq F(u, u) \tag{1.8}$$

This analysis shows that the Large-Eddy Simulation problem is very complex, since it depends explicitely on the exact solution, the computational grid and the numerical method, making each problem appearing as *unique*. Therefore, it is necessary to find some *mathematical models* for the Large-Eddy Simulation problem which will mimic its main features, the most important one being the removal of the small scales of the exact solution. A simplified heuristic view of this problem is illustrated in Fig. 1.5, where the effect of the Nyquist filter is represented.

Several mathematical models have been proposed to handle the true Large-Eddy Simulation problem. The most popular one (see [216, 440, 495, 619, 627]) relies on the representation of the removal of the small scales as the result of the application of a low-pass convolution filter (in terms of wave number) to the exact solution. The definition and the properties of this filtering operator are presented in Chap. 2. The application of this filter to the Navier–Stokes equations, described in Chap. 3, yields the corresponding constitutive mathematical model for the large-eddy simulation. Alternate mathematical models are detailed in Chap. 4.

The second question raised by the Large-Eddy Simulation approach deals with the search for the best approximate solution $u_\Pi \in \{u_d\}$ that will minimize the error $e(u, u_d)$. The short analysis given above shows that the projection error, $e_\Pi(u, u_d)$ cannot be avoided. Therefore, the best, *ideal* Large-Eddy Solution is such that

$$e(u, u_d) = e(u, u_\Pi) = e_\Pi(u, u_\Pi) \tag{1.9}$$

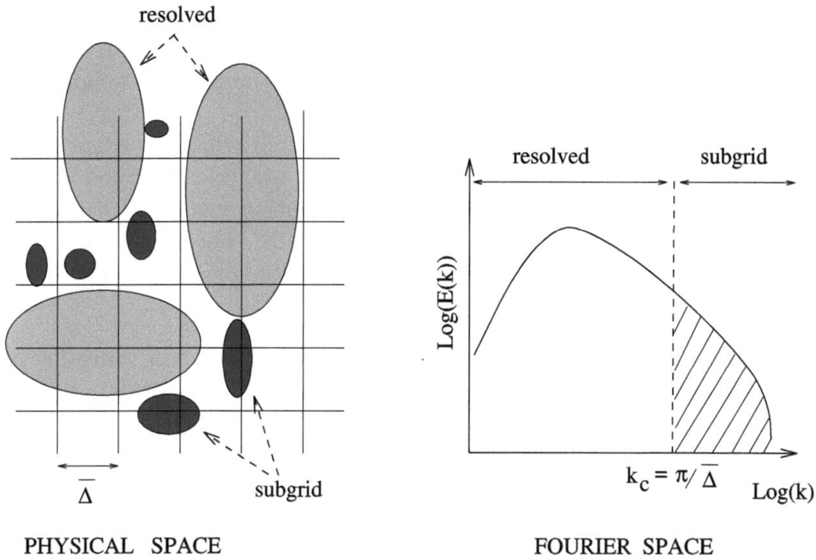

Fig. 1.5. Schematic view of the simplest scale separation operator: grid and theoretical filters are the same, yielding a sharp cutoff filtering in Fourier space between the resolved and subgrid scales. The associated cutoff wave number is denoted k_c, which is directly computed from the cutoff length $\overline{\Delta}$ in the physical space. Here, $\overline{\Delta}$ is assumed to be equal to the size of the computational mesh.

and, following relation (1.5), it is associated to

$$e_{\mathrm{d}}(u, u_{\Pi}) + e_{\mathrm{r}}(u, u_{\Pi}) = 0 \tag{1.10}$$

The best solution in sought in practice trying to enforce the sequel relation (1.10). Two basic different ways are identified for that purpose:

- The *explicit Large-Eddy Simulation approach*, in which an extra forcing term, referred to as a *subgrid model*, is introduced in the governing equation to cancel the *resolution error*. Two modeling approaches are discussed here: *functional modeling*, based on representing kinetic energy transfers (covered in Chaps. 5 and 6), and *structural modeling*, which aims to reproduce the eigenvectors of the statistical correlation tensors of the subgrid modes (presented in Chap. 7). The basic assumptions and the subgrid models corresponding to each of these approaches are presented. In the hypothetical case where a perfect subgrid model could be found, expression (1.10) shows that the *discretization error* $e_{\mathrm{d}}(u, u_{\mathrm{d}})$ must also be driven to zero to recover the ideal Large-Eddy Simulation solution. A perfect numerical method is obviously a natural candidate for that purpose, but reminding that the error measure is based on statistical moments, the much less stringent requirement that the numerical method must be *neutral* with respect to the error definition is sufficient. Chapter 8 is devoted to the theoretical

problems related to the effects of the numerical method used in the simulation. The representation of the numerical error in the form of an additional filter is introduced, along with the problem of the relative weight of the various filters used in the numerical simulation.
- The *implicit Large-Eddy Simulation approach*, in which no extra term is introduced in the governing equations, but the numerical method is chosen such that the numerical error and the resolution error will cancel each other, yielding a direct fulfilment of relation (1.10). This approach is briefly presented in this book in Sect. 5.3.4. The interested reader can refer to [276] for an exhaustive description.

The fact that the ideal solution u_{Π} is associated to a non-vanishing projection error $e_{\Pi}(u, u_{\Pi})$ raises the problem of the reliability of data obtained via Large-Eddy Simulation for practical purposes. Several theoretical and practical problems are met when addressing the issue of validating and exploiting Large-Eddy Simulation. The definition of the best solution being intrinsically based on the definition of the error functional (which is arbitrary), a universal answer seems to be meaningless. Questions concerning the analysis and validation of the large-eddy simulation calculations are dealt with in Chap. 9. The concept of statistically partially equivalent simulations is introduced, which is of major importance to interpret the nature of the data recovered from Large-Eddy Simulation. A short survey of available results dealing with the properties of filtered Navier–Stokes solutions (ideally u_{Π}) and Large-Eddy Simulation solutions (true u_{d} fields) is presented.

The discussions presented above deal with the definition of the Large-Eddy Simulation problem inside the computational domain. As all differential problems, it must be supplemented with ad hoc boundary conditions to yield a well-posed problem. Thus, the new question of defining discrete boundary conditions in a consistent way appears. The problem is similar to the previous one: what boundary conditions should be used to reach the best solution u_{Π}? A weaker constraint is to find boundary conditions which do not deteriorate the accuracy that could potentially be reached with the selected numerical scheme and closure. The boundary conditions used for large-eddy simulation are discussed in Chap. 10, where the main cases treated are solid walls and turbulent inflow conditions. In the solid wall case, the emphasis is put on the problem of defining wall stress models, which are subgrid models derived for the specific purpose of taking into account the dynamics of the inner layer of turbulent boundary layers. The issue of defining efficient turbulent inflow condtions raises from the need to truncate the computational domain, which leads to the requirement of finding a way to take into account upstream turbulent fluctuations in the boundary conditions.

Despite the fact that it yields very significant complexity reduction in terms of degrees of freedom with respect to Direct Numerical Simulation, Large-Eddy Simulation still requires considerable computational efforts to handle realistic applications. To obtain further complexity reduction, several hybridizations of the Large-Eddy Simulation technique have been proposed. Methods for reducing the cost of Large-Eddy Simulation by coupling it with multiresolution and multidomain techniques are presented in Chap. 11. Hybrid RANS/LES approaches are presented in Chap. 12. The definition of such multiresolution methods and/or hybrid RANS/LES techniques raises many practical and theoretical issues. Among the most important ones, the emphasis is put in the dedicated chapters on the coupling strategies and the fact that the instantaneous fields can be fully discontinuous (fully meaning here that the velocity field is not a priori continuous at the interfaces between domains with different resolution, but also that even the number of space dimension and the number of unknwons can be different).

Practical aspects concerning the implementation of subgrid models are described in Chap. 13. Lastly, the discussion is illustrated by examples of large-eddy simulation applications for different categories of flows, in Chap. 14.

Chapter 15 is devoted the the extension of concepts, methods and models presented in previous chapters to the case of a more complex physical system, in which an additional equation for a scalar is added to the Navier–Stokes equations. Two cases are considered: the passive scalar case, in which there is no feedback in the momentum equation and the new problem is restricted to closing the filtered scalar equation, and the active scalar case, which corresponds to a two-way coupling between the scalar field and the velocity field. In the latter, the definition of subgrid models for both the velocity and the scalar is a full problem. For the sake of clarity, the discussion is limited to stably stratified flows and buoyancy driven flows. Combustion and two-phase flows are not treated.

2. Formal Introduction to Scale Separation: Band-Pass Filtering

The idea of scale separation introduced in the preceding chapter will now be formalized on the mathematical level, to show how to handle the equations and derive the subgrid models.

This chapter is devoted to the representation of the filtering as a convolution product, which is the most common way to model the removal of small scales in the Larg-Eddy Simulation approach. Other definitions, such as partial statistical averaging or conditional averaging [251, 250, 465], will be presented in Chap. 4. The filtering approach is first presented in the ideal case of a filter of uniform cutoff length over an infinite domain (Sect. 2.1). Fundamental properties of filters and their approximation via differential operators is presented. Extensions to the cases of a bounded domain and a filter of variable cutoff length are then discussed (Sect. 2.2). The chapter is closed by discussing a few properties of the Eulerian time-domain filters (Sect. 2.3).

2.1 Definition and Properties of the Filter in the Homogeneous Case

The framework is restricted here to the case of homogeneous isotropic filters, for the sake of easier analysis, and to allow a better understanding of the physics of the phenomena. The filter considered is *isotropic*. This means that its properties are independent of the position and orientation of the frame of reference in space, which implies that it is applied to an unbounded domain and that the cutoff scale is constant and identical in all directions of space. This is the framework in which subgrid modeling developed historically. The extension to anisotropic and inhomogeneous[1] filters, which researchers have only more recently begun to look into, is described in Sect. 2.2.

2.1.1 Definition

Scales are separated by applying a scale high-pass filter, i.e. low-pass in frequency, to the exact solution. This filtering is represented mathematically in

[1] That is, whose characteristics, such as the mathematical form or cutoff frequency, are not invariant by translation or rotation of the frame of reference in which they are defined.

physical space as a convolution product. The resolved part $\overline{\phi}(\boldsymbol{x},t)$ of a space-time variable $\phi(\boldsymbol{x},t)$ is defined formally by the relation:

$$\overline{\phi}(\boldsymbol{x},t) = \int_{-\infty}^{+\infty}\int_{-\infty}^{+\infty} \phi(\boldsymbol{\xi},t')G(\boldsymbol{x}-\boldsymbol{\xi},t-t')dt'd^3\boldsymbol{\xi} \quad , \tag{2.1}$$

in which the convolution kernel G is characteristic of the filter used, which is associated with the cutoff scales in space and time, $\overline{\Delta}$ and $\overline{\tau}_{\mathrm{c}}$, respectively. This relation is denoted symbolically by:

$$\overline{\phi} = G \star \phi \quad . \tag{2.2}$$

The dual definition in the Fourier space is obtained by multiplying the spectrum $\widehat{\phi}(\boldsymbol{k},\omega)$ of $\phi(\boldsymbol{x},t)$ by the spectrum $\widehat{G}(\boldsymbol{k},\omega)$ of the kernel $G(\boldsymbol{x},t)$:

$$\widehat{\overline{\phi}}(\boldsymbol{k},\omega) = \widehat{\phi}(\boldsymbol{k},\omega)\widehat{G}(\boldsymbol{k},\omega) \quad , \tag{2.3}$$

or, in symbolic form:

$$\widehat{\overline{\phi}} = \widehat{G}\widehat{\phi} \quad , \tag{2.4}$$

where k and ω are the spatial wave number and time frequency, respectively.

The function $\widehat{G}$ is the transfer function associated with the kernel G. The spatial cutoff length $\overline{\Delta}$ is associated with the cutoff wave number k_{c} and time $\overline{\tau}_{\mathrm{c}}$ with the cutoff frequency ω_{c}. The unresolved part of $\phi(\boldsymbol{x},t)$, denoted $\phi'(\boldsymbol{x},t)$, is defined operationally as:

$$\begin{aligned} \phi'(\boldsymbol{x},t) &= \phi(\boldsymbol{x},t) - \overline{\phi}(\boldsymbol{x},t) & (2.5)\\ &= \phi(\boldsymbol{x},t) - \int_{-\infty}^{+\infty}\int_{-\infty}^{+\infty} \phi(\boldsymbol{\xi},t')G(\boldsymbol{x}-\boldsymbol{\xi},t-t')dt'd^3\boldsymbol{\xi}, & (2.6)\end{aligned}$$

or:

$$\phi' = (1-G) \star \phi \quad . \tag{2.7}$$

The corresponding form in spectral space is:

$$\widehat{\phi}'(\boldsymbol{k},\omega) = \widehat{\phi}(\boldsymbol{k},\omega) - \widehat{\overline{\phi}}(\boldsymbol{k},\omega) = \left(1-\widehat{G}(\boldsymbol{k},\omega)\right)\widehat{\phi}(\boldsymbol{k},\omega) \quad , \tag{2.8}$$

i.e.

$$\widehat{\phi}' = (1-\widehat{G})\widehat{\phi} \quad . \tag{2.9}$$

2.1.2 Fundamental Properties

In order to be able to manipulate the Navier–Stokes equations after applying a filter, we require that the filter verify the three following properties:

1. Conservation of constants
$$\overline{a} = a \Longleftrightarrow \int_{-\infty}^{+\infty} \int_{-\infty}^{+\infty} G(\boldsymbol{\xi}, t') d^3\boldsymbol{\xi} dt' = 1 \quad . \tag{2.10}$$
2. Linearity
$$\overline{\phi + \psi} = \overline{\phi} + \overline{\psi} \quad . \tag{2.11}$$
This property is automatically satisfied, since the product of convolution verifies it independently of the characteristics of the kernel G.
3. Commutation with derivation
$$\overline{\frac{\partial \phi}{\partial s}} = \frac{\partial \overline{\phi}}{\partial s}, \quad s = \boldsymbol{x}, t \quad . \tag{2.12}$$

Introducing the commutator $[f, g]$ of two operators f and g applied to the dummy variable ϕ

$$[f, g]\phi \equiv f \circ g(\phi) - g \circ f(\phi) = f(g(\phi)) - g(f(\phi)) \quad , \tag{2.13}$$

the relation (2.12) can be re-written symbolically

$$\left[G\star, \frac{\partial}{\partial s} \right] = 0 \quad . \tag{2.14}$$

The commutator defined by relation (2.13) has the following properties[2]:

$$[f, g] = -[g, f] \quad \text{Skew-symmetry} \quad , \tag{2.15}$$
$$[f \circ g, h] = [f, h] \circ g + f \circ [g, h] \quad \text{Leibniz identity} \quad , \tag{2.16}$$
$$[f, [g, h]] + [g, [h, f]] + [h, [f, g]] = 0 \quad \text{Jacobi's identity} \quad . \tag{2.17}$$

The filters that verify these three properties are not, in the general case, Reynolds operators (see Appendix A), i.e.

$$\overline{\overline{\phi}} = G \star G \star \phi = G^2 \star \phi \neq \overline{\phi} = G \star \phi \quad , \tag{2.18}$$
$$\overline{\phi'} = G \star (1 - G) \star \phi \neq 0 \quad , \tag{2.19}$$

[2] In the linear case, the commutator satisfies all the properties of the Poisson-bracket operator, as defined in classical mechanics.

which is equivalent to saying that G is not a projector (excluding the trivial case of the identity application). Let us recall that an application P is defined as being a projector if $P \circ P = P$. Such an application is idempotent because it verifies the relation

$$P^n \equiv \underbrace{P \circ P \circ ... \circ P}_{n \text{ times}} = P, \ \forall n \in \mathbb{N}^+ \quad . \tag{2.20}$$

When G is not a projector, the filtering can be interpreted as a change of variable, and can be inverted, so there is no loss of information[3] [243]. The kernel of the application is reduced to the null element, i.e. $\ker(G) = \{0\}$.

If the filter is a Reynolds operator, we get

$$G^2 = 1 \quad , \tag{2.21}$$

or, remembering the property of conservation of constants:

$$G = 1 \quad . \tag{2.22}$$

In the spectral space, the idempotency property implies that the transfer function takes the following form:

$$\widehat{G}(\boldsymbol{k}, \omega) = \left\{ \begin{array}{l} 0 \\ 1 \end{array} \right. \qquad \forall \boldsymbol{k}, \ \forall \omega \quad . \tag{2.23}$$

The convolution kernel $\widehat{G}$ therefore takes the form of a sum of Dirac functions and Heaviside functions associated with non-intersecting domains. The conservation of constants implies that $\widehat{G}$ is 1 for the modes that are constant in space and time. The application can no longer be inverted because its kernel $\ker(G) = \{\phi'\}$ is no longer reduced to the zero element; and consequently, the filtering induces an irremediable loss of information.

A filter is said to be positive if:

$$G(\boldsymbol{x}, t) > 0, \forall \boldsymbol{x} \, \text{and} \, \forall t \quad . \tag{2.24}$$

2.1.3 Characterization of Different Approximations

The various methods mentioned in the previous section for reducing the number of degrees of freedom will now be explained. We now assume that the

[3] The reduction of the number of degrees of freedom comes from the fact that the new variable, i.e. the filtered variable, is more regular than the original one in the sense that it contains fewer high frequencies. Its characteristic scale in space is therefore larger, which makes it possible to use a coarser solution to describe it, and therefore fewer degrees of freedom.

The result is a direct numerical simulation of the smoothed variable. As in all numerical simulations, a numerical cutoff is imposed by the use of a finite number of degrees of freedom. But in the case considered here the numerical cutoff is assumed to occur within the dissipative range of the spectrum, so that no active scales are missing.

space-time convolution kernel $G(\boldsymbol{x}-\boldsymbol{\xi}, t-t')$ in $\mathbb{R}^4$ is obtained by tensorizing mono-dimensional kernels:

$$G(\boldsymbol{x}-\boldsymbol{\xi}, t-t') = G(\boldsymbol{x}-\boldsymbol{\xi})G_t(t-t') = G_t(t-t') \prod_{i=1,3} G_i(x_i-\xi_i) \quad . \tag{2.25}$$

The Reynolds time average over a time interval T is found by taking:

$$G_t(t-t') = \frac{\mathcal{H}_T}{T}, \quad G_i(x_i-\xi_i) = \delta(x_i-\xi_i), \quad i=1,2,3 \quad , \tag{2.26}$$

in which δ is a Dirac function and $\mathcal{H}_T$ the Heaviside function corresponding to the interval chosen. This average is extended to the ith direction of space by letting $G_i(x_i-\xi_i) = \mathcal{H}_L/L$, in which L is the desired integration interval.

The phase average corresponding to the frequency ω_c is obtained by letting:

$$\widehat{G}_t(\omega) = \delta(\omega-\omega_c), \quad G_i(x_i-\xi_i) = \delta(x_i-\xi_i), \quad i=1,2,3 \quad . \tag{2.27}$$

In all of the following, the emphasis will be put on the large-eddy simulation technique based on spatial filtering, because it is the most employed approach, with very rare exceptions [160, 161, 603, 107]. This is expressed by:

$$G_t(t-t') = \delta(t-t') \quad . \tag{2.28}$$

Different forms of the kernel $G_i(x_i-\xi_i)$ in common use are described in the following section. It should nonetheless be noted that, when a spatial filtering is imposed, it automatically induces an implicit time filtering, since the dynamics of the Navier–Stokes equations makes it possible to associate a characteristic time with each characteristic length scale. This time scale is evaluated as follows. Let $\overline{\Delta}$ be the cutoff length associated with the filter, and $k_c = \pi/\overline{\Delta}$ the associated wave number. Let $E(k)$ be the energy spectrum of the exact solution (see Appendix A for a definition). The kinetic energy associated with the wave number k_c is $k_c E(k_c)$. The velocity scale v_c associated with this same wave number is estimated as:

$$v_c = \sqrt{k_c E(k_c)} \quad . \tag{2.29}$$

The characteristic time t_c associated with the length $\overline{\Delta}$ is calculated by dimensional arguments as follows:

$$t_c = \overline{\Delta}/v_c \quad . \tag{2.30}$$

The corresponding frequency is $\omega_c = 2\pi/t_c$. The physical analysis shows that, for spectrum forms $E(k)$ considered in the large-eddy simulation framework, v_c is a monotonic decreasing function of k_c (resp. monotonic increasing function of $\overline{\Delta}$), so that ω_c is a monotonic increasing function of k_c (resp. monotonic decreasing function of $\overline{\Delta}$). Suppressing the spatial scales corresponding to wave numbers higher than k_c induces the disappearance of the time frequencies higher than ω_c. We nonetheless assume that the description with a spatial filtering alone is relevant.

Eulerian time-domain filtering for spatial large-eddy simulation is recovered taking

$$G_i(x_i - \xi_i) = \delta(x_i - \xi_i) \quad . \tag{2.31}$$

A reasoning similar to the one given above shows that time filtering induces an implicit spatial filtering.

2.1.4 Differential Filters

A subset of the filters defined in the previous section is the set of *differential filters* [242, 243, 245, 248]. These filters are such that the kernel G is the Green's function associated to an inverse linear differential operator F:

$$\begin{aligned} \phi &= F(G \star \phi) = F(\overline{\phi}) \\ &= \overline{\phi} + \theta \frac{\partial \overline{\phi}}{\partial t} + \overline{\Delta}_l \frac{\partial \overline{\phi}}{\partial x_l} + \overline{\Delta}_{lm} \frac{\partial^2 \overline{\phi}}{\partial x_l \partial x_m} + \ldots \quad , \end{aligned} \tag{2.32}$$

where θ and $\overline{\Delta}_l$ are some time and space scales, respectively. Differential filters can be grouped into several classes: elliptic, parabolic or hyperbolic filters. In the framework of a generalized space-time filtering, Germano [242, 243, 245] recommends using a parabolic or hyperbolic time filter and an elliptic space filter, for reasons of physical consistency with the nature of the Navier–Stokes equations. It is recalled that a filter is said to be elliptic (resp. parabolic or hyperbolic) if F is an elliptic (resp. parabolic, hyperbolic) operator. Examples are given below [248].

Time Low-Pass Filter. A first example is the time low-pass filter. The associated inverse differential relation is:

$$\phi = \overline{\phi} + \theta \frac{\partial \overline{\phi}}{\partial t} \quad . \tag{2.33}$$

The corresponding convolution filter is:

$$\overline{\phi} = \frac{1}{\theta} \int_{-\infty}^{t} \phi(x, t') \exp\left(-\frac{t - t'}{\theta} \right) dt' \quad . \tag{2.34}$$

It is easily seen that this filter commutes with time and space derivatives. This filter is causal, because it incorporates no future information, and therefore is applicable to real-time or post-processing of the data.

Helmholtz Elliptic Filter. An elliptic filter is obtained by taking:

$$\phi = \overline{\phi} - \overline{\Delta}^2 \frac{\partial^2 \overline{\phi}}{\partial x_l^2} \quad . \tag{2.35}$$

It corresponds to a second-order elliptic operator, which depends only on space. The convolutional integral form is:

$$\overline{\phi} = \frac{1}{4\pi \overline{\Delta}^2} \int \frac{\phi(\xi, t)}{|x - \xi|} \exp\left(- \frac{|x - \xi|}{\overline{\Delta}} \right) d\xi \quad . \tag{2.36}$$

This filter satisfies the three previously mentioned basic properties.

Parabolic Filter. A parabolic filter is obtained taking

$$\phi = \overline{\phi} + \theta \frac{\partial \overline{\phi}}{\partial t} - \overline{\Delta}^2 \frac{\partial^2 \overline{\phi}}{\partial x_l^2} \quad , \tag{2.37}$$

yielding

$$\overline{\phi} = \frac{\sqrt{\theta}}{(4\pi)^{3/2} \overline{\Delta}^3} \int_{-\infty}^{t} \int \frac{\phi(\xi, t)}{(t - t')^{3/2}} \exp\left(- \frac{(x - \xi)^2 \theta}{4 \overline{\Delta}^2 (t - t')} - \frac{t - t'}{\theta} \right) d\xi dt' \quad . \tag{2.38}$$

It is easily verified that the parabolic filter satistifies the three required properties.

Convective and Lagrangian Filters. A convective filter is obtained by adding a convective part to the parabolic filter, leading to:

$$\phi = \overline{\phi} + \theta \frac{\partial \overline{\phi}}{\partial t} + \theta V_l \frac{\partial \overline{\phi}}{\partial x_l} - \overline{\Delta}^2 \frac{\partial^2 \overline{\phi}}{\partial x_l^2} \quad , \tag{2.39}$$

where $\boldsymbol{V}$ is an arbitrary velocity field. This filter is linear and constant preserving, but commutes with derivatives if and only if $\boldsymbol{V}$ is uniform. A Lagrangian filter is obtained when $\boldsymbol{V}$ is taken equal to $\boldsymbol{u}$, the velocity field. In this last case, the commutation property is obviously lost.

2.1.5 Three Classical Filters for Large-Eddy Simulation

Three convolution filters are ordinarily used for performing the spatial scale separation. For a cutoff length $\overline{\Delta}$, in the mono-dimensional case, these are written:

– Box or top-hat filter:

$$G(x - \xi) = \begin{cases} \dfrac{1}{\overline{\Delta}} & \text{if } |x - \xi| \leq \dfrac{\overline{\Delta}}{2} \\ 0 & \text{otherwise} \end{cases} \quad , \tag{2.40}$$

$$\widehat{G}(k) = \frac{\sin(k\overline{\Delta}/2)}{k\overline{\Delta}/2} \quad . \tag{2.41}$$

The convolution kernel G and the transfer function $\widehat{G}$ are represented in Figs. 2.1 and 2.2, respectively.

– Gaussian filter:

$$G(x-\xi) = \left(\frac{\gamma}{\pi\overline{\Delta}^2}\right)^{1/2} \exp\left(\frac{-\gamma|x-\xi|^2}{\overline{\Delta}^2}\right) \quad , \tag{2.42}$$

$$\widehat{G}(k) = \exp\left(\frac{-\overline{\Delta}^2 k^2}{4\gamma}\right) \quad , \tag{2.43}$$

in which γ is a constant generally taken to be equal to 6. The convolution kernel G and the transfer function $\widehat{G}$ are represented in Figs. 2.3 and 2.4, respectively.

– Spectral or sharp cutoff filter:

$$G(x-\xi) = \frac{\sin\left(k_c(x-\xi)\right)}{k_c(x-\xi)}, \text{ with } k_c = \frac{\pi}{\overline{\Delta}} \quad , \tag{2.44}$$

$$\widehat{G}(k) = \begin{cases} 1 & \text{if } |k| \leq k_c \\ 0 & \text{otherwise} \end{cases} \quad . \tag{2.45}$$

The convolution kernel G and the transfer function $\widehat{G}$ are represented in Figs. 2.5 and 2.6, respectively.

It is trivially verified that the first two filters are positive while the sharp cutoff filter is not. The top-hat filter is local in the physical space (its support is compact) and non-local in the Fourier space, inversely from the sharp cutoff filter, which is local in the spectral space and non-local in the physical space. As for the Gaussian filter, it is non-local both in the spectral and physical spaces. Of all the filters presented, only the sharp cutoff has the property:

$$\underbrace{\widehat{G}\cdot\widehat{G}...\cdot\widehat{G}}_{n \text{ times}} = \widehat{G}^n = \widehat{G} \quad ,$$

and is therefore idempotent in the spectral space. Lastly, the top-hat and Gaussian filters are said to be smooth because there is a frequency overlap between the quantities $\overline{\boldsymbol{u}}$ and $\boldsymbol{u}'$.

Modification of the exact solution spectrum by the filtering operator is illustrated in figure 2.7.

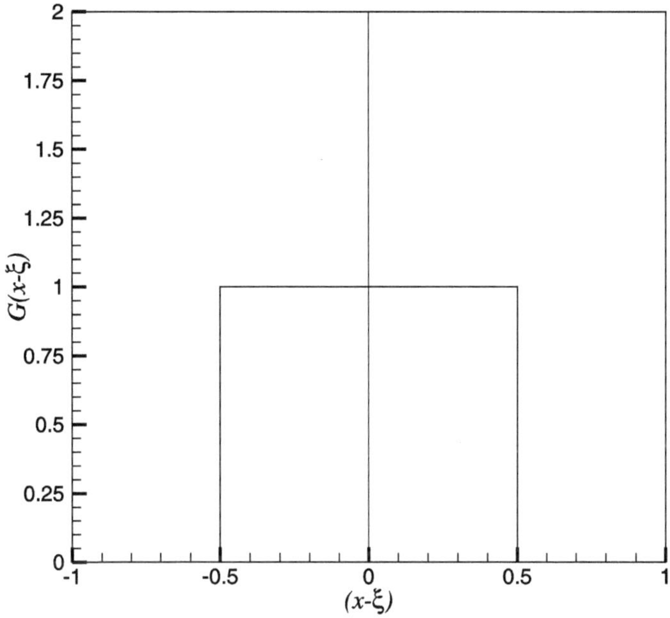

Fig. 2.1. Top-hat filter. Convolution kernel in the physical space normalized by $\overline{\Delta}$.

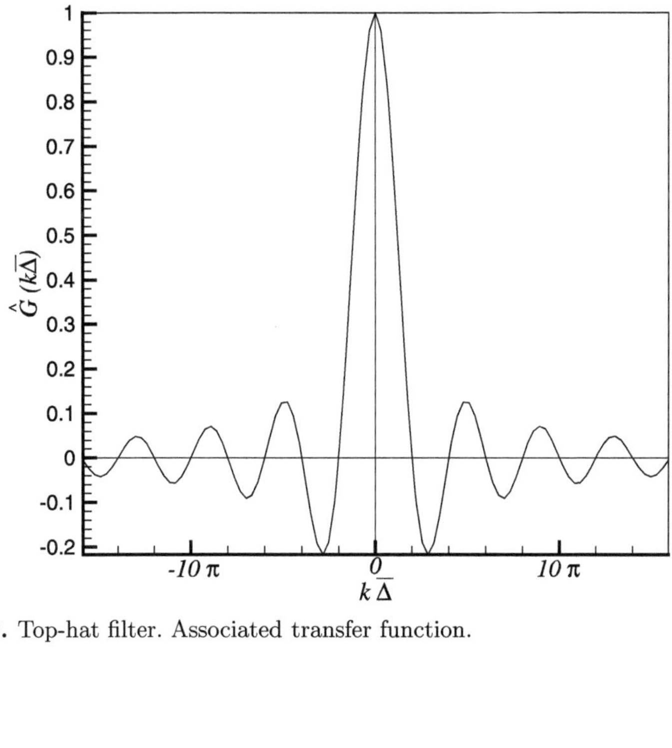

Fig. 2.2. Top-hat filter. Associated transfer function.

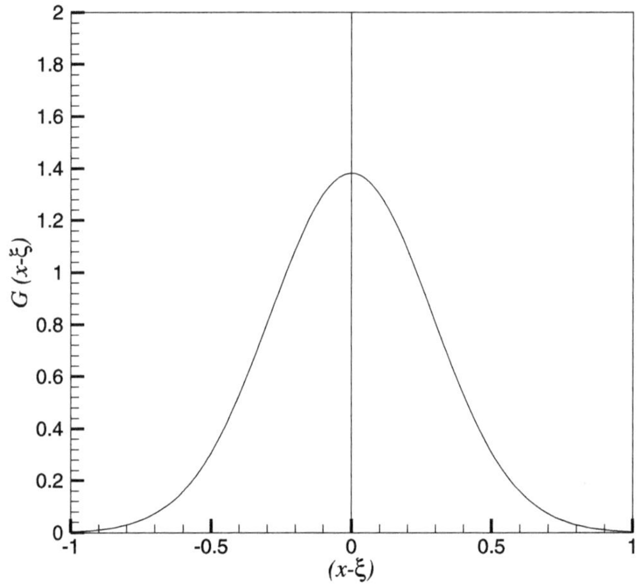

Fig. 2.3. Gaussian filter. Convolution kernel in the physical space normalized by $\overline{\Delta}$.

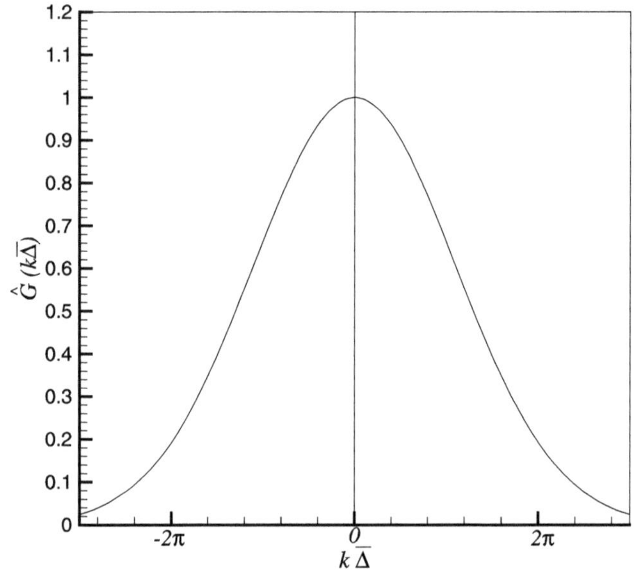

Fig. 2.4. Gaussian filter. Associated transfer function.

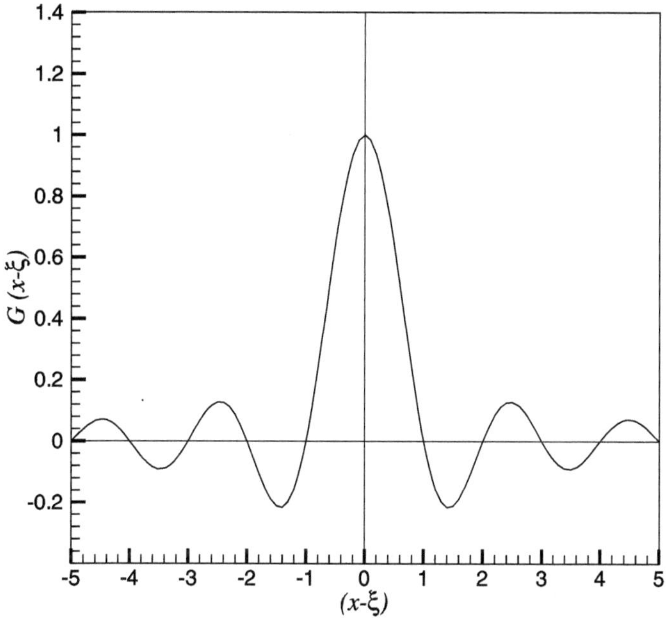

Fig. 2.5. Sharp cutoff filter. Convolution kernel in the physical space.

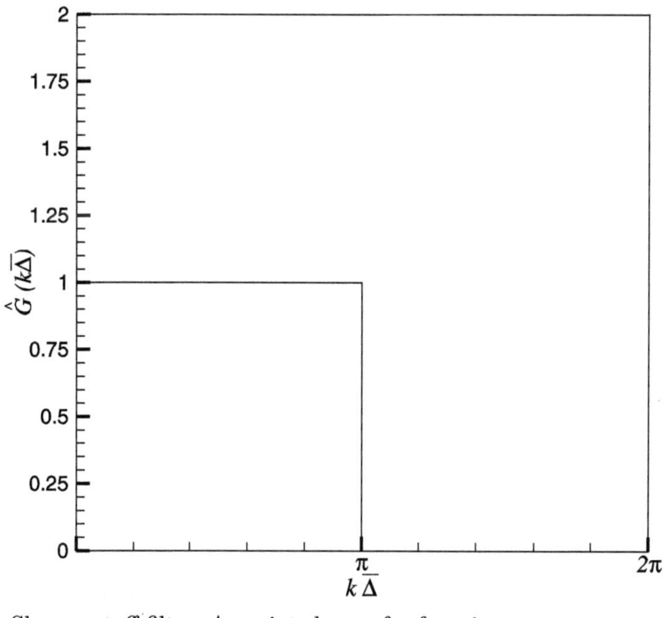

Fig. 2.6. Sharp cutoff filter. Associated transfer function.

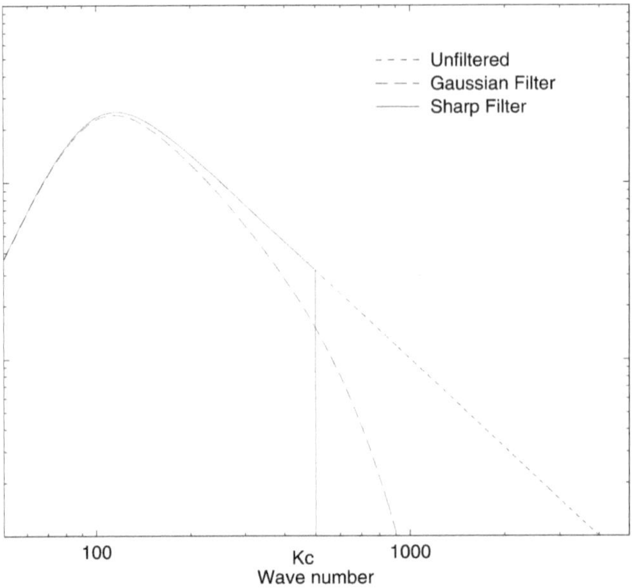

Fig. 2.7. Energy spectrum of the unfiltered and filtered solutions. Filters considered are a projective filter (sharp cutoff filter) and a smooth filter (Gaussian filter) with the same cutoff wave number $k_c = 500$.

2.1.6 Differential Interpretation of the Filters

General results. Convolution filters can be approximated as simple differential operators via a Taylor series expansion, if some additional constraints are fulfilled by the convolution kernel, thus yielding simplified and local filtering operators. Validation of the use of Taylor series expansions in the representation of the filtering operator and conditions for convergence will be discussed in the next section.

We first consider space filtering and recall its definition using a convolution product:

$$\overline{\phi}(x,t) = \int_{-\infty}^{+\infty} \phi(\xi,t) G(x-\xi) d\xi \quad . \tag{2.46}$$

To obtain a differential interpretation of the filter, we perform a Taylor series expansion of the $\phi(\xi,t)$ term at (x,t):

$$\phi(\xi,t) = \phi(x,t) + (\xi - x)\frac{\partial \phi(x,t)}{\partial x} + \frac{1}{2}(\xi - x)^2 \frac{\partial^2 \phi(x,t)}{\partial x^2} + ... \tag{2.47}$$

Introducing this expansion into (2.46), and considering the symmetry and conservation properties of the constants of the kernel G, we get:

$$\begin{aligned}
\overline{\phi}(x,t) &= \phi(x,t)\frac{1}{2}\frac{\partial^2\phi(x,t)}{\partial x^2}\int_{-\infty}^{+\infty} z^2 G(z)dz + ... \\
&+\frac{1}{n!}\frac{\partial^n\phi(x,t)}{\partial x^n}\int z^n G(z)dz + ... \\
&= \phi(x,t) + \sum_{l=1,\infty}\frac{\alpha^{(l)}}{l!}\frac{\partial^l\phi(x,t)}{\partial x^l} \quad ,
\end{aligned} \tag{2.48}$$

where $\alpha^{(l)}$ designates the lth-order moment of the convolution kernel:

$$\alpha^{(l)} = (-1)^l \int_{-\infty}^{+\infty} z^l G(z)dz \quad . \tag{2.49}$$

Assuming that the solution is 2π-periodic, the moments of the convolution kernel can be rewritten as follows [728, 607]:

$$\alpha^{(l)} = \overline{\Delta}^l M_l, \quad M_l = \int_{(\pi-x)/\overline{\Delta}}^{(-\pi-x)/\overline{\Delta}} \xi^l G(\xi)d\xi \quad , \tag{2.50}$$

leading to the following expression for the filtered variable $\overline{\phi}$:

$$\overline{\phi}(x) = \sum_{k=0}^{\infty}\frac{(-1)^k}{k!}\overline{\Delta}^k M_k(x)\frac{\partial^k\phi}{\partial x^k}(x) \quad . \tag{2.51}$$

With this relation, we can interpret the filtering as the application of a differential operator to the primitive variable ϕ. The subgrid part can also be rewritten using the following relation

$$\begin{aligned}
\phi'(x) &= \phi(x) - \overline{\phi}(x) \\
&= -\sum_{l=1,\infty}\frac{\alpha^{(l)}}{l!}\frac{\partial^l\phi(x)}{\partial x^l} \\
&= \sum_{k=1}^{\infty}\frac{(-1)^{k+1}}{k!}\overline{\Delta}^k M_k(x)\frac{\partial^k\phi}{\partial x^k}(x) \quad .
\end{aligned} \tag{2.52}$$

The filtered variable ϕ can also be expanded using derivatives of the transfer function $\hat{G}$ of the filter [607]. Assuming periodicity and differentiability of ϕ, we can write

$$\phi(x) = \sum_{k=-\infty}^{+\infty}\hat{\phi}_k e^{\imath kx} \quad , \quad \imath^2 = -1 \quad , \tag{2.53}$$

and

$$\frac{\partial^l\phi}{\partial x^l}(x) = \sum_{k=-\infty}^{+\infty}(\imath k)^l\hat{\phi}_k e^{\imath kx} \quad . \tag{2.54}$$

The filtered field is expanded as follows:

$$\overline{\phi}(x) = \sum_{k=-\infty}^{+\infty} \hat{G}(k)\hat{\phi}_k e^{\imath kx} \quad . \tag{2.55}$$

The filtered field can be expressed as a Taylor series expansion in the filter width $\overline{\Delta}$:

$$\overline{\phi}(x,\overline{\Delta}) = \overline{\phi}(x,0) + \overline{\Delta}\frac{\partial\overline{\phi}}{\partial\overline{\Delta}}(x,0) + \frac{\overline{\Delta}^2}{2}\frac{\partial^2\overline{\phi}}{\partial\overline{\Delta}^2}(x,0) + ... \quad . \tag{2.56}$$

By differentiating (2.55) with respect to $\overline{\Delta}$, we obtain

$$\frac{\partial^l \phi}{\partial\overline{\Delta}^l}(x,0) = l! a_l \frac{\partial^l \phi}{\partial x^l}(x) \quad , \tag{2.57}$$

with

$$a_l = \frac{1}{\imath^l l!}\frac{\partial^l \hat{G}}{\partial k^l}(0) \quad . \tag{2.58}$$

The resulting final expression of the filtered field is

$$\overline{\phi}(x) = \sum_{l=0}^{\infty}\sum_{k=-\infty}^{+\infty} \frac{(k\overline{\Delta})^l}{l!}\frac{\partial^l \hat{G}}{\partial k^l}(0)\hat{\phi}_k e^{\imath kx} \quad . \tag{2.59}$$

Time-domain filters defined as a convolution product can be expanded in an exactly similar way, yielding

$$\begin{aligned} \overline{\phi}(x,t) &= \int_{-\infty}^{+\infty} \phi(x,t')G(t-t')dt' \\ &= \phi(x,t) + \sum_{l=1,\infty} \frac{\alpha^{(l)}}{l!}\frac{\partial^l \phi(x,t)}{\partial t^l} \quad , \end{aligned} \tag{2.60}$$

The values of the first moments of the box and Gaussian filters are given in Table 2.1. It can be checked that the sharp cutoff filter leads to a divergent series, because of its non-localness.

For these two filters, we have the estimate

$$\alpha^{(n)} = O(\overline{\Delta}^n) \tag{2.61}$$

for space-domain filtering, and

$$\alpha^{(n)} = O(\overline{\tau_c}^n) \tag{2.62}$$

for time-domain filtering.

Table 2.1. Values of the first five non-zero moments for the box and Gaussian filters.

$\alpha^{(n)}$	$n=0$	$n=2$	$n=4$	$n=6$	$n=8$
box	1	$\overline{\Delta}^2/12$	$\overline{\Delta}^4/80$	$\overline{\Delta}^6/448$	$\overline{\Delta}^8/2304$
Gaussian	1	$\overline{\Delta}^2/12$	$\overline{\Delta}^4/48$	$5\overline{\Delta}^6/576$	$35\overline{\Delta}^8/6912$

For a general space–time filter, neglecting cross-derivatives of the kernel, this Taylor series expansion gives [160, 161]:

$$\begin{aligned}\overline{\phi}(x,t) &= \int_{-\infty}^{+\infty} \phi(\xi,t')G(x-\xi,t-t')d\xi dt' \\ &= \phi(x,t) + \sum_{l=1,\infty} \frac{\alpha_x^{(l)}}{l!}\frac{\partial^l \phi(x,t)}{\partial x^l} + \sum_{l=1,\infty} \frac{\alpha_t^{(l)}}{l!}\frac{\partial^l \phi(x,t)}{\partial t^l}\end{aligned} \quad , \tag{2.63}$$

with

$$\alpha_x^{(l)} = \int_{-\infty}^{+\infty}\int_{-\infty}^{+\infty} (\xi - x)^l G(x-\xi,t-t')d\xi dt' \quad , \tag{2.64}$$

and

$$\alpha_t^{(l)} = \int_{-\infty}^{+\infty}\int_{-\infty}^{+\infty} (t' - t)^l G(x-\xi,t-t')d\xi dt' \quad . \tag{2.65}$$

Conditions for Convergence of the Taylor Series Expansions. A first analysis of the convergence properties of the Taylor series expansions discussed above was provided by Vasilyev et al. [728], and is given below. Assuming that the periodic one-dimensional field ϕ does not contain wave numbers higher than $k_{\max}$, one can write the following Fourier integral:

$$\phi(x) = \int_{-k_{\max}}^{k_{\max}} \hat{\phi}(k)e^{-\imath kx}dk \quad , \tag{2.66}$$

where time-dependence has been omitted for the sake of simplicity. The total energy of ϕ, E_ϕ, is equal to

$$E_\phi = \int_{-k_{\max}}^{k_{\max}} |\hat{\phi}(k)|^2 dk \quad . \tag{2.67}$$

The mth derivative of ϕ can be written as

$$\frac{\partial^m \phi}{\partial x^m}(x) = (-\imath)^m \int_{-k_{\max}}^{k_{\max}} k^m \hat{\phi}(k)e^{-\imath kx}dk \quad . \tag{2.68}$$

From this expression, we get the following bounds for the derivative:

$$\begin{aligned}
\left|\frac{\partial^m \phi}{\partial x^m}\right| &\leq \int_{-k_{\max}}^{k_{\max}} |k|^{2m} |\hat{\phi}(k)|^2 dk \\
&\leq \left(\int_{-k_{\max}}^{k_{\max}} |k|^m dk\right)^{1/2} \left(\int_{-k_{\max}}^{k_{\max}} |\hat{\phi}(k)| dk\right)^{1/2} \\
&\leq \sqrt{\frac{2E_\phi k_{\max}}{2m+1}} k_{\max}^m \quad .
\end{aligned} \tag{2.69}$$

From relations (2.69) and (2.51) we obtain the following inequalities:

$$\begin{aligned}
\left|\sum_{l=0}^{\infty} \frac{(-1)^l}{l!} \overline{\Delta}^l M_l(x) \frac{\partial^l \phi}{\partial x^l}(x)\right| &\leq \sum_{l=0}^{\infty} \frac{1}{l!} \overline{\Delta}^l |M_l(x)| \left|\frac{\partial^l \phi}{\partial x^l}(x)\right| \\
&\leq \sqrt{2E_\phi k_{\max}} \sum_{l=0}^{\infty} \frac{\left(k_{\max}\overline{\Delta}\right)^l |M_l(x)|}{l!\sqrt{2l+1}} \; .
\end{aligned} \tag{2.70}$$

From this last inequality, it can easily be seen that the series (2.51) converges for any value of $\overline{\Delta}$ if the following constraint is satisfied:

$$\lim_{l \longrightarrow \infty} \frac{|M_{l+1}(x)|}{|M_l(x)|(l+1)} = 0 \quad . \tag{2.71}$$

For filters with compact support, the following criterion holds:

$$\lim_{l \longrightarrow \infty} \frac{(k_{\max}\overline{\Delta})|M_{l+1}(x)|}{|M_l(x)|(l+1)} < 1 \quad . \tag{2.72}$$

For symmetric filters, the analogous criterion is

$$\lim_{l \longrightarrow \infty} \frac{(k_{\max}\overline{\Delta})^2|M_{2l+2}(x)|}{|M_{2l}(x)|(2l+2)(2l+1)} < 1 \quad . \tag{2.73}$$

Pruett et al. [607] proved that all symmetric, non-negative[4] filters satisfy relation (2.73). This proof is now presented. If the filter is non-negative, following an integral mean value theorem, there exists a value c, $-2\pi \leq -\pi - x \leq c \leq \pi - x \leq 2\pi$, such that

$$\begin{aligned}
|M_{2l+2}| &= \left(\frac{|c|}{\overline{\Delta}}\right)^2 \left|\int_{(\pi-x)/\overline{\Delta}}^{(-\pi-x)/\overline{\Delta}} \xi^{2l} G(\xi) d\xi\right| \\
&= \left(\frac{|c|}{\overline{\Delta}}\right)^2 |M_{2l}| \quad ,
\end{aligned} \tag{2.74}$$

[4] It is recalled that the filter is said to be non-negative if $G(x) \geq 0, \forall x$.

whereby

$$\frac{|M_{2l+2}|}{|M_{2l}|} = \left(\frac{|c|}{\overline{\Delta}}\right)^2 \leq \left(\frac{2\pi}{\overline{\Delta}}\right)^2 \quad . \tag{2.75}$$

The expected result is trivially deduced from this last relation.

2.2 Spatial Filtering: Extension to the Inhomogeneous Case

2.2.1 General

In the above explanations, it was assumed that the filter is homogeneous and isotropic. These assumptions are at time too restrictive for the resulting conclusions to be usable. For example, the definition of bounded fluid domains forbids the use of filters that are non-local in space, since these would no longer be defined. The problem then arises of defining filters near the domain boundaries. At the same time, there may be some advantage in varying the filter cutoff length to adapt the structure of the solution better and thereby ensure optimum gain in terms of reducing the number of degrees of freedom in the system to be resolved.

From relation (2.1), we get the following general form of the commutation error for a convolution filter $G(y, \overline{\Delta}(x,t))$ on a domain Ω [230, 260]:

$$\left[\frac{\partial}{\partial x}, G\star\right]\phi = \frac{\partial}{\partial x}(G \star \phi) - G \star \frac{\partial \phi}{\partial x} \quad . \tag{2.76}$$

The first term of the right-hand side of (2.76) can be expanded as

$$\frac{\partial}{\partial x}(G \star \phi) = \frac{\partial}{\partial x}\int_\Omega G(x-\xi, \overline{\Delta}(x,t))\phi(\xi,t)d\xi \tag{2.77}$$

$$= \left(\frac{\partial G}{\partial \overline{\Delta}} \star \phi\right)\frac{\partial \overline{\Delta}}{\partial x} + \int_{\partial\Omega} G(x-\xi, \overline{\Delta}(x,t))\phi(\xi,t)n(\xi)ds + G \star \frac{\partial \phi}{\partial x} \quad , \tag{2.78}$$

where $n(\xi)$ is the outward unit normal vector to the boundary of Ω, $\partial\Omega$, yielding

$$\left[\frac{\partial}{\partial x}, G\star\right]\phi = \left(\frac{\partial G}{\partial \overline{\Delta}} \star \phi\right)\frac{\partial \overline{\Delta}}{\partial x} + \int_{\partial\Omega} G(x-\xi, \overline{\Delta}(x,t))\phi(\xi,t)n(\xi)ds. \tag{2.79}$$

The first term appearing in the right-hand side of relation (2.79) is due to spatial variation of the filtering length, while a domain boundary generates the second one. A similar development leads to:

$$\left[\frac{\partial}{\partial t}, G\star\right]\phi = \left(\frac{\partial G}{\partial \overline{\Delta}} \star \phi\right)\frac{\partial \overline{\Delta}}{\partial t} \quad . \tag{2.80}$$

These error terms must be eliminated, or bounded, in order to be able to define a controlled and consistent filtering process for large-eddy simulation. This is done by deriving new filtering operators. Several alternatives to the classical convolution products have been proposed, which are described in the following.

Franke and Frank [225] propose an extension of (2.79) to the case of a domain with *moving boundaries* and a uniform time depdendent filter length, i.e. $\partial\Omega = \partial\Omega(t)$ and $\overline{\Delta} = \overline{\Delta}(t)$. Limiting the analysis to the one dimensional-case for the sake of clarity, and taking $\Omega(t) = [a(t), b(t)]$, one first notices that the constraint dealing with the preservation of the constant

$$\int_{a(t)}^{b(t)} G(x-\xi, \overline{\Delta}(t))d\xi = 1 \quad , \tag{2.81}$$

yields the following *filter conservation law*

$$\int_{a(t)}^{b(t)} \frac{\partial G}{\partial \overline{\Delta}}\frac{d\overline{\Delta}(t)}{dt}d\xi = -\left(G(x-b(t), \overline{\Delta}(t))\frac{db(t)}{dt} - G(x-a(t), \overline{\Delta}(t))\frac{da(t)}{dt}\right) . \tag{2.82}$$

The commutation errors have the following forms:

$$\left[\frac{\partial}{\partial x}, G\star\right]\phi(x,t) = G(x-a(t), \overline{\Delta}(t))\phi(a(t),t) - G(x-b(t), \overline{\Delta}(t))\phi(b(t),t) \quad , \tag{2.83}$$

$$\begin{aligned}\left[\frac{\partial}{\partial t}, G\star\right]\phi(x,t) &= -\left[G(x-\xi, \overline{\Delta}(t))\phi(\xi,t)\frac{d\xi}{dt}\right]_{\xi=a(t)}^{\xi=b(t)} \\ &\quad + \int_{a(t)}^{b(t)} \phi(\xi,t)\frac{\partial G}{\partial \overline{\Delta}}\frac{d\overline{\Delta}(t)}{dt}d\xi \quad .\end{aligned} \tag{2.84}$$

2.2.2 Non-uniform Filtering Over an Arbitrary Domain

This section presents the findings concerning the extension of the filtering to the case where the filter cutoff length varies in space and where the domain over which it applies is bounded or infinite.

New Definition of Filters and Properties: Mono-dimensional Case. Alternative proposals in the homogeneous case. Ghosal and Moin [262] propose to define the filtering of a variable $\phi(\xi)$, defined over the interval

$]-\infty,+\infty[$, as

$$\overline{\phi}(\xi) = G \star \phi = \frac{1}{\overline{\Delta}} \int_{-\infty}^{+\infty} G\left(\frac{\xi - \eta}{\overline{\Delta}}\right) \phi(\eta) d\eta \quad , \tag{2.85}$$

in which the cutoff length $\overline{\Delta}$ is constant. The convolution kernel G is made to verify the following four properties:

1. Symmetry
$$G(-\xi) = G(\xi) \quad . \tag{2.86}$$
We note that this property was not explicitly required before, but that it is verified by the three filters described in Sect. 2.1.5.
2. Conservation of constants
$$\int_{-\infty}^{+\infty} a\ G(\xi) d\xi, \quad a = \text{const.} \tag{2.87}$$
3. Fast decay. $G(\xi) \rightarrow 0$ as $|\xi| \rightarrow \infty$ fast enough for all of its moments to be finite, i.e.
$$\int_{-\infty}^{+\infty} G(\xi)\xi^n d\xi < \infty, \quad \forall n \geq 0 \quad . \tag{2.88}$$
4. Quasi-local in physical space. $G(\xi)$ is localized (in a sense to be specified) in the interval $[-1/2, 1/2]$.

Extension of the Top-Hat Filter to the Inhomogeneous Case: Properties. Considering definition (2.85), the top-hat filter (2.40) is expressed:

$$G(\xi) = \begin{cases} 1 & \text{if } |\xi| \leq \frac{1}{2} \\ 0 & \text{otherwise} \end{cases} \quad . \tag{2.89}$$

There are a number of ways of extending this filter to the inhomogeneous case. The problem posed is strictly analogous to that of extending finite volume type schemes to the case of inhomogeneous structured grids: the control volumes can be defined directly on the computational grid or in a reference space carrying a uniform grid, after a change of variable. Two extensions of the box filter are discussed in the following, each based on a different approach.

Direct extension. If the cutoff length varies in space, one solution is to say:

$$\overline{\phi}(\xi) = \frac{1}{\Delta_+(\xi) + \Delta_-(\xi)} \int_{\xi - \Delta_-(\xi)}^{\xi + \Delta_+(\xi)} \phi(\eta) d\eta \quad , \tag{2.90}$$

in which $\Delta_+(\xi)$ and $\Delta_-(\xi)$ are positive functions and $(\Delta_+(\xi) + \Delta_-(\xi))$ is the cutoff length at point ξ. These different quantities are represented in Fig. 2.8.

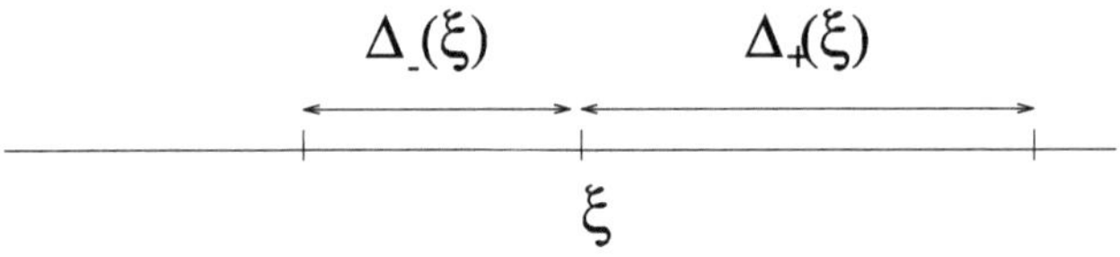

Fig. 2.8. Direct extension of the top-hat filter. Representation of the integration cell at point ξ.

If the domain is finite or semi- infinite, the functions $\Delta_+(\xi)$ and $\Delta_-(\xi)$ must decrease fast enough near the domain boundaries for the integration interval $[\xi-\Delta_-(\xi), \xi+\Delta_+(\xi)]$ to remain defined. The box filter is extended intuitively here, as an average over the control cell $[\xi-\Delta_-(\xi), \xi+\Delta_+(\xi)]$. This approach is similar to the finite volume techniques based on control volumes defined directly on the computational grid.

It is shown that this expression does not ensure the commutation property with derivation in space. Relation (2.12) becomes (the dependency of the functions Δ_+ and Δ_- as a function of ξ is not explicitly state, to streamline the notation):

$$\begin{aligned}\left[G\star, \frac{d}{d\xi}\right]\phi &= \frac{(d/d\xi)\,(\Delta_- + \Delta_+)}{\Delta_- + \Delta_+}\overline{\phi} \\ &- \frac{1}{\Delta_+ + \Delta_-}\left[\phi(\xi+\Delta_+)\frac{d\Delta_+}{d\xi} + \phi(\xi-\Delta_-)\frac{d\Delta_-}{d\xi}\right] \quad .\end{aligned} \tag{2.91}$$

The amplitude of the error committed cannot be evaluated a priori, and thus cannot be neglected. Also, when (2.90) is applied to the Navier–Stokes equations, all the terms, including the linear ones, will introduce unknown terms that will require a closure.

Extension by Variable Change. SOCF. To remedy this problem, a more general alternative description than relation (2.90) is proposed by Ghosal and Moin [262]. This new definition consists of defining filters that commute at the second order with the derivation in space (Second Order Commuting Filter, or SOCF). This is based on a change of variable that allows the use of a homogeneous filter. The function ϕ is assumed to be defined over a finite or infinite interval $[a, b]$. Any regular monotonic function defined over this interval can be related to a definite function over the interval $[-\infty, +\infty]$ by performing the variable change:

$$\xi = f(x) \quad , \tag{2.92}$$

in which f is a strictly monotonic differentiable function such that:

$$f(a) = -\infty, \quad f(b) = +\infty \quad . \tag{2.93}$$

The constant cutoff length $\overline{\Delta}$ defined over the reference space $[-\infty, +\infty]$ is associated with the variable cutoff length $\overline{\delta}(x)$ over the starting interval by the relation:

$$\overline{\delta}(x) = \frac{\overline{\Delta}}{f'(x)} \quad . \tag{2.94}$$

In the case of a finite or semi-infinite domain, the function f' takes infinite values at the bounds and the convolution kernel becomes a Dirac function. The filtering of a function $\psi(x)$ is defined in the inhomogeneous case in three steps:

1. We perform the variable change $x = f^{-1}(\xi)$, which leads to the definition of the function $\phi(\xi) = \psi(f^{-1}(\xi))$.
2. The function $\phi(\xi)$ is then filtered by the usual homogeneous filter (2.85):

$$\overline{\psi}(x) \equiv \overline{\phi}(\xi) = \frac{1}{\overline{\Delta}} \int_{-\infty}^{+\infty} G\left(\frac{f(x) - \eta}{\overline{\Delta}}\right) \phi(\eta) d\eta \quad . \tag{2.95}$$

3. The filtered quantity is then re-expressed in the original space:

$$\overline{\psi}(x) = \frac{1}{\overline{\Delta}} \int_a^b G\left(\frac{f(x) - f(y)}{\overline{\Delta}}\right) \psi(y) f'(y) dy \quad . \tag{2.96}$$

This new expression of the filter modifies the commutation error with the spatial derivation. Using (2.95) and integrating by parts, we get:

$$\begin{aligned} \frac{d\overline{\psi}}{dx} &= -\frac{f'(x)}{\overline{\Delta}} \left[G\left(\frac{f(x) - f(y)}{\overline{\Delta}}\right) \psi(y) \right]_{y=a}^{y=b} \\ &+ \frac{1}{\overline{\Delta}} \int_a^b G\left(\frac{f(x) - f(y)}{\overline{\Delta}}\right) f'(x) \psi'(y) dy \quad . \end{aligned} \tag{2.97}$$

The fast decay property of the kernel G makes it possible to cancel the first term of the rigth-hand side. The commutation error is:

$$\begin{aligned} \left[G\star, \frac{d}{d\xi}\right] \phi &= \frac{1}{\overline{\Delta}} \int_a^b G\left(\frac{f(x) - f(y)}{\overline{\Delta}}\right) f'(y) \psi'(y) \\ &\times \left[1 - \frac{f'(x)}{f'(y)}\right] dy \quad . \end{aligned} \tag{2.98}$$

In order to simplify this expression, we introduce a new variable ζ such that:

$$f(y) = f(x) + \overline{\Delta} \zeta \quad . \tag{2.99}$$

The variable y is then re-expressed as a series as a function of $\overline{\Delta}$:

$$y = y_0(\zeta) + \overline{\Delta} y_1(\zeta) + \overline{\Delta}^2 y_2(\zeta) + ... \tag{2.100}$$

Then, combining relations (2.99) and (2.100), we get (the dependence of the functions according to the variable x is not explicitly shown, to streamline the notation):

$$y = x + \frac{\overline{\Delta}\zeta}{f'} - \frac{\overline{\Delta}^2 f''\zeta}{2f'^3} + \ldots \quad , \tag{2.101}$$

which allows us to re-write relation (2.98) as:

$$\begin{aligned} \left[G\star, \frac{d}{d\xi}\right]\phi &= \int_{-\infty}^{+\infty} G(\zeta)\psi'(y(\zeta)) \left[1 - \frac{f'(x)}{f'(y(\zeta))}\right] d\zeta & (2.102) \\ &= C_1\overline{\Delta} + C_2\overline{\Delta}^2 + \ldots \quad , & (2.103) \end{aligned}$$

in which the coefficients C_1 and C_2 are expressed as:

$$\begin{aligned} C_1 &= \frac{f''\psi'}{f'^2} \int_{-\infty}^{+\infty} \zeta G(\zeta) d\zeta \quad , & (2.104) \\ C_2 &= \frac{2f'f''\psi'' + f'f'''\psi' - 3f''^2\psi'}{2f'^4} \int_{-\infty}^{+\infty} \zeta^2 G(\zeta) d\zeta \quad . & (2.105) \end{aligned}$$

The symmetry property of the kernel G implies $C_1 = 0$, which ensures that the filter commutation error with the spatial derivation is of the second order as a function of the cutoff length $\overline{\Delta}$. The authors call this Second-Order Commuting Filter (SOCF).

A study of the spectral distribution of the commutation error is available in reference [262]. Rather than detailing this analysis here, only the major results will be explained. Considering a function of the form:

$$\psi(x) = \widehat{\psi}_k e^{\imath kx}, \quad \imath^2 = -1 \quad , \tag{2.106}$$

the two derivation operations are written:

$$\overline{\frac{d\psi}{dx}} = \overline{\imath k\psi}, \quad \frac{d\overline{\psi}}{dx} = \imath k\overline{\psi} \quad . \tag{2.107}$$

The commutation error can be measured by comparing the two wave numbers k and k', the latter being such that $\overline{\imath k\psi} = \imath k'\overline{\psi}$. The commutation error is zero if $k = k'$. Algebraic manipulations lead to the relation:

$$\frac{k'}{k} = 1 - \imath\overline{\Delta}\frac{f''}{f'^2} \frac{\displaystyle\int_{-\infty}^{+\infty} \zeta G(\zeta) \sin(k\overline{\Delta}\zeta/f') d\zeta}{\displaystyle\int_{-\infty}^{+\infty} G(\zeta) \cos(k\overline{\Delta}\zeta/f') d\zeta} \quad . \tag{2.108}$$

Using the modal decomposition (2.106), the commutation error can be expressed in differential form. The calculations lead to:

$$\left[G\star,\frac{d}{dx}\right]\psi = \alpha^{(2)}\frac{f''}{f'^3}\overline{\Delta}^2\frac{d^2\overline{\psi}}{dx^2}+O(k\overline{\Delta})^4 \quad (2.109)$$

$$= -\alpha^{(2)}\overline{\delta}^2\left(\frac{\overline{\delta}'}{\overline{\delta}}\right)\frac{d^2\overline{\psi}}{dx^2}+O(k\overline{\delta})^4 \quad , \quad (2.110)$$

in which $\overline{\delta}(x)$ is the local cutoff length and $\alpha^{(2)}$ the second-order moment of G, i.e.

$$\alpha^{(2)}=\int_{-\infty}^{+\infty}\zeta^2 G(\zeta)d\zeta \quad . \quad (2.111)$$

Van der Ven's Filters. Commuting filters can be defined with the spatial derivation at an order higher than 2, at least in the case of an infinite domain. To obtain such filters, Van der Ven [725] proposes defining the filtering for the case of a variable cutoff length $\overline{\delta}(x)$ by direct extension of the form (2.85):

$$\overline{\phi}(x)=\frac{1}{\overline{\delta}(x)}\int_{-\infty}^{+\infty}G\left(\frac{x-y}{\overline{\delta}(x)}\right)\phi(y)dy \quad . \quad (2.112)$$

The function G is assumed here to be class C^1, symmetrical, and must conserve the constants. Also, the function $\overline{\delta}(x)$ is also assumed to be class C^1. This definition is achieved by linearizing the general formula (2.96) around x, that is by letting $\phi'(y)=\phi'(x)$ and $\phi(x)-\phi(y)=\phi'(x)(x-y)$ and including relation (2.94). This linearization operation is equivalent to considering that the function ϕ is linear in a neighbourhood of x containing the support of the convolution kernel. By introducing the variable change $y=x-\zeta\overline{\delta}(x)$, the corresponding commutation error is expressed:

$$\left[G\star,\frac{d}{dx}\right]\phi=\frac{\overline{\delta}'}{\overline{\delta}}\int\left(G(\zeta)+\zeta G'(\zeta)\right)\phi(x-\zeta\delta(x))d\zeta \quad . \quad (2.113)$$

To increase the order of the commutation error, we look for functions G that are solutions to the equation

$$G+\zeta G'=a\ G^{(n)}, \quad n>1 \quad , \quad (2.114)$$

in which a is a real and $G^{(n)}$ designates the nth derivative of the kernel G. For such functions, the commutation error becomes:

$$\left[G\star,\frac{d}{dx}\right]\phi = a\frac{\overline{\delta}'}{\overline{\delta}}(-1)^n\int G(\zeta)\left(\frac{\partial}{\partial\zeta}\right)^n\phi(x-\zeta\overline{\delta}(x))d\zeta \quad (2.115)$$

$$= a\overline{\delta}'(x)\overline{\delta}(x)^{n-1}\overline{\phi^{(n)}}(x) \quad , \quad (2.116)$$

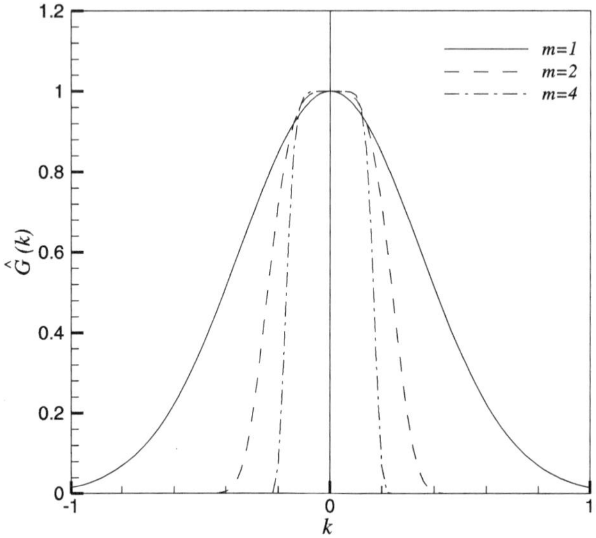

Fig. 2.9. High-order commuting filter. Graph of the associated transfer function for different values of the parameter m.

and is thus formally of order $n-1$. Simple analysis shows that the Fourier transform $\widehat{G}$ of the solution to problem (2.114) verifying the constant conservation property is of the form:

$$\widehat{G}(k) = \exp\left(\frac{-a\imath^n}{n} k^n\right) \quad . \tag{2.117}$$

The symmetry property of G implies that $n = 2m$ is even, and therefore:

$$\widehat{G}(k) = \exp\left(\frac{-a(-1)^m}{2m} k^{2m}\right) \quad . \tag{2.118}$$

The fast decay property is recovered for $a = b(-1)^m, b > 0$. It can be seen that the Gaussian filter then occurs again by letting $m = 1$. It is important to note that this analysis is valid only for infinite domains, because when the bounds of the fluid domain are included they bring out additional error terms with which it is no longer possible to be sure of the order of the commutation error. The transfer function obtained for various values of the parameter m is represented in Fig. 2.9.

High-Order Commuting Filters. Van der Ven's analysis has been generalized by Vasilyev et al. [728] so as to contain previous works (SOCF and Van der Ven's filters) as special cases. As for SOCF, the filtering process is defined thanks to the use of a reference space. We now consider that the physical domain $[a, b]$ is mapped into the domain $[\alpha, \beta]$. Ghosal and Moin used $\alpha = -\infty$ and $\beta = +\infty$. The correspondances between the two domains are summarized in Table 2.2.

Table 2.2. Correspondances for Vasilyev's high-order commuting filters.

Domain	$[a,b]$	$[\alpha,\beta]$
Coordinate	$x = f^{-1}(\xi)$	$\xi = f(x)$
Filter length	$\overline{\delta}(x) = \overline{\Delta}/f'(x)$	$\overline{\Delta}$
Function	$\psi(x)$	$\phi(\xi) = \psi(f^{-1}(\xi))$

Considering this new mapping, relation (2.95) is transformed as

$$\overline{\phi}(\xi) = \frac{1}{\overline{\Delta}} \int_{\beta}^{\alpha} G\left(\frac{\xi-\eta}{\overline{\Delta}}\right) \phi(\eta) d\eta \quad , \tag{2.119}$$

and using the change of variables (2.99), we get

$$\overline{\phi}(\xi) = \int_{\frac{\xi-\beta}{\overline{\Delta}}}^{\frac{\xi-\alpha}{\overline{\Delta}}} G(\zeta)\, \phi(\xi - \overline{\Delta}\zeta) d\zeta \quad . \tag{2.120}$$

The next step consists in performing a Taylor expansion of $\phi(\xi - \overline{\Delta}\zeta)$ in powers of $\overline{\Delta}$:

$$\phi(\xi - \overline{\Delta}\zeta) = \sum_{k=0,+\infty} \frac{(-1)^k}{k!} \overline{\Delta}^k \zeta^k \frac{\partial^k \phi}{\partial \xi^k}(\xi) \quad . \tag{2.121}$$

Substituting (2.121) into (2.120), we get

$$\overline{\phi}(\xi) = \sum_{k=0,+\infty} \frac{(-1)^k}{k!} \overline{\Delta}^k \alpha^{(k)}(\xi) \frac{\partial^k \phi}{\partial \xi^k}(\xi) \quad , \tag{2.122}$$

where the kth moment of the filter kernel is now defined as

$$\alpha^{(k)}(\xi) = \int_{\frac{\xi-\beta}{\overline{\Delta}}}^{\frac{\xi-\alpha}{\overline{\Delta}}} G(\zeta)\, \zeta^k d\zeta \quad . \tag{2.123}$$

Using the relation (2.122), the space derivative of the filtered variable expressed in the physical space can be evaluated as follows:

$$\frac{d\overline{\psi}}{dx}(x) = f'(x) \frac{d\overline{\phi}}{d\xi}(\xi) \tag{2.124}$$

$$= f'(x) \sum_{k=0,+\infty} \frac{(-1)^k}{k!} \overline{\Delta}^k \left(\frac{d\alpha^{(k)}}{d\xi}(\xi) \frac{\partial^k \phi}{\partial \xi^k}(\xi) + \alpha^{(k)}(\xi) \frac{\partial^{k+1} \phi}{\partial \xi^{k+1}}(\xi) \right) . \tag{2.125}$$

A similar procedure is used to evaluate the second part of the commutation error. Using (2.124) and the same change of variables, we get:

$$\overline{\frac{d\psi}{dx}}(x) = \frac{1}{\overline{\Delta}} \int_{\alpha}^{\beta} G\left(\frac{\xi - \eta}{\overline{\Delta}}\right) \frac{d\phi}{d\eta}(\eta) f'(f^{-1}(\eta)) d\eta \quad , \tag{2.126}$$

with

$$f'(f^{-1}(\eta)) = \sum_{l=1,+\infty} \frac{1}{(l-1)!} \left(\sum_{k=1,+\infty} \frac{(-1)^k}{k!} \overline{\Delta}^k \zeta^k \frac{\partial^k f^{-1}}{\partial \xi^k}(\xi) \right)^{l-1} \frac{\partial^l f}{\partial x^l}(x) \quad , \tag{2.127}$$

and

$$\frac{d\phi}{d\eta}(\eta) = \sum_{k=0,+\infty} \frac{(-1)^k}{k!} \overline{\Delta}^k \zeta^k \frac{\partial^{k+1} \phi}{\partial \xi^{k+1}}(\xi) \quad . \tag{2.128}$$

Making the assumptions that all the Taylor expansion series are convergent[5], the commutation error in the physical space is equal to

$$\left[G\star, \frac{d}{dx}\right]\psi = \sum_{k=1,+\infty} A_k \alpha^{(k)}(\xi) \overline{\Delta}^k + \sum_{k=0,+\infty} B_k \frac{d\alpha^{(k)}}{d\xi}(\xi) \overline{\Delta}^k \quad , \tag{2.129}$$

where A_k and B_k are real non-zero coefficients. It is easily seen from relation (2.129) that the commutation error is determined by the filter moments and the mapping function. The order of the commutation error can then be governed by chosing an adequate filter kernel. Vasilyev proposes to use a function G such that:

$$\alpha^{(0)} = 1 \quad \forall \xi \in [\alpha, \beta] \quad , \tag{2.130}$$

$$\alpha^{(k)} = 0 \quad 1 \leq k \leq n-1, \quad \forall \xi \in [\alpha, \beta] \quad , \tag{2.131}$$

$$\alpha^{(k)} < \infty \quad k \geq n, \quad \forall \xi \in [\alpha, \beta] \quad . \tag{2.132}$$

An immediate consequence is

$$\frac{d\alpha^{(k)}}{d\xi}(\xi) = 0, \quad 0 \leq k \leq n-1, \quad \forall \xi \in [\alpha, \beta] \quad , \tag{2.133}$$

leading to

$$\left[G\star, \frac{d}{dx}\right]\psi = O(\overline{\Delta}^n) \quad . \tag{2.134}$$

The commutation error can be controlled by choosing a kernel G with desired moment values. It is important noting that conditions (2.130) – (2.132) do not require that the filter kernel be symmetric. Discrete filters verifying theses properties will be discussed in Sect. 13.2.

[5] Vasilyev et al. [728] show that this is always true for practical numerical simulations.

Extension to the Multidimensional Case.

SOCF Filters. SOCF filters are extensible to the three-dimensional case for finite or infinite domains. Let (x_1, x_2, x_3) be a Cartesian system, and (X_1, X_2, X_3) the reference axis system associated with a uniform isotropic grid with a mesh size $\overline{\Delta}$. The two systems are related by the relations:

$$X_1 = H_1(x_1, x_2, x_3), \qquad x_1 = h_1(X_1, X_2, X_3) \quad , \tag{2.135}$$

$$X_2 = H_2(x_1, x_2, x_3), \qquad x_2 = h_2(X_1, X_2, X_3) \quad , \tag{2.136}$$

$$X_3 = H_3(x_1, x_2, x_3), \qquad x_3 = h_3(X_1, X_2, X_3) \quad , \tag{2.137}$$

or, in vectorial form:

$$\boldsymbol{X} = \boldsymbol{H}(\boldsymbol{x}), \quad \boldsymbol{x} = \boldsymbol{h}(\boldsymbol{X}), \quad \boldsymbol{h} = \boldsymbol{H}^{-1} \quad . \tag{2.138}$$

The filtering of a function $\psi(\boldsymbol{x})$ is defined analogously to the mono-dimensional case. We first make a variable change to work in the reference coordinate system, in which a homogeneous filter is applied, and then perform the inverse transformation. The three-dimensional convolution kernel is defined by tensorizing homogeneous mono-dimensional kernels.

After making the first change of variables, we get:

$$\overline{\psi}(\boldsymbol{h}(\boldsymbol{X})) = \frac{1}{\overline{\Delta}^3} \int \prod_{i=1,3} G\left(\frac{X_i - X_i'}{\overline{\Delta}}\right) \psi(\boldsymbol{h}(\boldsymbol{X}')) d^3\boldsymbol{X}' \quad , \tag{2.139}$$

or, in the original space:

$$\begin{aligned} \overline{\psi}(\boldsymbol{x}) &= \frac{1}{\overline{\Delta}^3} \int \prod_{i=1,3} G\left(\frac{H_i(\boldsymbol{x}) - X_i'}{\overline{\Delta}}\right) \psi(\boldsymbol{h}(\boldsymbol{X}')) d^3\boldsymbol{X}' & (2.140) \\ &= \frac{1}{\overline{\Delta}^3} \int \prod_{i=1,3} G\left(\frac{H_i(\boldsymbol{x}) - H_i(\boldsymbol{x}')}{\overline{\Delta}}\right) \psi(\boldsymbol{x}') J(\boldsymbol{x}') d^3\boldsymbol{x}', & (2.141) \end{aligned}$$

where $J(\boldsymbol{x})$ is the Jacobian of the transformation $\boldsymbol{X} = \boldsymbol{H}(\boldsymbol{x})$. Analysis of the error shows that, for filters defined this way, the commutation error with the derivation in space is always of the second order, i.e.

$$\overline{\frac{\partial \psi}{\partial x_k}} - \frac{\partial \overline{\psi}}{\partial x_k} = O(\overline{\Delta}^2) \quad , \tag{2.142}$$

where the second term of the left-hand side is written:

$$\begin{aligned} \frac{\partial \overline{\psi}}{\partial x_k} &= \frac{1}{\overline{\Delta}^3} \int \frac{1}{\overline{\Delta}} G' \left(\frac{H_j(\boldsymbol{x}) - X_j'}{\overline{\Delta}}\right) \\ &\times \prod_{i=1,3; i \neq j} G\left(\frac{H_i(\boldsymbol{x}) - X_i'}{\overline{\Delta}}\right) H_{j,k}(\boldsymbol{x}) \psi(\boldsymbol{h}(\boldsymbol{X}')) d^3\boldsymbol{X}', & (2.143) \end{aligned}$$

with the notation:

$$H_{j,k}(\boldsymbol{x}) = \frac{\partial H_j(\boldsymbol{x})}{\partial x_k} \quad . \tag{2.144}$$

Differential analysis of the commutation error is performed by considering the solutions of the form:

$$\psi(\boldsymbol{x}) = \widehat{\psi_{\boldsymbol{k}}} \exp(\imath \boldsymbol{k} \cdot \boldsymbol{x}) \quad . \tag{2.145}$$

An analogous approach to the one already made in the mono-dimensional case leads to the relation:

$$\left[G\star, \frac{\partial}{\partial x_k}\right] \psi = -\alpha^{(2)} \overline{\Delta}^2 \Gamma_{kmp} \frac{\partial^2 \overline{\psi}}{\partial x_m \partial x_p} + O(k\overline{\Delta})^4 \quad , \tag{2.146}$$

where the function Γ_{kmp} is defined as:

$$\Gamma_{kmp} = h_{m,jq}(\boldsymbol{H}(\boldsymbol{x})) h_{p,q}(\boldsymbol{H}(\boldsymbol{x})) H_{j,k}(\boldsymbol{x}) \quad . \tag{2.147}$$

Van der Ven's Filters. Van der Ven's simplified filtering naturally extends to the three-dimensional case in Cartesian coordinates by letting:

$$\overline{\phi}(\boldsymbol{x}) = \frac{1}{\prod_{i=1,3} \overline{\delta}_i(\boldsymbol{x})} \int_{R^3} \prod_{i=1,3} G\left(\frac{x_i - x_i'}{\overline{\delta}_i(\boldsymbol{x})}\right) \phi(\boldsymbol{x}') d^3\boldsymbol{x}' \quad , \tag{2.148}$$

in which $\overline{\delta}_i(\boldsymbol{x})$ is the filter cutoff length in the ith direction of space at point x. For a kernel G verifying (2.114), the commutation error is expressed:

$$\left[G\star, \frac{\partial}{\partial x_j}\right] \phi = a \sum_{i=1,3} \frac{\partial \overline{\delta}_i(\boldsymbol{x})}{\partial x_j} \overline{\delta}_i(\boldsymbol{x})^{n-1} \overline{\frac{\partial^n \phi(\boldsymbol{x})}{\partial x_i^n}} \quad , \tag{2.149}$$

and is formally of order $n-1$.

High-Order Commuting Filters. Vasilyev's filters are generalized to the multiple dimension case in the same way as SOCF.

2.2.3 Local Spectrum of Commutation Error

A spectral tool for analyzing the wavenumber sprectrum of the commutation error, referred to as the *local spectrum analysis*, was introduced by Vasilyev and Goldstein [727]. This tool enables an accurate understanding of the impact of the commutation error on the derivatives of the filtered quantities. Writing the convolution filter as

$$\overline{\phi}(x) = \frac{1}{\overline{\Delta}(x)} \int_{-\infty}^{+\infty} G\left(\frac{x-y}{\overline{\Delta}(x)}\right) \phi(y) dy \quad , \tag{2.150}$$

and introducing its local Fourier decomposition (which can be evaluated using a windowed Fourier transform on bounded domains)

$$\overline{\phi}(x) = \int_{-\infty}^{+\infty} \widehat{G}\left(k\overline{\Delta}(x)\right) \widehat{\phi}(k) e^{\imath kx} dk \quad , \tag{2.151}$$

one can identify the coefficients of the local Fourier transform of the filtered quantity $\overline{\phi}$

$$\widehat{\overline{\phi}}(k,x) = \widehat{G}\left(k\overline{\Delta}(x)\right) \widehat{\phi}(k) \quad . \tag{2.152}$$

The commutation error can be written as

$$\left[\frac{d}{dx}, G\star\right](\phi)(x) = \frac{1}{2\pi} \int_{-\infty}^{+\infty} \left(\frac{1}{\overline{\Delta}(x)} \frac{d\overline{\Delta}(x)}{dx} \widehat{\mathcal{K}}(k\overline{\Delta}(x))\widehat{\phi}(k) \right) e^{\imath kx} dk \quad , \tag{2.153}$$

where the transfer function $\widehat{\mathcal{K}}$ is defined as

$$\widehat{\mathcal{K}}(k) = -k \frac{d\widehat{G}(k)}{dk} \quad . \tag{2.154}$$

By analogy with the previous case, the local spectrum of the commutation error is defined as

$$\widehat{\left[\frac{d}{dx}, G\star\right]}(\phi)(k,x) = \frac{1}{\overline{\Delta}(x)} \frac{d\overline{\Delta}(x)}{dx} \widehat{\mathcal{K}}(k\overline{\Delta}(x))\widehat{\phi}(k) \quad . \tag{2.155}$$

This expresion shows that the gradient of the cutoff length $\overline{\Delta}(x)$ affects the amplitude of the commutation error, while the filter shape (more precisly the gradient of the transfer function in Fourier space) governs the spectral repartition of the error. Analyses carried out considering convolution kernels presented in Sect.2.1.5 reveal that the spectrum of commutation error is global for smooth filters like the Gaussian filter (i.e. error occurs at all scales) while it is much more local for sharp filters (i.e. the commutation error is concentrated on a narrow wavenumber range).

2.3 Time Filtering: a Few Properties

We consider here continuous causal filters of the form [603]:

$$\overline{\phi}(\boldsymbol{x},t) = G \star \phi(\boldsymbol{x},t) = \int_{-\infty}^{t} \phi(\boldsymbol{\xi},t')G(t-t')dt' \quad , \tag{2.156}$$

where the kernel G satisfies immediately two of the three fundamental properties given in Sect. 2.1.2, namely the linearity constraint and the constant-conservation constraint.

Due to the causality constraint, the time-domain support of these filters is bounded, i.e.

$$\lim_{t \longrightarrow -\infty} G(t) = 0 \quad . \tag{2.157}$$

Then, the following commutation properties hold:

$$\left[G\star, \frac{\partial}{\partial x_j}\right] \phi = 0 \quad , \tag{2.158}$$

$$\left[G\star, \frac{\partial}{\partial t}\right] = \phi(\boldsymbol{x}, t) G(0) - \left(\frac{\partial G}{\partial t}\right) \star \phi \quad . \tag{2.159}$$

It is observed that the use of spatially bounded domains does not introduce any commutation error terms, the filter being independent of the position in space.

Two examples [606] are the exponential filter

$$G(t) = \frac{1}{\overline{\Delta}} e^{\frac{t}{\overline{\Delta}}} \quad \longrightarrow \quad \overline{\phi}(\boldsymbol{x}, t) = \frac{1}{\overline{\Delta}} \int_{-\infty}^{t} \phi(\boldsymbol{\xi}, t') \exp\left(\frac{t - t'}{\overline{\Delta}}\right) dt' \quad , \tag{2.160}$$

where $\overline{\Delta}$ is the characteristic cutoff time, and the Heaviside filter

$$G(t) = \frac{1}{\overline{\Delta}} H(t + \overline{\Delta}) \quad , \tag{2.161}$$

where $H(t)$ is the Heaviside function, yielding

$$\overline{\phi}(\boldsymbol{x}, t) = \frac{1}{\overline{\Delta}} \int_{t-\overline{\Delta}}^{t} \phi(\boldsymbol{\xi}, t') dt' \quad . \tag{2.162}$$

An interesting property of Eulerian time-domain filtering is that *local* differential expression of the filters are easily derived, whose practical implementation is easier than those of their original counterparts. Differentiating relation (2.162), on obtains the differential form of the Heaviside filter:

$$\frac{\partial \overline{\phi}(\boldsymbol{x}, t)}{\partial t} = \frac{1}{\overline{\Delta}} \left(\phi(\boldsymbol{x}, t) - \phi(\boldsymbol{x}, t - \overline{\Delta})\right) \quad . \tag{2.163}$$

The differential exponential filter is expressed as

$$\frac{\partial \overline{\phi}(\boldsymbol{x}, t)}{\partial t} = \frac{1}{\overline{\Delta}} \left(\phi(\boldsymbol{x}, t) - \overline{\phi}(\boldsymbol{x}, t - \overline{\Delta})\right) \quad . \tag{2.164}$$

3. Application to Navier–Stokes Equations

This chapter is devoted to the derivation of the constitutive equations of the large-eddy simulation technique, which is to say the filtered Navier–Stokes equations. Our interest here is in the case of an incompressible viscous Newtonian fluid of uniform density and temperature.

We first describe the application of an isotropic spatial filter[1] to the Navier–Stokes equations expressed in Cartesian coordinates or in general coordinates.

The emphasis will be put on the Eulerian, velocity–pressure formulation.[2] An important point is that these two formulations lead to different commutation errors with the filtering operator, and thus yield different theoretical and practical problems. The main point is that, when curvilinear grids are considered, two possibilities arise for solving numerically the filtered governing equations:

- *Conventional Approach*: First the filter is applied to the Navier–Stokes equations written in Cartesian coordinates, and then the filtered equations are transformed in general coordinates. Here, the filter is applied in the physical space, and the filter kernels are developed within the usual Cartesian framework (see Chap. 2).
- *Alternate Approach*: First the Navier–Stokes equations are expressed in general coordinates, and then the filter is applied to the transformed equations. In this case, the transformed variables are filtered using uniform filter kernels, leading to vanishing commutation errors. If physical variables are filtered using a transformed filter kernel, some commutation errors appear and specific filters must be employed (see Chap. 2).

The differences originate from the fact that the transformation in general coordinates is a nonlinear operation, yielding different commutation errors between the two operations.

---

[1] Refer to the definition given in Chap. 2.

[2] A few works dealing with the velocity–vorticity form of the Navier–Stokes equations exist [163, 485, 486]. It is important to note that the Lagrangian framework is employed in [485, 486]. Results dealing with large-eddy simulation within the framework of lattice-Boltzmann methods are presented in the review by Chen and Doolen [127].

Most of the existing published works deal with the conventional approach. As a consequence, this chapter will be mostly devoted to this approach.

It should be noted that this ideal framework, which implies that the fluid domain is unbounded, is the one nearly all authors use because it is only in this framework that the theory on which the subgrid modeling is based can be fully developed. The commutation errors between the filter and the derivation in space are then ignored. Section 3.4 is on the application of an inhomogeneous filter to the basic equations written in Cartesian coordinates.

We begin by deriving the filtered Navier–Stokes equations following the conventional approach. The various decompositions of the nonlinear term as a function of the filtered quantities are then discussed. We lastly introduce the closure problem, i.e. the representation of the unknowns as a function of the variables in the filtered problem.

3.1 Navier–Stokes Equations

We recall here the equations governing the evolution of an incompressible Newtonian fluid, first in the physical space, in general coordinates, and then in the spectral space.

3.1.1 Formulation in Physical Space

In the physical space, the velocity field $\boldsymbol{u} = (u_1, u_2, u_3)$ expressed in a reference Cartesian coordinate system $\boldsymbol{x} = (x_1, x_2, x_3)$ is a solution of the system comprising the momentum and continuity equations:

$$\frac{\partial u_i}{\partial t} + \frac{\partial}{\partial x_j}(u_i u_j) = -\frac{\partial p}{\partial x_i} + \nu \frac{\partial}{\partial x_j}\left(\frac{\partial u_i}{\partial x_j} + \frac{\partial u_j}{\partial x_i}\right), \quad i = 1,2,3 \quad , \tag{3.1}$$

$$\frac{\partial u_i}{\partial x_i} = 0 \quad , \tag{3.2}$$

in which $p = P/\rho$ and ν are, respectively, the static pressure and the assumedly constant, uniform kinematic viscosity. To obtain a well-posed problem, we have to add initial and boundary conditions to this system.

3.1.2 Formulation in General Coordinates

The incompressible Navier–Stokes equations written in general coordinates in strong conservation-law form [733] read:

$$\frac{\partial}{\partial \xi^k}(J^{-1}\xi_i^k u_i) = 0 \quad , \tag{3.3}$$

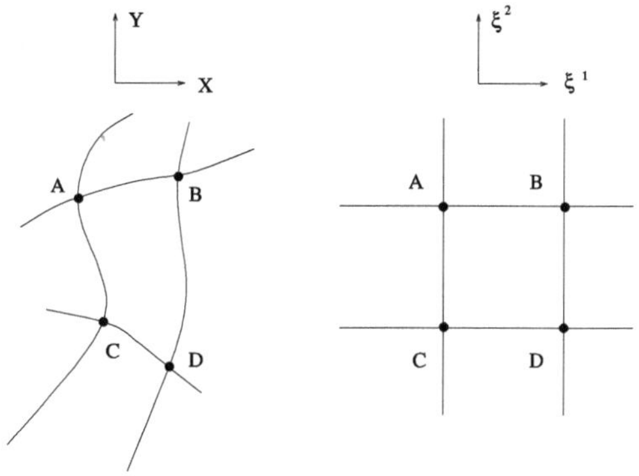

Fig. 3.1. Schematic of the coordinate transformation.

$$\frac{\partial}{\partial t}(J^{-1}u_i)+\frac{\partial}{\partial \xi^k}(U^k u_i)=-\frac{\partial}{\partial \xi^k}(J^{-1}\xi_i^k p)+\nu\frac{\partial}{\partial \xi^k}\left(J^{-1}G^{kl}\frac{\partial u_i}{\partial \xi^l}\right), \quad (3.4)$$

where ξ^k are the coordinate directions in the transformed space, $\xi_i^k = \partial\xi^k/\partial x_i$, J^{-1} is the Jacobian of the transformation, $G^{kl} = \xi_i^k \xi_i^l$ denotes the contravariant metric tensor, u_i the Cartesian components of the velocity field, and $U^k = J^{-1}\xi_i^k u_i$ the contravariant velocity component (see Fig. 3.1).

3.1.3 Formulation in Spectral Space

The dual system in spectral space is obtained by applying a Fourier transform to equations (3.1) and (3.2). By making use of the fact that the incompressibility constraint is reflected geometrically by the orthogonality[3] of the wave vector $\boldsymbol{k}$ and of the mode $\widehat{\boldsymbol{u}}(\boldsymbol{k})$, defined as (see Appendix A for greater detail on the spectral analysis of turbulence):

$$\widehat{\boldsymbol{u}}(\boldsymbol{k})=\frac{1}{(2\pi)^3}\int\int\int \boldsymbol{u}(\boldsymbol{x})e^{-\imath\boldsymbol{k}\cdot\boldsymbol{x}}d^3\boldsymbol{x}, \quad \imath^2=-1 \quad , \quad (3.5)$$

the system (3.1) - (3.2) can be reduced to a single equation:

$$\left(\frac{\partial}{\partial t}+\nu k^2\right)\widehat{u}_i(\boldsymbol{k})=T_i(\boldsymbol{k}) \quad , \quad (3.6)$$

in which the non-linear term $T_i(\boldsymbol{k})$ is of the form:

$$T_i(\boldsymbol{k})=M_{ijm}(\boldsymbol{k})\int\int \widehat{u}_j(\boldsymbol{p})\widehat{u}_m(\boldsymbol{q})\delta(\boldsymbol{k}-\boldsymbol{p}-\boldsymbol{q})d^3\boldsymbol{p}d^3\boldsymbol{q} \quad , \quad (3.7)$$

[3] This orthogonality relation is demonstrated by re-writing the incompressibility constraint of the velocity field in the spectral space as:

$$\frac{\partial u_i}{\partial x_i}=0 \iff k_i\widehat{u}_i(k)\equiv \boldsymbol{k}\cdot\widehat{\boldsymbol{u}}(\boldsymbol{k})=0 \quad .$$

with:

$$M_{ijm}(\boldsymbol{k}) = -\frac{\imath}{2}\left(k_m P_{ij}(\boldsymbol{k}) + k_j P_{im}(\boldsymbol{k})\right) \quad , \tag{3.8}$$

in which δ is the Kronecker symbol and $P_{ij}(\boldsymbol{k})$ is the projection operator on the plane orthogonal to the vector $\boldsymbol{k}$. This operator is written:

$$P_{ij}(\boldsymbol{k}) = \left(\delta_{ij} - \frac{k_i k_j}{k^2}\right) \quad . \tag{3.9}$$

3.2 Filtered Navier–Stokes Equations in Cartesian Coordinates (Homogeneous Case)

This section describes the equations of large-eddy simulation such as they are obtained by applying a homogeneous filter verifying the properties of linearity, conservation of constants, and commutation with derivation, to the Navier–Stokes equations. These are the equations that will be resolved in the numerical simulation.

3.2.1 Formulation in Physical Space

In light of the commutation with derivation property, the application of a filter to equations (3.1) and (3.2) is expressed:

$$\frac{\partial \overline{u}_i}{\partial t} + \frac{\partial}{\partial x_j}\left(\overline{u_i u_j}\right) = -\frac{\partial \overline{p}}{\partial x_i} + \nu \frac{\partial}{\partial x_j}\left(\frac{\partial \overline{u}_i}{\partial x_j} + \frac{\partial \overline{u}_j}{\partial x_i}\right) \quad , \tag{3.10}$$

$$\frac{\partial \overline{u}_i}{\partial x_i} = 0 \quad , \tag{3.11}$$

where $\overline{p}$ is the filtered pressure. The filtered momentum equation brings out the non-linear term $\overline{u_i u_j}$ which, in order for this equation to be usable, will have to be expressed as a function of $\overline{\boldsymbol{u}}$ and $\boldsymbol{u}'$, which are now the only unknowns left in the problem and where:

$$\boldsymbol{u}' = \boldsymbol{u} - \overline{\boldsymbol{u}} \quad . \tag{3.12}$$

This decomposition is not unique, and will be discussed in the following section.

3.2.2 Formulation in Spectral Space

Using the equivalence $\widehat{\overline{u}}_i(\boldsymbol{k}) = \widehat{G}(\boldsymbol{k})\widehat{u}_i(\boldsymbol{k})$, the momentum equation in the spectral space obtained by multiplying equation (3.6) by the transfer function

$\widehat{G}(\boldsymbol{k})$ is expressed:

$$\left(\frac{\partial}{\partial t} + 2\nu k^2\right) \widehat{G}(\boldsymbol{k})\widehat{u}_i(\boldsymbol{k}) = \widehat{G}(\boldsymbol{k})T_i(\boldsymbol{k}) \quad , \tag{3.13}$$

in which the filtered non-linear term $\widehat{G}(\boldsymbol{k})T_i(\boldsymbol{k})$ is written:

$$\widehat{G}(\boldsymbol{k})T_i(\boldsymbol{k}) = M_{ijm}(\boldsymbol{k}) \int\int \widehat{G}(\boldsymbol{k})\widehat{u}_j(\boldsymbol{p})\widehat{u}_m(\boldsymbol{q})\delta(\boldsymbol{k}-\boldsymbol{p}-\boldsymbol{q})d^3\boldsymbol{p}d^3\boldsymbol{q} \quad . \tag{3.14}$$

The filtered non-linear term (3.14) brings out the contributions of the modes $\widehat{\boldsymbol{u}}(\boldsymbol{p})$ and $\widehat{\boldsymbol{u}}(\boldsymbol{q})$. To complete the decomposition, these modes also have to be expressed as the sum of a filtered part and a fluctuation. This is the same problem as the one encountered when writing the equations in the physical space. This operation is described in the following section.

3.3 Decomposition of the Non-linear Term. Associated Equations for the Conventional Approach

This section details the various existing decompositions of the non-linear term and the associated equations written in Cartesian coordinates.

3.3.1 Leonard's Decomposition

Expression in Physical Space. Leonard [436] expresses the non-linear term in the form of a triple summation:

$$\begin{aligned} \overline{u_i u_j} &= \overline{(\overline{u}_i + u'_i)(\overline{u}_j + u'_j)} \qquad (3.15) \\ &= \overline{\overline{u}_i\overline{u}_j} + \overline{\overline{u}_i u'_j} + \overline{\overline{u}_j u'_i} + \overline{u'_i u'_j} \quad . \qquad (3.16) \end{aligned}$$

The non-linear term is now written entirely as a function of the filtered quantity $\overline{\boldsymbol{u}}$ and the fluctuation $\boldsymbol{u}'$. We then have two versions of this [762].

The first considers that all the terms appearing in the evolution equations of a filtered quantity must themselves be filtered quantities, because the simulation solution has to be the same for all the terms. The filtered momentum equation is then expressed:

$$\frac{\partial \overline{u}_i}{\partial t} + \frac{\partial}{\partial x_j}\left(\overline{\overline{u}_i\overline{u}_j}\right) = -\frac{\partial \overline{p}}{\partial x_i} + \nu\frac{\partial}{\partial x_j}\left(\frac{\partial \overline{u}_i}{\partial x_j} + \frac{\partial \overline{u}_j}{\partial x_i}\right) - \frac{\partial \tau_{ij}}{\partial x_j} \quad , \tag{3.17}$$

in which the subgrid tensor τ, grouping together all the terms that are not exclusively dependent on the large scales, is defined as:

$$\tau_{ij} = C_{ij} + R_{ij} = \overline{u_i u_j} - \overline{\overline{u}_i\overline{u}_j} \quad , \tag{3.18}$$

where the cross-stress tensor, C, which represents the interactions between large and small scales, and the Reynolds subgrid tensor, R, which reflects the interactions between subgrid scales, are expressed:

$$\begin{aligned} C_{ij} &= \overline{\overline{u}_i u'_j} + \overline{\overline{u}_j u'_i} \quad , & (3.19)\\ R_{ij} &= \overline{u'_i u'_j} \quad . & (3.20) \end{aligned}$$

In the following, this decomposition will be called double decomposition.

The other point of view consists of considering that it must be possible to evaluate the terms directly from the filtered variables. But the $\overline{\overline{u}_i \overline{u}_j}$ term cannot be calculated directly because it requires a second application of the filter. To remedy this, Leonard proposes a further decomposition:

$$\begin{aligned} \overline{\overline{u}_i \overline{u}_j} &= \left(\overline{\overline{u}_i \overline{u}_j} - \overline{u}_i \overline{u}_j\right) + \overline{u}_i \overline{u}_j \\ &= L_{ij} + \overline{u}_i \overline{u}_j \quad . & (3.21) \end{aligned}$$

The new L term, called Leonard tensor, represents interactions among the large scales. Using this new decomposition, the filtered momentum equation becomes:

$$\frac{\partial \overline{u}_i}{\partial t} + \frac{\partial}{\partial x_j}(\overline{u}_i \overline{u}_j) = -\frac{\partial \overline{p}}{\partial x_i} + \nu \frac{\partial}{\partial x_j}\left(\frac{\partial \overline{u}_i}{\partial x_j} + \frac{\partial \overline{u}_j}{\partial x_i}\right) - \frac{\partial \tau_{ij}}{\partial x_j} \quad . \quad (3.22)$$

The subgrid tensor τ, which now groups all the terms that are not expressed directly from $\overline{u}$, takes the form:

$$\tau_{ij} = L_{ij} + C_{ij} + R_{ij} = \overline{u_i u_j} - \overline{u}_i \overline{u}_j \quad . \quad (3.23)$$

This decomposition will be designated hereafter the Leonard or triple decomposition. Equation (3.22) and the subgrid term τ_{ij} defined by (3.23) can be obtained directly from the Navier–Stokes equations without using the Leonard decomposition for this. It should be noted that the term $\overline{u}_i \overline{u}_j$ is a quadratic term and that it contains frequencies that are in theory higher than each of the terms composing. So in order to represent it completely, more degrees of freedom are needed than for each of the terms $\overline{u}_i$ and $\overline{u}_j$[4].

We may point out that, if the filter is a Reynolds operator, then the tensors C_{ij} and L_{ij} are identically zero[5] and the two decompositions are then equivalent, since the subgrid tensor is reduced to the tensor R_{ij}.

---

[4] In practice, if the large-eddy simulation filter is associated with a given computational grid on which the Navier–Stokes equations are resolved, this means that the grid used for composing the $\overline{u}_i \overline{u}_j$ product has to be twice as fine (in each direction of space) as the one used to represent the velocity field. If the product is composed on the same grid, then only the $\overline{\overline{u}_i \overline{u}_j}$ term can be calculated.

[5] It is recalled that if the filter is a Reynolds operator, then we have the three following properties (see Appendix A):

$$\overline{\overline{\boldsymbol{u}}} = \overline{\boldsymbol{u}}, \ \overline{\boldsymbol{u}'} = 0, \ \overline{\overline{\boldsymbol{u}}\boldsymbol{u}} = \overline{\boldsymbol{u}}\,\overline{\boldsymbol{u}} \quad ,$$

Writing the Navier–Stokes equations (3.1) in the symbolic form

$$\frac{\partial \boldsymbol{u}}{\partial t} + \mathcal{NS}(\boldsymbol{u}) = 0 \quad , \tag{3.24}$$

the filtered Navier–Stokes equations are expressed

$$G \star \frac{\partial \boldsymbol{u}}{\partial t} = \frac{\partial \overline{\boldsymbol{u}}}{\partial t} = -G \star \mathcal{NS}(\boldsymbol{u}) \tag{3.25}$$

$$= -\mathcal{NS}(\overline{\boldsymbol{u}}) - [G\star, \mathcal{NS}](\boldsymbol{u}) \quad , \tag{3.26}$$

where $[.,.]$ is the commutator operator introduced in Sect. 2.1.2. We note that the subgrid tensor corresponds to the commutation error between the filter and the non-linear term. Introducing the bilinear form $\mathcal{B}(\cdot,\cdot)$:

$$\mathcal{B}(u_i, u_j) \equiv u_i u_j \quad , \tag{3.27}$$

we get

$$\tau_{ij} = [G\star, \mathcal{B}](u_i, u_j) \quad . \tag{3.28}$$

Double decomposition (3.18) leads to the following equation for the resolved kinetic energy $q_{\mathrm{r}}^2 = \overline{u}_i \overline{u}_i / 2$:

$$\begin{aligned} \frac{\partial q_{\mathrm{r}}^2}{\partial t} = & \underbrace{\overline{\overline{u}_i \overline{u}_j} \frac{\partial \overline{u}_i}{\partial x_j}}_{I} + \underbrace{\tau_{ij} \frac{\partial \overline{u}_i}{\partial x_j}}_{II} - \underbrace{\nu \frac{\partial \overline{u}_i}{\partial x_j} \frac{\partial \overline{u}_i}{\partial x_j}}_{III} \\ & - \underbrace{\frac{\partial}{\partial x_i} (\overline{u}_i \overline{p})}_{IV} + \underbrace{\frac{\partial}{\partial x_i} \left(\nu \frac{\partial q_{\mathrm{r}}^2}{\partial x_i} \right)}_{V} \\ & - \underbrace{\frac{\partial}{\partial x_j} (\overline{u}_i \overline{\overline{u}_i \overline{u}_j})}_{VI} - \underbrace{\frac{\partial}{\partial x_j} (\overline{u}_i \tau_{ij})}_{VII} \quad . \end{aligned} \tag{3.29}$$

This equation shows the existence of several mechanisms exchanging kinetic energy at the resolved scales:

---

whence

$$\begin{aligned} C_{ij} &= \overline{\overline{u}_i u'_j} + \overline{\overline{u}_j u'_i} \\ &= \overline{\overline{u}}_i \overline{u'}_j + \overline{\overline{u}}_j \overline{u'}_i \\ &= 0 \quad , \end{aligned}$$

$$\begin{aligned} L_{ij} &= \overline{\overline{\overline{u}_i \overline{u}_j}} - \overline{\overline{u}_i \overline{u}_j} \\ &= \overline{\overline{u}}_i \overline{\overline{u}}_j - \overline{u}_i \overline{u}_j \\ &= 0 \quad . \end{aligned}$$

- I - production
- II - subgrid dissipation
- III - dissipation by viscous effects
- IV - diffusion by pressure effect
- V - diffusion by viscous effects
- VI - diffusion by interaction among resolved scales
- VII - diffusion by interaction with subgrid modes.

Leonard's decomposition (3.23) can be used to obtain the similar form:

$$\begin{aligned}\frac{\partial q_{\rm r}^2}{\partial t} &= -\underbrace{\frac{\partial q_{\rm r}^2 \overline{u}_j}{\partial x_j}}_{VIII} + \underbrace{\tau_{ij}\frac{\partial \overline{u}_i}{\partial x_j}}_{IX} - \underbrace{\nu \frac{\partial \overline{u}_i}{\partial x_j}\frac{\partial \overline{u}_i}{\partial x_j}}_{X} \\ &- \underbrace{\frac{\partial}{\partial x_i}(\overline{u}_i \overline{p})}_{XI} + \underbrace{\frac{\partial}{\partial x_i}\left(\nu \frac{\partial q_{\rm r}^2}{\partial x_i}\right)}_{XII} \\ &+ \underbrace{\overline{u}_i \overline{u}_j \frac{\partial \overline{u}_i}{\partial x_j}}_{XIII} - \underbrace{\frac{\partial}{\partial x_j}(\overline{u}_i \tau_{ij})}_{XIV} \quad . \end{aligned} \tag{3.30}$$

This equation differs from the previous one only in the first and sixth terms of the right-hand side, and in the definition of tensor τ:

- VIII - advection
- IX - *idem* II
- X - *idem* III
- XI - *idem* IV
- XII - *idem* V
- XIII - production
- XIV - *idem* VII

The momentum equation for the small scales is obtained by subtracting the large scale equation from the unfiltered momentum equation (3.1), making, for the double decomposition:

$$\begin{aligned}\frac{\partial u_i'}{\partial t} + \frac{\partial}{\partial x_j}\left((\overline{u}_i + u_i')(\overline{u}_j + u_j') - \overline{\overline{u}_i \overline{u}_j}\right) &= -\frac{\partial p'}{\partial x_i} + \frac{\partial \tau_{ij}}{\partial x_i} \\ &+ \nu \frac{\partial}{\partial x_j}\left(\frac{\partial u_i'}{\partial x_j} + \frac{\partial u_j'}{\partial x_i}\right) , \end{aligned} \tag{3.31}$$

and, for the triple decomposition:

$$\begin{aligned}\frac{\partial u_i'}{\partial t} + \frac{\partial}{\partial x_j}\left((\overline{u}_i + u_i')(\overline{u}_j + u_j') - \overline{u}_i \overline{u}_j\right) &= -\frac{\partial p'}{\partial x_i} + \frac{\partial \tau_{ij}}{\partial x_i} \\ &+ \nu \frac{\partial}{\partial x_j}\left(\frac{\partial u_i'}{\partial x_j} + \frac{\partial u_j'}{\partial x_i}\right) . \end{aligned} \tag{3.32}$$

The *filtered subgrid kinetic energy* $q^2_{\mathrm{sgs}} = \overline{u'_k u'_k}/2$ equation obtained by multiplying (3.32) by u'_i and filtering the relation thus derived is expressed:

$$\begin{aligned}
\frac{\partial q^2_{\mathrm{sgs}}}{\partial t} &= \underbrace{-\frac{\partial}{\partial x_j}\left(q^2_{\mathrm{sgs}}\overline{u}_j\right)}_{XV} \underbrace{-\frac{1}{2}\frac{\partial}{\partial x_j}\left(\overline{u_i u_i u_j} - \overline{u}_j\overline{u_i u_i}\right)}_{XVI} \underbrace{-\frac{\partial}{\partial x_j}\left(\overline{p u_j} - \overline{p}\,\overline{u}_j\right)}_{XVII} \\
&+ \underbrace{\frac{\partial}{\partial x_j}\left(\nu \frac{\partial q^2_{\mathrm{sgs}}}{\partial x_j}\right)}_{XVIII} + \underbrace{\frac{\partial}{\partial x_j}\left(\tau_{ij}\overline{u}_i\right)}_{XIX} \\
&- \underbrace{\nu\left(\overline{\frac{\partial u_i}{\partial x_j}\frac{\partial u_i}{\partial x_j}} - \frac{\partial \overline{u}_i}{\partial x_j}\frac{\partial \overline{u}_i}{\partial x_j}\right)}_{XX} \underbrace{- \tau_{ij}\frac{\partial \overline{u}_i}{\partial x_j}}_{XXI} \quad . \qquad (3.33)
\end{aligned}$$

- XV - advection
- XVI - turbulent transport
- XVII - diffusion by pressure effects
- XVIII - diffusion by viscous effects
- XIX - diffusion by subgrid modes
- XX - dissipation by viscous effects
- XXI - subgrid dissipation.

For the double decomposition, equation (3.31) leads to:

$$\begin{aligned}
\frac{\partial q^2_{\mathrm{sgs}}}{\partial t} &= \underbrace{-\frac{\partial}{\partial x_j}\left(\overline{u_i u_i u_j} - \overline{\overline{u}_i\overline{\overline{u}_i\overline{u}_j}}\right)}_{XXII} + \underbrace{\overline{u_i u_j \frac{\partial u_i}{\partial x_j}} - \overline{\overline{\overline{u}_i\overline{u}_j}\frac{\partial \overline{u}_i}{\partial x_j}}}_{XXIII} \\
&- \underbrace{\frac{\partial}{\partial x_j}\left(\overline{p u_j} - \overline{\overline{p}\,\overline{u}_j}\right)}_{XXIV} + \underbrace{\nu\left(\overline{u_i\frac{\partial^2 u_i}{\partial x_j^2}} - \overline{\overline{u}_i\frac{\partial^2 \overline{u}_i}{\partial x_j^2}}\right)}_{XXV} \\
&+ \underbrace{\frac{\partial}{\partial x_j}\left(\tau_{ij}\overline{\overline{u}_i}\right) - \overline{\tau_{ij}\frac{\partial \overline{u}_i}{\partial x_j}}}_{XXVI} \quad , \qquad (3.34)
\end{aligned}$$

with:

- XXII - turbulent transport
- XXIII - production
- XXIV - diffusion by pressure effects
- XXV - viscous effects
- XXVI - subgrid dissipation and diffusion

It is recalled that, if the filter used is not positive, the generalized subgrid kinetic energy $q^2_{\rm gsgs}$ defined as the half-trace of the subgrid tensor,

$$q^2_{\rm gsgs} = \tau_{kk}/2 = \frac{1}{2}\overline{u'_i u'_i} + \overline{\overline{u}_i u'_i} \quad ,$$

can admit negative values locally (see Sect. 3.3.5). If the filter is a Reynolds operator, the subgrid tensor is then reduced to the subgrid Reynolds tensor and the generalized subgrid kinetic energy is equal to the subgrid kinetic energy, i.e.

$$q^2_{\rm sgs} \equiv \frac{1}{2}\overline{u'_i u'_i} = q^2_{\rm gsgs} \equiv \tau_{kk}/2 \quad . \tag{3.35}$$

Expression in Spectral Space. Both versions of the Leonard decomposition can be transcribed in the spectral space. Using the definition of the fluctuation $\widehat{\boldsymbol{u}}'(\boldsymbol{k})$ as

$$\widehat{u}'_i(\boldsymbol{k}) = (1 - \widehat{G}(\boldsymbol{k}))\widehat{u}_i(\boldsymbol{k}) \quad , \tag{3.36}$$

the filtered non-linear term $\widehat{G}(\boldsymbol{k})T_i(\boldsymbol{k})$ is expressed, for the triple decomposition:

$$\begin{aligned}
\widehat{G}(\boldsymbol{k})T_i(\boldsymbol{k}) &= M_{ijm}(\boldsymbol{k}) \int\int \widehat{G}(\boldsymbol{p})\widehat{G}(\boldsymbol{q})\widehat{u}_j(\boldsymbol{p})\widehat{u}_m(\boldsymbol{q})\delta(\boldsymbol{k}-\boldsymbol{p}-\boldsymbol{q})d^3\boldsymbol{p}d^3\boldsymbol{q} \\
&- M_{ijm}(\boldsymbol{k}) \int\int (1-\widehat{G}(\boldsymbol{k}))\widehat{G}(\boldsymbol{p})\widehat{G}(\boldsymbol{q}) \\
&\quad \times \widehat{u}_j(\boldsymbol{p})\widehat{u}_m(\boldsymbol{q})\delta(\boldsymbol{k}-\boldsymbol{p}-\boldsymbol{q})d^3\boldsymbol{p}d^3\boldsymbol{q} \\
&+ M_{ijm}(\boldsymbol{k}) \int\int \widehat{G}(\boldsymbol{k}) \left(\widehat{G}(\boldsymbol{p})(1-\widehat{G}(\boldsymbol{q})) + \widehat{G}(\boldsymbol{q})(1-\widehat{G}(\boldsymbol{p}))\right) \\
&\quad \times \widehat{u}_j(\boldsymbol{p})\widehat{u}_m(\boldsymbol{q})\delta(\boldsymbol{k}-\boldsymbol{p}-\boldsymbol{q})d^3\boldsymbol{p}d^3\boldsymbol{q} \\
&+ M_{ijm}(\boldsymbol{k}) \int\int \widehat{G}(\boldsymbol{k}) \left((1-\widehat{G}(\boldsymbol{q}))(1-\widehat{G}(\boldsymbol{p}))\right) \\
&\quad \times \widehat{u}_j(\boldsymbol{p})\widehat{u}_m(\boldsymbol{q})\delta(\boldsymbol{k}-\boldsymbol{p}-\boldsymbol{q})d^3\boldsymbol{p}d^3\boldsymbol{q} \quad .
\end{aligned} \tag{3.37}$$

The first term of the right-hand side corresponds to the contribution $\overline{\overline{u}_i\overline{u}_j}$, the second to the Leonard tensor L, the third to the cross stresses represented by the tensor C, and the fourth to the subgrid Reynolds tensor R. This is illustrated by Fig. 3.2.

The double decomposition is derived by combination of the first two terms of the right-hand side of (3.37):

$$\begin{aligned}
\widehat{G}(\boldsymbol{k})T_i(\boldsymbol{k}) &= M_{ijm}(\boldsymbol{k}) \int\int \widehat{G}(\boldsymbol{p})\widehat{G}(\boldsymbol{q})\widehat{G}(\boldsymbol{k}) \\
&\quad \times \widehat{u}_j(\boldsymbol{p})\widehat{u}_m(\boldsymbol{q})\delta(\boldsymbol{k}-\boldsymbol{p}-\boldsymbol{q})d^3\boldsymbol{p}d^3\boldsymbol{q}
\end{aligned}$$

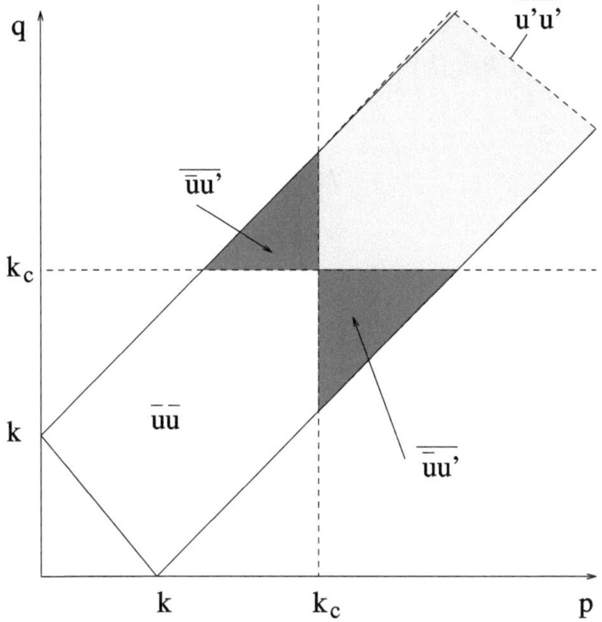

Fig. 3.2. Representation of the various Leonard decomposition terms in the spectral space, when using a sharp cutoff filter with a cutoff frequency k_c.

$$
\begin{aligned}
&+ \quad M_{ijm}(\boldsymbol{k}) \int\int \widehat{G}(\boldsymbol{k}) \left(\widehat{G}(\boldsymbol{p})(1-\widehat{G}(\boldsymbol{q})) + \widehat{G}(\boldsymbol{q})(1-\widehat{G}(\boldsymbol{p})) \right) \\
&\quad \times \widehat{u}_j(\boldsymbol{p})\widehat{u}_m(\boldsymbol{q})\delta(\boldsymbol{k}-\boldsymbol{p}-\boldsymbol{q}) d^3\boldsymbol{p} d^3\boldsymbol{q} \\
&+ \quad M_{ijm}(\boldsymbol{k}) \int\int \widehat{G}(\boldsymbol{k}) \left((1-\widehat{G}(\boldsymbol{q}))(1-\widehat{G}(\boldsymbol{p})) \right) \\
&\quad \times \widehat{u}_j(\boldsymbol{p})\widehat{u}_m(\boldsymbol{q})\delta(\boldsymbol{k}-\boldsymbol{p}-\boldsymbol{q}) d^3\boldsymbol{p} d^3\boldsymbol{q} \quad . \qquad (3.38)
\end{aligned}
$$

The first term of the right-hand side corresponds to the contribution $\overline{\overline{u}_i \overline{u}_j}$ in the physical space, and the last two remain unchanged with respect to the triple decomposition. Let us note that the sum of the contributions of the cross tensor and the subgrid Reynolds tensor simplifies to the form:

$$
\begin{aligned}
C_{ij} + R_{ij} \quad = \quad & M_{ijm}(\boldsymbol{k}) \int\int (1-\widehat{G}(\boldsymbol{p})\widehat{G}(\boldsymbol{q}))\widehat{G}(\boldsymbol{k}) \\
& \times \widehat{u}_j(\boldsymbol{p})\widehat{u}_m(\boldsymbol{q})\delta(\boldsymbol{k}-\boldsymbol{p}-\boldsymbol{q}) d^3\boldsymbol{p} d^3\boldsymbol{q} \quad . \qquad (3.39)
\end{aligned}
$$

The momentum equations corresponding to these two decompositions are found by replacing the right-hand side of equation (3.13) with the desired terms. For the double decomposition, we get:

$$\begin{aligned}\left(\frac{\partial}{\partial t}+\nu k^2\right)\widehat{G}(\boldsymbol{k})\widehat{u}_i(\boldsymbol{k}) &= M_{ijm}(\boldsymbol{k})\int\int \widehat{G}(\boldsymbol{p})\widehat{G}(\boldsymbol{q})\widehat{G}(\boldsymbol{k}) \\ &\quad \times\widehat{u}_j(\boldsymbol{p})\widehat{u}_m(\boldsymbol{q})\delta(\boldsymbol{k}-\boldsymbol{p}-\boldsymbol{q})d^3\boldsymbol{p}d^3\boldsymbol{q} \\ &+ M_{ijm}(\boldsymbol{k})\int\int (1-\widehat{G}(\boldsymbol{p})\widehat{G}(\boldsymbol{q}))\widehat{G}(\boldsymbol{k}) \\ &\quad \times\widehat{u}_j(\boldsymbol{p})\widehat{u}_m(\boldsymbol{q})\delta(\boldsymbol{k}-\boldsymbol{p}-\boldsymbol{q})d^3\boldsymbol{p}d^3\boldsymbol{q} \quad ,\end{aligned} \tag{3.40}$$

and for the triple decomposition:

$$\begin{aligned}\left(\frac{\partial}{\partial t}+\nu k^2\right)\widehat{G}(\boldsymbol{k})\widehat{u}_i(\boldsymbol{k}) &= M_{ijm}(\boldsymbol{k})\int\int \widehat{G}(\boldsymbol{p})\widehat{G}(\boldsymbol{q}) \\ &\quad \times\widehat{u}_j(\boldsymbol{p})\widehat{u}_m(\boldsymbol{q})\delta(\boldsymbol{k}-\boldsymbol{p}-\boldsymbol{q})d^3\boldsymbol{p}d^3\boldsymbol{q} \\ &- M_{ijm}(\boldsymbol{k})\int\int (1-\widehat{G}(\boldsymbol{k}))\widehat{G}(\boldsymbol{p})\widehat{G}(\boldsymbol{q}) \\ &\quad \times\widehat{u}_j(\boldsymbol{p})\widehat{u}_m(\boldsymbol{q})\delta(\boldsymbol{k}-\boldsymbol{p}-\boldsymbol{q})d^3\boldsymbol{p}d^3\boldsymbol{q} \\ &+ M_{ijm}(\boldsymbol{k})\int\int \widehat{G}(\boldsymbol{k})\Big(\widehat{G}(\boldsymbol{p})(1-\widehat{G}(\boldsymbol{q})) \\ &\quad +\widehat{G}(\boldsymbol{q})(1-\widehat{G}(\boldsymbol{p}))\Big) \\ &\quad \times\widehat{u}_j(\boldsymbol{p})\widehat{u}_m(\boldsymbol{q})\delta(\boldsymbol{k}-\boldsymbol{p}-\boldsymbol{q})d^3\boldsymbol{p}d^3\boldsymbol{q} \\ &+ M_{ijm}(\boldsymbol{k})\int\int \widehat{G}(\boldsymbol{k})\Big((1-\widehat{G}(\boldsymbol{q}))(1-\widehat{G}(\boldsymbol{p}))\Big) \\ &\quad \times\widehat{u}_j(\boldsymbol{p})\widehat{u}_m(\boldsymbol{q})\delta(\boldsymbol{k}-\boldsymbol{p}-\boldsymbol{q})d^3\boldsymbol{p}d^3\boldsymbol{q} \quad .\end{aligned} \tag{3.41}$$

For both decompositions, the momentum equation can be expressed in the symbolic form:

$$\left(\frac{\partial}{\partial t}+\nu k^2\right)\widehat{G}(\boldsymbol{k})\widehat{u}_i(\boldsymbol{k}) = T_{\rm r}(\boldsymbol{k}) + T_{\rm sgs}(\boldsymbol{k}) \quad , \tag{3.42}$$

in which $T_{\rm r}(\boldsymbol{k})$ designates the transfer terms calculated directly from the resolved modes, and is therefore equivalent to the contribution of the $\overline{u}_i\overline{u}_j$ term in the case of the triple decomposition, and that of the $\overline{\overline{u}_i\overline{u}_j}$ term for the double decomposition. The $T_{\rm sgs}(k)$ term designates the other non-linear terms, and therefore corresponds to the contribution of the subsidiary term such as defined above. Let $E(k)$ be the energy contained on the sphere of radius k. It is calculated as:

$$E(k) = \frac{1}{2}k^2\int \widehat{\boldsymbol{u}}(\boldsymbol{k})\cdot\widehat{\boldsymbol{u}}^*(\boldsymbol{k})dS(\boldsymbol{k}) \quad , \tag{3.43}$$

where $dS(\boldsymbol{k})$ is the surface element of the sphere, and where the asterisk designates a conjugate complex number. The kinetic energy of the resolved modes contained on this same sphere, denoted $\overline{E}_{\rm r}(k)$, is defined by the relation

$$\overline{E}_{\rm r}(k) = \frac{1}{2}k^2 \int \widehat{G}(\boldsymbol{k})\widehat{\boldsymbol{u}}(\boldsymbol{k}) \cdot \widehat{G}(\boldsymbol{k})\widehat{\boldsymbol{u}}^*(\boldsymbol{k}) dS(\boldsymbol{k}) \quad (3.44)$$

$$= \widehat{G}^2(k)E(k) \quad . \quad (3.45)$$

We return to the kinetic energy of the resolved modes, $q_{\rm r}^2 = \overline{u}_i\overline{u}_i/2$, by summation on all the wave numbers:

$$q_{\rm r}^2 = \int_0^\infty \overline{E}_{\rm r}(k)dk \quad . \quad (3.46)$$

It is important to note that $\overline{E}_{\rm r}(k)$ is related to the energy of the resolved modes, which is generally not equal to the filtered part of the kinetic energy which, for its part, is associated with the quantity denoted $\overline{E}(k)$, defined as

$$\overline{E}(k) = \widehat{G}(k)E(k) \quad . \quad (3.47)$$

The identity of these two quantities is verified when the transfer function is such that $\widehat{G}^2(k) = \widehat{G}(k)$, $\forall k$, i.e. when the filter used is a projector. The evolution equation for $\overline{E}_{\rm r}(k)$ is obtained by multiplying the filtered momentum equation (3.13) by $k^2\widehat{G}(\boldsymbol{k})\widehat{\boldsymbol{u}}^*(\boldsymbol{k})$, and then integrating the result on the sphere of radius k. Using the double decomposition we get the following equation:

$$\left(\frac{\partial}{\partial t} + 2\nu k^2\right)\overline{E}_{\rm r}(k) = \frac{1}{2}\int\int_{\Delta} \widehat{G}(\boldsymbol{p})\widehat{G}(\boldsymbol{q})\widehat{G}^2(\boldsymbol{k})S(\boldsymbol{k}|\boldsymbol{p},\boldsymbol{q})d\boldsymbol{p}d\boldsymbol{q}$$
$$+ \frac{1}{2}\int\int_{\Delta} (1-\widehat{G}(\boldsymbol{p})\widehat{G}(\boldsymbol{q}))\widehat{G}^2(\boldsymbol{k})S(\boldsymbol{k}|\boldsymbol{p},\boldsymbol{q})d\boldsymbol{p}d\boldsymbol{q} \quad , \quad (3.48)$$

and the triple decomposition:

$$\left(\frac{\partial}{\partial t} + 2\nu k^2\right)\overline{E}_{\rm r}(k) = \frac{1}{2}\int\int_{\Delta} \widehat{G}(\boldsymbol{p})\widehat{G}(\boldsymbol{q})\widehat{G}(\boldsymbol{k})S(\boldsymbol{k}|\boldsymbol{p},\boldsymbol{q})d\boldsymbol{p}d\boldsymbol{q}$$
$$- \frac{1}{2}\int\int \widehat{G}(\boldsymbol{k})(1-\widehat{G}(\boldsymbol{k}))\widehat{G}(\boldsymbol{p})\widehat{G}(\boldsymbol{q})S(\boldsymbol{k}|\boldsymbol{p},\boldsymbol{q})d\boldsymbol{p}d\boldsymbol{q}$$
$$+ \frac{1}{2}\int\int_{\Delta} \widehat{G}^2(\boldsymbol{k})\Big(\widehat{G}(\boldsymbol{p})$$
$$\times (1-\widehat{G}(\boldsymbol{q})) + \widehat{G}(\boldsymbol{q})(1-\widehat{G}(\boldsymbol{p}))\Big) S(\boldsymbol{k}|\boldsymbol{p},\boldsymbol{q})d\boldsymbol{p}d\boldsymbol{q}$$
$$+ \frac{1}{2}\int\int_{\Delta} \widehat{G}^2(\boldsymbol{k})\Big((1-\widehat{G}(\boldsymbol{q}))(1-\widehat{G}(\boldsymbol{p}))\Big)$$
$$\times S(\boldsymbol{k}|\boldsymbol{p},\boldsymbol{q})d\boldsymbol{p}d\boldsymbol{q} \quad , \quad (3.49)$$

in which

$$S(\boldsymbol{k}|\boldsymbol{p},\boldsymbol{q}) = 16\pi^2 kpq M_{ijm}(\boldsymbol{k})\widehat{u}_j(\boldsymbol{p})\widehat{u}_m(\boldsymbol{q})\widehat{u}_i(-\boldsymbol{k})\delta(\boldsymbol{k}-\boldsymbol{p}-\boldsymbol{q}) \quad , \tag{3.50}$$

and where the symbol $\int\int_\Delta$ designates integration over the interval $|k-p| < q < k+p$.

Following the example of what was done for the momentum equations, the kinetic energy evolution equation for the resolved modes can be expressed in the abbreviated form

$$\left(\frac{\partial}{\partial t} + 2\nu k^2\right)\overline{E}_{\mathrm{r}}(\boldsymbol{k}) = T_{\mathrm{r}}^{\mathrm{e}}(\boldsymbol{k}) + T_{\mathrm{sgs}}^{\mathrm{e}}(\boldsymbol{k}) \quad . \tag{3.51}$$

The terms $T_{\mathrm{r}}^{\mathrm{e}}(\boldsymbol{k})$ and $T_{\mathrm{sgs}}^{\mathrm{e}}(\boldsymbol{k})$ represent, respectively, the energy transfers of mode $\boldsymbol{k}$ with all the other modes associated with the terms that can be calculated directly from the resolved modes, and the subgrid terms.The kinetic energy conservation property for inviscid fluids, i.e. in the case of zero viscosity, implies:

$$\int (T_{\mathrm{r}}^{\mathrm{e}}(\boldsymbol{k}) + T_{\mathrm{sgs}}^{\mathrm{e}}(\boldsymbol{k}))d^3\boldsymbol{k} = 0 \quad . \tag{3.52}$$

The momentum equations for the unresolved scales are obtained by algebraic manipulations strictly analogous to those used for obtaining the equations for the resolved scales, except that this time equation (3.6) is multiplied by $(1-\widehat{G}(\boldsymbol{k}))$ instead of $\widehat{G}(\boldsymbol{k})$. These equations are written:

$$\left(\frac{\partial}{\partial t} + \nu k^2\right)(1-\widehat{G}(\boldsymbol{k}))\widehat{u}_i(\boldsymbol{k}) = (1-\widehat{G}(\boldsymbol{k}))T_i(\boldsymbol{k}) \quad . \tag{3.53}$$

Calculations similar to those explained above lead to:

$$\begin{aligned}
\left(\frac{\partial}{\partial t} + \nu k^2\right)\widehat{u}_i'(\boldsymbol{k}) &= M_{ijm}(\boldsymbol{k})\int\int \widehat{G}(\boldsymbol{p})\widehat{G}(\boldsymbol{q})(1-\widehat{G}(\boldsymbol{k})) \\
&\quad \times \widehat{u}_j(\boldsymbol{p})\widehat{u}_m(\boldsymbol{q})\delta(\boldsymbol{k}-\boldsymbol{p}-\boldsymbol{q})d^3\boldsymbol{p}d^3\boldsymbol{q} \\
&+ M_{ijm}(\boldsymbol{k})\int\int (1-\widehat{G}(\boldsymbol{k}))\Big(\widehat{G}(\boldsymbol{p})(1-\widehat{G}(\boldsymbol{q})) \\
&\quad + \widehat{G}(\boldsymbol{q})(1-\widehat{G}(\boldsymbol{p}))\Big) \\
&\quad \times \widehat{u}_j(\boldsymbol{p})\widehat{u}_m(\boldsymbol{q})\delta(\boldsymbol{k}-\boldsymbol{p}-\boldsymbol{q})d^3\boldsymbol{p}d^3\boldsymbol{q} \\
&+ M_{ijm}(\boldsymbol{k})\int\int (1-\widehat{G}(\boldsymbol{k}))\Big((1-\widehat{G}(\boldsymbol{q}))(1-\widehat{G}(\boldsymbol{p}))\Big) \\
&\quad \times \widehat{u}_j(\boldsymbol{p})\widehat{u}_m(\boldsymbol{q})\delta(\boldsymbol{k}-\boldsymbol{p}-\boldsymbol{q})d^3\boldsymbol{p}d^3\boldsymbol{q} \quad .
\end{aligned} \tag{3.54}$$

The first term of the right-hand side represents the contribution of the interactions between large scale modes, the second the contribution of the cross interactions, and the last the interactions among the subgrid modes.

Let $\overline{E}_{\rm sgs}$ be the energy contained in the subgrid modes. This energy is defined as:

$$\overline{E}_{\rm sgs}(k) = \frac{1}{2}k^2 \int (1-\widehat{G}(\boldsymbol{k}))\widehat{\boldsymbol{u}}(\boldsymbol{k})\cdot(1-\widehat{G}(\boldsymbol{k}))\widehat{\boldsymbol{u}}^*(\boldsymbol{k})dS(\boldsymbol{k}) \quad (3.55)$$

$$= (1-\widehat{G})^2(k)E(k) \quad , \quad (3.56)$$

and is different, in the general case, from the kinetic energy fluctuation $E'(k) = (1-\widehat{G})(k)E(k)$, though the equality of these two quantities is verified when the filter is a Reynolds operator. Simple calculations give us the following evolution equation for $\overline{E}_{\rm sgs}(k)$:

$$\left(\frac{\partial}{\partial t} + 2\nu k^2\right)\overline{E}_{\rm sgs}(k) = \frac{1}{2}\int\int_\Delta \widehat{G}(\boldsymbol{p})\widehat{G}(\boldsymbol{q})(1-\widehat{G}(\boldsymbol{k}))^2 S(\boldsymbol{k}|\boldsymbol{p},\boldsymbol{q})d\boldsymbol{p}d\boldsymbol{q}$$
$$+ \frac{1}{2}\int\int_\Delta (1-\widehat{G}(\boldsymbol{k}))\Big(\widehat{G}(\boldsymbol{p})(1-\widehat{G}(\boldsymbol{q}))$$
$$+\widehat{G}(\boldsymbol{q})(1-\widehat{G}(\boldsymbol{p}))\Big) S(\boldsymbol{k}|\boldsymbol{p},\boldsymbol{q})d\boldsymbol{p}d\boldsymbol{q}$$
$$+ \frac{1}{2}\int\int_\Delta (1-\widehat{G}(\boldsymbol{k}))^2(1-\widehat{G}(\boldsymbol{q}))(1-\widehat{G}(\boldsymbol{p}))$$
$$\times S(\boldsymbol{k}|\boldsymbol{p},\boldsymbol{q})d\boldsymbol{p}d\boldsymbol{q} \quad , \quad (3.57)$$

where the notation used is the same as for the kinetic energy evolution equation of the resolved modes. The subgrid kinetic energy $q^2_{\rm sgs}$ is obtained by summation over the entire spectrum:

$$q^2_{\rm sgs} = \int_0^\infty \overline{E}_{\rm sgs}(k)dk \quad . \quad (3.58)$$

3.3.2 Germano Consistent Decomposition

This section presents the Germano consistent decomposition, which is a generalization of the Leonard decomposition.

Definition and Properties of Generalized Central Moments. For convenience, we use $[\phi]_{\rm G}$ to denote the resolved part of the field ϕ, defined as in the first chapter, where G is the convolution kernel, i.e.:

$$[\phi]_{\rm G}(\boldsymbol{x}) \equiv G \star \phi(\boldsymbol{x}) \equiv \int_{-\infty}^{+\infty} G(\boldsymbol{x}-\boldsymbol{\xi})\phi(\boldsymbol{\xi})d^3\boldsymbol{\xi} \quad . \quad (3.59)$$

We define the generalized central moments with the filter G, denoted $\tau_{\rm G}$, as [244, 246, 247, 248]:

$$\tau_{\rm G}(\phi_1,\phi_2) = [\phi_1\phi_2]_{\rm G} - [\phi_1]_{\rm G}[\phi_2]_{\rm G} \quad , \quad (3.60)$$

$$\tau_{\rm G}(\phi_1,\phi_2,\phi_3) = [\phi_1\phi_2\phi_3]_{\rm G} - [\phi_1]_{\rm G}\tau_{\rm G}(\phi_2,\phi_3) - [\phi_2]_{\rm G}\tau_{\rm G}(\phi_1,\phi_3)$$
$$-[\phi_3]_{\rm G}\tau_{\rm G}(\phi_1,\phi_2) - [\phi_1]_{\rm G}[\phi_2]_{\rm G}[\phi_3]_{\rm G} \quad , \quad (3.61)$$

$$\tau_{\rm G}(\phi_1,\phi_2,\phi_3,\phi_4) = \ldots \quad (3.62)$$

The generalized central moments thus defined verify the following properties:

$$\tau_G(\phi, \psi) = \tau_G(\psi, \phi) \quad , \tag{3.63}$$

$$\tau_G(\phi, a) = 0, \quad \text{for a = const.} \quad , \tag{3.64}$$

$$\tau_G(\phi, \psi, a) = 0, \quad \text{for a = const.} \quad , \tag{3.65}$$

$$\partial\tau_G(\phi, \psi)/\partial s = \tau_G(\partial\phi/\partial s, \psi) + \tau_G(\phi, \partial\psi/\partial s), \ s = \boldsymbol{x}, t \quad . \tag{3.66}$$

If we perform the decomposition $\phi = \phi_1 + \phi_2, \psi = \psi_1 + \psi_2$, we get:

$$\begin{aligned} \tau_G(\psi_1 + \psi_2, \phi_1 + \phi_2) &= \tau_G(\psi_1, \phi_1) + \tau_G(\psi_1, \phi_2) \\ &+ \tau_G(\psi_2, \phi_1) + \tau_G(\psi_2, \phi_2) \quad . \end{aligned} \tag{3.67}$$

The generalized central moments also appear as the coefficients of the following formal Taylor expansion [247]:

$$\begin{aligned} [\phi(a_1, ..., a_n)]_G &= \phi([a_1]_G, ..., [a_n]_G) + \sum_{l,m} \frac{\tau_G(a_l, a_m)}{2!} y_{lm} \\ &+ \sum_{l,m,k} \frac{\tau_G(a_l, a_m, a_k)}{3!} y_{lmk} + ... \quad , \end{aligned} \tag{3.68}$$

with

$$y_{lm} = \frac{\partial^2 \phi([a_1]_G, ..., [a_n]_G)}{\partial [a_l]_G \partial [a_m]_G}, \quad y_{lmk} = \frac{\partial^3 \phi([a_1]_G, ..., [a_n]_G)}{\partial [a_l]_G \partial [a_m]_G \partial [a_k]_G} \quad ,$$

and where the a_i are generic turbulent quantities. The relation (3.68) establishes a link between the filtered value of the functional ϕ and its unfiltered counterpart applied to the filtered variables $[a_i]_G$.

Consistent Decomposition: Associated Equations. By applying the property (3.67) to the decomposition $\phi = [\phi]_G + \phi', \psi = [\psi]_G + \psi'$, we get:

$$\begin{aligned} \tau_G([\phi]_G + \phi', [\psi]_G + \psi') &= \tau_G([\phi]_G, [\psi]_G) + \tau_G(\phi', [\psi]_G) \\ &+ \tau_G([\phi]_G, \psi') + \tau_G(\phi', \psi') \quad . \end{aligned} \tag{3.69}$$

This decomposition is said to be *consistent* because it is consistent with the definition of the generalized central moments, ensuring that all the terms in it are of the same form, which is not true of the Leonard decomposition. The various terms of the right-hand side of equation (3.69) can be interpreted as generalizations of the terms of the Leonard triple decomposition. By applying this definition to the components of the velocity fields, the subgrid tensor (3.23) appears in a double form:

$$\begin{aligned} \tau_G(u_i, u_j) &= [u_i u_j]_G - [u_i]_G [u_j]_G \\ &= L_{ij} + C_{ij} + R_{ij} \\ &= \mathcal{L}_{ij} + \mathcal{C}_{ij} + \mathcal{R}_{ij} \quad , \end{aligned} \tag{3.70}$$

in which the tensors $\mathcal{L}$, $\mathcal{C}$ and $\mathcal{R}$ are defined as:

$$\mathcal{L}_{ij} = \tau_{\mathrm{G}}([u_i]_{\mathrm{G}}, [u_j]_{\mathrm{G}}) \quad , \tag{3.71}$$

$$\mathcal{C}_{ij} = \tau_{\mathrm{G}}([u_i]_{\mathrm{G}}, u'_j) + \tau_{\mathrm{G}}(u'_i, [u_j]_{\mathrm{G}}) \quad , \tag{3.72}$$

$$\mathcal{R}_{ij} = \tau_{\mathrm{G}}(u'_i, u'_j) \quad , \tag{3.73}$$

and represent, respectively, the interactions between the large scales, the cross interactions, and the interactions among subgrid scales. They therefore represent tensors defined by Leonard, but are not the same as them in the general case.

By bringing out the generalized central moments, the filtered momentum equations are written in the form:

$$\frac{\partial [u_i]_{\mathrm{G}}}{\partial t} + \frac{\partial}{\partial x_j}([u_i]_{\mathrm{G}}[u_j]_{\mathrm{G}}) = -\frac{\partial [p]_{\mathrm{G}}}{\partial x_i} + \nu \frac{\partial}{\partial x_j}\left(\frac{\partial [u_i]_{\mathrm{G}}}{\partial x_j} + \frac{\partial [u_j]_{\mathrm{G}}}{\partial x_i}\right) - \frac{\partial \tau_{\mathrm{G}}(u_i, u_j)}{\partial x_j} \quad . \tag{3.74}$$

This equation is equivalent to the one derived from the triple Leonard decomposition. Similarly, the subgrid kinetic energy evolution equation (3.33) is re-written as:

$$\frac{\partial q^2_{\mathrm{sgs}}}{\partial t} = \frac{\partial}{\partial x_j}\left(\frac{1}{2}\tau_{\mathrm{G}}(u_i, u_i, u_j) + \tau_{\mathrm{G}}(p, u_j) - \nu \frac{\partial q^2_{\mathrm{sgs}}}{\partial x_j}\right)$$

$$- \nu \tau_{\mathrm{G}}(\partial u_i/\partial x_j, \partial u_i/\partial x_j) - \tau_{\mathrm{G}}(u_i, u_j)\frac{\partial [u_i]_{\mathrm{G}}}{\partial x_j} \quad . \tag{3.75}$$

It is easy to check that the structure of the filtered equations is, in terms of generalized central moments, independent of the filter used. This is called the *filtering (or averaging) invariance property*.

3.3.3 Germano Identity

Basic Germano Identity. Subgrid tensors corresponding to two different filtering levels can be related by an exact relation derived by Germano [246].

A sequential application of two filters, F and G, is denoted:

$$[u_i]_{\mathrm{FG}} = [[u_i]_{\mathrm{F}}]_{\mathrm{G}} = [[u_i]_{\mathrm{G}}]_{\mathrm{F}} \quad , \tag{3.76}$$

or equivalently:

$$[u_i]_{\mathrm{FG}}(\boldsymbol{x}) = \int_{-\infty}^{+\infty} G(\boldsymbol{x} - \boldsymbol{y}) d^3\boldsymbol{y} \int_{-\infty}^{+\infty} F(\boldsymbol{y} - \boldsymbol{\xi}) u_i(\boldsymbol{\xi}) d^3\boldsymbol{\xi} \quad . \tag{3.77}$$

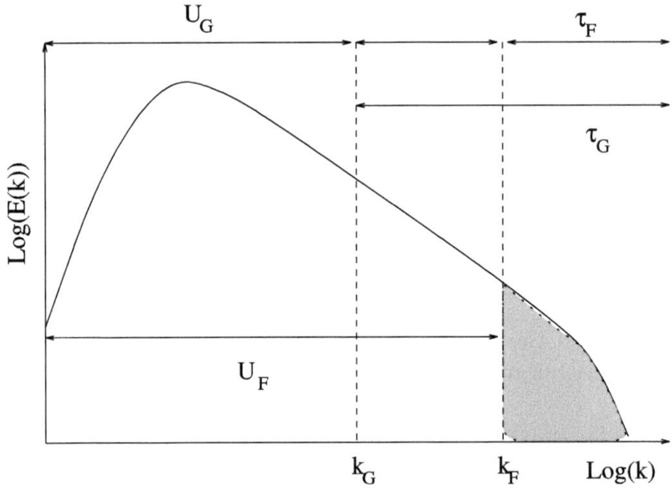

Fig. 3.3. Illustration of the two filtering levels, F and G, involved in the Germano identity. Associated cutoff wave numbers are denoted $k_{\rm F}$ and $k_{\rm G}$, respectively. Resolved velocity fields are $\boldsymbol{u}_{\rm F}$ and $\boldsymbol{u}_{\rm G}$, and the associated subgrid tensors are $\tau_{\rm F}$ and $\tau_{\rm G}$, respectively.

Here, $[u_i]_{\rm FG}$ corresponds to the resolved field for the double filtering FG. The two filtering levels are illustrated in Fig. 3.3.

The subgrid tensor associated with the level FG is defined as the following generalized central moment:

$$\tau_{\rm FG}(u_i, u_j) = [u_i u_j]_{\rm FG} - [u_i]_{\rm FG}[u_j]_{\rm FG} \quad . \tag{3.78}$$

This expression is a trivial extension of the definition of the subgrid tensor associated with the G filtering level. By definition, the subgrid tensor $\tau_{\rm G}([u_i]_{\rm F}, [u_j]_{\rm F})$ calculated from the scales resolved for the F filtering level, is written:

$$\tau_{\rm G}([u_i]_{\rm F}, [u_j]_{\rm F}) = [[u_i]_{\rm F}[u_j]_{\rm F}]_{\rm G} - [u_i]_{\rm FG}[u_j]_{\rm FG} \quad . \tag{3.79}$$

These two subgrid tensors are related by the following exact relation, called the Germano identity:

$$\tau_{\rm FG}(u_i, u_j) = [\tau_{\rm F}(u_i, u_j)]_{\rm G} + \tau_{\rm G}([u_i]_{\rm F}, [u_j]_{\rm F}) \quad . \tag{3.80}$$

This relation can be interpreted physically as follows. The subgrid tensor at the FG filtering level is equal to the sum of the subgrid tensor at the F level filtered at the G level and the subgrid tensor at the G level calculated from the field resolved at the F level. This relation is local in space and time and is independent of the filter used.

It is interesting noting that re-writing the subgrid tensor as

$$\tau_{FG}(u_i, u_j) = [F \star G\star, B](u_i, u_j) \quad ,$$

where $[.,.]$ is the commutator operator (see equation (2.13)) and $B(.,.)$ the bilinear form defined by relation (3.27), the Germano identity (3.80) is strictly equivalent to relation (2.16):

$$[F \star G\star, B](u_i, u_j) = [F\star, B] \circ (G\star)(u_i, u_j) + (F\star) \circ [G\star, B](u_i, u_j). \quad (3.81)$$

The previous Germano identity can be referred to as the *multiplicative Germano identity* [251], because it is based on a sequential application of the two filters. An *additive Germano identity* can also be defined considering that the second filtering level is defined by the operator $(F + G)\star$ and not by the operator $(G\star) \circ (F\star)$. The equivalent relation for (3.80) is

$$\tau_{\mathrm{F+G}}(u_i, u_j) = \tau_{\mathrm{F}}(u_i, u_j) + \tau_{\mathrm{G}}(u_i, u_j) - ([u_i]_{\mathrm{F}}[u_j]_{\mathrm{G}} + [u_i]_{\mathrm{G}}[u_j]_{\mathrm{F}}) \quad . \quad (3.82)$$

Multilevel Germano Identity. The Multiplicative Germano Identity can be extended to the case of N filtering levels, $G_i, i = 1, N$, with associated characteristic lengths $\overline{\Delta}_1 \leq \overline{\Delta}_2 \leq ... \leq \overline{\Delta}_N$ [248, 710, 633].

We define the nth level filtered variable $\overline{\phi}^n$ as

$$\overline{\phi}^n = G_n \star G_{n-1} \star ... \star G_1 \star \phi = \mathcal{G}_1^n \star \phi \quad , \quad (3.83)$$

with

$$\mathcal{G}_m^n \equiv G_n \star G_{n-1} \star ... \star G_m, \mathcal{G}_n^n = Id, \quad \forall m \in [1, n] \quad . \quad (3.84)$$

Let $\tau_{ij}^n = \overline{u_i u_j}^n - \overline{u}_i^n \overline{u}_j^n$ be the subgrid tensor associated to the nth filtering level. The classical two-level Germano identity (3.80) reads

$$\tau_{ij}^{n+1} = \overline{\tau_{ij}^n}^{n+1} + L_{ij}^{n+1}, \quad L_{ij}^{n+1} = \overline{\overline{u}_i^n \overline{u}_j^n}^{n+1} - \overline{u}_i^{n+1} \overline{u}_j^{n+1} \quad . \quad (3.85)$$

Simple algebraic developments lead to the following relation between two filtering levels n and m, with $m < n$:

$$\tau_{ij}^n = L_{ij}^n + \sum_{k=m+1,n-1} \mathcal{G}_{k+1}^n L_{ij}^k + \mathcal{G}_{m+1}^n \tau_{ij}^m \quad , \quad (3.86)$$

resulting in a fully general mutilevel identity.

Generalized Germano Identity. A more general multiplicative identity is obtained by applying an arbitrary operator $\mathcal{L}$ to the basic identity (3.81) [629], yielding

$$\begin{aligned} \mathcal{L}\{[F \star G\star, B](u_i, u_j)]\} &= \mathcal{L}\{[F\star, B] \circ (G\star)(u_i, u_j) \\ &+ (F\star) \circ [G\star, B](u_i, u_j)\} \quad . \end{aligned} \quad (3.87)$$

For linear operators, we get

$$\begin{aligned} \mathcal{L}\{[F \star G\star, B](u_i, u_j)]\} &= \mathcal{L}\{[F\star, B] \circ (G\star)(u_i, u_j)\} \\ &+ \mathcal{L}\{(F\star) \circ [G\star, B](u_i, u_j)\} \quad . \end{aligned} \quad (3.88)$$

Application to the multilevel identity (3.86) is straightforward.

3.3.4 Invariance Properties

One of the basic principles of modeling in mechanics is to conserve the generic properties of the starting equations [681, 230, 260, 572, 326, 398].

We consider in the present section the analysis of some invariance/symmetry properties of the filtered Navier–Stokes equations, and the resulting constraints for subgrid models. It is remembered that a differential equation will be said to be invariant under a transformation if it is left unchanged by this transformation. It is important to note that these properties are not shared by the boundary conditions. It is shown that properties of the filtered Navier–Stokes equations depend on the filter used to operate the scale separation. The preservation of the symmetry properties of the original Navier–Stokes equations will then lead to the definition of specific requirements for the filter kernel[6] $G(x, \xi)$. The properties considered below are:

- Galilean invariance for the spatial filtering approach (p. 64).
- Galilean invariance for the time-domain filtering approach (p. 66).
- General frame-invariance properties for the time-domain filtering approach (p. 67).
- Time invariance for the spatial filtering approach (p. 69).
- Rotation invariance for the spatial filtering approach (p. 70).
- Reflection invariance for the spatial filtering approach (p. 70).
- Asymptotic Material Frame Indifference for the spatial filtering approach (p. 71).

Galilean Invariance for Spatial filter. This section is devoted to the analysis by Speziale [681] of the preservation of the Galilean invariance property for translations of the Navier–Stokes equations, first by applying a spatial filter, then by using the different decompositions presented above.

Let us take the Galilean transformation (translation):

$$\boldsymbol{x}^\bullet = \boldsymbol{x} + \boldsymbol{V}t + \boldsymbol{b}, \quad t^\bullet = t \quad , \tag{3.89}$$

in which $\boldsymbol{V}$ and $\boldsymbol{b}$ are arbitrary uniform vectors in space and constant in time. If the $(\boldsymbol{x}, t)$ frame of reference is associated with an inertial frame, then so is $(\boldsymbol{x}^\bullet, t^\bullet)$. Let $\boldsymbol{u}$ and $\boldsymbol{u}^\bullet$ be the velocity vectors expressed in the base frame of reference and the new translated one, respectively. The passage from one system to the other is defined by the relations:

$$\boldsymbol{u}^\bullet = \boldsymbol{u} + \boldsymbol{V} \quad , \tag{3.90}$$

$$\frac{\partial}{\partial x_i^\bullet} = \frac{\partial}{\partial x_i} \quad , \tag{3.91}$$

$$\frac{\partial}{\partial t^\bullet} = \frac{\partial}{\partial t} - V_i \frac{\partial}{\partial x_i} \quad . \tag{3.92}$$

[6] We will only consider filters with constant and uniform cutoff length, i.e. $\overline{\Delta}$ is independent on both space and time. Variable length filters are anisotropic or nonhomogeneous, and violate the following properties in the most general case.

The proof of the invariance of the Navier–Stokes equations for the transformation (3.89) is trivial and is not reproduced here. With this property in hand, what remains to be shown in order to prove the invariance of the filtered equations by such a transformation is that the filtering process preserves this property.

Let there be a variable ϕ such that

$$\phi^\bullet = \phi \quad . \tag{3.93}$$

The filtering in the translated frame of reference is expressed:

$$\overline{\phi^\bullet} = \int G(\boldsymbol{x}^\bullet - \boldsymbol{x}^{\bullet\prime})\phi^\bullet(\boldsymbol{x}^{\bullet\prime})d^3\boldsymbol{x}^{\bullet\prime} \quad . \tag{3.94}$$

By using the previous relations, we get:

$$\boldsymbol{x}^\bullet - \boldsymbol{x}^{\bullet\prime} = (\boldsymbol{x} + \boldsymbol{V}t + \boldsymbol{b}) - (\boldsymbol{x}' + \boldsymbol{V}t + \boldsymbol{b}) = \boldsymbol{x} - \boldsymbol{x}' \quad , \tag{3.95}$$

$$d^3\boldsymbol{x}^{\bullet\prime} = \left| \frac{\partial x^{\bullet\prime}_i}{\partial x'_j} \right| d^3\boldsymbol{x}' = d^3\boldsymbol{x}' \quad , \tag{3.96}$$

and thus, by substitution, the equality:

$$\overline{\phi^\bullet} = \int G(\boldsymbol{x} - \boldsymbol{x}')\phi(\boldsymbol{x}')d^3\boldsymbol{x}' = \overline{\phi} \quad , \tag{3.97}$$

which completes the proof[7]. The invariance of the Navier–Stokes equations for the transformation (3.89) implies that the sum of the subgrid terms and the convection term, calculated directly from the large scales, is also invariant, but not that each term taken individually is invariant. In the following, we study the properties of each term arising from the Leonard and Germano decompositions.

The above relations imply:

$$\overline{\boldsymbol{u}^\bullet} = \overline{\boldsymbol{u}} + \boldsymbol{V}, \quad \boldsymbol{u}^{\bullet\prime} = \boldsymbol{u}', \quad \overline{\boldsymbol{u}^{\bullet\prime}} = \overline{\boldsymbol{u}'} \quad , \tag{3.98}$$

which reflects the fact that the velocity fluctuations are invariant by Galilean transformation, while the total velocity is not. In the spectral space, this corresponds to the fact that only the constant mode does not remain invariant by this type of transformation since, with the V field being uniform, it alone is affected by the change of coordinate system[8].

[7] A sufficient condition is that the filter kernel appears as a function of $\boldsymbol{x} - \boldsymbol{x}'$.

[8] This is expressed:

$$\boldsymbol{V} = \text{cste} \Longrightarrow \widehat{\boldsymbol{V}}(\boldsymbol{k}) = 0 \ \forall \boldsymbol{k} \neq 0 \quad ,$$

and thus

$$\widehat{\boldsymbol{u}^\bullet}(\boldsymbol{k}) = \widehat{\boldsymbol{u}}(\boldsymbol{k}) \ \forall \boldsymbol{k} \neq 0 \quad ,$$

$$\widehat{\boldsymbol{u}^\bullet}(0) = \widehat{\boldsymbol{u}}(0) + \widehat{\boldsymbol{V}}(0) \quad .$$

In the translated frame, the Leonard tensor takes the form:

$$
\begin{aligned}
L_{ij}^{\bullet} &= \overline{\overline{u_i^{\bullet}}\,\overline{u_j^{\bullet}}} - \overline{u_i^{\bullet}}\,\overline{u_j^{\bullet}} & (3.99)\\
&= L_{ij} + \left(\overline{V_i\overline{u}_j + V_j\overline{u}_i}\right) - \left(V_i\overline{u}_j + V_j\overline{u}_i\right) & (3.100)\\
&= L_{ij} - \left(V_i\overline{u}_j' + V_j\overline{u}_i'\right) \quad . & (3.101)
\end{aligned}
$$

So this tensor is not invariant. Similar analyses show that:

$$
\begin{aligned}
C_{ij}^{\bullet} &= C_{ij} + \left(V_i\overline{u}_j' + V_j\overline{u}_i'\right) \quad , & (3.102)\\
R_{ij}^{\bullet} &= R_{ij} \quad , & (3.103)\\
L_{ij}^{\bullet} + C_{ij}^{\bullet} &= L_{ij} + C_{ij} \quad . & (3.104)
\end{aligned}
$$

The tensor C is thus not invariant in the general case, while the tensor R and the groups $L{+}C$ and $L{+}C{+}R$ are. A difference can be seen to appear here between the double and triple decompositions: the double retains groups of terms (subgrid tensor and terms computed directly) that are not individually invariant, while the groups in the triple decomposition are.

The generalized central moments are invariant by construction. That is, by combining relations (3.67) and (3.63), we immediately get:

$$
\tau_{\mathrm{G}}^{\bullet}(u_i^{\bullet}, u_j^{\bullet}) = \tau_{\mathrm{G}}(u_i, u_j) \quad . \tag{3.105}
$$

This property results in all the terms in Germano's consistent decomposition being invariant by Galilean transformation, which is all the more true for the tensors $\mathcal{L}$, $\mathcal{C}$ and $\mathcal{R}$.

Galilean Invariance and Doppler Effect for Time-Domain Filters. Pruett [603] extended the above analysis to the case of Eulerian

time-domain filtering. Using the properties of the Eulerian time-domain filtering and the fact that Navier–Stokes equations are form-invariant under Galilean transformations, one can easily prove that time-domain filtered Navier–Stokes equations are also form-invariant under these transformations:

$$
\frac{\partial \overline{u}_i^{\bullet}}{\partial t^{\bullet}} + \frac{\partial}{\partial x_j^{\bullet}}\left(\overline{u_i^{\bullet} u_j^{\bullet}}\right) = -\frac{\partial \overline{p}^{\bullet}}{\partial x_i^{\bullet}} + \nu \frac{\partial}{\partial x_j^{\bullet}}\left(\frac{\partial \overline{u}_i^{\bullet}}{\partial x_j^{\bullet}} + \frac{\partial \overline{u}_j^{\bullet}}{\partial x_i^{\bullet}}\right) \quad , \tag{3.106}
$$

$$
\frac{\partial \overline{u}_i^{\bullet}}{\partial x_i^{\bullet}} = 0 \quad . \tag{3.107}
$$

It was shown in the preceding section that the spatially filtered part of a Galilean-invariant function is itself Galilean invariant, i.e.

$$
\boldsymbol{u}^{\bullet} = \boldsymbol{u} + \boldsymbol{V} \Longrightarrow \overline{\boldsymbol{u}^{\bullet}} = \overline{\boldsymbol{u}} + \boldsymbol{V}, \quad p^{\bullet} = p \Longrightarrow \overline{p^{\bullet}} = \overline{p} \quad . \tag{3.108}
$$

A fundamental difference between spatial- and time-domain filtering is that what applies to the former does not apply to the latter. Writing the definition of the filtered velocity in the translated frame, we have:

$$\begin{aligned}
\overline{u_i^\bullet} &\equiv \overline{u_i(t^\bullet, \boldsymbol{x}^\bullet - Vt^\bullet) + V_i} \\
&= \int_{-\infty}^{t^\bullet} u_i(t^{\bullet\prime}, \boldsymbol{x}^\bullet - \boldsymbol{V}t^{\bullet\prime}) G(t^{\bullet\prime} - t^\bullet) dt^{\bullet\prime} + V_i \\
&= \int_{-\infty}^{t} u_i(t', \boldsymbol{x}^\bullet - \boldsymbol{V}t') G(t' - t) dt' + V_i \\
&\neq \int_{-\infty}^{t} u_i(t', \boldsymbol{x}^\bullet - \boldsymbol{V}t) G(t' - t) dt' + V_i \\
&= \int_{-\infty}^{t} u_i(t', \boldsymbol{x}) G(t' - t) dt' + V_i \\
&\equiv \overline{u}_i(\boldsymbol{x}, t) + V_i \quad .
\end{aligned} \tag{3.109}$$

It is seen from these equations that Eulerian temporally filtered quantities experience a Doppler shift in the direction of the translational velocity $\boldsymbol{V}$, and that equalities presented in (3.108) are recovered only when $\boldsymbol{V} = 0$. The equations remain invariant under Galilean transformations because each term is subjected to the same shift.

This Doppler shift results in a wave number-dependent frequency shift between the two frames [603], indicating that Eulerian time-domain filtering may be inadequate for flows in which structures are convected at very different characteristic velocities, such as boundary layers [289]. On the contrary, free shear flows (mixing layers, jets, and wakes) seem to be better adapted.

General Investigation of Frame-Invariance Properties of Time-Domain Filtered Navier–Stokes Equations. Previous analysis of Galilean invariance properties of Eulerian time-domain filter was extended to the more general case of Euclidean group of transformations by Pruett et al. [606].

In the case the observer is fixed in the inertial frame (referred to as case I below), the spatial coordinates in the noninertial frame vary with time, while those in the inertial frame are fixed. Thus, the general change of reference frame is expressed as

$$x_i^\bullet(t^\bullet) = Q_{ij}\left[x_i + V_i\right] \quad , \tag{3.110}$$

where $x_i^\bullet$ is refers to the coordinates of a point in a frame of reference in arbitrary time-dependent motion (rotation and translation) relative to an inertial frame tied to coordinates x_i, and $Q = Q(t)$ is a time-dependent orthogonal tensor. The vector $\boldsymbol{V}$ is also time-dependent, i.e. $\boldsymbol{V} = \boldsymbol{V}(t)$. The time in the new reference frame is obtained considering a shift $t_0 : t^\bullet = t + t_0$. The velocity in the moving frame is

$$u_i^\bullet(t^\bullet, \boldsymbol{x}^\bullet) = \dot{Q}_{ij}\left[x_i + V_i\right] + Q_{ij}\left[u_i + \dot{V}_i\right] \quad . \tag{3.111}$$

The filtering cutoff Δ being frame-invariant, it will not be explicited in the following.

In the opposite case (case II) where the observer is fixed in the noninertial frame, one obtains

$$x_i(t) = Q_{ji} x_j^\bullet - V_i \quad , \tag{3.112}$$

$$u_i(t, \boldsymbol{x}) = \dot{Q}_{ji} x_j^\bullet + Q_{ji} u_j^\bullet - \dot{V}_i \quad . \tag{3.113}$$

The spatial coordinates in the inertial frame are now time-dependent, and those in the noninertial frame are fixed.

These two different cases must be treated separately when analyzing the properties of the time-filetered Navier–Stokes equations.

Application of the Eulerian time filter in case I yields

$$\overline{x_i^\bullet}(t^\bullet) = \overline{Q}_{ij} x_i + \overline{Q_{ij} V_i} \quad , \tag{3.114}$$

$$\overline{u_i^\bullet}(t^\bullet, \boldsymbol{x}^\bullet) = \overline{\dot{Q}_{ij} \left[x_i + V_i \right]} + \overline{Q_{ij} \left[u_i + \dot{V}_i \right]} \quad . \tag{3.115}$$

The subgrid velocity field $\boldsymbol{u}^{\bullet\prime} = \boldsymbol{u}^\bullet - \overline{\boldsymbol{u}^\bullet}$ is then equal to

$$u_i^{\bullet\prime} = \left(Q_{ik} u_k - \overline{Q_{ik} u_k} \right) + \left(\dot{Q}_{ik} - \overline{\dot{Q}}_{ik} \right) x_k + \left((Q_{ik} \dot{b}_k) - \overline{(Q_{ik} \dot{b}_k)} \right) \quad . \tag{3.116}$$

In case II, the filtered and subgrid quantities are expressed as follows

$$\overline{x}_i(t) = \overline{Q}_{ji} x_j^\bullet - \overline{V}_i \quad , \tag{3.117}$$

$$\overline{u}_i(t, \boldsymbol{x}) = \overline{\dot{Q}_{ji} x_j^\bullet} + \overline{Q}_{ji} u_j^\bullet - \overline{\dot{V}}_i \quad , \tag{3.118}$$

$$u_i' = \left(Q_{ki} u_k^\bullet - \overline{Q_{ki} u_k^\bullet} \right) + \left(\dot{Q}_{kj} - \overline{\dot{Q}}_{kj} \right) x_k^\bullet - \left(\dot{b}_i - \overline{\dot{b}}_i \right) \quad . \tag{3.119}$$

A look at equations (3.115) and (3.118) show that the velocity is not frame invariant under general Euclidean transformations. The same conclusion apply for the subgrid velocity field. A noticeable difference between time- and space-filtering is that, because $Q(t)$ is a time-dependent parameter, filtered and unfiltered velocity fields do not transform in the same manner in the time-filtering approach.

The Navier–Stokes equations are known to be not frame-invariant under the Euclidean group of transformation, and can be expressed as

$$\frac{\partial u_i^\bullet}{\partial t^\bullet} + \partial u_k^\bullet \frac{\partial u_i^\bullet}{\partial x_k^\bullet} = -\frac{\partial P^\bullet}{\partial x_i^\bullet} + \nu \frac{\partial^2 u_i^\bullet}{\partial x_k^\bullet \partial x_k^\bullet} + 2\Omega_{ik} u_k^\bullet + \dot{\Omega}_{ik} x_k^\bullet \quad , \tag{3.120}$$

where the modified pressure $P^\bullet$ and the rotation rate tensor are defined as

$$P^\bullet = p^\bullet + \frac{1}{2}\Omega_{kl}\Omega_{ln}x_n^\bullet x_k^\bullet - Q_{nk}\ddot{V}_k^\bullet x_n^\bullet \quad , \tag{3.121}$$

$$\Omega_{ik} \equiv \dot{Q}_{il}Q_{kl} \quad , \tag{3.122}$$

with $\ddot{V}_k^\bullet \equiv Q_{kn}\ddot{V}_n$.

The Navier–Stokes equations under the Euclidean transformation group for an observer fixed in the noninertial reference frame are obtained by first taking the material derivative of (3.111) and applying the filter, leading to

$$\overline{\frac{Du_i^\bullet}{Dt^\bullet}} = \overline{(\dot{\Omega}_{ik} - \Omega_{il}\Omega_{lk})x_k^\bullet} + 2\overline{\Omega_{ik}u_k^\bullet} + \overline{Q_{ij}\ddot{V}_j} + \overline{Q_{ij}\frac{Du_j}{Dt}} \quad . \tag{3.123}$$

and then inserting the following expression deduced from the Navier–Stokes equations written in the inertial frame:

$$\overline{Q_{ij}\frac{Du_j}{Dt}} = -\frac{\overline{p^\bullet}}{\partial x_i^\bullet} + \nu\frac{\partial^2\overline{u_i^\bullet}}{\partial x_k^\bullet \partial x_k^\bullet} \quad , \tag{3.124}$$

yielding

$$\frac{\partial\overline{u_i^\bullet}}{\partial t^\bullet} + \overline{u_k^\bullet}\frac{\partial\overline{u^\bullet}_i}{\partial x_k^\bullet} = -\frac{\partial\overline{P^\bullet}}{\partial x_i^\bullet} + \nu\frac{\partial^2\overline{u_i^\bullet}}{\partial x_k^\bullet\partial x_k^\bullet} + 2\overline{\Omega_{ik}u_k^\bullet} + \overline{\dot{\Omega}_{ik}x_k^\bullet} - \frac{\partial\tau_{ik}^\bullet}{\partial x_k^\bullet} \quad , \tag{3.125}$$

where the filtered pressure $\overline{P^\bullet}$ and the subgrid scale tensor $\tau^\bullet$ are defined as

$$\overline{P^\bullet} = \overline{p^\bullet} + \frac{1}{2}\overline{\Omega_{kl}\Omega_{ln}x_n^\bullet x_k^\bullet} - \overline{\ddot{V}_k^\bullet x_k^\bullet} \tag{3.126}$$

$$\tau_{ik}^\bullet = \overline{u_i^\bullet u_k^\bullet} - \overline{u^\bullet}_i\overline{u_k^\bullet} \quad . \tag{3.127}$$

A comparison of equations (3.125) and (3.120) shows that the time-filtered Navier–Stokes equations do not retain the same form in the most general case, to the contrary of the spatially filtered ones. The differences appear in the Coriolis terms, the centrifugal and the rotational acceleration terms.

Time Invariance (Spatial Filters). A time shift of the amount t_0 yields the following change of coordinates:

$$t^\bullet = t + t_0, \quad \boldsymbol{x}^\bullet = \boldsymbol{x}, \quad \boldsymbol{u}^\bullet = \boldsymbol{u} \quad . \tag{3.128}$$

Since we are considering space dependent filters only, the filtered Navier–Stokes equations are automatically time-invariant, without any restriction on the filter kernel. We have:

$$\overline{\boldsymbol{u}}^\bullet = \overline{\boldsymbol{u}}, \quad \boldsymbol{u}^{\bullet\prime} = \boldsymbol{u}' \quad , \tag{3.129}$$

and

$$\tau_{ik}^\bullet = \tau_{ik} \quad , \tag{3.130}$$
$$L_{ik}^\bullet = L_{ik} \quad , \tag{3.131}$$
$$R_{ik}^\bullet = R_{ik} \quad , \tag{3.132}$$
$$C_{ik}^\bullet = C_{ik} \quad . \tag{3.133}$$

All the subgrid terms are invariant.

Rotation Invariance (Spatial Filters). We now consider the following change of reference system:

$$t^\bullet = t, \quad \boldsymbol{x}^\bullet = A\boldsymbol{x}, \quad \boldsymbol{u}^\bullet = A\boldsymbol{u} \quad , \tag{3.134}$$

where A is the rotation matrix with $A^T A = AA^T = Id$ and $|A| = 1$. Simple calculations similar to those shown for in the section devoted to Galilean invariance lead to the following relations:

$$\overline{\boldsymbol{u}}^\bullet = A\overline{\boldsymbol{u}}, \quad \boldsymbol{u}^{\bullet\prime} = A\boldsymbol{u}' \quad , \tag{3.135}$$

if and only if the filter kernel $G(x, \xi)$ satisfies

$$G(A(x - \xi)) = G(x - \xi) \Longrightarrow G(x, \xi) = G(|x - \xi|) \quad , \tag{3.136}$$

meaning that the filter must be spherically symmetric. The subgrid terms are transformed as:

$$\tau_{ik}^\bullet = A_{im} A_{kn} \tau_{mn} \quad , \tag{3.137}$$
$$L_{ik}^\bullet = A_{im} A_{kn} L_{mn} \quad , \tag{3.138}$$
$$R_{ik}^\bullet = A_{im} A_{kn} R_{mn} \quad , \tag{3.139}$$
$$C_{ik}^\bullet = A_{im} A_{kn} C_{mn} \quad , \tag{3.140}$$

and are seen to be invariant.

Reflection Invariance (Spatial Filters). We now consider a reflection in the lth direction of space:

$$t^\bullet = t;\ x_l^\bullet = -x_l; x_i^\bullet = x_i, i \neq l; u_l^\bullet = -u_l; u_i^\bullet = u_i, i \neq l \quad . \tag{3.141}$$

If the filter is such that $G(x - \xi) = G(-x + \xi)$, i.e. is symmetric, then

$$\overline{u_l^\bullet} = -\overline{u}_l; \overline{u_i^\bullet} = \overline{u}_i, i \neq l; u_l^{\bullet\prime} = -u'_l; u_i^{\bullet\prime} = u'_i, i \neq l \quad , \tag{3.142}$$

yielding

$$\tau^{\bullet}_{ik} = \beta\tau_{ik} \quad , \qquad (3.143)$$
$$L^{\bullet}_{ik} = \beta L_{ik} \quad , \qquad (3.144)$$
$$R^{\bullet}_{ik} = \beta R_{ik} \quad , \qquad (3.145)$$
$$C^{\bullet}_{ik} = \beta C_{ik} \quad , \qquad (3.146)$$

with $\beta = -1$ if $i = l$ or $k = l$ and $i \neq l$, and $\beta = 1$ otherwise. We can see that the subgrid tensor and all the terms appearing in both the double and triple decomposition are invariant.

Asymptotic Material Frame Indifference (Spatial Filters). The last symmetry considered in the present section is the asymptotic material frame indifference, which is a generalization of the preceding cases. The change of frame is expressed as:

$$t^{\bullet} = t, \quad \boldsymbol{x}^{\bullet} = A(t)\boldsymbol{x} + c(t), \quad \boldsymbol{u}^{\bullet} = A\boldsymbol{u} + d(t), \quad d(t) = \dot{c} + \dot{A}\boldsymbol{x} \quad , \qquad (3.147)$$

where the rotation matrix A is such that $A^T A = AA^T = Id$, $|A| = 1$ and $c(t)$ is a vector. The Navier–Stokes equations are not form-invariant under this group of transformation in the general case. Form invariance is recovered in the asymptotic limit of two-dimensional flows.

The resulting changes of the subgrid and resolved velocity field are:

$$\overline{\boldsymbol{u}}^{\bullet} = A\overline{\boldsymbol{u}} + d, \quad \boldsymbol{u}^{\bullet\prime} = A\boldsymbol{u}' \quad , \qquad (3.148)$$

yielding

$$\tau^{\bullet}_{ik} = A_{im}\tau_{mn}A_{kn} \quad , \qquad (3.149)$$
$$L^{\bullet}_{ik} = A_{im}\tau_{mn}L_{kn} - B_{ik} \quad , \qquad (3.150)$$
$$C^{\bullet}_{ik} = A_{im}\tau_{mn}C_{kn} + B_{ik} \quad , \qquad (3.151)$$
$$R^{\bullet}_{ik} = A_{im}R_{mn}A_{kn} \quad , \qquad (3.152)$$

with

$$B_{ij} = u'_i d_j + u'_j d_i \quad .$$

These properties are subjected to the condition $G(x,\xi) = G(|x-\xi|)$. We can see that the properties of the subgrid tensors are the same as in the case of the Galilean invariance case.

Table 3.1 summarizes the results dealing with the symmetry properties.

Table 3.1. Invariance properties of spatial convolution filters and subgrid tensors.

Symmetry	$G(x,\xi)$	L	C	$L+C$	R
Galilean translation	$G(x-\xi)$	no	no	yes	yes
Time shift	$G(x,\xi)$	yes	yes	yes	yes
Rotation	$G(\|x-\xi\|)$	yes	yes	yes	yes
Reflection	$G(x-\xi)=G(\xi-x)$	yes	yes	yes	yes
Asymptotic material indifference	$G(\|x-\xi\|)$	no	no	yes	yes

3.3.5 Realizability Conditions

A second-rank tensor τ is realizable or semi-positive definite, if the following inequalities are verified (without summation on the repeated greek indices) [746, 260]:

$$\tau_{\alpha\alpha} \geq 0, \qquad \alpha = 1,2,3 \quad , \tag{3.153}$$

$$|\tau_{\alpha\beta}| \leq \sqrt{\tau_{\alpha\alpha}\tau_{\beta\beta}}, \qquad \alpha,\beta = 1,2,3 \quad , \tag{3.154}$$

$$\det(\tau) \geq 0 \quad . \tag{3.155}$$

These conditions can be written in several equivalent forms [260]. Some of these are listed below.

1. The quadratic form
$$Q = x_i \tau_{ij} x_j \tag{3.156}$$
is positive semidefinite.
2. The three principal invariants of τ are nonnegative:

$$I_1 = \sum \tau_{\alpha\beta} \geq 0 \quad , \tag{3.157}$$

$$I_2 = \sum_{\alpha\neq\beta} (\tau_{\alpha\alpha}\tau_{\beta\beta} - \tau_{\alpha\beta}^2) \geq 0 \quad , \tag{3.158}$$

$$I_3 = \det(\tau) \geq 0 \quad . \tag{3.159}$$

The positiveness of the filter as defined by relation (2.24) is a necessary and sufficient condition to ensure the realizability of the subgrid tensor τ. Below, we reproduce the demonstration given by Vreman et al. [746], which is limited to the case of a spatial filter $G(\boldsymbol{x}-\boldsymbol{\xi})$ without restricting the general applicability of the result.

Let us first assume that $G \geq 0$. To prove that the tensor τ is realizable at any position $\boldsymbol{x}$ of the fluid domain Ω, we define the sub-domain Ω_x representing the support of the application $\boldsymbol{\xi} \rightarrow G(\boldsymbol{x}-\boldsymbol{\xi})$. Let F_x be the space of real functions defined on Ω_x. Since G is positive, for $\phi, \psi \in F_x$, the application

$$(\phi, \psi)_x = \int_{\Omega_x} G(\boldsymbol{x} - \boldsymbol{\xi})\phi(\boldsymbol{\xi})\psi(\boldsymbol{\xi})d\boldsymbol{\xi} \tag{3.160}$$

defines an inner product on F_x. Using the definition of the filtering, the subgrid tensor can be re-written in the form:

$$\begin{aligned}
\tau_{ij}(\boldsymbol{x}) &= \overline{u_i u_j}(\boldsymbol{x}) - \overline{u}_i(\boldsymbol{x})\overline{u}_j(\boldsymbol{x}) \\
&= \overline{u_i u_j}(\boldsymbol{x}) - \overline{u}_i(\boldsymbol{x})\overline{u}_j(\boldsymbol{x}) - \overline{u}_j(\boldsymbol{x})\overline{u}_i(\boldsymbol{x}) + \overline{u}_i(\boldsymbol{x})\overline{u}_j(\boldsymbol{x}) \\
&= \int_{\Omega_x} G(\boldsymbol{x} - \boldsymbol{\xi})u_i(\boldsymbol{\xi})u_j(\boldsymbol{\xi})d^3\boldsymbol{\xi} - \overline{u}_i(\boldsymbol{x}) \int_{\Omega_x} G(\boldsymbol{x} - \boldsymbol{\xi})u_j(\boldsymbol{\xi})d^3\boldsymbol{\xi} \\
&- \overline{u}_j(\boldsymbol{x}) \int_{\Omega_x} G(\boldsymbol{x} - \boldsymbol{\xi})u_i(\boldsymbol{\xi})d^3\boldsymbol{\xi} - \overline{u}_i(\boldsymbol{x})\overline{u}_j(\boldsymbol{x}) \int_{\Omega_x} G(\boldsymbol{x} - \boldsymbol{\xi})d^3\boldsymbol{\xi} \\
&= \int_{\Omega_x} G(\boldsymbol{x} - \boldsymbol{\xi}) \left(u_i(\boldsymbol{\xi}) - \overline{u}_i(\boldsymbol{x})\right) \left(u_j(\boldsymbol{\xi}) - \overline{u}_j(\boldsymbol{x})\right) \\
&= (u_i^x, u_j^x)_x \quad ,
\end{aligned} \tag{3.161}$$

where the difference $u_i^x(\boldsymbol{\xi}) = u_i(\boldsymbol{\xi}) - \overline{u}_i(\boldsymbol{x})$ is defined on Ω_x. The tensor τ thus appears as a Grammian 3×3 matrix of inner products, and is consequently always defined as semi-positive. This shows that the stated condition is sufficient.

Let us now assume that the condition $G \geq 0$ is not verified for a piecewise continuous kernel. There then exists a pair $(\boldsymbol{x}, \boldsymbol{y}) \in \Omega \times \Omega$, an $\epsilon \in \mathbb{R}^+$, $\epsilon > 0$, and a neighbourhood $V = \{\boldsymbol{\xi} \in \Omega, |\boldsymbol{\xi} - \boldsymbol{y}| < \epsilon\}$, such that $G(\boldsymbol{x} - \boldsymbol{\xi}) < 0$, $\forall \boldsymbol{\xi} \in V$. For a function u_1 defined on Ω such that $u_1(\boldsymbol{\xi}) \neq 0$ if $\boldsymbol{\xi} \in V$ et $u_1(\boldsymbol{\xi}) = 0$ everywhere else, then the component τ_{11} is negative:

$$\tau_{11}(x) = \overline{u_1^2}(x) - (\overline{u}_1(x))^2 \leq \int_V G(\boldsymbol{x} - \boldsymbol{\xi}) \left(u_1(\boldsymbol{\xi})\right)^2 d^3\boldsymbol{\xi} < 0 \quad . \tag{3.162}$$

The tensor τ is thus not semi-positive definite, which concludes the demonstration. The properties of the three usual analytical filter presented in Sect. 2.1.5 are summarized in Table 3.2.

Table 3.2. Positiveness property of convolution filters.

Filter	Eq.	Positiveness
Box	(2.40)	yes
Gaussian	(2.42)	yes
Sharp cutoff	(2.44)	no

3.4 Extension to the Inhomogeneous Case for the Conventional Approach

The results of the previous sections were obtained by applying isotropic homogeneous filters on an unbounded domain to Navier–Stokes equations written in Cartesian coordinates. What is presented here are the equations obtained by applying non-homogeneous convolution filters on bounded domains to these equations. Using the commutator (2.13), the most general form of the filtered Navier–Stokes equations is:

$$\begin{aligned}\frac{\partial \overline{u}_i}{\partial t} &+ \frac{\partial}{\partial x_j}(\overline{u}_i\overline{u}_j) + \frac{\partial \overline{p}}{\partial x_i} - \nu\frac{\partial}{\partial x_j}\left(\frac{\partial \overline{u}_i}{\partial x_j} + \frac{\partial \overline{u}_j}{\partial x_i}\right) = -\frac{\partial \tau_{ij}}{\partial x_j} \\ &- \left[G\star, \frac{\partial}{\partial t}\right](u_i) - \left[G\star, \frac{\partial}{\partial x_j}\right](u_iu_j) \\ &- \left[G\star, \frac{\partial}{\partial x_i}\right](p) + \nu\left[G\star, \frac{\partial^2}{\partial x_k \partial x_k}\right](u_i) \quad , \end{aligned} \qquad (3.163)$$

$$\frac{\partial \overline{u}_i}{\partial x_i} = -\left[G\star, \frac{\partial}{\partial x_i}\right](u_i) \quad . \qquad (3.164)$$

All the terms appearing in the right-hand side of equations (3.163) and (3.164) are commutation errors. The first term of the right-hand side of the filtered momentum equation is the subgrid force, and is subject to modeling. The other terms are artefacts due to the filter, and escape subgrid modeling.

An interesting remark drawn from equation (3.164) is that the filtered field is not divergence-free if some commutation errors arise. An analysis of the breakdown of continuity constraint in large-eddy simulation is provided by Langford and Moser [420], which shows that for many common large-eddy simulation representations, there is no exact continuity constraint on the filtered velocity field. But for mean-preserving representations a bulk continuity constraint holds.

The governing equations obtained using second-order commuting filters (SOCF), as well as the techniques proposed by Ghosal and Moin [262] and Iovenio and Tordella [343] to reduce the commutation error and Vasilyev's high-order commuting filters [728], are presented in the following.

3.4.1 Second-Order Commuting Filter

Here we propose to generalize Leonard's approach by applying SOCF filters. The decomposition of the non-linear term considered here as an example is the triple decomposition; but the double decomposition is also usable. For convenience in writing the filtered equations, we introduce the operator $\mathcal{D}_i$

such that:

$$\overline{\frac{\partial \psi}{\partial x_i}} = \mathcal{D}_i \overline{\psi} \quad . \tag{3.165}$$

According to the results of Sect. 2.2.2, the operator $\mathcal{D}_i$ is of the form:

$$\mathcal{D}_i = \frac{\partial}{\partial x_i} - \alpha^{(2)} \overline{\Delta}^2 \Gamma_{ijk} \frac{\partial^2}{\partial x_i^2} + O(\overline{\Delta}^4) \quad , \tag{3.166}$$

in which the term Γ is defined by the relation (2.147). By applying the filter and bringing out the subgrid tensor $\tau_{ij} = \overline{u_i u_j} - \overline{u}_i \overline{u}_j$, we get for the momentum equation:

$$\frac{\partial \overline{u}_i}{\partial t} + \mathcal{D}_j(\overline{u}_i \overline{u}_j) = -\mathcal{D}_i \overline{p} + \nu \mathcal{D}_j \mathcal{D}_j \overline{u}_i - \mathcal{D}_j \tau_{ij} \quad . \tag{3.167}$$

To measure the errors, we introduce the expansion as a function of $\overline{\Delta}$:

$$\overline{p} = \overline{p}^{(0)} + \overline{\Delta}^2 \overline{p}^{(1)} + \dots \quad , \tag{3.168}$$

$$\overline{\boldsymbol{u}} = \overline{\boldsymbol{u}}^{(0)} + \overline{\Delta}^2 \overline{\boldsymbol{u}}^{(1)} + \dots \tag{3.169}$$

The terms corresponding to the odd powers of $\overline{\Delta}$ are identically zero because of the symmetry of the convolution kernel. By substituting this decomposition in (3.167), at the first order we get:

$$\frac{\partial \overline{u}_i^{(0)}}{\partial t} + \frac{\partial}{\partial x_j}\left(\overline{u}_i^{(0)} \overline{u}_j^{(0)}\right) = -\frac{\partial \overline{p}^{(0)}}{\partial x_i} + \nu \frac{\partial}{\partial x_j}\left(\frac{\partial \overline{u}_i^{(0)}}{\partial x_j} + \frac{\partial \overline{u}_j^{(0)}}{\partial x_i}\right) - \frac{\partial \tau_{ij}^{(0)}}{\partial x_j} \quad , \tag{3.170}$$

in which $\tau_{ij}^{(0)}$ is the subgrid term calculated from the field $\overline{\boldsymbol{u}}^{(0)}$. The associated continuity equation is:

$$\frac{\partial \overline{u}_i^{(0)}}{\partial x_i} = 0 \quad . \tag{3.171}$$

These equations are identical to those obtained in the homogeneous case, but relate to a variable containing an error in $O(\overline{\Delta}^2)$ with respect to the exact solution.

To reduce the error, the problem of the term in $\overline{\Delta}^2$ has to be resolved, i.e. solve the equations that use the variables $\overline{\boldsymbol{u}}^{(1)}$ and $\overline{p}^{(1)}$. Simple expansions lead to the system:

$$\begin{aligned} \frac{\partial \overline{u}_i^{(1)}}{\partial t} + \frac{\partial}{\partial x_j}\left(\overline{u}_i^{(1)} \overline{u}_j^{(0)} + \overline{u}_i^{(0)} \overline{u}_j^{(1)}\right) &= -\frac{\partial \overline{p}^{(1)}}{\partial x_i} + \nu \frac{\partial}{\partial x_j}\left(\frac{\partial \overline{u}_i^{(1)}}{\partial x_j} + \frac{\partial \overline{u}_j^{(1)}}{\partial x_i}\right) \\ & \quad - \frac{\partial \tau_{ij}^{(1)}}{\partial x_j} + \alpha^{(2)} f_i^{(1)} \quad , \end{aligned} \tag{3.172}$$

in which the coupling term $f_i^{(1)}$ defined as:

$$
\begin{aligned}
f_i^{(1)} &= \Gamma_{jmn}\frac{\partial^2(\overline{u}_i^{(0)}\overline{u}_j^{(0)})}{\partial x_m \partial x_n} + \Gamma_{imn}\frac{\partial^2 \overline{p}^{(0)}}{\partial x_m \partial x_n} + \Gamma_{jmn}\frac{\partial^2 \tau_{ij}^{(0)}}{\partial x_m \partial x_n} \\
&- \nu \frac{\partial \Gamma_{kmn}}{\partial x_k}\frac{\partial^2 \overline{u}_i^{(0)}}{\partial x_m \partial x_n} - 2\Gamma_{kmn}\frac{\partial^3 \overline{u}_i^{(0)}}{\partial x_k \partial x_m \partial x_n} \quad ,
\end{aligned} \tag{3.173}
$$

$$
\frac{\partial \overline{u}_i^{(1)}}{\partial x_i} = 0 \quad . \tag{3.174}
$$

By solving this second problem, we can ensure the accuracy of the solution up to the order $O(\overline{\Delta}^4)$.

Another procedure aiming at removing the commutation error was proposed by Iovenio and Tordella [343]. It relies on an approximation of the commutation error terms up to the fourth order in terms of $\overline{\Delta}$ which is based on the use of several filtering levels. Reminding that the commutation error between the filtering operator and the first-order spatial derivative can be expressed as

$$
\left[G\star, \frac{d}{dx}\right](\phi) = -\frac{d\overline{\Delta}(x)}{dx}\frac{\partial}{\partial \overline{\Delta}}\overline{\phi}^{\overline{\Delta}}(x) \quad , \tag{3.175}
$$

where $\overline{\phi}^{\overline{\Delta}}$ denotes the filtered quantity obtained applying a filter with length $\overline{\Delta}$ on the variable ϕ, and introducing the central second order finite-difference approximation for the gradient of the filtered quantity with respect to the filter width:

$$
\frac{\partial}{\partial \overline{\Delta}}\overline{\phi}^{\overline{\Delta}} = \frac{1}{2h}\left(\overline{\phi}^{\overline{\Delta}+h} - \overline{\phi}^{\overline{\Delta}-h}\right) + O(h^2) \tag{3.176}
$$

one obtains the following explicit, two filtering level approximation for the commutation error

$$
\left[G\star, \frac{d}{dx}\right](\phi) \simeq -\frac{d\overline{\Delta}(x)}{dx}\frac{1}{2\overline{\Delta}}\left(\overline{\left(\overline{\phi}^{\overline{\Delta}}\right)}^{2\overline{\Delta}} - \overline{\phi}^{\overline{\Delta}}\right) \quad , \tag{3.177}
$$

This evaluation is independent of the exact filter shape, and makes it possible to cancel the leading error term in each part of the filtered Navier-Stokes equations (3.163) - (3.164). It just involves the definition of an auxiliary filtering level with a cutoff length equal to $2\overline{\Delta}$.

3.4.2 High-Order Commuting Filters

The use of Vasilyev's filters (see Sect. 2.2.2) instead of SOCF yields a set of governing filtered equations formally equivalent to (3.167), but with:

$$\mathcal{D}_i = \frac{\partial}{\partial x_i} + O(\overline{\Delta}^n) \quad , \tag{3.178}$$

where the order of accuracy n is fixed by the number of vanishing moments of the filter kernel. The classical filtered equations, without extra-terms accounting for the commutation errors, relate to a variable containing an error scaling as $O(\overline{\Delta}^n)$ with respect to the exact filtered solution.

3.5 Filtered Navier–Stokes Equations in General Coordinates

3.5.1 Basic Form of the Filtered Equations

Jordan [358, 359], followed by other researchers [780, 18], proposed operating the filtering in the transformed plane, following the alternate approach, as defined at the beginning of this chapter. Assuming that the filter width and local grid spacing are equal, the resolved and filtered flowfields are identical. It is recalled that the filtering operation is applied along the curvilinear lines:

$$\overline{\psi(\xi^k)\phi(\xi^k)} = \int G(\xi^k - \xi'^k)\psi(\xi'^k)\phi(\xi'^k)d\xi'^k \quad , \tag{3.179}$$

where ψ is a metric coefficient or a group of metric coefficients, ϕ a physical variable (velocity component, pressure), G a homogeneous filter kernel, and ξ^k the coordinate along the considered line. It is easily deduced from the results presented in Sect. 2.2 that the commutation error vanishes in the present case, thanks to the homogeneity of the kernel: $\partial G/\partial\overline{\Delta} = 0$. But it is worth noting that the error term coming from the boundary of the domain will not cancel in the general case.[9]

Application of the filter to the Navier–Stokes equations written in generalized coordinates (3.3) and (3.4) leads to the following set of governing equations for large-eddy simulation:

$$\frac{\partial}{\partial \xi^k}(\overline{J^{-1}\xi_i^k u_i}) = 0 \quad , \tag{3.180}$$

$$\frac{\partial}{\partial t}(\overline{J^{-1}u_i}) + \frac{\partial}{\partial \xi^k}(\overline{U^k u_i}) = -\frac{\partial}{\partial \xi^k}(\overline{J^{-1}\xi_i^k p}) + \nu\frac{\partial}{\partial \xi^k}\left(\overline{J^{-1}G^{kl}\frac{\partial}{\partial \xi^l}(u_i)}\right) \quad . \tag{3.181}$$

[9] This point is extensively discussed in Chap. 10.

3.5.2 Simplified Form of the Equations – Non-linear Terms Decomposition

It is seen that many filtered nonlinear terms appear in (3.180) and (3.181) which originate from the coordinate transformation. In order to uncouple geometrical quantities, such as metrics and Jacobian, from quantities related to the flow, like velocity, and to obtain a simpler problem, further assumptions are required. The metrics being computed by a finite difference approximation in practice, they can be considered as filtered quantities, yielding:

$$\overline{U^k} = \overline{J^{-1}\xi_j^k u_j} \simeq \overline{J^{-1}}\ \overline{\xi_j^k}\overline{u_j} \quad . \tag{3.182}$$

All the terms appearing in the filtered equations can be simplified similarly. As for the conventional approach, convective nonlinear terms need to be decomposed in order to allow us to use them for practical purpose. The resulting equations are:

$$\frac{\partial}{\partial \xi^k}(\overline{U^k}) = 0 \quad , \tag{3.183}$$

$$\begin{aligned} \frac{\partial}{\partial t}(\overline{J^{-1}u_i}) + \frac{\partial}{\partial \xi^k}(\overline{U^k}\,\overline{u_i}) \quad &= \quad -\frac{\partial}{\partial \xi^k}(\overline{J^{-1}\xi_i^k p}) \\ &\quad +\nu\frac{\partial}{\partial \xi^k}\left(\overline{J^{-1}}\ \overline{G^{kl}}\frac{\partial}{\partial \xi^l}(\overline{u_i})\right) - \frac{\partial}{\partial \xi^k}(\sigma_i^k) \quad , \end{aligned} \tag{3.184}$$

where the contravariant counterpart of the subgrid tensor is defined as

$$\sigma_i^k = \overline{J^{-1}\xi_j^k u_i u_j} - \overline{J^{-1}\xi_j^k u_j}\,\overline{u_i} = \overline{U^k u_i} - \overline{U^k}\,\overline{u_i} \quad . \tag{3.185}$$

Taking into account the fact that the metrics are assumed to be smooth filtered quantities, the contravariant subgrid tensor can be tied to the subgrid tensor defined in Cartesian coordinates:

$$\sigma_i^k = \overline{J^{-1}\xi_j^k u_i u_j} - \overline{J^{-1}\xi_j^k u_j}\ \overline{u_i} = \overline{J^{-1}}\ \overline{\xi_j^k}\ (\overline{u_i u_j} - \overline{u}_i\overline{u}_j) = \overline{J^{-1}}\ \overline{\xi_j^k}\tau_{ij} \quad . \tag{3.186}$$

3.6 Closure Problem

3.6.1 Statement of the Problem

As was already said in the first chapter, large-eddy simulation is a technique for reducing the number of degrees of freedom of the solution. This is done by separating the scales in the exact solution into two categories: resolved scales and subgrid scales. The selection is made by the filtering technique described above.

The complexity of the solution is reduced by retaining only the large scales in the numerical solution process, which entails reducing the number of degrees of freedom in the solution in space and time. The information concerning the small scales is consequently lost, and none of the terms that use these scales, i.e. the terms in $\boldsymbol{u}'$ in the physical space and in $(1 - \widehat{G})$ in the spectral space, can be calculated directly. They are grouped into the subgrid tensor τ. This scale selection determines the level of resolution of the mathematical model.

Nonetheless, in order for the dynamics of the resolved scales to remain correct, the subgrid terms have to be taken into consideration, and thus have to be modeled. The modeling consists of approximating the coupling terms on the basis of the information contained in the resolved scales alone. The modeling problem is twofold:

1. Since the subgrid scales are lacking in the simulation, their existence is unknown and cannot be decided locally in space and time. The problem thus arises of knowing if the exact solution contains, at each point in space and time, any smaller scales than the resolution established by the filter. In order to answer this question, additional information has to be introduced, in either of two ways. The first is to use additional assumptions derived from acquired knowledge in fluid mechanics to link the existence of subgrid modes to certain properties of the resolved scales. The second way is to enrich the simulation by introducing new unknowns directly related to the subgrid modes, such as their kinetic energy, for example.
2. Once the existence of the subgrid modes is determined, their interactions with the resolved scales have to be reflected. The quality of the simulation will depend on the fidelity with which the subgrid model reflects these interactions.

Various modeling strategies and models that have been developed are presented in the following.

An important remark, somewhat tautological, is that the modeling process should take into account the filtering operator [597, 171, 604]. This can be seen by remarking that filtered and subgrid fields are *defined* by the filtering operator, and that a change in the filter will automatically lead to a new definition of these quantities and modify their properties.

3.6.2 Postulates

So far, we have assumed nothing concerning the type of flow at hand, aside from those assumptions that allowed us to demonstrate the momentum and continuity equations. Subgrid modeling usually assumes the following hypothesis

Hypothesis 3.1 *If subgrid scales exist, then the flow is locally (in space and time) turbulent.*

Consequently, the subgrid models will be built on the known properties of turbulent flows.

It should be noted that theories exist that use other basic hypotheses. We may mention as an example the description of suspensions in the form of a fluid with modified properties [423]: the solid particles are assumed to have predefined characteristics (mass, form, spatial distribution, and so forth) and have a characteristic size very much less than the filter cutoff length, i.e. at the scale at which we want to describe the flow dynamics directly. Their actions are taken into account globally, which means a very high saving compared with an individual description of each particle. The different descriptions obtained by homogenization techniques also enter into this framework.

3.6.3 Functional and Structural Modeling

Preliminary Remarks. Before discussing the various ways of modeling the subgrid terms, we have to set some constraints in order to orient the choices [627]. The subgrid modeling must be done in compliance with two constraints:

1. Physical constraint. The model must be consistent from the viewpoint of the phenomenon being modeled, i.e.:
   - Conserve the basic properties of the starting equation, such as Galilean invariance and asymptotic behaviors;
   - Be zero wherever the exact solution exhibits no small scales corresponding to the subgrid scales;
   - Induce an effect of the same kind (dispersive or dissipative, for example) as the modeled terms;
   - Not destroy the dynamics of the solve scales, and thus especially not inhibit the flow driving mechanisms.
2. Numerical constraint. A subgrid model can only be thought of as included in a numerical simulation method, and must consequently:
   - Be of acceptable algorithmic cost, and especially be local in time and space;
   - Not destabilize the numerical simulation;
   - Be insensitive to discretization, i.e. the physical effects induced theoretically by the model must not be inhibited by the discretization.

Modeling Strategies. The problem of subgrid modeling consists in taking the interaction with the fluctuating field $\boldsymbol{u}'$, represented by the term $\nabla \cdot \tau$, into account in the evolution equation of the filtered field $\overline{\boldsymbol{u}}$. Two modeling strategies exist [627]:

- *Structural modeling* of the subgrid term, which consists in making the best approximation of the tensor τ by constructing it from an evaluation of $\overline{\boldsymbol{u}}$ or a formal series expansion. The modeling assumption therefore consists in using a relation of the form $\boldsymbol{u}' = \mathcal{H}(\overline{\boldsymbol{u}})$ or $\tau = \mathcal{H}(\overline{\boldsymbol{u}})$.

– *Functional modeling*, which consists in modeling the action of the subgrid terms on the quantity $\overline{\boldsymbol{u}}$ and not the tensor τ itself, i.e. introducing a dissipative or dispersive term, for example, that has a similar effect but not necessarily the same structure (not the same proper axes, for example). The closure hypothesis can then be expressed in the form $\nabla \cdot \tau = \mathcal{H}(\overline{\boldsymbol{u}})$.

These two modeling approaches do not require the same foreknowledge of the dynamics of the equations treated and theoretically do not offer the same potential in terms of the quality of results obtained.

The structural approach requires no knowledge of the nature of the inter-scale interaction, but does require enough knowledge of the structure of the small scales of the solution in order to be able to determine one of the relations $u' = \mathcal{H}(\overline{\boldsymbol{u}})$ or $\tau = \mathcal{H}(\overline{\boldsymbol{u}})$ to be possible, one of the two following conditions has to be met:

– The dynamics of the equation being computed leads to a universal form of the small scales (and therefore to their total structural independence from the resolved motion, as all that remains to be determined is their energy level).
– The dynamics of the equation induces a sufficiently strong and simple inter-scale correlation for the structure of the subgrid scales to be deduced from the information contained in the resolved field.

As concerns the modeling of the inter-scale interaction by just taking its effect into account, this requires no foreknowledge of the subgrid scale structure, but does require knowing the nature of the interaction [184] [383]. Moreover, in order for such an approach to be practical, the effect of the small scales on the large must be universal in character, and therefore independent of the large scales of the flow.

4. Other Mathematical Models for the Large-Eddy Simulation Problem

The two preceding chapters are devoted to the convolution filtering mathematical model for Large-Eddy simulation. Others approaches are now described, that can be gouped in two classes:

- Mathematical models which rely on a statistical average (Sect. 4.1), recovering this way some interesting features of the Reynolds-Averaged Navier–Stokes model by precluding some drawbacks of the convolution filter approach in general domains.
- Models derived from regularized versions of the Navier–Stokes equations (Sect. 4.2), that were proposed to alleviate some theoretical problems dealing with the existence, the uniqueness and the regularity of the general solution of the three-dimensional, unsteady, incompressible Navier–Stokes equations. These regularized models have smooth solutions, in the sense that their gradients remain controlled, and are re-interpreted within the Large-Eddy Simulation framework as good candidates to account for the removal of small scales.

4.1 Ensemble-Averaged Models

4.1.1 Yoshizawa's Partial Statistical Average Model

Yoshizawa [791] proposes to combine scale decomposition and statistical average to define an *ad hoc* mathematical model for Large-Eddy Simulation, referred to as the *partial statistical average procedure*. Writing the generalized Fourier decomposition of a dummy variable $\phi(x,t)$ as

$$\phi(x,t) = \sum_{k=1,+\infty} \phi_k(t)\psi_k(x) \tag{4.1}$$

where $\phi_k(t)$ and $\psi_k(x)$ are the coefficients of the decomposition and the basis functions, respectively, the filtered part of $\phi(x,t)$ is defined as

$$\overline{\phi}(x,t) = \sum_{k=1,k_c} \phi_k(t)\psi_k(x) + \sum_{k=k_c,+\infty} \langle \phi_k(t) \rangle \psi_k(x) \tag{4.2}$$

where $\langle\cdot\rangle$ denotes a statistical average operator and k_c is related to the cutoff index of the decomposition. The partial statistical averaging method appears then as the restriction of the usual ensemble average to scales which correspond to modes higher than k_c. The cutoff length $\overline{\Delta}$ is deduced from the characteristic lengthscale associated to ψ_{k_c}.

Since it relies on an ensemble average operator, this procedure does not suffer the drawbacks of the convolution filtering approach and can be applied on curvilinear grids on bounded domains in a straightforward manner. But it requires the computation of the coefficients $\langle\phi_k(t)\rangle$, and therefore several realizations of the flow are necessary, rendering its practical implementation very expensive from the computational viewpoint. In the simple case of homogeneous flows, the statistical average can be transformed into a spatial average invoking the ergodic theorem (see Appendix A for a brief discussion).

4.1.2 McComb's Conditional Mode Elimination Procedure

Another procedure was proposed independently by McComb and coworkers [465], which is referred to as *conditional mode elimination*. These authors based their approach on the *local chaos hypothesis*, which states that in a fully turbulent flow the small scales are more uncertain than the large ones. This assumption is compatible with Kolmogorov's local isotropy hypothesis (see Sect. A.5.1 for a discussion) dealing with the universality of the small scales and their increasing (as a function of the wavenumber) statistical decoupling from the large ones. More precisely, McComb's interpretation says that uncertainty in the high-wavenumber modes originates in the amplification of some degree of uncertainty in low-wavenumber modes by the non-linear chaotic nature of turbulence. This scheme is illustrated in Fig. 4.1.

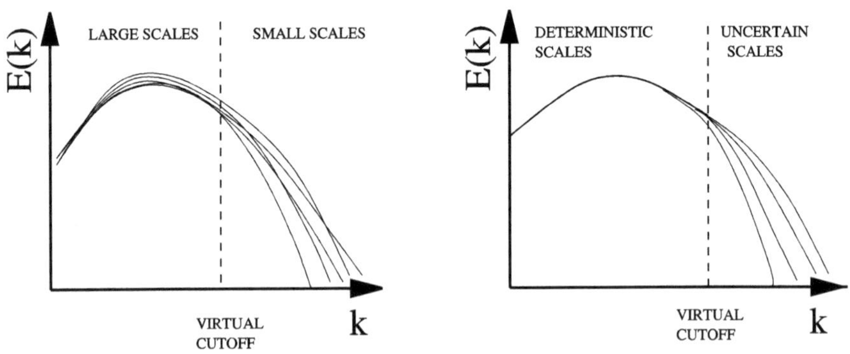

Fig. 4.1. Schematic view of the local chaos hypothesis proposed by McComb in the Fourier space. *Left*: several instantaneous spectra are shown, in which increasing uncertainty is observed. *Right*: ideal view, where wave numbers smaller than k_c are strictly deterministic, while higher wave number exhibit a fully chaotic behavior.

The scale separation with a cutoff length $\overline{\Delta}$ is achieved carying out a conditional statistical average of scales smaller than $\overline{\Delta}$, $\phi^{<}$, based on fixed realizations of scales larger than $\overline{\Delta}$, $\phi^{>}$. The former are assumed to be uncertain and to exhibit and infinite number of different realizations for each realization of the large scales.

The filtered part of $\phi(x,t)$ is then expressed as

$$\overline{\phi}(x,t) = \phi^{>}(x,t) + \langle \phi^{<} | \phi^{>} \rangle (x,t) \tag{4.3}$$

where $\langle f|g \rangle$ denotes the conditional statistical average of f with respect to g.

As Yoshizawa's procedure, the conditional mode elimination does not suffer the drawbacks of the filtering approach. These two ensemble-average based models for Large-Eddy Simulation are equivalent in many cases.

4.2 Regularized Navier–Stokes Models

The mathematical models discussed in this section were not originally proposed to represent the properties of the Large-Eddy Simulation technique. They are surrogates to the Navier–Stokes equations, which have better properties from the pure mathematical point of view: while the question of the existence, uniqueness and regulatity of the solution of the three-dimensional, unsteady and incompressible Navier–Stokes equations is still an open problem, these new models allow for a complete mathematical analysis. One of the main obstacle faced in the mathematical analysis of the Navier–Stokes equations is that it cannot yet be proven that its solutions remain smooth for arbitrarily long times. More precisely, no *a priori* estimates has been found which guarantees that the enstrophy remains finite everywhere in the domain filled by the fluid (but it can be proven that it is bounded in the mean). The physical interpretation associated with this picture is that some very intermittent vorticity bursts can occur, injecting kinetic energy at scales much smaller than the Kolmorogov scale, resulting in quasi-infinite local values of the enstrophy. Such events correspond to finite-time singularities of the solution, and violate the axiom of continuum mechanics.

A large number of mathematical results dealing with these problems have been published, which will not be further discussed here. The important point is that some systems, which are very close to the Navier–Stokes equations, have been proposed. A common feature is that they are well-posed from the mathematical point of view, meaning that their solutions are proved to be regular. As a consequence, they appear as regularized systems derived from the original Navier–Stokes equations, the regularization being associated to the disappearance of singularities. From a physical point of view, these new systems do not allow the occurance of local infinite gradient thanks to an extra damping of the smallest scales. This smoothing property originates their interpretation as models for Large-Eddy Simulation.

4.2.1 Leray's Model

The first model was proposed by Leray in 1934, who suggested to regularize the Navier–Stokes equations as follows:

$$\frac{\partial u_i}{\partial t} + \frac{\partial \overline{u}_k u_i}{\partial x_k} = -\frac{\partial p}{\partial x_i} + \nu \frac{\partial^2 u_i}{\partial x_k \partial x_k} \tag{4.4}$$

$$\frac{\partial u_i}{\partial x_i} = 0 \tag{4.5}$$

where the regularized (i.e. filtered) velocity field is defined as

$$\overline{\boldsymbol{u}}(\boldsymbol{x}, t) = \phi_\epsilon \star \boldsymbol{u}(\boldsymbol{x}, t) \tag{4.6}$$

where the mollifying function (i.e. the filter kernel) ϕ_ϵ is assumed to have a compact support, to be C^∞ and to have an integral equal to one. It can be proved under these assumptions that the solution of the regularized system (4.4)–(4.5) is unique and C^∞. A main drawback is that it does not share all the frame-invariance properties of the Navier–Stokes equations. As quoted by Geurts and Holm [258], the system proposed by Leray can be rewritten in the usual Large-Eddy Simulation framework applying the filter a second times, leading to

$$\frac{\partial \overline{u}_i}{\partial t} + \frac{\partial \overline{u}_k \overline{u}_i}{\partial x_k} = -\frac{\partial \overline{p}}{\partial x_i} + \nu \frac{\partial^2 \overline{u}_i}{\partial x_k \partial x_k} - \frac{\partial \tau_{ij}^{\text{Leray}}}{\partial x_j} \tag{4.7}$$

$$\frac{\partial \overline{u}_i}{\partial x_i} = 0 \tag{4.8}$$

where the subgrid tensor ansatz is defined as

$$\tau_{ij}^{\text{Leray}} = \overline{u}_i \overline{u}_j - \overline{u_i \overline{u}_j} \tag{4.9}$$

An important difference with the usual definition of the subgrid tensor τ_{ij} is that this new tensor is not symmetric. Leray's regularized model makes it possible to carry out a complete mathematical analysis, but suffers the same problem when dealing with curvilinear grids on bounded domains as the original convolution filter model described in the preceding chapters.

4.2.2 Holm's Navier–Stokes-α Model

The second regularized model presented in this chapter is the Navier–Stokes-α proposed by Holm (see [221, 282, 182, 221, 258, 281]). The regularization is achieved by imposing an energy penalty which damps the scales smaller than the threshold scale α (to be interpreted as $\overline{\Delta}$ within the usual large-eddy simulation framework)[1], while still allowing for non linear sweeping of the small scales by the largest ones. The regularization appears as a non-linearly dispersive modification of the convection term in the Navier–Stokes equations.

The system of the Navier–Stokes-α (also referred to as the Camassa-Holm equations) can be derived in two different ways, which are now presented.

Method 1: Kelvin-filtered Navier–Stokes equations. The first way to obtain the Navier–Stokes-α model is to introduce the Kelvin-filtering. The Navier–Stokes equations satisfy Kelvin's circulation theorem

$$\frac{d}{dt} \oint_{\Gamma(\boldsymbol{u})} \boldsymbol{u} \cdot d\boldsymbol{x} = \oint_{\Gamma(\boldsymbol{u})} (\nu \nabla^2) \cdot d\boldsymbol{x} \tag{4.10}$$

where $\Gamma(\boldsymbol{u})$ is a closed fluid loop that moves with velocity $\boldsymbol{u}$. The original set of equations is regularized by modifying the fluid loop along which the circulation is integrated: instead of using a fluid loop moving at velocity $\boldsymbol{u}$, an new fluid loop moving at the regularized velocity $\overline{\boldsymbol{u}}$ is considered. The exact definition of $\overline{\boldsymbol{u}}$ is not necessary at this point and will be given later. The new circulation relationship is

$$\frac{d}{dt} \oint_{\Gamma(\overline{\boldsymbol{u}})} \boldsymbol{u} \cdot d\boldsymbol{x} = \oint_{\Gamma(\overline{\boldsymbol{u}})} (\nu \nabla^2) \cdot d\boldsymbol{x} \tag{4.11}$$

and corresponds to the following modified momentum equation:

$$\frac{\partial \boldsymbol{u}}{\partial t} + \overline{\boldsymbol{u}} \cdot \nabla \boldsymbol{u} + \nabla^T \overline{\boldsymbol{u}} \cdot \boldsymbol{u} = -\nabla p + \nu \nabla^2 \boldsymbol{u} \tag{4.12}$$

with

$$\nabla \cdot \overline{\boldsymbol{u}} = 0 \quad . \tag{4.13}$$

This set of equations describes the Kelvin-filtered Navier–Stokes equations. The Navier–Stokes-α equations are recovered specifying the regularized field $\overline{\boldsymbol{u}}$ as the result of the application of the Helmholtz filter (2.35) to the original field $\boldsymbol{u}$:

$$\boldsymbol{u} = (1 - \alpha^2 \nabla^2) \overline{\boldsymbol{u}} \quad . \tag{4.14}$$

It can be proved that the kinetic energy E_α defined as

$$E_\alpha = \frac{1}{2} \int \boldsymbol{u} \cdot \overline{\boldsymbol{u}} d\boldsymbol{x} = \int \left[\frac{1}{2} |\overline{\boldsymbol{u}}|^2 + \frac{\alpha^2}{2} |\nabla^2 \overline{\boldsymbol{u}}|^2 \right] d\boldsymbol{x} \quad , \tag{4.15}$$

[1] It can be shown that in the case of three-dimensional fully developed turbulence, the solution of the Navier–Stokes-α exhibits the usual $k^{-5/3}$ behavior for scales larger than α and a k^{-3} behavior for scales smaller than α.

is bounded, showing that the filtered field $\overline{\boldsymbol{u}}$ remains regular. The equation (4.12) can be rewritten under the usual form in Large-Eddy Simulation as a momentum equation for the filtered velocity field $\overline{\boldsymbol{u}}$ (formally identical to (4.4)). The corresponding definition of the subgrid tensor is

$$\tau_{ij}^{\mathrm{NS}\alpha} = (\overline{\overline{u}_i\,\overline{u}_j} - \overline{u}_i\,\overline{u}_j) - \alpha^2 \left(\overline{\frac{\partial \overline{u}_k}{\partial x_i}\frac{\partial \overline{u}_k}{\partial x_j}} + \overline{\overline{u}_j \nabla^2 \overline{u}_i} \right) \quad . \tag{4.16}$$

Method 2: Modified Leray's Model. Guermond, Oden and Prudhomme [282] observe that the Navier–Stokes-α system can be interpreted as a frame-invariant modification of original Leray's regularized model. Starting from the rotational form of the Navier–Stokes equations

$$\frac{\partial \boldsymbol{u}}{\partial t} + (\nabla \times \boldsymbol{u}) \times \boldsymbol{u} = -\nabla \pi + \nu \nabla^2 \boldsymbol{u}, \quad \pi = p + \frac{1}{2}u^2 \quad , \tag{4.17}$$

$$\nabla \cdot \boldsymbol{u} = 0 \quad , \tag{4.18}$$

and regularizing it using the technique proposed by Leray, one obtains

$$\frac{\partial \boldsymbol{u}}{\partial t} + (\nabla \times \boldsymbol{u}) \times \overline{\boldsymbol{u}} = -\nabla \pi + \nu \nabla^2 \boldsymbol{u}, \quad \pi = p + \frac{1}{2}u^2 \quad , \tag{4.19}$$

$$\nabla \cdot \boldsymbol{u} = 0 \quad . \tag{4.20}$$

Now using the relations

$$(\nabla \times \boldsymbol{u}) \times \overline{\boldsymbol{u}} = \overline{\boldsymbol{u}} \cdot \nabla \boldsymbol{u} - (\nabla^T \boldsymbol{u})\overline{\boldsymbol{u}}, \quad \nabla(\boldsymbol{u} \cdot (\nabla^T \boldsymbol{u})) = (\nabla^T \boldsymbol{u})\overline{\boldsymbol{u}} + (\nabla^T \overline{\boldsymbol{u}})\boldsymbol{u} \quad , \tag{4.21}$$

the following form of the regularized system is recovered

$$\frac{\partial \boldsymbol{u}}{\partial t} + \overline{\boldsymbol{u}} \cdot \nabla \boldsymbol{u} + (\nabla^T \overline{\boldsymbol{u}}) \cdot \boldsymbol{u} = -\nabla \pi' + \nu \nabla^2 \boldsymbol{u}, \quad \pi' = \pi - \boldsymbol{u} \cdot \overline{\boldsymbol{u}} \quad , \tag{4.22}$$

$$\nabla \cdot \overline{\boldsymbol{u}} = 0 \quad . \tag{4.23}$$

The Navier–Stokes-α model is obatined using the Helmholtz filter (4.14). The corresponding equation for $\overline{\boldsymbol{u}}$ is

$$\frac{\partial \overline{\boldsymbol{u}}}{\partial t} + \overline{\boldsymbol{u}} \cdot \nabla \overline{\boldsymbol{u}} = \nabla \cdot \mathcal{T} \quad , \tag{4.24}$$

with

$$\mathcal{T} = -\overline{p} Id + 2\nu(1 - \alpha^2 \nabla^2)\overline{S} + 2\alpha^2 \overline{S}^{\circ} \quad , \tag{4.25}$$

where $\overline{S}^{\circ}$ is related to the Jaumann derivative of the regularized strain rate tensor:

$$\overline{S}^{\circ} = \frac{\partial \overline{S}}{\partial t} + \overline{\boldsymbol{u}} \cdot \nabla \overline{S} + \overline{S\Omega} - \overline{\Omega S}, \quad \overline{\Omega} = \frac{1}{2}\left(\nabla \overline{u} - \nabla^T \overline{u}\right) \quad . \tag{4.26}$$

This system is formally similar to the constitutive law of a rate-dependent incompressible fluid of second grade with slightly modified dissipation, and it is frame-invariant. It is equivalent to the Leray model in which the term which is responsible for the failure in the frame preservation, i.e. $\alpha^2(\nabla^T \overline{\boldsymbol{u}} \nabla^2 \overline{\boldsymbol{u}})$, has been removed. Therefore, the Navier–Stokes-α equations appear as a perturbation of order α^2 (i.e. $\overline{\Delta}^2$) of the original Leray model.

4.2.3 Ladyzenskaja's Model

Another regularized version of the Navier–Stokes equations was proposed by Ladyzenskaja and Kaniel [417, 418, 377], who introduced a non-linear modification of the stress tensor which is expected to be more relevant than the linear relationship for Newtonian fluids when velocity gradients are large. The equation for the regularized field $\overline{\boldsymbol{u}}$ is

$$\frac{\partial \overline{\boldsymbol{u}}}{\partial t} + \overline{\boldsymbol{u}} \cdot \nabla \overline{\boldsymbol{u}} = -\nabla \overline{p} + \nu \nabla^2 \overline{\boldsymbol{u}} - \varepsilon \nabla \cdot T(\nabla \overline{\boldsymbol{u}}) \quad , \tag{4.27}$$

$$\nabla \cdot \overline{\boldsymbol{u}} = 0 \quad , \tag{4.28}$$

where ε is a strictly arbitrary constant and the non-linear stress tensor T is defined as

$$T(\nabla \overline{\boldsymbol{u}}) = \nu_T(\|\nabla \overline{\boldsymbol{u}}\|^2) \nabla \overline{\boldsymbol{u}} \quad , \tag{4.29}$$

where the non-linear viscosity $\nu_T(\tau)$ is a positive monotonically-increasing function of $\tau \geq 0$ that obeys the following law for large values of τ:

$$c\tau^{\mu} \leq \nu_T(\tau) \leq c'\tau^{\mu}, \quad 0 < c < c', \quad \mu \geq \frac{1}{4} \quad . \tag{4.30}$$

The equivalent expression for the subgrid stress tensor is

$$\tau_{ij}^{\text{Ladyzenskaja}} = \varepsilon T_{ij}(\|\nabla \overline{\boldsymbol{u}}\|^2) \quad . \tag{4.31}$$

Since T depends only on the gradient of the resolved field $\overline{\boldsymbol{u}}$, Ladyzenskaja's model is closed and does not require further modeling work.

5. Functional Modeling (Isotropic Case)

It would be illusory to try to describe the structure of the scales of motion and the interactions in all imaginable configurations, in light of the very large disparity of physical phenomena encountered. So we have to restrict this description to cases which by nature include scales that are too small for today's computer facilities to solve them entirely, and which are at the same time accessible to theoretical analysis. This description will therefore be centered on the inter-scale interactions in the case of fully developed isotropic homogeneous turbulence[1], which is moreover the only case accessible by theoretical analysis and is consequently the only theoretical framework used today for developing subgrid models. Attempts to extend this theory to anisotropic and/or inhomogeneous cases are discussed in Chap. 6. The text will mainly be oriented toward the large-eddy simulation aspects. For a detailed description of the isotropic homogeneous turbulence properties, which are reviewed in Appendix A, the reader may refer to the works of Lesieur [439] and Batchelor [45].

5.1 Phenomenology of Inter-Scale Interactions

It is important to note here the framework of restrictions that apply to the results we will be presenting. These results concern three-dimensional flows and thus do not cover the physics of two-dimensional flows (in the sense of flows with two directions[2], and not two-component[3] flows), which have a totally different dynamics [403, 404, 405, 438, 481]. The modeling in the two-dimensional case leads to specific models [42, 624, 625] which will not

---

[1] That is, whose statistical properties are invariant by translation, rotation, or symmetry.

[2] These are flows such that there exists a direction x for which we have the property:

$$\frac{\partial u}{\partial x} \equiv 0 \quad .$$

[3] These are flows such that there exists a framework in which the velocity field has an identically zero component.

be presented. For details on two-dimensional turbulence, the reader may also refer to [439].

5.1.1 Local Isotropy Assumption: Consequences

In the case of fully developed turbulence, Kolmogorov's statistical description of the small scales of the flow, based on the assumption of local isotropy, has been the one most used for a very long time.

By introducing the idea of local isotropy, Kolmogorov assumes that the small scales belonging to the inertial range of the energy spectrum of a fully developed inhomogeneous turbulent flow are:

- *Statistically isotropic*, and therefore entirely characterized by a characteristic velocity and time;
- *Without time memory*, therefore in energy equilibrium with the large scales of the flow by instantaneous re-adjustment.

This isotropy of the small scales implies that they are statistically independent of the large energetic scales, which are characteristic of each flow and are therefore anisotropic. Experimental work [512] has shown that this assumption is not valid in shear flows for all the scales belonging to the inertial range, but only for those whose size is of the order of the Kolmogorov scale. Numerical experiments [32] show that turbulent stresses are nearly isotropic for wave numbers k such that $kL_\varepsilon > 50$, where L_ε is the integral dissipation length[4]. These experiments have also shown that the existence of an inertial region does not depend on the local isotropy hypothesis. The causes of this persistence of the anisotropy in the inertial range due to interactions existing between the various scales of the flow will be mentioned in Chap. 6. Works based on direct numerical simulations have also shown that the assumption of equilibrium between the resolved and subgrid scales may be faulted, at least temporarily, when the flow is subject to unsteady forcing [594, 570, 454, 504]. This is due to the fact that the relaxation times of these two scale ranges are different. In the case of impulsively accelerated flows (plane channel, boundary layer, axisymmetric straining) the subgrid scales react more quickly than the resolved ones, and then also relax more quickly toward an equilibrium solution.

The existence of a zone of the spectrum, corresponding to the higher frequencies, where the scales of motion are statistically isotropic, justifies the study of the inter-modal interactions in the ideal case of isotropic homogeneous turbulence. Strictly speaking, the results can be used for determining subgrid models only if the cutoff associated with the filter is in this region,

[4] The integral dissipation length is defined as

$$L_\varepsilon = \frac{\langle u_i u_i \rangle^{3/2}}{\varepsilon} \quad ,$$

where ε is the energy dissipation rate.

because the dynamics of the unresolved scales then corresponds well to that of the isotropic homogeneous turbulence. It should be noted that this last condition implies that the representation of the dynamics, while incomplete, is nonetheless very fine, which theoretically limits the gain in complexity that can be expected from large-eddy simulation technique.

Another point is that the local isotropy hypothesis is formulated for fully developed turbulent flows at very high Reynolds numbers. As it affirms the universal character of the small scales' behavior for these flows, it ensures the possibility using the large-eddy simulation technique strictly, if the filter cutoff frequency is set sufficiently high. There is no theoretical justification, though, for applying the results of this analysis to other flows, such as transitional flows.

5.1.2 Interactions Between Resolved and Subgrid Scales

In order to study the interactions between the resolved and subgrid scales, we adopt an isotropic filter by a cutoff wave number k_c. The subgrid scales are those represented by the $\boldsymbol{k}$ modes such that $k \geq k_c$.

In the case of fully developed isotropic homogeneous turbulence, the statistical description of the inter-scale interactions is reduced to that of the kinetic energy transfers. Consequently, only the information associated with the amplitude of the fluctuations is conserved, and none concerning the phase is taken into account.

These transfers are analyzed using several tools:

– Analytical theories of turbulence, also called two-point closures, which describe triadic interactions on the basis of certain assumptions. They will therefore express the non-linear term $S(\boldsymbol{k}|\boldsymbol{p},\boldsymbol{q})$, defined by relation (3.50) completely. For a description of these theories, the reader may refer to Lesieur's book [439], and we also mention Waleffe's analysis [748, 749], certain conclusions of which are presented in the following.
– Direct numerical simulations, which provide a complete description of the dynamics.
– Renormalization Group Theory [622, 328, 812, 809, 464, 775, 804, 805, 813, 802, 803, 810], with several variants.

Typology of the Triadic Interactions. It appears from the developments of Sect. 3.1.3 (also see Appendix A) that the $\widehat{\boldsymbol{u}}(\boldsymbol{k})$ mode interacts only with those modes whose wave vectors $\boldsymbol{p}$ and $\boldsymbol{q}$ form a closed triangle with $\boldsymbol{k}$. The wave vector triads $(\boldsymbol{k},\boldsymbol{p},\boldsymbol{q})$ thus defined are classified in several groups [805] which are represented in Fig. 5.1:

– *Local triads* for which

$$\frac{1}{a} \leq \max\left\{\frac{p}{k},\frac{q}{k}\right\} \leq a, \qquad a = O(1) \quad ,$$

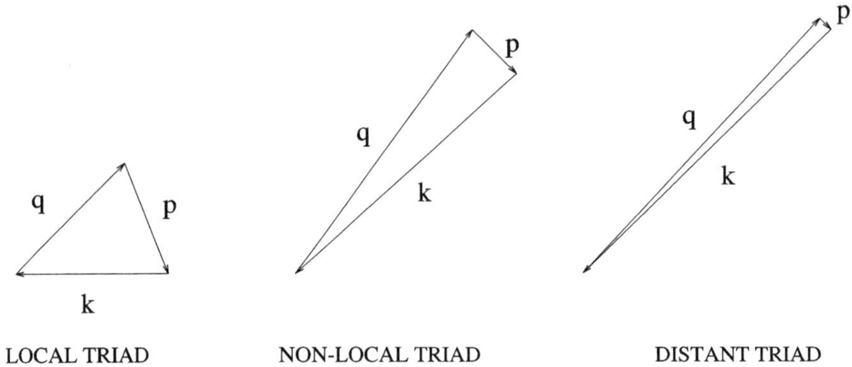

Fig. 5.1. Different types of triads.

which correspond to interactions among wave vectors of neighboring modules, and therefore to interactions among scales of slightly different sizes;
- *Non-local triads*, which are all those interactions that do not fall within the first category, i.e. interactions among scales of widely differing sizes. Here, we adopt the terminology proposed in [74], which distinguishes between two sub-classes of non-local triads, one being distant triads of interactions in which $k \ll p \sim q$ or $k \sim q \gg p$. It should be noted that these terms are not unequivocal, as certain authors [439, 442] refer to these "distant" triads as being just "non-local".

By extension, a phenomenon will be called local if it involves wave vectors $\boldsymbol{k}$ and $\boldsymbol{p}$ such that $1/a \leq p/k \leq a$, and otherwise non-local or distant.

Canonical Analysis. This section presents the results from analysis of the simplest theoretical case, which we call here canonical analysis. This consists of assuming the following two hypotheses:

1. Hypothesis concerning the flow. The energy spectrum $E(\boldsymbol{k})$ of the exact solution is a Kolmogorov spectrum, i.e.
$$E(\boldsymbol{k}) = K_0 \varepsilon^{2/3} k^{-5/3}, \qquad k \in [0, \infty] \quad , \tag{5.1}$$
where K_0 is the Komogorov constant and ε the kinetic energy dissipation rate. We point out that this spectrum is not integrable since its corresponds to an infinite kinetic energy.
2. Hypothesis concerning the filter. The filter is a sharp cutoff type. The subgrid tensor is thus reduced to the subgrid Reynolds tensor.

In analyzing the energy transfers $T^{\mathrm{e}}_{\mathrm{sgs}}(k)$ (see relation (3.51)) between the modes to either side of a cutoff wave number k_{c} located in the inertial range of the spectrum, Kraichnan [405] uses the Test Field Model (TFM) to bring out the existence of two spectral bands (see Fig. 5.2) for which the interactions with the small scales (p and/or $q \geq k_{\mathrm{c}}$) are of different kinds.

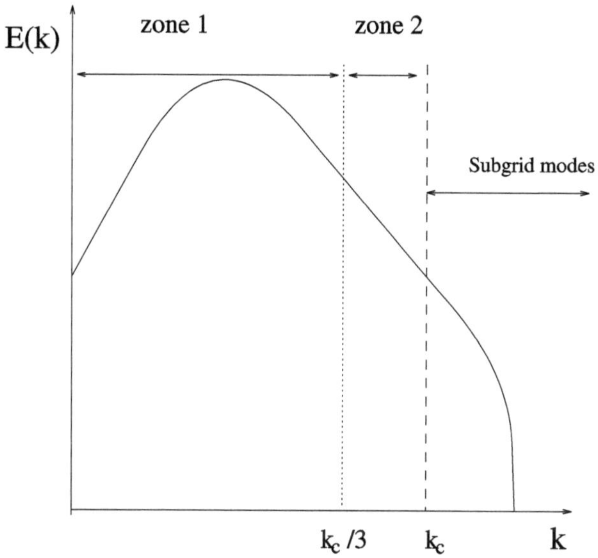

Fig. 5.2. Interaction regions between resolved and subgrid scales.

1. In the first region (1 in Fig. 5.2), which corresponds to the modes such that $k \ll k_c$, the dominant dynamic mechanism is a random displacement of the momentum associated with $\boldsymbol{k}$ by disturbances associated with $\boldsymbol{p}$ and $\boldsymbol{q}$. This phenomenon, analogous to the effects of the molecular viscosity, entails a kinetic energy decay associated with $\boldsymbol{k}$ and, since the total kinetic energy is conserved, a resulting increase of it associated with $\boldsymbol{p}$ and $\boldsymbol{q}$. So here it is a matter of a non-local transfer of energy associated with non-local triadic interactions. These transfers, which induce a damping of the fluctuations, are associated with what Waleffe [748, 749] classifies as type F triads (represented in Fig. 5.3).

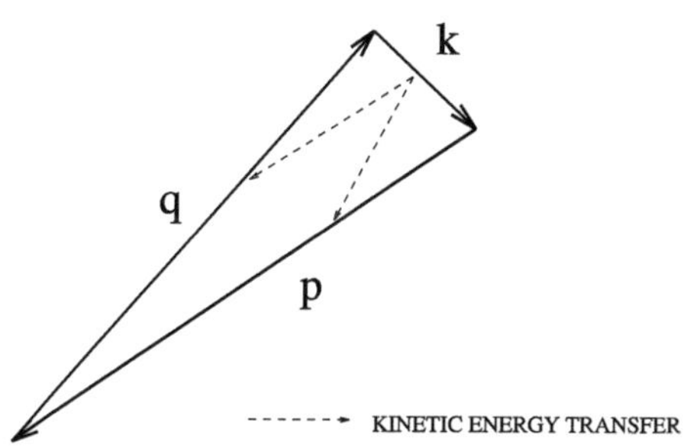

Fig. 5.3. Non-local triad $(\boldsymbol{k}, \boldsymbol{p}, \boldsymbol{q})$ of the F type according to Waleffe's classification, and the associated non-local energy transfers. The kinetic energy of the mode corresponding to the smallest wave vector $\boldsymbol{k}$ is distributed to the other two modes $\boldsymbol{p}$ and $\boldsymbol{q}$, creating a forward energy cascade in the region where $k \ll k_c$.

Subsequent analyses using the Direct Interaction Approximation (DIA) and the Eddy Damped Quasi-Normal Markovian (EDQNM) models [120, 136, 442, 443, 647] or Waleffe's analyses [748, 749] have refined this representation by showing the existence of two competitive mechanisms in the region where $k \ll k_c$. The first region is where the energy of the large scales is drained by the small ones, as already shown by Kraichnan. The second mechanism, of much lesser intensity, is a return of energy from the small scales $\boldsymbol{p}$ and $\boldsymbol{q}$ to the large scale $\boldsymbol{k}$. This mechanism also corresponds to a non-local energy transfer associated with non-local triadic interactions that Waleffe classifies as type R (see Fig. 5.4). It represents a backward stochastic energy cascade associated with an energy spectrum in k^4 for very small wave numbers. This phenomenon has been predicted analytically [442] and verified by numerical experimentation [441, 120]. The analytical studies and numerical simulations show that this backward cascade process is dominant for very small wave numbers. On the average, these modes receive more energy from the subgrid modes than they give to them.

2. In the second region (region 2 in Fig. 5.2), which corresponds to the $\boldsymbol{k}$ modes such that $(k_c - k) \ll k_c$, the mechanisms already present in region 1 persist. The energy transfer to the small scales is at the origin of the forward kinetic energy cascade.
   Moreover, another mechanism appears involving triads such that p or $q \ll k_c$, which is that the interactions between the scales of this region and the subgrid scales are much more intense than in the first. Let us take $q \ll k_c$. This mechanism is a coherent straining of the small scales $\boldsymbol{k}$ and $\boldsymbol{p}$ by the shear associated with $\boldsymbol{q}$, resulting in a wave number diffusion process between $\boldsymbol{k}$ and $\boldsymbol{p}$ through the cutoff, with one of the structures being stretched (vortex stretching phenomenon) and the other unstretched. What we are observing here is a local energy transfer between $\boldsymbol{k}$ and $\boldsymbol{p}$ associated with non-local triadic interactions due to the type R triads (see Fig. 5.4). Waleffe refines the analysis of this phenomenon: a very large part of the energy is transferred locally from the intermediate wave number located just ahead of the cutoff toward the larger wave number just after it, and the remaining fraction of energy is transferred to the smaller wave number. These findings have been corroborated by numerical data [120, 185, 189] and other theoretical analyses [136, 443].

The energy transfers $T_{\text{sgs}}^{\text{e}}(k)$ (see relation (3.51)) between mode $\boldsymbol{k}$ and the subgrid modes can be represented in a form analogous to molecular dissipation. To do this, by following Heisenberg (see [688] for a description of Heisenberg's theory), we define an effective viscosity $\nu_{\text{e}}(k|k_c)$, which represents the energy transfers between the $\boldsymbol{k}$ mode and the modes located beyond the k_c cutoff such that:

$$T_{\text{sgs}}^{\text{e}}(k) = -2\nu_{\text{e}}(k|k_c)k^2 E(k) \quad . \tag{5.2}$$

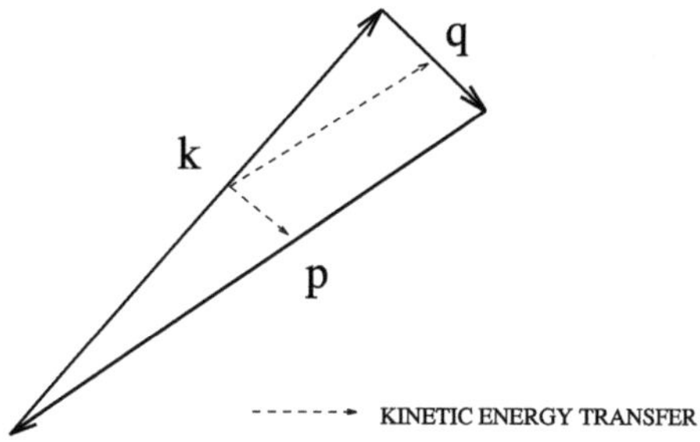

Fig. 5.4. Non-local $(\boldsymbol{k}, \boldsymbol{p}, \boldsymbol{q})$ triad of the R type according to Waleffe's classification, and the associated energy transfers in the case $q \ll k_c$. The kinetic energy of the mode corresponding to the intermediate wave vector $\boldsymbol{k}$ is distributed locally to the largest wave vector $\boldsymbol{p}$ and non-locally to the smallest wave vector, $\boldsymbol{q}$. The former transfer originates the intensification of the coupling in the $(k_c - k) \ll k_c$ spectral band, while the latter originates the backward kinetic energy cascade.

It should be pointed out that this viscosity is real, i.e. $\nu_e(k|k_c) \in \mathbb{R}$, and that if any information related to the phase were included, it would lead the definition of a complex term having an a priori non-zero imaginary part, which may seem to be more natural for representing a dispersive type of coupling. Such a term is obtained not by starting with the kinetic energy equation, but with the momentum equation[5].

The two energy cascades, forward and backward, can be introduced separately by introducing distinct effective viscosities, constructed in such a way as to ensure energy transfers equivalent to those of these cascades. We get the following two forms:

$$\nu_e^+(k|k_c,t) = -\frac{T_{sgs}^+(k|k_c,t)}{2k^2E(k,t)} \quad , \tag{5.3}$$

$$\nu_e^-(k|k_c,t) = -\frac{T_{sgs}^-(k|k_c,t)}{2k^2E(k,t)} \quad , \tag{5.4}$$

in which $T_{sgs}^+(k|k_c,t)$ (resp. $T_{sgs}^-(k|k_c,t)$) is the energy transfer term from the k mode to the subgrid modes (resp. from the subgrid modes to the k mode). This leads to the decomposition:

$$T_{sgs}^e(k) = T_{sgs}^+(k|k_c,t) + T_{sgs}^-(k|k_c,t) \tag{5.5}$$

$$= -2k^2E(k,t)\left(\nu_e^+(k|k_c,t) + \nu_e^-(k|k_c,t)\right) \quad . \tag{5.6}$$

These two viscosities depend explicitly on the wave number k and the cutoff wave vector k_c, as well as the shape of the spectrum. The result of

[5] This possibility is only mentioned here, because no works have been published on it to date.

these dependencies on the flow is that the viscosities are not, because they characterize the flow and not the fluid. They are of opposite sign: $\nu_e^+(k|k_c,t)$ ensures a loss of energy of the resolved scales and is consequently positive, like the molecular viscosity, whereas $\nu_e^-(k|k_c,t)$, which represents an energy gain in the resolved scales, is negative.

The conclusions of the theoretical analyses [405, 443] and numerical studies [120] are in agreement on the form of these two viscosities. Their behavior is presented in Fig. 5.5 in the canonical case.

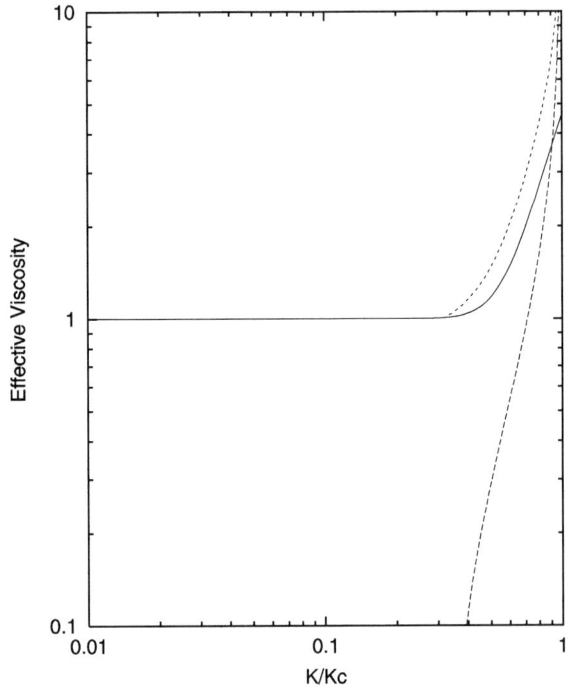

Fig. 5.5. Representation of effective viscosities in the canonical case. Short dashes: $\nu_e^+(k|k_c,t)$; long dashes: $-\nu_e^-(k|k_c,t)$; solid: $\nu_e^+(k|k_c,t)+\nu_e^-(k|k_c,t)$.

We may note that these two viscosities become very high for wave numbers close to the cutoff. These two effective viscosities diverge as $(k_c - k)^{-2/3}$ as k tends toward k_c. However, their sum $\nu_e(k|k_c,t)$ remains finite and Leslie et al. [443] proposes the estimation:

$$\nu_e(k_c|k_c,t) = 5.24\nu_e^+(0|k_c,t) \quad . \tag{5.7}$$

The interactions with the subgrid scales is therefore especially important in the dynamics of the smallest resolved scales. More precisely, Kraichnan's theoretical analysis leads to the conclusion that about 75% of the energy transfers of a k mode occur with the modes located in the $[k/2, 2k]$ spectral band[6]. No transfers outside this spectral band have been observed in direct numerical simulations at low Reynolds numbers [185, 804, 805]. The difference with the theoretical analysis stems from the fact that this analysis is performed in the limit of the infinite Reynolds numbers.

In the limit of the very small wave numbers, we have the asymptotic behaviors:

$$2k^2 E(k,t)\nu_e^+(k|k_c,t) \propto k^{1/3} \quad , \tag{5.8}$$

$$2k^2 E(k,t)\nu_e^-(k|k_c,t) \propto k^4 \quad . \tag{5.9}$$

The effective viscosity associated with the energy cascade takes the constant asymptotic value:

$$\nu_e^+(0|k_c,t) = 0.292\varepsilon^{1/3}k_c^{-4/3} \quad . \tag{5.10}$$

We put the emphasis on the fact that the effective viscosity discussed here is defined considering the kinetic enery transfer between unresolved and resolved modes. As quoted by McComb et al. [465], it is possible to define different effective viscosities by considering other balance equations, such as the enstrophy transfer. An important consequence is that kinetic-energy-based effective viscosities are efficient surrogates of true transfer terms in the kinetic energy equation, but may be very bad representations of subgrid effects for other physical mechanisms.

Dependency According to the Filter. Leslie and Quarini [443] extended the above analysis to the case of the Gaussian filter. The spectrum considered is always of the Kolmogorov type. The Leonard term is now non-zero. The results of the analysis show very pronounced differences from the canonical analysis. Two regions of the spectrum are still distinguishable, though, with regard to the variation of the effective viscosities ν_e^+ and ν_e^-, which are shown in Fig. 5.6:

- In the first region, where $k \ll k_c$, the transfer terms still observe a constant asymptotic behavior, independent of the wave number considered, as in the canonical case. The backward cascade term is negligible compared with the forward cascade term.
- In the second region, on the other hand, when approaching cutoff, the two transfer terms do not have divergent behavior, contrary to what is observed in the canonical case. The forward cascade term decreases monotonically and cancels out after the cutoff for wave numbers more than a decade beyond it. The backward cascade term increases up to cutoff and exhibits a decreasing behavior analogous to that of the forward cascade term. The maximum intensity of the backward cascade is encountered for modes just after the cutoff.

[6] The same local character of kinetic energy transfer is observed in non-homogeneous flow, such has the plane channel flow [186].

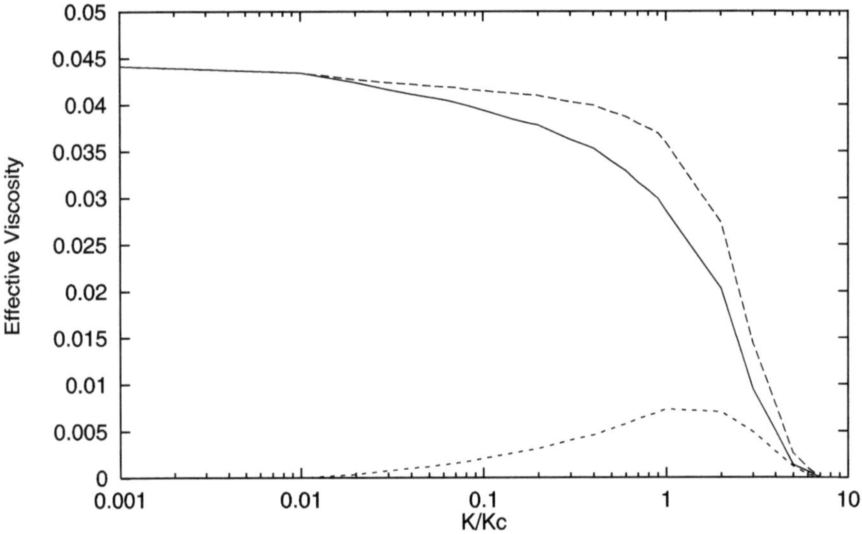

Fig. 5.6. Effective viscosities in the application of a Gaussian filter to a Kolmogorov spectrum. Long dots $\nu_e^+(k|k_c,t)$; dots: $-\nu_e^-(k|k_c,t)$; solid: $\nu_e^+(k|k_c,t) + \nu_e^-(k|k_c,t)$.

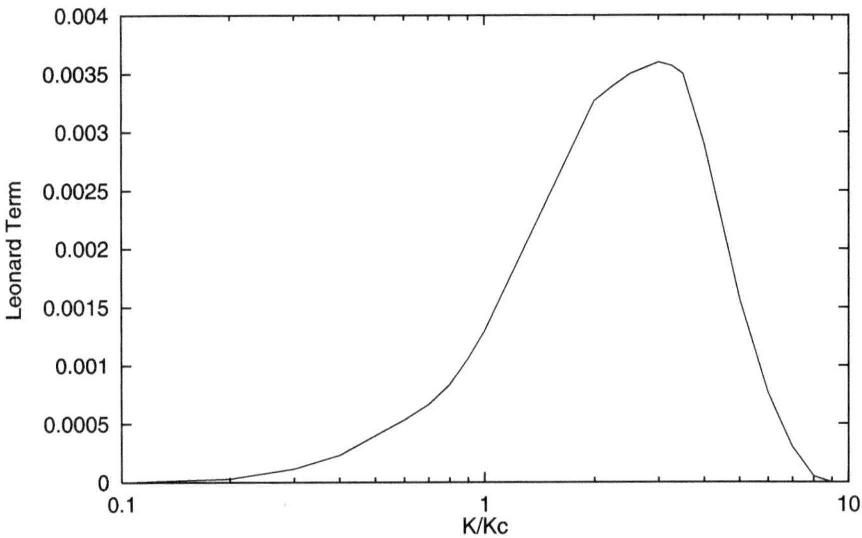

Fig. 5.7. Effective viscosity corresponding to the Leonard term in the case of the application of a Gaussian filter to a Kolmogorov spectrum.

In contrast to the sharp cutoff filter used for the canonical analysis, the Gaussian filter makes it possible to define Leonard terms and non-identically zero cross terms. The effective viscosity associated with these terms is shown in Fig. 5.7, where it can be seen that it is negligible for all the modes more than a decade away from the cutoff. In the same way as for the backward cascade term, the maximum amplitude is observed for modes located just after the cutoff. This term remains smaller than the forward and backward cascade terms for all the wave numbers.

Dependency According to Spectrum Shape. The results of the canonical analysis are also dependent on the shape of the spectrum considered. The analysis is repeated for the case of the application of the sharp cutoff filter to a production spectrum of the form:

$$E(\boldsymbol{k}) = A_s(k/k_p) K_0 \varepsilon^{2/3} k^{-5/3} \quad , \tag{5.11}$$

with

$$A_s(x) = \frac{x^{s+5/3}}{1 + x^{s+5/3}} \quad , \tag{5.12}$$

and where k_p is the wave number that corresponds to the maximum of the energy spectrum [443]. The shape of the spectrum thus defined is illustrated in Fig. 5.8 for several values of the s parameter.

The variation of the total effective viscosity ν_e for different values of the quotient k_c/k_p is diagrammed in Fig. 5.9. For low values of this quotient, i.e. when the cutoff is located at the beginning of the inertial range, we observe

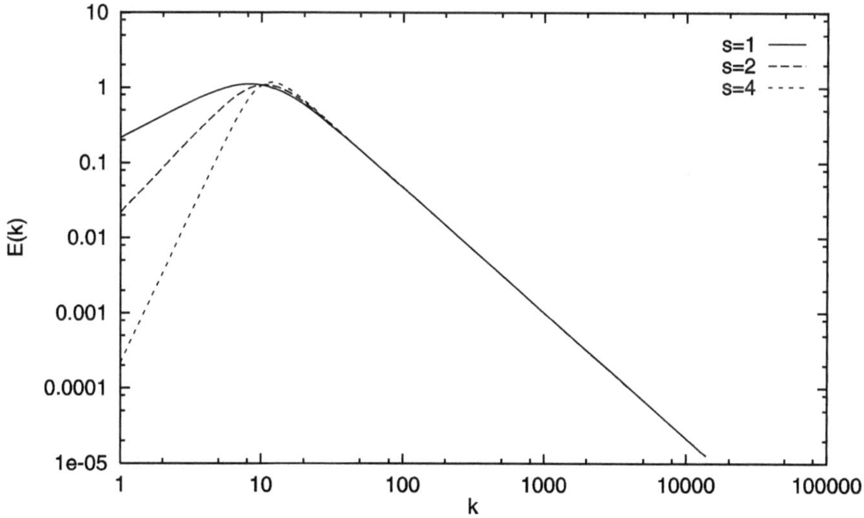

Fig. 5.8. Production spectrum for different values of the shape parameter s.

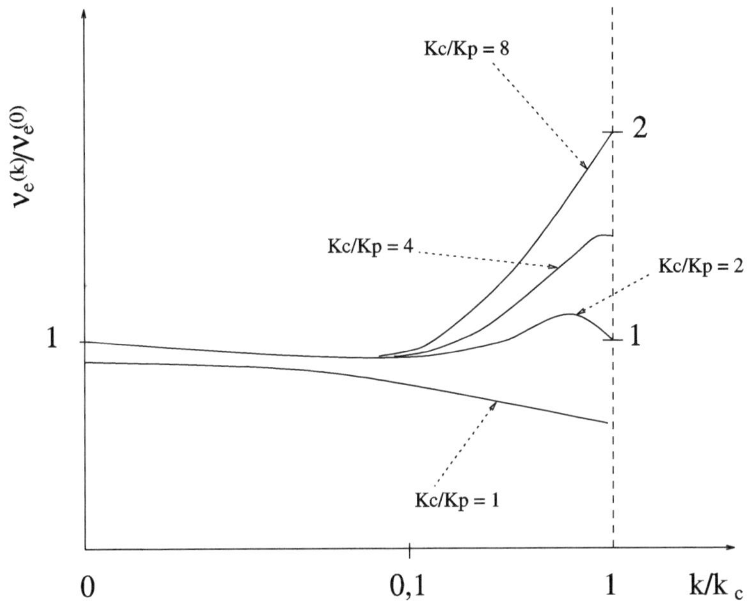

Fig. 5.9. Total effective viscosity $\nu_e(k|k_c)$ in the case of the application of a sharp cutoff filter to a production spectrum for different values of the quotient k_c/k_p, normalized by its value at the origin.

that the viscosity may decrease at the approach to the cutoff, while it is strictly increasing in the canonical case. This difference is due to the fact that the asymptotic reasoning that was applicable in the canonical case is no longer valid, because the non-localness of the triadic interactions involved relay the difference in spectrum shape to the whole of it. For higher values of this quotient, i.e. when the cutoff is located *sufficiently far* into the inertial range (for large values of the ratio k_c/k_p), a behavior that is qualitatively similar to that observed in the canonical case is once again found[7].

For $k_c = k_p$, no increase is observed in the energy transfers as k tends toward k_c. The behavior approximates that observed for the canonical analysis as the ratio k_p/k_c decreases.

5.1.3 A View in Physical Space

Analyses described in the preceding section were all performed in the Fourier space, and do not give any information about the location of the subgrid transfer in the physical space and its correlation with the resolved scale features[8]. Complementary informations on the subgrid transfer in the physical space have been found by several authors using direct numerical simulation.

[7] In practice, k_c/k_p=8 seems appropriate.

[8] This is a prerequisite for designing a functional subgrid model in physical space.

Kerr et al. [383] propose to use the rotational form of the non-linear term of the momentum equation:

$$N(x) = u(x) \times \omega(x) - \nabla p_h(x) \quad , \tag{5.13}$$

where $\omega = \nabla \times u$ and p_h the pressure term. By splitting the velocity and vorticity field into a resolved and a subgrid contribution, we get:

$$\underbrace{u \times \omega - \overline{u} \times \overline{\omega}}_{I} = \underbrace{\overline{u} \times \omega'}_{II} + \underbrace{u' \times \overline{\omega}}_{III} + \underbrace{u' \times \omega'}_{IV} \quad . \tag{5.14}$$

The four terms represent different coupling mechanisms between the resolved motion and the subgrid scales:

- I - exact subgrid term,
- II - interaction between resolved velocity and subgrid vorticity,
- III - interaction between subgrid velocity and resolved vorticity,
- IV - interaction between subgrid velocity and subgrid vorticity.

The corresponding complete non-linear terms $N^I, ..., N^{IV}$ are built by adding the specific pressure term. The associated subgrid kinetic energy transfer terms are computed as $\varepsilon^l = \overline{u} \cdot N^l$. The authors made three significant observations for isotropic turbulence:

- Subgrid kinetic energy transfer is strongly correlated with the boundaries of regions of large vorticity production (stretching), i.e. regions where $\overline{\omega}_i \overline{S}_{ij} \overline{\omega}_j$ is large;
- Term II, $\overline{u} \times \omega'$, has a correlation with subgrid non-linear term I up to 0.9. This term dominates the backward energy cascade;
- Up to 90% of the subgrid kinetic energy transfer comes from term III, i.e. from the interaction of subgrid velocity with resolved vorticity. This term mostly contributes to the forward energy cascade.

Additional results of Borue and Orszag [72] show that the subgrid transfer takes place in regions where the vorticity stretching term is positive or in regions with negative skewness of the resolved strain rate tensor, $\mathrm{Tr}(\overline{S}^3)$. These authors also found that there is only a very poor local correlation between the subgrid transfer $\tau_{ij}\overline{S}_{ij}$ and the local strain $\overline{S}_{ij}\overline{S}_{ij}$, where $\overline{S}_{ij}$ is the resolved strain rate tensor.

Horiuti [325, 327] decomposed the subgrid tensor into several contributions,[9] and used direct numerical simulation data of isotropic turbulence to analyze their contributions. A first remark is that the eigenvectors of the total subgrid tensor have a preferred orientation of 42° relative to those of $\overline{S}$. Eigenvectors of $(\overline{S}_{ik}\overline{S}_{kj} - \overline{\Omega}_{ik}\overline{\Omega}_{kj})$ are highly aligned with those of $\overline{S}$, while

[9] This decomposition is discussed in the section devoted to nonlinear models, p. 223.

those of $(\overline{S}_{ik}\overline{\Omega}_{kj} - \overline{S}_{ik}\overline{\Omega}_{kj})$ exhibit a 42° angle, from which stems the global observed difference. The first term is associated with the forward energy cascade. The second one makes no contribution to the total production of subgrid kinetic energy, but is relevant to the vortex stretching and the backward energy cascade process.[10] Similar results were obtained by Meneveau and coworkers [704, 703].

The role of coherent structures in interscale transfer is of major importance in shear flows. Da Silva and Métais [158] carried out an exhaustive study in the plane jet case: the most intense forward cascade events occur near these coherent structures and not randomly in space. The local equilibrium assumption is observed to hold globally but not locally as most viscous dissipation of subgrid kinetic energy takes place within coherent structure cores, while forward and backward cascade occur at different locations.

5.1.4 Summary

The different analyses performed in the framework of fully developed isotropic turbulence show that:

1. Interactions between the small and large scales is reflected by two main mechanisms:
   - A drainage of energy from the resolved scales by the subgrid scales (forward energy cascade phenomenon);
   - A weak feedback of energy, proportional to k^4 to the resolved scales (backward energy cascade phenomenon).
2. The interactions between the subgrid scales and the smallest of the resolved scales depend on the filter used and on the shape of the spectrum. In certain cases, the coupling with the subgrid scales is strengthened for wave numbers close to the cutoff and the energy toward the subgrid modes is intensified.
3. These cascade mechanisms are associated to specific features of the velocity and vorticity field in physical space.

5.2 Basic Functional Modeling Hypothesis

All the subgrid models entering into this category make more or less implicit use of the following hypothesis:

Hypothesis 5.1 *The action of the subgrid scales on the resolved scales is essentially an energetic action, so that the balance of the energy transfers alone between the two scale ranges is sufficient to describe the action of the subgrid scales.*

[10] This is an indication that the backward energy cascade is not associated with negative subgrid viscosity from the theoretical point of view.

Using this hypothesis as a basis for modeling, then, we neglect a part of the information contained in the small scales, such as the structural information related to the anisotropy. As was seen above, the energy transfers between subgrid scales and resolved scales mainly exhibit two mechanisms: a forward energy transfer toward the subgrid scales and a backward transfer to the resolved scales which, it seems, is much weaker in intensity. All the approaches existing today for numerical simulation at high Reynolds numbers consider the energy lost by the resolved scales, while only a few rare attempts have been made to consider the backward energy cascade.

Once hypothesis 5.1 is assumed, the modeling consists in modifying the different evolution equations of the system in such a way as to integrate the desired dissipation or energy production effects into them. To do this, two different approaches can be found in today's works:

- *Explicit modeling* of the desired effects, i.e. including them by adding additional terms to the equations: the actual subgrid models;
- *Implicit inclusion* by the numerical scheme used, by arranging it so the truncation error induces the desired effects.

Let us note that while the explicit approach is what would have to be called the classical modeling approach, the implicit one appears generally only as an a posteriori interpretation of dissipative properties for certain numerical methods used.

5.3 Modeling of the Forward Energy Cascade Process

This section describes the main functional models of the energy cascade mechanism. Those derived in the Fourier space, conceived for simulations based on spectral numerical methods, and models derived in the physical space, suited to the other numerical methods, are presented separately.

5.3.1 Spectral Models

The models belonging to this category are all effective viscosity models drawing upon the analyses of Kraichnan for the canonical case presented above. The following models are described:

1. The Chollet–Lesieur model (p. 106) which, based on the results of the canonical analysis (inertial range of the spectrum with a slope of -5/3, sharp cutoff filter, no effects associated with a production type spectrum) yields an analytical expression for the effective viscosity as a function of the wave number considered and the cutoff wave number. It will reflect the local effects at the cutoff, i.e. the increase in the energy transfer toward the subgrid scales. This model explicitly brings out a dependency

of the effective viscosity as a function of the kinetic energy at the cutoff. This guarantees that, when all the modes of the exact solution are resolved, the subgrid model automatically cancels out. The fact that this information is local in frequency allows the model to consider (at least partially) the spectral disequilibrium phenomena that occur at the level of the resolved scales[11], though without relaxing the hypotheses underlying the canonical analysis. Only the amplitude of the transfers is variable, and not their pre-supposed shape.

2. The effective viscosity model (p. 107), which is a simplification of the previous one and is based on the same assumptions. The effective viscosity is then independent of the wave number and is calculated so as to ensure the same average value as the Chollet–Lesieur model. It is simpler to compute, but does not reflect the local effects at the cutoff.
3. The dynamic spectral model (p. 107), which is an extension of the Chollet–Lesieur model for spectra having a slope different from that of the canonical case (i.e. - 5/3). Richer information is considered here: while the Chollet–Lesieur model is based only on the energy level at the cutoff, the dynamic spectral model also incorporates the spectrum slope at the cutoff. With this improvement, we can cancel the subgrid model in certain cases for which the kinetic energy at cutoff is non-zero but where the kinetic energy transfer to the subgrid modes is zero[12]. This model also reflects the local effects at the cutoff. The other basic assumptions underlying the Chollet–Lesieur model are maintained.
4. The Lesieur–Rogallo model (p. 108), which computes the intensity of the transfers by a dynamic procedure. This is an extension of the Chollet–Lesieur model for flows in spectral disequilibrium, as modifications in the nature of the transfers to the subgrid scales can be considered. The dynamic procedure consists in including in the model information relative to the energy transfers at play with the highest resolved frequencies. The assumptions concerning the filter are not relaxed, though.
5. Models based on the analytical theories of turbulence (p. 108), which compute the effective viscosity without assuming anything about the spectrum shape of the resolved scales, are thus very general. On the other hand, the spectrum shape of the subgrid scales is assumed to be that of a canonical inertial range. These models, which are capable of including very complex physical phenomena, require very much more implementation and computation effort than the previous models. The assumptions concerning the filter are the same as for the previous models.

Chollet–Lesieur Model. Subsequent to Kraichnan's investigations, Chollet and Lesieur [136] proposed an effective viscosity model using the results of the EDQNM closure on the canonical case. The full subgrid transfer term

[11] This is by their action on the transfers between resolved scales and the variations induced on the energy level at cutoff.

[12] As is the case, for example, for two-dimensional flows.

including the backward cascade is written:

$$T^{\mathrm{e}}_{\mathrm{sgs}}(k|k_{\mathrm{c}}) = -2k^2 E(k)\nu_{\mathrm{e}}(k|k_{\mathrm{c}}) \quad , \tag{5.15}$$

in which the effective viscosity $\nu_{\mathrm{e}}(k|k_{\mathrm{c}})$ is defined as the product

$$\nu_{\mathrm{e}}(k|k_{\mathrm{c}}) = \nu^{+}_{\mathrm{e}}(k|k_{\mathrm{c}})\nu^{\infty}_{\mathrm{e}} \quad . \tag{5.16}$$

The constant term $\nu^{\infty}_{\mathrm{e}}$, independent of k, corresponds the asymptotic value of the effective viscosity for wave numbers that are small compared with the cutoff wave number k_{c}. This value is evaluated using the cutoff energy $E(k_{\mathrm{c}})$:

$$\nu^{\infty}_{\mathrm{e}} = 0.441 K_0^{-3/2} \sqrt{\frac{E(k_{\mathrm{c}})}{k_{\mathrm{c}}}} \quad . \tag{5.17}$$

The function $\nu_{\mathrm{e}}(k|k_{\mathrm{c}})$ reflects the variations of the effective viscosity in the proximity of the cutoff. The authors propose the following form, which is obtained by approximating the exact solution with a law of exponential form:

$$\nu^{+}_{\mathrm{e}}(k|k_{\mathrm{c}}) = 1 + 34.59 \exp(-3.03 k_{\mathrm{c}}/k) \quad . \tag{5.18}$$

This form makes it possible to obtain an effective viscosity that is nearly independent of k for wave numbers that are small compared with k_{c}, with a finite increase near the cutoff. There is a limited inclusion of the backward cascade with this model: the effective viscosity remains strictly positive for all wave numbers, while the backward cascade is dominant for very small wave numbers, which would correspond to negative values of the effective viscosity.

Constant Effective Viscosity Model. A simplified form of the effective viscosity of (5.16) can be derived independently of the wave number k [440]. By averaging the effective viscosity along k and assuming that the subgrid modes are in a state of energy balance, we get:

$$\nu_{\mathrm{e}}(k|k_{\mathrm{c}}) = \nu_{\mathrm{e}} = \frac{2}{3} K_0^{-3/2} \sqrt{\frac{E(k_{\mathrm{c}})}{k_{\mathrm{c}}}} \quad . \tag{5.19}$$

Dynamic Spectral Model. The asymptotic value of the effective viscosity (5.17) has been extended to the case of spectra of slope $-m$ by Métais and Lesieur [514] using the EDQNM closure. For a spectrum proportional to $k^{-m}, m \leq 3$, we get:

$$\nu^{\infty}_{\mathrm{e}}(m) = 0.31 \frac{5-m}{m+1} \sqrt{3-m} K_0^{-3/2} \sqrt{\frac{E(k_{\mathrm{c}})}{k_{\mathrm{c}}}} \quad . \tag{5.20}$$

For $m > 3$, the energy transfer cancels out, inducing zero effective viscosity. Here, we find a behavior similar to that of two-dimensional turbulence. Extension of this idea in physical space has been derived by Lamballais and his coworkers [422, 675].

Lesieur–Rogallo Model. By introducing a new filtering level corresponding to the wave number $k_m < k_c$, Lesieur and Rogallo [441] propose a dynamic algorithm for adapting the Chollet–Lesieur model. The contribution to the transfer $T(\boldsymbol{k}), k < k_c$, corresponding to the $(\boldsymbol{k}, \boldsymbol{p}, \boldsymbol{q})$ triads such that p and/or q are in the interval $[k_m, k_c]$, can be computed explicitly by Fourier transforms. This contribution is denoted $T_{\rm sub}(k|k_m, k_c)$ and is associated with the effective viscosity:

$$\nu_{\rm e}(k|k_m, k_c) = -\frac{T_{\rm sub}(k|k_m, k_c)}{2k^2 E(k)} \quad . \tag{5.21}$$

The effective viscosity corresponding to the interactions with wave numbers located beyond k_m is the sum:

$$\nu_{\rm e}(k|k_m) = \nu_{\rm e}(k|k_m, k_c) + \nu_{\rm e}(k|k_c) \quad . \tag{5.22}$$

This relation corresponds exactly to Germano's identity and was previously derived by the authors. The two terms $\nu_{\rm e}(k|k_m)$ and $\nu_{\rm e}(k|k_c)$ are then modeled by the Chollet–Lesieur model. We adopt the hypothesis that when $k < k_m$, then $k \ll k_c$, which leads to $\nu_{\rm e}^+(k|k_c) = \nu_{\rm e}^+(0)$. Relation (5.22) then leads to the equation:

$$\nu_{\rm e}^+(k|k_{\rm m}) = \nu_{\rm e}(k|k_m, k_c)\sqrt{\frac{k_{\rm m}}{E(k_{\rm m})}} + \nu_{\rm e}^+(0)\left(\frac{k_m}{k_c}\right)^{4/3} \quad . \tag{5.23}$$

The factor $\nu_{\rm e}^+(0)$ is evaluated by considering that we have the relations

$$\nu_{\rm e}^+(k|k_m) \approx \nu_{\rm e}^+(0), \quad \nu_{\rm e}(k|k_m, k_c) \approx \nu_{\rm e}(0|k_m, k_c) \quad , \tag{5.24}$$

for $k \ll k_m$, which leads to:

$$\nu_{\rm e}^+(0) = \nu_{\rm e}(0|k_m, k_c)\sqrt{\frac{k_{\rm m}}{E(k_{\rm m})}}\left[1 - \left(\frac{k_m}{k_c}\right)^{4/3}\right]^{-1} \quad . \tag{5.25}$$

Models Based on Analytical Theories of Turbulence. The effective viscosity models presented above are all based on an approximation of the effective viscosity profile obtained in the canonical case, and are therefore intrinsically linked to the underlying hypotheses, especially those concerning the shape of the energy spectrum. One way of relaxing this constraint is to compute the effective viscosity directly from the computed spectrum using analytical theories of turbulence. This approach has been used by Aupoix [24], Chollet [132, 133], and Bertoglio [56, 57, 58].

More recently, following the recommendations of Leslie and Quarini, which are to model the forward and backward cascade mechanisms separately,

Chasnov [120] in 1991 proposed an effective viscosity model considering only the energy draining effects, with the backward cascade being modeled separately (see Sect. 5.4). Starting with an EDQNM analysis, Chasnov proposes computing the effective viscosity $\nu_e(k|k_c)$ as:

$$\nu_e(k|k_c) = \frac{1}{2k^2} \int_{k_c}^{\infty} dp \int_{p-k}^{p} dq \Theta_{kpq} \left(\frac{p^2}{q}(xy + z^3)E(q) + \frac{q^2}{p}(xz + y^3)E(p) \right) , \quad (5.26)$$

in which x, y and z are geometric factors associated with the $(\boldsymbol{k}, \boldsymbol{p}, \boldsymbol{q})$ triads and Θ_{kpq} a relaxation time. These terms are explained in Appendix B. To compute this integral, the shape of the energy spectrum beyond the cutoff k_c must be known. As it is not known *a priori*, it must be specified elsewhere. In practice, Chasnov uses a Kolmogorov spectrum extending from the cutoff to infinity. To simplify the computations, the relation (5.26) is not used outside the interval $[k_c \leq p \leq 3k_c]$. For wave numbers $p > 3k_c$, the following simplified asymptotic form already proposed by Kraichnan is used:

$$\nu_e(k|k_c) = \frac{1}{15} \int_{k_c}^{\infty} dp \Theta_{kpq} \left(5E(p) + p\frac{\partial E(p)}{\partial p} \right) \quad . \quad (5.27)$$

5.3.2 Physical Space Models

Subgrid Viscosity Concept. The forward energy cascade mechanism to the subgrid scales is modeled explicitly using the following hypothesis:

Hypothesis 5.2 *The energy transfer mechanism from the resolved to the subgrid scales is analogous to the molecular mechanisms represented by the diffusion term, in which the viscosity ν appears.*

This hypothesis is equivalent to assuming that the behavior of the subgrid scales is analogous to the Brownian motion superimposed on the motion of the resolved scales. In gaskinetics theory, molecular agitation draws energy from the flow by way of molecular viscosity. So the energy cascade mechanism will be modeled by a term having a mathematical structure similar to that of molecular diffusion, but in which the molecular viscosity will be replaced by a subgrid viscosity denoted ν_{sgs}. As Boussinesq proposed, this choice of mathematical form of the subgrid model is written:

$$-\nabla \cdot \tau^d = \nabla \cdot \left(\nu_{sgs}(\nabla \overline{\boldsymbol{u}} + \nabla^T \overline{\boldsymbol{u}}) \right) \quad , \quad (5.28)$$

in which τ^d is the deviator of τ, i.e.:

$$\tau_{ij}^d \equiv \tau_{ij} - \frac{1}{3}\tau_{kk}\delta_{ij} \quad . \quad (5.29)$$

The complementary spherical tensor $\frac{1}{3}\tau_{kk}\delta_{ij}$ is added to the filtered static pressure term and consequently requires no modeling. This decomposition is necessary since the tensor $(\nabla\overline{\boldsymbol{u}} + \nabla^T\overline{\boldsymbol{u}})$ has a zero trace, and we can only model a tensor that also has a zero trace. This leads to the definition of the modified pressure Π:

$$\Pi = \overline{p} + \frac{1}{3}\tau_{kk} \quad . \tag{5.30}$$

It is important to note that the modified pressure and filtered pressure $\overline{p}$ may take very different values when the generalized subgrid kinetic energy becomes large [374]. The closure thus now consists in determining the relation:

$$\nu_{\text{sgs}} = \mathcal{N}(\overline{\boldsymbol{u}}) \quad . \tag{5.31}$$

The use of hypothesis (5.2) and of a model structured as above calls for a few comments.

Obtaining a scalar subgrid viscosity requires the adoption of the following hypothesis:

Hypothesis 5.3 *A characteristic length l_0 and a characteristic time t_0 are sufficient for describing the subgrid scales.*

Then, by dimensional reasoning similar to Prandtl's, we arrive at:

$$\nu_{\text{sgs}} \propto \frac{l_0^2}{t_0} \quad . \tag{5.32}$$

Models of the form (5.28) are local in space and time, which is a necessity if they are to be used in practice. This local character, similar to that of the molecular diffusion terms, implies [26, 405, 813]:

Hypothesis 5.4 (Scale Separation Hypothesis) *There exists a total separation between the subgrid and resolved scales.*

A spectrum verifying this hypothesis is presented in Fig. 5.10.

Using L_0 and T_0 to denote the characteristic scales, respectively, of the resolved field in space and time, this hypothesis can be reformulated as:

$$\frac{l_0}{L_0} \ll 1, \quad \frac{t_0}{T_0} \ll 1 \quad . \tag{5.33}$$

This hypothesis is verified in the case of molecular viscosity. The ratio between the size of the smallest dynamically active scale, η_{K}, and the mean free path ξ_{fp} of the molecules of a gas is evaluated as:

$$\frac{\xi_{\text{fp}}}{\eta_{\text{K}}} \simeq \frac{\text{Ma}}{Re^{1/4}} \quad , \tag{5.34}$$

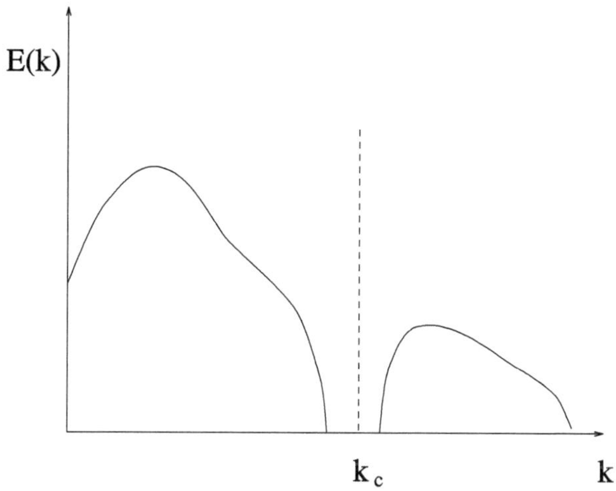

Fig. 5.10. Energy spectrum corresponding to a total scale separation for cutoff wave number k_c.

where Ma is the Mach number, defined as the ratio of the fluid velocity to the speed of sound, and Re the Reynolds number [708]. In most of the cases encountered, this ratio is less than 10^3, which ensures the pertinence of using a continuum model. For applications involving rarefied gases, this ratio can take on much higher values of the order of unity, and the Navier–Stokes equations are then no longer an adequate model for describing the fluid dynamics.

Filtering associated to large-eddy simulation does not introduce such a separation between resolved and subgrid scales because the turbulent energy spectrum is continuous. The characteristic scales of the smallest resolved scales are consequently very close to those of the largest subgrid scales[13]. This continuity originates the existence of the spectrum region located near the cutoff, in which the effective viscosity varies rapidly as a function of the wave number. The result of this difference in nature with the molecular viscosity is that the subgrid viscosity is not a characteristic of the fluid but of the flow. Let us not that Yoshizawa [786, 788], using a re-normalization technique, has shown that the subgrid viscosity is defined as a fourth-order non-local tensor in space and time, in the most general case. The use of the scale separation hypothesis therefore turns out to be indispensable for constructing local models, although it is contrary to the scale similar hypothesis of Bardina et al. [40], which is discussed in Chap. 7.

It is worth noting that subgrid-viscosity based models for the forward energy cascace induce a spurious alignment of the eigenvectors for resolved strain rate tensor and subgrid-scale tensor, because they are expressed as

[13] This is all the more true for smooth filters such as the Gaussian and box filters, which allow a frequency overlapping between the resolved and subgrid scales.

$\tau^{\mathrm{d}} \propto (\nabla\overline{\boldsymbol{u}} + \nabla^T\overline{\boldsymbol{u}})$.[14] Tao et al. [703, 704] and Horiuti [325] have shown that this alignment is unphysical: the eigenvectors for the subgrid tensor have a strongly preferred relative orientation of 35 to 45 degrees with the resolved strain rate eigenvectors.

The modeling problem consists in determining the characteristic scales l_0 and t_0.

Model Types. The subgrid viscosity models can be classified in three categories according to the quantities they bring into play [26]:

1. Models based on the resolved scales (p. 113): the subgrid viscosity is evaluated using global quantities related to the resolved scales. The existence of subgrid scales at a given point in space and time will therefore be deduced from the global characteristics of the resolved scales, which requires the introduction of assumptions.
2. Models based on the energy at the cutoff (p. 116): the subgrid viscosity is calculated from the energy of the highest resolved frequency. Here, it is a matter of information contained in the resolved field, but localized in frequency and therefore theoretically more pertinent for describing the phenomena at cutoff than the quantities that are global and thus not localized in frequency, which enter into the models of the previous class. The existence of subgrid scales is associated with a non-zero value of the energy at cutoff[15].
3. Models based on the subgrid scales (p. 116), which use information directly related to the subgrid scales. The existence of the subgrid scales is no longer determined on the basis of assumptions concerning the characteristics of the resolved scales as it is in the previous cases, but rather directly from this additional information. These models, because they are richer, also theoretically allow a better description of these scales than the previous models.

These model classes are presented in the following. All the developments are based on the analysis of the energy transfers in the canonical case. In order to be able to apply the models formulated from these analyses to more realistic flows, such as the homogeneous isotropic flows associated with a production type spectrum, we adopt the assumption that the filter cutoff fre-

[14] But it must be also remembered that the purpose of these functional models is not to predict the subgrid tensor, but just to enforce the correct resolved kinetic energy balance. This reconstruction of the subgrid tensor is nothing but an a posteriori interpretation. This fact is used by Germano to derive new subgrid viscosity models [252].

[15] This hypothesis is based on the fact that the energy spectrum $E(k)$ of an isotropic turbulent flow in spectral equilibrium corresponding to a Kolmogorov spectrum is a monotonic continuous decreasing function of the wave number k. If there exists a wave number k^* such that $E(k^*) = 0$, then $E(k) = 0, \forall k > k^*$. Also, if the energy is non-zero at the cutoff, then subgrid modes exist, i.e. if $E(k_{\mathrm{c}}) \neq 0$, then there exists a neighbourhood $\Omega_{k_{\mathrm{c}}} = [k_{\mathrm{c}}, k_{\mathrm{c}} + \epsilon_{\mathrm{c}}]$, $\epsilon_{\mathrm{c}} > 0$ such that $E(k_{\mathrm{c}}) \geq E(k) \geq 0 \, \forall k \in \Omega_{k_{\mathrm{c}}}$.

quency is located sufficiently far into the inertial range for these analyses to remain valid (refer to Sect. 5.1.2). The use of these subgrid models for arbitrary developed turbulent flows (anisotropic, inhomogeneous) is justified by the local isotropy hypothesis: we assume then that the cutoff occurs in the scale range that verifies this hypothesis.

The case corresponding to an isotropic homogeneous flow associated with a production spectrum is represented in Fig. 5.11. Three energy fluxes are defined: the injection rate of turbulent kinetic energy into the flow by the driving mechanisms (forcing, instabilities), denoted ε_I; the kinetic energy transfer rate through the cutoff, denoted $\tilde{\varepsilon}$; and the kinetic energy dissipation rate by the viscous effects, denoted ε.

Models Based on the Resolved Scales. These models are of the generic form:

$$\nu_{\mathrm{sgs}} = \nu_{\mathrm{sgs}}\left(\overline{\Delta}, \tilde{\varepsilon}\right) \quad , \tag{5.35}$$

in which $\overline{\Delta}$ is the characteristic cutoff length of the filter and $\tilde{\varepsilon}$ the instantaneous energy flux through the cutoff. We implicitly adopt the assumption here, then, that the subgrid modes exist, i.e. that the exact solution is not entirely represented by the filtered field when this flux is non-zero.

First Method. Simple dimensional analysis shows that:

$$\nu_{\mathrm{sgs}} \propto \tilde{\varepsilon}^{1/3}\,\overline{\Delta}^{4/3} \quad . \tag{5.36}$$

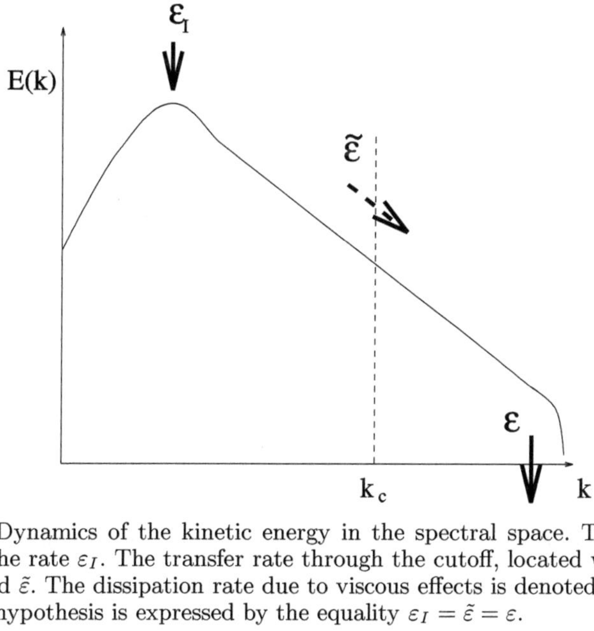

Fig. 5.11. Dynamics of the kinetic energy in the spectral space. The energy is injected at the rate ε_I. The transfer rate through the cutoff, located wave number k_c, is denoted $\tilde{\varepsilon}$. The dissipation rate due to viscous effects is denoted ε. The local equilibrium hypothesis is expressed by the equality $\varepsilon_I = \tilde{\varepsilon} = \varepsilon$.

Reasoning as in the framework of Kolmogorov's hypotheses for isotropic homogeneous turbulence, for the case of an infinite inertial spectrum of the form

$$E(k) = K_0 \langle\varepsilon\rangle^{2/3} k^{-5/3}, \quad K_0 \sim 1.4 \quad , \tag{5.37}$$

in which ε is the kinetic energy dissipation rate, we get the equation:

$$\langle\nu_{\rm sgs}\rangle = \frac{A}{K_0 \pi^{4/3}} \langle\tilde{\varepsilon}\rangle^{1/3} \overline{\Delta}^{4/3} \quad , \tag{5.38}$$

in which the constant A is evaluated as $A = 0.438$ with the TFM model and as $A = 0.441$ by the EDQNM theory [26]. The angle brackets operator $\langle\,\rangle$, designates a statistical average. This statistical averaging operation is intrinsically associated with a spatial mean by the fact of the flow's spatial homogeneity and isotropy hypotheses. This notation is used in the following to symbolize the fact that the reasoning followed in the framework of isotropic homogeneous turbulence applies only to the statistical averages and not to the local values in the physical space. The problem is then to evaluate the average flux $\langle\tilde{\varepsilon}\rangle$. In the isotropic homogeneous case, we have:

$$\langle 2|\overline{S}|^2\rangle = \langle 2\overline{S}_{ij}\overline{S}_{ij}\rangle = \int_0^{k_c} 2k^2 E(k) dk, \quad k_c = \frac{\pi}{\overline{\Delta}} \quad . \tag{5.39}$$

If the cutoff k_c is located far enough into the inertial range, the above relation can be expressed solely as a function of this region's characteristic quantities. Using a spectrum of the shape (5.37), we get:

$$\langle 2|\overline{S}|^2\rangle = \pi^{4/3} K_0 \frac{3}{2} \langle\varepsilon\rangle^{2/3} \overline{\Delta}^{-4/3} \quad . \tag{5.40}$$

Using the hypothesis[16] [447]:

$$\langle|\overline{S}|^{3/2}\rangle \simeq \langle|\overline{S}|\rangle^{3/2} \quad , \tag{5.41}$$

we get the equality:

$$\langle\varepsilon\rangle = \frac{1}{\pi^2} \left(\frac{3K_0}{2}\right)^{-3/2} \overline{\Delta}^2 \langle 2|\overline{S}|^2\rangle^{3/2} \quad . \tag{5.42}$$

In order to evaluate the dissipation rate $\langle\varepsilon\rangle$ from the information contained in the resolved scales, we assume the following:

Hypothesis 5.5 (Local Equilibrium Hypothesis) *The flow is in constant spectral equilibrium, so there is no accumulation of energy at any frequency and the shape of the energy spectrum remains invariant with time.*

[16] The error margin measured in direct numerical simulations of isotropic homogeneous turbulence is of the order of 20% [507].

This implies an instantaneous adjustment of all the scales of the solution to the turbulent kinetic energy production mechanism, and therefore equality between the production, dissipation, and energy flux through the cutoff:

$$\langle \varepsilon_I \rangle = \langle \tilde{\varepsilon} \rangle = \langle \varepsilon \rangle \quad . \tag{5.43}$$

Using this equality and relations (5.38) and (5.42), we get the closure relation:

$$\langle \nu_{\mathrm{sgs}} \rangle = \left(C\overline{\Delta} \right)^2 \langle 2|\overline{S}|^2 \rangle^{1/2} \quad , \tag{5.44}$$

where the constant C is evaluated as:

$$C = \frac{\sqrt{A}}{\pi\sqrt{K_0}} \left(\frac{3K_0}{2} \right)^{-1/4} \sim 0.148 \quad . \tag{5.45}$$

Second Method. The local equilibrium hypothesis allows:

$$\langle \varepsilon \rangle = \langle \tilde{\varepsilon} \rangle \equiv \langle -\overline{S}_{ij}\tau_{ij} \rangle = \langle \nu_{\mathrm{sgs}} 2\overline{S}_{ij}\overline{S}_{ij} \rangle \quad . \tag{5.46}$$

The idea is then to assume that:

$$\langle \nu_{\mathrm{sgs}} 2\overline{S}_{ij}\overline{S}_{ij} \rangle = \langle \nu_{\mathrm{sgs}} \rangle \langle 2\overline{S}_{ij}\overline{S}_{ij} \rangle \quad . \tag{5.47}$$

By stating at the outset that the subgrid viscosity is of the form (5.44) and using relation (5.40), a new value is found for the constant C:

$$C = \frac{1}{\pi} \left(\frac{3K_0}{2} \right)^{-3/4} \sim 0.18 \quad . \tag{5.48}$$

We note that the value of this constant is independent of the cutoff wave number k_c, but because of the way it is calculated, we can expect a dependency as a function of the spectrum shape.

Alternate Form. This modeling induces a dependency as a function of the cutoff length $\overline{\Delta}$ and the strain rate tensor $\overline{S}$ of the resolved velocity field. In the isotropic homogeneous case, we have the equality:

$$\langle 2|\overline{S}|^2 \rangle = \langle \overline{\omega} \cdot \overline{\omega} \rangle, \quad \overline{\omega} = \nabla \times \overline{\boldsymbol{u}} \quad . \tag{5.49}$$

By substitution, we get the equivalent form [487]:

$$\langle \nu_{\mathrm{sgs}} \rangle = \left(C\overline{\Delta} \right)^2 \langle \overline{\omega} \cdot \overline{\omega} \rangle^{1/2} \quad . \tag{5.50}$$

These two versions bring in the gradients of the resolved velocity field. This poses a problem of physical consistency since the subgrid viscosity is non-zero as soon as the velocity field exhibits spatial variations, even if it is laminar and all the scales are resolved. The hypothesis that links the existence of the subgrid modes to that of the mean field gradients therefore prevents us from considering the large scale intermittency and thereby requires us to develop models which by nature can only be effective for dealing with flows that are completely turbulent and under-resolved everywhere[17]. Poor behavior can therefore be expected when treating intermittent or weakly developed turbulent flows (i.e. in which the inertial range does not appear in the spectrum) due to too strong an action by the model.

Models Based on the Energy at Cutoff. The models of this category are based on the intrinsic hypothesis that if the energy at the cutoff is non-zero, then subgrid modes exist.

First Method. Using relation (5.38) and supposing that the cutoff occurs within an inertial region, i.e.:

$$E(k_c) = K_0 \langle \varepsilon \rangle^{2/3} k_c^{-5/3} \quad , \tag{5.51}$$

by substitution, we get:

$$\langle \nu_{sgs} \rangle = \frac{A}{\sqrt{K_0}} \sqrt{\frac{E(k_c)}{k_c}}, \ k_c = \pi/\overline{\Delta} \quad . \tag{5.52}$$

This model raises the problem of determining the energy at the cutoff in the physical space, but on the other hand ensures that the subgrid viscosity will be null if the flow is well resolved, i.e. if the highest-frequency mode captured by the grid is zero. This type of model thus ensures a better physical consistency than those models based on the large scales. It should be noted that it is equivalent to the spectral model of constant effective viscosity.

Second Method. As in the case of models based on the large scales, there is a second way of determining the model constant. By combining relations (5.46) and (5.51), we get:

$$\langle \nu_{sgs} \rangle = \frac{2}{3K_0^{3/2}} \sqrt{\frac{E(k_c)}{k_c}} \quad . \tag{5.53}$$

Models Based on Subgrid Scales. Here we considers models of the form:

$$\langle \nu_{sgs} \rangle = \langle \nu_{sgs} \rangle \left(\overline{\Delta}, \langle q_{sgs}^2 \rangle, \langle \varepsilon \rangle \right) \quad , \tag{5.54}$$

in which $\langle q_{sgs}^2 \rangle$ is the kinetic energy of the subgrid scales and $\langle \varepsilon \rangle$ the kinetic energy dissipation rate[18]. These models contain more information about the

[17] In the sense that the subgrid modes exist at each point in space and at each time step.

[18] Other models are of course possible using other subgrid scale quantities like a length or time scale, but we limit ourselves here to classes of models for which practical results exist.

subgrid modes than those belonging to the two categories described above, and thereby make it possible to do without the local equilibrium hypothesis (5.5) by introducing characteristic scales specific to the subgrid modes by way of $\langle q_{\rm sgs}^2 \rangle$ and $\langle \varepsilon \rangle$. This capacity to handle the energy disequilibrium is expressed by the relation:

$$\langle \widetilde{\varepsilon} \rangle \equiv \langle -\tau_{ij} \overline{S}_{ij} \rangle \neq \langle \varepsilon \rangle \quad , \tag{5.55}$$

which should be compared with (5.43). In the case of an inertial range extending to infinity beyond the cutoff, we have the relation:

$$\langle q_{\rm sgs}^2 \rangle \equiv \langle \frac{1}{2} \overline{u_i' u_i'} \rangle = \int_{k_{\rm c}}^{\infty} E(k) dk = \frac{3}{2} K_0 \langle \varepsilon \rangle^{2/3} k_{\rm c}^{-2/3} \quad , \tag{5.56}$$

from which we deduce:

$$\langle \varepsilon \rangle = \frac{k_{\rm c}}{(3K_0/2)^{3/2}} \langle q_{\rm sgs}^2 \rangle^{3/2} \quad . \tag{5.57}$$

By introducing this last equation into relation (5.38), we come to the general form:

$$\langle \nu_{\rm sgs} \rangle = C_\alpha \langle \varepsilon \rangle^{\alpha/3} \langle q_{\rm sgs}^2 \rangle^{(1-\alpha)/2} \overline{\Delta}^{1+\alpha/3} \quad , \tag{5.58}$$

in which

$$C_\alpha = \frac{A}{K_0 \pi^{4/3}} \left(\frac{3K_0}{2} \right)^{(\alpha-1)/2} \pi^{(1-\alpha)/3} \quad , \tag{5.59}$$

and in which α is a real weighting parameter. Interesting forms of $\nu_{\rm sgs}$ have been found for certain values:

- For $\alpha = 1$, we get

$$\langle \nu_{\rm sgs} \rangle = \frac{A}{K_0 \pi^{4/3}} \overline{\Delta}^{4/3} \langle \varepsilon \rangle^{1/3} \quad . \tag{5.60}$$

  This form uses only the dissipation and is analogous to that of the models based on the resolved scales. If the local equilibrium hypothesis is used, these two types of models are formally equivalent.
- For $\alpha = 0$, we get

$$\langle \nu_{\rm sgs} \rangle = \sqrt{\frac{2}{3}} \frac{A}{\pi K_0^{3/2}} \overline{\Delta} \left(\langle q_{\rm sgs}^2 \rangle \right)^{1/2} \quad . \tag{5.61}$$

  This model uses only the kinetic energy of the subgrid scales. As such, it is formally analogous to the definition of the diffusion coefficient of an ideal gas in the framework of gaskinetics theory. In the case of an inertial spectrum extending to infinity beyond the cutoff, this model is strictly equivalent to the model based on the energy at cutoff, since in this precise case we have the relation:

$$\frac{3}{2}k_c E(k_c) = \langle q_{sgs}^2 \rangle \quad . \tag{5.62}$$

– For $\alpha = -3$, we have:

$$\langle \nu_{sgs} \rangle = \frac{4A}{9K_0^3} \frac{\left(\langle q_{sgs}^2 \rangle\right)^2}{\langle \varepsilon \rangle} \quad . \tag{5.63}$$

This model is formally analogous to the $k-\varepsilon$ statistical model of turbulence for the Reynolds Averaged Navier–Stokes equations, and does not bring in the filter cutoff length explicitly.

The closure problem consists in determining the quantities $\langle \varepsilon \rangle$ and $\langle q_{sgs}^2 \rangle$. To do this, we can introduce one or more equations for the evolution of these quantities or we can deduce them from the information contained in the resolved field. As these quantities represent the subgrid scales, we are justified in thinking that, if they are correctly evaluated, the subgrid viscosity will be negligible when the flow is well resolved numerically. However, it should be noted that these models in principle require more computation than those based on the resolved scales, because they produce more information concerning the subgrid scales.

Extension to Other Spectrum Shapes. The above developments are based on a Kolmogorov spectrum, which reflects only the existence of a region of similarity of the real spectra. This approach can be extended to other more realistic spectrum shapes, mainly including the viscous effects. Several extensions of the models based on the large scales were proposed by Voke [737] for this. The total dissipation $\langle \varepsilon \rangle$ can be decomposed into the sum of the dissipation associated with the large scales, denoted $\langle \varepsilon_r \rangle$, and the dissipation associated with the subgrid scales, denoted $\langle \varepsilon_{sgs} \rangle$, (see Fig. 5.12):

$$\langle \varepsilon \rangle = \langle \varepsilon_r \rangle + \langle \varepsilon_{sgs} \rangle \quad . \tag{5.64}$$

These three quantities can be evaluated as:

$$\langle \varepsilon \rangle = \langle 2(\nu_{sgs} + \nu)|\overline{S}|^2 \rangle \quad , \tag{5.65}$$

$$\langle \varepsilon_r \rangle = 2\nu \langle |\overline{S}|^2 \rangle = 2\nu \int_0^{k_c} k^2 E(k) dk \quad , \tag{5.66}$$

$$\langle \varepsilon_{sgs} \rangle = 2\langle \nu_{sgs} |\overline{S}|^2 \rangle = C_s \overline{\Delta}^2 \left(2\langle |\overline{S}|^2 \rangle\right)^{3/2} \quad , \tag{5.67}$$

from which we get:

$$\frac{\langle \varepsilon_r \rangle}{\langle \varepsilon \rangle} = \frac{1}{1+\tilde{\nu}}, \quad \tilde{\nu} = \frac{\langle \nu_{sgs} \rangle}{\nu} \quad . \tag{5.68}$$

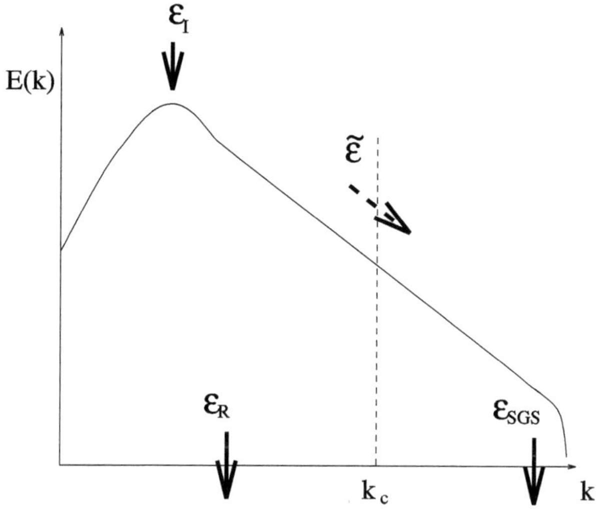

Fig. 5.12. Kinetic energy dynamics in the spectral space. The energy is injected at the rate ε_I. The transfer rate through the cutoff located at the wave number k_c is denoted $\tilde{\varepsilon}$. The dissipation rate in the form of heat by the viscous effects associated with the scales located before and after the cutoff k_c are denoted ε_r and ε_{sgs}, respectively.

This ratio is evaluated by calculating the $\langle \varepsilon_{\text{r}} \rangle$ term analytically from the chosen spectrum shapes, which provides a way of then computing the subgrid viscosity $\langle \nu_{\text{sgs}} \rangle$.

We define the three following parameters:

$$\kappa = \frac{k}{k_{\text{d}}} = k \left(\frac{\nu^3}{\langle \varepsilon \rangle} \right)^{1/4} \quad , \quad \kappa_{\text{c}} = \frac{k_{\text{c}}}{k_{\text{d}}} \quad , \tag{5.69}$$

$$Re_{\overline{\Delta}} = \frac{\overline{\Delta}^2 \sqrt{2|\overline{S}|^2}}{\nu} \quad , \tag{5.70}$$

in which k_{d} is the wave number associated with the Kolmogorov scale (see Appendix A), and $Re_{\overline{\Delta}}$ is the mesh-Reynolds number. Algebraic substitutions lead to:

$$\kappa = \pi Re_{\overline{\Delta}}^{-1/2} (1 + \tilde{\nu})^{-1/4} \quad . \tag{5.71}$$

The spectra studied here are of the generic form:

$$E(k) = K_0 \varepsilon^{2/3} k^{-5/3} f(\kappa) \quad , \tag{5.72}$$

in which f is a damping function for large wave numbers. The following are some commonly used forms of this function:

– Heisenberg–Chandrasekhar spectrum:

$$f(\kappa) = \left[1 + \left(\frac{3K_0}{2}\right)^3 \kappa^4\right]^{-4/3} \quad . \tag{5.73}$$

– Kovasznay spectrum:

$$f(\kappa) = \left(1 - \frac{K_0}{2}\kappa^{4/3}\right)^2 \quad . \tag{5.74}$$

Note that this function cancels out for $\kappa = (2/K_0)^{3/4}$, which requires that the spectrum be forced to zero for wave numbers beyond this limit.
– Pao spectrum:

$$f(\kappa) = \exp\left(-\frac{3K_0}{2}\kappa^{4/3}\right) \quad . \tag{5.75}$$

These three spectrum shapes are graphed in Fig. 5.13. An analytical integration leads to:

– For the Heisenberg–Chandrasekhar spectrum:

$$\frac{\langle \varepsilon_{\mathrm{r}} \rangle}{\langle \varepsilon \rangle} = \kappa_{\mathrm{c}}^{4/3} \left[\left(\frac{2}{3K_0}\right)^3 + \kappa_{\mathrm{c}}^4\right]^{-1/3} \quad , \tag{5.76}$$

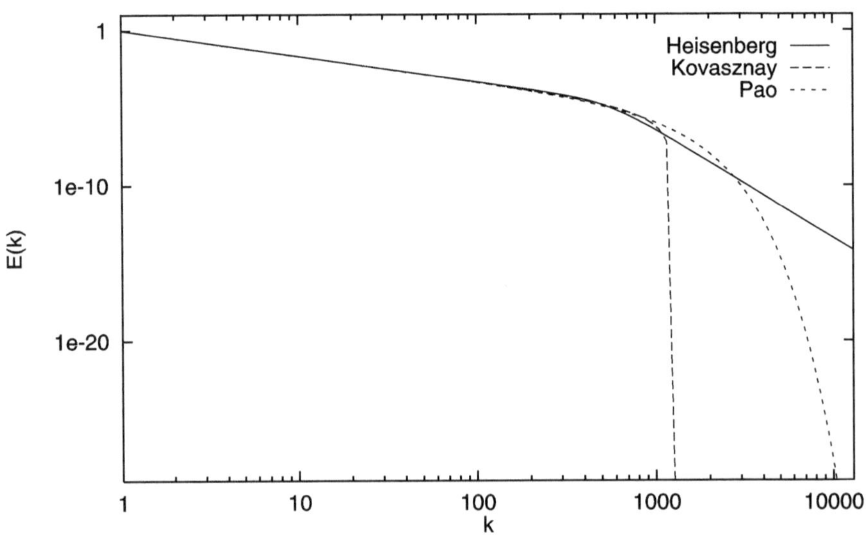

Fig. 5.13. Graph of Heisenberg–Chandrasekhar, Kovasznay, and Pao spectra, for $k_d = 1000$.

or:

$$\langle \nu_{\rm sgs} \rangle = \nu \left\{ \kappa_{\rm c}^{-4/3} \left[\left(\frac{2}{3K_0} \right)^3 + \kappa_{\rm c}^4 \right]^{1/3} - 1 \right\} \quad . \tag{5.77}$$

– For the Kovazsnay spectrum:

$$\frac{\langle \varepsilon_{\rm r} \rangle}{\langle \varepsilon \rangle} = 1 - \left(1 - \frac{K_0}{2} \kappa_{\rm c}^{4/3} \right)^3 \quad , \tag{5.78}$$

or:

$$\langle \nu_{\rm sgs} \rangle = \nu \left\{ \left[1 - \left(1 - \frac{K_0}{2} \kappa_{\rm c}^{4/3} \right)^3 \right]^{-1} - 1 \right\} \quad . \tag{5.79}$$

– For the Pao spectrum:

$$\frac{\langle \varepsilon_{\rm r} \rangle}{\langle \varepsilon \rangle} = 1 - \exp \left(\frac{3K_0}{2} \kappa_{\rm c}^{4/3} \right)^3 \quad , \tag{5.80}$$

or:

$$\langle \nu_{\rm sgs} \rangle = \nu \left\{ \left[1 - \exp \left(\frac{3K_0}{2} \kappa_{\rm c}^{4/3} \right)^3 \right]^{-1} - 1 \right\} \quad . \tag{5.81}$$

These new estimates of the subgrid viscosity $\langle \nu_{\rm sgs} \rangle$ make it possible to take the viscous effects into account, but requires that the spectrum shape be set *a priori*, as well as the value of the ratio $\kappa_{\rm c}$ between the cutoff wave number $k_{\rm c}$ and the wave number $k_{\rm d}$ associated with the Kolmogorov scale.

Inclusion of the Local Effects at Cutoff. The subgrid viscosity models in the physical space, such as they have been developed, do not reflect the increase in the coupling intensity with the subgrid modes when we consider modes near the cutoff. These models are therefore analogous to that of constant effective viscosity. To bring out these effects in the proximity of the cutoff, Chollet [134], Ferziger [217], Lesieur and Métais [440], Deschamps [176], Borue and Orszag [68, 69, 70, 71], Winckelmans and co-workers [762, 163, 759] and Layton [427] propose introducing high-order dissipation terms that will have a strong effect on the high frequencies of the resolved field without affecting the low frequencies.

Chollet [134], when pointing out that the energy transfer term $T^{\rm e}_{\rm sgs}$ can be written in the general form

$$T^{\rm e}_{\rm sgs}(k|k_{\rm c}) = -2\nu_{\rm e}^{(n)}(k|k_{\rm c}) k^{2n} E(k) \quad , \tag{5.82}$$

in which $\nu_e^{(n)}(k|k_c)$ is a hyper-viscosity, proposes modeling the subgrid term in the physical space as the sum of an ordinary subgrid viscosity model and a sixth-order dissipation. This new model is expressed:

$$\nabla \cdot \tau = -\langle \nu_{\rm sgs} \rangle \left(C_1 \nabla^2 + C_2 \nabla^6 \right) \overline{\boldsymbol{u}} \quad , \tag{5.83}$$

in which C_1 and C_2 are constants. Ferziger proposes introducing a fourth-order dissipation by adding to the subgrid tensor τ the tensor $\tau^{(4)}$, defined as:

$$\tau_{ij}^{(4)} = \frac{\partial}{\partial x_j} \left(\nu_{\rm sgs}^{(4)} \frac{\partial^2 \overline{u}_i}{\partial x_k \partial x_k} \right) + \frac{\partial}{\partial x_i} \left(\nu_{\rm sgs}^{(4)} \frac{\partial^2 \overline{u}_j}{\partial x_k \partial x_k} \right) \quad , \tag{5.84}$$

or as

$$\tau_{ij}^{(4)} = \frac{\partial^2}{\partial x_k \partial x_k} \left(\nu_{\rm sgs}^{(4)} \left(\frac{\partial \overline{u}_i}{\partial x_j} + \frac{\partial \overline{u}_j}{\partial x_i} \right) \right) \quad , \tag{5.85}$$

in which the hyper-viscosity $\nu_{\rm sgs}^{(4)}$ is defined by dimensional arguments as

$$\nu_{\rm sgs}^{(4)} = C_{\rm m} \overline{\Delta}^4 |\overline{S}| \quad . \tag{5.86}$$

The full subgrid term that appears in the momentum equations is then written:

$$\tau_{ij} = \tau_{ij}^{(2)} + \tau_{ij}^{(4)} \quad , \tag{5.87}$$

in which $\tau_{ij}^{(2)}$ is a subgrid viscosity model described above. A similar form is proposed by Lesieur and Métais: after defining the velocity field $\boldsymbol{u}^\diamond$ as

$$\boldsymbol{u}^\diamond = \nabla^{2p} \overline{\boldsymbol{u}} \quad , \tag{5.88}$$

the two authors propose the composite form:

$$\tau_{ij} = -\nu_{\rm sgs} \overline{S}_{ij} + (-1)^{p+1} \nu_{\rm sgs}^\diamond S_{ij}^\diamond \quad , \tag{5.89}$$

in which $\nu_{\rm sgs}^\diamond$ hyper-viscosity obtained by applying a subgrid viscosity model to the $\boldsymbol{u}^\diamond$ field, and $S^\diamond$ the strain rate tensor computed from this same field. The constant of the subgrid model used should be modified to verify the local equilibrium relation, which is

$$\langle -\tau_{ij} \overline{S}_{ij} \rangle = \langle \varepsilon \rangle \quad .$$

This composite form of the subgrid dissipation has been validated experimentaly by Cerutti et al. [115], who computed the spectral distribution of dissipation and the corresponding spectral viscosity from experimental data.

It is worth noting that subgrid dissipations defined thusly, as the sum of second- and fourth-order dissipations, are similar in form to certain numerical schemes designed for capturing strong gradients, like that of Jameson et al. [346].

Borue and Orszag [68, 69, 70, 71] propose to eliminate the molecular and the subgrid viscosities by replacing them by a higher power of the Laplacian operator. Numerical tests show that three-dimensional inertial-range dynamics is relatively independent of the form of the hyperviscosity. It was also shown that for a given numerical resolution, hyperviscous dissipations increase the extent of the inertial range of three-dimensional turbulence by an order of magnitude. It is worth noting that this type of iterated Laplacian is commonly used for two-dimensional simulations. Borue and Orszag used a height-time iterated Laplacian to get these conclusions. Such operators are easily defined when using spectral methods, but are of poor interest when dealing with finite difference of finite volume techniques.

Subgrid-Viscosity Models. Various subgrid viscosity models belonging to the three categories defined above will now be described. These are the following:

1. The Smagorinsky model (p. 124), which is based on the resolved scales. This model, though very simple to implant, suffers from the defects already mentioned for the models based on the large scales.
2. The second-order Structure Function model developed by Métais and Lesieur (p. 124), which is an extension into physical space of the models based on the energy at cutoff. Theoretically based on local frequency information, this model should be capable of treating large-scale intermittency better than the Smagorinsky model. However, the impossibility of localizing the information in both space and frequency (see discussion further on) reduces its efficiency.
3. The third-order Structure Function models developed by Shao (p. 126), which can be interpreted as an extension of the previous model based on the second-order structure function. The use of the Kolmogorov–Meneveau equation [510] for the filtered third-order structure function enables the definition of several models which do not contain arbitrary constants and have improved potentiality for non-equilibrium flows.
4. A model based on the kinetic energy of the subgrid modes (p. 128). This energy is considered as an additional variable of the problem, and is evaluated by solving an evolution equation. Since it contains information relative to the subgrid scales, it is theoretically more capable of handling large-scale intermittency than the previous model. Moreover, the local equilibrium hypothesis can be relaxed, so that the spectral nonequilibrium can be integrated better. The model requires additional hypotheses, though (modeling, boundary conditions).
5. The Yoshizawa model (p. 129), which includes an additional evolution equation for a quantity linked to a characteristic subgrid scale, by which it can be classed among models based on the subgrid scales. It has the same advantages and disadvantages as the previous model.
6. The Mixed Scale Model (p. 130), which uses information related both to the subgrid modes and to the resolved scales, though without incorpo-

rating additional evolution equations. The subgrid scale data is deduced from that contained in the resolved scales by extrapolation in the frequency domain. This model is of intermediate level of complexity (and quality) between those based on the large scales and those that use additional variables.

Smagorinsky Model. The Smagorinsky model [676] is based on the large scales. It is generally used in a local form the physical space, i.e. variable in space, in order to be more adaptable to the flow being calculated. It is obtained by space and time localization of the statistical relations given in the previous section. There is no particular justification for this local use of relations that are on average true for the whole, since they only ensure that the energy transfers through the cutoff are expressed correctly on the average, and not locally.

This model is expressed:

$$\nu_{\mathrm{sgs}}(\boldsymbol{x},t) = \left(C_{\mathrm{s}}\overline{\Delta}\right)^2 (2|\overline{S}(\boldsymbol{x},t)|^2)^{1/2} \quad . \tag{5.90}$$

The constant theoretical value C_{s} is evaluated by the relations (5.45) or (5.48). It should nonetheless be noted that the value of this constant is, in practice, adjusted to improve the results. Clark et al. [143] use $C_{\mathrm{s}} = 0.2$ for a case of isotropic homogeneous turbulence, while Deardorff [172] uses $C_{\mathrm{s}} = 0.1$ for a plane channel flow. Studies of shear flows using experimental data yield similar evaluations ($C_{\mathrm{s}} \simeq 0.1 - 0.12$) [503, 570, 724]. This decrease in the value of the constant with respect to its theoretical value is due to the fact that the field gradient is now non-zero and that it contributes to the $|\overline{S}(\boldsymbol{x},t)|$ term. To enforce the local equilibrium relation, the value of the constant has to be reduced. It should be noted that this new value ensures only that the right quantity of resolved kinetic energy will be dissipated on the whole throughout the field, but that the quality of the level of local dissipation is uncontrolled. [19]

Second-Order Structure Function Model. This model is a transposition of Métais and Lesieur's constant effective viscosity model into the physical space, and can consequently be interpreted as a model based on the energy at cutoff, expressed in physical space. The authors [514] propose evaluating the energy at cutoff $E(k_{\mathrm{c}})$ by means of the second-order velocity structure

---

[19] Canuto and Cheng [99] derived a more general expression for the constant C_{S}, which appears as an explicit function of the subgrid kinetic energy and the local shear:

$$C_{\mathrm{S}} \propto \left(\frac{q^2_{\mathrm{sgs}}}{\varepsilon|\overline{S}|\overline{\Delta}^2}\right)^{1/2} \quad ,$$

which is effectively a decreasing function of the local shear $|\overline{S}|$. That demonstrates the limited theoretical range of application of the usual Smagorinsky model.

function. This is defined as:

$$D_{\mathrm{LL}}(\boldsymbol{x},r,t)=\int_{|\boldsymbol{x}'|=r}\left[\boldsymbol{u}(\boldsymbol{x},t)-\boldsymbol{u}(\boldsymbol{x}+\boldsymbol{x}',t)\right]^2 d^3\boldsymbol{x}' \quad . \tag{5.91}$$

In the case of isotropic homogeneous turbulence, we have the relation:

$$D_{\mathrm{LL}}(r,t)=\int D_{\mathrm{LL}}(\boldsymbol{x},r,t)d^3\boldsymbol{x}=4\int_0^\infty E(k,t)\left(1-\frac{\sin(kr)}{kr}\right)dk \quad . \tag{5.92}$$

Using a Kolmogorov spectrum, the calculation of (5.92) leads to:

$$D_{\mathrm{LL}}(r,t)=\frac{9}{5}\Gamma(1/3)K_0\varepsilon^{2/3}r^{2/3} \quad , \tag{5.93}$$

or, expressing the dissipation ε, as a function of $D_{\mathrm{LL}}(r,t)$ in the expression for the Kolmogorov spectrum:

$$E(k)=\frac{5}{9\Gamma(1/3)}D_{\mathrm{LL}}(r,t)r^{-2/3}k^{-5/3} \quad . \tag{5.94}$$

To derive a subgrid model, we now have to evaluate the second-order structure function from the resolved scales alone. To do this, we decompose by:

$$D_{\mathrm{LL}}(r,t)=\overline{D}_{\mathrm{LL}}(r,t)+C_0(r,t) \quad , \tag{5.95}$$

in which $\overline{D}_{\mathrm{LL}}(r,t)$ is computed from the resolved scales and $C_0(r,t)$ corresponds to the contribution of the subgrid scales:

$$C_0(r,t)=4\int_{k_c}^\infty E(k,t)\left(1-\frac{\sin(kr)}{kr}\right)dk \quad . \tag{5.96}$$

By replacing the quantity $E(k,t)$ in equation (5.96) by its value (5.94), we get:

$$C_0(r,t)=D_{\mathrm{LL}}(r,t)\left(\frac{r}{\overline{\Delta}}\right)^{-2/3}H_{\mathrm{sf}}(r/\overline{\Delta}) \quad , \tag{5.97}$$

in which H_{sf} is the function

$$H_{\mathrm{sf}}(x)=\frac{20}{9\Gamma(1/3)}\left[\frac{3}{2\pi^{2/3}}+x^{2/3}\mathrm{Im}\left\{\exp(\imath 5\pi/6)\Gamma(-5/3,\imath\pi x)\right\}\right] \quad . \tag{5.98}$$

Once it is substituted in (5.95), this equation makes it possible to evaluate the energy at the cutoff. The second-order Structure Function model takes the form:

$$\langle\nu_{\mathrm{sgs}}(r)\rangle=A(r/\overline{\Delta})\overline{\Delta}\sqrt{\overline{D}_{\mathrm{LL}}(r,t)} \quad , \tag{5.99}$$

in which

$$A(x) = \frac{2K_0^{-3/2}}{3\pi^{4/3}\sqrt{(9/5)\Gamma(1/3)}} x^{-4/3}\left(1 - x^{-2/3}H_{\rm sf}(x)\right)^{-1/2} \quad . \tag{5.100}$$

In the same way as for the Smagorinsky model, a local model in space can be had by using relation (5.94) locally in order to include the local intermittency of the turbulence. The model is then written:

$$\nu_{\rm sgs}(\boldsymbol{x}, r) = A(r/\overline{\Delta})\overline{\Delta}\sqrt{\overline{D}_{\rm LL}(\boldsymbol{x}, r, t)} \quad . \tag{5.101}$$

In the case where $r = \overline{\Delta}$, the model takes the simplified form:

$$\nu_{\rm sgs}(\boldsymbol{x}, \overline{\Delta}, t) = 0.105\overline{\Delta}\sqrt{\overline{D}_{\rm LL}(\boldsymbol{x}, \overline{\Delta}, t)} \quad . \tag{5.102}$$

A link can be established with the models based on the resolved scale gradients by noting that:

$$\boldsymbol{u}(\boldsymbol{x}, t) - \boldsymbol{u}(\boldsymbol{x} + \boldsymbol{x}', t) = -\boldsymbol{x}' \cdot \nabla \boldsymbol{u}(\boldsymbol{x}, t) + O(|\boldsymbol{x}'|^2) \quad . \tag{5.103}$$

This last relation shows that the function $\overline{F}_2$ is homogeneous to a norm of the resolved velocity field gradient. If this function is evaluated in the simulation in a way similar to how the resolved strain rate tensor is computed for the Smagorinsky model, we can in theory expect the Structure Function model to suffer some of the same weaknesses: the information contained in the model will be local in space, therefore non-local in frequency, which induces a poor estimation of the kinetic energy at cutoff and a loss of precision of the model in the treatment of large-scale intermittency and spectral nonequilibrium.

Third-Order Structure Functions Models. Shao et al. [669, 153] defined a more general class of structure function-based subgrid viscosities considering the Kolmogorov–Meneveau equation for the third-order velocity structure function in isotropic turbulence [510]

$$-\frac{4}{5} r\varepsilon = \overline{D}_{\rm LLL} - 6G_{\rm LLL} \quad , \tag{5.104}$$

where $\overline{D}_{\rm LLL}$ is the third-order longitudinal velocity correlation of the filtered field

$$\overline{D}_{\rm LLL}(r) = \langle [\overline{\boldsymbol{u}}(\boldsymbol{x} + \boldsymbol{r}) - \overline{\boldsymbol{u}}(\boldsymbol{x})]^3 \rangle \quad , \tag{5.105}$$

where $\langle \cdot \rangle$ denotes the statistical average operator (which is equivalent to the integral sequence introduced in the presentation of the second-order structure function model). The two other quantities are the longitudinal velocity-stress correlation tensor

$$G_{\rm LLL}(r) = \langle \overline{u}_1(\boldsymbol{x})\tau_{11}(\boldsymbol{x} + \boldsymbol{r}) \rangle \quad , \tag{5.106}$$

and the average subgrid dissipation

$$\varepsilon = -\tau_{ij}\overline{S}_{ij} \quad . \tag{5.107}$$

Shao's procedure consists in using relation (5.104) to compute the subgrid viscosity. To this end, he assumes that the velocity-stress correlation obeys the following scale-similarity hypothesis

$$G_{\rm LLL}(r) \propto r^p \quad , \tag{5.108}$$

where $p = -1/3$ corresponds to the Kolmogorov local isotropy hypotheses. Using that assumption, one obtains the following relationship for two space increments r_1 and r_2:

$$\frac{0.8 r_1 \varepsilon + \overline{D}_{\rm LLL}(r_1)}{0.8 r_2 \varepsilon + \overline{D}_{\rm LLL}(r_2)} = \left(\frac{r_1}{r_2}\right)^{-1/3} \quad . \tag{5.109}$$

Now introducing a subgrid viscosity model with subgrid viscosity $\nu_{\rm sgs}$, several possibilities exist for the evaluation of the subgrid viscosity. The three following models have been proposed by Shao and his coworkers:

- The *one-scale, constant subgrid viscosity model.* Assuming that the subgrid viscosity is constant, one obtains

$$\tau_{ij} = -2\nu_{\rm sgs}\overline{S}_{ij}, \quad \varepsilon = \nu_{\rm sgs}|\overline{S}|^2, \quad G_{\rm LLL} = \nu_{\rm sgs}\overline{D}_{\rm LL,r} \quad , \tag{5.110}$$

  where comma separated indices denote derivatives. Inserting these relations into (5.104), an expression for the eddy viscosity is recovered:

$$\nu_{\rm sgs} = \frac{-S_k}{0.8\frac{|\overline{S}|^2}{\overline{D}_{\rm LL}} - 4} r \sqrt{\overline{D}_{\rm LL}} \quad , \tag{5.111}$$

  where the skewness of the longitudinal filtered velocity increment is defined as

$$S_k = \frac{\overline{D}_{\rm LLL}}{\overline{D}_{\rm LL}^{3/2}} \quad . \tag{5.112}$$

  The Métais-Lesieur model is recovered taking $r = \overline{\Delta}$ and using properties of isotropic turbulence to evaluate the different terms appearing in (5.111).
- The *one-scale, variable subgrid viscosity model.* Relaxing the previous constraint dealing with the constant appearing in the structure-function model, one obtains the following asymptotic expression

$$\nu_{\rm sgs} = \frac{-S_k}{8} r \sqrt{\overline{D}_{\rm LL}} \quad . \tag{5.113}$$

The constant is observed to be self-adpative, depending on the computed value of the skewness parameter S_k.

– *The multiscale structure function model.* The last model derived by Shao is more general and is based on the two-scale relation (5.109). Using the relation $\varepsilon = \nu_{\mathrm{sgs}}|\overline{S}|^2$ and considering two diffrent separation distances r_1 and r_2, one obtain the following expression

$$\nu_{\mathrm{sgs}} = \frac{\overline{D}_{\mathrm{LLL}}(r_1) - \left(\frac{r_1}{r_2}\right)^{1/3} \overline{D}_{\mathrm{LLL}}(r_2)}{-0.4|\overline{S}|^2 \left(1 - \left(\frac{r_1}{r_2}\right)^{4/3}\right) r_1} \quad . \tag{5.114}$$

These expressions are made local in the physical space by using local values of the different parameters involving the resolved velocity field $\overline{\boldsymbol{u}}$.

Model Based on the Subgrid Kinetic Energy. One model, of the form (5.61), based on the subgrid scales, was developed independently by a number of authors [318, 653, 791, 792, 525, 690, 391]. The subgrid viscosity is computed from the kinetic energy of the subgrid modes q_{sgs}^2:

$$\nu_{\mathrm{sgs}}(\boldsymbol{x}, t) = C_{\mathrm{m}} \overline{\Delta} \sqrt{q_{\mathrm{sgs}}^2(\boldsymbol{x}, t)} \quad , \tag{5.115}$$

where, for reference:

$$q_{\mathrm{sgs}}^2(\boldsymbol{x}, t) = \frac{1}{2} \overline{(u_i(\boldsymbol{x}, t) - \overline{u}_i(\boldsymbol{x}, t))^2} \quad . \tag{5.116}$$

The constant C_{m} is evaluated by the relation (5.61). This energy constitutes another variable of the problem and is evaluated by solving an evolution equation. This equation is obtained from the exact evolution equation (3.33), whose unknown terms are modeled according to Lilly's proposals [447], or by a re-normalization method. The various terms are modeled as follows (refer to the work of McComb [464], for example):

– The diffusion term is modeled by a gradient hypothesis, by stating that the non-linear term is proportional to the kinetic energy q_{sgs}^2 gradient (Kolmogorov-Prandtl relation):

$$\frac{\partial}{\partial x_j} \left(\frac{1}{2} \overline{u_i' u_i' u_j'} + \overline{u_j' p} \right) = C_2 \frac{\partial}{\partial x_j} \left(\overline{\Delta} \sqrt{q_{\mathrm{sgs}}^2} \frac{\partial q_{\mathrm{sgs}}^2}{\partial x_j} \right) \quad . \tag{5.117}$$

– The dissipation term is modeled using dimensional reasoning, by:

$$\varepsilon = \frac{\nu}{2} \overline{\frac{\partial u_i'}{\partial x_j} \frac{\partial u_i'}{\partial x_j}} = C_1 \frac{\left(q_{\mathrm{sgs}}^2\right)^{3/2}}{\overline{\Delta}} \quad . \tag{5.118}$$

The resulting evolution equation is:

$$\underbrace{\frac{\partial q_{\text{sgs}}^2}{\partial t} + \frac{\partial \overline{u}_j q_{\text{sgs}}^2}{\partial x_j}}_{I} = \underbrace{-\tau_{ij}\overline{S}_{ij}}_{II} - \underbrace{C_1 \frac{{q_{\text{sgs}}^2}^{3/2}}{\overline{\Delta}}}_{III} + \underbrace{C_2 \frac{\partial}{\partial x_j}\left(\overline{\Delta}\sqrt{q_{\text{sgs}}^2}\frac{\partial q_{\text{sgs}}^2}{\partial x_j}\right)}_{IV} + \underbrace{\nu \frac{\partial^2 q_{\text{sgs}}^2}{\partial x_j \partial x_j}}_{V} , \quad (5.119)$$

in which C_1 and C_2 are two positive constants and the various terms represent:

- *I* - advection by the resolved modes,
- *II* - production by the resolved modes,
- *III* - turbulent dissipation,
- *IV* - turbulent diffusion,
- *V* - viscous dissipation.

Using an analytical theory of turbulence, Yoshizawa [791, 792] and Horiuti [318] propose $C_1 = 1$ and $C_2 = 0.1$.

Yoshizawa Model. The filter cutoff length, $\overline{\Delta}$, is the only length scale used in deriving models based on the large scales, as this derivation has been explained above. The characteristic length associated with the subgrid scales, denoted Δ_{f}, is assumed to be proportional to this length, and the developments of Sect. 5.3.2 show that:

$$\Delta_{\text{f}} = C_{\text{s}}\overline{\Delta} \quad . \quad (5.120)$$

The variations in the structure of the subgrid modes cannot be included by setting a constant value for the coefficient C_{s}, as is done in the case of the Smagorinsky model, for example. To remedy this, Yoshizawa [787, 790] proposes differentiating these two characteristic scales and introducing an additional evolution equation to evaluate Δ_{f}. This length can be evaluated from the dissipation ε and the subgrid kinetic energy q_{sgs}^2 by the relation:

$$\Delta_{\text{f}} = C_1 \frac{(q_{\text{sgs}}^2)^{3/2}}{\varepsilon} + C_2 \frac{(q_{\text{sgs}}^2)^{3/2}}{\varepsilon^2}\frac{Dq_{\text{sgs}}^2}{Dt} - C_3 \frac{(q_{\text{sgs}}^2)^{5/2}}{\varepsilon^3}\frac{D\varepsilon}{Dt} \quad , \quad (5.121)$$

in which D/Dt is the material derivative associated with the resolved velocity field. The values of the constants appearing in equation (5.121) can be determined by an analysis conducted with the TSDIA technique [790]: $C_1 = 1.84$, $C_2 = 4.95$ et $C_3 = 2.91$. We now express the proportionality relation between the two lengths as:

$$\Delta_{\text{f}} = (1 + r(\boldsymbol{x}, t))\overline{\Delta} \quad . \quad (5.122)$$

By evaluating the subgrid kinetic energy as:

$$q_{\rm sgs}^2 = \left(\overline{\Delta}\varepsilon / C_1\right)^{2/3} \quad , \tag{5.123}$$

relations (5.121) and (5.122) lead to:

$$r = C_4 \overline{\Delta}^{2/3} \varepsilon^{-4/3} \frac{D\varepsilon}{Dt} \quad , \tag{5.124}$$

with $C_4 = 0.04$. Using the local equilibrium hypothesis, we get:

$$\varepsilon = -\tau_{ij}\overline{S}_{ij} \simeq C_5 \Delta_{\rm f}^2 |\overline{S}|^3 \quad , \tag{5.125}$$

in which $C_5 = 6.52.10^{-3}$. This definition completes the calculation of the factor r and the length $\Delta_{\rm f}$. This variation of the characteristic length $\Delta_{\rm f}$ can be re-interpreted as a variation of the constant in the Smagorinsky model:

$$C_{\rm s} = C_{\rm s0} \left(1 - C_{\rm a} |\overline{S}|^{-2} \frac{D|\overline{S}|}{Dt} + C_{\rm b} \overline{\Delta}^2 |\overline{S}|^{-2} \frac{\partial}{\partial x_j} \left(|\overline{S}|^{-2} \frac{\partial |\overline{S}|}{\partial x_j} \right) \right) \quad . \tag{5.126}$$

The constants $C_{\rm s0}$, $C_{\rm a}$ and $C_{\rm b}$ are evaluated at 0.16, 1.8, and 0.047, respectively, by Yoshizawa [790] and Murakami [554]. In practice, $C_{\rm b}$ is taken to be equal to zero and the constant $C_{\rm s}$ is bounded in order to ensure the stability of the simulation: $0.1 \leq C_{\rm s} \leq 0.27$. Morinishi and Kobayashi [541] recommend using the values $C_{\rm a} = 32$ and $C_{\rm s0} = 0.1$.

Mixed Scale Model. Ta Phuoc Loc and Sagaut [627, 626] defined models having a triple dependency on the large and small structures of the resolved field as a function of the cutoff length. These models, which make up the one-parameter Mixed Scale Model family, are derived by taking a weighted geometric average of the models based on the large scales and those based on the energy at cutoff:

$$\nu_{\rm sgs}(\alpha)(\boldsymbol{x},t) = C_{\rm m} \left|\mathcal{F}(\overline{\boldsymbol{u}}(\boldsymbol{x},t))\right|^{\alpha} \left(q_{\rm c}^2\right)^{\frac{1-\alpha}{2}} (\boldsymbol{x},t)\, \overline{\Delta}^{1+\alpha} \quad , \tag{5.127}$$

with

$$\mathcal{F}(\overline{\boldsymbol{u}}(\boldsymbol{x},t)) = \overline{S}(\boldsymbol{x},t) \text{ or } \nabla \times \overline{\boldsymbol{u}}(\boldsymbol{x},t) \quad . \tag{5.128}$$

It should be noted that localized versions of the models are used here, so that any flows that do not verify the spatial homogeneity property can be processed better. The kinetic energy $q_{\rm c}^2$ can be evaluated using any method presented in Sect. 9.2.3. In the original formulation of the model, it is evaluated in the physical space by the formula:

$$q_{\rm c}^2(\boldsymbol{x},t) = \frac{1}{2} (\overline{u}_i(\boldsymbol{x},t))' \, (\overline{u}_i(\boldsymbol{x},t))' \quad . \tag{5.129}$$

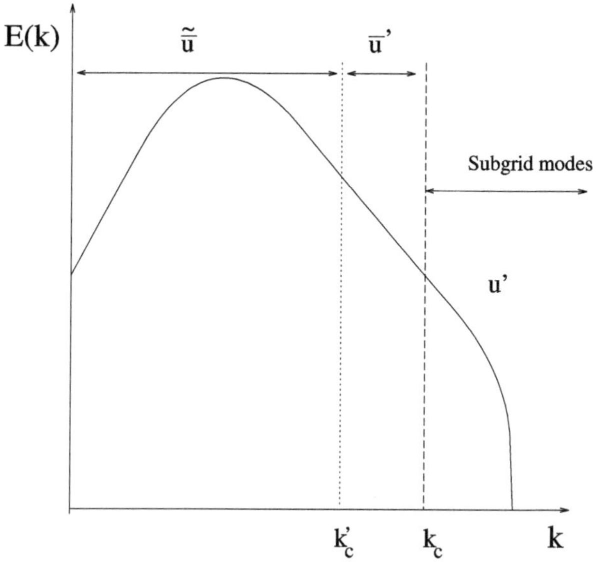

Fig. 5.14. Spectral subdivisions for double sharp-cutoff filtering. $\widetilde{\overline{u}}$ is the resolved field in the sense of the test filter, $(\overline{u})'$ the test field, and u' the unresolved scales in the sense of the initial filter.

The *test field* $(\overline{\boldsymbol{u}})'$ represents the high-frequency part of the resolved velocity field, defined using a second filter, referred to as the *test filter*, designated by the *tilde* symbol and associated with the cutoff length $\widetilde{\overline{\Delta}} > \overline{\Delta}$ (see Fig. 5.14):

$$(\overline{\boldsymbol{u}})' = \overline{\boldsymbol{u}} - \widetilde{\overline{\boldsymbol{u}}} \quad . \tag{5.130}$$

The resulting model can be interpreted in two ways:

- As a model based on the kinetic energy of the subgrid scales, i.e. the second form of the models based on the subgrid scales in Sect. 5.3.2, if we use Bardina's hypothesis of scale similarity (described in Chap. 7), which allows us to set:

$$q_{\mathrm{c}}^2 \simeq q_{\mathrm{sgs}}^2 \quad , \tag{5.131}$$

  in which q_{sgs}^2 is the kinetic energy of the subgrid scales. This assumption can be refined in the framework of the canonical analysis. Assuming that the two cutoffs occur in the inertial range of the spectrum, we get:

$$q_{\mathrm{c}}^2 = \int_{k_{\mathrm{c}}'}^{k_{\mathrm{c}}} E(k)dk = \frac{3}{2} K_0 \varepsilon^{2/3} \left({k_{\mathrm{c}}'}^{-2/3} - {k_{\mathrm{c}}}^{-2/3} \right) \quad , \tag{5.132}$$

  in which k_{c} and k_{c}' are wave numbers associated with $\overline{\Delta}$ and $\widetilde{\overline{\Delta}}$, respectively.

We then define the relation:

$$q_c^2 = \beta q_{sgs}^2, \quad \beta = \left[\left(\frac{k_c'}{k_c} \right)^{-2/3} - 1 \right] \quad . \tag{5.133}$$

It can be seen that the approximation is exact if $\beta = 1$, i.e. if:

$$k_c' = \frac{1}{\sqrt{8}} k_c \quad . \tag{5.134}$$

This approximation is also used by Bardina et al. [40] and Yoshizawa et al. [794] to derive models based on the subgrid kinetic energy without using any additional transport equation.

– As a model based on the energy at cutoff, and therefore as a generalization of the spectral model of constant effective viscosity into the physical space. That is, using the same assumptions as before, we get:

$$q_c^2 = \frac{3}{2} \beta k_c E(k_c) \quad . \tag{5.135}$$

Here, the approximation is exact if $k_c' = k_c/\sqrt{8}$.

It is important to note that the Mixed Scale Model makes no use of any commutation property between the test filter and the derivation operators. Also, we note that for $\alpha \in [0, 1]$ the subgrid viscosity $\nu_{sgs}(\alpha)$ is always defined, whereas the model appears in the form of a quotient for other values of α can then raise problems of numerical stability once it is discretized, because the denominator may cancel out.

The model constant can be evaluated theoretically by analytical theories of turbulence in the canonical case. Resuming the results of Sect. 5.3.2, we get:

$$C_m = C_q^{1-\alpha} \, C_s^{2\alpha} \quad , \tag{5.136}$$

in which $C_s \sim 0.18$ or $C_s \sim 0.148$ and $C_q \sim 0.20$.

Some other particular cases of the Mixed Scale Model can be found. Wong and Lilly [766], Carati [103] and Tsubokura [717] proposed using $\alpha = -1$, yielding a model independent of the cutoff length $\overline{\Delta}$. Yoshizawa et al. [793] used $\alpha = 0$, but introduced an exponential damping term in order to enforce a satisfactory asymptotic near-wall behavior (see p. 159):

$$\nu_{sgs} = 0.03 (q_c^2)^{1/2} \, \overline{\Delta} \left[1 - \exp \left(-21 \frac{q_c^2}{\overline{\Delta}^2 |\overline{S}|^2} \right) \right] \quad . \tag{5.137}$$

Mathematical analysis in the case $\alpha = 0$ was provided by Iliescu and Layton [342] and Layton and Lewandowski [430].

5.3.3 Improvement of Models in the Physical Space

Statement of the Problem. Experience shows that the various models yield good results when they are applied to homogeneous turbulent flows and that the cutoff is placed sufficiently far into the inertial range of the spectrum, i.e. when a large part of the total kinetic energy is contained in the resolved scales[20].

In other cases, as in transitional flows, highly anisotropic flows, highly under-resolved flows, or those in high energetic disequilibrium, the subgrid models behave much less satisfactorily. Aside from the problem stemming from numerical errors, there are mainly two reasons for this:

1. The characteristics of these flows does not correspond to the hypotheses on which the models are derived, which means that the models are at fault. We then have two possibilities: deriving models from new physical hypotheses or adjusting existing ones, more or less empirically. The first choice is theoretically more exact, but there is a lack of descriptions of turbulence for frameworks other than that of isotropic homogeneous turbulence. Still, a few attempts have been made to consider the anisotropy appearing in this category. These are discussed in Chap. 6. The other solution, if the physics of the models is put to fault, consists in reducing their importance, i.e. increasing the cutoff frequency to capture a larger part of the flow physics directly. This means increasing the number of degrees of freedom and striking a compromise between the grid enrichment techniques and subgrid modeling efforts.
2. Deriving models based on the energy at cutoff or the subgrid scales (with no additional evolution equation) for simulations in the physical space runs up against Gabor-Heisenberg's generalized principle of uncertainty [204, 627], which stipulates that the precision of the information cannot be improved in space and in frequency at the same time. This is illustrated by Fig. 5.15. Very good frequency localization implies high non-localization in space, which reduces the possibilities of taking the intermittency[21] into account and precludes the treatment of highly inhomogeneous flows. Inversely, very good localization of the information in space prevents any good spectral resolution, which leads to high errors, e.g. in computing the energy at the cutoff. Yet this frequency localization is very important, since it alone can be used to detect the presence of the subgrid scales. It is important to recall here that large-eddy simulation is based on a selection in frequency of modes making up the exact

[20] Certain authors estimate this share to be between 80% and 90% [119]. Another criterion sometimes mentioned is that the cutoff scale should be of the order of Taylor's microscale. Bagget et al. [32] propose to define the cutoff length in such a way that the subgrid scales will be quasi-isotropic, leading to $\overline{\Delta} \approx L_\varepsilon/10$, where L_ε is the integral dissipation length.

[21] Direct numerical simulations and experimental data show that the true subgrid dissipation and its surrogates do not have the same scaling laws [114, 510].

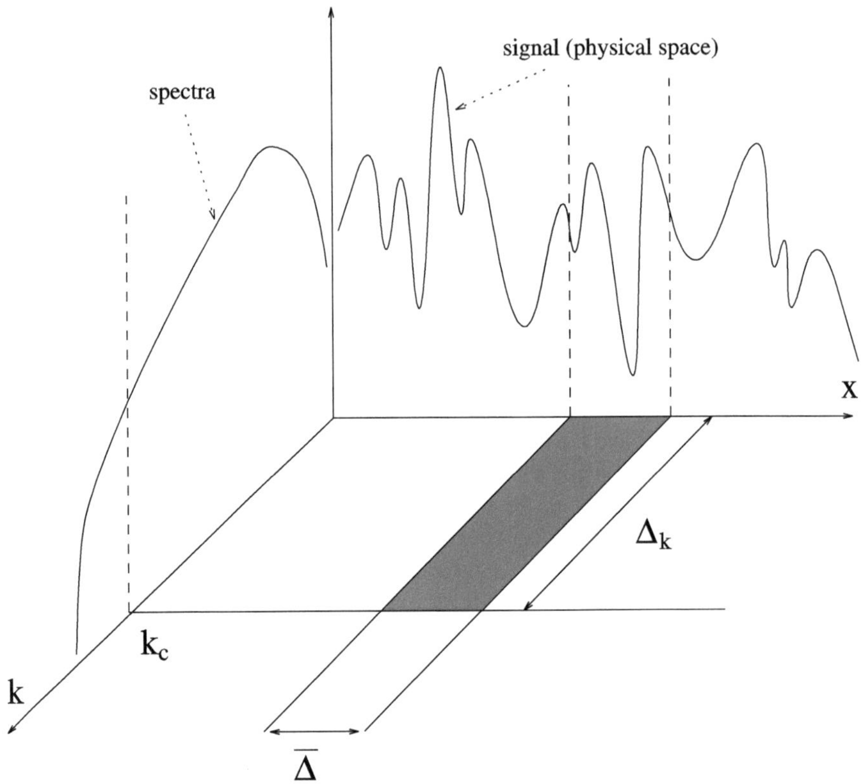

Fig. 5.15. Representation of the resolution in the space-frequency plane. The spat ial resolution $\overline{\Delta}$ is associated with frequency resolution Δ_k. Gabor-Heisenberg's uncertainty principle stipulates that the product $\overline{\Delta} \times \Delta_k$ remains constant, i.e. that the area of the gray domain keeps the same value (from [204], courtesy of F. Ducros).

solution. Problems arise here, induced by the localization of statistical average relations that are exact, as this localization may correspond to a statistical average. Two solutions may be considered: developing an acceptable compromise between the precision in space and frequency, or enriching the information contained in the simulation, which is done either by adding variables to it as in the case of models based on the kinetic energy of the subgrid modes, or by assuming further hypotheses when deriving models.

In the following, we present techniques developed to improve the simulation results, though without modifying the structure of the subgrid models deeply. The purpose of all these modifications is to adapt the subgrid model better to the local state of the flow and remedy the lack of frequency localization of the information.

We will be describing:

1. Dynamic procedures for computing subgrid model constants (p. 137). These constants are computed in such a way as to reduce an a priori estimate of the error committed with the model considered, locally in space and time, in the least squares sense. This estimation is made using the Germano identity, and requires the use of an analytical filter. It should be noted that the dynamic procedures do not change the model in the sense that its form (e.g. subgrid viscosity) remains the same. All that is done here is to minimize a norm of the error associated with the form of the model considered. The errors committed intrinsically[22] by adopting an a priori form of the subgrid tensor are not modified. These procedures, while theoretically very attractive, do pose problems of numerical stability and can induce non-negligible extra computational costs. This variation of the constant at each point and each time step makes it possible to minimize the error locally for each degree of freedom, while determining a constant value offers only the less efficient possibility of an overall minimization. This is illustrated by the above discussion of the constant in the Smagorinsky model.
2. Dynamic procedures that are not directly based on the Germano identity (p. 152): the multilevel dynamic procedure by Terracol and Sagaut and the multiscale structure function method by Shao. These procedures have basically the same capability to monitor the constant in the subgrid model as the previous dynamic procedures. They also involve the definition of a test filter level, and are both based on considerations dealing with the dissipation scaling as a function of the resolution. Like the procedures based on the Germano identity, they are based on the implicit assumption that some degree of self-similarity exists in the computed flow.
3. Structural sensors (p. 154), which condition the existence of the subgrid scales to the verification of certain constraints by the highest frequencies of the resolved scales. More precisely, we consider here that the subgrid scales exist if the highest resolved frequencies verify topological properties that are expected in the case of isotropic homogeneous turbulence. When these criteria are verified, we adopt the hypothesis that the highest resolved frequencies have a dynamics close to that of the scales contained in the inertial range. On the basis of energy spectrum continuity (see the note of page p. 112), we then deduce that unresolved scales exist, and the subgrid model is then used, but is otherwise canceled.
4. The accentuation technique (p. 156), which consists in artificially increasing the contribution of the highest resolved frequencies when evaluating

[22] For example, the subgrid viscosity models described above all induce a linear dependency between the subgrid tensor and the resolved-scale tensor:

$$\tau_{ij}^{\mathrm{d}} = -\nu_{\mathrm{sgs}}\overline{S}_{ij} \quad .$$

the subgrid viscosity. This technique allows a better frequency localization of the information included in the model, and therefore a better treatment of the intermittence phenomena, as the model is sensitive only to the higher resolved frequencies. This result is obtained by applying a frequency high-pass filter to the resolved field.

5. The damping functions for the near-wall region (p. 159), by which certain modifications in the turbulence dynamics and characteristic scales of the subgrid modes in the boundary layers can be taken into account. These functions are established in such a way as to cancel the subgrid viscosity models in the near-wall region so that they will not inhibit the driving mechanisms occurring in this area. These models are of limited generality as they presuppose a particular form of the flow dynamics in the region considered. They also require that the relative position of each point with respect to the solid wall be known, which can raise problems in practice such as when using multidomain techniques or when several surfaces exist. And lastly, they constitute only an amplitude correction of the subgrid viscosity models for the forward energy cascade: they are not able to include any changes in the form of this mechanism, or the emergence of new mechanisms.

The three "generalist" techniques (dynamic procedure, structural sensor, accentuation) for adapting the subgrid viscosity models are all based on extracting a test field from the resolved scales by applying a test filter to these scales. This field corresponds to the highest frequencies catured by the simulation, so we can see that all these techniques are based on a frequency localization of the information contained in the subgrid models. The loss of localness in space is reflected by the fact that the number of neighbors involved in the subgrid model computation is increased by using the test filter.

Dynamic Procedures for Computing the Constants.

Dynamic Models. Many dynamic procedures have been proposed to evaluate the parameters in the subgrid models. The following methods are presented

1. The original method proposed by Germano, and its modification proposed by Lilly to improve its robustness (p. 137). Its recent improvements for complex kinetic energy spectrum shapes are also discussed.
2. The Lagrangian dynamic procedure (p. 144), which is well suited for fully non-homogeneous flows.
3. The constrained localized dynamic procedure (p. 146), which relax some strong assumptions used in the Germano–Lilly approach. q
4. The approximate localized dynamic procedure (p. 148), which is a simplification of the constrained localized dynamic procedure that do nor requires to solve an integral problem to compute the dynamic constant.
5. The generalized dynamic procedures (p. 149), which aim at optimizing the approximation of the subgrid acceleration and make it possible to account for discretization errors.

6. The dynamic inverse procedure (p. 150), which is designed to improve the dynamic procedure when the cutoff is located at the very begining of the inertial range of the kinetic energy spectrum.
7. The Taylor series expansion based dynamic procedure (p. 151), which results in a differential expression for the dynamic constant, the test filter being replaced by its differential approximation.
8. The dynamic procedure based on dimensional parameters (p. 151), which yields a very simple expression.

Germano–Lilly Dynamic Procedure. In order to adapt the models better to the local structure of the flow, Germano et al. [253] proposed an algorithm for adapting the Smagorinsky model by automatically adjusting the constant at each point in space and at each time step. This procedure, described below, is applicable to any model that makes explicit use of an arbitrary constant $C_{\rm d}$, such that the constant now becomes time- and space-dependent: $C_{\rm d}$ becomes $C_{\rm d}(\boldsymbol{x}, t)$.

The dynamic procedure is based on the multiplicative Germano identity (3.80) , now written in the form:

$$L_{ij} = T_{ij} - \tilde{\tau}_{ij} \quad , \tag{5.138}$$

in which

$$\tau_{ij} \equiv L_{ij} + C_{ij} + R_{ij} = \overline{u_i u_j} - \overline{u}_i \overline{u}_j \quad , \tag{5.139}$$

$$T_{ij} \equiv \widetilde{\overline{u_i u_j}} - \widetilde{\overline{u}}_i \widetilde{\overline{u}}_j \quad , \tag{5.140}$$

$$L_{ij} \equiv \widetilde{\overline{u}_i \overline{u}_j} - \widetilde{\overline{u}}_i \widetilde{\overline{u}}_j \quad , \tag{5.141}$$

in which the tilde symbol *tilde* designates the test filter. The tensors τ and T are the subgrid tensors corresponding, respectively, to the first and second filtering levels. The latter filtering level is associated with the characteristic length $\widetilde{\overline{\Delta}}$, with $\widetilde{\overline{\Delta}} > \overline{\Delta}$. Numerical tests show that an optimal value is $\widetilde{\overline{\Delta}} = 2\overline{\Delta}$. The tensor L can be computed directly from the resolved field.

We then assume that the two subgrid tensors τ and T can be modeled by the same constant $C_{\rm d}$ for both filtering levels. Formally, this is expressed:

$$\tau_{ij} - \frac{1}{3}\tau_{kk}\delta_{ij} = C_{\rm d}\beta_{ij} \quad , \tag{5.142}$$

$$T_{ij} - \frac{1}{3}T_{kk}\delta_{ij} = C_{\rm d}\alpha_{ij} \quad , \tag{5.143}$$

in which the tensors α and β designate the deviators of the subgrid tensors obtained using the subgrid model deprived of its constant. It is important noting that the use of the same subgrid model with the same constant is equivalent to a scale-invariance assumption on both the subgrid fluxes and the filter, to be discussed in the following.

Table 5.1. Examples of subgrid model kernels for the dynamic procedure.

Model	β_{ij}	α_{ij}
(5.90)	$-2\overline{\Delta}^2\lvert\overline{S}\rvert\overline{S}_{ij}$	$-2\widetilde{\overline{\Delta}}^2\lvert\widetilde{\overline{S}}\rvert\widetilde{\overline{S}}_{ij}$
(5.102)	$-2\overline{\Delta}\sqrt{\overline{F}(\overline{\Delta})}\overline{S}_{ij}$	$-2\widetilde{\overline{\Delta}}\sqrt{\widetilde{\overline{F}}(\widetilde{\overline{\Delta}})}\widetilde{\overline{S}}_{ij}$
(5.127)	$-2\overline{\Delta}^{1+\alpha}\lvert\mathcal{F}(\overline{\boldsymbol{u}})\rvert^\alpha(q_c^2)^{\frac{1-\alpha}{2}}\overline{S}_{ij}$	$-2\widetilde{\overline{\Delta}}^{1+\alpha}\lvert\mathcal{F}(\widetilde{\overline{\boldsymbol{u}}})\rvert^\alpha(\tilde{q}_c^2)^{\frac{1-\alpha}{2}}\widetilde{\overline{S}}_{ij}$

Some examples of subgrid model kernels for α_{ij} and β_{ij} are given in Table 5.1.

Introducing the above two formulas in the relation (5.138), we get[23]:

$$L_{ij} - \frac{1}{3}L_{kk}\delta_{ij} \equiv L_{ij}^{\mathrm{d}} = C_{\mathrm{d}}\alpha_{ij} - \widetilde{C_{\mathrm{d}}\beta_{ij}} \quad . \tag{5.144}$$

We cannot use this equation directly for determining the constant C_{d} because the second term uses the constant only through a filtered product [621]. In order to continue modeling, we need to make the approximation:

$$\widetilde{C_{\mathrm{d}}\beta_{ij}} = C_{\mathrm{d}}\widetilde{\beta}_{ij} \quad , \tag{5.145}$$

which is equivalent to considering that C_{d} is constant over an interval at least equal to the test filter cutoff length. The parameter C_{d} will thus be computed in such a way as to minimize the error committed[24], which is evaluated using the residual E_{ij}:

$$E_{ij} = L_{ij} - \frac{1}{3}L_{kk}\delta_{ij} - C_{\mathrm{d}}\alpha_{ij} + C_{\mathrm{d}}\widetilde{\beta}_{ij} \quad . \tag{5.146}$$

This definition consists of six independent relations, which in theory makes it possible to determine six values of the constant[25]. In order to conserve a single relation and thereby determine a single value of the constant, Germano et al. propose contracting the relation (5.146) with the resolved strain rate tensor. The value sought for the constant is a solution of the problem:

$$\frac{\partial E_{ij}\overline{S}_{ij}}{\partial C_{\mathrm{d}}} = 0 \quad . \tag{5.147}$$

[23] It is important to note that, for the present case, the tensor L_{ij} is replaced by its deviatoric part L_{ij}^{d}, because we are dealing with a zero-trace subgrid viscosity modeling.

[24] Meneveau and Katz [505] propose to use the dynamic procedure to rank the subgrid models, the best one being associated with the lowest value of the residual.

[25] Which would lead to the definition of a tensorial subgrid viscosity model.

This method can be efficient, but does raise the problem of indetermination when the tensor $\overline{S}_{ij}$ cancels out. To remedy this problem, Lilly [448] proposes calculating the constant C_d by a least-squares method, by which the constant C_d now becomes a solution of the problem:

$$\frac{\partial E_{ij} E_{ij}}{\partial C_\mathrm{d}} = 0 \quad , \tag{5.148}$$

or

$$C_\mathrm{d} = \frac{m_{ij} L_{ij}^\mathrm{d}}{m_{kl} m_{kl}} \quad , \tag{5.149}$$

in which

$$m_{ij} = \alpha_{ij} - \widetilde{\beta}_{ij} \quad . \tag{5.150}$$

The constant C_d thus computed has the following properties:

- It can take negative values, so the model can have an anti-dissipative effect locally. This is a characteristic that is often interpreted as a modeling of the backward energy cascade mechanism. This point is detailed in Sect. 5.4.
- It is not bounded, since it appears in the form of a fraction whose denominator can cancel out[26].

These two properties have important practical consequences on the numerical solution because they are both potentially destructive of the stability of the simulation. Numerical tests have shown that the constant can remain negative over long time intervals, causing an exponential growth in the high frequency fluctuations of the resolved field. The constant therefore needs an ad hoc process to ensure the model's good numerical properties. There are a number of different ways of performing this process on the constant: statistical average in the directions of statistical homogeneity [253, 779], in time or local in space [799]; limitation using arbitrary bounds [799] (clipping); or by a combination of these methods [779, 799]. Let us note that the averaging procedures can be defined in two non-equivalent ways [801]: by averaging the denominator and numerator separately, which is denoted symbolically:

$$C_\mathrm{d} = \frac{\langle m_{ij} L_{ij}^\mathrm{d} \rangle}{\langle m_{kl} m_{kl} \rangle} \quad , \tag{5.151}$$

or by averaging the quotient, i.e. on the constant itself:

$$C_\mathrm{d} = \langle C_\mathrm{d} \rangle = \left\langle \frac{m_{ij} L_{ij}^\mathrm{d}}{m_{kl} m_{kl}} \right\rangle \quad . \tag{5.152}$$

[26] This problem is linked to the implementation of the model in the simulation. In the continuous case, if the denominator tends toward zero, then the numerator cancels out too. These are calculation errors that lead to a problem of division by zero.

The ensemble average can be performed over homogeneous directions of the simulation (if they exist) or over different realizations, i.e. over several statistically equivalent simulations carried out in parallel [102, 108].

The time average process is expressed:

$$C_{\rm d}(\boldsymbol{x},(n+1)\Delta t) = a_1 C_{\rm d}(\boldsymbol{x},(n+1)\Delta t) + (1-a_1)C_{\rm d}(\boldsymbol{x},n\Delta t) \quad , \tag{5.153}$$

in which Δt is the time step used for the simulation and $a_1 \leq 1$ a constant. Lastly, the constant clipping process is intended to ensure that the following two conditions are verified:

$$\nu + \nu_{\rm sgs} \geq 0 \quad , \tag{5.154}$$

$$C_{\rm d} \leq C_{\rm max} \quad . \tag{5.155}$$

The first condition ensures that the total resolved dissipation $\varepsilon = \nu \overline{S}_{ij}\overline{S}_{ij} - \tau_{ij}\overline{S}_{ij}$ remains positive or zero. The second establishes an upper bound. In practice, $C_{\rm max}$ is of the order of the theoretical value of the Smagorinsky constant, i.e. $C_{\rm max} \simeq (0.2)^2$.

The models in which the constant is computed by this procedure are called "dynamic" because they automatically adapt to the local state of the flow. When the Smagorinsky model is coupled with this procedure, it is habitually called the dynamic model, because this combination was the first to be put to the test and is still the one most extensively used among the dynamic models.

The dynamic character of the constant $C_{\rm d}$ is illustrated in Fig. 5.16, which displays the time history of the square root of the dynamic constant in freely decaying isotropic turbulence. It is observed that during the first stage of the computation the constant is smaller than the theoretical value of the Smagorinsky constant $C_{\rm d} \sim 0.18$ given by equation (5.48), because the spectrum is not fully developed. In the second stage, when a self-similar state is reached, the theoretical value is automatically recovered.

The use of the same value of the constant for the subgrid model at the two filtering levels appearing in equation (5.138) implicitely relies on the two following self-similarity assumptions:

- The two cutoff wave numbers are located in the inertial range of the kinetic energy spectrum;
- The filter kernels associated to the two filtering levels are themselves self-similar.

These two constraints are not automatically satisfied, and the validity of the dynamic procedure for computing the constant requires a careful analysis.

Meneveau and Lund [507] propose an extension of the dynamic procedure for a cutoff located in the viscous range of the spectrum. Writing the constant

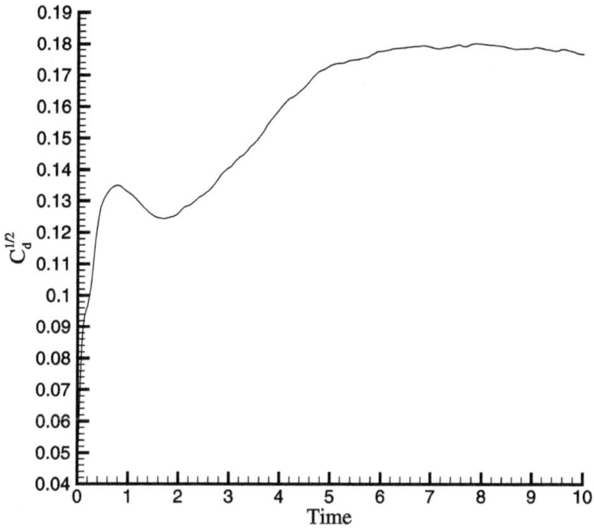

Fig. 5.16. Time history of the square root of the dynamic constant in large-eddy simulation of freely decaying isotropic turbulence (dynamic Smagorinsky model). Courtesy of E. Garnier, ONERA.

of the subgrid-scale model C as an explicit function of the filter characteristic length, the Germano–Lilly procedure leads to

$$C(\overline{\Delta}) = C(\widetilde{\overline{\Delta}}) = C_{\mathrm{d}} \quad . \tag{5.156}$$

Let η be the Kolmogorov length scale. It was said in the introduction that the flow is fully resolved if $\overline{\Delta} = \eta$. Therefore, the dynamic procedure is consistent if, and only if

$$\lim_{\overline{\Delta} \to \eta} C_{\mathrm{d}} = C(\eta) = 0 \quad . \tag{5.157}$$

Numerical experiments carried out by the two authors show that the Germano–Lilly procedure is not consistent, because it returns the value of the constant associated to the test filter level

$$C_{\mathrm{d}} = C(\widetilde{\overline{\Delta}}) \quad , \tag{5.158}$$

yielding

$$\lim_{\overline{\Delta} \to \eta} C_{\mathrm{d}} = C(r\eta) \neq 0, \quad r = \widetilde{\overline{\Delta}}/\overline{\Delta} \quad . \tag{5.159}$$

Numerical tests also showed that taking the limit $r \to 1$ or computing the two values $C(\overline{\Delta})$ and $C(r\overline{\Delta})$ using least-square-error minimization without assuming them to be equal yield inconsistent or ill-behaved solutions. A solution is to use prior knowledge to compute the dynamic constant. A robust algorithm is obtained by rewriting equation (5.146) as follows:

$$E_{ij} = L^{\mathrm{d}}_{ij} - C(\overline{\Delta}) \left(f(\overline{\Delta}, r)\alpha_{ij} - \widetilde{\beta}_{ij} \right) \quad , \tag{5.160}$$

where $f(\overline{\Delta}, r) = C(r\overline{\Delta})/C(\overline{\Delta})$ is evaluated by calculations similar to those of Voke (see page 118). A simple analytical fitting is obtained in the case $r = 2$:

$$f(\overline{\Delta}, 2) \approx \max(100, 10^{-x}), \ x = 3.23(Re_{2\overline{\Delta}}^{-0.92} - Re_{\overline{\Delta}}^{-0.92}) \quad , \tag{5.161}$$

where the mesh-Reynolds numbers are evaluated as (see equation (5.70)):

$$Re_{\overline{\Delta}} = \frac{\overline{\Delta}^2 |\overline{S}|}{\nu}, \quad Re_{2\overline{\Delta}} = \frac{4\overline{\Delta}^2 |\widetilde{\overline{S}}|}{\nu} \quad .$$

Other cases can be considered where the similarity hypothesis between the subgrid stresses at different resolution levels may be violated, leading to different values of the constant [601]. Among them:

- The case of a very coarse resolution, with a cutoff located at the very beginning of the inertial range or in the production range.
- The case of a turbulence undergoing rapid strains, where a transition length $\Delta_{\mathrm{T}} \propto S^{-3/2}\varepsilon^{1/2}$ appears. Here, S and ε are the strain magnitude and the dissipation rate, respectively. Dimensional arguments show that, roughly speaking, scales larger than Δ_{T} are rapidly distorted but have no time to adjust dynamically, while scales smaller than Δ_{T} can relax faster via nonlinear interactions.

For each of these cases, scale dependence of the model near the critical length scale is expected, which leads to a possible loss of efficiency of the classical Germano–Lilly dynamic procedure.

A more general dynamic procedure, which does not rely on the assumption of scale similarity or location of the cutoff in the dissipation range, was proposed by Porté-Agel et al. [601]. This new scale-dependent dynamic procedure is obtained by considering a third filtering level (i.e. a second test-filtering level) with a characteristic cutoff length scale $\widehat{\overline{\Delta}} > \widetilde{\overline{\Delta}}$. Filtered variables at this new level are denoted by a caret.

Writing the Germano identity between level $\overline{\Delta}$ and level $\widehat{\overline{\Delta}}$ leads to

$$Q_{ij} - \frac{1}{3}Q_{kk}\delta_{ij} \equiv \widehat{\overline{u}_i\overline{u}_j} - \widehat{\overline{u}}_i\widehat{\overline{u}}_j = C(\widehat{\overline{\Delta}})\gamma_{ij} - C(\overline{\Delta})\widehat{\beta_{ij}} \quad , \tag{5.162}$$

where γ_{ij} and β_{ij} denote the expression of the subgrid model at levels $\widehat{\overline{\Delta}}$ and $\overline{\Delta}$, respectively. By taking

$$n_{ij} = \left(\Lambda(\widehat{\overline{\Delta}}, \overline{\Delta})\gamma_{ij} - \widehat{\beta_{ij}} \right) \quad , \tag{5.163}$$

with

$$\Lambda(\widehat{\overline{\Delta}}, \overline{\Delta}) = \frac{C(\widehat{\overline{\Delta}})}{C(\overline{\Delta})} \quad , \tag{5.164}$$

we obtain the following value for the constant at level $\overline{\Delta}$:

$$C(\overline{\Delta}) = \frac{n_{ij}Q_{ij}}{n_{ij}n_{ij}} \quad . \tag{5.165}$$

By now considering relation (5.149), which expresses the Germano identity between the first two filtering levels, where m_{ij} is now written as

$$m_{ij} = \left(\Lambda(\widetilde{\overline{\Delta}}, \overline{\Delta})\alpha_{ij} - \widetilde{\beta_{ij}} \right) \quad , \tag{5.166}$$

and by equating the values of $C(\overline{\Delta})$ obtained using the two test-filtering levels, we obtain the following relation:

$$(L_{ij}m_{ij})(n_{ij}n_{ij}) - (Q_{ij}n_{ij})(m_{ij}m_{ij}) = 0 \quad , \tag{5.167}$$

which has two unknowns, $\Lambda(\widetilde{\overline{\Delta}}, \overline{\Delta})$ and $\Lambda(\widehat{\overline{\Delta}}, \overline{\Delta})$. In order to obtain a closed system, some additional assumptions are needed. It is proposed in [601] to assume a power-law scaling of the dynamic constant, $C(x) \propto x^p$, leading to

$$C(a\overline{\Delta}) = C(\overline{\Delta})a^r \quad . \tag{5.168}$$

For this power-law behavior, the function $\Lambda(.,.)$ does not depend on the scales but only on the ratio of the scales, i.e. $\Lambda(x,y) = (x/y)^r$. Using this simplification, (5.167) appears as a fifth-order polynomial in $C(\overline{\Delta})$. The dynamic constant is taken equal to the largest root.

We now consider the problem of the filter self-similarity. Let G_1 and G_2 be the filter kernels associated with the first and second filtering level. For the sake of simplicity, we use the notations $\overline{\Delta} = \Delta_1$ and $\widetilde{\overline{\Delta}} = \Delta_2$. We assume that the filter kernels are rewritten in a form such that:

$$\overline{u}(x) = G_1 \star u(x) = \int G_1 \left(\frac{|x-\xi|}{\Delta_1} \right) u(\xi) d\xi \quad , \tag{5.169}$$

$$\widetilde{\overline{u}}(x) = G_2 \star u(x) = \int G_2 \left(\frac{|x-\xi|}{\Delta_2} \right) u(\xi) d\xi \quad . \tag{5.170}$$

We also introduce the test filter G_t, which is defined such that

$$\widetilde{\overline{u}} = G_2 \star u = G_t \star \overline{u} = G_t \star G_1 \star u \quad . \tag{5.171}$$

The filters G_1 and G_2 are self-similar if and only if

$$G_1(y) = \frac{1}{r^d} G_2 \left(\frac{y}{r} \right), \quad r = \Delta_2/\Delta_1 \quad . \tag{5.172}$$

Hence, the two filters must have identical shapes and may only differ by their associated characteristic length. The problem is that in practice only G_t is known, and the self-similarity property might not be *a priori* verified. Carati and Vanden Eijnden [104] show that the interpretation of the resolved field is fully determined by the choice of the test filter G_t, and that the use of the same model for the two levels of filtering is fully justified. This is demonstrated by re-interpreting previous filters in the following way. Let us consider an infinite set of self-similar filters $\{F_n \equiv F(l_n)\}$ defined as

$$F_n(x) = \frac{1}{r^n}\mathcal{F}\left(\frac{x}{l_n}\right), l_n = r^n l_0 \quad , \tag{5.173}$$

where $\mathcal{F}$, $r > 1$ and l_0 are the filter kernel, an arbitrary parameter and a reference length, respectively. Let us introduce a second set $\{F_n^* \equiv F^*(l_n^*)\}$ defined by

$$F_n^* \equiv F_n \star F_{n-1} \star ... \star F_{-\infty} \quad . \tag{5.174}$$

For positive kernel $\mathcal{F}$, we get the following properties:

– The length l_n^* obeys the same geometrical law as l_n:

$$l_n^* = r l_{n-1}^*, \quad \text{and} \ l_n^* = \frac{r}{\sqrt{r^2-1}} l_n \quad . \tag{5.175}$$

– $\{F_n^*\}$ constitute a set of self-similar filters.

Using these two set of filters, the classical filters involved in the dynamic procedure can be defined as self-similar filters:

$$G_t(\Delta_t) = F_n(l_n) \quad , \tag{5.176}$$
$$G_1(\Delta_1) = F_{n-1}^*(l_{n-1}^*) \quad , \tag{5.177}$$
$$G_2(\Delta_2) = F_n^*(l_n^*) \quad . \tag{5.178}$$

For any test-filter G_t and any value of r, the first filter operator can be constructed explicitly:

$$G_1 = G_t(\Delta_t/r) \star G_t(\Delta_t/r^2) \star ... \star G_t(\Delta_t/r^\infty) \quad . \tag{5.179}$$

This relation shows that for any test filter of the form (5.176), the two filtering operators can be rewritten as self-similar ones, justifying the use of the same model at all the filtering levels.

Lagrangian Dynamic Procedure. The constant regularization procedures based on averages in the homogeneous directions have the drawback of not being usable in complex configurations, which are totally inhomogeneous. One technique for remedying this problem is to take this average along the fluid particle trajectories. This new procedure [508], called the dynamic Lagrangian procedure, has the advantage of being applicable in all configurations.

The trajectory of a fluid particle located at position $\boldsymbol{x}$ at time t is, for times t' previous to t, denoted as:

$$\boldsymbol{z}(t') = \boldsymbol{x} - \int_{t'}^{t} \overline{\boldsymbol{u}}[\boldsymbol{z}(t''), t'']dt'' \quad . \tag{5.180}$$

The residual (5.146) is written in the following Lagrangian form:

$$E_{ij}(\boldsymbol{z}, t') = L_{ij}(\boldsymbol{z}, t') - C_{\mathrm{d}}(\boldsymbol{x}, t) m_{ij}(\boldsymbol{z}, t') \quad . \tag{5.181}$$

We see that the value of the constant is fixed at point $\boldsymbol{x}$ at time t, which is equivalent to the same linearization operation as for the Germano–Lilly procedure. The value of the constant that should be used for computing the subgrid model at $\boldsymbol{x}$ at time t is determined by minimizing the error along the fluid particle trajectories. Here too, we reduce to a well-posed problem by defining a scalar residual E_{lag}, which is defined as the weighted integral along the trajectories of the residual proposed by Lilly:

$$E_{\mathrm{lag}} = \int_{-\infty}^{t} E_{ij}(\boldsymbol{z}(t'), t') E_{ij}(\boldsymbol{z}(t'), t') W(t - t') dt' \quad , \tag{5.182}$$

in which the weighting function $W(t-t')$ is introduced to control the memory effect. The constant is a solution of the problem:

$$\frac{\partial E_{\mathrm{lag}}}{\partial C_{\mathrm{d}}} = \int_{-\infty}^{t} 2E_{ij}(\boldsymbol{z}(t'), t') \frac{\partial E_{ij}(\boldsymbol{z}(t'), t')}{\partial C_{\mathrm{d}}} W(t - t') dt' = 0 \quad , \tag{5.183}$$

or:

$$C_{\mathrm{d}}(\boldsymbol{x}, t) = \frac{\mathcal{J}_{\mathrm{LM}}}{\mathcal{J}_{\mathrm{MM}}} \quad , \tag{5.184}$$

in which

$$\mathcal{J}_{\mathrm{LM}}(\boldsymbol{x}, t) = \int_{-\infty}^{t} L_{ij} m_{ij}(\boldsymbol{z}(t'), t') W(t - t') dt' \quad , \tag{5.185}$$

$$\mathcal{J}_{\mathrm{MM}}(\boldsymbol{x}, t) = \int_{-\infty}^{t} m_{ij} m_{ij}(\boldsymbol{z}(t'), t') W(t - t') dt' \quad . \tag{5.186}$$

These expressions are non-local in time, which makes them unusable for the simulation, because they require that the entire history of the simulation be kept in memory, which exceeds the storage capacities of today's supercomputers. To remedy this, we choose a fast-decay memory function W:

$$W(t - t') = \frac{1}{T_{\mathrm{lag}}} \exp\left(-\frac{t - t'}{T_{\mathrm{lag}}}\right) \quad , \tag{5.187}$$

in which T_{lag} is the Lagrangian correlation time. With the memory function in this form, we can get the following equations:

$$\frac{D\mathcal{J}_{\text{LM}}}{Dt} \equiv \frac{\partial \mathcal{J}_{\text{LM}}}{\partial t} + \overline{u}_i \frac{\partial \mathcal{J}_{\text{LM}}}{\partial x_i} = \frac{1}{T_{\text{lag}}} \left(L_{ij} m_{ij} - \mathcal{J}_{\text{LM}} \right) \quad , \tag{5.188}$$

$$\frac{D\mathcal{J}_{\text{MM}}}{Dt} \equiv \frac{\partial \mathcal{J}_{\text{MM}}}{\partial t} + \overline{u}_i \frac{\partial \mathcal{J}_{\text{MM}}}{\partial x_i} = \frac{1}{T_{\text{lag}}} \left(m_{ij} m_{ij} - \mathcal{J}_{\text{MM}} \right) \quad , \tag{5.189}$$

the solution of which can be used to compute the subgrid model constant at each point and at each time step. The correlation time T_{lag} is estimated by tests in isotropic homogeneous turbulence at:

$$T_{\text{lag}}(\boldsymbol{x}, t) = 1.5 \, \overline{\Delta} \left(\mathcal{J}_{\text{MM}} \mathcal{J}_{\text{LM}} \right)^{-1/8} \quad , \tag{5.190}$$

which comes down to considering that the correlation time is reduced in the high-shear regions where $\mathcal{J}_{\text{MM}}$ is large, and in those regions where the non-linear transfers are high, i.e. where $\mathcal{J}_{\text{LM}}$ is large.

This procedure does not guarantee that the constant will be positive, and must therefore be coupled with a regularization procedure. Meneveau et al. [508] recommend a clipping procedure.

Solving equations (5.188) and (5.189) yields a large amount of additional numerical work, resulting in a very expensive subgrid model. To alleviate this problem, the solution to these two equations may be approximated using the following Lagrangian tracking technique [596]:

$$\begin{aligned} \mathcal{J}_{\text{LM}}(\boldsymbol{x}, n\Delta t) &= a \, L_{ij}(\boldsymbol{x}, n\Delta t) m_{ij}(\boldsymbol{x}, n\Delta t) \\ &+ (1-a) \mathcal{J}_{\text{LM}}(\boldsymbol{x} - \Delta t \overline{\boldsymbol{u}}(\boldsymbol{x}, n\Delta t), (n-1)\Delta t) \, , \quad (5.191) \\ \mathcal{J}_{\text{MM}}(\boldsymbol{x}, n\Delta t) &= a \, m_{ij}(\boldsymbol{x}, n\Delta t) m_{ij}(\boldsymbol{x}, n\Delta t) \\ &+ (1-a) \mathcal{J}_{\text{MM}}(\boldsymbol{x} - \Delta t \overline{\boldsymbol{u}}(\boldsymbol{x}, n\Delta t), (n-1)\Delta t) \, , \quad (5.192) \end{aligned}$$

where

$$a = \frac{\Delta t / T_{\text{lag}}}{1 + \Delta t / T_{\text{lag}}} \quad . \tag{5.193}$$

This new procedure requires only the storage of the two parameters $\mathcal{J}_{\text{LM}}$ and $\mathcal{J}_{\text{MM}}$ at the previous time step and the use of an interpolation procedure. The authors indicate that a linear interpolation is acceptable.

Constrained Localized Dynamic Procedure. Another generalization of the Germano–Lilly dynamic procedure was proposed for inhomogeneous cases by Ghosal et al. [261]. This new procedure is based on the idea of minimizing an integral problem rather than a local one in space, as is done in the Germano–Lilly procedure, which avoids the need to linearize the constant when applying the test filter. We now look for the constant $C_{\rm d}$ that will minimize the function $\mathcal{F}[C_{\rm d}]$, with

$$\mathcal{F}[C_{\rm d}] = \int E_{ij}(\boldsymbol{x})E_{ij}(\boldsymbol{x})d^3\boldsymbol{x} \quad , \tag{5.194}$$

in which E_{ij} is defined from relation (5.144) and not by (5.146) as was the case for the previously explained versions of the dynamic procedure. The constant sought is such that the variation of $\mathcal{F}[C_{\rm d}]$ is zero:

$$\delta\mathcal{F}[C_{\rm d}] = 2\int E_{ij}(\boldsymbol{x})\delta E_{ij}(\boldsymbol{x})d^3\boldsymbol{x} = 0 \quad , \tag{5.195}$$

or, by replacing E_{ij} with its value:

$$\int \left(-\alpha_{ij}E_{ij}\delta C_{\rm d} + E_{ij}\beta_{ij}\widetilde{\delta C_{\rm d}}\right) d^3\boldsymbol{x} = 0 \quad . \tag{5.196}$$

Expressing the convolution product associated with the test filter, we get:

$$\int \left(-\alpha_{ij}E_{ij} + \beta_{ij}\int E_{ij}(\boldsymbol{y})G(\boldsymbol{x}-\boldsymbol{y})d^3\boldsymbol{y}\right)\delta C_{\rm d}(\boldsymbol{x})d^3\boldsymbol{x} = 0 \quad , \tag{5.197}$$

from which we deduce the following Euler-Lagrange equation:

$$-\alpha_{ij}E_{ij} + \beta_{ij}\int E_{ij}(\boldsymbol{y})G(\boldsymbol{x}-\boldsymbol{y})d^3\boldsymbol{y} = 0 \quad . \tag{5.198}$$

This equation can be re-written in the form of a Fredholm's integral equation of the second kind for the constant $C_{\rm d}$:

$$f(\boldsymbol{x}) = C_{\rm d}(\boldsymbol{x}) - \int \mathcal{K}(\boldsymbol{x},\boldsymbol{y})C_{\rm d}(\boldsymbol{y})d^3\boldsymbol{y} \quad , \tag{5.199}$$

where

$$f(\boldsymbol{x}) = \frac{1}{\alpha_{kl}(\boldsymbol{x})\alpha_{kl}(\boldsymbol{x})}\left(\alpha_{ij}(\boldsymbol{x})L_{ij}(\boldsymbol{x}) - \beta_{ij}(\boldsymbol{x})\int L_{ij}(\boldsymbol{y})G(\boldsymbol{x}-\boldsymbol{y})d^3\boldsymbol{y}\right) \quad , \tag{5.200}$$

$$\mathcal{K}(\boldsymbol{x},\boldsymbol{y}) = \frac{\mathcal{K}_{\mathcal{A}}(\boldsymbol{x},\boldsymbol{y}) + \mathcal{K}_{\mathcal{A}}(\boldsymbol{y},\boldsymbol{x}) + \mathcal{K}_{\mathcal{S}}(\boldsymbol{x},\boldsymbol{y})}{\alpha_{kl}(\boldsymbol{x})\alpha_{kl}(\boldsymbol{x})} \quad , \tag{5.201}$$

and

$$\mathcal{K}_{\mathcal{A}}(\boldsymbol{x},\boldsymbol{y}) = \alpha_{ij}(\boldsymbol{x})\beta_{ij}(\boldsymbol{y})G(\boldsymbol{x}-\boldsymbol{y}) \quad , \tag{5.202}$$

$$\mathcal{K}_{\mathcal{S}}(\boldsymbol{x},\boldsymbol{y}) = \beta_{ij}(\boldsymbol{x})\beta_{ij}(\boldsymbol{y})\int G(\boldsymbol{z}-\boldsymbol{x})G(\boldsymbol{z}-\boldsymbol{y})d^3\boldsymbol{z} \quad . \tag{5.203}$$

This new formulation raises no problems concerning the linearization of the constant, but does not solve the instability problems stemming from the negative values it may take. This procedure is called the localized dynamic procedure.

To remedy the instability problem, the authors propose constraining the constant to remain positive. The constant $C_{\rm d}(\boldsymbol{x})$ is then expressed as the square of a new real variable $\xi(\boldsymbol{x})$. Replacing the constant with its decomposition as a function of ξ, the Euler-Lagrange equation (5.198) becomes:

$$\left(-\alpha_{ij}E_{ij} + \beta_{ij}\int E_{ij}(\boldsymbol{y})G(\boldsymbol{x}-\boldsymbol{y})d^3\boldsymbol{y}\right)\xi(\boldsymbol{x}) = 0 \quad . \tag{5.204}$$

This equality is true if either of the factors is zero, i.e. if $\xi(\boldsymbol{x}) = 0$ or if the relation (5.198) is verified, which is denoted symbolically $C_{\rm d}(\boldsymbol{x}) = \mathcal{G}[C_{\rm d}(\boldsymbol{x})]$. In the first case, the constant is also zero. To make sure it remains positive, the constant is computed by an iterative procedure:

$$C_{\rm d}^{(n+1)}(\boldsymbol{x}) = \begin{cases} \mathcal{G}[C_{\rm d}^{(n)}(\boldsymbol{x})] & \text{if } \mathcal{G}[C_{\rm d}^{(n)}(\boldsymbol{x})] \geq 0 \\ 0 & \text{otherwise} \end{cases} \quad , \tag{5.205}$$

in which

$$\mathcal{G}[C_{\rm d}(\boldsymbol{x})] = f(\boldsymbol{x}) - \int \mathcal{K}(\boldsymbol{x},\boldsymbol{y})C_{\rm d}(\boldsymbol{y})d^3\boldsymbol{y} \quad . \tag{5.206}$$

This completes the description of the constrained localized dynamic procedure. It is applicable to all configurations and ensures that the subgrid model constant remains positive. This solution is denoted symbolically:

$$C_{\rm d}(\boldsymbol{x}) = \left[f(\boldsymbol{x}) + \int \mathcal{K}(\boldsymbol{x},\boldsymbol{y})C_{\rm d}(\boldsymbol{y})d^3\boldsymbol{y}\right]_+ \quad , \tag{5.207}$$

in which + designates the positive part.

Approximate Localized Dynamic Procedure. The localized dynamic procedure decribed in the preceding paragraph makes it possible to regularize the dynamic procedure in fully non-homogeneous flows, and removes the mathematical inconsistency of the Germano–Lilly procedure. But it requires to solve an integral equation, and thus induces a significant overhead.

To alleviate this problem, Piomelli and Liu [596] propose an Approximate Localized Dynamic Procedure, which is not based on a variational approach but on a time extrapolation process. Equation (5.144) is recast in the form

$$L_{ij}^d = C_{\rm d}\alpha_{ij} - \widetilde{C_{\rm d}^*\beta_{ij}} \quad , \tag{5.208}$$

where $C_{\rm d}^*$ is an estimate of the dynamic constant $C_{\rm d}$, which is assumed to be known. Writing the new formulation of the residual E_{ij}, the dynamic constant is now evaluated as

$$C_{\rm d} = \frac{\alpha_{ij}(L_{ij}^d + \widetilde{C_{\rm d}^* \beta_{ij}})}{\alpha_{ij}\alpha_{ij}} \quad . \tag{5.209}$$

The authors propose to evaluate the estimate $C_{\rm d}^*$ by a time extrapolation:

$$C_{\rm d}^* = C_{\rm d}^{(n-1)} + \Delta t \left. \frac{\partial C_{\rm d}}{\partial t} \right|^{(n-1)} + \ldots \quad , \tag{5.210}$$

where the superscript $(n-1)$ is related to the value of the variable at the $(n-1)$th time step, and Δt is the value of the time step. In practice, Piomelli and Liu consider first- and second-order extrapolation schemes. The resulting dynamic procedure is fully local, and does not induce large extra computational effort as the original localized procedure does. Numerical experiments carried out by these authors demonstrate that it still requires clipping to yield a well-behaved algorithm.

Generalized Dynamic Procedure. It is also possible to derive a dynamic procedure using the generalized Germano identity (3.87) [629]. We assume that the operator $\mathcal{L}$ appearing in equation (3.88) is linear, and that there exists a linear operator $\mathcal{L}'$ such that

$$\mathcal{L}(a\ N) = a\mathcal{L}(N) + \mathcal{L}'(a, N) \quad , \tag{5.211}$$

where a is a scalar real function and N an arbitrary second rank tensor. The computation of the dynamic constant $C_{\rm d}$ is now based on the minimization of the residual E_{ij}

$$E_{ij} = \mathcal{L}(L_{ij}^{\rm d}) - C_{\rm d}\mathcal{L}(m_{ij}) \quad , \tag{5.212}$$

where $L_{ij}^{\rm d}$ and m_{ij} are defined by equations (5.144) and (5.150). A least-square minimizations yields:

$$C_{\rm d}' = \frac{\mathcal{L}(L_{ij}^{\rm d})\mathcal{L}(m_{ij})}{\mathcal{L}(m_{ij})\mathcal{L}(m_{ij})} \quad . \tag{5.213}$$

The reduction of the residual obtained using this new relation with respect to the classical one is analyzed by evaluating the difference:

$$\delta E_{ij} = E_{ij}' - \mathcal{L}(E_{ij}) \quad , \tag{5.214}$$

where E_{ij}' is given by relation (5.212) and E_{ij} by (5.146). Inserting the two dynamic constants $C_{\rm d}'$ and $C_{\rm d}$, defined respectively by relations (5.213) and (5.149), we get:

$$\delta E_{ij} = (C_{\rm d} - C_{\rm d}')\mathcal{L}(m_{ij}) + \mathcal{L}'(C_{\rm d}, m_{ij}) \quad . \tag{5.215}$$

An obvious example for the linear operator $\mathcal{L}$ is the divergence operator. The associated $\mathcal{L}'$ is the gradient operator.

An alternative consisting in minimizing a different form of the residual has been proposed by Morinishi and Vasilyev [542, 544] and Mossi [550]:

$$E_{ij} = \mathcal{L}(L_{ij}^{\mathrm{d}}) - \mathcal{L}(C_{\mathrm{d}} m_{ij}) \tag{5.216}$$

$$= \mathcal{L}(L_{ij}^{\mathrm{d}}) - C_{\mathrm{d}} \mathcal{L}(m_{ij}) - \mathcal{L}'(C_{\mathrm{d}}, m_{ij}) \quad . \tag{5.217}$$

The use of this new form of the residual generally requires solving a differential equation, and then yields a more complex procedure than the form (5.212).

These two procedures theoretically more accurate results than the classical one, because they provide reduce the error committed on the subgrid force term itself, rather than on the subgrid tensor. They also take into account for the numerical error associated to the discrete form of $\mathcal{L}$.

Dynamic Inverse Procedure. We have already seen that the use of the dynamic procedure may induce some problems if the cutoff is not located in the inertial range of the spectra, but in the viscous one. A similar problem arises if the cutoff wave number associated to the test filter occurs at the very beginning of the inertial range, or in the production range of the spectrum. In order to compensate inaccuracies arising from the use of a large filter length associated with the test filter, Kuerten et al. [415] developed a new approach, referred to as the Dynamic Inverse Procedure. It relies on the idea that if a dynamic procedure is developed involving only length scales comparable to the basic filter length, self-similarity properties will be preserved and consistent modeling may result. Such a procedure is obtained by defining the first filtering operator G and the second one F by

$$G = H^{-1} \circ L, \quad F = H \quad , \tag{5.218}$$

where L is the classical filter level and H an explicit test filter, whose inverse H^{-1} is assumed to be known explicitly. Inserting these definitions into the Germano identity (3.80), we get a direct evaluation of the subgrid tensor τ:

$$[F \star G\star, B](u_i, u_j) = [L\star, B](u_i, u_j) \tag{5.219}$$

$$= \overline{u_i u_j} - \overline{u}_i \overline{u}_j \tag{5.220}$$

$$\equiv \tau_{ij} \tag{5.221}$$

$$= [H\star, B] \circ (H^{-1} \star L\star)(u_i, u_j) + (H\star) \circ [H^{-1} \star L\star, B](u_i, u_j) \quad . \tag{5.222}$$

This new identity can be recast in a form similar to the original Germano identity

$$L_{ij}^{\diamond} = \tau_{ij} - H \star T_{ij}^{\diamond} \quad , \tag{5.223}$$

with

$$L^{\diamond}_{ij} = H \star ((H^{-1} \star \overline{u}_i)(H^{-1} \star \overline{u}_j)) - \overline{u}_i\overline{u}_j \quad ,$$

$$T^{\diamond}_{ij} = H^{-1} \star \overline{u_i u_j} - (H^{-1} \star \overline{u}_i)(H^{-1} \star \overline{u}_j) \quad .$$

The term $L^{\diamond}_{ij}$ is explicitly known in practice, and does not require any modeling. Using the same notation as in the section dedicated to the Germano–Lilly procedure, we get, for the Smagorinsky model:

$$\tau_{ij} = C_{\rm d}\beta_{ij},\ \beta_{ij} = -2\overline{\Delta}^2|\overline{S}|\overline{S}_{ij} \quad , \tag{5.224}$$

$$T^{\diamond}_{ij} = C_{\rm d}\alpha_{ij},\ \alpha_{ij} = -2\widehat{\overline{\Delta}}^2|\widehat{\overline{S}}|\widehat{\overline{S}}_{ij} \quad , \tag{5.225}$$

where $\widehat{\overline{\Delta}}$ and $\widehat{\overline{S}}$ are the characteristic length and the strain rate tensor associated to the $H^{-1} \circ L$ filtering level, respectively. Building the residual E_{ij} as

$$E_{ij} = L^{\diamond}_{ij} - C_{\rm d}(\beta_{ij} - H \star \alpha_{ij}) = L^{\diamond}_{ij} - C_{\rm d}m_{ij} \quad , \tag{5.226}$$

the least-square-error minimization procedure yields:

$$C_{\rm d} = \frac{L^{\diamond}_{ij}m_{ij}}{m_{ij}m_{ij}} \quad . \tag{5.227}$$

In this new procedure, the two lengths involved are $\overline{\Delta}$ and $\widehat{\overline{\Delta}}$. Since the latter is associated to an inverse filtering operator, we get $\widehat{\overline{\Delta}} \leq \overline{\Delta}$, ensuring that the dynamic procedure will not bring in lengths associated to the production range of the spectrum. In practice, this procedure is observed to suffer the same stability problems than the Germano–Lilly procedure, and needs to be used together with a stabilization procedure (averaging, clipping, etc.).

Taylor Series Expansion Based Dynamic Models. The dynamic procedures presented above rely on the use of a discrete test filter. Chester et al. [129] proposed a new formulation for the dynamic procedure based on the differential approximation of the test filter. All quantities appearing at the test filter level can therefore be rewritten as sums and products of partial derivatives of the resolved velocity field, leading to a new expression of dynamic constants which involves only higher-order derivatives of the velocity field.

Dynamic Procedure with Dimensional Constants. The dynamic procedures described in the preceding paragraphs are designed to find the best values of non-dimensional constants in subgrid scale models. Wong and Lilly [766], followed by Carati and his co-workers [103] propose to extend this procedure to evaluate dimensional parameters which appear in some models. They applied this idea to the so-called Kolmogov formulation for the subgrid viscosity:

$$\nu_{\mathrm{sgs}} = C\overline{\Delta}^{4/3}\varepsilon^{1/3} = C_\varepsilon \overline{\Delta}^{4/3} \quad , \tag{5.228}$$

where the parameter $C_\varepsilon = C\varepsilon^{1/3}$ has the dimension of the cubic root of the subgrid dissipation rate ε. Introducing this closure at both grid and test filtering levels, one obtains (the *tilde* symbol is related to the test filter level):

$$\tau_{ij} = -2C_\varepsilon \overline{\Delta}^{4/3} \overline{S}_{ij} \tag{5.229}$$

$$T_{ij} = -2C_\varepsilon \widetilde{\overline{\Delta}}^{4/3} \widetilde{\overline{S}}_{ij} \tag{5.230}$$

leading to the following expression of the residual

$$E_{ij} = L^{\mathrm{d}}_{ij} - C_\varepsilon \widetilde{\overline{\Delta}}^{4/3} \widetilde{\overline{S}}_{ij} + \widetilde{\left(C_\varepsilon \overline{\Delta}^{4/3} \overline{S}_{ij} \right)} \quad . \tag{5.231}$$

This expression can be used to generate integral expressions for C_ε. A very simple local definition is recovered further assuming that $C_\varepsilon \overline{\Delta}^{4/3}$ is almost constant over distance of the order of $\widetilde{\overline{\Delta}}$. Using the additional property that the test filter perfectly commutes with spatial derivatives, relation (5.231) simplifies as

$$E_{ij} = L^{\mathrm{d}}_{ij} - 2C_\varepsilon \left(\overline{\Delta}^{4/3} - \widetilde{\overline{\Delta}}^{4/3} \right) \widetilde{\overline{S}}_{ij} \quad . \tag{5.232}$$

The least-square optimization method therefore yields the following formula for the dynamic C_ε:

$$C_\varepsilon = \frac{1}{2\left(\overline{\Delta}^{4/3} - \widetilde{\overline{\Delta}}^{4/3} \right)} \frac{L^{\mathrm{d}}_{ij} \widetilde{\overline{S}}_{ij}}{\widetilde{\overline{S}}_{ij} \widetilde{\overline{S}}_{ij}} \quad . \tag{5.233}$$

As the original Germano-Lilly procedure for non-dimensional parameters, this procedure suffers some numerical instability problems and must therefore be regularized using clipping and/or averaging.

Dynamic Procedures Without the Germano Identity.

Multilevel Procedure by Terracol and Sagaut. This method proposed by Terracol and Sagaut [709] relies on the hypothesis that the computed resolved kinetic energy spectrum obeys a power-law like

$$E(k) = E_0 k^\alpha \quad , \tag{5.234}$$

where α is the scaling parameter. It is worth noting that Barenblatt [41] suggests that both E_0 and α might be Reynolds-number dependent. A more accurate expression for the kinetic energy spectrum is

$$E(k) = K_0 \varepsilon^{2/3} k^{-5/3} (k\Lambda)^{\zeta} \quad , \tag{5.235}$$

where $K_0 = 1.4$ is the Kolmogorov constant, Λ a length scale and ζ an intermittency factor. Under this assumption, the mean subgrid dissipation rate across a cutoff wave number k_c, $\varepsilon(k_c)$, scales like

$$\varepsilon(k_c) = \varepsilon_0 k_c^{\gamma}, \quad \gamma = \frac{3\alpha + 5}{2} = \frac{3}{2}\zeta \quad . \tag{5.236}$$

where ε_0 is a k_c-independent parameter. It is observed that in the Kolmogorov case ($\alpha = -5/3$), one obtains $\gamma = 0$, leading to a constant dissipation rate.

Let us now introduce a set of cutoff wave numbers k_n, with $k_1 > k_2 > ...$ The following recursive law is straithgforwardly derived from (5.236)

$$\frac{\varepsilon(k_n)}{\varepsilon(k_{n+1})} = R_{n,n+1}^{\gamma}, \quad R_{n,n+1} = \frac{k_n}{k_{n+1}} \quad , \tag{5.237}$$

leading to the following two-level evaluation of the parameter γ:

$$\gamma = \frac{\log(\varepsilon(k_n)/\varepsilon(k_{n+1}))}{\log(R_{n,n+1})} \quad . \tag{5.238}$$

Now introducing a generic subgrid model for the nth cutoff level

$$\tau_{ij}^n = C f_{ij}(\overline{\boldsymbol{u}}^n, \overline{\Delta}^n) \quad , \tag{5.239}$$

where C is the constant of the model to be dynamically computed, $\overline{\boldsymbol{u}}^n$ the resolved field at the considered cutoff level and $\overline{\Delta}^n \equiv \pi/k_n$ the current cutoff length, the dissipation rate can also be expressed as

$$\varepsilon(k_n) = -\tau_{ij}^n \overline{S}_{ij}^n = -C f_{ij}(\overline{\boldsymbol{u}}^n, \overline{\Delta}^n) \overline{S}_{ij}^n \quad , \tag{5.240}$$

where $\overline{S}^n$ is the resolved strain rate at level n. Equation (5.237) shows that the ratio $\varepsilon(k_n)/\varepsilon(k_{n+1})$ is independent of the model constant C. Using this property, Terracol and Sagaut propose to introduce two test filter levels k_2 and k_3 (k_1 being the grid filter level where the equations must be closed, i.e. $\overline{\Delta}^1 = \overline{\Delta}$). The intermittency factor γ is then computed using relation (5.238), and one obtains the following evaluation for the subgrid dissipation rate at the grid filter level:

$$\varepsilon(k_1) = R_{1,2}^{\gamma} \varepsilon'(k_2) \quad , \tag{5.241}$$

where $\varepsilon'(k_2)$ is evaluated using a reliable approximation of the subgrid tensor to close the sequence (in practice, a scale-similarity model is used in Ref. [709]). The corresponding value of C for the model at the grid level is then deduced from (5.240):

$$C = R_{1,2}^{\gamma} \frac{\varepsilon'(k_2)}{-f_{ij}(\overline{\boldsymbol{u}}, \overline{\Delta}) \overline{S}_{ij}} \quad . \tag{5.242}$$

Multiscale Method Based on the Kolmogorov-Meneveau Equation. Another procedure was developed by Shao [669, 153] starting from the Kolmogorov-Meneveau equation for filtered third-order velocity structure function:

$$-\frac{4}{5}r\varepsilon = \overline{D}_{\mathrm{LLL}} - 6G_{\mathrm{LLL}} \quad , \tag{5.243}$$

where $\overline{D}_{\mathrm{LLL}}$ is the third-order longitudinal velocity correlation of the filtered field, $G_{\mathrm{LLL}}(r)$ the longitudinal velocity-stress correlation tensor and $\varepsilon = -\tau_{ij}\overline{S}_{ij}$ the average subgrid dissipation (see p. 126 for additional details).

Now assuming that the following self-similarity law is valid

$$G_{\mathrm{LLL}}(r) \propto r^p \quad , \tag{5.244}$$

where $p = -1/3$ corresponds to the Kolmogorov local isotropy hypotheses, one obtains the following relationship for two space increments r_1 and r_2:

$$\frac{0.8r_1\varepsilon + \overline{D}_{\mathrm{LLL}}(r_1)}{0.8r_2\varepsilon + \overline{D}_{\mathrm{LLL}}(r_2)} = \left(\frac{r_1}{r_2}\right)^{-1/3} \quad . \tag{5.245}$$

Now introducing the same generic subgrid closure as for the Terracol–Sagaut procedure

$$\tau_{ij} = Cf_{ij}(\overline{\boldsymbol{u}}, \overline{\Delta}) \quad , \tag{5.246}$$

and inserting it into (5.245) to evaluate $\varepsilon = -\tau_{ij}\overline{S}_{ij}$, taking $r_1 = \overline{\Delta}$ and $r_2 > r_1$, the dynamic value of the constant C is

$$C = \frac{\left(\frac{\overline{\Delta}}{r_2}\right)^{-1/3} \overline{D}_{\mathrm{LLL}}(r_2) - \overline{D}_{\mathrm{LLL}}(\overline{\Delta})}{0.8 f_{ij}(\overline{\boldsymbol{u}}, \overline{\Delta})\overline{S}_{ij}\left(\overline{\Delta} - \left(\frac{\overline{\Delta}}{r_2}\right)^{-1/3} r_2\right)} \quad . \tag{5.247}$$

The only fixed parameter in the Shao procedure is the scaling parameter p in (5.244). This parameter can be computed dynamically introducing a third space increment r_3, leading to the definition of a dynamic procedure with the same properties as the one proposed by Terracol and Sagaut. The proposal of Shao can also be extended to subgrid models with several adjustable constant by introducing an additional space increment for each new constant and solving a linear algebra problem.

Structural Sensors. Selective Models. In order to improve the prediction of intermittent phenomena, we introduce a sensor based on structural information. This is done by incorporating a selection function in the model, based on the local angular fluctuations of the vorticity, developed by David [166, 440].

The idea here is to modulate the subgrid model in such a way as to apply it only when the assumptions underlying the modeling are verified, i.e. when all the scales of the exact solution are not resolved and the flow is of the fully developed turbulence type. The problem therefore consists in determining if these two hypotheses are verified at each point and each time step. David's structural sensor tests the second hypothesis. To do this, we assume that, if the flow is turbulent and developed, the highest resolved frequencies have certain characteristics specific to isotropic homogeneous turbulence, and particularly structural properties.

So the properties specific to isotropic homogeneous turbulence need to be identified. David, taking direct numerical simulations as a base, observed that the probability density function of the local angular fluctuation of the vorticity vector exhibit a peak around the value of 20°. Consequently, he proposes identifying the flow as being locally under-resolved and turbulent at those points for which the local angular fluctuations of the vorticity vector corresponding to the highest resolved frequencies are greater than or equal to a threshold value θ_0.

The selection criterion will therefore be based on an estimation of the angle θ between the instantaneous vorticity vector $\boldsymbol{\omega}$ and the local average vortcity vector $\tilde{\boldsymbol{\omega}}$ (see Fig. 5.17), which is computed by applying a test filter to the vorticity vector.

The angle θ is given by the following relation:

$$\theta(\boldsymbol{x}) = \arcsin\left(\frac{\|\tilde{\boldsymbol{\omega}}(\boldsymbol{x}) \times \boldsymbol{\omega}(\boldsymbol{x})\|}{\|\tilde{\boldsymbol{\omega}}(\boldsymbol{x})\|.\|\boldsymbol{\omega}(\boldsymbol{x})\|}\right) \quad . \tag{5.248}$$

We define a selection function to damp the subgrid model when the angle θ is less than a threshold angle θ_0.

In the original version developed by David, the selection function f_{θ_0} is a Boolean operator:

$$f_{\theta_0}(\theta) = \begin{cases} 1 & \text{if } \theta \geq \theta_0 \\ 0 & \text{otherwise} \end{cases} \quad . \tag{5.249}$$

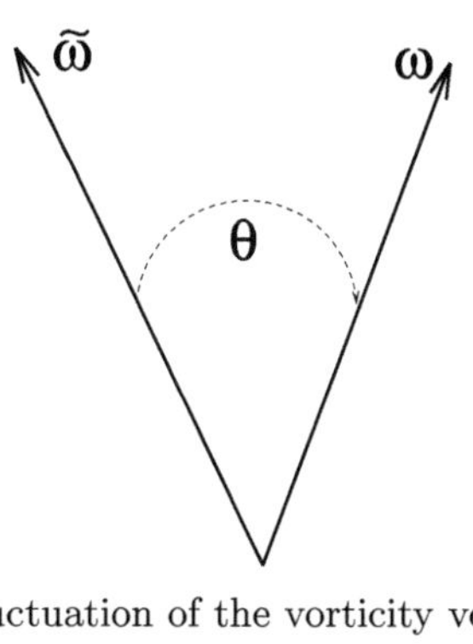

Fig. 5.17. Local angular fluctuation of the vorticity vector.

This function is discontinuous, which may pose problems in the numerical solution. One variant of it that exhibits no discontinuity for the threshold value is defined as follows [636]:

$$f_{\theta_0}(\theta) = \begin{cases} 1 & \text{if } \theta \geq \theta_0 \\ r(\theta)^n & \text{otherwise} \end{cases} \quad , \tag{5.250}$$

in which θ_0 is the chosen threshold value and r the function:

$$r(\theta) = \frac{\tan^2(\theta/2)}{\tan^2(\theta_0/2)} \quad , \tag{5.251}$$

where the exponent n is positive. In practice, it is taken to be equal to 2. Considering the fact that we can express the angle θ as a function of the norms of the vorticity vector $\boldsymbol{\omega}$, the average vorticity vector $\tilde{\boldsymbol{\omega}}$, and the norm ω' of the fluctuating vorticity vector defined as $\boldsymbol{\omega}' = \boldsymbol{\omega} - \tilde{\boldsymbol{\omega}}$, by the relation:

$$\omega'^2 = \tilde{\omega}^2 + \omega^2 - 2\tilde{\omega}\omega\cos\theta \quad ,$$

and the trigonometric relation:

$$\tan^2(\theta/2) = \frac{1-\cos\theta}{1+\cos\theta} \quad ,$$

the quantity $\tan^2(\theta/2)$ is estimated using the relation:

$$\tan^2(\theta/2) = \frac{2\tilde{\omega}\omega - \tilde{\omega}^2 - \omega^2 + \omega'^2}{2\tilde{\omega}\omega + \tilde{\omega}^2 + \omega^2 - \omega'^2} \quad . \tag{5.252}$$

The selection function is used as a multiplicative factor of the subgrid viscosity, leading to the definition of selective models:

$$\nu_{\text{sgs}} = \nu_{\text{sgs}}(\boldsymbol{x}, t) f_{\theta_0}(\theta(\boldsymbol{x})) \quad , \tag{5.253}$$

in which ν_{sgs} is calculated by an arbitrary subgrid viscosity model. It should be noted that, in order to keep the same average subgrid viscosity value over the entire fluid domain, the constant that appears in the subgrid model has to be multiplied by a factor of 1.65. This factor is evaluated on the basis of isotropic homogeneous turbulence simulations.

Accentuation Technique. Filtered Models.

Accentuation Technique. Since large-eddy simulation is based on a frequency selection, improving the subgrid models in the physical space requires a better diagnostic concerning the spectral distribution of the energy in the calculated solution. More precisely, what we want to do here is to determine if the exact solution is entirely resolved, in which case the subgrid model should be reduced to zero, or if there exist subgrid scales that have to be taken into account by means of a model. When models expressed in the physical space do not use additional variables, they suffer from imprecision due to Gabor-Heisenberg's principle of uncertainty already mentioned above, because the contribution of the low frequencies precludes any precise determination of the energy at the cutoff. Let us recall that, if this energy is zero, the exact solution is completely represented and, if otherwise, then subgrid modes exist. In order to be able to detect the existence of the subgrid modes better, Ducros [204, 205] proposes an accentuation technique which consists in applying the subgrid models to a modified velocity field obtained by applying a frequency high-pass filter to the resolved velocity field. This filter, denoted HP^n, is defined recursively as:

$$\mathrm{HP}^1(\overline{\boldsymbol{u}}) \simeq \overline{\Delta}^2 \nabla^2 \overline{\boldsymbol{u}} \quad , \tag{5.254}$$

$$\mathrm{HP}^n(\overline{\boldsymbol{u}}) = \mathrm{HP}(\mathrm{HP}^{n-1}(\overline{\boldsymbol{u}})) \quad . \tag{5.255}$$

We note that the application of this filter in the discrete case results in a loss of localness in the physical space, which is in conformity with Gabor-Heisenberg's principle of uncertainty. We use $E_{\mathrm{HP}^n}(k)$ to denote the energy spectrum of the field thus obtained. This spectrum is related to the initial spectrum $\overline{E}(k)$ of the resolved scales by:

$$\overline{E}_{\mathrm{HP}^n}(k) = T_{\mathrm{HP}^n}(k)\overline{E}(k) \quad , \tag{5.256}$$

in which $T_{\mathrm{HP}^n}(k)$ is a transfer function which Ducros evaluates in the form:

$$T_{\mathrm{HP}^n}(k) = b^n \left(\frac{k}{k_c}\right)^{\gamma n} \quad . \tag{5.257}$$

Here, b and γ are positive constants that depend on the discrete filter used in the numerical simulation[27]. The shape of the spectrum obtained by the transfer function to a Kolmogorov spectrum is graphed in Fig. 5.18 for several values of the parameter n. This type of filter modifies the spectrum of the initial solution by emphasizing the contribution of the highest frequencies.

The resulting field therefore represents mainly the high frequencies of the initial field and serves to compute the subgrid model. To remain consistent, the subgrid model has to be modified. Such models are called filtered models. The case of the Structure Function model is given as an example. Filtered versions of the Smagorinsky and Mixed Scale models have been developped by Sagaut, Comte and Ducros [628].

Filtered Second-Order Structure Function Model. We define the second-order structure function of the filtered field:

$$\overline{D}_{\mathrm{LL}}^{\mathrm{HP}^n}(\boldsymbol{x}, r, t) = \int_{|\boldsymbol{x}'|=r} \left[\mathrm{HP}^n(\overline{u})(\boldsymbol{x}, t) - \mathrm{HP}^n(\overline{u})(\boldsymbol{x}+\boldsymbol{x}', t)\right]^2 d^3\boldsymbol{x}' \quad , \tag{5.258}$$

[27] For a Laplacian type filter discretized by second-order accurate finite difference scheme iterated three times ($n = 3$), Ducros finds $b^3 = 64,000$ and $3\gamma = 9.16$.

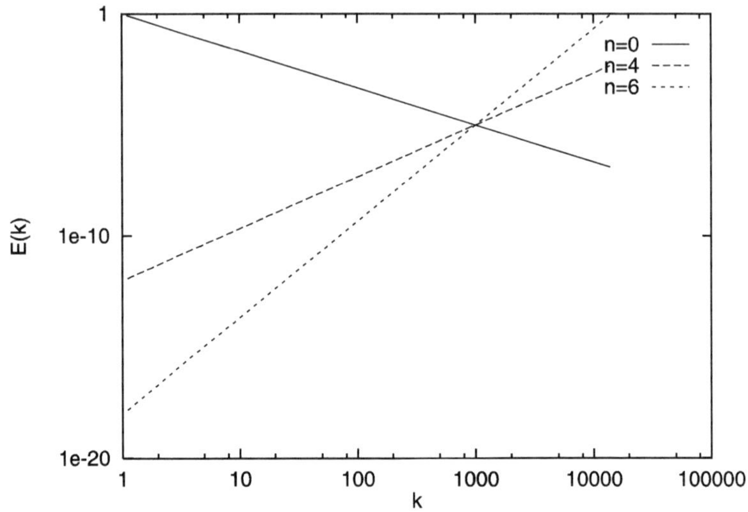

Fig. 5.18. Energy spectrum of the accentuation solution for different values of the parameter n ($b = \gamma = 1, k_c = 1000$).

for which the statistical average over the entire fluid domain, denoted $\langle \overline{D}_{\rm LL}^{{\rm HP}^n} \rangle(r,t)$, is related to the kinetic energy spectrum by the relation:

$$\langle \overline{D}_{\rm LL}^{{\rm HP}^n} \rangle(r,t) = 4 \int_0^{k_{\rm c}} \overline{E}_{{\rm HP}^n}(k) \left(1 - \frac{\sin(k\overline{\Delta})}{k\overline{\Delta}} \right) dk \quad . \tag{5.259}$$

According to the theorem of averages, there exists a wave number $k_* \in [0, k_{\rm c}]$ such that:

$$\overline{E}_{{\rm HP}^n}(k_*) = \frac{\langle \overline{D}_{\rm LL}^{{\rm HP}^n} \rangle(r,t)}{4(\pi/k_{\rm c}) \int_0^{\pi} (1 - \sin(\xi)/\xi)\, d\xi} \quad . \tag{5.260}$$

Using a Kolmogorov spectrum, we can state the equality:

$$\frac{\overline{E}(k_{\rm c})}{k_{\rm c}^{-5/3}} = \frac{\overline{E}_{{\rm HP}^n}(k_*)}{k_*} \quad . \tag{5.261}$$

Considering this last relation, along with (5.256) and (5.257), the subgrid viscosity models based on the energy at cutoff are expressed:

$$\langle \nu_{\rm sgs} \rangle = \frac{2}{3} \frac{K_0^{-3/2}}{k_{\rm c}^{1/2}} \sqrt{ \left(\frac{k_*}{k_{\rm c}} \right)^{5/3 - \gamma n} \frac{1}{b^n} \overline{E}_{{\rm HP}^n}(k_*) } \quad , \tag{5.262}$$

in which:

$$\left(\frac{k_*}{k_c}\right)^{-5/3+\gamma n} = \frac{1}{\pi^{-5/3+\gamma n}} \frac{\displaystyle\int_0^\pi \xi^{-5/3+\gamma n}\left(1-\sin(\xi)/\xi\right) d\xi}{\displaystyle\int_0^\pi \left(1-\sin(\xi)/\xi\right) d\xi} \quad . \tag{5.263}$$

By localizing these relations in the physical space, we deduce the filtered structure function model:

$$\begin{aligned}\nu_{\mathrm{sgs}}(\boldsymbol{x},\overline{\Delta},t) &= \frac{2}{3}\frac{K_0^{-3/2}}{\pi^{4/3}}\frac{\overline{\Delta}}{2}\left(\frac{\pi^{\gamma n}}{b^n}\right)^{1/2} \frac{\left(\overline{D}_{\mathrm{LL}}^{\mathrm{HP}^n}(\boldsymbol{x},\overline{\Delta},t)\right)^{1/2}}{\left(\displaystyle\int_0^\pi \xi^{-5/3+\gamma n}\left(1-\sin(\xi)/\xi\right) d\xi\right)^{1/2}} \\ &= C^{(n)}\overline{\Delta}\sqrt{\overline{D}_{\mathrm{LL}}^{\mathrm{HP}^n}(\boldsymbol{x},\overline{\Delta},t)} \quad .\end{aligned} \tag{5.264}$$

The values of the constant $C^{(n)}$ are given in the following Table: In practice, Ducros recommends using $n = 3$.

Table 5.2. Values of the Structure Function model constant for different iterations of the high-pass filter.

n	0	1	2	3	4
$C^{(n)}$	0.0637	0.020	0.0043	0.000841	$1.57 \cdot 10^{-4}$

Damping Functions for the Near-wall Region. The presence of a solid wall modifies the turbulence dynamics in several ways, which are discussed in Chap. 10. The only fact concerning us here is that the presence of a wall inhibits the growth of the small scales. This phenomenon implies that the characteristic mixing length of the subgrid modes Δ_{f} has to be reduced in the near-surface region, which corresponds to a reduction in the intensity of the subgrid viscosity. To represent the dynamics in the near-wall region correctly, it is important to make sure that the subgrid models verify the good properties in this region. In the case of a canonical boundary layer (see Chap. 10), the statistical asymptotic behavior of the velocity components and subgrid tensions can be determined analytically. Let u be the main velocity component in the x direction, v the transverse component in the y direction, and w the velocity component normal to the wall, in the z direction. Using the incompressibility constraint, a Taylor series expansion of the velocity component in the region very near the wall yields:

$$\langle u\rangle \propto z, \ \langle v\rangle \propto z, \ \langle w\rangle \propto z^2, \tag{5.265}$$

$$\begin{aligned}&\langle\tau_{11}\rangle \propto z^2, \quad \langle\tau_{22}\rangle \propto z^2, \quad \langle\tau_{33}\rangle \propto z^4, \\ &\langle\tau_{13}\rangle \propto z^3, \quad \langle\tau_{12}\rangle \propto z^2, \quad \langle\tau_{23}\rangle \propto z^3.\end{aligned} \tag{5.266}$$

Experience shows that it is important to reproduce the behavior of the component τ_{13} in order to ensure the quality of the simulation results. It is generally assumed that the most important stress in the near-wall region is τ_{13}, because it is directly linked to the mean turbulence production term, P, which is evaluated as

$$P \propto \langle \tau_{13} \rangle \frac{d\langle u \rangle}{dz} \quad . \tag{5.267}$$

Thus, it is expected that subgrid-viscosity models will be such that

$$\langle \nu_{\rm sgs} \rangle \frac{d\langle u \rangle}{dz} \propto \langle \tau_{13} \rangle \propto z^3 \quad . \tag{5.268}$$

We deduce the following law from relations (5.265) and (5.268):

$$\langle \nu_{\rm sgs} \rangle \propto z^3 \quad . \tag{5.269}$$

We verify that the subgrid-viscosity models based on the large scales alone do not verify this asymptotic behavior. This is understood by looking at the wall value of the subgrid viscosity associated with the mean velocity field $\langle u \rangle$. A second-order Taylor series expansion of some zero-equation subgrid viscosity models presented in the preceding section yields:

$$\begin{aligned} \nu_{\rm sgs}|_{\rm w} &\propto \overline{\Delta}|_{\rm w}^2 \left| \frac{\partial \langle u \rangle}{\partial z} \right|_{\rm w} \quad \text{(Smagorinsky)} \quad , \\ \nu_{\rm sgs}|_{\rm w} &\propto \overline{\Delta}|_{\rm w} \Delta z_1 \left| \frac{\partial \langle u \rangle}{\partial z} \right|_{\rm w} \quad \text{(2nd order Structure Function)} \quad , \\ \nu_{\rm sgs}|_{\rm w} &\propto \overline{\Delta}|_{\rm w}^{3/2} \Delta z_1 \left| \frac{\partial \langle u \rangle}{\partial z} \right|_{\rm w}^{1/2} \left| \frac{\partial^2 \langle u \rangle}{\partial z^2} \right|_{\rm w}^{1/2} \quad \text{(Mixed Scale)} \quad , \end{aligned} \tag{5.270}$$

where the w subscript denotes values taken at the wall, and Δz_1 is the distance to the wall at which the model is evaluated. Because $\partial \langle u \rangle / \partial z$ is not zero at the wall,[28], we have the following asymptotic scalings of the modeled subgrid viscosity at solid walls:

$$\begin{aligned} \nu_{\rm sgs}|_{\rm w} &= O(\overline{\Delta}|_{\rm w}^2) \quad \text{(Smagorinsky)} \quad , \\ \nu_{\rm sgs}|_{\rm w} &= O(\overline{\Delta}|_{\rm w} \Delta z_1) \quad \text{(2nd order Structure Function)} \quad , \\ \nu_{\rm sgs}|_{\rm w} &= 0 \quad \text{(Mixed Scale)} \quad . \end{aligned} \tag{5.271}$$

In practice, the Mixed Scale model can predict a zero subgrid viscosity at the wall if the computational grid is fine enough to make it possible to

[28] Or, equivalently, the skin friction is not zero.

evaluate correctly the second-order wall–normal velocity derivative, i.e. if at least three grid points are located within the region where the mean velocity profile obeys a linear law. Consequently, the first two models must be modified in the near-wall region in order to enforce a correct asymptotic behavior of the subgrid terms in that region.

This is done by introducing damping functions. The usual relation:

$$\Delta_{\mathrm{f}} = C\overline{\Delta} \quad , \tag{5.272}$$

is replaced by:

$$\Delta_{\mathrm{f}} = C\overline{\Delta} f_{\mathrm{w}}(z) \quad , \tag{5.273}$$

in which $f_{\mathrm{w}}(z)$ is the damping function and z the distance to the wall. From Van Driest's results, we define:

$$f_{\mathrm{w}}(z) = 1 - \exp\left(-z u_\tau / 25\nu\right) \quad , \tag{5.274}$$

in which the friction velocity u_τ is defined in Sect. 10.2.1. Piomelli et al. [600] propose the alternate form:

$$f_{\mathrm{w}}(z) = \left(1 - \exp\left(-(z u_\tau / 25\nu)^3\right)\right)^{1/2} \quad . \tag{5.275}$$

From this last form we can get a correct asymptotic behavior of the subgrid viscosity, i.e. a decrease in z^{+3} in the near-wall region, contrary to the Van Driest function. Experience shows that we can avoid recourse to these functions by using a dynamic procedure, a filtered model, a selective model, or the Yoshizawa model. It is worth noting that subgrid viscosity models can be designed, which automatically follow the correct behavior in the near-wall region. An example is the WALE model, developed by Nicoud and Ducros [567].

5.3.4 Implicit Diffusion: the ILES Concept

Large-eddy simulation approaches using a numerical viscosity with no explicit modeling are all based implicitly on the hypothesis:

Hypothesis 5.6 *The action of subgrid scales on the resolved scales is equivalent to a strictly dissipative action.*

This approach is referred to as *Implicit Large-Eddy Simulation* (ILES). Simulations belonging to this category use dissipation terms introduced either in the framework of upwind schemes for the convection or explicit artificial dissipation term, or by the use of implicit [716] or explicit [210] frequency low-pass filters. The approach most used is doubtless the use of upwind schemes for the convective term. The diffusive term introduced then varies both in degree and order, depending on the scheme used (QUICK [437], Godunov [776], PPM [145], TVD [150], FCT [66], MPDATA [489, 488], among others) and the dissipation induced can in certain cases be very close[29] to that introduced by a physical model [275]. Let us note that most of the schemes introduce dissipations of the second and/or fourth order and, in so doing, are very close to subgrid models. This point is discussed more precisely in Chap. 8. This approach is widely used in cases where the other modeling approaches become difficult for one of the two following reasons:

- The dynamic mechanisms escape the physical modeling because they are unknown or too complex to be modeled exactly and explicitly, which is true when complex thermodynamic mechanisms, for example, interact strongly with the hydrodynamic mechanisms (e.g. in cases of combustion [135] or shock/turbulence interaction [435]).
- Explicit modeling offers *no a priori* guarantee of certain realizability constraints related to the quantities studied (such as the temperature [125] or molar concentrations of pollutants [474]). This point is illustrated in Fig. 5.19, which displays the probability density function of a passive scalar computed by Large-Eddy Simulation with different numerical schemes for the convection term.

In cases belonging to one of these two classes, the error committed by using an implicit viscosity may in theory have no more harmful consequence on the quality of the result obtained than that which would be introduced by using an explicit model based on inadequate physical considerations. This approach is used essentially for dealing with very complex configurations or those harboring numerical difficulties, because it allows the use of robust numerical methods. Nonetheless, high-resolution simulations of flows are beginning to make their appearance [756, 602, 776, 274].

A large number of stabilized numerical methods have been used for large-eddy simulation, but only a few of them have been designed for this specific purpose or more simply have been analyzed in that sense. A few general approaches for designing stabilized methods which mimic functional subgrid modeling are discussed below:

1. The MILES (Monotone Integrated Large Eddy Simulation) approach within the framework of flux-limiting finite volume methods, as discussed by Grinstein and Fureby (p. 163).
2. The adaptive flux reconstruction technique within the framework of non-limited finite volume methods, proposed by (p. 165).
3. Finite element schemes with embedded subgrid stabilization (p. 166).
4. The use of Spectral Vanishing Viscosities (p. 169) which are well suited for numerical methods with spectral-like accuracy.
5. The high-order filtering technique (p. 170), originally developed within the finite-difference framework, and wich is equivalent to some approximate deconvolution based structural models.

[29] In the sense where these dissipations are localized at the same points and are of the same order of magnitude.

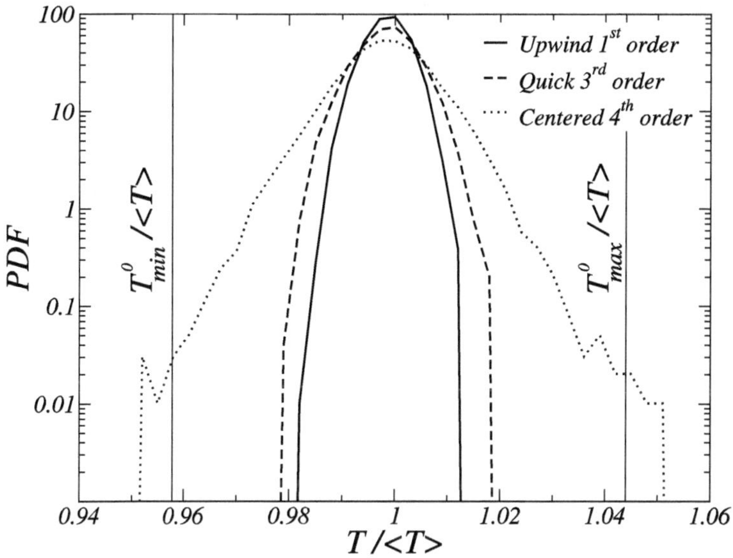

Fig. 5.19. Probability of the density function of the temperature (modeled as a passive scalar) in a channel flow obtained via Large-Eddy Simulation. Vertical lines denote physical bounds. It is observed that the simulation carried out with centered fourth-order accurate scheme admits non-physical values, while the use of the stabilized schemes to solve the passive scalar equation cures this problem. Courtesy of F. Chatelain (CEA).

MILES Approach. A theoretical analysis of the MILES approach within the framework of flux-limiting finite volume discretizations has been carried out by Fureby and Grinstein [273, 228, 231, 274, 229], which puts the emphasis on the existing relationship between leading numerical error terms and tensorial subgrid viscosities.

Defining a control cell Ω of face-normal unit vector $\boldsymbol{n}$, the convective fluxes are usually discretized using Green's theorem

$$\int_{\Omega} \nabla \cdot (\overline{\boldsymbol{u}} \otimes \overline{\boldsymbol{u}}) d\Omega = \int_{\partial\Omega} (\overline{\boldsymbol{u}} \cdot \boldsymbol{n}) \overline{\boldsymbol{u}} dS \quad , \tag{5.276}$$

where $\otimes$ denotes the tensorial product and $\partial\Omega$ is the boundary of Ω. The associated discrete relation is

$$\int_{\partial\Omega} (\overline{\boldsymbol{u}} \cdot \boldsymbol{n}) \overline{\boldsymbol{u}} dS \approx \sum_{\mathrm{f}} F_{\mathrm{f}}^{\mathrm{C}}(\overline{\boldsymbol{u}}) \quad , \tag{5.277}$$

where f are the faces of Ω and the discrete flux function is expressed as

$$F_{\mathrm{f}}^{\mathrm{C}}(\overline{\boldsymbol{u}}) = ((\overline{\boldsymbol{u}} \cdot d\boldsymbol{A})\overline{\boldsymbol{u}})_{\mathrm{f}} \quad , \tag{5.278}$$

where $d\boldsymbol{A}$ is the face-area vector of face f of $\partial\Omega$, and $()_\mathrm{f}$ is the integrated value on face f. For flux-limiting methods, the numerical flux is decomposed as the weighted sum of a high-order flux function F_f^H that works well in smooth regions and a low-order flux function F_f^L:

$$F_\mathrm{f}^\mathrm{C}(\overline{\boldsymbol{u}}) = F_\mathrm{f}^\mathrm{H}(\overline{\boldsymbol{u}}) + (1 - \Gamma(\overline{\boldsymbol{u}}))\left[F_\mathrm{f}^\mathrm{H}(\overline{\boldsymbol{u}}) - F_\mathrm{f}^\mathrm{L}(\overline{\boldsymbol{u}})\right] \quad , \tag{5.279}$$

where $\Gamma(\overline{\boldsymbol{u}})$ is the flux limiter.[30] Fureby and Grinstein analyzed the leading error term using the following assumptions: (i) time integration is performed using a three-point backward scheme, (ii) the high-order flux functions use first-order functional reconstruction, and (iii) the low-order flux functions use upwind differencing. Retaining the leading dissipative error term, the continuous equivalent formulation for the discretized fluxes is

$$\underbrace{\nabla\cdot(\overline{\boldsymbol{u}}\otimes\overline{\boldsymbol{u}})}_{\text{exact}} - \underbrace{\nabla\cdot(\overline{\boldsymbol{u}}\otimes\boldsymbol{r} + \boldsymbol{r}\otimes\overline{\boldsymbol{u}} + \boldsymbol{r}\otimes\boldsymbol{r})}_{\text{dissipative error}} \quad , \tag{5.280}$$

with

$$\boldsymbol{r} = \beta(\nabla\overline{\boldsymbol{u}})\boldsymbol{d}, \quad \beta = \frac{1}{2}(1 - \Gamma(\overline{\boldsymbol{u}}))\mathrm{sgn}\left(\frac{\overline{\boldsymbol{u}}\cdot\boldsymbol{d}}{|\boldsymbol{d}|}\right) \quad , \tag{5.281}$$

where $\boldsymbol{d}$ is the topology vector connecting neighboring control volumes. Comparison of the error term in (5.280) and the usual subgrid term appearing in filtered Navier–Stokes equations (3.17) yields the following identification for the MILES subgrid tensor:

$$\begin{aligned}\tau_\mathrm{MILES} &= -(\boldsymbol{u}\otimes\boldsymbol{r} + \boldsymbol{r}\otimes\boldsymbol{u} + \boldsymbol{r}\otimes\boldsymbol{r}) \\ &= -\underbrace{\beta\left[(\overline{\boldsymbol{u}}\otimes\boldsymbol{d})\nabla^\mathrm{T}\overline{\boldsymbol{u}} + \nabla\overline{\boldsymbol{u}}(\overline{\boldsymbol{u}}\otimes\boldsymbol{d})^\mathrm{T}\right]}_{I} \\ &\quad + \underbrace{\beta^2(\nabla\overline{\boldsymbol{u}})\boldsymbol{d}\otimes(\nabla\overline{\boldsymbol{u}})\boldsymbol{d}}_{II} \quad .\end{aligned} \tag{5.282}$$

Term I appears as a general subgrid-viscosity model with a tensorial diffusivity $\beta(\overline{\boldsymbol{u}}\otimes\boldsymbol{d})$, while term II mimics the Leonard tensor, leading to the definition of an implicit mixed model (see Sect. 7.4).[31] A scalar-valued measure of the viscosity is $\sqrt{2}/8|\overline{\boldsymbol{u}}|\overline{\Delta}_\mathrm{MILES}$, where the characteristic length associated with the grid is $\overline{\Delta}_\mathrm{MILES} = \sqrt{\mathrm{tr}[(\nabla^\mathrm{T}\boldsymbol{d})(\boldsymbol{d}\otimes\boldsymbol{d})(\nabla\boldsymbol{d})]}$.

The authors remarked that these error terms are invariant under the Galilean group of transformations, but are not frame indifferent. Realizability and non-negative dissipation of subgrid kinetic energy may be enforced for some choice of the limiter.

[30] Many flux limiters can be found in the literature: minmod, superbee, FCT limiter, ... The reader is referred to specialized reference books [307] for a detailed discussion of these functions.

[31] MILES can also be interpreted as an implicit deconvolution model, using the analogy discussed in Sect. 7.3.3.

Adaptive Flux Reconstruction. Adams [2] established a theoretical bridge between high-order adaptive flux reconstruction used in certain finite-volume schemes and the use of a subgrid viscosity. In the simplified case of the following one-dimensional conservation law

$$\frac{\partial u}{\partial t} + \frac{\partial F(u)}{\partial x} = 0 \quad , \tag{5.283}$$

the finite volume technique leads to the computation of the cell-averaged variable:

$$\overline{u}_j = \int_{x_{j-1/2}}^{x_{j+1/2}} u(\xi) d\xi \quad , \tag{5.284}$$

with $\Delta x = x_{j+1/2} - x_{j-1/2}$ the cell spacing of the jth cell. High-order finite volume methods rely on the reconstruction of the unfiltered value $u_{j+1/2}$ on both sides of the cell face $x_{j+1/2}$ on each cell j. This is achieved by defiltering the variable $\overline{u}$ and defining a high-order polynomial interpolant.

The defiltering step is similar to the deconvolution approach, whose related results are presented in Sect. 7.2.1 and will not be repeated here. Sticking to Adams' demonstration, a second-order deconvolution is employed:

$$u_j = \overline{u}_j - \frac{\Delta x^2}{24} \frac{\partial^2 \overline{u}_j}{\partial x^2} \quad . \tag{5.285}$$

Following the WENO (Weightest Essentially Non-Oscillatory) concept [350], a hierarchical family of left-hand-side interpolants of increasing order is:

$$P^{+,(0)}(x)_j = u_j \quad , \tag{5.286}$$

$$P_j^{+,(1)}(x) = P_j^{+,(0)}(x) + \alpha_{1,1}^{+}(x - x_j)\Delta_j^{(1)} \quad , \tag{5.287}$$

$$P_j^{+,(2)}(x) = P_j^{+,(1)} + (x-x_j)(x-x_{j+1})\left(\alpha_{1,2}^{+}\Delta_{j-1}^{(2)} + \alpha_{2,2}^{+}\Delta_j^{(2)}\right) \quad , \tag{5.288}$$

$$\ldots = \ldots$$

where $P_j^{+,(k)}(x)$ is the kth-order interpolant, $\Delta_j^{(p)}$ the divided difference of degree p of the variable, and $\alpha_{m,n}^{+}$ some weigthing parameters. Right-hand-sides are defined in the same way, except that $P^{-,(0)}(x)_j = u_{j+1}$ and weights are noted $\alpha_{m,n}^{-}$. A kth-order interpolation is obtained under the following constraints:

$$\sum_m \alpha_{m,n}^{\pm} = 1, \quad \alpha_{m,n}^{\pm} > 0, \quad n = 1,.., k-1 \quad . \tag{5.289}$$

Applying this procedure to (5.283), the numerical convection term can be expressed as

$$\frac{\partial F(u)}{\partial x} \approx \frac{1}{\Delta x}\left[f_{j+1/2}(x_{j+1/2}) - f_{j-1/2}(x_{j-1/2})\right] \quad , \tag{5.290}$$

where $f_{j\pm 1/2}$ is a numerical flux function. Adams' analysis is based on the local Lax–Friedrichs flux:

$$f_{j+1/2}(x) = \left[f(P_j^+(x)) + f(P_{j+1}^-(x))\right] - \beta_{j+1/2}(P_{j+1}^-(x) - P_j^+(x)) \quad , \tag{5.291}$$

with

$$\beta_{j+1/2} = \max_{u_j, u_{j+1}} |f'(u)| \quad . \tag{5.292}$$

In the simplified case of the Burgers equation, i.e. $F(u) = u^2/2$, the leading error term is:

$$\begin{aligned} \mathcal{E} &= \left(\frac{1}{8} - \frac{\gamma_1}{16}\right) \Delta x^2 \left(\overline{u}_j \frac{\partial^3 \overline{u}_j}{\partial x^3} + \frac{\partial \overline{u}_j}{\partial x}\frac{\partial^2 \overline{u}_j}{\partial x^2}\right) \\ &\quad + \frac{1}{8}(\beta_{j+1/2}\delta_1 - \beta_{j-1/2}\delta_2)\Delta x^2 \frac{\partial^3 \overline{u}_j}{\partial x^3} \\ &\quad + (\beta_{j+1/2} - \beta_{j-1/2})\gamma_2 \Delta x \frac{\partial^2 \overline{u}_j}{\partial x^2} \quad , \end{aligned} \tag{5.293}$$

where the coefficients are defined as

$$\gamma_1 = (\alpha_{1,2}^+ + \alpha_{2,2}^+ + \alpha_{1,2}^- + \alpha_{2,2}^-), \quad \gamma_2 = (\alpha_{1,2}^+ + \alpha_{2,2}^+ - \alpha_{1,2}^- - \alpha_{2,2}^-) \quad , \tag{5.294}$$

$$\delta_1 = (\alpha_{2,2}^- - \alpha_{2,2}^+), \quad \delta_2 = (\alpha_{1,2}^- - \alpha_{1,2}^+) \quad . \tag{5.295}$$

Subgrid-viscosity models can be recovered by chosing adequately the values of the constants appearing in (5.293). The Smagorinsky model with length scale $\overline{\Delta} = \Delta x$ and constant C_S is obtained by taking:

$$\gamma_1 = 2, \ \ \gamma_2 = C_S, \ \ \delta_1 = \delta_2 = 0, \ \ \beta_{j\pm 1/2} = \pm|\overline{u}_{j+1/2} - \overline{u}_{j-1/2}| \quad . \tag{5.296}$$

Variational Schemes with Embedded Subgrid Stabilization. We now present finite element methods with some built-in subgrid stabilization [280, 79, 144, 299, 623, 331, 332, 336]. The presentation will be limited to the main ideas for a simple linear advection–diffusion equation. The reader is referred to original articles for detailed mathematical results and extension to Navier–Stokes equations. These methods are all based on the variational formulation of the problem. For a passive scalar ϕ, we have:

$$\begin{aligned} \int_\Omega \frac{\partial \phi}{\partial t}\psi dV + \int_\Omega u\frac{\partial \phi}{\partial x}\psi dV &= \nu\int_\Omega \frac{\partial^2 \phi}{\partial x^2}\psi dV + \int_\Omega f\psi dV \\ &= -\nu\int_\Omega \frac{\partial \phi}{\partial x}\frac{\partial \psi}{\partial x} dV + \int_\Omega f\psi dV \end{aligned} \quad (5.297)$$

where Ω is the fluid domain, ψ a weighting function, u the advection velocity and f a source term. Boundary terms are assumed to vanish for the sake of simplicity. Let $\mathcal{L}$ be the time-dependent advection–diffusion operator:

$$\mathcal{L} = \frac{\partial}{\partial t} + u\frac{\partial}{\partial x} - \nu\frac{\partial}{\partial x^2} \quad . \quad (5.298)$$

Using this operator, (5.297) can be recast under the symbolic compact form

$$(\psi, \mathcal{L}\phi)_\Omega = (\mathcal{L}^*\psi, \phi)_\Omega = a(\psi, \phi) = (\psi, f)_\Omega \quad , \quad (5.299)$$

where $(.,.)_\Omega$ is a scalar product, $a(.,.)$ the bilinear form deduced from the preceding equations and $\mathcal{L}^*$ the adjoint operator:

$$\mathcal{L}^* = -\frac{\partial}{\partial t} - u\frac{\partial}{\partial x} - \nu\frac{\partial}{\partial x^2} \quad . \quad (5.300)$$

We now split the trial and weighting functions as the sum of a resolved and a subgrid function, i.e. $\phi = \overline{\phi} + \phi'$ and $\psi = \overline{\psi} + \psi'$. Inserting these decompositions into relation (5.299), we obtain

$$a(\psi, \phi) = a(\overline{\psi} + \psi', \overline{\phi} + \phi') = (\overline{\psi} + \psi', f)_\Omega \quad , \quad (5.301)$$

and, assuming that $\overline{\psi}$ and ψ' are linearly independent, we get the two following subproblems:

$$a(\overline{\psi}, \overline{\phi}) + a(\overline{\psi}, \phi') = (\overline{\psi}, f)_\Omega \quad , \quad (5.302)$$

and

$$a(\psi', \overline{\phi}) + a(\psi', \phi') = (\psi', f)_\Omega \quad , \quad (5.303)$$

or, equivalently,

$$a(\overline{\psi}, \overline{\phi}) + (\mathcal{L}^*\overline{\psi}, \phi')_\Omega = (\overline{\psi}, f)_\Omega \quad , \quad (5.304)$$

and

$$(\psi', \mathcal{L}\overline{\phi})_\Omega + (\psi', \mathcal{L}\phi')_\Omega = (\psi', f)_\Omega \quad . \quad (5.305)$$

A first solution consists of discretizing (5.304) and (5.305) using standard shape functions for the resolved scales and oscillatory *bubble* functions for the

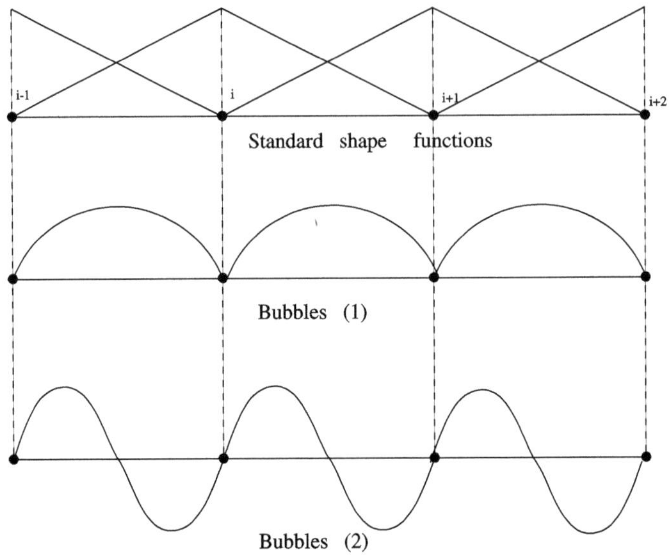

Fig. 5.20. Schematic of the embedded subgrid stabilization approach: linear finite element shape functions in one dimension plus typical bubbles

subgrid scales (see Fig. 5.20). The resulting method is a two-scale method, with embedded subgrid stabilization. It is important to note that degrees of freedom associated with bubble functions are eliminated by static condensation, i.e. are expressed as functions of the resolved scales, and do not require the solution of additional evolution equations.

Hughes and Stewart [336] proposed regularizing (5.304), which governs the motion of resolved scales, yielding

$$a(\overline{\psi},\overline{\phi}) + (\mathcal{L}^*\overline{\psi}, M(\mathcal{L}\overline{\phi} - f))_\Omega = (\overline{\psi}, f)_\Omega \quad , \tag{5.306}$$

where $(\mathcal{L}\overline{\phi} - f)$ is the residual of the resolved scales and M an operator originating from an elliptic regularization, which can be evaluated using bubble functions.

Usual stabilized methods also rely on the regularization of (5.304) without considering the subgrid scale equation. A general form of the stabilized problem is

$$a(\overline{\psi},\overline{\phi}) + (\mathbb{L}\overline{\psi}, \tau_{\text{stab}}(\mathcal{L}\overline{\phi} - f))_\Omega = (\overline{\psi}, f)_\Omega \quad , \tag{5.307}$$

where, typically, $\mathbb{L}$ is a differential operator and τ_{stab} is an algebraic operator which approximates the integral operator $(-M)$ of (5.306).

Classical examples are:

– Standard Galerkin method, which does not introduce any stabilizing term:

$$\mathbb{L} = 0 \quad . \tag{5.308}$$

– Streamwise upwind Petrov–Galerkin method, which introduces a stabilizing term based on the advection operator:

$$\mathbb{L} = u\frac{\partial}{\partial x} \quad . \tag{5.309}$$

– Galerkin/least-squares method, which extends the preceding method by including the whole differential operator:

$$\mathbb{L} = \mathcal{L} \quad . \tag{5.310}$$

– Subgrid Stabilization (Bubbles), which recovers the method of Hughes:

$$\mathbb{L} = -\mathcal{L}^* \quad . \tag{5.311}$$

The amount of numerical dissipation is governed by the parameter τ_{stab}, which can assume either tensorial or scalar expression. Many definitions can be found, most of them yielding $|\tau_{\text{stab}}| \propto \Delta x^2$, which is the right scale for a subgrid dissipation. Results dealing with the tuning of this parameter for turbulent flow simulations are almost nonexistent.

Spectral Vanishing Viscosities. Karniadakis et al. [379, 395] propose to adapt the Tadmor spectral viscosity [700] for large-eddy simulation purpose. This approach will be presented using a simplified non-linear conservation law for the sake of clarity. Considering the model Burgers equation

$$\frac{\partial u}{\partial t} + \frac{\partial}{\partial x}\left(\frac{u^2}{2}\right) = 0 \quad , \tag{5.312}$$

which can develop singularities, Tadmor proposes to regularize it for numerical purpose as

$$\frac{\partial \overline{u}}{\partial t} + \frac{\partial}{\partial x}\left(\frac{\overline{u}^2}{2}\right) = \epsilon\frac{\partial}{\partial x}\left(Q_\epsilon \frac{\partial \overline{u}}{\partial x}\right) \quad , \tag{5.313}$$

where ϵ, Q_ϵ and $\overline{u}$ are the articial viscosity parameter, the artificial viscosity kernel and the regularized solution (interpreted as the resolved field in large-eddy simulation), respectively. The original formulation of the regularization proposed by Tadmor is expressed in the spectral space as

$$\epsilon\frac{\partial}{\partial x}\left(Q_\epsilon \frac{\partial \overline{u}}{\partial x}\right) = -\epsilon \sum_{M \leq |k| \leq N} k^2 \widehat{Q}(k) \widehat{\overline{u}}_k e^{ikx} \quad , \tag{5.314}$$

where k is the wave number, N the number of Fourier modes, and M the wave number above which the artificial viscosity is activated. Several forms for the viscosity kernel have been suggested, among which the continuous kernel of

Maday [475] for pseudo-spectral methods based on Legendre polynomials:

$$\widehat{Q}(k) = \exp\left(-\frac{(k-N)^2}{(k-M)^2}\right), \quad k > M \quad , \tag{5.315}$$

with $M \simeq 5\sqrt{N}$ and $\epsilon \sim 1/N$. A major difference with spectral functional model in Fourier space is that Tadmor-type regularizations vanish at low wave numbers, while functional subgrid viscosities don't. The extension of the method to multidimensional curvilinear grids has been extensivly studied by Pasquetti and Xu [773, 582, 581].

Karniadakis et al. [395] further modify this model by defining a dynamic version of the artificial viscosity in which the parameter ϵ is tuned regarding the local state of the flow. To this end, the regularization term is redefined as

$$c(x,t) Q \frac{\partial^2 Q \overline{u}}{\partial x^2} \quad , \tag{5.316}$$

where the self-adaptive amplitude parameter $c(x,t)$ can be computed considering either the gradient of the solution

$$c(x,t) = \frac{\kappa}{N} \frac{|\nabla \overline{u}|}{\|\nabla \overline{u}\|_\infty} \quad , \tag{5.317}$$

where κ is an adjustable arbitrary parameter, or the strain tensor

$$c(x,t) = \frac{|\overline{S}|}{\|\overline{S}\|_\infty} \quad . \tag{5.318}$$

To prevent a too high dissipation near solid walls, Kirby and Karniadakis multiply Q by a damping function

$$g(y^+) = \frac{2}{\pi} \tan^{-1}\left(\frac{2ky^+}{\pi}\right) \left[1 - \exp\left(-\frac{y^+}{C}\right)\right]^2 \quad , \tag{5.319}$$

where y is he distance to the wall, the superscript $+$ refers to quantities expressed in wall units and C is a parameter.

High-Order Filtered Methods. Visbal and Rizetta define [734] another procedure to perform large-eddy simulation based on numerical stabilization without explicit physical subgrid model. Their procedure is based on the application, at the end of each time, of an high-order low-pass filter to the solution. This stabilizing procedure and the associated corresponding method originating in the works by Visbal and Gaitonde is discussed in references given therein. In practice, they use a symmetric compact finite difference filter with the followng properties:

1. it is non-dispersive, i.e. it is strictly dissipative,
2. it does not amplify any waves,
3. it preserves constant functions,
4. it completely eliminates the odd-even mode.

A tenth-order compact filter is observed to yield satisfactory results in simple cases (decaying isotropic turbulence). It is important to notice that this procedure is formally equivalent to the filtering-form of the full deconvolution procedure proposed by Mathew et al. [499] (see p. 220 for details). Therefore, this implicit procedure can be completely rewritten within the structural modeling framework. Other authors [63] develop similar strategies, based on the use of very-high order accurate finite difference schemes.

5.4 Modeling the Backward Energy Cascade Process

5.4.1 Preliminary Remarks

The above models reflect only the forward cascade process, i.e. the dominant average effect of the subgrid scales. The second energy transfer mechanism, the backward energy cascade, is much less often taken into account in simulations. We may mention two reasons for this. Firstly, the intensity of this return is very weak compared with that of the forward cascade toward the small scales (at least on the average in the isotropic homogeneous case) and its role in the flow dynamics is still very poorly understood. Secondly, modeling it requires the addition of an energy source term to the equations being computed, which is potentially a generator of numerical problems.

Two methods are used for modeling the backward energy cascade:

- Adding a stochastic forcing term constructed from random variables and the information contained in the resolved field. This approach makes it possible to include a random character of the subgrid scales, and each simulation can be considered a particular realization. The space-time correlations characteristic of the scales originating the backward cascade cannot be represented by this approach, though, which limits its physical representativeness.
- Modifying the viscosity associated with the forward cascade mechanism defined in the previous section, so as to take the energy injected at the large scales into account. The backward cascade is then represented by a negative viscosity, which is added to that of the cascade model. This approach is statistical and deterministic, and also subject to caution because it is not based on a physical description of the backward cascade phenomenon and, in particular, possesses no spectral distribution in k^4 predicted by the analytical theories like EDQNM (see also footnote p. 104). Its advantage resides mainly in the fact that it allows a reduction of the total dissipation

of the simulation, which is generally too high. Certain dynamic procedures for automatically computing the constants can generate negative values of them, inducing an energy injection in the resolved field. This property is sometimes interpreted as the capacity of the dynamic procedure to reflect the backward cascade process. This approach can therefore be classed in the category of statistical deterministic backward cascade models.

Representing the backward cascade by way of a negative viscosity is controversial because the theoretical analyses, such as by the EDQNM model, distinguish very clearly between the cascade and backward cascade terms, both in their intensity and in their mathematical form [443, 442]. This representation is therefore to be linked to other statistical deterministic descriptions of the backward cascade, which take into account only an average reduction of the effective viscosity, such as the Chollet–Lesieur effective viscosity spectral model.

The main backward cascade models belonging to these two categories are described in the following.

5.4.2 Deterministic Statistical Models

This section describes the deterministic models for the backward cascade. These models, which are based on a modification of the subgrid viscosity associated with the forward cascade process, are:

1. The spectral model based on the theories of turbulence proposed by Chasnov (p. 172). A negative subgrid viscosity is computed directly from the EDQNM theory. No hypothesis is adopted concerning the spectrum shape of the resolved scales, so that the spectral disequilibrium mechanisms can be taken into account at the level of these scales, but the spectrum shape of the subgrid scales is set arbitrarily. Also, the filter is assumed to be of the sharp cutoff type.
2. The dynamic model with an equation for the subgrid kinetic energy (p. 173), to make sure this energy remains positive. This ensures that the backward cascade process is represented physically, in the sense that a limited quantity of energy can be restored to the resolved scales by the subgrid modes. However, this approach does not allow a correct representation of the spectral distribution of the backward cascade. Only the quantity of restored energy is controlled.

Chasnov's Spectral Model. Chasnov [120] adds a model for the backward cascade, also based on an EDQNM analysis, to the forward cascade model already described (see Sect. 5.3.1). The backward cascade process is represented deterministically by a negative effective viscosity term $\nu_e^-(k|k_c)$, which is of the form:

$$\nu_e^-(k|k_c, t) = -\frac{F^-(k|k_c, t)}{2k^2 E(k, t)} \quad . \tag{5.320}$$

The stochastic forcing term is computed as:

$$F^{-}(k|k_{\rm c},t)=\int_{k_{\rm c}}^{\infty}dp\int_{p-k}^{p}dq\Theta_{kpq}\frac{k^{3}}{pq}(1-2x^{2}z^{2}-xyz)E(q,t)E(p,t), \quad (5.321)$$

in which x, y, and z are geometric factors associated with the triad $(\boldsymbol{k},\boldsymbol{p},\boldsymbol{q})$, and Θ_{kpq} is a relaxation time described in Appendix B. As is done when computing the draining term (see Chasnov's effective viscosity model in Sect. 5.3.1), we assume that the spectrum takes the Kolmogorov form beyond the cutoff $k_{\rm c}$. To simplify the computations, formula (5.321) is not used for wave numbers $k_{\rm c} \leq p \leq 3k_{\rm c}$. For the other wave numbers, we use the asymptotic form

$$F^{-}(k|k_{\rm c},t)=\frac{14}{15}k^{4}\int_{k_{\rm c}}^{\infty}dp\Theta_{kpp}(t)\frac{E^{2}(p,t)}{p^{2}} \quad . \quad (5.322)$$

This expression complete Chasnov's spectral subgrid model which, though quite close to the Kraichnan type effective viscosity models, makes it possible to take into account the backward cascade effects that are dominant for very small wave numbers.

Localized Dynamic Model with Energy Equation. The Germano–Lilly dynamic procedure and the localized dynamic procedure lead to the definition of subgrid models that raise numerical stability problems because the model constant can take negative values over long time intervals, leading to exponential growth of the disturbances.

This excessive duration of the dynamic constant in the negative state corresponds to too large a return of kinetic energy toward the large scales [101]. This phenomenon can be interpreted as a violation of the spectrum realizability constraint: when the backward cascade is over-estimated, a negative kinetic energy is implicitly defined in the subgrid scales. A simple idea for limiting the backward cascade consists in guaranteeing spectrum realizability[32]. The subgrid scales cannot then restore more energy than they contain. To verify this constraint, local information is needed on the subgrid kinetic energy, which naturally means defining this as an additional variable in the simulation.

A localized dynamic model including an energy equation is proposed by Ghosal et al. [261]. Similar models have been proposed independently by Ronchi et al. [621, 511], Wong [765] and Kim and Menon [391, 392, 583]. The subgrid model used is based on the kinetic energy of the subgrid modes. Using the same notation as in Sect. (5.3.3), we get:

$$\alpha_{ij}=-2\widetilde{\overline{\Delta}}\sqrt{Q_{\rm sgs}^{2}}\widetilde{\overline{S}}_{ij} \quad , \quad (5.323)$$

$$\beta_{ij}=-2\overline{\Delta}\sqrt{q_{\rm sgs}^{2}}\overline{S}_{ij} \quad , \quad (5.324)$$

[32] The spectrum $E(k)$ is said to be realizable if $E(k)\geq 0$, $\forall k$.

in which the energies $Q^2_{\rm sgs}$ and $q^2_{\rm sgs}$ are defined as:

$$Q^2_{\rm sgs} = \frac{1}{2}\left(\widetilde{\overline{u_i u_i}} - \widetilde{\overline{u}}_i \widetilde{\overline{u}}_i\right) = \frac{1}{2} T_{ii} \quad , \tag{5.325}$$

$$q^2_{\rm sgs} = \frac{1}{2}\left(\overline{u_i u_i} - \overline{u}_i \overline{u}_i\right) = \frac{1}{2} \tau_{ii} \quad . \tag{5.326}$$

Germano's identity (5.138) is written:

$$Q^2_{\rm sgs} = \widetilde{q^2_{\rm sgs}} + \frac{1}{2} L_{ii} \quad . \tag{5.327}$$

The model is completed by calculating $q^2_{\rm sgs}$ by means of an additional evolution equation. We use the equation already used by Schumann, Horiuti, and Yoshizawa, among others (see Sect. 5.3.1):

$$\begin{aligned} \frac{\partial q^2_{\rm sgs}}{\partial t} + \frac{\partial \overline{u}_j q^2_{\rm sgs}}{\partial x_j} &= -\tau_{ij}\overline{S}_{ij} - C_1 \frac{\left(q^2_{\rm sgs}\right)^{3/2}}{\overline{\Delta}} \\ &+ C_2 \frac{\partial}{\partial x_j}\left(\overline{\Delta}\sqrt{q^2_{\rm sgs}}\frac{\partial q^2_{\rm sgs}}{\partial x_j}\right) + \nu \frac{\partial^2 q^2_{\rm sgs}}{\partial x_j \partial x_j}, \end{aligned} \tag{5.328}$$

in which the constants C_1 and C_2 are computed by a constrained localized dynamic procedure described above. The dynamic constant $C_{\rm d}$ is computed by a localized dynamic procedure.

This model ensures that the kinetic energy $q^2_{\rm sgs}$ will remain positive, i.e. that the subgrid scale spectrum will be realizable. This property ensures that the dynamic constant cannot remain negative too long and thereby destabilize the simulation. However, finer analysis shows that the realizability conditions concerning the subgrid tensor τ (see Sect. 3.3.5) are verified only on the condition:

$$-\frac{\sqrt{q^2_{\rm sgs}}}{3\overline{\Delta}|s_\gamma|} \le C_{\rm d} \le \frac{\sqrt{q^2_{\rm sgs}}}{3\overline{\Delta} s_\alpha} \quad , \tag{5.329}$$

where s_α and s_γ are, respectively, the largest and smallest eigenvalues of the strain rate tensor $\overline{S}$. The model proposed therefore does not ensure the realizability of the subgrid tensor.

The two constants C_1 and C_2 are computed using an extension of the constrained localized dynamic procedure. To do this, we express the kinetic energy $Q^2_{\rm sgs}$ evolution equation as:

$$\begin{aligned} \frac{\partial Q^2_{\rm sgs}}{\partial t} + \frac{\partial \widetilde{\overline{u}}_j Q^2_{\rm sgs}}{\partial x_j} &= -T_{ij}\widetilde{\overline{S}}_{ij} - C_1 \frac{\left(Q^2_{\rm sgs}\right)^{3/2}}{\widetilde{\overline{\Delta}}} \\ &+ \; C_2 \frac{\partial}{\partial x_j}\left(\sqrt{Q^2_{\rm sgs}}\frac{\partial Q^2_{\rm sgs}}{\partial x_j}\right) + \nu \frac{\partial^2 Q^2_{\rm sgs}}{\partial x_j \partial x_j} \; . \end{aligned} \tag{5.330}$$

One variant of the Germano's relation relates the subgrid kinetic energy flux f_j to its analog at the level of the test filter F_j:

$$F_j - \widetilde{f}_j = Z_j \equiv \widetilde{\overline{u}}_j(\overline{p} + \widetilde{q^2_{\rm sgs} + \overline{u}_i\overline{u}_i/2}) - \overline{u}_j(\overline{p} + \widetilde{q^2_{\rm sgs} + \overline{u}_i\overline{u}_i/2}) \quad , \qquad (5.331)$$

in which $\overline{p}$ is the resolved pressure.

To determine the constant C_2, we substitute in this relation the modeled fluxes:

$$f_j = C_2\overline{\Delta}\sqrt{q^2_{\rm sgs}}\frac{\partial q^2_{\rm sgs}}{\partial x_j} \quad , \qquad (5.332)$$

$$F_j = C_2\widetilde{\overline{\Delta}}\sqrt{Q^2_{\rm sgs}}\frac{\partial Q^2_{\rm sgs}}{\partial x_j} \quad , \qquad (5.333)$$

which leads to:

$$Z_j = X_jC_2 - \widetilde{Y_jC_2} \quad , \qquad (5.334)$$

in which

$$X_j = \widetilde{\overline{\Delta}}\sqrt{Q^2_{\rm sgs}}\frac{\partial Q^2_{\rm sgs}}{\partial x_j} \quad , \qquad (5.335)$$

$$Y_j = \overline{\Delta}\sqrt{q^2_{\rm sgs}}\frac{\partial q^2_{\rm sgs}}{\partial x_j} \quad . \qquad (5.336)$$

Using the same method as was explained for the localized dynamic procedure, the constant C_2 is evaluated by minimizing the quantity:

$$\int \left(Z_j - X_jC_2 + \widetilde{Y_jC_2}\right)\left(Z_j - X_jC_2 + \widetilde{Y_jC_2}\right) \quad . \qquad (5.337)$$

By analogy with the preceding developments, the solution is obtained in the form:

$$C_2(\boldsymbol{x}) = \left[f_{C_2}(\boldsymbol{x}) + \int \mathcal{K}_{C_2}(\boldsymbol{x},\boldsymbol{y})C_2(\boldsymbol{y})d^3\boldsymbol{y}\right]_+ \quad , \qquad (5.338)$$

in which:

$$f_{C_2}(\boldsymbol{x}) = \frac{1}{X_j(\boldsymbol{x})X_j(\boldsymbol{x})}\left(X_j(\boldsymbol{x})Z_j(\boldsymbol{x}) - Y_j(\boldsymbol{x})\int Z_j(\boldsymbol{y})G(\boldsymbol{x}-\boldsymbol{y})d^3\boldsymbol{y}\right) \quad , \qquad (5.339)$$

$$\mathcal{K}_{C_2}(\boldsymbol{x},\boldsymbol{y}) = \frac{\mathcal{K}^{C_2}_{\mathcal{A}}(\boldsymbol{x},\boldsymbol{y}) + \mathcal{K}^{C_2}_{\mathcal{A}}(\boldsymbol{y},\boldsymbol{x}) - \mathcal{K}^{C_2}_{\mathcal{S}}(\boldsymbol{x},\boldsymbol{y})}{X_j(\boldsymbol{x})X_j(\boldsymbol{x})} \quad , \qquad (5.340)$$

in which

$$\mathcal{K}_{\mathcal{A}}^{C_2}(\boldsymbol{x},\boldsymbol{y}) = X_j(\boldsymbol{x})Y_j(\boldsymbol{y})G(\boldsymbol{x}-\boldsymbol{y}) \quad , \tag{5.341}$$

$$\mathcal{K}_{\mathcal{S}}^{C_2}(\boldsymbol{x},\boldsymbol{y}) = Y_j(\boldsymbol{x})Y_j(\boldsymbol{y}) \int G(\boldsymbol{z}-\boldsymbol{x})G(\boldsymbol{z}-\boldsymbol{y})d^3\boldsymbol{z} \quad . \tag{5.342}$$

This completes the computation of constant C_2. To determine the constant C_1, we substitute (5.327) in (5.330) and get:

$$\frac{\partial \widetilde{q^2_{\text{sgs}}}}{\partial t} + \frac{\partial \widetilde{\overline{u}}_j \widetilde{q^2_{\text{sgs}}}}{\partial x_j} = -E \frac{\partial F_j}{\partial x_j} + \nu \frac{\partial^2 \widetilde{q^2_{\text{sgs}}}}{\partial x_j \partial x_j} \quad , \tag{5.343}$$

in which E is defined as:

$$E = T_{ij}\widetilde{\overline{S}}_{ij} + \frac{C_1 (Q^2_{\text{sgs}})^{3/2}}{\widetilde{\overline{\Delta}}} - \nu \frac{1}{2}\frac{\partial^2 L_{ii}}{\partial x_j \partial x_j} + \frac{1}{2}\left(\frac{\partial L_{ii}}{\partial t} + \frac{\partial \widetilde{\overline{u}}_j L_{ii}}{\partial x_j}\right) \quad . \tag{5.344}$$

Applying the test filter to relation (5.328), we get:

$$\frac{\partial \widetilde{q^2_{\text{sgs}}}}{\partial t} + \frac{\partial \widetilde{\overline{u}}_j \widetilde{q^2_{\text{sgs}}}}{\partial x_j} = -\widetilde{\tau_{ij}\overline{S}_{ij}} - \left(C_1 \frac{\widetilde{(q^2_{\text{sgs}})^{3/2}}}{\overline{\Delta}}\right) + \frac{\partial \widetilde{f}_j}{\partial x_j} + \nu \frac{\partial^2 \widetilde{q^2_{\text{sgs}}}}{\partial x_j \partial x_j} \quad . \tag{5.345}$$

By eliminating the term $\partial \widetilde{q^2_{\text{sgs}}}/\partial t$ between relations (5.343) and (5.345), then replacing the quantity $F_j - \widetilde{f}_j$ by its expression (5.331) and the quantity T_{ij} by its value as provided by the Germano identity, we get:

$$\chi = \phi C_1 - \widetilde{\psi C_1} \quad , \tag{5.346}$$

in which

$$\chi = \widetilde{\tau_{ij}\overline{S}_{ij}} - \widetilde{\tau}_{ij}\widetilde{\overline{S}}_{ij} - L_{ij}\widetilde{\overline{S}}_{ij} + \frac{\partial \rho_j}{\partial x_j} - \frac{1}{2}D_t L_{ii} + \frac{1}{2}\nu \frac{\partial^2 L_{ii}}{\partial x_j \partial x_j} \quad , \tag{5.347}$$

$$\phi = (Q^2_{\text{sgs}})^{3/2}/\widetilde{\overline{\Delta}} \quad , \tag{5.348}$$

$$\psi = (q^2_{\text{sgs}})^{3/2}/\overline{\Delta} \quad , \tag{5.349}$$

and

$$\rho_j = \widetilde{\overline{u}}_j(\overline{p} + \widetilde{\overline{u}_i\overline{u}_i}/2) - \overline{u}_j(\widetilde{\overline{p} + \overline{u}_i\overline{u}_i}/2) \quad . \tag{5.350}$$

The symbol D_t designates the material derivative $\partial/\partial t + \widetilde{\overline{u}}_j \partial/\partial x_j$. The constant C_1 is computed by minimizing the quantity

$$\int \left(\chi - \phi C_1 + \widetilde{\psi C_1}\right)\left(\chi - \phi C_1 + \widetilde{\psi C_1}\right) \quad , \tag{5.351}$$

by a constrained localized dynamic procedure, which is written:

$$C_1(\boldsymbol{x}) = \left[f_{C_1}(\boldsymbol{x}) + \int \mathcal{K}_{C_1}(\boldsymbol{x}, \boldsymbol{y}) C_1(\boldsymbol{y}) d^3\boldsymbol{y} \right]_+ \quad , \tag{5.352}$$

in which

$$f_{C_1}(\boldsymbol{x}) = \frac{1}{\phi(\boldsymbol{x})\phi(\boldsymbol{x})} \left(\phi(\boldsymbol{x})\chi(\boldsymbol{x}) - \psi(\boldsymbol{x}) \int \chi(\boldsymbol{y}) G(\boldsymbol{x} - \boldsymbol{y}) d^3\boldsymbol{y} \right) \quad , \tag{5.353}$$

$$\mathcal{K}_{C_1}(\boldsymbol{x}, \boldsymbol{y}) = \frac{\mathcal{K}_{\mathcal{A}}^{C_1}(\boldsymbol{x}, \boldsymbol{y}) + \mathcal{K}_{\mathcal{A}}^{C_1}(\boldsymbol{y}, \boldsymbol{x}) - \mathcal{K}_{\mathcal{S}}^{C_1}(\boldsymbol{x}, \boldsymbol{y})}{\phi(\boldsymbol{x})\phi(\boldsymbol{x})} \quad , \tag{5.354}$$

in which

$$\mathcal{K}_{\mathcal{A}}^{C_1}(\boldsymbol{x}, \boldsymbol{y}) = \phi(\boldsymbol{x})\psi(\boldsymbol{y}) G(\boldsymbol{x} - \boldsymbol{y}) \quad , \tag{5.355}$$

$$\mathcal{K}_{\mathcal{S}}^{C_1}(\boldsymbol{x}, \boldsymbol{y}) = \psi(\boldsymbol{x})\psi(\boldsymbol{y}) \int G(\boldsymbol{z} - \boldsymbol{x}) G(\boldsymbol{z} - \boldsymbol{y}) d^3\boldsymbol{z} \quad , \tag{5.356}$$

which completes the computation of the constant C_1.

The version by Menon et al. [391, 392, 583], also extensively used by Davidson and his coworkers [576, 577] is much simpler as far as the practical implementation is addressed. This simplified formulation is defined as follows:

$$C_1 = \frac{\widetilde{\overline{\Delta}} \varepsilon_{\mathrm{test}}}{(Q_{\mathrm{sgs}}^2)^{3/2}} \quad , \tag{5.357}$$

and

$$C_2 = C_{\mathrm{d}} = \frac{1}{2} \frac{L_{ij}^{\mathrm{d}} \widetilde{\overline{S}}_{ij}}{\widetilde{\overline{S}}_{ij} \widetilde{\overline{S}}_{ij}} \quad , \tag{5.358}$$

where Q_{sgs}^2 is computed using relation (5.327) and the dissipation at the test filter level is evaluated using a scale-similarity hypothesis, yielding

$$\varepsilon_{\mathrm{test}} = (\nu + \nu_{\mathrm{sgs}}) \left[\left(\widetilde{\frac{\partial \overline{u}_i}{\partial x_j} \frac{\partial \overline{u}_i}{\partial x_j}} \right) - \left(\frac{\partial \widetilde{\overline{u}}_i}{\partial x_j} \frac{\partial \widetilde{\overline{u}}_i}{\partial x_j} \right) \right] \quad . \tag{5.359}$$

Since it is strictly local in the sense that no integral problem is involved, this new formulation is much less demanding than the previous one in terms of computational effort.

Another simplified local one-equation dynamic model was proposed by Fureby [231], which is defined by the following relations:

$$C_1 = \frac{\zeta m}{mm} \quad , \tag{5.360}$$

where

$$m = \frac{\left(Q_{\rm sgs}^2\right)^{3/2}}{\widetilde{\overline{\Delta}}} - \frac{\widetilde{\left(q_{\rm sgs}^2\right)^{3/2}}}{\overline{\Delta}} \quad , \tag{5.361}$$

$$\zeta = \widetilde{\tau_{ij}\overline{S}_{ij}} - T_{ij}\widetilde{\overline{S}}_{ij} - \frac{\partial}{\partial t}\left(\frac{1}{2}L_{kk}\right) - \frac{\partial}{\partial x_j}\left(\frac{1}{2}L_{kk}\widetilde{\overline{u}}_j\right) \quad . \tag{5.362}$$

The remaining parameter is computed as follows:

$$C_2 = C_{\rm d} = \frac{L_{ij}^{\rm d} M_{ij}}{M_{ij}M_{ij}} \quad , \tag{5.363}$$

where

$$M_{ij} = \frac{1}{2}\left(\alpha_{ij} - \widetilde{\beta_{ij}}\right) \quad . \tag{5.364}$$

A more complex model is proposed by Krajnovic and Davidson [406], who use a linear-combination model (see Sect. 7.4) to close both the momentum equations and the prognostic equation for the subgrid kinetic energy.

5.4.3 Stochastic Models

Models belonging to this category are based on introducing a random forcing term into the momentum equations. It should be noted that this random character does not reflect the space-time correlation scales of the subgrid fluctuations, which limits the physical validity of this approach and can raise numerical stability problems. It does, however, obtain forcing term formulations at low algorithmic cost. The models described here are:

1. Bertoglio's model in the spectral space (p. 179). The forcing term is constructed using a stochastic process, which is designed in order to induce the desired backward energy flux and to possess a finite correlation time scale. This is the only random model for the backward cascade derived in the spectral space.

2. Leith's model (p. 180). The forcing term is represented by an acceleration vector deriving from a vector potential, whose amplitude is evaluated by simple dimensional arguments. The backward cascade is completely decoupled from forward cascade here: there is no control on the realizability of the subgrid scales.
3. Mason–Thomson model (p. 182), which can be considered as an improvement of the preceding model. The evaluations of the vector potential amplitude and subgrid viscosity modeling the forward cascade are coupled, so as to ensure that the local equilibrium hypothesis is verified. This ensures that the subgrid kinetic energy remains positive.
4. Schumann model (p. 183), in which the backward cascade is represented not as a force deriving from a vector potential but rather as the divergence of a tensor constructed from a random solenoidal velocity field whose kinetic energy is equal to the subgrid kinetic energy.
5. Stochastic dynamic model (p. 184), which makes it possible to calculate the subgrid viscosity and a random forcing term simultaneously and dynamically. This coupling guarantees that the subgrid scales are realizable, but at the cost of a considerable increase in the algorithmic complexity of the model.

Bertoglio Model. Bertoglio and Mathieu [57, 58] propose a spectral stochastic subgrid model based on the EDQNM analysis. This model appears as a new source term $f_i(\boldsymbol{k},t)$ in the filtered mometum equations, and is evaluated as a stochastic process. The following constraints are enforced:

- f must not modify the velocity field incompressibility, i.e. $k_i f_i(\boldsymbol{k},t) = 0$;
- f will have a Gaussian probability density function;
- The correlation time of f, noted t_f, is finite;
- f must induce the desired effect on the statistical second-order moments of the resolved velocity field:

$$\langle f_i(\boldsymbol{k},t)\overline{\widehat{u}}_j^*(\boldsymbol{k},t) + f_j(\boldsymbol{k},t)\overline{\widehat{u}}_i(\boldsymbol{k},t)\rangle = T_{ij}^-(\boldsymbol{k},t)\left(\frac{2\pi}{L}\right)^3 \quad , \tag{5.365}$$

where $T_{ij}^-(\boldsymbol{k},t)$ is the exact backward transfer term appearing in the variation equation for $\langle\overline{\widehat{u}}_i(\boldsymbol{k},t)\overline{\widehat{u}}_j^*(\boldsymbol{k},t)\rangle$ and L the size of the computational domain in physical space.

Assuming that the response function of the simulated field is isotropic and independent of f, and that the time correlations exhibit an exponential decay, we get the following velocity-independent relation:

$$\langle f_i(\boldsymbol{k},t)f_j^*(\boldsymbol{k},t) + f_i^*(\boldsymbol{k},t)f_j(\boldsymbol{k},t)\rangle = T_{ij}^-(\boldsymbol{k},t)\left(\frac{2\pi}{L}\right)^3\left(\frac{1}{\theta(\boldsymbol{k},t)} + \frac{1}{t_f}\right) \quad , \tag{5.366}$$

where $\theta(\boldsymbol{k}, t)$ is a relaxation time evaluated from the resolved scales. We now have to compute the stochastic variable f_i. The authors propose the following algorithm, which is based on three random variables a, b and c:

$$\begin{aligned} f_1^{(n+1)} &= \left(1 - \frac{\Delta t}{t_f}\right) f_1^{(n)} + \sqrt{h_{11}^{(n)}} \beta_{11}^{(n+1)} \sqrt{\frac{\Delta t}{t_f}} \exp(\imath 2\pi a^{(n+1)}) \\ &+ \sqrt{h_{22}^{(n)}} \beta_{12}^{(n+1)} \sqrt{\frac{\Delta t}{t_f}} \exp(\imath 2\pi c^{(n+1)}) \quad , \end{aligned} \tag{5.367}$$

$$\begin{aligned} f_2^{(n+1)} &= \left(1 - \frac{\Delta t}{t_f}\right) f_2^{(n)} + \sqrt{h_{22}^{(n)}} \beta_{22}^{(n+1)} \sqrt{\frac{\Delta t}{t_f}} \exp(\imath 2\pi b^{(n+1)}) \\ &+ \sqrt{h_{11}^{(n)}} \beta_{21}^{(n+1)} \sqrt{\frac{\Delta t}{t_f}} \exp(\imath 2\pi c^{(n+1)}) \quad , \end{aligned} \tag{5.368}$$

where the superscript (n) denotes the value at the nth time step, Δt is the value of the time step, and $h_{ij}(\boldsymbol{k}, t) = \langle f_i(\boldsymbol{k}, t) f_j^*(\boldsymbol{k}, t)\rangle$. Moreover, we get the complementary set of equations, which close the system:

$$\begin{aligned} (\beta_{11}^{(n+1)})^2 &= \frac{1}{h_{11}^{(n)}} \left((h_{11}^{(n+1)} - h_{11}^{(n)}) \frac{t_f}{\Delta t} - h_{22}^{(n)} (\beta_{12}^{(n+1)})^2 \right) \\ &+2 - \frac{\Delta t}{t_f} \quad , \end{aligned} \tag{5.369}$$

$$\begin{aligned} (\beta_{22}^{(n+1)})^2 &= \frac{1}{h_{22}^{(n)}} \left((h_{22}^{(n+1)} - h_{22}^{(n)}) \frac{t_f}{\Delta t} - h_{11}^{(n)} (\beta_{21}^{(n+1)})^2 \right) \\ &+2 - \frac{\Delta t}{t_f} \quad , \end{aligned} \tag{5.370}$$

$$\begin{aligned} \beta_{12}^{(n+1)} \beta_{12}^{(n+1)} &= \frac{1}{\sqrt{h_{11}^{(n)} h_{11}^{(n)}}} \left((h_{12}^{(n+1)} - h_{12}^{(n)}) \frac{t_f}{\Delta t} \right. \\ &\left. + h_{12}^{(n)} \left(2 - \frac{\Delta t}{t_f}\right)\right) \quad , \end{aligned} \tag{5.371}$$

$$(\beta_{21}^{(n+1)})^2 = (\beta_{12}^{(n+1)})^2 \quad , \tag{5.372}$$

which completes the description of the model. The resulting random force satisfies all the cited constraints, but it requires the foreknowledge of the h_{ij} tensor. This tensor is evaluated using the EDQNM theory, which requires the spectrum of the subgrid scales to be known. To alleviate this problem, arbitrary form of the spectrum can be employed.

Leith Model. A stochastic backward cascade model expressed in the physical space was derived by Leith in 1990 [434]. This model takes the form of a random forcing term that is added to the momentum equations. This term is computed at each point in space and each time step with the introduction of a vector potential ϕ^{b} for the acceleration, in the form of a white isotropic noise in space and time. The random forcing term with null divergence $\boldsymbol{f}^{\mathrm{b}}$ is deduced from this vector potential.

We first assume that the space and time auto-correlation scales of the subgrid modes are small compared with the cutoff lengths in space $\overline{\Delta}$ and in time Δt associated with the filter[33]. This way, the subgrid modes appear to be de-correlated in space and time. The correlation at two points and two times of the vector potential ϕ^{b} is then expressed:

$$\langle \phi_i^{\mathrm{b}}(\boldsymbol{x},t)\phi_k^{\mathrm{b}}(\boldsymbol{x}',t')\rangle = \sigma(\boldsymbol{x},t)\delta(\boldsymbol{x}-\boldsymbol{x}')\delta(t-t')\delta_{ik} \quad , \tag{5.373}$$

in which σ is the variance. This is computed as:

$$\sigma(\boldsymbol{x},t) = \frac{1}{3}\int dt' \int d^3\boldsymbol{x}' \langle \phi_k^{\mathrm{b}}(\boldsymbol{x},t)\phi_k^{\mathrm{b}}(\boldsymbol{x}',t')\rangle \quad . \tag{5.374}$$

Simple dimensional reasoning shows that:

$$\sigma(\boldsymbol{x},t) \approx |\overline{S}|^3\overline{\Delta}^7 \quad . \tag{5.375}$$

Also, as the vector potential appears as a white noise in space and time at the fixed resolution level, the integral (5.374) is written:

$$\sigma(\boldsymbol{x},t) = \frac{1}{3}\langle \phi_k^{\mathrm{b}}(\boldsymbol{x},t)\phi_k^{\mathrm{b}}(\boldsymbol{x},t)\rangle\overline{\Delta}^3\Delta t \quad . \tag{5.376}$$

Considering relations (5.375) and (5.376), we get:

$$\langle \phi_k^{\mathrm{b}}(\boldsymbol{x},t)\phi_k^{\mathrm{b}}(\boldsymbol{x},t)\rangle \approx |\overline{S}|^3\overline{\Delta}^4\frac{1}{\Delta t} \quad . \tag{5.377}$$

The shape proposed for the kth component of the vector potential is:

$$\phi_k^{\mathrm{b}} = C_{\mathrm{b}}|\overline{S}|^{3/2}\overline{\Delta}^2\Delta t^{-1/2}g \quad , \tag{5.378}$$

in which C_{b} is a constant of the order of unity, Δt the simulation time cutoff length (i.e. the time step), and g the random Gaussian variable of zero average and variance equal to unity. The vector $\boldsymbol{f}^{\mathrm{b}}$ is then computed by taking the rotational of the vector potential, which guarantees that it is solenoidal.

In practice, Leith sets the value of the constant C_{b} at 0.4 and applies a spatial filter with a cutoff length of $2\overline{\Delta}$, so as to ensure better algorithm stability.

[33] We again find here a total scale separation hypothesis that is not verified in reality.

Mason–Thomson Model. A similar model is proposed by Mason and Thomson [498]. The difference from the Leith model resides in the scaling of the vector potential. By calling $\Delta_{\rm f}$ and $\overline{\Delta}$ the characteristic lengths of the subgrid scales and spatial filter, respectively, the variants of the resolved stresses due to the subgrid fluctuations is, if $\Delta_{\rm f} \ll \overline{\Delta}$, of the order of $(\Delta_{\rm f}/\overline{\Delta})^3 u_{\rm e}^4$, in which $u_{\rm e}$ is the characteristic subgrid velocity. The amplitude a of the fluctuations in the gradients of the stresses is:

$$a \approx \frac{\Delta_{\rm f}^{3/2}}{\overline{\Delta}^{5/2}} u_{\rm e}^2 \quad , \tag{5.379}$$

which is also the amplitude of the associated acceleration. The corresponding kinetic energy variation rate of the resolved scales, $q_{\rm r}^2$, is estimated as:

$$\frac{\partial q_{\rm r}^2}{\partial t} \approx a^2 t_{\rm e} \approx \frac{\Delta_{\rm f}^3}{\overline{\Delta}^5} u_{\rm e}^4 t_{\rm e} \quad , \tag{5.380}$$

in which $t_{\rm e}$ is the characteristic time of the subgrid scales. As $t_{\rm e} \approx \Delta_{\rm f}/u_{\rm e}$ and the dissipation rate is evaluated by dimensional arguments as $\varepsilon \approx u_{\rm e}^3/\Delta_{\rm f}$, we can say:

$$\frac{\partial q_{\rm r}^2}{\partial t} = C_{\rm b} \frac{\Delta_{\rm f}^5}{\overline{\Delta}^5} \varepsilon \quad . \tag{5.381}$$

The ratio $\Delta_{\rm f}/\overline{\Delta}$ is evaluated as the ratio of the subgrid scale mixing length to the filter cutoff length, and is thus equal to the constant of the subgrid viscosity models discussed in Sect. 5.3.2. Previous developments have shown that this constant is not unequivocally determinate, but that it is close to 0.2. The constant $C_{\rm b}$ is evaluated at 1.4 by an EDQNM analysis.

The dissipation rate that appears in equation (5.381) is evaluated in light of the backward cascade. The local subgrid scale equilibrium hypothesis is expressed by:

$$-\tau_{ij}\overline{S}_{ij} = \varepsilon + C_{\rm b} \frac{\Delta_{\rm f}^5}{\overline{\Delta}^5} \varepsilon \quad , \tag{5.382}$$

in which τ_{ij} is the subgrid tensor. The term on the left represents the subgrid kinetic energy production, the first term in the right-hand side the dissipation, and the last term the energy loss to the resolved scales by the backward cascade. The dissipation rate is evaluated using this last relation:

$$\varepsilon = \frac{-\tau_{ij}\overline{S}_{ij}}{1 + (\Delta_{\rm f}/\overline{\Delta})^5} \quad , \tag{5.383}$$

which completes the computation of the right-hand side of equation (5.381), with the tensor τ_{ij} being evaluated using a subgrid viscosity model.

This equation can be re-written as:

$$\frac{\partial q_{\mathrm{r}}^2}{\partial t} = \sigma_{\mathrm{a}}^2 \Delta t \quad , \tag{5.384}$$

in which σ_{a}^2 is the sum of the variances of the acceleration component amplitudes. From the equality of the two relations (5.381) and (5.384), we can say:

$$\sigma_{\mathrm{a}}^2 = C_{\mathrm{b}} \frac{\Delta_{\mathrm{f}}^5}{\overline{\Delta}^5} \frac{\varepsilon}{\Delta t} \quad . \tag{5.385}$$

The vector potential scaling factor a and σ_{a}^2 are related by:

$$a = \sqrt{\sigma_{\mathrm{a}}^2 \frac{\Delta t}{t_{\mathrm{e}}}} \quad . \tag{5.386}$$

To complete the model, we now have to evaluate the ratio of the subgrid scale characteristic time to the time resolution scale. This is done simply by evaluating the characteristic time t_{e} from the subgrid viscosity ν_{sgs} computed by the model used, to reflect the cascade:

$$t_{\mathrm{e}} = \frac{\Delta_{\mathrm{f}}^2}{\nu_{\mathrm{sgs}}} \quad , \tag{5.387}$$

which completes the description of the model, since the rest of the procedure is the same as what Leith defined.

Schumann Model. Schumann proposed a stochastic model for subgrid tensor fluctuations that originate the backward cascade of kinetic energy [654]. The subgrid tensor τ is represented as the sum of a turbulent viscosity model and a stochastic part R^{st}:

$$\tau_{ij} = \nu_{\mathrm{sgs}} \overline{S}_{ij} + \frac{2}{3} q_{\mathrm{sgs}}^2 \delta_{ij} + R_{ij}^{\mathrm{st}} \quad . \tag{5.388}$$

The average random stresses R_{ij}^{st} are zero:

$$\langle R_{ij}^{\mathrm{st}} \rangle = 0 \quad . \tag{5.389}$$

They are defined as:

$$R_{ij}^{\mathrm{st}} = \gamma_m \left(v_i v_j - \frac{2}{3} q_{\mathrm{sgs}}^2 \delta_{ij} \right) \quad , \tag{5.390}$$

in which γ_m is a parameter and v_i a random velocity. From dimensional arguments, we can define this as:

$$v_i = \sqrt{\frac{2 q_{\mathrm{sgs}}^2}{3}} g_i \quad , \tag{5.391}$$

in which g_i is a white random number in space and has a characteristic correlation time τ_v:

$$\langle g_i \rangle = 0 \quad , \tag{5.392}$$

$$\langle g_i(\boldsymbol{x},t) g_j(\boldsymbol{x}',t') \rangle = \delta_{ij}\delta(\boldsymbol{x}-\boldsymbol{x}') \exp(|t-t'|/\tau_v) \quad . \tag{5.393}$$

The v_i field is made solenoidal by applying a projection step. We note that the time scale τ_v is such that:

$$\tau_v \sqrt{q_{\rm sgs}^2}/\overline{\Delta} \approx 1 \quad . \tag{5.394}$$

The parameter γ_m determines the portion of random stresses that generate the backward cascade. Assuming that only the scales belonging to the interval $[k_{\rm c}, nk_{\rm c}]$ are active, for a spectrum of slope of $-m$ we get:

$$\gamma_m^2 = \frac{\displaystyle\int_{k_{\rm c}}^{nk_{\rm c}} k^{-2m} dk}{\displaystyle\int_{k_{\rm c}}^{\infty} k^{-2m} dk} = 1 - n^{1-2m} \quad . \tag{5.395}$$

For $n = 2$ and $m = 5/3$, we get $\gamma_m = 0.90$. The subgrid kinetic energy $q_{\rm sgs}^2$ is evaluated from the subgrid viscosity model.

Stochastic Localized Dynamic Model. A localized dynamic procedure including a stochastic forcing term was proposed by Carati et al. [101]. The contribution of the subgrid terms in the momentum equation appears here as the sum of a subgrid viscosity model, denoted $C_{\rm d}\beta_{ij}$ using the notation of Sect. 5.3.3, which models the energy cascade, and a forcing term denoted $\boldsymbol{f}$:

$$\frac{\partial \tau_{ij}}{\partial x_j} = \frac{\partial C_{\rm d}\beta_{ij}}{\partial x_j} + f_i \quad . \tag{5.396}$$

The β_{ij} term can be computed using any subgrid viscosity model. The force $\boldsymbol{f}$ is chosen in the form of a white noise in time with null divergence in space. The correlation of this term at two points in space and two times is therefore expressed:

$$\langle f_i(\boldsymbol{x},t) f_j(\boldsymbol{x}',t') \rangle = A^2(\boldsymbol{x},t) H_{ij}(\boldsymbol{x}-\boldsymbol{x}')\delta(t-t') \quad . \tag{5.397}$$

The statistical average here is an average over all the realizations of f conditioned by a given velocity field $\boldsymbol{u}(\boldsymbol{x},t)$. The factor A^2 is such that $H_{ii}(0) = 1$. Since a stochastic term has been introduced into the subgrid model, the residual E_{ij} on which the dynamic procedure for computing the constant $C_{\rm d}$ is founded also possesses a stochastic nature. This property will therefore be shared by the dynamically computed constant, which is not acceptable. To find the original properties of the dynamic constant, we take a statistical average of the residual, denoted $\langle E_{ij} \rangle$, which gets rid of the random terms. The constant of the subgrid viscosity model is computed by a localized dynamic procedure based on the statistical average of the residual, which is written:

$$\langle E_{ij} \rangle = L_{ij} + \widetilde{C_{\rm d}\beta_{ij}} - C_{\rm d}\alpha_{ij} \quad . \tag{5.398}$$

The amplitude of the random forcing term can also be computed dynamically. To bring out the non-zero contribution of the stochastic term in the statistical average, we base this new procedure on the resolved kinetic energy balance at the level of the test filter $Q_{\rm r}^2 = \widetilde{\overline{u}}_i \widetilde{\overline{u}}_i / 2$. The evolution equation of this quantity is obtained in two different forms (only the pertinent terms are detailed, the others are symbolized):

$$\frac{\partial Q_{\rm r}^2}{\partial t} = ... - \widetilde{\overline{u}}_i \frac{\partial}{\partial x_j} (C_{\rm d}\alpha_{ij} + P\delta_{ij}) + \mathcal{E}_F \quad , \tag{5.399}$$

$$\frac{\partial Q_{\rm r}^2}{\partial t} = ... - \widetilde{\overline{u}}_i \frac{\partial}{\partial x_j} \left(\widetilde{C_{\rm d}\beta_{ij}} + L_{ij} + \widetilde{p}\delta_{ij} \right) + \mathcal{E}_{\widetilde{f}} \quad . \tag{5.400}$$

The pressure terms P and p are in equilibrium with the velocity fields $\widetilde{\overline{\boldsymbol{u}}}$ and $\overline{\boldsymbol{u}}$, respectively. The quantities $\mathcal{E}_F$ and $\mathcal{E}_{\widetilde{f}}$ are the backward cascade energy injections associated, respectively, with the forcing term $\boldsymbol{F}$ computed directly at the level of the test filter, and with the forcing term $\widetilde{\boldsymbol{f}}$ computed at the first level and then filtered. The difference between equations (5.399) and (5.400) leads to:

$$Z \equiv \mathcal{E}_F - \mathcal{E}_{\widetilde{f}} - g \neq 0 \quad , \tag{5.401}$$

in which the fully known term g is of the form:

$$g = \widetilde{\overline{u}}_i \frac{\partial}{\partial x_j} \left(C_{\rm d}\alpha_{ij} + P\delta_{ij} - \widetilde{C_{\rm d}\beta_{ij}} - L_{ij} - \widetilde{p}\delta_{ij} \right) \quad . \tag{5.402}$$

The quantity Z plays a role for the kinetic energy that is analogous to the residual E_{ij} for the momentum. Minimizing the quantity

$$\mathcal{Z} = \int \langle Z \rangle^2 \tag{5.403}$$

can thus serve as a basis for defining a dynamic procedure for evaluating the stochastic forcing.

To go any further, the shape of the $\boldsymbol{f}$ term has to be specified. To simplify the use, we assume that the correlation length of $\boldsymbol{f}$ is small compared with the cutoff length $\overline{\Delta}$. The function $\boldsymbol{f}$ thus appears as de-correlated in space, which is reflected by:

$$\langle \mathcal{E}_{\rm f} \rangle = \frac{1}{2} A^2(\boldsymbol{x}, t) \quad . \tag{5.404}$$

In order to be able to calculate $\mathcal{E}_{\mathrm{f}}$ dynamically, we assume that the backward cascade is of equal intensity at the two filtering levels considered, i.e.

$$\langle \mathcal{E}_{\mathrm{f}} \rangle = \langle \mathcal{E}_F \rangle \quad . \tag{5.405}$$

Also, since $\boldsymbol{f}$ is de-correlated at the $\widetilde{\overline{\Delta}}$ scale, we assume:

$$\mathcal{E}_{\widetilde{f}} \ll \langle \mathcal{E}_{\mathrm{f}} \rangle = \langle \mathcal{E}_F \rangle \quad , \tag{5.406}$$

which makes it possible to change relation (5.401) to become

$$\langle Z \rangle = \langle \mathcal{E}_F \rangle - g \quad . \tag{5.407}$$

We now choose $\boldsymbol{f}$ in the form:

$$f_i = P_{ij}(\mathcal{A} e_j) \quad , \tag{5.408}$$

in which e_j is a random isotropic Gaussian function, $\mathcal{A}$ a dimensioned constant that will play the same role as the subgrid viscosity model constant, and P_{ij} the projection operator on a space of zero divergence. We have the relations:

$$\langle e_i(\boldsymbol{x}, t) \rangle = 0 \quad , \tag{5.409}$$

$$\langle e_i(\boldsymbol{x}, t) e_i(\boldsymbol{x}', t') \rangle = \frac{1}{3} \delta_{ij} \delta(t - t') \delta(\boldsymbol{x} - \boldsymbol{x}') \quad . \tag{5.410}$$

Considering (5.408), (5.410) and (5.404), we get:

$$\langle \mathcal{E}_{\mathrm{f}} \rangle = \frac{1}{2} A^2 = \frac{1}{3} \mathcal{A}^2 \quad . \tag{5.411}$$

The computation of the model is completed by evaluating the constant $\mathcal{A}$ by a constrained localized dynamic procedure based on minimizing the functional (5.403), which can be re-written in the form:

$$\mathcal{Z}[\mathcal{A}] = \int \left(\frac{\mathcal{A}^2}{3} - g \right)^2 \quad . \tag{5.412}$$

6. Functional Modeling: Extension to Anisotropic Cases

6.1 Statement of the Problem

The developments of the previous chapters are all conducted in the isotropic framework, which implies that both the filter used and the flow are isotropic. They can be extended to anisotropic or inhomogeneous cases only by localizing the statistical relations in space and time and introducing heuristic procedures for adjusting the models. But when large-eddy simulation is applied to inhomogeneous flows, we very often have to use anisotropic grids, which correspond to using a anisotropic filter. So there are two factors contributing to the violation of the hypotheses underlying the models presented so far: filter anisotropy (respectively inhomogeneity) and flow anisotropy (respectively inhomogeneity).

This chapter is devoted to extensions of the modeling to anisotropic cases. Two situations are considered: application of a anisotropic homogeneous filter to an isotropic homogeneous turbulent flow (Sect. 6.2), and application of an isotropic filter to an anisotropic flow (Sect. 6.3).

6.2 Application of Anisotropic Filter to Isotropic Flow

The filters considered in the following are anisotropic in the sense that the filter cutoff length is different in each direction of space. The different types of anisotropy possible for Cartesian filtering cells are represented in Fig. 6.1.

In order to use an anisotropic filter to describe an isotropic flow, we are first required to modify the subgrid models, because theoretical work and numerical experiments have shown that the resolved fields and the subgrid thus defined are anisotropic [368]. For example, for a mesh cell with an aspect ratio $\overline{\Delta}_2/\overline{\Delta}_1 = 8, \overline{\Delta}_3/\overline{\Delta}_1 = 4$, the subgrid stresses will differ from their values obtained with an isotropic filter by about ten percent. It is very important to note, though, that this anisotropy is an artifact due to the filter but that the dynamic of the subgrid scales still corresponds that of isotropic homogeneous turbulence.

On the functional modeling level, the problem is in determining the characteristic length that has to be used to compute the model.

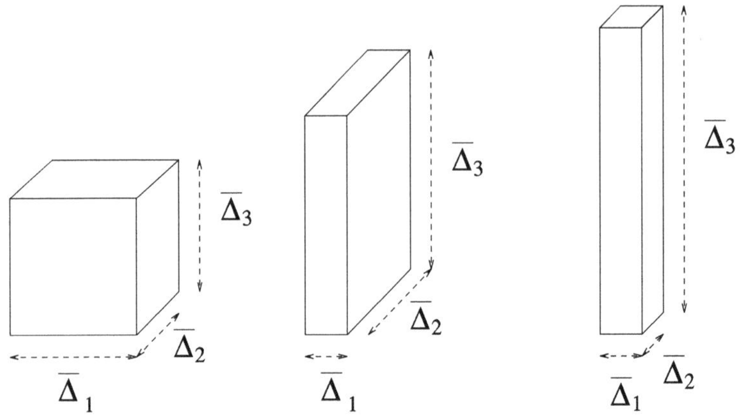

Fig. 6.1. Different types of filtering cells. Isotropic cell (on the left): $\overline{\Delta}_1 = \overline{\Delta}_2 = \overline{\Delta}_3$; pancake-type anisotropic cell (center): $\overline{\Delta}_1 \ll \overline{\Delta}_2 \approx \overline{\Delta}_3$; cigar-type anisotropic cell (right): $\overline{\Delta}_1 \approx \overline{\Delta}_2 \ll \overline{\Delta}_3$.

Two approaches are available:

- The first consists in defining a single length scale for representing the filter. This lets us keep models analogous to those defined in the isotropic case, using for example scalar subgrid viscosities for representing the forward cascade process. This involves only a minor modification of the subgrid models since only the computation of the characteristic cutoff scale is modified. But it should be noted that such an approach can in theory be valid only for cases of low anisotropy, for which the different cutoff lengths are of the same order of magnitude.
- The second approach is based on the introduction of several characteristic length scales in the model. This sometimes yields major modifications in the isotropic models, such as the definition of tensorial subgrid viscosities to represent the forward cascade process. In theory, this approach takes the filter anisotropy better into account, but complicates the modeling stage.

6.2.1 Scalar Models

These models are all of the generic form $\overline{\Delta} = \overline{\Delta}(\overline{\Delta}_1, \overline{\Delta}_2, \overline{\Delta}_3)$. We present here:

1. Deardorff's original model and its variants (p. 189). These forms are empirical and have no theoretical basis. All we do is simply to show that they are consistent with the isotropic case, i.e. $\overline{\Delta} = \overline{\Delta}_1$ when $\overline{\Delta}_1 = \overline{\Delta}_2 = \overline{\Delta}_3$.
2. The model of Scotti et al. (p. 189), which is based on a theoretical analysis considering a Kolmogorov spectrum with an anisotropic homogeneous filter. This model makes a complex evaluation possible of the filter cutoff length, but is limited to the case of Cartesian filtering cells.

Deardorff's Proposal. The method most widely used today is without doubt the one proposed by Deardorff [172], which consists in evaluating the filter cutoff length as the cube root of the volume V_Ω of the filtering cell Ω. Or, in the Cartesian case:

$$\overline{\Delta}(\boldsymbol{x}) = \left(\overline{\Delta}_1(\boldsymbol{x})\overline{\Delta}_2(\boldsymbol{x})\overline{\Delta}_3(\boldsymbol{x})\right)^{1/3} \quad , \tag{6.1}$$

in which $\overline{\Delta}_i(\boldsymbol{x})$ is the filter cutoff length in the ith direction of space at position $\boldsymbol{x}$.

Extensions of Deardorff's Proposal. Simple extensions of definition (6.1) are often used, but are limited to the case of Cartesian filtering cells:

$$\overline{\Delta}(\boldsymbol{x}) = \sqrt{(\overline{\Delta}_1^2(\boldsymbol{x}) + \overline{\Delta}_2^2(\boldsymbol{x}) + \overline{\Delta}_3^2(\boldsymbol{x}))/3} \quad , \tag{6.2}$$

$$\overline{\Delta}(\boldsymbol{x}) = \max\left(\overline{\Delta}_1(\boldsymbol{x}), \overline{\Delta}_2(\boldsymbol{x}), \overline{\Delta}_3(\boldsymbol{x})\right) \quad . \tag{6.3}$$

Another way to compute the characteristic length on a strongly non-uniform mesh, which prevents the occurrence of large values of subgrid viscosity, was proposed by Arad [16]. It relies on the use of the harmonic mean of the usual length scales:

$$\overline{\Delta}(\boldsymbol{x}) = \left(\hat{\Delta}_1(\boldsymbol{x})\hat{\Delta}_2(\boldsymbol{x})\hat{\Delta}_3(\boldsymbol{x})\right)^{1/3} \quad , \tag{6.4}$$

with

$$\hat{\Delta}_i(\boldsymbol{x}) = \left((\overline{\Delta}_i(\boldsymbol{x}))^{-\omega} + (\overline{\Delta}_i^M)^{-\omega}\right)^{-1/\omega} \quad , \quad i = 1, 2, 3 \quad , \tag{6.5}$$

where $\overline{\Delta}_i^M$ is a prescribed bound for $\overline{\Delta}_i(\boldsymbol{x})$, and $\omega > 0$.

Proposal of Scotti et al. More recently, Scotti, Meneveau, and Lilly [664] proposed a new definition of $\overline{\Delta}$ based on an improved estimate of the dissipation rate ε in the anisotropic case. The filter is assumed to be anisotropic but homogeneous, i.e. the cutoff length is constant in each direction of space.

We define $\overline{\Delta}_{\max} = \max(\overline{\Delta}_1, \overline{\Delta}_2, \overline{\Delta}_3)$. Aspect ratios of less than unity, constructed from the other two cutoff lengths with respect to $\Delta_{\max}$, are denoted a_1 and a_2[1]. The form physically sought for the anisotropy correction is:

$$\overline{\Delta} = \overline{\Delta}_{\text{iso}} f(a_1, a_2) \quad , \tag{6.6}$$

in which $\overline{\Delta}_{\text{iso}}$ is Deardorff's isotropic evaluation computed by relation (6.1).

[1] For example, by taking $\Delta_{\max} = \overline{\Delta}_1$, we get $a_1 = \overline{\Delta}_2/\overline{\Delta}_1$ and $a_2 = \overline{\Delta}_3/\overline{\Delta}_1$.

Using the approximation:

$$\langle \varepsilon \rangle = \overline{\Delta}^2 \langle 2\overline{S}_{ij}\overline{S}_{ij}\rangle^{3/2} \quad , \tag{6.7}$$

and the following equality, which is valid for a Kolmogorov spectrum,

$$\langle \overline{S}_{ij}\overline{S}_{ij}\rangle = \langle \varepsilon \rangle^{2/3} \frac{K_0}{2\pi} \int |\widehat{G}(\boldsymbol{k})|^2 k^{-5/3} d^3\boldsymbol{k} \quad , \tag{6.8}$$

where $\widehat{G}(\boldsymbol{k})$ is the kernel of the anisotropic filter considered, after calculation we get:

$$\overline{\Delta} = \left(\frac{K_0}{2\pi} \int |\widehat{G}(\boldsymbol{k})|^2 k^{-5/3} d^3\boldsymbol{k} \right)^{-3/4} \quad . \tag{6.9}$$

Considering a sharp cutoff filter, we get the following approximate relation by integrating equation (6.9):

$$f(a_1, a_2) = \cosh \sqrt{\frac{4}{27} \left[(\ln a_1)^2 - \ln a_1 \ln a_2 + (\ln a_2)^2 \right]} \quad . \tag{6.10}$$

It is interesting to note that the dynamic procedure (see Sect. 5.3.3) for the computation of the Smagorinsky constant can be interpreted as an implicit way to compute the $f(a_1, a_2)$ function [663]. Introducing the subgrid mixing length Δ_f, the Smagorinsky model reads

$$\nu_{\rm sgs} = \Delta_f^2 |\overline{S}| \tag{6.11}$$

$$= C_{\rm d} \overline{\Delta}_{\rm iso}^2 |\overline{S}| \tag{6.12}$$

$$= (C_{\rm S} \overline{\Delta}_{\rm iso} f(a_1, a_2))^2 |\overline{S}| \quad , \tag{6.13}$$

where $C_{\rm d}$ is the value of the constant computed using a dynamic procedure, and $C_{\rm S}$ the theoretical value of the Smagorinsky constant evaluated through the canonical analysis. A trivial identification leads to:

$$f(a_1, a_2) = \sqrt{C_{\rm d}} / C_{\rm S} \quad . \tag{6.14}$$

This interpretation is meaningful for positive values of the dynamic constant. A variant can be derived by using the anisotropy measure $f(a_1, a_2)$ instead of the isotropic one inside the dynamic procedure (see equation (6.12)), yielding new definitions of the tensors α_{ij} and β_{ij} appearing in the dynamic procedure (see Table 5.1). Taking the Smagorinsky model as an example, we get:

$$\beta_{ij} = -2(\overline{\Delta}_{\rm iso} f(\overline{a}_1, \overline{a}_2))^2 |\overline{S}| \overline{S}_{ij}, \quad \alpha_{ij} = -2(\widetilde{\overline{\Delta}}_{\rm iso} f(\widetilde{\overline{a}}_1, \widetilde{\overline{a}}_2))^2 |\widetilde{\overline{S}}| \widetilde{\overline{S}}_{ij} \quad , \tag{6.15}$$

where $f(\overline{a}_1, \overline{a}_2)$ and $f(\widetilde{\overline{a}}_1, \widetilde{\overline{a}}_2)$ are the anisotropy measures associated to the first and second filtering levels, respectively. The corresponding formulation of the f function is now:

$$f(a_1, a_2) = \sqrt{C_{\rm d}} / (C_{\rm S} \overline{\Delta}_{\rm iso}) \quad . \tag{6.16}$$

6.2.2 Batten's Mixed Space-Time Scalar Estimator

It was shown in Sect. 2.1.3 that spatial filtering induces a time filtering. In a reciprocal manner, enforcing a time-frequency cutoff leads to the definition of an intrinsic spatial cutoff length. To account for that phenomenon, Batten [49] defines the cutoff length as

$$\overline{\Delta}(\boldsymbol{x}) = 2 \max \left(\overline{\Delta}_1(\boldsymbol{x}), \overline{\Delta}_2(\boldsymbol{x}), \overline{\Delta}_3(\boldsymbol{x}), \sqrt{q_{\text{sgs}}^2} \Delta t \right) \quad , \tag{6.17}$$

where Δt and q_{sgs}^2 are the time step of the simulation and the subgrid kinetic energy, respectively. The subgrid kinetic energy can be computed solving an prognostic transport equation (p. 128) or one of the methods discussed in Sect. 9.2.3.

6.2.3 Tensorial Models

The tensorial models presented in the following are constructed empirically, with no physical basis. They are justified only by intuition and only for highly anisotropic filtering cells of the cigar type, for example (see Fig. 6.1). Representing the filter by a single and unique characteristic length is no longer relevant. The filter's characteristic scales and their inclusion in the subgrid viscosity model are determined intuitively. Two such models are described:

1. The model of Bardina et al. (p. 191), which describes the geometry of the filtering cell by means of six characteristic lengths calculated from the inertia tensor of the filtering cell. This approach is completely general and is a applicable to all possible types of filtering cells (Cartesian, curvilinear, and other), but entrains a high complexification in the subgrid models.
2. The model of Zahrai et al. (p. 192), which is applicable only to Cartesian cells and is simple to include in the subgrid viscosity models.

Proposal of Bardina et al.

Definition of a Characteristic Tensor. These authors [39] propose replacing the isotropic scalar evaluation of the cutoff length associated with the grid by an anisotropic tensorial evaluation linked directly to the filtering cell geometry: $V(\boldsymbol{x}) = (\overline{\Delta}_1(\boldsymbol{x})\overline{\Delta}_2(\boldsymbol{x})\overline{\Delta}_3(\boldsymbol{x}))$. To do this, we introduce the moments of the inertia tensor $\mathcal{I}$ associated at each point $\boldsymbol{x}$:

$$\mathcal{I}_{ij}(\boldsymbol{x}) = \frac{1}{V(\boldsymbol{x})} \int_V x_i x_j dV \quad . \tag{6.18}$$

Since the components of the inertia tensor are homogeneous at the square of a length, the tensor of characteristic lengths is obtained by taking the square root of them. In the case of a pancake filtering cell aligned with the axes of the Cartesian coordinate system, we get the diagonal matrix:

$$\mathcal{I}_{ij} = \frac{2}{3}\begin{pmatrix} \overline{\Delta}_1^2 & 0 & 0 \\ 0 & \overline{\Delta}_2^2 & 0 \\ 0 & 0 & \overline{\Delta}_3^2 \end{pmatrix} \quad . \tag{6.19}$$

Application to the Smagorinsky Model. As we model only the anisotropic part of the subgrid tensor, the tensor $\mathcal{I}$ is decomposed into the sum of a spherical term $\mathcal{I}^{\mathrm{i}}$ and an anisotropic term $\mathcal{I}^{\mathrm{d}}$:

$$\mathcal{I}_{ij} = \mathcal{I}^{\mathrm{i}}\delta_{ij} + (\mathcal{I}_{ij} - \mathcal{I}^{\mathrm{i}}\delta_{ij}) = \mathcal{I}^{\mathrm{i}}\delta_{ij} + \mathcal{I}_{ij}^{\mathrm{d}} \quad , \tag{6.20}$$

with

$$\mathcal{I}^{\mathrm{i}} = \frac{1}{3}\mathcal{I}_{kk} = \frac{1}{3}(\overline{\Delta}_1^2 + \overline{\Delta}_2^2 + \overline{\Delta}_3^2) \quad . \tag{6.21}$$

Modifying the usual Smagorinsky model, the authors finally propose the following anisotropic tensorial model for deviator of the subgrid tensor τ:

$$\begin{aligned} \tau_{ij} - \frac{1}{3}\tau_{kk}\delta_{ij} &= C_1\mathcal{I}^{\mathrm{i}}|\overline{S}|\overline{S}_{ij} \\ &+ C_2|\overline{S}|\left(\mathcal{I}_{ik}\overline{S}_{kj} + \mathcal{I}_{jk}\overline{S}_{ki} - \frac{1}{3}\mathcal{I}_{lk}\overline{S}_{kl}\delta_{ij}\right) \\ &+ C_3\frac{|\overline{S}|}{\mathcal{I}^{\mathrm{i}}}\left(\mathcal{I}_{ik}\mathcal{I}_{jl}\overline{S}_{kl} - \frac{1}{3}\mathcal{I}_{mk}\mathcal{I}_{ml}\overline{S}_{kl}\delta_{ij}\right) \quad , \end{aligned} \tag{6.22}$$

in which C_1, C_2 and C_3 are constants to be evaluated.

Proposal of Zahrai et al.

Principle. Zahrai et al. [795] proposed conserving the isotropic evaluation of the dissipation rate determined by Deardorff and further considering that this quantity is constant over each mesh cell:

$$\langle\varepsilon\rangle = \left(\overline{\Delta}_1(\boldsymbol{x})\overline{\Delta}_2(\boldsymbol{x})\overline{\Delta}_3(\boldsymbol{x})\right)^{2/3}\langle 2\overline{S}_{ij}\overline{S}_{ij}\rangle^{3/2} \quad . \tag{6.23}$$

On the other hand, when deriving the subgrid model, we consider that the filter's characteristic length in each direction is equal to the cutoff length in that direction. This procedure calls for the definition of a tensorial model for the subgrid viscosity.

Application to the Smagorinsky model. In the case of the Smagorinsky model, we get for component k:

$$(\nu_{\mathrm{sgs}})_k = C_1(\overline{\Delta}_1\overline{\Delta}_2\overline{\Delta}_3)^{2/9}(\overline{\Delta}_k)^{4/3}\langle 2\overline{S}_{ij}\overline{S}_{ij}\rangle^{3/2} \quad , \tag{6.24}$$

where C_1 is a constant.

6.3 Application of an Isotropic Filter to a Shear Flow

We will now be examining the inclusion of subgrid scale anisotropy in the functional models.

The first part of this section presents theoretical results concerning subgrid scale anisotropy and the interaction mechanisms between the large and small scales in this case. These results are obtained either by the EDQNM theory or by asymptotic analysis of the triadic interactions.

The second part of the section describes the modifications that have been proposed for functional type subgrid models. Only models for the forward energy cascade will be presented, because no model for the backward cascade has yet been proposed in the anisotropic case.

6.3.1 Phenomenology of Inter-Scale Interactions

Anisotropic EDQNM Analysis. Aupoix [24] proposes a basic analysis of the effects of anisotropy in the homogeneous case using Cambon's anisotropic EDQNM model. The essential details of this model are given in Appendix B.

The velocity field $\boldsymbol{u}$ is decomposed as usual into average part $\langle \boldsymbol{u} \rangle$ and a fluctuating part $\boldsymbol{u}'$:

$$\boldsymbol{u} = \langle \boldsymbol{u} \rangle + \boldsymbol{u}' \quad . \tag{6.25}$$

To study anisotropic homogeneous flows, we define the spectral tensor

$$\Phi_{ij}(\boldsymbol{k}) = \langle \widehat{u}_i'^{*}(\boldsymbol{k}) \widehat{u}_j'(\boldsymbol{k}) \rangle \quad , \tag{6.26}$$

which is related to the double correlations in the physical space by the relation:

$$\langle u_i' u_j' \rangle(\boldsymbol{x}) = \int\int\int \Phi_{ij}(\boldsymbol{k}) d^3\boldsymbol{k} \quad . \tag{6.27}$$

Starting with the Navier–Stokes equations, we obtain the evolution equation (see Appendices A and B):

$$\begin{aligned} \left(\frac{\partial}{\partial t} + 2\nu k^2\right) \Phi_{ij}(\boldsymbol{k}) \quad &+ \quad \frac{\partial \langle u_i \rangle}{\partial x_l} \Phi_{jl}(\boldsymbol{k}) + \frac{\partial \langle u_j \rangle}{\partial x_l} \Phi_{il}(\boldsymbol{k}) \\ &- \quad 2\frac{\partial \langle u_l \rangle}{\partial x_m} \left(k_i \Phi_{jm}(\boldsymbol{k}) + k_j \Phi_{mi}(\boldsymbol{k})\right) \\ &- \quad \frac{\partial \langle u_l \rangle}{\partial x_m} \frac{\partial}{\partial k_m} \left(k_l \Phi_{ij}(\boldsymbol{k})\right) \\ &= \quad P_{il}(\boldsymbol{k}) T_{lj}(\boldsymbol{k}) + P_{jl}(\boldsymbol{k}) T_{li}^{*}(\boldsymbol{k}) \quad , \end{aligned} \tag{6.28}$$

where

$$T_{ij}(\boldsymbol{k}) = k_l \int\int\int \langle u_i(\boldsymbol{k}) u_l(\boldsymbol{p}) u_j(-\boldsymbol{k}-\boldsymbol{p})\rangle d^3\boldsymbol{p} \quad , \tag{6.29}$$

and

$$P_{ij}(\boldsymbol{k}) = \left(\delta_{ij} - \frac{k_i k_j}{k^2}\right) \quad , \tag{6.30}$$

and where the * designates the complex conjugate number. We then simplify the equations by integrating the tensor Φ on spheres of radius k=cste:

$$\phi_{ij}(k) = \int \Phi_{ij}(\boldsymbol{k}) dA(\boldsymbol{k}) \quad , \tag{6.31}$$

and obtain the evolution equations:

$$\begin{aligned} \left(\frac{\partial}{\partial t} + 2\nu k^2\right)\phi_{ij}(k) &= -\frac{\partial\langle u_i\rangle}{\partial x_k}\phi_{jl}(k) - \frac{\partial\langle u_j\rangle}{\partial x_l}\phi_{il}(k) \\ &+ P^{\mathrm{l}}_{ij}(k) + S^{\mathrm{l}}_{ij}(k) + P^{\mathrm{nl}}_{ij}(k) + S^{\mathrm{nl}}_{ij}(k) \quad , \end{aligned} \tag{6.32}$$

where the terms $P^{\mathrm{l}}, S^{\mathrm{l}}, P^{\mathrm{nl}}$ and S^{nl} are the linear pressure, linear transfer, non-linear pressure, and non-linear transfer contributions, respectively. The linear terms are associated with the action of the average velocity gradient, and the non-linear terms with the action of the turbulence on itself.

The expression of these terms and their closure by the anisotropic EDQNM approximation are given in Appendix B. Using these relations, Aupoix derives an expression for the interaction between the modes corresponding to wave numbers greater than a given cutoff wave number k_{c} (i.e. the small or subgrid scales) and those associated with small wave numbers such that $k \leq k_{\mathrm{c}}$ (i.e. the large or resolved scales). To obtain a simple expression for the coupling among the different scales by the non-linear terms P^{nl} and S^{nl}, we adopt the hypothesis that there exists a total separation of scales (in the sense defined in Sect. 5.3.2) between the subgrid and resolved modes, so that we can obtain the following two asymptotic forms:

$$\begin{aligned} P^{\mathrm{nl}}_{ij}(k) &= -\frac{32}{175}k^4 \int_{k_{\mathrm{c}}}^{\infty} \Theta_{0pp}\left[10 + a(p)\right] \frac{E^2(p)H_{ij}(p)}{p^2} dp \\ &+ \frac{16}{105}k^2 E(k) \int_{k_{\mathrm{c}}}^{\infty} \Theta_{0pp}\left[(a(p)+3)p\frac{\partial}{\partial p}\left(E(p)H_{ij}(p)\right)\right. \\ &\left. + E(p)H_{ij}(p)\left(5\left\{a(p)+3\right\} + p\frac{\partial a(p)}{\partial p}\right)\right] dp \quad , \end{aligned} \tag{6.33}$$

$$\begin{aligned} S^{\rm nl}_{ij}(k) &= 2k^4 \int_{k_C}^{\infty} \Theta_{0pp} \frac{E^2(p)}{p^2} \left[\frac{14}{15} \left(\frac{1}{3}\delta_{ij} + 2H_{ij}(p) \right) + \frac{8}{25} a(p) H_{ij}(p) \right] dp \\ &- 2k^2 \phi_{ij}(k) \frac{1}{15} \int_{k_C}^{\infty} \Theta_{0pp} \left[5E(p) + p\frac{\partial E(p)}{\partial p} \right] dp \\ &- 2k^2 E(k) \int_{k_C}^{\infty} \Theta_{0pp} \left[\frac{2}{15} \left\{ 5E(p)H_{ij}(p) + p\frac{\partial}{\partial p}(E(p)H_{ij}(p)) \right\} \right. \\ & \left. + E(p)H_{ij}(p) \left\{ \frac{8}{15}(a(p)+3) + \frac{8}{25}a(p) \right\} \right] dp \quad , \end{aligned} \tag{6.34}$$

where $E(k)$ is the energy spectrum, defined as:

$$E(k) = \frac{1}{2}\phi_{ll}(k) \quad , \tag{6.35}$$

and $H_{ij}(k)$ the anisotropy spectrum:

$$H_{ij}(k) = \frac{\phi_{ij}(k)}{2E(k)} - \frac{1}{3}\delta_{ij} \quad . \tag{6.36}$$

It is easily verified that, in the isotropic case, H_{ij} cancels out by construction. The function $a(k)$ is a structural parameter that represents the anisotropic distribution on the sphere of radius k, and Θ_{kpq} the characteristic relaxation time evaluated by the EDQNM hypotheses. The expression of this term is given in Appendix B.

These equations can be simplified by using the asymptotic value of the structural parameter $a(k)$. By taking $a(k) = -4.5$, we get:

$$\begin{aligned} P^{\rm nl}_{ij}(k) + S^{\rm nl}_{ij}(k) &= k^4 \int_{k_C}^{\infty} \Theta_{0pp} \frac{E^2(p)}{p^2} \left[\frac{28}{45}\delta_{ij} - \frac{368}{175} H_{ij}(p) \right] dp \\ &- 2k^2 \phi_{ij}(k) \frac{1}{15} \int_{k_C}^{\infty} \Theta_{0pp} \left[5E(p) + p\frac{\partial E(p)}{\partial p} \right] dp \\ &+ k^2 E(k) \int_{k_C}^{\infty} \Theta_{0pp} \left[\frac{1052}{525} E(p)H_{ij}(p) \right. \\ & \left. - \frac{52}{105}\frac{\partial}{\partial p}(E(p)H_{ij}(p)) \right] dp \quad . \end{aligned} \tag{6.37}$$

From this equation, it can be seen that the anisotropy of the small scales takes on a certain importance. In a case where the anisotropic spectrum has the same (resp. opposite) sign for the small scales as it does for the large, the term in k^4 constitutes a return of energy that has the effect of a return toward isotropy (resp. departure from isotropy), and the term in $k^2E(k)$ represents an backward energy cascade associated with an increasing anisotropy (resp. a return to isotropy). Lastly, the term in $k^2\phi_{ij}(k)$ is a term of isotropic drainage of energy to the large scales by the small, and represents here the energy cascade phenomenon modeled by the isotropic subgrid models.

Asymptotic Analysis of Triadic Interactions. Another analysis of inter-scale interactions in the isotropic case is the asymptotic analysis of triadic interactions [74, 785].

The evolution equation of the Fourier mode $\widehat{\boldsymbol{u}}(\boldsymbol{k})$ is written in the symbolic form:

$$\frac{\partial \widehat{\boldsymbol{u}}(\boldsymbol{k})}{\partial t} = \dot{\boldsymbol{u}}(\boldsymbol{k}) = [\dot{\boldsymbol{u}}(\boldsymbol{k})]_{\rm nl} + [\dot{\boldsymbol{u}}(\boldsymbol{k})]_{\rm vis} \quad , \tag{6.38}$$

where $[\dot{\boldsymbol{u}}(\boldsymbol{k})]_{\rm nl}$ and $[\dot{\boldsymbol{u}}(\boldsymbol{k})]_{\rm vis}$ represent, respectively, the non-linear terms associated with the convection and pressure, and the linear term associated with the viscous effects, defined as:

$$[\dot{\boldsymbol{u}}(\boldsymbol{k})]_{\rm nl} = -{\rm i} \sum_{p} \widehat{\boldsymbol{u}}(\boldsymbol{p})_{\perp k} \left(\boldsymbol{k} \cdot \widehat{\boldsymbol{u}}(\boldsymbol{k}-\boldsymbol{p})\right) \quad , \tag{6.39}$$

with

$$\widehat{u}_i(\boldsymbol{p})_{\perp k} = \left(\delta_{ij} - \frac{k_i k_j}{k^2}\right) \widehat{u}_j(\boldsymbol{p}) \quad , \tag{6.40}$$

$$[\dot{\boldsymbol{u}}(\boldsymbol{k})]_{\rm vis} = -\nu k^2 \widehat{\boldsymbol{u}}(\boldsymbol{k}) \quad . \tag{6.41}$$

The evolution equation of the modal energy, $e(\boldsymbol{k}) = \widehat{\boldsymbol{u}}(\boldsymbol{k}) \cdot \widehat{\boldsymbol{u}}^*(\boldsymbol{k})$, is of the form:

$$\frac{\partial e(\boldsymbol{k})}{\partial t} = \widehat{\boldsymbol{u}}(\boldsymbol{k}) \cdot \dot{\boldsymbol{u}}^*(\boldsymbol{k}) + cc = [\dot{e}(\boldsymbol{k})]_{\rm nl} + [\dot{e}(\boldsymbol{k})]_{\rm vis} \quad , \tag{6.42}$$

with

$$[\dot{e}(\boldsymbol{k})]_{\rm nl} = -{\rm i} \sum_{p} \left[\widehat{\boldsymbol{u}}^*(\boldsymbol{k}) \cdot \widehat{\boldsymbol{u}}(\boldsymbol{p})\right] \left[\boldsymbol{k} \cdot \widehat{\boldsymbol{u}}(\boldsymbol{k}-\boldsymbol{p})\right] + cc \quad , \tag{6.43}$$

$$[\dot{e}(\boldsymbol{k})]_{\rm vis} = -2\nu k^2 e(\boldsymbol{k}) \quad , \tag{6.44}$$

where the symbol cc designates the complex conjugate number of the term that precedes it. The non-linear energy transfer term brings in three wave vectors $(\boldsymbol{k}, \boldsymbol{p}, \boldsymbol{q} = \boldsymbol{k}-\boldsymbol{p})$ and is consequently a linear sum of non-linear triadic interactions. We recall (see Sect. 5.1.2) that the interactions can be classified into various categories ranging from local interactions, for which the norms of the three wave vectors are similar (i.e. $k \sim p \sim q$), to distant interactions for which the norm of one of the wave vectors is very small compared with the other two (for example $k \ll p \sim q$). The local interactions therefore correspond to the inter-scale interactions of the same size and the distant interactions to the interactions between a large scale and two small scales.

Also, any interaction that introduces a $(\boldsymbol{k}, \boldsymbol{p}, \boldsymbol{q})$ triad that does not verify the relation $k \sim p \sim q$ is called a non-local interaction.

In the following, we will be analyzing an isolated distant triadic interaction associated with three modes: $\boldsymbol{k}, \boldsymbol{p}$ and $\boldsymbol{q}$. We adopt the configuration $k \ll p \sim q$ and assume that $\boldsymbol{k}$ is large scale located in the energetic portion of the spectrum. An asymptotic analysis shows that:

$$[\dot{\boldsymbol{u}}(\boldsymbol{k})]_{\mathrm{nl}} = O(\delta) \quad , \tag{6.45}$$

$$[\dot{\boldsymbol{u}}(\boldsymbol{p})]_{\mathrm{nl}} = -\mathrm{i}\left(\widehat{\boldsymbol{u}}^*(\boldsymbol{q})\left[\boldsymbol{p} \cdot \widehat{\boldsymbol{u}}^*(\boldsymbol{k})\right]\right) + O(\delta) \quad , \tag{6.46}$$

$$[\dot{\boldsymbol{u}}(\boldsymbol{q})]_{\mathrm{nl}} = -\mathrm{i}\left(\widehat{\boldsymbol{u}}^*(\boldsymbol{p})\left[\boldsymbol{p} \cdot \widehat{\boldsymbol{u}}^*(\boldsymbol{k})\right]\right) + O(\delta) \quad , \tag{6.47}$$

where δ is the small parameter defined as

$$\delta = \frac{k}{p} \ll 1 \quad .$$

The corresponding energy transfer analysis leads to the following relations:

$$[\dot{e}(\boldsymbol{k})]_{\mathrm{nl}} = O(\delta) \quad , \tag{6.48}$$

$$[\dot{e}(\boldsymbol{p})]_{\mathrm{nl}} = -\left[\dot{e}(\boldsymbol{q})\right]_{\mathrm{nl}} = \mathrm{i}\left\{\widehat{\boldsymbol{u}}(\boldsymbol{p}) \cdot \widehat{\boldsymbol{u}}(\boldsymbol{q})\left[\boldsymbol{p} \cdot \widehat{\boldsymbol{u}}(\boldsymbol{k})\right] + cc\right\} + O(\delta) \quad . \tag{6.49}$$

Several remarks can be made:

- The interaction between large and small scales persists in the limit of the infinite Reynolds numbers. Consistently with the Kolmogorov hypotheses, these interactions occur with no energy transfer between the large and small scales. Numerical simulations have shown that the energy transfers are negligible between two modes separated by more than two decades.
- The variation rate of the high frequencies $\widehat{\boldsymbol{u}}(\boldsymbol{p})$ and $\widehat{\boldsymbol{u}}(\boldsymbol{q})$ is directly proportional to the amplitude of the low-frequency mode $\widehat{\boldsymbol{u}}(\boldsymbol{k})$. This implies that the strength of the coupling with the low-frequency modes increases with the energy of the modes.

Moreover, complementary analysis shows that, for modes whose wavelength is of the order of the Taylor micro-scale λ defined as (see Appendix A):

$$\lambda = \sqrt{\frac{\overline{u'^2}}{\overline{\left(\dfrac{\partial u'}{\partial x}\right)^2}}} \quad , \tag{6.50}$$

the ratio between the energy transfers due to the distant interactions and those due to the local interactions vary as:

$$\frac{[\dot{e}(k_\lambda)]_{\text{distant}}}{[\dot{e}(k_\lambda)]_{\text{local}}} \sim Re_\lambda^{11/6} \quad , \tag{6.51}$$

where Re_λ is the Reynolds number referenced to the Taylor micro-scale and the velocity fluctuation u'. This relation shows that the coupling increases with the Reynolds number, with the result that an anisotropic distribution of the energy at the low frequencies creates an anisotropic forcing of the high frequencies, leading to a deviation from isotropy of these high frequencies.

A competitive mechanism exists that has an isotropy reduction effect at the small scales. This is the energy cascade associated with non-local triadic interactions that do not enter into the asymptotic limit of the distant interactions.

For a wave vector of norm k, the ratio of the characteristic times $\tau(k)_{\text{cascade}}$ and $\tau(k)_{\text{distant}}$, associated respectively with the energy transfer of the cascade mechanism and that due to the distant interactions, is evaluated as:

$$\frac{\tau(k)_{\text{cascade}}}{\tau(k)_{\text{distant}}} \sim \text{constant} \times (k/k_{\text{injection}})^{11/6} \quad , \tag{6.52}$$

where $k_{\text{injection}}$ is the mode in which the energy injection occurs in the spectrum. So we see that the distant interactions are much faster than the energy cascade. Also, the first effect of a sudden imposition of large scale anisotropy will be to anisotropize the small scales, followed by competition between the two mechanisms. The dominance of one of the two depends on a number of factors, such as the separation between the k and $k_{\text{injection}}$ scales, or the intensity and coherence of the anisotropy at the large scale.

Numerical simulations [785] performed in the framework of homogeneous turbulence have shown a persistence of anisotropy at the small scales. However, it should be noted that this anisotropy is detected only on statistical moments of the velocity field of order three or more, with first- and second-order moments being isotropic.

6.3.2 Anisotropic Models: Scalar Subgrid Viscosities

The subgrid viscosity models presented in this section have been designed to alleviate the problem observed with basic subgrid viscosities, i.e. to prevent the occurance of too high dissipation levels in shear flows, which are known to have disastrous effects in near-wall regions.[2] The models presented below are:

---

[2] Problems encountered in free shear flows are usually less important, since large scales are often driven by inviscid instabilities, while the existence of a critical Reynolds number may lead to relaminarization if the subgrid viscosity is too high.

1. WALE model by Nicoud and Ducros (p. 199), which is built to recover the expected asymptotic behavior in the near-wall region in equilibrium turbulent boundary layers on fine grids, without any additional damping function.
2. Casalino–Jacob Weighted Gradient Model (p. 199), which is based on a modification of the of the Smagorinsky constant to make it sensitive to the mean shear stresses, rendering it more local in terms of wave number.
3. Models based on the idea of separating the field into an isotropic part and inhomogeneous part (p. 200), in order to be able to isolate the contribution of the mean field in the computation of the subgrid viscosity, for models based on the large scales, and thereby better localize the information contained in these models by frequency. This technique, however, is applicable only to flows whose mean velocity profile is known or can be computed on the fly.

Wall-Adapting Local Eddy-Viscosity Model. It has been seen before (p. 159) that most subgrid viscosity models do not exhibit the correct behavior in the vicinity of solid walls in equilibrium boundary layers on fine grids, resulting in a too high damping of fluctuations in that region and to a wrong prediction of the skin friction. The common way to alleviate this problem is to add a damping function, which requires the distance to the wall and the skin friction as input parameters, leading to complex implementation issues. Another possibility is to use self-adpative models, which involve a larger algorithmic complexity.

An elegant solution to solve the near-wall region problem on fine grids is proposed by Nicoud and Ducros [567], who found a combination of resolved velocity spatial derivatives that exhibits the expected asymptotic behavior $\nu_{\rm sgs} \propto {z^+}^3$, where z^+ is the distance to the wall expressed in wall units.

The subgrid viscosity is defined as

$$\nu_{\rm sgs} = (C_w \overline{\Delta})^2 \frac{\left(\mathcal{S}^d_{ij}\mathcal{S}^d_{ij}\right)^{3/2}}{\left(\overline{S}_{ij}\overline{S}_{ij}\right)^{5/2} + \left(\mathcal{S}^d_{ij}\mathcal{S}^d_{ij}\right)^{5/4}} \quad , \tag{6.53}$$

with $C_w = 0.55 - 0.60$ and

$$\mathcal{S}^d_{ij} = \overline{S}_{ik}\overline{S}_{kj} + \overline{\Omega}_{ik}\overline{\Omega}_{kj} - \frac{1}{3}\left(\overline{S}_{mn}\overline{S}_{mn} - \overline{\Omega}_{mn}\overline{\Omega}_{mn}\right)\delta_{ij} \quad . \tag{6.54}$$

This model also possesses the interesting property that the subgrid viscosity vanishes when the flow is two-dimensional, in agreement with the physical analysis.

Weighted Gradient Subgrid Viscosity Model. Casalino, Boudet and Jacob [112] introduced a modification in the evaluation of the Smagorinsky constant to render it sensitive to resolved gradients, with the purpose of recovering a better accuracy in shear flows.

The weighted gradient subgrid viscosity model is written as

$$\nu_{\text{sgs}} = (C(\boldsymbol{x},t)\overline{\Delta})^2 |\overline{S}_{ij}| \quad , \tag{6.55}$$

where the self-adaptive constant is equal to

$$C(\boldsymbol{x},t) = \gamma C_S \left(\frac{S^*_{ij} S^*_{ij}}{\overline{S}_{ij}\overline{S}_{ij}} \right)^{m/4} \quad , \tag{6.56}$$

where $C_S = 0.18$ is the conventional Smagorinsky constant, γ and m are free parameters and the weighted strain tensor S^* is defined as (without summation over repeated greek indices)

$$S^*_{\alpha\beta} = W_{\alpha\beta} \overline{S}_{\alpha\beta} \quad . \tag{6.57}$$

The weighting matrix coefficients are inversely proportional to the third moment of corresponding strain rate tensor coefficients:

$$W_{\alpha\beta} = \frac{\left| \left\langle \overline{S}^3_{\alpha\beta} \right\rangle \right|}{\sum_{\alpha,\beta=1,3} \left| \left\langle \overline{S}^3_{\alpha\beta} \right\rangle^{-1} \right|} \quad . \tag{6.58}$$

The bracket operator is related to a local average. Numerical tests have shown that $(m, \gamma) = (1,3)$ and $(3,10)$ yield satisfactory results in free shear flows, the last values yielding the recovery of the theoretical behavior of the subgrid viscosity in the vicinity of solid walls on fine meshes.

Models Based on a Splitting Technique. Subgrid viscosity models are mostly developed in the framework of the hypotheses of the canonical analysis, i.e. for homogeneous turbulent flows. Experience shows that the performance of these models declines when they are used in an inhomogeneous framework, which corresponds to a non-uniform average flow. One simple idea initially proposed by Schumann [653] is to split the velocity field into inhomogeneous and isotropic parts and to compute a specific subgrid term for each of these parts.

In practice, Schumann proposes an anisotropic subgrid viscosity model for dealing with flows whose average gradient is non-zero, and in particular any flow regions close to solid walls. The model is obtained by splitting the deviator part of the subgrid tensor τ^{d} into one locally isotropic part and one inhomogeneous:

$$\tau^{\text{d}}_{ij} = -2\nu_{\text{sgs}} \left(\overline{S}_{ij} - \langle \overline{S}_{ij} \rangle \right) - 2\nu^{\text{a}}_{\text{sgs}} \langle \overline{S}_{ij} \rangle \quad , \tag{6.59}$$

where the angle brackets $\langle . \rangle$ designate an statistical average, which in practice is a spatial average in the directions of homogeneity in the solution. The coefficients ν_{sgs} and $\nu^{\text{a}}_{\text{sgs}}$ are the scalar subgrid viscosities representing a locally isotropic turbulence and an inhomogeneous turbulence, respectively. Moin and Kim [537] and Horiuti [319] give the following definitions:

$$\nu_{\mathrm{sgs}} = \left(C_1\overline{\Delta}\right)^2 \sqrt{2\left(\overline{S}_{ij} - \langle\overline{S}_{ij}\rangle\right)\left(\overline{S}_{ij} - \langle\overline{S}_{ij}\rangle\right)} \quad , \tag{6.60}$$

$$\nu_{\mathrm{sgs}}^{\mathrm{a}} = \left(C_2\overline{\Delta}_{\mathrm{z}}\right)^2 \sqrt{2\langle\overline{S}_{ij}\rangle\langle\overline{S}_{ij}\rangle} \quad , \tag{6.61}$$

where C_1 and C_2 are two constants. Horiuti recommends $C_1 = 0.1$ and $C_2 = 0.254$, while Moin and Kim use $C_1 = C_2 = 0.254$. The isotropic part is a function of the fluctuation of the viscosity gradients, so as to make sure that the extra-diagonal components thus predicted for the subgrid tensor cancel out on the average over time. This is consistent with the isotropic hypothesis.

The two characteristic lengths $\overline{\Delta}$ and $\overline{\Delta}_{\mathrm{z}}$ represent the cutoff lengths for the two types of structures, and are evaluated as:

$$\overline{\Delta}(z) = (\overline{\Delta}_1\overline{\Delta}_2\overline{\Delta}_3)^{1/3}(1 - \exp(zu_\tau/A\nu)) \quad , \tag{6.62}$$

$$\overline{\Delta}_{\mathrm{z}}(z) = \overline{\Delta}_3(1 - \exp([zu_\tau/A\nu]^2)) \quad , \tag{6.63}$$

where z is the distance to the solid wall, $\overline{\Delta}_3$ the cutoff length in the direction normal to the surface, and u_τ the friction velocity at the surface (see Sect. 10.2.1). The constant A is taken to be equal to 25.

This model was initially designed for the case of a plane channel flow. It requires being able to compute the statistical average of the velocity field, and thus can be extended only to sheared flows exhibiting at least one direction of homogeneity, or requires the use of several statistically equivalent simulations to perform the ensemble average [102, 108].

Sullivan et al. [698] propose a variant of it that incorporates an anisotropy factor (so that the model constant can be varied to represent the field anisotropy better):

$$\tau_{ij}^{\mathrm{d}} = -2\nu_{\mathrm{sgs}}\gamma\overline{S}_{ij} - 2\nu_{\mathrm{sgs}}^{\mathrm{a}}\langle\overline{S}_{ij}\rangle \quad . \tag{6.64}$$

The authors propose computing the viscosity $\nu_{\mathrm{sgs}}^{\mathrm{a}}$ as before. The ν_{sgs} term, on the other hand, is now calculated by a model with one evolution equation for the subgrid kinetic energy (see equation (5.119) in Chap. 5). Only the subgrid kinetic energy production by the isotropic part is included, which is equivalent to replacing the II term in equation (5.119) with

$$2\nu_{\mathrm{sgs}}\gamma\left(\overline{S}_{ij} - \langle\overline{S}_{ij}\rangle\right)\left(\overline{S}_{ij} - \langle\overline{S}_{ij}\rangle\right) \quad . \tag{6.65}$$

The authors evaluate the anisotropy factor from the shearing rates of the large and small scales. The average per plane of fluctuation homogeneity of the resolved strain rate tensor, calculated by

$$S' = \sqrt{2\langle\left(\overline{S}_{ij} - \langle\overline{S}_{ij}\rangle\right)\left(\overline{S}_{ij} - \langle\overline{S}_{ij}\rangle\right)\rangle} \quad , \tag{6.66}$$

is used for evaluating the shear of the small scales. The shear of the large scales is estimated as

$$S^{\diamond} = \sqrt{2\langle \overline{S}_{ij} \rangle \langle \overline{S}_{ij} \rangle} \quad . \tag{6.67}$$

The isotropy factor is evaluated as:

$$\gamma = \frac{S'}{S' + S^{\diamond}} \quad . \tag{6.68}$$

6.3.3 Anisotropic Models: Tensorial Subgrid Viscosities

Here we describe the main models proposed in the anisotropic framework. Except for Aupoix's spectral model, none of these take explicit account of the backward cascade mechanism. They are:

1. Aupoix's spectral model (p. 203), which is based on the anisotropic EDQNM analysis. The interaction terms are evaluated by adopting a pre-set shape of the energy spectra and subgrid mode anisotropy. This model, which requires a great deal of computation, has the advantage of including all the coupling mechanisms between large and small scales.
2. Horiuti's model (p. 204), which is based on an evaluation of the anisotropy tensor of the subgrid modes from the equivalent tensor constructed from the highest frequencies in the resolved field. This tensor is then used to modulate the subgrid viscosity empirically in each direction of space. This is equivalent to considering several characteristic velocity scales for representing the subgrid modes. This model can only modulate the subgrid dissipation differently for each velocity component and each direction of space, but does not include the more complex anisotropic transfer mechanisms through the cutoff.
3. The model of Carati and Cabot (p. 205), who propose a general form of the subgrid viscosity in the form of a fourth-rank tensor. The components of this tensor are determined on the basis of symmetry relations. However, this model is a applicable only when the flow statistically exhibits an axial symmetry, which restricts its field of validity.
4. The model of Abba et al. (p. 207) which, as in the previous example, considers the subgrid viscosity in the form of a fourth-rank tensor. The model is based on the choice of a local adapted reference system for representing the subgrid modes, and which is chosen empirically when the flow possesses no obvious symmetries.
5. The model proposed by Carati (p. 207), which is based on the evaluation of the statistical anisotropy tensor. The requirement (mean velocity profile) is the same as in the Schumann splitted model.

Aupoix Spectral Model. In order to take the anisotropy of the subgrid scales into account, Aupoix [24] proposes adopting preset shapes of the energy spectra and anisotropy so that the relations stemming from the previously described EDQNM analysis of anisotropy can be used. Aupoix proposes the following model for the energy spectrum:

$$E(k) = K_0 \varepsilon^{2/3} k^{-5/3} \exp\{f(k/k_{\rm d})\} \quad , \tag{6.69}$$

where

$$f(x) = \exp\left[-3.5x^2\left(1 - \exp\left\{6x + 1.2 - \sqrt{196x^2 - 33.6x + 1.4532}\right\}\right)\right] \quad . \tag{6.70}$$

This spectrum is illustrated in Fig. 6.2. The anisotropy spectrum is modeled by:

$$\begin{aligned} H_{ij}(k) &= b_{ij}\left[5 + \frac{k}{E(k)}\frac{\partial E(k)}{\partial k}\right] \\ &\times \left[1 + \mathcal{H}\left(\frac{k}{k_{\rm max}} - 1\right)\mathcal{H}\left(|\mathcal{F}(\overline{\boldsymbol{u}})|\right)\left\{\left(\frac{k}{k_{\rm max}}\right)^{-2/3} - 1\right\}\right] , \end{aligned} \tag{6.71}$$

where $\mathcal{F}(\overline{\boldsymbol{u}}) = \nabla \times \overline{\boldsymbol{u}}$, $k_{\rm max}$ is the wave number corresponding to the energy spectrum maximum, and $\mathcal{H}$ the Heaviside function defined by:

$$\mathcal{H}(x) = \begin{cases} 0 & \text{if } x \leq 0 \\ 1 & \text{otherwise} \end{cases} \quad ,$$

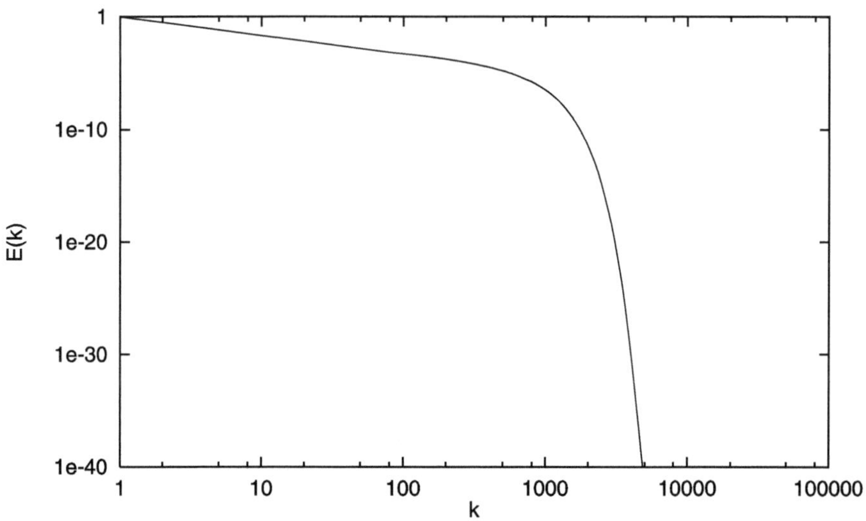

Fig. 6.2. Aupoix spectrum ($k_d = 1000$).

and where b_{ij} is the anisotropy tensor defined as:

$$b_{ij} = \frac{\overline{u_i' u_j'}}{q_{\rm sgs}^2} - \frac{1}{3}\delta_{ij} \quad . \tag{6.72}$$

Horiuti's Model. Horiuti [322] proposes extending the Smagorinsky model to the isotropic case by choosing a different velocity scale for characterizing each component of the subgrid tensor.

Starting with an ordinary dimensional analysis, the subgrid viscosity $\nu_{\rm sgs}$ is expressed as a function of the subgrid kinetic energy $q_{\rm sgs}^2$ and the dissipation rate ε:

$$\nu_{\rm sgs} = C_1 \frac{(q_{\rm sgs}^2)^2}{\varepsilon} \quad . \tag{6.73}$$

To make a better adjustment of the dissipation induced by the subgrid model to the local state of the flow, Horiuti proposes replacing equation (6.73) by:

$$\nu_{\rm sgs} = C_1 \frac{(q_{\rm sgs}^2)^2}{\varepsilon} \Upsilon \quad , \tag{6.74}$$

in which Υ is a dimensionless parameter whose function is to regulate the dissipation rate as a function of the anisotropy of the resolved field. The proposed form for Υ is:

$$\Upsilon = \frac{3E^{\rm s}}{2q_{\rm sgs}^2} \quad , \tag{6.75}$$

where $E^{\rm s}$ is the square of a characteristic velocity scale of the subgrid modes. For example, near solid walls, Horiuti proposes using the fluctuation of the velocity component normal to the wall, which makes it possible for the model to cancel out automatically. To generalize this approach, we associate a characteristic velocity $E_{ij}^{\rm s}$ with each subgrid stress τ_{ij}.

In practice, the author proposes evaluating these characteristic velocities by the scale similarity hypothesis by means of a test filter indicated by a *tilde*:

$$E_{ij}^{\rm s} = (\overline{u}_i - \tilde{\overline{u}}_i)(\overline{u}_j - \tilde{\overline{u}}_j) \quad , \tag{6.76}$$

which makes it possible to define a tensorial parameter Υ_{ij} as:

$$\Upsilon_{ij} = \frac{3(\overline{u}_i - \tilde{\overline{u}}_i)(\overline{u}_j - \tilde{\overline{u}}_j)}{\sum_{l=1,3}(\overline{u}_l - \tilde{\overline{u}}_l)^2} \quad . \tag{6.77}$$

This tensorial parameter characterizes the anisotropy of the test field $(\overline{\boldsymbol{u}} - \tilde{\overline{\boldsymbol{u}}})$ and can be considered as an approximation of the anisotropy tensor associated with this velocity field (to within a coefficient of $1/3\ \delta_{ij}$). Using a model based on the large scales, Horiuti derives the tensorial subgrid viscosity $\nu_{e_{ij}}$:

$$\nu_{e_{ij}} = \left(C_1 \overline{\Delta}\right)^2 |\mathcal{F}(\overline{\boldsymbol{u}})| \Upsilon_{ij} \quad , \tag{6.78}$$

with

$$\mathcal{F}(\overline{\boldsymbol{u}}) = \nabla \times \overline{\boldsymbol{u}}, \quad \text{or} \quad \left(\nabla \overline{\boldsymbol{u}} + \nabla^T \overline{\boldsymbol{u}}\right) \quad ,$$

where the constant C_1 is evaluated as it is for the scalar models. He proposes a model of the general form for the subgrid tensor τ:

$$\tau_{ij} = \delta_{ij} \left(\frac{2}{3} K + \frac{2}{3} P \right) - \nu_{e_{il}} \frac{\partial \overline{u}_j}{\partial x_l} - \nu_{e_{jl}} \frac{\partial \overline{u}_i}{\partial x_l} \quad , \tag{6.79}$$

where

$$K = \frac{1}{2} \sum_{l=1,3} (\overline{u}_l - \tilde{\overline{u}}_l)^2, \qquad P = \nu_{e_{lm}} \frac{\partial \overline{u}_m}{\partial x_l} \quad .$$

It is important to note that this is a model for the entire subgrid tensor and not for its deviatoric part alone, as is the case for the isotropic models.

Carati and Cabot Model. Carati and Cabot [100] propose a tensorial anisotropic extension of the subgrid viscosity models. Generally, the deviator τ^{d} of the subgrid tensor τ is modeled as:

$$\tau_{ij}^{\mathrm{d}} = \nu_{ijkl}^{(1)} \overline{S}_{kl} + \nu_{ijkl}^{(2)} \overline{\Omega}_{kl} \quad , \tag{6.80}$$

where the tensors $\overline{S}$ and $\overline{\Omega}$ are defined as:

$$\overline{S} = \frac{1}{2} \left(\nabla \overline{\boldsymbol{u}} + \nabla^T \overline{\boldsymbol{u}} \right), \qquad \overline{\Omega} = \frac{1}{2} \left(\nabla \overline{\boldsymbol{u}} - \nabla^T \overline{\boldsymbol{u}} \right) \quad .$$

The two viscosities $\nu^{(1)}$ and $\nu^{(2)}$ are fourth-rank tensors theoretically defined by 81 independent coefficients. However, the properties of the tensors τ^{d}, $\overline{S}$ and $\overline{\Omega}$ make it possible to reduce the number of these parameters.

The tensors τ^{d} and $\overline{S}$ are symmetrical and have zero trace, which entails:

$$\begin{aligned} \nu_{ijkl}^{(1)} &= \nu_{jikl}^{(1)} \quad , \\ \nu_{ijkl}^{(1)} &= \nu_{jilk}^{(1)} \quad , \\ \nu_{iikl}^{(1)} &= 0 \quad , \\ \nu_{ijkk}^{(1)} &= 0 \quad . \end{aligned}$$

The tensor $\nu^{(1)}$ therefore contains 25 independent coefficients. By a similar analysis, we can say:

$$\begin{aligned} \nu^{(2)}_{ijkl} &= \nu^{(2)}_{jikl} \quad , \\ \nu^{(2)}_{ijkl} &= -\nu^{(2)}_{jilk} \quad , \\ \nu^{(2)}_{iikl} &= 0 \quad . \end{aligned} \tag{6.81}$$

The tensor $\nu^{(2)}$ therefore contains 15 independent coefficients, which raises the number of coefficients to be determined to 40.

Further reductions can be made using the symmetry properties of the flow. For the case of symmetry about the axis defined by the vector $\boldsymbol{n} = (n_1, n_2, n_3)$, the authors show that the model takes a reduced form that now uses only four coefficients, $C_1, ..., C_4$:

$$\begin{aligned} \tau^{\mathrm{d}}_{ij} &= -2C_1\overline{S}_{ij} - 2C_2\left(n_i\overline{s}_j + \overline{s}_i n_j - \frac{2}{3}\overline{s}_k n_k \delta_{ij}\right) \\ &- C_3\left(n_i n_j - \frac{1}{3}n^2\delta_{ij}\right)\overline{s}_k n_k - 2C_4\left(\overline{r}_i n_j + n_i \overline{r}_j\right) \quad , \end{aligned} \tag{6.82}$$

where $\overline{s}_i = \overline{S}_{ik}n_k$ and $\overline{r}_i = \overline{\Omega}_{ik}n_k$.

Adopting the additional hypothesis that the tensors $\nu^{(1)}$ and $\nu^{(2)}$ verify the Onsager symmetry relations for the covariant vector $\boldsymbol{n}$ and the contravariant vector $\boldsymbol{p}$:

$$\begin{aligned} \nu^{(1)}_{ijkl}(\boldsymbol{n}) &= \nu^{(1)}_{klij}(\boldsymbol{n}) \quad , \\ \nu^{(2)}_{ijkl}(\boldsymbol{n}) &= \nu^{(2)}_{klij}(\boldsymbol{n}) \quad , \\ \nu^{(1)}_{ijkl}(\boldsymbol{p}) &= \nu^{(1)}_{klij}(-\boldsymbol{p}) \quad , \\ \nu^{(2)}_{ijkl}(\boldsymbol{p}) &= \nu^{(2)}_{klij}(-\boldsymbol{p}) \quad , \end{aligned} \tag{6.83}$$

we get the following reduced form:

$$\tau^{\mathrm{d}}_{ij} = -2\nu_1\overline{S}^{\parallel}_{ij} - 2\nu_2 n^2 \overline{S}^{\perp}_{ij} \quad , \tag{6.84}$$

where ν_1 and ν_2 are two scalar viscosities and

$$\overline{S}^{\parallel}_{ij} = \frac{1}{n^2}\left(n_i\overline{s}_j + \overline{s}_i n_j\right) - \frac{1}{3n^2}\overline{s}_k n_k \delta_{ij}, \quad \overline{S}^{\perp}_{ij} = \overline{S}_{ij} - \overline{S}^{\parallel}_{ij} \quad .$$

Carati then proposes determining the two parameters ν_1 and ν_2 by an ordinary dynamic procedure.

Model of Abba et al. Another tensor formulation was proposed by Abba et al. [1]. These authors propose defining the subgrid viscosity in the form of the fourth-rank tensor denoted ν_{ijkl}. This tensor is defined as the product of a scalar isotropic subgrid viscosity $\nu_{\rm iso}$ and an fourth-rank tensor denoted $\mathcal{C}$, whose components are dimensionless constants which will play the role of the scalar constants ordinarily used. The tensor subgrid viscosity ν_{ijkl} thus defined is expressed:

$$\nu_{ijkl} = \mathcal{C}_{ijkl}\nu_{\rm iso} = \left(\sum_{\alpha,\beta} C_{\alpha\beta} a_{i\alpha} a_{j\beta} a_{k\alpha} a_{l\beta}\right) \nu_{\rm iso} \quad , \tag{6.85}$$

where $a_{i\alpha}$ designates the ith component of the unit vector $\boldsymbol{a}_\alpha$ $(\alpha = 1,2,3)$, $C_{\alpha\beta}$ is a symmetrical 3×3 matrix that replaces the scalar Smagorinsky constant. The three vectors $\boldsymbol{a}_\alpha$ are arbitrary and have to be defined as a function of some foreknowledge of the flow topology and its symmetries. When this information is not known, the authors propose using the local framework defined by the following three vectors:

$$\boldsymbol{a}_1 = \frac{\boldsymbol{u}}{u}, \quad \boldsymbol{a}_3 = \frac{\nabla(|u|^2) \times \boldsymbol{u}}{|\nabla(|u|^2) \times \boldsymbol{u}|}, \quad \boldsymbol{a}_2 = \boldsymbol{a}_3 \times \boldsymbol{a}_1 \quad . \tag{6.86}$$

The authors apply this modification to the Smagorinsky model. The scalar viscosity is thus evaluated by the formula:

$$\nu_{\rm iso} = \overline{\Delta}^2 |\overline{S}| \quad . \tag{6.87}$$

The subgrid tensor deviator is then modeled as:

$$\tau_{ij}^{\rm d} = -2\sum_{k,l} \mathcal{C}_{ijkl}\overline{\Delta}^2 |\overline{S}|\overline{S}_{kl} + \frac{2}{3}\delta_{ij}\overline{\Delta}^2 \mathcal{C}_{mmkl}|\overline{S}|\overline{S}_{kl} \quad . \tag{6.88}$$

The model constants are then evaluated by means of a dynamic procedure.

Models Based on a Splitting Technique. An anisotropic tensorial subgrid viscosity model is proposed by Carati et al. [103]. The resulting form of the subgrid stress tensor is

$$\tau_{ij} = \nu_{\rm scalar}\gamma_{ik}\gamma_{jl}\overline{S}_{kl} \quad , \tag{6.89}$$

where $\nu_{\rm scalar}$ plays the role of a scalar subgrid viscosity to be computed using an arbitrary functional model and

$$\gamma_{ij} = 3\frac{\langle u_i'' u_j'' \rangle}{\langle u_k'' u_k'' \rangle} \quad , \tag{6.90}$$

where $\boldsymbol{u}''(\boldsymbol{x},t) = \boldsymbol{u}(\boldsymbol{x},t) - \langle \boldsymbol{u}(\boldsymbol{x},t) \rangle$ is the local instantaneous fluctuation of the resolved field around its statistical mean value. In practice, the mean flow values can be obtained performing statistical averages over homogeneous directions of the flow, over several realizations computed in parallel, or by performing a steady RANS computation.

6.4 Remarks on Flows Submitted to Strong Rotation Effects

All developements presented above deal with shear flows. Rotation is known to also lead to isotropy breakdown and to very complex changes in the interscale dynamics [97]:

- The turbulent kinetic energy dissipation rate is observed to diminish, due to a scrambling in the non-linear triadic interactions. These phenomena have been extensively described using the weak wave turbulence framework and advanced EDQNM closures.
- The Kolmogorov spectrum is no longer valid when strong rotational effects are present.
- Anisotropy is seen to rise, even starting from initially isotropic state. This trend is observed on all statistical moments, including Reynolds stresses, integral length scales, ... Spectral analysis reveals that the induced anisotropy escapes the classical description in the physical space, and that special representations in the Fourier space must be used to describe it acuurately.
- Very complex coupling of rotation with strain is observed, leading to a very large class of dynamical regimes.

The question thus arises of the validity of the subgrid models in the presence of dominant rotational effects, since they do not account for this change in the interscale transfers. In fact, none of the subgrid model presented above is able to account for rotation effects. But numerical experiments [183, 702, 557, 577, 152, 398, 422, 576, 596, 777, 215] show that good results in flows driven by strong rotation are obtained using models that are local in terms of wave numbers on fine grids: dynamic models (and other self-adaptive models), approximate deconvolution models, ... The explanation for this success is that rotation effects are already taken into account by resolved scales, and that local subgrid models will be built on scales that are already modified by the rotation.

7. Structural Modeling

7.1 Introduction and Motivations

This chapter describes some of the family of structural models. As has already been said, these are established with no prior knowledge of the nature of the interactions between the subgrid scales and those that make up the resolved field.

These models can be grouped into several categories:

- Those derived by formal series expansions (Sect. 7.2). These models make no use of any foreknowledge of the physics of the flows, and are based only on series expansions of the various terms that appear in the filtered Navier–Stokes equations. This group of model encompasses models based on deconvolution procedures, nonlinear models and those based on the homogenization technique.
- Those that use the physical hypothesis of scale similarity (Sect. 7.3). These models are based on the scale similarity hypothesis, which establishes a correspondence between the statistical structure of the flow at different filtering levels. Despite the fact that these models are formally equivalent to deconvolution-type models, I chose to present them in a separate section, because they were originally derived on the grounds of physical assumptions rather than on mathematical considerations. The link between the two classes of model is explicitly discussed in Sect. 7.3.3.
- The mixed models, which are based on linear combinations of the functional and structural types, are presented in Sect. 7.4. These models have historically been developed within the framework of the scale-similarity hypothesis, but recent developments dealing with the deconvolution approach show that they are a natural part of deconvolution-based subgrid models. Here again, I chose to fit these to the classical presentation, in order to allow the reader to establish the link with published references more easily. The theoretical equivalence with full deconvolution models is discussed at the end of this section.
- Those based on transport equations for the subgrid tensor components (Sect. 7.5). These models, though they require no information concerning the way the subgrid modes act on the resolved scales, require a very

complex level of modeling since all the unknown terms in the transport equations for the subgrid tensor components have to be evaluated.
- Those constructed from deterministic models for the subgrid structures (Sect. 7.6). They assume that preferential directions of alignments are known for the subgrid structures.
- Those based on an explicit reconstruction of the subgrid velocity fluctuations on an auxiliary grid (Sect. 7.7). These models are the only ones which aim at reconstructing the subgrid motion directly. They can be interpreted as solutions for the full deconvolution problem, as defined in Sect. 7.2. The main difference with deconvolution-like models is that they require the definition of a finer auxiliary grid, on which the solution of the hard deconvolution problem is explicitly reconstructed.
- Those based on a direct identification of subgrid terms using advanced mathematical tools, such as linear stochastic estimation or neural networks (Sect. 7.8).
- Those based on specific numerical algorithms, whose errors are designed to mimic the subgrid forces (Sect. 7.9).

7.2 Formal Series Expansions

The structural models presented in this section belong to one of the three following families:

1. Models based on approximate deconvolution (Sect. 7.2.1). They rely on an attempt to recover, at least partially, the original unfiltered velocity field by inverting the filtering operator. The full deconvolution being impossible to compute in practice, only approximate deconvolution is used.
2. Nonlinear models (Sect. 7.2.2), which rely on the formal derivation of a surrogate of the subgrid stress tensor τ as a function of the gradient of the resolved velocity field.
3. Models based on the homogenization technique (Sect. 7.2.3).

7.2.1 Models Based on Approximate Deconvolution

General Statement of the Deconvolution Problem. The deconvolution approach, also sometimes referred to as the defiltering approach, aims at reconstructing the unfiltered field from the filtered one. The subgrid modes are no longer modeled, but reconstructed using an ad hoc mathematical procedure [181].

We recall that, writing the Navier–Stokes equations symbolically as

$$\frac{\partial \boldsymbol{u}}{\partial t} + \mathcal{NS}(\boldsymbol{u}) = 0 \quad , \tag{7.1}$$

we get the following for the filtered field evolution equation (see Chap. 3)

$$\frac{\partial \overline{\boldsymbol{u}}}{\partial t} + \mathcal{NS}(\overline{\boldsymbol{u}}) = [\mathcal{NS}, G\star](\boldsymbol{u}) \quad , \tag{7.2}$$

where G is the filter kernel, and $[\cdot,\cdot]$ is the commutator operator. The exact subgrid term, which corresponds to the right-hand side of relation (7.2), appears as a function of the exact nonfiltered field $\boldsymbol{u}$. This field being unknown during the computation, the idea here is to approximate it using a deconvolution procedure:

$$\boldsymbol{u} \approx \boldsymbol{u}^{\bullet} \equiv G_l^{-1} \star \overline{\boldsymbol{u}} = G_l^{-1} \star G \star \boldsymbol{u} \quad , \tag{7.3}$$

where G_l^{-1} is an lth-order approximate inverse of the filter G

$$G_l^{-1} \star G = Id + O(\overline{\Delta}^l) \quad .$$

The subgrid term is then approximated as

$$[\mathcal{NS}, G\star](\boldsymbol{u}) \simeq [\mathcal{NS}, G\star](\boldsymbol{u}^{\bullet}) = [\mathcal{NS}, G\star](G_l^{-1} \star \overline{\boldsymbol{u}}) \quad , \tag{7.4}$$

achieving the description of the procedure. Combining the right-hand side and the left-hand side of the resulting equation, we get:

$$\frac{\partial \overline{\boldsymbol{u}}}{\partial t} + G \star \mathcal{NS}(G_l^{-1} \star \overline{\boldsymbol{u}}) = 0 \quad . \tag{7.5}$$

The nonlinear term appears as $(G\star) \circ (\mathcal{NS}) \circ (G_l^{-1})$, i.e. as the sequential application of: (i) the approximate deconvolution operator, (ii) the Navier–Stokes operator, and (iii) a regularization operator, referred to as primary regularization in the parlance of Adams [691, 4, 692, 693, 694]. It is worth noting that the efficiency of the present strategy will be conditioned by our capability of finding the approximate inverse operator, the two others being a priori known.

The deconvolution procedure, in the general presentation given above, calls for several important remarks:

1. It is efficient for invertible filters only, i.e. non-projective filters. Projective filters induce an irreversible loss of information, which cannot be recovered (see Sect. 2.1.2). For smooth filters, the unfiltered field can theoretically be reconstructed [106, 807].
2. In practice, the grid filter is always present, because of the finite number of degrees of freedom used to compute the solution. As a consequence of the Nyquist theorem, a projective filter with space and time cutoffs equal to $2\Delta x$ and $2\Delta t$, respectively, is always present. Here, Δx is the mesh size of the computational grid used for the large-eddy simulation, and Δt the time step employed for the numerical time integration. As a consequence, the filter to be considered in a practical simulation, referred to as

the *effective filter*, is a combination of the convolution filter with cutoff length scale $\overline{\Delta}$ and the projective grid filter. The latter can be modeled as a sharp cutoff filter with cutoff wave number $k_c = \pi/\Delta x$. A complete discussion about the effective filter is given in Sect. 8.2. This implies that the deconvolution procedure cannot reconstruct structures smaller than $2\Delta x$, and, following the terminology of Adams, two problems can be identified in the deconvolution approach:

a) The *soft deconvolution problem*, which corresponds to the reconstruction of the unfiltered field for wave numbers $k \in [0, \pi/\Delta x]$. The corresponding reconstruction procedures described below are:
   - Procedures relying on an iterative reconstruction of the inverse of the filtering operator (p. 212);
   - Procedures based on a truncated Taylor series expansion of the filtering operator (p. 213).
b) The *hard deconvolution problem*: solving the soft deconvolution problem does not suffice for closing the reconstruction problem, because interactions with scales smaller than $2\Delta x$ are not taken into account. In order to alleviate this problem, the soft deconvolution procedure must be supplemented with a secondary procedure, sometimes referred to as secondary regularization. These scales being definitively lost, they must be modeled (and not reconstructed) using a functional model. All the models presented in Chap. 5 can be used. The specific penalization procedure developed by Stolz and Adams and other techniques are discussed in a dedicated section (p. 218).

These two processes are illustrated in Fig. 7.1.

Solving the Soft Deconvolution Problem: Iterative Deconvolution. Adams et al. [691, 692, 4, 694, 693, 743, 744, 742] developed an iterative deconvolution procedure based on the Van Cittert method. If the filter kernel G has an inverse G^{-1}, the latter can be obtained using the following expansion:

$$G^{-1} = (Id - (Id - G))^{-1} \quad , \tag{7.6}$$

$$= \sum_{p=0,\infty} (Id - G)^p \quad , \tag{7.7}$$

yielding the following reconstruction for the defiltered variable ϕ:

$$\phi = \overline{\phi} + (\overline{\phi} - \overline{\overline{\phi}}) + (\overline{\phi} - 2\overline{\overline{\phi}} + \overline{\overline{\overline{\phi}}}) + ... \quad , \tag{7.8}$$

or equivalently

$$\phi' = (\overline{\phi} - \overline{\overline{\phi}}) + (\overline{\phi} - 2\overline{\overline{\phi}} + \overline{\overline{\overline{\phi}}}) + ... \tag{7.9}$$

The series are known to be convergent if $\|Id - G\| < 1$. A practical model is obtained by truncating the expansion at a given power. Stolz and Adams [691] recommend using a fifth-order ($p = 5$) expansion.

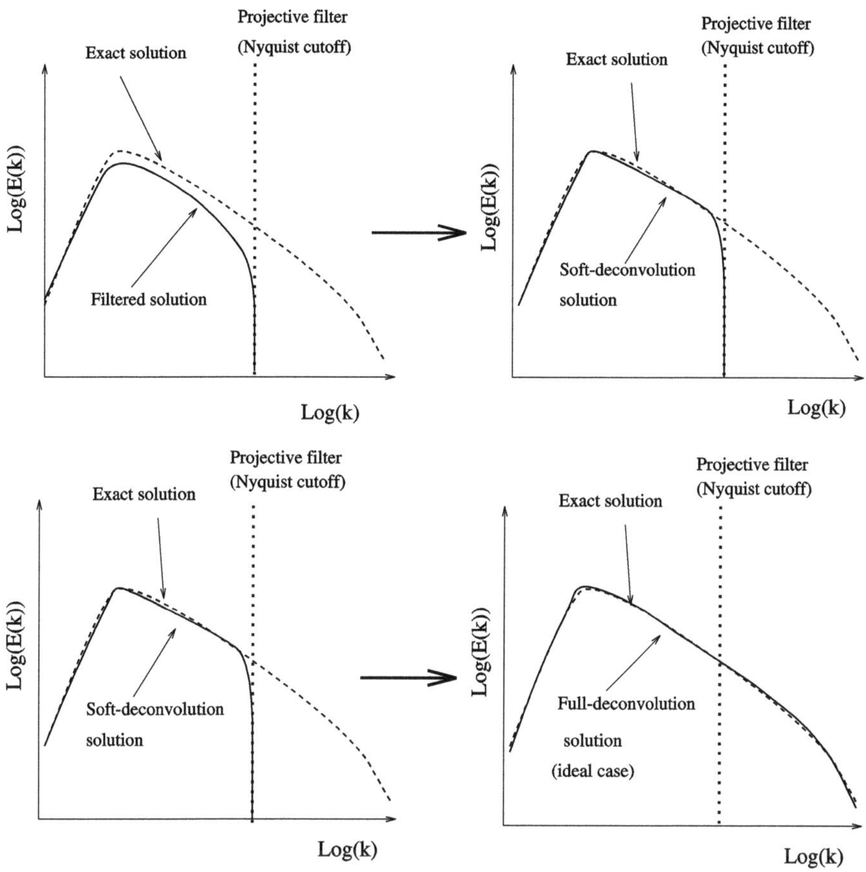

Fig. 7.1. Schematic of the full deconvolution problem. *Top*: soft deconvolution problem; *Bottom*: ideal solution of the full (soft+hard) deconvolution problem.

Other possible reconstruction techniques, such as the Tikhonov regularization, the singular value decomposition or the conjugate gradient method, are discussed in reference [4].

Solving the Soft Deconvolution Problem: Truncated Taylor Series-Expansion-Based Deconvolution Procedures. The deconvolution procedures presented in this section are all based on the representation of the filtering operator in the form of a Taylor series expansion (see Sect. 2.1.6). We recall the general form of such an expansion for the dummy variable ϕ:

$$\overline{\phi}(x) = \sum_{k=0}^{\infty} \frac{(-1)^k}{k!} \overline{\Delta}^k M_k(x) \frac{\partial^k \phi}{\partial x^k}(x) \quad , \tag{7.10}$$

where M_k is the kth-order moment of the filter kernel G.

The common idea shared by all these models is to approximate the filtering operator by truncating the Taylor series expansion, defining a low-order differential operator:

$$\overline{\phi}(x) \approx \sum_{k=0}^{N} \frac{(-1)^k}{k!} \overline{\Delta}^k M_k(x) \frac{\partial^k \phi}{\partial x^k}(x) \quad , \tag{7.11}$$

with $N \leq 4$ in practice. The accuracy of the reconstruction obviously depends on the convergence rate of the Taylor series expansion for the filtering operator. Pruett et al. [607] proved that it is fastly converging for some typical filters, such as Gaussian and top-hat filters. The unfiltered field ϕ can formally be expressed as the solution of the following inverse problem:

$$\phi(x) \approx \left(\sum_{k=0}^{N} \frac{(-1)^k}{k!} \overline{\Delta}^k M_k(x) \frac{\partial^k}{\partial x^k} \right)^{-1} \overline{\phi}(x) \quad . \tag{7.12}$$

The last step consists of inverting this operator, $\overline{\Delta}$ being considered as a small parameter. Two possibilities arise at this stage of the soft deconvolution procedure.

The first one consists of using an implicit method to solve (7.12). This procedure is rarely used, because of its algorithmic cost.

The second one, considered by a large number of authors, relies on the use of an explicit approximation of the solution of (7.12), which is obtained using once again a Taylor series expansion. Recalling that for a small parameter ϵ we have

$$(1+\epsilon)^{-1} = 1 - \epsilon + \epsilon^2 - \epsilon^3 + \epsilon^4 - ... \quad , \tag{7.13}$$

and assuming that $\overline{\Delta}$ can play the role of a small parameter, we get the following approximate inverse relation, which is valid for symmetric filters:

$$\phi(x) \approx \left(Id - \frac{1}{2} \overline{\Delta}^2 M_2(x) \frac{\partial^2}{\partial x^2} + ... \right) \overline{\phi}(x) \quad . \tag{7.14}$$

This last form can be computed immediately from the resolved field. Limiting the expansion to the second order, the subgrid part is expressed as:

$$\begin{aligned} \phi'(x) &= \frac{1}{2} \overline{\Delta}^2 M_2 \frac{\partial^2 \phi(x)}{\partial x^2} + O(\overline{\Delta}^4) \\ &= \frac{1}{2} \overline{\Delta}^2 M_2 \frac{\partial^2}{\partial x^2} \left(\overline{\phi} + O(\overline{\Delta}^2) \right) \\ &= \overline{\phi}'(x) + O(\overline{\Delta}^2) \quad . \end{aligned} \tag{7.15}$$

This can be used to express all the contributions as a function of the resolved field, with second-order accuracy. The various terms of the Leonard decomposition are approximated to the second order as:

$$L_{ij} \equiv \overline{\overline{u}_i \, \overline{u}_j} - \overline{u}_i \, \overline{u}_j = \frac{1}{2} \overline{\Delta}^2 M_2 \frac{\partial^2}{\partial x^2} (\overline{u}_i \, \overline{u}_j) + O(\overline{\Delta}^4) \quad , \tag{7.16}$$

$$C_{ij} \equiv \overline{\overline{u}_i \, u'_j} + \overline{\overline{u}_j \, u'_i} = -\frac{1}{2} \overline{\Delta}^2 M_2 \left(\overline{u}_i \frac{\partial^2}{\partial x^2} \overline{u}_j + \overline{u}_j \frac{\partial^2}{\partial x^2} \overline{u}_i \right) + O(\overline{\Delta}^4) \quad . \tag{7.17}$$

The combination of these two terms leads to:

$$L_{ij} + C_{ij} = \overline{\Delta}^2 M_2 \frac{\partial \overline{u}_i}{\partial x} \frac{\partial \overline{u}_j}{\partial x} + O(\overline{\Delta}^4) \quad . \tag{7.18}$$

As for the subgrid Reynolds tensor, it appears only as a fourth-order term:

$$R_{ij} \equiv \overline{u'_i u'_j} = \frac{1}{4} \left(\overline{\Delta}^2 M_2 \right)^2 \frac{\partial^2 \overline{u}_i}{\partial x^2} \frac{\partial^2 \overline{u}_j}{\partial x^2} + O(\overline{\Delta}^6) \quad , \tag{7.19}$$

so that it disappears in a second-order expansion of the full subgrid tensor. The resulting model (7.18) is referred to as the gradient model, tensor-diffusivity model or Clark's model. It can also be rewritten using undivided differences, leading to the definition of the increment model proposed by Brun and Friedrich [82]. In practice, this approach is used only to derive models for the tensors L and C, which escape functional modeling [98, 143, 160, 161, 459]. Certain authors also use these evaluations to neglect these tensors when the numerical scheme produces errors of the same order, which is the case for second-order accurate schemes.

Finer analysis allows a better evaluation of the order of magnitude of the subgrid tensor. By using a subgrid viscosity model, i.e.

$$\tau_{ij} = -2 \nu_{\mathrm{sgs}} \overline{S}_{ij} \quad , \tag{7.20}$$

and using the local equilibrium hypothesis:

$$\varepsilon = -\tau_{ij} \overline{S}_{ij} = \nu_{\mathrm{sgs}} |\overline{S}|^2 \quad , \tag{7.21}$$

the amplitude of the subgrid tensor can be evaluated as:

$$|\tau_{ij}| \approx \nu_{\mathrm{sgs}} |\overline{S}| \approx \sqrt{\varepsilon \nu_{\mathrm{sgs}}} \quad . \tag{7.22}$$

By basing the computation of the subgrid viscosity on the subgrid kinetic energy:

$$\nu_{\mathrm{sgs}} \approx \overline{\Delta} \sqrt{q^2_{\mathrm{sgs}}} \quad , \tag{7.23}$$

and computing this energy from a Kolmogorov spectrum:

$$\sqrt{q^2_{\mathrm{sgs}}} \quad = \quad \left(\int_{k_c}^{\infty} E(k) dk \right)^{1/2}$$

$$\propto \left(\int_{k_c}^{\infty} k^{-5/3} dk \right)^{1/2}$$
$$\propto (k_c)^{-1/3}$$
$$\propto \overline{\Delta}^{1/3} \quad ,$$

we get for the subgrid viscosity:

$$\nu_{sgs} \propto \overline{\Delta}\,\overline{\Delta}^{1/3} = \overline{\Delta}^{4/3} \quad . \tag{7.24}$$

The order of magnitude of the corresponding subgrid tensor is:

$$|\tau_{ij}| \propto \sqrt{\varepsilon \nu_{sgs}} \propto \overline{\Delta}^{2/3} \quad . \tag{7.25}$$

This estimation is clearly different from those given previously, and shows that the subgrid tensor is theoretically dominant compared with the terms in $\overline{\Delta}^2$. This last evaluation is usually interpreted as being that of the subgrid Reynolds tensor R_{ij}, while the estimations of the tensors C_{ij} and L_{ij} given above are generally considered to be correct.

A generalized expansion for the whole subgrid tensor is proposed by Carati et al. [105]. These authors have proved that the differential expansion

$$\overline{\phi\psi} = \sum_{l,m=0}^{\infty} C_{lm}(G) \frac{\partial^l \overline{\phi}}{\partial x^l} \frac{\partial^m \overline{\psi}}{\partial x^m} \quad , \tag{7.26}$$

where $\phi(x)$ and $\psi(x)$ are two C^∞ real functions, and $C_{lm}(G)$ are some real coefficients which depend explicitly on the filter, is valid for all the filter kernels G such that:

$$\frac{G(-\mathrm{i}(\phi + \psi))}{G(-\mathrm{i}\phi)G(-\mathrm{i}\psi)} \in \mathbb{R}, \quad \mathrm{i}^2 = -1 \quad . \tag{7.27}$$

This is particularly true of all symmetric kernels. The resulting general form of the gradient model deduced from (7.26) is:

$$\tau_{ij} = \overline{u_i u_j} - \overline{u}_i \overline{u}_j = \sum_{l,m=0,\infty;(l,m)\neq(0,0)} C_{lm}(G) \frac{\partial^l \overline{u}_i}{\partial x^l} \frac{\partial^m \overline{u}_j}{\partial x^m} \quad . \tag{7.28}$$

The use of the Gaussian filter yields the following simplified form:

$$\tau_{ij} = \sum_{m=0,\infty} \left(\frac{\overline{\Delta}^2}{16} \right)^m \frac{\partial^m \overline{u}_i}{\partial x^m} \frac{\partial^m \overline{u}_j}{\partial x^m} \quad . \tag{7.29}$$

By retaining only the first term in (7.29), one recovers the gradient model (7.18).

Expression (7.28) can be recast in a general tensor-diffusivity form, as demonstrated by Adams and Leonard [3]. These authors have shown that, for discontinuous, but otherwise smooth, functions ψ and ϕ the series (7.26) can be summed up, leading to

$$\overline{\phi\psi} = \overline{\phi}\,\overline{\psi} + \frac{(\overline{\Delta}/4)^2}{2} R(\zeta_\psi) R(\zeta_\phi) \frac{\partial \overline{\phi}}{\partial x} \frac{\partial \overline{\psi}}{\partial x} \quad , \tag{7.30}$$

where a non-dimensional eddy-diffusivity $R(\zeta)$ is introduced. It is defined by the following relation:

$$\zeta^2 R^2(\zeta) = \frac{1}{2} - 2 \left(\int_0^{G^{-1}(\zeta)} G(\zeta') d\zeta' \right)^2 \quad . \tag{7.31}$$

A consistency constraint is $R(\zeta) \longrightarrow 1$ for $\zeta \longrightarrow 0$. This constraint is satisfied for the top-hat filter, but is violated for the Gaussian filter. Numerical experiments show that this form is ill-conditionned and leads to unstable results, and must be supplemented by a secondary regularization. To this end, most of the authors explicitly add another dissipative term to the momentum equation (see p. 218).

Layton et al. [341, 206, 234] proposed regularizing the gradient model by applying a smoothing operator. By choosing a convolution filter with a kernel G_2, the resulting model, referred to as the rational approximation model, is written as

$$\tau_{ij} = G_2 \star \left(\frac{\overline{\Delta}^2}{16} \frac{\partial \overline{u}_i}{\partial x} \frac{\partial \overline{u}_j}{\partial x} \right) \quad . \tag{7.32}$$

Iliescu et al. [341] discuss both explicit and implicit implementation of (7.32), which was applied to channel flows [219, 340, 339].

The need for a secondary regularization to get a stable computation using the gradient-like models may be understood by analysing their dissipative properties. The link with functional models of subgrid viscosity type is established by looking at the corresponding subgrid force term appearing in the momentum equations [761]. In three dimensions, we have:

$$\overline{\Delta}^2 M_2 \frac{\partial}{\partial x_j} \left(\frac{\partial \overline{u}_i}{\partial x_k} \frac{\partial \overline{u}_j}{\partial x_k} \right) = \overline{\Delta}^2 M_2 \overline{S}_{jk} \frac{\partial}{\partial x_j} \frac{\partial}{\partial x_k} \overline{u}_i \quad , \tag{7.33}$$

showing that $\overline{\Delta}^2 M_2 \overline{S}_{jk}$ plays the role of a tensorial subgrid viscosity. Since the tensor $\overline{S}_{jk}$ is not positive-definite, antidissipation occurs along stretching directions, which are associated with negative eigenvalues.

Full three-dimensional expressions up to the fourth-order terms of the subgrid tensor have been derived by several authors [81, 607], but these are very cumbersome and will not be presented here.

Taylor series expansion has also been used by several authors [81, 717] to derive an equivalent differential expression for the test filter and some tensorial quantities appearing in Germano's dynamic procedure for evaluation of the constants (see p. 137).

A worthy remark can be made dealing with the evaluation of the constants appearing in the models derived from truncated Taylor series expansions, which are considered as truncated polynomials in $\overline{\Delta}$. These parameters are commonly taken equal to the one appearing in the full (untruncated) expansions, and do not correspond to the coefficients of the best truncated polynomial approximation of the full expansions. Considering the best approximation, different coefficients are usually found, which may lead to different properties with respect to the preservation of symmetries of the filtered Navier–Stokes equations discussed in Sect. 3.3.4.

Connection between Taylor Expansion-Based Deconvolution and Iterative Deconvolution. Stolz, Adams and Kleiser [693] have demonstrated the practical equivalence between the gradient-type models (or tensor diffusivity models) and the use of the Van Cittert iterative technique for the soft deconvolution problem.

In the particular case of the Gaussian filter, an exact fourth-order defiltered variable is

$$\phi \approx G_l^{-1} \star \overline{\phi} = \overline{\phi} + \frac{\overline{\Delta}^2}{24} \frac{\partial^2 \overline{\phi}}{\partial x^2} + \frac{\overline{\Delta}^4}{1152} \frac{\partial^4 \overline{\phi}}{\partial x^4} + O(\overline{\Delta}^6) \quad . \tag{7.34}$$

The use of this relation yields the following expression for the approximated subgrid tensor:

$$\begin{aligned} \overline{u_i u_j} - \overline{u}_i \overline{u}_j &\approx \overline{(G_l^{-1} \star \overline{u}_i)(G_l^{-1} \star \overline{u}_j)} - \overline{u}_i \overline{u}_j \\ &= \frac{\overline{\Delta}^2}{24} \frac{\partial \overline{u}_i}{\partial x} \frac{\partial \overline{u}_j}{\partial x} + \frac{\overline{\Delta}^4}{288} \frac{\partial^2 \overline{u}_i}{\partial x^2} \frac{\partial^2 \overline{u}_j}{\partial x^2} + O(\overline{\Delta}^6) \quad . \end{aligned} \tag{7.35}$$

It is observed that each iterative deconvolution procedure yields the definition of a particular tensor-diffusivity model. Practical differences may arise from the discretization of the continuous derivate operators.

Solving the Hard Deconvolution Problem: Secondary Regularization. The secondary regularization is needed to obtain stable numerical simulation. This can be seen by two different ways:

1. The soft deconvolution procedure is restricted to the reconstruction of scales which are resolvable on the considered computational grid. Interactions with unresolved scales need to be taken into account from a theoretical point of view.
2. It has been demonstrated (at least in the particular case of the tensor-diffusivity model) that negative dissipation can occur, yielding possible numerical troubles. The net drain of resolved kinetic energy by unresolved scales needs to be taken into account (see Chap. 5 for a detailed discussion).

In practice, this secondary regularization is achieved by adding a dissipative term to the defiltered equations. Starting from relation (7.5), we arrive at the following formal evolution equation:

$$\frac{\partial \overline{\boldsymbol{u}}}{\partial t} + G \star \mathcal{NS}(G_l^{-1} \star \overline{\boldsymbol{u}}) = \mathcal{S}(\overline{\boldsymbol{u}}, \overline{\Delta}) \quad , \tag{7.36}$$

where the dissipative source term $\mathcal{S}(\overline{\boldsymbol{u}}, \overline{\Delta})$ can a priori be considered as a function of $\overline{\boldsymbol{u}}$ and $\overline{\Delta}$.

The Stolz–Adams Penalty Term. In order to account for kinetic energy transfer with scales which are not recovered by the deconvolution procedure, Stolz and Adams [691, 692, 694, 4] introduced a relaxation term, yielding

$$\mathcal{S}(\overline{\boldsymbol{u}}, \overline{\Delta}) = -\chi(Id - G_l^{-1} \star G) \star \overline{\boldsymbol{u}} \quad , \tag{7.37}$$

where χ is an empirical relaxation time. Since $(Id - G_l^{-1} \star G)$ is positive semidefinite, this relaxation term is purely dissipative. The use of this term can be interpreted as applying a second filtering operator $(G_l^{-1}\star) \circ (G\star)$ to $\overline{\boldsymbol{u}}$ every $1/\chi\Delta t$ time steps, Δt being the time step selected to perform the numerical time integration. It can also be interpreted as a penalization of the filtered solution. An important point is that this relaxation regularization is not equivalent to the use of subgrid viscosity type dissipation, since the associated spectral distributions of the dissipation are very different.

The relaxation time χ can be empiricaly chosen or dynamically evaluated. Numerical simulations show that the results may not be very sensitive to the value of this parameter. Stolz et al. [692, 693] reported no important changes for the channel flow case in the range $12.5u_\tau/h \leq \chi \leq 100u_\tau/h$, where u_τ is the skin friction and h the channel height. This relaxation time can also be evaluated dynamically in order to enforce a constant kinetic energy of the resolved subfilter modes, i.e. of the modes which are filtered out of the exact solution by the convolution filter but which are resolved on the computational grid. These modes correspond to the spectral band $[\pi/\overline{\Delta}, \pi/\Delta x]$. A measure of this energy is

$$E_{\mathrm{HF}} = \frac{1}{2}\left((Id - G_l^{-1} \star G) \star \overline{\boldsymbol{u}}\right)^2 \quad . \tag{7.38}$$

The equilibrium hypothesis can be expressed as

$$\frac{\partial E_{\mathrm{HF}}}{\partial t} = 0 \quad . \tag{7.39}$$

By combining relations (7.36) and (7.37), we get

$$\frac{\partial E_{\mathrm{HF}}}{\partial t} = \left[(Id - G_l^{-1} \star G) \star \overline{\boldsymbol{u}}\right] \cdot \left[-G \star \mathcal{NS}(G_l^{-1} \star \overline{\boldsymbol{u}}) - \chi(Id - G_l^{-1} \star G) \star \overline{\boldsymbol{u}}\right] \; , \tag{7.40}$$

leading to

$$\chi = -\frac{\left[(Id - G_l^{-1} \star G) \star \overline{\boldsymbol{u}}\right] \cdot \left[G \star \mathcal{NS}(G_l^{-1} \star \overline{\boldsymbol{u}})\right]}{\left[(Id - G_l^{-1} \star G) \star \overline{\boldsymbol{u}}\right] \cdot \left[(Id - G_l^{-1} \star G) \star \overline{\boldsymbol{u}}\right]} \quad . \tag{7.41}$$

Other Possibilities. The secondary regularization is achieved by many authors using subgrid-viscosity models, which provide the desired drain of resolved kinetic energy. All the subgrid-viscosity models can be used. The most employed ones are the Smagorinsky model and the dynamic Smagorinsky model. But it is worth noting that the secondary regularization can be achieved by means of the Implicit Large-Eddy Simulation approach (Sect. 5.3.4). As an example, Pasquetti and Xu used the Spectral Vanishing Viscosity approach in Refs. [582, 581].

Full Deconvolution Model Examples. As seen at the beginning of this section, the deconvolution approach necessitates the use of two models. An a priori infinite number of combinations between models for the soft and the hard deconvolution problems can be defined. Some examples are listed in Table 7.1. These two-part models for the full deconvolution problem can also be recast within the framework of mixed modeling. This point is discussed in Sect. 7.4.

Table 7.1. Examples of solutions to the Full Deconvolution Problem.

Ref.	Soft Deconvolution	Hard Deconvolution
[4, 694, 693, 692, 691]	Van Cittert (7.8)	relaxation (7.37)
[759]	Van Cittert (7.8)	Smagorinsky (5.90)
[143, 759]	tensor-diffusivity (7.18)	Smagorinsky (5.90)
[763, 764, 12]	tensor-diffusivity (7.18)	dynamic Smagorinsky
[341]	rational model (7.32)	Smagorinsky (5.90)

Single-Step Filtering Implementation of the Full Approximate Deconvolution Model. The practical implementation of the full approximate deconvolution model has been shown by Mathew *et al.* [499] to simplify as the application of a single filtering operator to the solution at the end of each time step of the numerical computation. Observing that relation (7.5) can be recast as

$$G \star \left[\frac{\partial \boldsymbol{u}^\bullet}{\partial t} + \mathcal{NS}(\boldsymbol{u}^\bullet)\right] = 0 \quad , \tag{7.42}$$

under the assumption that $\boldsymbol{u}^\bullet$ is sufficient close to $\boldsymbol{u}$, one obtains the following relation:

$$\frac{\partial \overline{\boldsymbol{u}}}{\partial t} = G \star \frac{\partial \boldsymbol{u}^\bullet}{\partial t} \quad . \tag{7.43}$$

Therefore, the basic implementation requires three steps to obtain the solution at time $(n+1)\Delta t$ (noticed $\overline{\boldsymbol{u}}(n+1)$ below) starting from the solution at the previous time step, $\overline{\boldsymbol{u}}(n)$

1. Evaluation of the approximate unfiltered field at time $n\Delta t$:

$$\boldsymbol{u}^\bullet(n) = G_l^{-1} \star \overline{\boldsymbol{u}}(n) \quad .$$

2. Time advancement of the approximate unfiltered solution :

$$\boldsymbol{u}^\bullet(n+1) = \boldsymbol{u}^\bullet(n) + \Delta t \frac{\partial \boldsymbol{u}^\bullet}{\partial t} + O(\Delta t^2) \quad .$$

3. Restriction of the new approximate unfiltered solution :

$$\overline{\boldsymbol{u}}(n+1) = G \star \boldsymbol{u}^\bullet(n+1) \quad .$$

Looking at this sequence, Mathew observes that the first and third steps can be combined in a unique one, leading to the following two-step algorithm

1. Time advancement of the approximate unfiltered solution :

$$\boldsymbol{u}^\circ = \boldsymbol{u}^\bullet(n) + \Delta t \frac{\partial \boldsymbol{u}^\bullet}{\partial t} + O(\Delta t^2) \quad .$$

2. Restriction of the new approximate unfiltered solution :

$$\boldsymbol{u}^\bullet(n+1) = (G_l^{-1} \star G) \star \boldsymbol{u}^\circ = Q_l \star \boldsymbol{u}^\circ \quad .$$

It is seen that the sole filter involved in practice is the filter Q_l, which is an lth order perturbation of the identity whose effect is mostly concentrated on high resolved wavenumbers. This two-step method accounts for the soft deconvolution model only. The hard deconvolution problem, or equivalently the secondary regularization, is reintroduced within the same framework using the fact that it is equivalent (at least in the case of the penalty term proposed by Stolz and Adams) to the application of the filter $(G_l^{-1}\star)\circ(G\star) = Q_l$ to the solution every $1/\chi\Delta t$ time steps. The full deconvolution problem can thus be implemented in a simple two-step procedure:

1. Time advancement of the approximate unfiltered solution:

$$\boldsymbol{u}^\circ = \boldsymbol{u}^\bullet(n) + \Delta t \frac{\partial \boldsymbol{u}^\bullet}{\partial t} + O(\Delta t^2) \quad .$$

2. Restriction of the new approximate unfiltered solution :

$$\boldsymbol{u}^\bullet(n+1) = \mathcal{G} \star \boldsymbol{u}^\circ \quad .$$

where the basic value $\mathcal{G} = Q_l$ is changed into $\mathcal{G} = Q_l \star Q_l = Q_l^2$ every $1/\chi\Delta t$ time steps.

Toward Higher-Order Deconvolution Models. The approximate deconvolution approach presented above can be seen as the lowest-order member of a general class of deconvolution approaches. This fact is emphasized by Mathew and his coworkers [499], who carried out the development at the next order. Restricting for the sake of simplicity the discussion to the formal, one-dimensional scalar conservation law :

$$\frac{\partial u}{\partial t} + \frac{\partial f(u)}{\partial x} = 0 \quad , \tag{7.44}$$

the deconvolution approach can be written as

$$\frac{\partial \overline{u}}{\partial t} + \frac{\partial f(\overline{u})}{\partial x} = \frac{\partial f(\overline{u})}{\partial x} - G \star \frac{\partial f(u)}{\partial x} = \mathcal{R} \quad . \tag{7.45}$$

The low-order approximate deconvolution procedure presented above corresponds to the following closure

$$\mathcal{R} \simeq \mathcal{R}_1 = \frac{\partial f(\overline{u})}{\partial x} - G \star \frac{\partial f(u^\bullet)}{\partial x} \quad . \tag{7.46}$$

The higher-order method is derived writing the remainder in this closure relation as

$$\mathcal{R} = \mathcal{R}_1 + \mathcal{R}_2, \quad \mathcal{R}_2 = G \star \left(\frac{\partial f(u^\bullet)}{\partial x} - \frac{\partial f(u)}{\partial x} \right) \quad , \tag{7.47}$$

and finding a computable approximation for $\mathcal{R}_2$. Such an expansion is found introducing the Taylor series expansion

$$\mathcal{R}_2 = G \star \frac{\partial}{\partial x} \left(\left. \frac{\partial f}{\partial u} \right|_u (u^\bullet - u) + O(u^\bullet - u)^2 \right) \quad . \tag{7.48}$$

An estimation of the leading term of this expansion is

$$\mathcal{R}_2 = G \star \frac{\partial}{\partial x} \left(\left. \frac{\partial f}{\partial u} \right|_u (G_l^{-1} \star G - Id) \star u \right) \quad , \tag{7.49}$$

which can be approximated as

$$\mathcal{R}_2 = G \star \frac{\partial}{\partial x} \left(\left. \frac{\partial f}{\partial u} \right|_{u=u^\bullet} (G_l^{-1} \star G - Id) \star u^\bullet \right) \quad . \tag{7.50}$$

The resulting higher-order formulation of the approximate deconvolution approach is

$$\frac{\partial \overline{u}}{\partial t} + G \star \frac{\partial f(u^\bullet)}{\partial x} = G \star \frac{\partial}{\partial x} \left(\left. \frac{\partial f}{\partial u} \right|_{u=u^\bullet} (G_l^{-1} \star G - Id) \star u^\bullet \right) \quad . \tag{7.51}$$

It is worth noticing that this refined modeling makes a term appearing in the right hand side of equation (7.51) that is similar to the empirical penalty term for secondary regularization (7.37) introduced by Stolz and Adams.

7.2.2 Non-linear Models

There are a number of ways of deriving nonlinear models: Horiuti [321], Speziale [683], Yoshizawa [788], and Wong [765] start with an expansion in a small parameter, while Lund and Novikov [462] use the mathematical properties of the tensors considered. It is this last approach that will be described first, because it is the one that best reveals the difference with the functional models. Kosovic's simplified model [401] and Wong's dynamic model [765] are then described.

Generic Model of Lund and Novikov. We assume that the deviator of the subgrid tensor can be expressed as a function of the resolved velocity field gradients (and not the velocity field itself, to ensure the Galilean invariance property), the unit tensor, and the square of the cutoff length $\overline{\Delta}$:

$$\tau_{ij} - \frac{1}{3}\tau_{kk}\delta_{ij} \equiv \tau_{ij}^{\mathrm{d}} = \mathcal{F}(\overline{S}_{ij}, \overline{\Omega}_{ij}, \delta_{ij}, \overline{\Delta}^2) \quad . \tag{7.52}$$

The isotropic part of τ is not taken into account, and is integrated in the pressure term because $\overline{S}$ and $\overline{\Omega}$ have zero traces. To simplify the expansions in the following, we use the reduced notation:

$$\overline{S}\,\overline{\Omega} = \overline{S}_{ik}\overline{\Omega}_{kj}, \quad \mathrm{tr}(\overline{S}\,\overline{\Omega}^2) = \overline{S}_{ij}\overline{\Omega}_{jk}\overline{\Omega}_{ki} \quad .$$

The most general form for relation (7.52) is a polynomial of infinite degree of tensors whose terms are of the form $\overline{S}^{a_1}\overline{\Omega}^{a_2}\overline{S}^{a_3}\overline{\Omega}^{a_4}$..., where the a_i are positive integers. Each terms in the series is multiplied by a coefficient, which is itself a function of the invariants of $\overline{S}$ and $\overline{\Omega}$. This series can be reduced to a finite number of linearly independent terms by the Cayley- Hamilton theorem. Since the tensor τ^{d} is symmetrical, we retain only the symmetrical terms here. The computations lead to the definition of eleven tensors, $m_1, ..., m_{11}$, with which $I_1, ..., I_6$ are associated:

$$\begin{array}{lllllll} m_1 &=& \overline{S}, & m_2 &=& \overline{S}^2, \\ m_3 &=& \overline{\Omega}^2, & m_4 &=& \overline{S}\,\overline{\Omega} - \overline{\Omega}\,\overline{S}, \\ m_5 &=& \overline{S}^2\overline{\Omega} - \overline{\Omega}\overline{S}^2, & m_6 &=& Id, \end{array}$$

$$\begin{array}{lllllll} m_7 &=& \overline{S}\,\overline{\Omega}^2 + \overline{\Omega}^2\overline{S}, & m_8 &=& \overline{\Omega}\,\overline{S}\,\overline{\Omega}^2 - \overline{\Omega}^2\overline{S}\,\overline{\Omega}, \\ m_9 &=& \overline{S}\,\overline{\Omega}\,\overline{S}^2 - \overline{S}^2\overline{\Omega}\,\overline{S}, & m_{10} &=& \overline{S}^2\overline{\Omega}^2 + \overline{\Omega}^2\overline{S}^2, \\ m_{11} &=& \overline{\Omega}\,\overline{S}^2\overline{\Omega}^2 - \overline{\Omega}^2\overline{S}^2\overline{\Omega}, \end{array} \tag{7.53}$$

$$\begin{array}{lllllll} I_1 &=& \mathrm{tr}(\overline{S}^2), & I_2 &=& \mathrm{tr}(\overline{\Omega}^2), \\ I_3 &=& \mathrm{tr}(\overline{S}^3), & I_4 &=& \mathrm{tr}(\overline{S}\,\overline{\Omega}^2), \\ I_5 &=& \mathrm{tr}(\overline{S}^2\overline{\Omega}^2), & I_6 &=& \mathrm{tr}(\overline{S}^2\overline{\Omega}^2\overline{S}\,\overline{\Omega}), \end{array} \tag{7.54}$$

where Id designates the identity tensor.

These tensors are independent in the sense that none can be decomposed into a linear sum of the ten others, if the coefficients are constrained to appear as polynomials of the six invariants defined above. If we relax this last constraint by considering the polynomial quotients of the invariants too, then only six of the eleven tensors are linearly independent. The tensors defined above are no longer linearly independent in two cases: when the tensor $\overline{S}$ has a double eigenvalue and when two components of the vorticity disappear when expressed in the specific reference of $\overline{S}$. The first case corresponds to an axisymmetrical shear and the second to a situation where the rotation is about a single axis aligned with one of the eigenvectors of $\overline{S}$. Assuming that neither of these conditions is verified, six of the terms of (7.53) are sufficient for representing the tensor τ, and five for representing its deviator part, which is consistent with the fact that a second-order symmetrical tensor with zero trace has only five degrees of freedom in the third dimension. We then obtain the generic polynomial form:

$$\begin{aligned}\tau^{\mathrm{d}} &= C_1\overline{\Delta}^2|\overline{S}|\overline{S} + C_2\overline{\Delta}^2(\overline{S}^2)^{\mathrm{d}} + C_3\overline{\Delta}^2(\overline{\Omega}^2)^{\mathrm{d}} \\ &+ C_4\overline{\Delta}^2(\overline{S}\,\overline{\Omega} - \overline{\Omega}\,\overline{S}) + C_5\overline{\Delta}^2\frac{1}{|\overline{S}|}(\overline{S}^2\overline{\Omega} - \overline{S}\,\overline{\Omega}^2) \quad , \end{aligned} \qquad (7.55)$$

where the C_i, $i = 1,5$ are constants to be determined. This type of model is analogous in form to the non-linear statistical turbulence models [682, 683]. Numerical experiments performed by the authors on cases of isotropic homogeneous turbulence have shown that this modeling, while yielding good results, is very costly. Also, computing the different constants raises problems because their dependence as a function of the tensor invariants involved is complex. Meneveau et al. [509] attempted to compute these components by statistical techniques, but achieved no significant improvement over the linear model in the prediction of the subgrid tensor eigenvectors. A priori tests carried out by Horiuti [325] have shown that the $(\overline{\Omega S} - \overline{S\Omega})$ term is responsible for a significant improvement of the correlation coefficient with the true subgrid tensor.

We note that the first term of the expansion corresponds to subgrid viscosity models for the forward energy cascade based on large scales, which makes it possible to interpret this type of expansion as a sequence of departures from symmetry: the isotropic part of the tensor is represented by a spherical tensor, and the first term represents a first departure from symmetry but prevents the inclusion of the inequality of the normal subgrid stresses[1]. The anisotropy of the normal stresses is included by the following terms, which therefore represent a new departure from symmetry.

[1] This is true for all modeling of the form $\tau = (\boldsymbol{V} \otimes \boldsymbol{V})$ in which $\boldsymbol{V}$ is an arbitrary vector. It is trivially verified that the tensor $(\boldsymbol{V} \otimes \boldsymbol{V})$ admits only a single non-zero eigenvalue $\lambda = (V_1^2 + V_2^2 + V_3^2)$, while the subgrid tensor in the most general case has three distinct eigenvalues.

Kosovic's Simplified Non-Linear Model. In order to reduce the algorithmic cost of the subgrid model, Kosovic [401] proposes neglecting certain terms in the generic model presented above. After neglecting the high-order terms on the basis of an analysis of their orders of magnitude, the author proposes the following model:

$$\begin{aligned}\tau_{ij} &= -(C_s\overline{\Delta})^2\left[2(2|\overline{S}|^2)^{1/2}\overline{S}_{ij} + C_1\left(\overline{S}_{ik}\overline{S}_{kj} - \frac{1}{3}\overline{S}_{mn}\overline{S}_{mn}\delta_{ij}\right)\right.\\ &+ \left. C_2\left(\overline{S}_{ik}\overline{\Omega}_{kj} - \overline{\Omega}_{ik}\overline{S}_{kj}\right)\right] \quad , \qquad (7.56)\end{aligned}$$

where C_s is the constant of the subgrid viscosity model based on the large scales (see Sect. 5.3.2) and C_1 and C_2 two constants to be determined. After computation, the local equilibrium hypothesis is expressed:

$$\begin{aligned}\langle\varepsilon\rangle &= -\langle\tau_{ij}\overline{S}_{ij}\rangle\\ &= (C_s\overline{\Delta})^2\langle 2\left[(2|\overline{S}|^2)^{1/2}\overline{S}_{ij}\overline{S}_{ij} + C_1\overline{S}_{ik}\overline{S}_{kj}\overline{S}_{ji}\right]\rangle \quad . \qquad (7.57)\end{aligned}$$

In the framework of the canonical case (isotropic turbulence, infinite inertial range, sharp cutoff filter), we get (see [45]):

$$\begin{aligned}\langle\overline{S}_{ij}\overline{S}_{ij}\rangle &= \frac{30}{4}\langle\left(\frac{\partial\overline{u}_1}{\partial x_1}\right)^2\rangle\\ &= \frac{3}{4}K_0\langle\varepsilon\rangle^{2/3}k_c^{4/3} \quad , \qquad (7.58)\end{aligned}$$

$$\begin{aligned}\langle\overline{S}_{ik}\overline{S}_{kj}\overline{S}_{ji}\rangle &= \frac{105}{8}\langle\left(\frac{\partial\overline{u}_1}{\partial x_1}\right)^3\rangle\\ &= -\frac{105}{8}\mathcal{S}(k_c)\left(\frac{1}{10}K_0\right)^{3/2}\langle\varepsilon\rangle k_c^2 \quad , \qquad (7.59)\end{aligned}$$

where coefficient $\mathcal{S}(k_c)$ is defined as:

$$\mathcal{S}(k_c) = -\langle\left(\frac{\partial\overline{u}_1}{\partial x_1}\right)^3\rangle / \langle\left(\frac{\partial\overline{u}_1}{\partial x_1}\right)^2\rangle^{3/2} \quad . \qquad (7.60)$$

Substituting these expressions in relation (7.57) yields:

$$\langle\varepsilon\rangle = (C_s\overline{\Delta})^2\left[1 - \frac{7}{\sqrt{960}}C_1\mathcal{S}(k_c)\right]\left(\frac{3}{2}K_0\right)^{3/2}k_c^2\langle\varepsilon\rangle \quad . \qquad (7.61)$$

This relation provides a way of relating the constants C_s and C_1 and thereby computing C_1 once C_s is determined by reasoning similar to that explained in the chapter on functional models. The asymptotic value of $\mathcal{S}(k_c)$ is evaluated by theory and experimental observation at between 0.4 and 0.8, as $k_c \to \infty$. The constant C_2 cannot be determined this way, since the contribution of the anti-symmetrical of the velocity gradient to the energy transfer is null[2].

On the basis of simple examples of anisotropic homogeneous turbulence, Kosovic proposes:

$$C_2 \approx C_1 \quad , \tag{7.62}$$

which completes the description of the model.

Dynamic Non-Linear Model. Kosovic's approach uses some hypotheses intrinsic to the subgrid modes, for example the existence of a theoretical the spectrum shape and the local equilibrium hypothesis. To relax these constraints, Wong [765] proposes computing the constants of the non-linear models by means of a dynamic procedure.

To do this, the author proposes a model of the form (we use the same notation here as in the description of the dynamic model with one equation for the kinetic energy, in Sect. 5.4.2):

$$\tau_{ij} = \frac{2}{3} q_{\rm sgs}^2 \delta_{ij} - 2C_1 \overline{\Delta} \sqrt{q_{\rm sgs}^2}\, \overline{S}_{ij} - C_2 \overline{N}_{ij} \quad , \tag{7.63}$$

where C_1 and C_2 are constants and $q_{\rm sgs}^2$ the subgrid kinetic energy, and

$$\overline{N}_{ij} = \overline{S}_{ik}\overline{S}_{kj} - \frac{1}{3}\overline{S}_{mn}\overline{S}_{mn}\delta_{ij} + \dot{\overline{S}}_{ij} - \frac{1}{3}\dot{\overline{S}}_{mm}\delta_{ij} \quad , \tag{7.64}$$

where $\dot{\overline{S}}_{ij}$ is the Oldroyd[3] derivative of $\overline{S}_{ij}$:

$$\dot{\overline{S}}_{ij} = \frac{D\overline{S}_{ij}}{Dt} - \frac{\partial \overline{u}_i}{\partial x_k}\overline{S}_{kj} - \frac{\partial \overline{u}_j}{\partial x_k}\overline{S}_{ki} \quad , \tag{7.65}$$

where D/Dt is the material derivative associated with the velocity field $\overline{\boldsymbol{u}}$. The isotropic part of this model is based on the kinetic energy of the subgrid modes (see Sect. 5.3.2). Usually, we introduce a test filter symbolized by a *tilde*, the cutoff length of which is denoted $\widetilde{\overline{\Delta}}$. Using the same model, the subgrid tensor corresponding to the test filter is expressed:

$$T_{ij} = \frac{2}{3} Q_{\rm sgs}^2 \delta_{ij} - 2C_1 \widetilde{\overline{\Delta}} \sqrt{Q_{\rm sgs}^2}\, \widetilde{\overline{S}}_{ij} - C_2 \widetilde{\overline{H}}_{ij} \quad , \tag{7.66}$$

---

[2] This is because we have the relation

$$\overline{\Omega}_{ij}\overline{S}_{ij} \equiv 0 \quad ,$$

since the tensors $\overline{\Omega}$ and $\overline{S}$ are anti-symmetrical and symmetrical, respectively.

[3] This derivative responds to the principle of objectivity, i.e. it is invariant if the frame of reference in which the motion is observed is changed.

where Q^2_{sgs} is the subgrid kinetic energy corresponding to the test filter, and $\widetilde{\overline{H}}_{ij}$ the tensor analogous to $\overline{N}_{ij}$, constructed from the velocity field $\widetilde{\overline{\boldsymbol{u}}}$. Using the two expressions (7.63) and (7.66), the Germano identity (5.138) is expressed:

$$\begin{aligned} L_{ij} &= T_{ij} - \widetilde{\tau}_{ij} \\ &\simeq \frac{2}{3}(Q^2_{\text{sgs}} - \widetilde{q^2_{\text{sgs}}})\delta_{ij} + 2C_1\overline{\Delta}A_{ij} + C_2\overline{\Delta}^2 B_{ij} \quad , \end{aligned} \tag{7.67}$$

in which

$$A_{ij} = \overline{S}_{ij}\sqrt{q^2_{\text{sgs}}} - \frac{\widetilde{\overline{\Delta}}}{\overline{\Delta}}\widetilde{\overline{S}}_{ij}\sqrt{Q^2_{\text{sgs}}} \quad , \tag{7.68}$$

$$B_{ij} = \widetilde{\overline{N}}_{ij} - \left(\frac{\widetilde{\overline{\Delta}}}{\overline{\Delta}}\right)^2 \widetilde{\overline{H}}_{ij} \quad . \tag{7.69}$$

We then define the residual E_{ij}:

$$E_{ij} = L_{ij} - \frac{2}{3}(Q^2_{\text{sgs}} - \widetilde{q^2_{\text{sgs}}})\delta_{ij} + 2C_1\overline{\Delta}A_{ij} + C_2\overline{\Delta}^2 B_{ij} \quad . \tag{7.70}$$

The two constants C_1 and C_2 are then computed in such a way as to minimize the scalar residual $E_{ij}E_{ij}$, i.e.

$$\frac{\partial E_{ij}E_{ij}}{\partial C_1} = \frac{\partial E_{ij}E_{ij}}{\partial C_2} = 0 \quad . \tag{7.71}$$

A simultaneous evaluation of these two parameters leads to:

$$2\overline{\Delta}C_1 \approx \frac{L_{mn}(A_{mn}B_{pq}B_{pq} - B_{mn}A_{pq}B_{pq})}{A_{kl}A_{kl}B_{ij}B_{ij} - (A_{ij}B_{ij})^2} \quad , \tag{7.72}$$

$$\overline{\Delta}^2 C_2 \approx \frac{L_{mn}(B_{mn}A_{pq}A_{pq} - A_{mn}A_{pq}B_{pq})}{A_{kl}A_{kl}B_{ij}B_{ij} - (A_{ij}B_{ij})^2} \quad . \tag{7.73}$$

The quantities q^2_{sgs} and Q^2_{sgs} are obtained by solving the corresponding evolution equations, which are described in the chapter on functional models. This completes computation of the subgrid model.

One variant that does not require the use of additional evolution equations is derived using a model based on the gradient of the resolved scales instead of one based on the subgrid kinetic energy, to describe the isotropic term. The subgrid tensor deviator is now modeled as:

$$\tau_{ij} - \frac{1}{3}\tau_{kk}\delta_{ij} = -2C_1\overline{\Delta}^2|\overline{S}|\overline{S}_{ij} - C_2\overline{N}_{ij} \quad . \tag{7.74}$$

The two parameters computed by the dynamic procedure are now $\overline{\Delta}^2 C_1$ and $\overline{\Delta}^2 C_2$. The expressions obtained are identical in form to relations (7.72) and (7.73), where the tensor A_{ij} is defined as:

$$A_{ij} = |\overline{S}|\overline{S}_{ij} - |\widetilde{\overline{S}}|\widetilde{\overline{S}}_{ij} \left(\frac{\widetilde{\overline{\Delta}}}{\overline{\Delta}} \right)^2 \quad . \tag{7.75}$$

7.2.3 Homogenization-Technique-Based Models

General Description. The theory of homogenization is a two-scale expansion technique originally developed in structural mechanics to model inhomogeneous materials with a periodic microstructure. If the slow (i.e. large) scale and the rapid (i.e. small) scale are very different (i.e. if the microstructure is very fine compared with the large scale variations of the material), the composite material can be represented by an homogeneous material whose characteristics can be computed theoretically through the two-scale expansion and an averaging step. Using the scale separation assumption between the resolved scales of motion and the subgrid scales, the homogenization was introduced by Perrier and Pironneau [588] to obtain an theoretical evaluation for the subgrid viscosity. This approach was resurrected twenty years later by Persson, Fureby and Svanstedt [589] who derived a new homogenization-based tensorial subgrid viscosity model and performed the first simulations with this class of models.

The homogenization approach, which consists in solving the evolution equations of the filtered field separately from those of the subgrid modes, is based on the assumption that the cutoff is located within the inertial range at each point. The resolved field $\overline{\boldsymbol{u}}$ and the subgrid field $\boldsymbol{u}'$ are computed on two different grids by a coupling algorithm. In all of the following, we adopt the hypothesis that $\overline{\boldsymbol{u}'} = 0$. The subgrid modes $\boldsymbol{u}'$ are then represented by a random process $\boldsymbol{v}^\delta$, which depends on the dissipation ε, and the viscosity ν, and which is transported by the resolved field $\overline{\boldsymbol{u}}$. This modeling is denoted symbolically:

$$\boldsymbol{u}' = \boldsymbol{v}^\delta \left(\varepsilon, \frac{\boldsymbol{x} - \overline{\boldsymbol{u}}t}{\delta}, \frac{t}{\delta^2} \right) \quad , \tag{7.76}$$

in which δ^{-1} is the largest wave number in the inertial range and δ^{-2} the highest frequency considered. As the inertial range is assumed to extend to the high wave numbers, δ is taken as small parameter. Let $\boldsymbol{u}^\delta$ be the solution to the problem:

$$\frac{\partial u_i^\delta}{\partial t} + \frac{\partial (u_i^\delta + v_i^\delta)(u_j^\delta + v_j^\delta)}{\partial x_j} - \nu \frac{\partial^2 u_i^\delta}{\partial x_k \partial x_k} = - \frac{\partial p^\delta}{\partial x_i} - \frac{\partial v_i^\delta}{\partial t} + \nu \frac{\partial^2 v_i^\delta}{\partial x_k \partial x_k} \quad . \tag{7.77}$$

If $\boldsymbol{v}^\delta$ is close to $\boldsymbol{u}'$, then $\overline{\boldsymbol{u}^\delta}$ is close to $\overline{\boldsymbol{u}}$. More precisely, we have:

$$\boldsymbol{u}^\delta = \overline{\boldsymbol{u}} + \delta \boldsymbol{u}^1 + \delta^2 \boldsymbol{u}^2 + ... \qquad (7.78)$$

The lowest-order terms in the filtered expansion yields a homogenized problem for $\overline{\boldsymbol{u}}$ which involves a subgrid stress tensor $\overline{\boldsymbol{v}^\delta \cdot \nabla \boldsymbol{u}^1 + \boldsymbol{u}^1 \cdot \nabla \boldsymbol{v}^\delta}$. The homogenization technique relies on the computation of $\boldsymbol{u}^1$ and the definition of stochastic model for $\boldsymbol{v}^\delta$.

A modeling of this kind, while satisfactory on the theoretical level, is not so in practice because the function $\boldsymbol{v}^\delta$ oscillates very quickly in space and time, and the number of degrees of freedom needed in the discrete system to describe its variations remains very high. To reduce the size of the discrete system significantly, other hypotheses are needed, leading to the definition of simplified models which are described in the following. The description of the models is limited to the minimum amount of details for the sake of simplicity. The reader is referred to Ref. [589] for a detailed presentation of the two-scale expansion of the Navier–Stokes equations.

Perrier–Pironneau Models. The first simplification introduced by Perrier and Pironneau [588] consists in choosing the random process in the form:

$$\boldsymbol{v}^\delta(\boldsymbol{x}, t) = \frac{1}{\delta} \boldsymbol{v}(\boldsymbol{x}, t, \boldsymbol{x}', t') \quad , \qquad (7.79)$$

in the space and time scales $\boldsymbol{x}'$ and t', respectively, of the subgrid modes are defined as:

$$\boldsymbol{x}' = \frac{\boldsymbol{x} - \overline{\boldsymbol{u}} t}{\delta}, \; t' = \frac{t}{\delta^2} \quad . \qquad (7.80)$$

The new variable $\boldsymbol{v}(\boldsymbol{x}, t, \boldsymbol{x}', t')$ oscillates slowly and can thus be represented with fewer degrees of freedom. Assuming that $\boldsymbol{v}$ is periodical depending on the variables $\boldsymbol{x}'$ and t' on a domain $\Omega_v = Z \times]0, T'[$, and that the average of $\boldsymbol{v}$ is null over this domain[4], it is demonstrated that the subgrid tensor is expressed in the form:

$$\tau = B \nabla \overline{\boldsymbol{u}} \quad , \qquad (7.81)$$

where the term $B\nabla\overline{\boldsymbol{u}}$ is computed by taking the average on the cell of periodicity Ω_v of the term $(\boldsymbol{v} \cdot \nabla \boldsymbol{u}^1 + \boldsymbol{u}^1 \cdot \nabla \boldsymbol{v})$, where $\boldsymbol{u}^1$ is the a solution on this cell of the problem:

$$\frac{\partial \boldsymbol{u}^1}{\partial t'} - \nu \nabla^2_{\boldsymbol{x}'} \boldsymbol{u}^1 + \boldsymbol{v} \cdot \nabla_{\boldsymbol{x}'} \boldsymbol{u}^1 + \boldsymbol{u}^1 \cdot \nabla_{\boldsymbol{x}'} \boldsymbol{v} = \nabla q - \boldsymbol{v} \cdot \nabla \overline{\boldsymbol{u}} - \overline{\boldsymbol{u}} \cdot \nabla \boldsymbol{v} \quad , \qquad (7.82)$$

$$\nabla_{\boldsymbol{x}'} \cdot \boldsymbol{u}^1 = 0 \quad , \qquad (7.83)$$

[4] This is equivalent to considering that $\boldsymbol{v}(\boldsymbol{x}, t, \boldsymbol{x}', t')$ is statistically homogeneous and isotropic, which is theoretically justifiable by the physical hypothesis of local isotropy.

where $\nabla_{\boldsymbol{x}'}$ designates the gradient with respect to the $\boldsymbol{x}'$ variables and q the Lagrange multiplier that enforces the constraint (7.83). This model, though simpler, is still difficult to use because the variable $(\boldsymbol{x} - \overline{\boldsymbol{u}}t)$ is difficult to manipulate. So other simplifications are needed.

To arrive at a usable model, the authors propose neglecting the transport of the random variable by the filtered field in the field's evolution equation. This way, the random variable can be chosen in the form:

$$\boldsymbol{v}^{\delta}(\boldsymbol{x},t) = \frac{1}{\delta}\boldsymbol{v}(\boldsymbol{x},t,\boldsymbol{x}'',t') \quad , \tag{7.84}$$

with

$$\boldsymbol{x}'' = \frac{\boldsymbol{x}}{\delta} \quad , \tag{7.85}$$

and where the time t' is defined as before. Assuming that $\boldsymbol{v}$ is periodic along $\boldsymbol{x}''$ and t' on the domain Ω_v and has an average of zero over this interval, the subgrid term takes the form:

$$\tau = A\nabla\overline{\boldsymbol{u}} \quad , \tag{7.86}$$

where A is a definite positive tensor such that the term $A\nabla\overline{\boldsymbol{u}}$ is equal to the average of the term $(\boldsymbol{v} \otimes \boldsymbol{u}^1)$ over Ω_v, in which $\boldsymbol{u}^1$ is a solution on Ω_v of the problem:

$$\frac{\partial \boldsymbol{u}^1}{\partial t'} - \nu\nabla^2_{\boldsymbol{x}''}\boldsymbol{u}^1 + \boldsymbol{v}\cdot\nabla_{\boldsymbol{x}''}\boldsymbol{u}^1 = \nabla q + \boldsymbol{v}\cdot\nabla\overline{\boldsymbol{u}} \quad , \tag{7.87}$$

$$\nabla_{\boldsymbol{x}''}\cdot\boldsymbol{u}^1 = 0 \quad . \tag{7.88}$$

Persson Tensorial Subgrid Viscosity. Persson, Fureby and Svanstedt [589, 227] followed a similar procedure to derive a fourth-rank tensorial subgrid viscosity. They use the following equation for $\boldsymbol{u}^1$ as a starting point (the rapid scale system coordinates is the same as in the second model of Perrier and Pironneau):

$$\frac{\partial \boldsymbol{u}^1}{\partial t'} - \nu\nabla^2_{\boldsymbol{x}''}\boldsymbol{u}^1 + \nabla_{\boldsymbol{x}''}\cdot(\boldsymbol{u}^1\otimes\boldsymbol{v} + \boldsymbol{v}\otimes\boldsymbol{u}^1) - \nabla_{\boldsymbol{x}''}q = \nabla\cdot(\boldsymbol{v}\otimes\overline{\boldsymbol{u}}) \quad , \tag{7.89}$$

$$\nabla_{\boldsymbol{x}''}\cdot\boldsymbol{u}^1 = 0 \quad . \tag{7.90}$$

Based on this system, the authors derived the following expression for the subgrid tensor:

$$\tau_{ij} = A_{ijkl}\frac{\partial \overline{u}_k}{\partial x_l} \quad , \tag{7.91}$$

where the anisotropic fourth-rank tensorial subgrid viscosity is defined as

$$A_{hjkl} = \frac{K_0 C_1 q_{\mathrm{sgs}}^2 \overline{\Delta}}{(2\pi)^{11/3} \nu} \widetilde{A}_{hjkl} \quad , \tag{7.92}$$

where the constant parameters are $K_0 = 1.4$ (Kolmogorov constant) and $C_1 = 1.05$. The subgrid kinetic energy q_{sgs}^2 is evaluated solving an additional evolution equation deduced from (5.119) that includes the new definition of the subgrid viscosity as a tensorial quantity instead of a scalar one. The non-dimensional parameter is given by the following spectral summation over the wave vectors $\boldsymbol{m}$

$$\begin{aligned} \widetilde{A}_{hjkl} \quad = \quad & - \sum_{\boldsymbol{m} \in \mathbb{Z}} R_m^{-17/3} \left[\left(1 - \frac{(m_k/\widetilde{\Delta}_k)^2 + (m_l/\widetilde{\Delta}_l)^2}{\widetilde{R}_m^2} \right. \right. \\ & \left. + \frac{(m_k/\widetilde{\Delta}_k)^2 (m_l/\widetilde{\Delta}_l)^2 (2 - \delta_{kl})}{\widetilde{R}_m^4} \right) (\delta_{hl}\delta_{jk} + \delta_{jl}\delta_{hk}) \\ & \left. + 2 \frac{(m_h/\widetilde{\Delta}_h)^2 (m_k/\widetilde{\Delta}_k)^2 (1 - \delta_{jl})}{\widetilde{R}_m^4} \right] \quad , \end{aligned} \tag{7.93}$$

where

$$\widetilde{R}_m = \sqrt{\sum_{n=1,3} \left(\frac{m_n}{\widetilde{\Delta}_n} \right)^2}, \quad R_m = \frac{\delta}{\overline{\Delta}} \widetilde{R}_m \quad , \tag{7.94}$$

where the scalar cutoff length on a Cartesian grid is defined as $\overline{\Delta} = (\overline{\Delta}_1 \overline{\Delta}_2 \overline{\Delta}_3)^{1/3}$, and δ is the expansion parameter. The cell aspect ratio are defined as $\widetilde{\Delta}_l = \overline{\Delta}_l / \overline{\Delta}$.

7.3 Scale Similarity Hypotheses and Models Using Them

7.3.1 Scale Similarity Hypotheses

Basic Hypothesis. The scale similarity hypothesis such as proposed by Bardina et al. [39, 40] consists in assuming that the statistical structure of the tensors constructed on the basis of the subgrid scales is similar to that of their equivalents evaluated on the basis of the smallest resolved scales. The spectrum of the solution based on this hypothesis is therefore broken down into three bands: the largest resolved scales, the smallest resolved scales (i.e. the test field), and the unresolved scales (see Fig. 5.14).

This statistical consistency can be interpreted in two complementary ways. The first uses the energy cascade idea. That is, the unresolved scales and the smallest resolved scales have a common history due to their interactions with the largest resolved scales. The classical representation of the cascade has it that the effect of the largest resolved scales is exerted on the smallest resolved scales, which in turn influence the subgrid scales, which are therefore indirectly forced by the largest resolved scales, but similarly to the smallest. The second interpretation is based on the idea of coherent structures. These structures have a non-local frequency signature[5], i.e. they have a contribution on the three spectral bands considered. Scale similarity is therefore associated with the fact that certain structures appear in each of the three bands, inducing a strong correlation of the field among the various levels of decomposition.

Extended Hypothesis. This hypothesis was generalized by Liu et al. [455] (see [506] for a more complete discussion) to a spectrum split into an arbitrary number of bands, as illustrated in Fig. 7.2. The scale similarity hypothesis is then re-formulated for two consecutive spectrum bands, with the consistent forcing being associated with the low frequency band closest to those considered. Thus the specific elements of the tensors constructed from the velocity field u^n and their analogous elements constructed from u^{n+1} are assumed to be the same. This hypothesis has been successfully verified in experiments in the case of a jet turbulence [455] and plane wake turbulence [570]. Liu et al. have also demonstrated that scale similarity persists during rapid straining [454].

7.3.2 Scale Similarity Models

This section presents the structural models constructed on the basis of the scale similarity hypothesis. All of them make use of a frequency extrapolation technique: the subgrid tensor is a approximated by an analogous tensor computed from the highest resolved frequencies. The following are described:

1. Bardina's model (p. 233) in which the subgrid tensor is computed by applying the analytical filter a second time and thereby evaluating the fluctuation of the resolved scales. This model is therefore inoperative when the filter is idempotent, because this fluctuation is then null.
2. Filtered Bardina model (p. 234), which is an improvement on the previous one. By construction, the subgrid tensor is a filtered quantity, which results in the application of a convolution product and is therefore nonlocal in the sense that it incorporates all the information contained in

[5] This is due to the fact that the variations of the velocity components associated with a vortex cannot be represented by a monochromatic wave. For example, the Lamb-Oseen vortex tangential velocity radial distribution is:

$$U_\theta = \frac{q}{r}\left(1 - e^{-r^2}\right) \quad ,$$

where r is the distance to the center and q the maximum velocity.

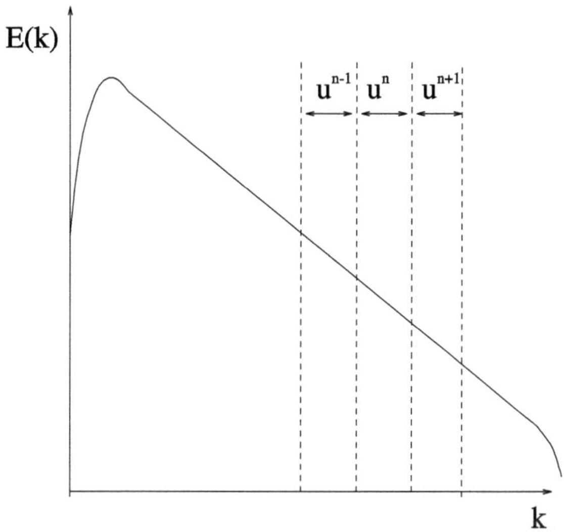

Fig. 7.2. Spectral decomposition based on the extended scale similarity hypothesis.

the support of the filter convolution kernel. It is proposed in this model to recover this non-local character by applying the filter to the modeled subgrid tensor.

3. Liu–Meneveau–Katz model (p. 234), which generalizes the Bardina model to the use of two consecutive filters of different shapes and cutoff frequencies, for computing the small scale fluctuations. This model can therefore be used for any type of filter.
4. The dynamic similarity model (p. 236), which can be used to compute the intensity of the modeled subgrid stresses by a dynamic procedure, whereas in the previous cases this intensity is prescribed by hypotheses on the form of the energy spectrum.

Bardina Model. Starting with the hypothesis, Bardina, Ferziger, and Reynolds [40] proposed modeling the C and R terms of the Leonard decomposition by a second application of the filter that was used to separate the scales. We furthermore have the approximation:

$$\overline{\phi\psi} \simeq \overline{\phi}\,\overline{\psi} \quad , \tag{7.95}$$

which allows us to say:

$$R_{ij} = (\overline{u}_i - \overline{\overline{u}}_i)(\overline{u}_j - \overline{\overline{u}}_j) \quad , \tag{7.96}$$

$$C_{ij} = (\overline{u}_i - \overline{\overline{u}}_i)\overline{\overline{u}}_j + (\overline{u}_j - \overline{\overline{u}}_j)\overline{\overline{u}}_i \quad , \tag{7.97}$$

or

$$R_{ij} + C_{ij} = (\overline{u}_i\overline{u}_j - \overline{\overline{u}}_i\overline{\overline{u}}_j) \quad . \tag{7.98}$$

Adding Leonard's term, which is computed directly from the resolved scales, we get:

$$\tau_{ij} = L_{ij} + R_{ij} + C_{ij} = (\overline{\overline{u}_i \overline{u}_j} - \overline{\overline{u}}_i \overline{\overline{u}}_j) \quad . \tag{7.99}$$

This can be re-written in another using the generalized central moments proposed by Germano [244]:

$$\tau_{ij} = \tau_{\mathrm{G}}([u_i]_{\mathrm{G}}, [u_j]_{\mathrm{G}}) \equiv \mathcal{L}_{ij} \quad . \tag{7.100}$$

The subgrid tensor is therefore approximated by the generalized central moment of the filtered field defined like the tensor $\mathcal{L}_{ij}$ in the Germano decomposition (see Sect. 3.3.2). Experience shows that this model is not effective when the filter is a Reynolds operator, because the contribution thus computed then cancels out. Contrary to the subgrid viscosity models, this one does not induce an alignment of the proper axis system of the subgrid tensor on those of the strain rate tensor. Tests performed on databases generated by direct numerical simulation have shown that this model leads to a very good level of correlation with the true subgrid tensor, including when the flow is anisotropic [320].

Despite its very good level of correlation[6], experience shows that this model is only slightly dissipative and that it underestimates the energy cascade. It does, on the other hand, include the backward cascade mechanism.

Filtered Bardina Model. The Bardina model (7.99) is local in space in the sense that it appears as a product of local values. This local character is in contradiction with the non-local nature of the subgrid tensor, so that each component appears in the form of a convolution product. To remedy this problem, Horiuti [323] and Layton and his colleagues [429, 428] propose the filtered Bardina model:

$$\tau_{ij} = \overline{(\overline{\overline{u}_i \overline{u}_j} - \overline{\overline{u}}_i \overline{\overline{u}}_j)} = \overline{\mathcal{L}}_{ij} \quad . \tag{7.101}$$

With this additional filtering operation, we recover the non-local character of the subgrid tensor.

Liu–Meneveau–Katz Model. The Bardina model uses a second application of the same filter, and therefore a single cutoff scale. This model is generalized to the case of two cutoff levels as [455]:

$$\tau_{ij} = C_1(\widetilde{\overline{u}_i \overline{u}_j} - \widetilde{\overline{u}}_i \widetilde{\overline{u}}_j) = C_1 \mathcal{L}_{ij}^{\mathrm{m}} \quad , \tag{7.102}$$

where the tensor $\mathcal{L}_{ij}^{\mathrm{m}}$ is now defined by two different levels of filtering. The test filter cutoff length designated by the *tilde* is larger than that of the first level.

[6] The correlation coefficient at the scalar level is generally higher than 0.8.

The constant C_l can be evaluated theoretically to ensure that the average value of the modeled generalized subgrid kinetic energy is equal to its exact counterpart [148]. This leads to the relation:

$$C_l = \frac{\langle \overline{u_k u_k} - \overline{u}_k \overline{u}_k \rangle}{\langle \widetilde{\overline{u}_k \overline{u}_k} - \widetilde{\overline{u}}_k \widetilde{\overline{u}}_k \rangle} \quad . \tag{7.103}$$

Let $\widehat{F}(k)$ and $\widehat{G}(k)$ be transfer functions associated with the grid filter and test filter, respectively, and let $E(k)$ be the energy spectrum of the exact solution. Relation (7.103) can then be re-written as:

$$C_l = \frac{\int_0^\infty (1 - \widehat{F}^2(k)) E(k) dk}{\int_0^\infty (1 - \widehat{G}^2(k)) \widehat{F}^2(k) E(k) dk} \quad . \tag{7.104}$$

Evaluations made using experimental data come to $C_l \simeq 1$ [570, 455][7]. Shah and Ferziger [668] propose extending this model to the case of non-symmetric filters.

To control the amplitude of the backward cascade induced by the model, especially near solid walls, Liu et al. [455] propose the modified form:

$$\tau_{ij} = C_l f(I_{\rm LS}) \mathcal{L}_{ij}^{\rm m} \quad , \tag{7.105}$$

where the dimensionless invariant $I_{\rm LS}$, defined as

$$I_{\rm LS} = \frac{\mathcal{L}_{lk}^{\rm m} \overline{S}_{lk}}{\sqrt{\mathcal{L}_{lk}^{\rm m} \mathcal{L}_{lk}^{\rm m}} \sqrt{\overline{S}_{lk} \overline{S}_{lk}}} \tag{7.106}$$

measures the alignment of the proper axes of the tensors $\mathcal{L}^{\rm m}$ and $\overline{S}$. As the kinetic energy dissipated by the subgrid model is expressed

$$\varepsilon = -\tau_{ij} \overline{S}_{ij} \quad , \tag{7.107}$$

we get, using model (7.105):

$$\varepsilon = -C_l f(I_{\rm LS}) I_{\rm LS} \quad . \tag{7.108}$$

The backward energy cascade is modulated by controlling the sign and amplitude of the product $f(I_{\rm LS}) I_{\rm LS}$. The authors considered a number of choices. The first is:

$$f(I_{\rm LS}) = \begin{cases} 1 & \text{if} \quad I_{\rm LS} \geq 0 \\ 0 & \text{otherwise} \end{cases} \quad . \tag{7.109}$$

This solution makes it possible to cancel out the representation of the backward cascade completely by forcing the model to be strictly dissipative.

[7] The initial value of 0.45 ± 0.15 given in [455] does not take the backward cascade into account.

One drawback to this is that the function f is discontinuous, which can generate numerical problems. A second solution that is continuous consists in taking:

$$f(I_{\mathrm{LS}}) = \begin{cases} I_{\mathrm{LS}} & \text{if} \quad I_{\mathrm{LS}} \geq 0 \\ 0 & \text{otherwise} \end{cases} \quad . \tag{7.110}$$

One last positive, continuous, upper-bounded solution is of the form:

$$f(I_{\mathrm{LS}}) = \begin{cases} (1 - \exp(-\gamma I_{\mathrm{LS}}^2)) & \text{if} \quad I_{\mathrm{LS}} \geq 0 \\ 0 & \text{otherwise} \end{cases} \quad , \tag{7.111}$$

in which $\gamma = 10$.

Dynamic Similarity Model. A dynamic version of the Liu–Meneveau–Katz model (7.102) was also proposed [455] for which the constant C_{l} will no longer be set arbitrarily. To compute this model, we introduce a third level of filtering identified by $\widehat{\cdot}$. The Q analogous to tensor $\mathcal{L}^{\mathrm{m}}$ for this new level of filtering is expressed:

$$Q_{ij} = (\widehat{\overline{\widetilde{u}}_i \overline{\widetilde{u}}_j} - \widehat{\overline{\widetilde{u}}}_i \widehat{\overline{\widetilde{u}}}_j) \quad . \tag{7.112}$$

The Germano–Lilly dynamic procedure, based here on the difference:

$$M_{ij} = f(I_{\mathrm{QS}}) Q_{ij} - \widetilde{f(I_{\mathrm{LS}}) \mathcal{L}_{ij}^{\mathrm{m}}} \quad , \tag{7.113}$$

where

$$I_{\mathrm{QS}} = \frac{Q_{mn} \widetilde{\overline{S}}_{mn}}{|Q| \, |\widetilde{\overline{S}}|} \quad , \tag{7.114}$$

yields:

$$C_{\mathrm{l}} = \frac{\mathcal{L}_{lk}^{\mathrm{m}} M_{lk}}{M_{pq} M_{pq}} \quad . \tag{7.115}$$

7.3.3 A Bridge Between Scale Similarity and Approximate Deconvolution Models. Generalized Similarity Models

The Bardina model can be interpreted as a particular case of the approximate soft deconvolution based models described in Sect. 7.2.

Using the second order differential approximation

$$\overline{\phi} = \phi + \frac{\alpha^{(2)}}{2} \frac{\partial^2 \phi}{\partial x^2} \quad , \tag{7.116}$$

the Bardina model (7.99) is strictly equivalent to the second order gradient model given by relations (7.18) and (7.19).

It can also be derived using the Van Cittert deconvolution procedure: a zeroth-order truncation in (7.7) is used to recover relation (7.95), while a first-order expansion is employed to derive (7.96).

The Bardina model then appears as a low-order formal expansion model for the subgrid tensor. Generalized scale similarity models can then be defined using higher-order truncations for the formal expansion [254]. They are formulated as

$$\tau_{ij} = \overline{(G_l^{-1} \star \overline{u}_i)(G_l^{-1} \star \overline{u}_j)} - \overline{(G_l^{-1} \star \overline{u})_i} - \overline{(G_l^{-1} \star \overline{u})_j} \quad , \tag{7.117}$$

where $G_l^{-1}\star$ designates the approximate soft deconvolution operator, defined in Sect. 7.2.1.

The same key idea can be extended to close the Navier–Stokes on a truncated wavelet basis: this was achieved by Hoffman for several simplified systems [315, 311, 309, 308].

7.4 Mixed Modeling

7.4.1 Motivations

The structural models based on the scale similarity idea and the soft deconvolution models/techniques on the one hand, and the functional models on the other hand, each have their advantages and disadvantages that make them seem complementary:

- The functional models, generally, correctly take into account the level of the energy transfers between the resolved scales and the subgrid modes. However, their prediction of the subgrid tensor structure, i.e. its eigenvectors, is very poor.
- The models based on the scale-similarity hypothesis or an approximate deconvolution procedure generally predict well the structure of the subgrid tensor better (and then are able to capture anisotropic effects and disequilibrium), but are less efficient for dealing with the level of the energy transfers. It is also observed that these models yield a poor prediction of the subgrid vorticity production [491], a fact coherent with their under-dissipative character.

Dubois et al [202] also observed that these two parts yield different correlations with reference data. Tests have shown that mixed models are able to capture disequilibrium and anisotropy effects [8, 454, 504, 506, 594, 12].

Shao et al. [670] propose a splitting of the kinetic energy transfer across the cutoff that enlights the role of each one of these two model classes. These authors combine the classical large-eddy simulation convolution filter to the ensemble average, yielding the following decompositions:

$$\begin{aligned} \boldsymbol{u} &= \langle \boldsymbol{u} \rangle + \boldsymbol{u}'^{e} && (7.118) \\ &= \overline{\boldsymbol{u}} + \boldsymbol{u}' && (7.119) \\ &= \langle \overline{\boldsymbol{u}} \rangle + \langle \boldsymbol{u}' \rangle + \overline{\boldsymbol{u}'^{e}} + \boldsymbol{u}'' \quad . && (7.120) \end{aligned}$$

Using this hybrid decomposition, the subgrid tensor splits into

$$\tau_{ij} = \tau_{ij}^{\text{rapid}} + \tau_{ij}^{\text{slow}} \quad , \tag{7.121}$$

with

$$\tau_{ij}^{\text{slow}} = \overline{u_i'^e u_j'^e} - \overline{u'^e}_i \overline{u'^e}_j \quad , \tag{7.122}$$

$$\begin{aligned} \tau_{ij}^{\text{rapid}} = & \overline{\langle u_i \rangle \langle u_j \rangle} - \langle \overline{u}_i \rangle \langle \overline{u}_j \rangle + \overline{u_i'^e \langle u_j \rangle} - \overline{u'^e}_i \langle \overline{u}_j \rangle \\ & + \overline{u_j'^e \langle u_i \rangle} - \overline{u'^e}_j \langle \overline{u}_i \rangle \quad . \end{aligned} \tag{7.123}$$

These two parts can be analysed as follows:

- The rapid part explicitly depends on the mean flow. This contribution arises only if the convolution filter is applied in directions where the mean flow gradients are non-zero. It is referred to as rapid because the time scale of its response to variations of the mean flow is small. Numerical experiments show that this part plays an important role when the turbulence is in a desiquilibrium state when: (i) production of kinetic energy is much larger than dissipation or (ii) the filter length is of the same order as the integral scale of turbulence. Subgrid stresses anisotropy is observed to be due to the interaction of this rapid part and the mean shear. Numerical simulations have shown that the rapid part escapes the functional modeling, but scale-similarity models and soft deconvolution models succeed in representing anisotropic energy transfer (both forward and backward cascades) associated to the rapid part.
- The slow part is always present in large-eddy simulation, because it does not depend on the mean flow gradients. It corresponds to the subgrid tensor analyzed through the previously described canonical analysis. It is referred to as slow because its relaxation time is long with respect to rapid part. Numerical tests show that subgrid viscosity model correctly capture the associated kinetic energy transfer.

One simple idea for generating subgrid models possessing good qualities on both the structural and energy levels is to combine a functional with a structural model, making what is called mixed models. This is generally done by combining a subgrid viscosity model for representing the energy cascade mechanism with a scale similarity. The stochastic backward cascade models are usually not included because the structural models are capable of including this phenomenon.

The resulting form is

$$\tau_{ij} - \frac{1}{3}\tau_{kk}\delta_{ij} = -2\nu_{\text{sgs}}\overline{S}_{ij} + (L_{ij} - \frac{1}{3}L_{kk}\delta_{ij}) \quad , \tag{7.124}$$

where ν_{sgs} is the subgrid viscosity (evaluated using one of the previously described model), and L_{ij} the evaluation obtained using one of the structural model[8].

Another argument for using a mixed model originates from the splitting of the full deconvolution model as the sum of the soft deconvolution model and the hard deconvolution model (see Sect. 7.2.1): the scale-similarity models are formally equivalent to truncated Taylor-series-expansion-based soft deconvolution models, and thus are not able to account for interactions between resolved scales and scales smaller than the mesh size on which the equations are solved. Consequently, they must be supplemented by another model, which will play the role of the secondary regularization within the framework of the deconvolution approach.

Examples of such models are described in the following.

7.4.2 Examples of Mixed Models

We present several examples of mixed models here:

1. The Smagorinsky–Bardina model (p. 240), for which the respective weights of each of the contributions are preset. This model is limited by the hypotheses underlying each of the two parts constituting it: the subgrid viscosity is still based on arguments of the infinite inertial range type. Experience shows, though, that combining the two models reduces the importance of the constraints associated with these underlying hypotheses, which improves the results.
2. A one-parameter mixed model whose subgrid viscosity is computed by a dynamic procedure of the Germano–Lilly type (p. 240). With this procedure, the respective weights of the structural and functional parts of the model can be modified, so that the subgrid viscosity model is now computed as a complement to the scale similarity model, which allows a better control of the dissipation induced. It can be said, though, that this procedure innately prefers the structural part.
3. The general form of N-parameter dynamic mixed model, as derived by Sagaut et al. (p. 241). This procedure is an extension of the previous one: the weights of the different parts of the model are dynamically computed, resulting in a possibly better approximation of the true subgrid stresses. The case of two-parameter dynamic mixed model is emphasized.

[8] Only scale-similarity models or approximate deconvolution models are used in practice to derive mixed models, because they are very easy to implement.

Other mixed models have already been presented within the framework of the deconvolution approach (see Sect. 7.2.1, p. 220), which will not be repeated in the present section. Application of the results dealing with the dynamic evaluation of the constants presented below to the Taylor series expansion based deconvolution models is straightforward.

Mixed Smagorinsky–Bardina Model. The first example is proposed by Bardina et al. [40] in the form of a linear combination of the Smagorinsky model (5.90) and the scale similarity model (7.99). The subgrid tensor deviator is then written:

$$\tau_{ij} - \frac{1}{3}\tau_{kk}\delta_{ij} = \frac{1}{2}\left(-2\nu_{\rm sgs}\overline{S}_{ij} + \mathcal{L}_{ij} - \frac{1}{3}\mathcal{L}_{kk}\delta_{ij}\right) \quad , \tag{7.125}$$

in which

$$\mathcal{L}_{ij} = \left(\overline{\overline{u}_i\,\overline{u}_j} - \overline{\overline{u}}_i\,\overline{\overline{u}}_j\right) \quad , \tag{7.126}$$

and

$$\nu_{\rm sgs} = C_{\rm s}\overline{\Delta}^2|\overline{S}| \quad . \tag{7.127}$$

Variants are obtained either by changing the subgrid viscosity model used or by replacing the tensor $\mathcal{L}$ with the tensor $\mathcal{L}^{\rm m}$ (7.105) or the tensor $\overline{\mathcal{L}}$ (7.101).

One-Parameter Mixed Dynamic Model. A mixed dynamic model was proposed by Zang, Street, and Koseff [799]. This is based initially on the Bardina model coupled with the Smagorinsky model, but the latter can be replaced by any other subgrid viscosity model. The subgrid viscosity model constant is computed by a dynamic procedure. The subgrid tensors corresponding to the two filtering levels are modeled by a mixed model:

$$\tau_{ij} - \frac{1}{3}\tau_{kk}\delta_{ij} = -2\nu_{\rm sgs}\overline{S}_{ij} + \mathcal{L}_{ij}^{\rm m} - \frac{1}{3}\mathcal{L}_{kk}^{\rm m}\delta_{ij} \quad , \tag{7.128}$$

$$T_{ij} - \frac{1}{3}T_{kk}\delta_{ij} = -2\nu_{\rm sgs}\widetilde{\overline{S}}_{ij} + \mathcal{Q}_{ij} - \frac{1}{3}\mathcal{Q}_{kk}\delta_{ij} \quad , \tag{7.129}$$

in which

$$\mathcal{Q}_{ij} = \widetilde{\overline{\overline{u}_i\overline{u}_j}} - \widetilde{\overline{\overline{u}}}_i\widetilde{\overline{\overline{u}}}_j \quad , \tag{7.130}$$

and

$$\nu_{\rm sgs} = C_{\rm d}\overline{\Delta}|\overline{S}| \quad . \tag{7.131}$$

The residual E_{ij} is now of the form:

$$E_{ij} = \mathcal{L}_{ij}^{\mathrm{m}} - \mathcal{H}_{ij} - \left(-2C_{\mathrm{d}}\overline{\Delta}^2 m_{ij} + \delta_{ij} P_{kk} \right) \quad , \tag{7.132}$$

in which

$$\mathcal{H}_{ij} = \widetilde{\overline{\overline{u}}_i \overline{\overline{u}}_j} - \widetilde{\overline{\overline{u}}}_i \widetilde{\overline{\overline{u}}}_j \quad , \tag{7.133}$$

$$\mathcal{L}_{ij}^{\mathrm{m}} = \widetilde{\overline{u}_i \overline{u}_j} - \widetilde{\overline{u}}_i \widetilde{\overline{u}}_j \quad , \tag{7.134}$$

$$m_{ij} = \left(\frac{\widetilde{\overline{\Delta}}}{\overline{\overline{\Delta}}} \right) |\widetilde{\overline{S}}| \widetilde{\overline{S}}_{ij} - \widetilde{|\overline{S}| \overline{S}_{ij}} \quad , \tag{7.135}$$

and where P_{kk} represents the trace of the subgrid tensor. The Germano–Lilly dynamic procedure leads to:

$$C_{\mathrm{d}} = \frac{(\mathcal{L}_{ij}^{\mathrm{m}} - \mathcal{H}_{ij}) m_{ij}}{m_{ij} m_{ij}} \quad . \tag{7.136}$$

In simulations performed with this model, the authors observed a reduction in the value of the dynamic constant with respect to that predicted by the usual dynamic model (i.e. based on the Smagorinsky model alone). This can be explained by the fact that the difference between the $\mathcal{L}^{\mathrm{m}}$ and $\mathcal{H}$ terms appears in the numerator of the fraction (7.136) and that this difference is small because these terms are very similar. This shows that the subgrid viscosity model serves only to model a residual part of the full subgrid tensor and not its entirety, as in the usual dynamic model.

Vreman et al. [747] propose a variant of this model. For the sake of mathematical consistency, by making the model for the tensor T_{ij} dependent only on the velocity field that corresponds to the same level of filtering, i.e. $\widetilde{\overline{\boldsymbol{u}}}$, these authors propose the following alternate form for the tensor $\mathcal{Q}_{ij}$:

$$\mathcal{Q}_{ij} = \widetilde{\overline{\overline{\widetilde{u}}_i \overline{\widetilde{u}}_j}} - \widetilde{\overline{\overline{\widetilde{u}}}}_i \widetilde{\overline{\overline{\widetilde{u}}}}_j \quad . \tag{7.137}$$

N-Parameter Dynamic Mixed Model.

General Formulation and Formal Resolution. A general form of multiparameter dynamic model was derived by Sagaut et al. [630]. Considering a formal N-part parametrization of the subgrid tensor, each term being associated to a real constant $C_l, k = 1, .., N$

$$\tau_{ij} = \sum_{l=1,N} C_l f_{ij}^l (\overline{\boldsymbol{u}}, \overline{\Delta}) \quad , \tag{7.138}$$

where the functions f_{ij}^l are the kernels of the different parts of the complete model. The equivalent formulation obtained at the test filter level is

$$T_{ij} = \sum_{l=1,N} C_l f_{ij}^l (\widetilde{\overline{\boldsymbol{u}}}, \widetilde{\overline{\Delta}}) \quad . \tag{7.139}$$

Inserting (7.138) and (7.139) into the Germano identity (5.138), we get the following definition of the residual E_{ij}:

$$E_{ij} = L_{ij} - \sum_{l=1,N} C_l m_{ij}^l, \quad m_{ij}^l = f_{ij}^l(\widetilde{\overline{\boldsymbol{u}}}, \widetilde{\overline{\Delta}}) - \widetilde{f_{ij}^l(\overline{\boldsymbol{u}}, \overline{\Delta})} \quad . \tag{7.140}$$

In order to obtain N linearly independent relations to compute the constants C_l, a first solution is to operate the contraction of the residual (7.140) with N independent tensors A_{ij}^l. The constants will then appear as the solutions of the following linear algebraic problem of rank N:

$$\sum_{l=1,N} C_l m_{ij}^l A_{ji}^k = L_{ij} A_{ji}^k, \quad k = 1, N \quad . \tag{7.141}$$

It is worth noting that the N constants are coupled, resulting in a global self-adaption of each constant. The particular case of the least-square minimization is recovered by taking $A_{ij}^k = m_{ij}^k, k = 1, N$.

In the case where some constants are not computed dynamically but are arbitrarily set, the linear system (7.141) corresponds to a ill-posed problem containing more constraints than degrees of freedom. Assuming that the N' first constants are arbitrarily chosen, we recover a well-posed problem of rank $N - N'$ by replacing L_{ij} with L'_{ij}, where

$$L'_{ij} = L_{ij} - \sum_{l=1,N'} C_l m_{ij}^l \quad . \tag{7.142}$$

Two-Parameter Dynamic Models. Mixed models have also been proposed by Salvetti [640, 643] and Horiuti [324] with two dynamic constants (one for the subgrid-viscosity part and one for the scale-similarity), corresponding to the $N = 2$ case in the previous section. These models have the advantage of avoiding any a priori preference for the contribution of one or the other model component. An extensive study of dynamic mixed model has been carried out by Sarghini et al. [646] for the plane channel case.

Numerical simulations show that two-paramater mixed models may yield disappointing results, because of a too low dissipation level. This is due to the fact that the coupled dynamic procedure described in the previous section gives a heavy weigth to the scale-similarity part of the model, because its correlation coefficient with the exact subgrid tensor is much higher than the one of the subgrid-viscosity model. This conclusion was confirmed by Anderson and Meneveau [12] in isotropic turbulence. These authors also observed a serious lack of robustness in regard to the dynamic Smagorinsky model when the test filter cutoff exceeds the integral scale of turbulence: negative values of the dynamic constant are returned in this case. To relieve this problem, Morinishi [542, 543] proposes to uncouple the computation of the dynamic constants. The modified algorithm for the dynamic procedure is:

1. Compute the constant associated to the subgrid-viscosity part of the model using a classical dynamic procedure, without taking the scale similarity part into account. This corresponds to the $N = 1$ case in the previous section. The resulting constant will ensure a correct level of dissipation.
2. Compute the constant associated to the scale-similarity part using a two-parameter dynamic procedure, but considering that the constant of the subgrid-viscosity part is fixed. This corresponds to $N = 2$ and $N' = 1$ in the previous section.

7.5 Differential Subgrid Stress Models

A natural way to find an expression for the subgrid stress tensor is to solve a prognostic equation for each component. This approach leads to the definition of six additional equations, and thus a very significant increase in both the model complexity and the computational cost. But the expected advantage is that such a model would a priori be able to account for a large class of physical mechanisms, yielding the definition of a very robust model. Three proposal have been published by different research groups:

1. The pioneering model of Deardorff (p. 243).
2. The model by Fureby et al. (p. 244), which is an improved version of the Deardorff model with better theoretical properties, such as realizability.
3. The models based of the use of transport equations for the velocity filtered probability density function (p. 245), which do not rely on an explicit model for the triple correlations.

7.5.1 Deardorff Model

Another approach for obtaining a model for the subgrid tensor consists in solving an evolution equation for each of its components. This approach proposed by Deardorff [173] is analogous in form to two-point statistical modeling. Here, we adopt the case where the filter is a Reynolds operator. The subgrid tensor τ_{ij} is thus reduced to the subgrid Reynolds tensor R_{ij}. We deduce the evolution equation of the subgrid tensor components from that of the subgrid modes (3.31)[9]:

$$\frac{\partial \tau_{ij}}{\partial t} = -\frac{\partial}{\partial x_k}\left(\overline{u}_k \tau_{ij}\right) - \tau_{ik}\frac{\partial \overline{u}_j}{\partial x_k} - \tau_{jk}\frac{\partial \overline{u}_i}{\partial x_k}$$

---

[9] This is done by applying the filter to the relation obtained by multiplying (3.31) by u'_j and taking the half-sum with the relation obtained by inverting the subscripts i and j.

$$-\frac{\partial}{\partial x_k}\overline{u_i'u_j'u_k'} + \overline{p'\left(\frac{\partial u_i'}{\partial x_j} + \frac{\partial u_j'}{\partial x_i}\right)}$$
$$-\frac{\partial}{\partial x_j}\overline{u_i'p'} - \frac{\partial}{\partial x_i}\overline{u_j'p'} - 2\nu\overline{\frac{\partial u_i'}{\partial x_k}\frac{\partial u_j'}{\partial x_k}} \quad . \tag{7.143}$$

The various terms in this equation have to be modeled. The models Deardorff proposes are:

– For the pressure–strain correlation term:

$$\overline{p'\left(\frac{\partial u_i'}{\partial x_j} + \frac{\partial u_j'}{\partial x_i}\right)} = -C_{\rm m}\frac{\sqrt{q^2_{\rm sgs}}}{\overline{\Delta}}\left(\tau_{ij} - \frac{2}{3}q^2_{\rm sgs}\delta_{ij}\right) + \frac{2}{5}q^2_{\rm sgs}\overline{S}_{ij} \quad , \tag{7.144}$$

where $C_{\rm m}$ is a constant, $q^2_{\rm sgs}$ the subgrid kinetic energy, and $\overline{S}_{ij}$ the strain rate tensor of the resolved field.

– For the dissipation term:

$$\nu\overline{\frac{\partial u_i'}{\partial x_k}\frac{\partial u_j'}{\partial x_k}} = \delta_{ij}C_{\rm e}\frac{(q^2_{\rm sgs})^{3/2}}{\overline{\Delta}} \quad , \tag{7.145}$$

where $C_{\rm e}$ is a constant.

– For the triple correlations:

$$\overline{u_i'u_j'u_k'} = -C_{3\rm m}\overline{\Delta}\sqrt{q^2_{\rm sgs}}\left(\frac{\partial}{\partial x_i}\tau_{jk} + \frac{\partial}{\partial x_j}\tau_{ik} + \frac{\partial}{\partial x_k}\tau_{ij}\right) \quad . \tag{7.146}$$

The pressure–velocity correlation terms $\overline{p'u_i'}$ are neglected. The values of the constants are determined in the case of isotropic homogeneous turbulence:

$$C_{\rm m} = 4.13, \ C_{\rm e} = 0.70, \ C_{3\rm m} = 0.2 \quad . \tag{7.147}$$

Lastly, the subgrid kinetic energy is determined using evolution equation (5.119).

7.5.2 Fureby Differential Subgrid Stress Model

An alternate form of the differential stress model is proposed by Fureby et al. [232], which has better symmetry preservation properties than the original Deardorff model. The triple correlation term is approaximated as

$$-\overline{u_i'u_j'u_k'} = \underbrace{c_q\sqrt{q^2_{\rm sgs}}\overline{\Delta}}_{\nu_{\rm sgs}}\frac{\partial \tau_{ij}}{\partial x_k} \quad , \tag{7.148}$$

with $c_q = 0.07$. Other terms are kept unchanged. Each term in the closed subgrid stress equations is now frame indifferent. The set of closed equations is realizable, and has the same transformation properties as the exact Navier–Stokes equations under a change of frame.

An alternate form of the subgrid viscosity in (7.148) which allows the representation of backscatter is also defined by Fureby:

$$\nu_{\text{sgs}} = -\frac{\tau_{ij}\overline{S}_{ij}}{\overline{S}_{ij}\overline{S}_{ij}} \quad . \tag{7.149}$$

7.5.3 Velocity-Filtered-Density-Function-Based Subgrid Stress Models

A methodology referred to as the Velocity Filtered Density Function to close the filtered Navier–Stokes equations is proposed by Gicquel et al. [264] on the grounds of the previous works by Pope and Givi. The subgrid scale stresses are reconstructed by considering the joint probability function of all of the components of the velocity vector. To this end, an exact evolution equation is derived for the velocity filtered density function in which unclosed terms are modeled. Two implementations have been proposed: a first one consists in discretizing transport equations for the subgrid stresses associated to the velocity filtered density function (this is the one emphasized below); the second one consists in solving it via a Lagrangian Monte Carlo scheme, which leads to the definition of an equivalent stochastic system. This last form should be classified as a structural model based on a stochastic reconstruction of subgrid scales, and will be mentioned in Sect. 7.7.

Definitions. The first step consists in defining the velocity filtered density function P_L

$$P_L(\boldsymbol{v};\boldsymbol{x},t) = \int_{-\infty}^{+\infty} \rho[\boldsymbol{v},\boldsymbol{u}(\boldsymbol{x}',t)]G(\boldsymbol{x}'-\boldsymbol{x})d\boldsymbol{x}' \quad , \tag{7.150}$$

where G is the convolution filter kernel and $\rho[\boldsymbol{v},\boldsymbol{u}(\boldsymbol{x}',t)]$ is the fine-grained density

$$\rho[\boldsymbol{v},\boldsymbol{u}(\boldsymbol{x}',t)] = \delta(\boldsymbol{v}-\boldsymbol{u}(\boldsymbol{x},t)) \quad . \tag{7.151}$$

It is observed that P_L has all the properties of the probability density function when the filter kernel is positive. The conditional filtered value of a dummy variable $\phi(\boldsymbol{x},t)$ is therefore defined as

$$\langle \phi(\boldsymbol{x},t)|\boldsymbol{u}(\boldsymbol{x},t)=\boldsymbol{v}\rangle_L = \langle \phi|\boldsymbol{v}\rangle_L \, \frac{\int_{-\infty}^{+\infty}\phi(\boldsymbol{x}',t)\rho[\boldsymbol{v},\boldsymbol{u}(\boldsymbol{x}',t)]G(\boldsymbol{x}'-\boldsymbol{x})d\boldsymbol{x}'}{P_L(\boldsymbol{v};\boldsymbol{x},t)} \quad . \tag{7.152}$$

where $\langle\alpha|\beta\rangle_L$ denotes the filtered value of α conditioned on β.

First Model. The starting point of the procedure is to write an expression for the time-derivative of P_L. Such an expression is found by applying the time-derivative operator to Eq. (7.150), yielding

$$\begin{aligned}\frac{\partial P_L(\boldsymbol{v};\boldsymbol{x},t)}{\partial t} &= \int_{-\infty}^{+\infty} \frac{\partial u_i(\boldsymbol{x}',t)}{\partial t}\frac{\partial \rho[\boldsymbol{v},\boldsymbol{u}(\boldsymbol{x}',t)]}{\partial v_i} G(\boldsymbol{x}'-\boldsymbol{x})d\boldsymbol{x}' \\ &= -\frac{\partial}{\partial v_i}\left[\left\langle \frac{\partial u_i}{\partial t}\middle| \boldsymbol{v}\right\rangle_L P_L(\boldsymbol{v};\boldsymbol{x},t)\right] \end{aligned} \quad (7.153)$$

The next step consists in eliminating the velocity time derivative using the momentum equation, leading to

$$\begin{aligned}\frac{\partial P_L}{\partial t} &+ \overline{u}_k\frac{\partial P_L}{\partial x_k} = -\frac{\partial}{\partial x_k}\left[(v_k-\overline{u}_k)P_L\right] + \frac{\partial \overline{p}}{\partial x_i}\frac{\partial P_L}{\partial v_i} \\ &- 2\nu\frac{\partial \overline{S}_{ik}}{\partial x_k}\frac{\partial P_L}{\partial v_i} + \frac{\partial}{\partial v_i}\left[\left(\left\langle \frac{\partial p}{\partial x_i}\middle|\boldsymbol{v}\right\rangle_L - \frac{\partial \overline{p}}{\partial x_i}\right)P_L\right] \\ &- 2\nu\frac{\partial}{\partial v_i}\left[\left(\left\langle \frac{\partial S_{ik}}{\partial x_i}\middle|\boldsymbol{v}\right\rangle_L - \frac{\partial \overline{S}_{ik}}{\partial x_i}\right)P_L\right] \quad , \end{aligned} \quad (7.154)$$

where the relation

$$\overline{\phi} \equiv G\star\phi = \int_{-\infty}^{+\infty} \langle\phi|\boldsymbol{v}\rangle_L P_L(\boldsymbol{v};\boldsymbol{x},t)d\boldsymbol{v} \quad , \quad (7.155)$$

was used. The last two terms in the right hand side of (7.154) are unclosed terms which require the definition of ad hoc subgrid models. The first term in the right hand side is related to the subgrid advection and is closed. The sum of the two unknown terms is modeled as follows

$$\begin{aligned}\frac{\partial}{\partial v_i}\left[\left(\left\langle \frac{\partial p}{\partial x_i}\middle|\boldsymbol{v}\right\rangle_L - \frac{\partial \overline{p}}{\partial x_i}\right)P_L\right] - 2\nu\frac{\partial}{\partial v_i}\left[\left(\left\langle \frac{\partial S_{ik}}{\partial x_i}\middle|\boldsymbol{v}\right\rangle_L - \frac{\partial \overline{S}_{ik}}{\partial x_i}\right)P_L\right] \\ \simeq -\frac{\partial}{\partial v_i}\left[G_{ij}(v_j-\overline{u}_i)P_L\right]\frac{1}{2}C_0\varepsilon\frac{\partial^2 P_L}{\partial v_i \partial v_i} \end{aligned} \quad (7.156)$$

with

$$G_{ij} = -\omega\left(\frac{1}{2}+\frac{3}{4}C_0\right)\delta_{ij} \quad , \quad (7.157)$$

with $C_0 = 2.1$ and where the subgrid dissipation rate ε and the subgrid mixing frequency ω are evaluated as

$$\varepsilon = \frac{\left(q^2_{\rm sgs}\right)^{3/2}}{\overline{\Delta}}, \quad \omega = \frac{\varepsilon}{q^2_{\rm sgs}} \quad . \quad (7.158)$$

The subgrid kinetic energy q^2_{sgs} is taken equal to half the trace of the subgrid stress tensor. This closure is equivalent the the one proposed by Rotta for the Reynolds-averaged Navier–Stokes equations.

The final stage consists in writing the corresponding equations for the subgrid stresses, which are defined here using the generalized central moment framework of Germano (Sect. 3.3.2), i.e. $\tau_{ij} = \tau_{\mathrm{G}}(u_i, u_j)$. Evolution equations consistant with the previous closed form of the velocity filtered density function are

$$\begin{aligned}\frac{\partial \tau_{ij}}{\partial t} &+ \frac{\partial}{\partial x_k}\left(\overline{u}_k \tau_{ij}\right) = -\frac{\partial \tau_{ijk}}{\partial x_k} + G_{ik}\tau_{jk} + G_{jk}\tau_{ik} \\ &-\tau_{ik}\frac{\partial \overline{u}_j}{\partial x_k} - \tau_{jk}\frac{\partial \overline{u}_i}{\partial x_k} + C_0 \varepsilon \delta_{ij} \quad . \end{aligned} \tag{7.159}$$

The last unknown term is the third-order generalized central moment $\tau_{ijk} = \tau_{\mathrm{G}}(u_i, u_j, u_k)$. Its value are taken from the Lagrangian Monte Carlo solver used to generate a prognostic stochastic velocity field whose pdf P_L satisfies equation (7.154). Details of the numerical implementation can be found in Ref. [264] and will not be reproduced here.

Second Model. An alternative form for relation (7.154) is

$$\begin{aligned}\frac{\partial P_L}{\partial t} &+ \overline{u}_k \frac{\partial P_L}{\partial x_k} = -\frac{\partial}{\partial x_k}\left[\left(v_k - \overline{u}_k\right) P_L\right] + \frac{\partial \overline{p}}{\partial x_i}\frac{\partial P_L}{\partial v_i} \\ &+ \frac{\partial}{\partial v_i}\left[\left(\left\langle \frac{\partial p}{\partial x_i}\middle| \boldsymbol{v}\right\rangle_L - \frac{\partial \overline{p}}{\partial x_i}\right) P_L\right] \\ &+ \nu \frac{\partial^2 P_L}{\partial x_k \partial x_k} - \frac{\partial^2}{\partial v_i \partial v_i}\left[\left\langle \nu \frac{\partial u_i}{\partial x_k}\frac{\partial u_j}{\partial x_k}\middle| \boldsymbol{v}\right\rangle_L P_L\right] \quad . \end{aligned} \tag{7.160}$$

Using the same procedure as in the previous case, the following closed form is obtained

$$\begin{aligned}\frac{\partial P_L}{\partial t} &+ \overline{u}_k \frac{\partial P_L}{\partial x_k} = -\frac{\partial}{\partial x_k}\left[\left(v_k - \overline{u}_k\right) P_L\right] + \frac{\partial \overline{p}}{\partial x_i}\frac{\partial P_L}{\partial v_i} \\ &+\nu \frac{\partial \overline{u}_i}{\partial x_k}\frac{\partial \overline{u}_j}{\partial x_k}\frac{\partial^2 P_L}{\partial v_i \partial v_j} + 2\nu \frac{\partial \overline{u}_i}{\partial x_k}\frac{\partial^2 P_L}{\partial v_i \partial v_k} \\ &-\frac{\partial}{\partial v_i}\left(G_{ij}(v_j - \overline{u}_j) P_L\right) + \frac{1}{2} C_0 \varepsilon \frac{\partial^2 P_L}{\partial v_i \partial v_j} \quad . \end{aligned} \tag{7.161}$$

The consistent subgrid stress equations are now

$$\begin{aligned}\frac{\partial \tau_{ij}}{\partial t} &+ \frac{\partial}{\partial x_k}\left(\overline{u}_k \tau_{ij}\right) = -\frac{\partial \tau_{ijk}}{\partial x_k} + G_{ik}\tau_{jk} + G_{jk}\tau_{ik} \\ &+\nu \frac{\partial^2 \tau_{ij}}{\partial x_k \partial x_k} - \tau_{ik}\frac{\partial \overline{u}_j}{\partial x_k} - \tau_{jk}\frac{\partial \overline{u}_i}{\partial x_k} + C_0 \varepsilon \delta_{ij} \quad . \end{aligned} \tag{7.162}$$

The difference with the previous model is that viscous diffusion is taken into account in the subgrid stress transport equations, leading to a better accuracy in flows where viscous effects are influencial. The numerical implementation relies on the same Monte Carlo approach as the previous model.

7.5.4 Link with the Subgrid Viscosity Models

We reach the functional subgrid viscosity models again starting with a model with transport equations for the subgrid stresses, at the cost of additional assumptions. For example, Yoshizawa et al. [794] proposed neglecting all the terms of equation (7.143), except those of production. The evolution equation thus reduced comes to:

$$\frac{\partial \tau_{ij}}{\partial t} = -\tau_{ik}\frac{\partial \overline{u}_j}{\partial x_k} - \tau_{jk}\frac{\partial \overline{u}_i}{\partial x_k} \quad . \tag{7.163}$$

Assuming that the subgrid modes are isotropic or quasi-isotropic, i.e. that the extra-diagonal elements of the subgrid tensor are very small compared with the diagonal elements, and that the latter are almost mutually equal, the right-hand side of the reduced equation (7.163) comes down to the simplified form:

$$-q_{\rm sgs}^2 \overline{S}_{ij} \quad , \tag{7.164}$$

in which $q_{\rm sgs}^2 = \overline{u_k' u_k'}/2$ is the subgrid kinetic energy. Let t_0 be the characteristic time of the subgrid modes. Considering the relations (7.163) and (7.164), and assuming that the relaxation time of the subgrid modes is much shorter than that of the resolved scales[10], we get

$$\tau_{ij} - \frac{1}{3}\tau_{kk}\delta_{ij} \approx -t_0 q_{\rm sgs}^2 \overline{S}_{ij} \quad . \tag{7.165}$$

The time t_0 can be evaluated by dimensional argument using the cutoff length $\overline{\Delta}$ and the subgrid kinetic energy:

$$t_0 \approx \frac{\overline{\Delta}}{\sqrt{q_{\rm sgs}^2}} \quad . \tag{7.166}$$

By entering this estimate into equation (7.165), we get an expression analogous to the one used in the functional modeling framework:

$$\tau_{ij} - \frac{1}{3}\tau_{kk}\delta_{ij} \approx -\overline{\Delta}\sqrt{q_{\rm sgs}^2}\,\overline{S}_{ij} \quad . \tag{7.167}$$

[10] We again find here the total scale-separation hypothesis 5.4.

7.6 Stretched-Vortex Subgrid Stress Models

7.6.1 General

Misra and Pullin [517, 608, 735], following on the works of Pullin and Saffman [609], proposed subgrid models using the assumption that the subgrid modes can be represented by stretched vortices whose orientation is governed by the resolved scales.

Supposing that the subgrid modes can be linked to a random superimposition of fields generated by axisymmetrical vortices, the subgrid tensor can be written in the form:

$$\tau_{ij} = 2 \int_{k_c}^{\infty} E(k) dk \langle E_{pi} Z_{pq} E_{qj} \rangle \quad , \tag{7.168}$$

in which $E(k)$ is the energy spectrum, E_{lm} the rotation matrix used to switch from the vortex coordinate system to the reference system, Z_{ij} the diagonal tensor whose main elements are $(1/2, 1/2, 0)$ and $\langle E_{pi} Z_{pq} E_{qj} \rangle$, the moment of the probability density function $P(\alpha, \beta)$ of the Euler angles α and β giving the orientation of the vortex axis with respect to the frame of reference. The statistical average performed on the Euler angles of a function f is defined as:

$$\langle f(E_{ij}) \rangle = \frac{1}{4\pi} \int_0^{\pi} \int_0^{2\pi} f(E_{ij}) P(\alpha, \beta) \sin(\alpha) d\alpha d\beta \quad . \tag{7.169}$$

Two pieces of information are therefore needed to compute the subgrid term: the shape of the energy spectrum for the subgrid modes and the subgrid structure orientation distribution function. As the use of an evolution equation for the probability density function yielded no satisfactory results, Misra and Pullin propose modeling this function as a product of Dirac functions or a linear combination of such products. These are of the general form:

$$P(\alpha, \beta) = \frac{4\pi}{\sin(\alpha)} \delta(\alpha - \theta) \delta(\beta - \phi) \quad , \tag{7.170}$$

where $\theta(\boldsymbol{x}, t)$ and $\phi(\boldsymbol{x}, t)$ determine the specific orientation considered. Defining the two unit vectors $\boldsymbol{e}$ and $\boldsymbol{e}^v$:

$$e_1 = \sin(\alpha)\cos(\beta), \; e_2 = \sin(\alpha)\sin(\beta), \; e_3 = \cos(\alpha), \tag{7.171}$$

$$e_1^v = \sin(\theta)\cos(\phi), \; e_2^v = \sin(\theta)\sin(\phi), \; e_3^v = \cos(\theta), \tag{7.172}$$

the subgrid tensor can be re-written in the form:

$$\tau_{ij} = \left(\delta_{ij} - e_i^v e_j^v\right) \int_{k_c}^{\infty} E(k) dk = \left(\delta_{ij} - e_i^v e_j^v\right) q_{\text{sgs}}^2 \quad . \tag{7.173}$$

The various models must thus specify the specific orientation directions of the subgrid structures. Three models are presented in the following. The subgrid kinetic energy $q^2_{\rm sgs}$ can be computed in different ways (see Sect. 9.2.3), for example by solving an additional evolution equation, or by using a double filtering technique. A local evaluation procedure in the physical space based on the second-order velocity structure function is proposed by Voekl et al. [735].

7.6.2 S3/S2 Alignment Model

A first hypothesis is to assume that the subgrid structures are oriented along the eigenvectors of the resolved strain rate tensor $\overline{S}_{ij}$ that corresponds to its two largest eigenvalues. This is equivalent to assuming that they respond instantaneously to the forcing of the large scales. Using $\boldsymbol{e}^{s2}$ and $\boldsymbol{e}^{s3}$ to denote these two vectors, and λ_2 and $\lambda_3 \geq \lambda_2$ the associated eigenvalues, we get

$$\tau_{ij} = q^2_{\rm sgs} \left[\lambda \left(\delta_{ij} - e^{s3}_i e^{s3}_j \right) + (1 - \lambda) \left(\delta_{ij} - e^{s2}_i e^{s2}_j \right) \right] \quad , \tag{7.174}$$

where the weighting coefficient is taken proportional to the norms of the eigenvalues:

$$\lambda = \frac{\lambda_3}{\lambda_3 + |\lambda_2|} \quad . \tag{7.175}$$

7.6.3 S3/ω Alignment Model

The second model is derived on the assumption that the subgrid structures are oriented along the third eigenvector of the tensor $\overline{S}_{ij}$, denoted $\boldsymbol{e}^{s3}$ as before, and the vorticity vector of the resolved field. The unit vector it carries is denoted $\boldsymbol{e}^{\omega}$ and is computed as:

$$\boldsymbol{e}^{\omega} = \frac{\nabla \times \overline{\boldsymbol{u}}}{|\nabla \times \overline{\boldsymbol{u}}|} \quad . \tag{7.176}$$

The subgrid tensor is evaluated as:

$$\tau_{ij} = q^2_{\rm sgs} \left[\lambda \left(\delta_{ij} - e^{s3}_i e^{s3}_j \right) + (1 - \lambda) \left(\delta_{ij} - e^{\omega}_i e^{\omega}_j \right) \right] \quad . \tag{7.177}$$

The weighting parameter λ is chosen arbitrarily. The authors performed tests considering the three values 0, 0.5, and 1.

7.6.4 Kinematic Model

Starting with the kinematics of a vortex filament entrained by a fixed velocity field, Misra and Pullin propose a third model, for which the vector $\boldsymbol{e}^v$ is obtained by solving an evolution equation. The equation for the ith component of this vector is:

$$\frac{\partial e^v_i}{\partial t} = e^v_j \frac{\partial \overline{u}_i}{\partial x_j} - e^v_i e^v_k e^v_j \frac{\partial \overline{u}_k}{\partial x_j} \quad . \tag{7.178}$$

The subgrid tensor is then evaluated by inserting the vector $\boldsymbol{e}^v$ thus computed into the expression (7.173).

7.7 Explicit Evaluation of Subgrid Scales

The models described in the present section are all based on an explicit evaluation of the subgrid scales $\boldsymbol{u}' \equiv (Id - G) \star \boldsymbol{u}$. Because the subgrid modes correspond to scales of motion that can not be represented at the considered filtering level (i.e. in practice on the computational grid), a new higher-resolution filtering level is introduced. Numerically, this is done by introducing an auxiliary computational grid (or a set of embedded auxiliary grids), whose mesh size is smaller than the original one. The subgrid field $\boldsymbol{u}'$ is evaluated on that grid using one of the model presented below, and then the non-linear $G \star ((\overline{\boldsymbol{u}} + \boldsymbol{u}') \otimes (\overline{\boldsymbol{u}} + \boldsymbol{u}'))$ is computed. The corresponding general algorithmic frame is

1. $\overline{\boldsymbol{u}}$ is known from a previous calculation, on the computational grid, i.e. at the G filtering level, whose characteristic length is $\overline{\Delta}$.
2. Define an auxiliary grid, associated to a new filtering level F with characteristic length $\tilde{\Delta} < \overline{\Delta}$, and interpolate $\overline{\boldsymbol{u}}$ on the auxiliary grid.
3. Compute the approximate subgrid field $\boldsymbol{u}'_a = (F - G) \star \boldsymbol{u}$ using a model on the auxiliary grid.
4. Compute the approximate non-filtered non-linear term at the F level on the auxiliary grid:
$$(\overline{\boldsymbol{u}} + \boldsymbol{u}'_a) \otimes (\overline{\boldsymbol{u}} + \boldsymbol{u}'_a) \quad .$$
5. Compute the approximate filtered non-linear term at the G level on the computational grid:
$$G \star ((\overline{\boldsymbol{u}} + \boldsymbol{u}'_a) \otimes (\overline{\boldsymbol{u}} + \boldsymbol{u}'_a)) \quad ,$$
and use it to compute the evolution of $\overline{\boldsymbol{u}}$.

It is worth noting that this class of models can be interpreted as a generalization of deconvolution-based models (see Sect. 7.2). Classical deconvolution models require the use of a second regularization to take into account the interactions with modes that cannot be reconstructed on the mesh. In the present case, these unresolved scales are explicitly reconstructed on a finer grid, rendering the approach more general. Looking at the algorithmic framework presented above, we can see that (i) the second step (interpolation) is equivalent to the soft deconvolution, the third step is an extension of the hard deconvolution and, (iii) the fifth step is associated to the primary regularization.

Several ways to compute the subgrid motion on the auxiliary grid have been proposed by different authors. They are classified by increasing order of complexity (computational cost):

1. Fractal Interpolation Procedure of the fluctuations, as proposed by Scotti and Meneveau (p. 253). The subgrid fluctuations are reconstructed in a deterministic way on the fine grid using an iterative fractal interpolation technique (several similar fractal reconstruction techniques can be found in [364]). This model is based on geometrical considerations only, and does not take into account any information dealing with the flow dynamics such as disequilibrium, anisotropy, ... But it provides an estimate of the subgrid motion at a very low cost.
2. Chaotic Map Model of McDonough et al. (p. 254). The subgrid fluctuations are approximated in a deterministic way using a very simple chaotic dynamical system, which is chosen in order to mimic some properties of the real turbulent fluctuations (amplitude, autocorrelation, distribution of velocity fluctuations, ...).
   This model is the easiest to implement, and induces a very small overhead. A problem is that it requires the definition of a realistic dynamical system, and then a complete knowledge of the turbulent motion characteristics at each point of the numerical simulation.
3. One-Dimensional Turbulence model for the fluctuations, as proposed by Kerstein and his co-workers (p. 257). This approach relies on the definition of a simplified model for the three subgrid velocity component along lines located inside the computational cells. This model can be seen as an improvement of the Chaotic Map Model, since the non linear cascade effect are taken into account via the use of chaotic map, but diffusion effects and resolved pressure coupling are incorporated by solving a differential equation.
4. Reconstruction of the subgrid velocity field using kinematic simulations (p. 259). This approach, proposed by Flohr and Vassilicos, provides an incompressible, random, statistically steady, isotropic turbulent velocity field with prescribed energy spectrum. This model does not require that we solve any differential equation, and thus has a very low algorithmic cost.
5. The Velocity Filtered Density Function proposed by Gicquel et al. (p. 260) is another model belonging to this family, based on the use of a stochastic model for the subgrid fluctuations. Its equivalent differential formulation being presented in Sect. 7.5.3, the emphasis is put here on the equivalent stochastic system. In this form, it can be interepreted as the most advanced stochastic reconstruction technique for the subgrid scales, since it retains all the complexity of the Navier–Stokes dynamics.
6. Subgrid Scale Estimation Procedure proposed by Domaradzki and his coworkers (p. 261). The subgrid fluctuation are now deduced from a simplified advection equation, deduced from the filtered Navier–Stokes operator. An evaluation of the subgrid motion production term is derived, and integrated over a time interval associated to characteristic relaxation time of the subgrid scale. This model makes it possible to evaluate the

subgrid motion at a very low computational cost, but requires the computation of an approximate inverse filter.

7. Multilevel Simulations (p. 263), which are based on the use of the exact Navier–Stokes equations on a set of embedded computational grids. The reduction of the computational effort with respect to the Direct Numerical Simulation is obtained by freezing (quasi-static approximation) the high-frequencies represented on fine grids for some time interval, leading to the definition of a cyclic strategy. These methods can be interpreted as a time-consistent extension of the classical multigrid procedures for steady computations. They correspond to the maximal computational effort, but also to the most realistic approach.

7.7.1 Fractal Interpolation Procedure

Scotti and Meneveau [661, 662] propose to reconstruct the subgrid velocity field using two informations: (i) the resolved velocity field, which is known on the coarsest grid, and (ii) the fractality of the velocity field. The fluctuations are evaluated by interpolating the resolved coarse-grid velocity field on the fine grid using a fractal interpolation technique.

We first describe this interpolation technique in the monodimensional case. It is based on an iterative mapping procedure. The fluctuating field $\boldsymbol{u}'_a$ is reconstructed within each interval of the coarse grid by introducing a local coordinate $\xi \in [0,1]$. Let us consider the interval $[x_{i-1}, x_{i+1}]$, where $i-1$ and $i+1$ are related to the grid index on the coarse grid. We have $\xi = (x - x_{i-1})/2\overline{\Delta}$. The proposed map kernel W for a function ϕ to interpolated on the considered interval is:

$$W[\phi](\xi) = \begin{cases} d_{i,1}\phi(2\xi) + q_{i,1}(2\xi) & \text{if } \xi \in [0,1/2] \\ d_{i,2}\phi(2\xi) + q_{i,2}(2\xi) & \text{if } \xi \in]1/2,1] \end{cases} , \tag{7.179}$$

where $q_{i,j}$ are polynomials and $d_{i,j}$ are stretching parameters. The authors propose to use the following linear polynomials:

$$\begin{aligned} q_{i,1}(\xi) &= (\phi(x_i) - \phi(x_{i-1}) - d_{i,1}(\phi(x_{i+1}) - \phi(x_{i-1}))\xi \\ &\quad +\phi(x_{i-1})(1 - d_{i,1}) \quad , \end{aligned} \tag{7.180}$$

$$\begin{aligned} q_{i,2}(\xi) &= (\phi(x_{i+1}) - \phi(x_i) - d_{i,2}(\phi(x_{i+1}) - \phi(x_{i-1}))\xi \\ &\quad -\phi(x_{i-1})d_{i,2} \quad . \end{aligned} \tag{7.181}$$

The fluctuation is defined as

$$\boldsymbol{u}'_a = \lim_{n\to\infty} W^n[\overline{\boldsymbol{u}}] = \underbrace{W \circ W \circ ... \circ W[u]}_{n \text{ times}} \quad . \tag{7.182}$$

The stretching parameters are such that the Hausdorff dimension D of the synthetic signal is equal to

$$D = \begin{cases} 1 + \frac{\log(|d_{i,1}|+|d_{i,2}|)}{\log(2)} & \text{if } 1 < |d_{i,1}| + |d_{i,2}| < 2 \\ 1 & \text{if } |d_{i,1}| + |d_{i,2}| \leq 1 \end{cases} \quad . \tag{7.183}$$

In order to conserve then mean value of the signal over the considered interval, we have $d_{i,1} = -d_{i,2} = d$. For three-dimensional isotropic turbulence, we have $D = 5/3$, yielding $d = \mp 2^{1/3}$.

This procedure theoretically requires an infinite number of iterations to build the fluctuating field. In practice, a finite number of iterations is used. The statistical convergence rate of process being exponential, it still remains a good approximation. A limited number of iterations can also be seen as a way to account for viscous effects.

The extension to the multidimensional case is straightforward, each direction of space being treated sequentially.

This procedure also makes it possible to compute analytically the subgrid tensor. The resulting model will not be presented here (see [662] for a complete description).

7.7.2 Chaotic Map Model

McDonough and his coworkers [552, 338, 469] propose an estimation procedure based on the definition of a chaotic dynamical system. The resulting model generates a contravariant subgrid-scale velocity field, represented at discrete time intervals on the computational grid:

$$\boldsymbol{u}'_a = A_u \boldsymbol{\zeta} \odot \boldsymbol{V} \quad , \tag{7.184}$$

where A is an amplitude coefficient evaluated from canonical analysis, $\boldsymbol{\zeta}$ an anisotropy correction vector consisting mainly of first-order structure function of high-pass filtered resolved scales, and $\boldsymbol{V}$ is a vector of chaotic algebraic maps. It is important noting that the two vectors are multiplied using a vector Hadamard product, defined for two vectors and a unit vector $\boldsymbol{i}$ according to:

$$(\boldsymbol{\zeta} \odot \boldsymbol{V}) \cdot \boldsymbol{i} \equiv (\boldsymbol{\zeta} \cdot \boldsymbol{i})(\boldsymbol{V} \cdot \boldsymbol{i}) \quad . \tag{7.185}$$

The amplitude factor is chosen such that the kinetic energy of the synthetic subgrid motion is equal to the energy contained in all the scales not resolved by the simulation. It is given by the expression:

$$A_u = C_u u_* Re_{\overline{\Delta}}^{1/6} \quad , \tag{7.186}$$

with

$$u_* = (\nu |\nabla \boldsymbol{u}|)^{1/2}, \quad Re_{\overline{\Delta}} = \frac{\overline{\Delta}^2 |\nabla \boldsymbol{u}|}{\nu} \quad ,$$

where ν is the molecular viscosity. The scalar coefficient C_u is evaluated from classical inertial range arguments. The suggested value is $C_u = 0.62$.

The anisotropy vector $\boldsymbol{\zeta}$ is computed making the assumption that the flow anisotropy is smoothly varying in wave-number. In a way similar to the one proposed by Horiuti (see Sect. 6.3.3), the first step consists in evaluating the anisotropy vector from the highest resolved frequency. In order to account for the anisotropy of the filter, the resolved contravariant velocity field $\boldsymbol{u}_c$ is considered. The resulting expression for $\boldsymbol{\zeta}$ is:

$$\boldsymbol{\zeta} = \sqrt{3}\frac{\boldsymbol{s}}{|J^{-1} \cdot \boldsymbol{s}|} \quad , \tag{7.187}$$

where J^{-1} is the inverse of the coordinate transformation matrix associated to the computational grid (and to the filter). The vector $\boldsymbol{s}$ is defined according to

$$\boldsymbol{s} \cdot \boldsymbol{i} = \sqrt{3}\frac{|\nabla(\widetilde{\boldsymbol{u}_c}' \cdot \boldsymbol{i})|}{|\nabla\widetilde{\boldsymbol{u}_c}'|} \quad , \tag{7.188}$$

where the $\widetilde{\boldsymbol{u}_c}'$ is related to the test field computed thanks to the use of the test filter of characteristic length $\widetilde{\overline{\Delta}} > \overline{\Delta}$.

We now describe the estimation procedure for the stochastic vector $\boldsymbol{V}$. In order to recover the desired cross-correlation between the subgrid velocity component, the vector $\boldsymbol{V}$ is defined as:

$$\boldsymbol{V} = A\boldsymbol{M} \quad , \tag{7.189}$$

where A is a tensor such that $R = A \cdot A^T$, where R is the correlation tensor of the subgrid scale velocity. In practice, McDonough proposes to use the evaluation:

$$A_{ij} = \frac{(\nabla\widetilde{u}_i')_j}{|\nabla\widetilde{u}_i'|} \quad . \tag{7.190}$$

Each component $M_i, i = 1, 2, 3$ of the vector $\boldsymbol{M}$ is of the form:

$$M_i = \sigma \sum_{l=0,N} a_l \sum_{m=1,N_l} M'_{lm} \quad , \tag{7.191}$$

where N_l is the binomial coefficient

$$N_l \equiv \begin{pmatrix} N \\ l \end{pmatrix} \quad ,$$

and $\sigma = 1.67$ is the standard deviation for the variable, and the weights a_l are given by

$$a_l = \sqrt{3}\left(p^l(1-p)^{(N-l)}\right)^{1/2}, \quad p = 0.7 \quad . \tag{7.192}$$

The maps M'_{lm} are all independent instances of one of the three following normalized maps:

– The *tent map*:

$$m^{(n+1)} = \begin{cases} R(-2-3m^{(n)}) & \text{if } m^{(n)} < -1/3 \\ R(3m^{(n)}) & \text{if } -1/3 \leq m^{(n)} \leq 1/3 \\ R(-2-3m^{(n)}) & \text{if } m^{(n)} > 1/3 \end{cases} , \qquad (7.193)$$

where $m^{(n)}$ is the nth instance of the discrete dynamical system, and $R \in [-1, 1]$.

– The *logistic map*:

$$m^{(n+1)} = RA_R m^{(n)} (1 - |m^{(n)}| A_m) \quad , \qquad (7.194)$$

with

$$A_R = 2 + 2\sqrt{2}, \quad A_m = \left(1 + \frac{1}{A_R}\right) \sqrt{\frac{3}{2}} \quad .$$

– The *sawtooth map*:

$$m^{(n+1)} = \begin{cases} R(2+3m^{(n)}) & \text{if } m^{(n)} < -1/3 \\ R(3m^{(n)}) & \text{if } -1/3 \leq m^{(n)} \leq 1/3 \\ R(-2+3m^{(n)}) & \text{if } m^{(n)} > 1/3 \end{cases} . \qquad (7.195)$$

The map parameter R is related to some physical flow parameter, since the bifurcation and autocorrelation behaviors of the map are governed by R. An ad hoc choice for R will make it possible to model some of the local history effects in a turbulent flow in a way that is quantitatively and qualitatively correct. It is chosen here to set the bifurcation parameter R on the basis of local flow values, rather than on global values such as the Reynolds number. That choice allows us to account for large-scale intermittency effects. Selecting the ratio of the Taylor λ and Kolmogorov η scales, a possible choice is:

$$R = \tanh \left\{ \left[\frac{(\lambda/\eta)}{(\lambda/\eta)_c} \right]^r \tanh^{-1}(R_c) \right\} \quad , \qquad (7.196)$$

where r is a scaling exponent empirically assumed to lie in the range $[4, 6]$, and $(\lambda/\eta)_c$ is a critical value of the microscale ratio that is mapped onto R_c, the critical value of R. Suggested values are given in Table 7.2.

The last point is related to the time scale of the subgrid scales. Let t_e be the characteristic relaxation time of the subgrid scales, to be evaluated using inertial range considerations. If this time scale is smaller than the time step Δt of the simulation (*the characteristic filter time*), then the stochastic

Table 7.2. Parameters of the Chaotic Map Model.

Map	R_c	$(\lambda/\eta)_c$	r
Logistic	$-(2+2\sqrt{2})^{1/2}$	26	5
Tent	-1/3	28.6	5
Sawtooth	-1/3	28.6	5

variables M_i' must be updated n_u times per time step, with

$$n_u \approx \frac{\Delta t}{t_e} = \left(\frac{\Delta t|\nabla \overline{\boldsymbol{u}}|}{f_M}\right) Re_{\overline{\Delta}}^{-1/3} \quad , \tag{7.197}$$

where f_M is a fundamental frequency associated with the chaotic maps used to generate the variables. It is defined as:

$$f_M = \frac{C}{\theta} \quad , \tag{7.198}$$

where C is some positive constant and θ the *integral iteration scale*

$$\theta = \frac{1}{2}\rho(0) + \sum_{l=1,\infty} \rho(l), \quad \rho(l) = \frac{\langle m^{(n)} m^{(n+l)} \rangle}{\langle m^{(n)} m^{(n)} \rangle} \quad , \tag{7.199}$$

which completes the description of the model. This model is Galilean- and frame-invariant, and automatically generates realizable Reynolds stresses. It reproduces the desired root-mean-square amplitude of subgrid fluctuations, along with the probability density function for this amplitude. Finally, the proper temporal auto-correlation function can be enforced.

7.7.3 Kerstein's ODT-Based Method

A more complex chaotic map model based on Kerstein's One-Dimensional Turbulence (ODT) approach[11] was also proposed [650, 387, 389, 390, 388, 305, 306, 208, 198, 386, 771]. This is a method for simulating turbulent fluctuations along one-dimensional lines of sight through a three-dimensional turbulent flows. The velocity fluctuations evolve by two mechanisms, namely the molecular diffusion and turbulent stirring. The latter mechanisms is taken into account by a sequence of fractal transformations denoted eddy events. An eddy event may be interpreted as a model of an individual eddy, whose location, length scale and frequency are determined using a non-linear probabilistic model.

[11] It is worth noting that ODT originates in the Linear Eddy Model [384, 385, 117, 118, 177, 696, 414, 413, 473].

The diffusive step consists in solving the following one-dimensional advection-diffusion equation for each subgrid velocity component along the line

$$\frac{\partial u_i'}{\partial t} + \frac{\partial}{\partial x_j}(V_j u_i') = \frac{\partial \overline{p}}{\partial x_i} + \nu \frac{\partial^2 u_i'}{\partial x^2} \quad , \tag{7.200}$$

where ν is the molecular viscosity and $\boldsymbol{V}$ is the local advective field such that

$$\overline{u}_i = \int_{\Omega} V_i(\xi) d\xi \quad , \tag{7.201}$$

where Ω is the volume based on the cutoff length $\overline{\Delta}$.

The second step, which accounts for non-linear effects, is more complex and consists in two mathematical operations. The first one is a measure-preserving map representing the turbulent stirring, while the second one is a modification of the velocity profiles in order to implement energy transfers among velocity components. These two steps can be expressed as

$$u_i'(x) \longleftarrow u_i'(f(x)) + c_i K(x) \quad , \tag{7.202}$$

where the stirring-related mapping $f(x)$ is defined as

$$f(x) = x_0 + \begin{cases} 3(x - x_0) & \text{if } x_0 \leq x \leq x_0 + l/3 \\ 2l - 3(x - x_0) & \text{if } x_0 + l/3 \leq x \leq x_0 + 2l/3 \\ 3(x - x_0) - 2l & \text{if } x_0 + 2l/3 \leq x \leq x_0 + l \\ x - x_0 & \text{otherwise} \end{cases} \tag{7.203}$$

where l is the length of the segment affected by the eddy event. The second term in the right hand side of (7.202) is implemented to capture pressure-induced energy redistribution between velocity components and therefore makes it possible to account for the return to isotropy of subgrid fluctuations. The kernel K is defined as

$$K(x) = x - f(x) \tag{7.204}$$

The amplitude coefficients c_i are determined for each eddy to enforce the two following constraints: (i) the total subgrid kinetic energy $E = \sum_i E_i = \sum_i \frac{1}{2} \int u_i'(x) u_i'(x) dx$ remains constant, and (ii) the subgrid scale spectrum must be realizable, i.e. the energy extracted from a velocity component cannot exceed the available energy in this component. The resulting definition of the coefficients is

$$c_i = \frac{27}{4l} \left(-w_i + \text{sign}(w_i) \sqrt{(1 - \alpha) w_i^2 + \frac{\alpha}{2} \sum_{j \neq i} w_j^2} \right) \quad , \tag{7.205}$$

where

$$w_i = \frac{1}{l^2} \int u'_i(f(x)) K(x) dx = \frac{4}{9l^2} \int_{x_0}^{x_0+l} u'_i(x)(l - 2(x - x_0)) dx \quad . \quad (7.206)$$

The degree of energy redistribution is governed by the parameter α, which is taken equal to 2/3 in Ref. [650] (corresponding to equipartition of the available energy among the velocity components).

The last element of the method is the eddy selection step, which give access to the time sequence of eddy events. All events are implemented instantaneously, but occur with frequencies comparable to turnover frequencies of associated turbulent structures. At each time step, the event-rate distribution is obtained by first associating a time scale $\tau(x_0, l)$ with every eddy event. Using l/τ and l^3/τ^2 as an eddy velocity scale and a measure of the energy of the eddy motion, respetively, the time scale τ is computed using the following relation

$$\left(\frac{l}{\tau}\right)^2 \sim (1-\alpha) w_1^2 + \frac{\alpha}{2}(w_2^2 + w_3^2) - Z\frac{\nu^2}{l^2} \quad , \quad (7.207)$$

where Z is the amplitude of the viscous penalty term that governs the size of the smallest eddies for given local strain conditions. A probabilistic model can be derived defining an event-rate distribution λ

$$\lambda(x_0, l, t) = \frac{C}{l^2 \tau(x_0, l, t)} \quad , \quad (7.208)$$

where C is an arbitrary parameter which determines the relative strength of turbulent stirring.

7.7.4 Kinematic-Simulation-Based Reconstruction

Following Flohr and Vassilicos [220], the incompressible, turbulent-like subgrid velocity field is generated by summing different Fourier modes

$$\boldsymbol{u}'(\boldsymbol{x}, t) = \sum_{n=1,N} \left(\boldsymbol{a}_n \cos(\boldsymbol{k}_n \cdot \boldsymbol{x} + \omega_n t) + \boldsymbol{b}_n \sin(\boldsymbol{k}_n \cdot \boldsymbol{x} + \omega_n t) \right) \quad , \quad (7.209)$$

where N is the number of Fourier modes, $\boldsymbol{a}_n$ and $\boldsymbol{b}_n$ are the amplitudes corresponding to wave vector $\boldsymbol{k}_n$, and ω_n is a time frequency. The wave vectors are randomly distributed in spherical shells:

$$\boldsymbol{k}_n = k_n(\sin\theta\cos\phi, \sin\theta\sin\phi, \cos\theta) \quad , \quad (7.210)$$

where θ and ϕ are uniformly distributed random angles within $[0, 2\pi[$ and $[0, \pi]$, respectively. The random uncorrelated amplitude vectors $\boldsymbol{a}_n$ and $\boldsymbol{b}_n$

are chosen such that

$$\boldsymbol{a}_n \cdot \boldsymbol{k}_n = \boldsymbol{b}_n \cdot \boldsymbol{k}_n = 0 \quad , \tag{7.211}$$

to ensure incompressibility, and

$$|\boldsymbol{a}_n|^2 = |\boldsymbol{b}_n|^n = 2E(k_n)\Delta k_n \quad , \tag{7.212}$$

where $E(k)$ is the prescribed energy spectrum, and Δk_n is the wave number increment between the shells. Recommended shell distributions in the spectral space are:

– Linear distribution

$$k_n = k_1 + \frac{k_N - k_1}{N-1}(n-1) \quad ; \tag{7.213}$$

– Geometric distribution

$$k_n = k_1 \left(\frac{k_N}{k_1}\right)^{(n-1)/(N-1)} \quad ; \tag{7.214}$$

– Algebraic distribution

$$k_n = k_1 n^{\log(k_N/k_1)/\log N} \quad . \tag{7.215}$$

The time frequency ω_n is arbitrary. Possible choices are $\omega_n = U_c k_n$ if all the modes are advected with a constant velocity U_c, and $\omega_n = \sqrt{k_n^3 E(k_n)}$ if it is proportional to the eddy-turnover time of mode n.

In practice, Flohr and Vassilicos use this model to evaluate the dynamics of a passive tracer, but do not couple it with the momentum equations. Nevertheless, it could be used to close the momentum equation too.

7.7.5 Velocity Filtered Density Function Approach

The reconstruction of the subgrid motion via a stochastic system which obeys the required probability density function is proposed by Gicquel et al. [264]. This method is also equivalent (up to the second order) to solving the differential equations for the subgrid stresses presented in Sect. 7.5.3. The bases of the method are presented in this section, and will not be repeated here.

The key of the present method is the definition of a Lagrangian Monte Carlo method, which is used to evaluate both the position $\mathcal{X}_i$ (in space) and the value of a surrogate of the subgrid velocity, $\mathcal{U}_i$, associated to a set of virtual particules. The value of the subgrid velocity in each cell of the Large-Eddy Simulation grid is defined as the statistical average over all the virtual particules that cross the cell during a fixed time interval.

The stochastic differential equations equivalent to the first model presented in Sect. 7.5.3 is

$$d\mathcal{X}_i(t) = \mathcal{U}_i(t)dt \quad , \tag{7.216}$$

$$\begin{aligned} d\mathcal{U}_i(t) &= \left(-\frac{\partial \overline{p}}{\partial x_i} + 2\nu \frac{\partial \overline{s}_{ik}}{\partial x_k} + G_{ij}(\mathcal{U}_j(t) - \overline{u}_j(t)) \right) dt \\ &\quad + \sqrt{C_0 \varepsilon}\, dW_i^v(t) \quad , \end{aligned} \tag{7.217}$$

where G_{ij}, C_0 and ε are defined in Sect. 7.5.3, and W_i^v denotes and independent Wiener–Levy process.

The second model proposed by Gicquel accounts for viscous diffusion and is expressed as

$$d\mathcal{X}_i(t) = \mathcal{U}_i(t)dt + \sqrt{2\nu}\, dW_i^x(t) \quad , \tag{7.218}$$

$$\begin{aligned} d\mathcal{U}_i(t) &= \left(-\frac{\partial \overline{p}}{\partial x_i} + 2\nu \frac{\partial \overline{s}_{ik}}{\partial x_k} + G_{ij}(\mathcal{U}_j(t) - \overline{u}_j(t)) \right) dt \\ &\quad + \sqrt{C_0 \varepsilon}\, dW_i^v(t) + \sqrt{2\nu} \frac{\partial \overline{u}_i}{\partial x_j} dW_j^x(t) \quad , \end{aligned} \tag{7.219}$$

where ν is the molecular viscosity and W_i^x is another independent Wiener–Levy process. In practice, convergence of the statistical average over the particules within each cell must be carefully checked to recover reliable results.

7.7.6 Subgrid Scale Estimation Procedure

A two-step subgrid scale estimation procedure in the physical space[12] is proposed by Domaradzki and his coworkers [458, 187, 394, 188]. In the first (kinematic) step, an approximate inversion of the filtering operator is performed, providing the value of the defiltered velocity field on the auxiliary grid. In the second (non-linear dynamic) step, scales smaller than the filter length associated to the primary grid are generated, resulting in an approximation of the full solution.

Let $\overline{\boldsymbol{u}}$ be the filtered field obtained on the primary computational grid, and $\boldsymbol{u}^\bullet$ the defiltered field on the secondary grid. That secondary grid is chosen such that the associated mesh size is twice as fine as the mesh size of the primary grid. We introduce the discrete filtering operator $G_{\rm d}$, defined such that

$$G_{\rm d} \boldsymbol{u}^\bullet = \overline{\boldsymbol{u}} \quad . \tag{7.220}$$

[12] A corresponding procedure in the spectral space is described in reference [190].

It is important to note that in this two-grid implementation, the right-hand side of equation (7.220) must first be interpolated on the auxiliary grid to recover a well-posed linear algebra problem. To avoid this interpolation step, Domaradzki proposes to solve directly the filtered Navier–Stokes equations on the finest grid, and to define formally the G filtering level by taking $\overline{\Delta} = 2\tilde{\Delta}$. The defiltered field $\boldsymbol{u}^\bullet$ is obtained by solving the inverse problem

$$\boldsymbol{u}^\bullet = (G_{\mathrm{d}})^{-1}\overline{\boldsymbol{u}} \quad . \tag{7.221}$$

This is done in practice by solving the corresponding linear system. In practice, the authors use an three-point discrete approximation of the box filter for G_{d} (see Sect. 13.2 for a description of discrete test filters). This step corresponds to an implicit deconvolution procedure (the previous ones were explicit procedures, based on the construction of the inverse operator *via* Taylor expansions or iterative procedures), and can be interpreted as an interpolation step of the filtered field on the auxiliary grid.

The $\boldsymbol{u}'_a$ subgrid velocity field is then evaluated using an approximation of its associated non-linear production term:

$$\boldsymbol{u}'_a = \theta_a \, N' \quad , \tag{7.222}$$

where θ_a and N' are a characteristic time scale and N' the production rate. These terms are evaluated as follows. The full convection term on the auxiliary grid is

$$-u_j^\bullet \frac{\partial u_i^\bullet}{\partial x_j}, j = 1,2,3 \quad . \tag{7.223}$$

This term accounts for the production of all the frequencies resolved on the auxiliary grid. Since we are interested in the production of the small scales only, we must remove the advection by the large scales, and restrict the resulting term tho the desired frequency range. The resulting term N'_i is

$$N'_i = (Id - G) \star \left(-(u_j^\bullet - \overline{u}_j) \frac{\partial u_i^\bullet}{\partial x_j} \right) \quad . \tag{7.224}$$

In practice, the convolution filter G is replaced by the discrete operator G_{d}. The production time θ_a is evaluated making the assumption that the subgrid kinetic energy is equal to the kinetic energy contained in the smallest resolved scales:

$$|\boldsymbol{u}'_a|^2 = \theta_a^2 |N'|^2 = \alpha^2 |\boldsymbol{u}^\bullet - \overline{\boldsymbol{u}}|^2 \implies \theta_a = \alpha \frac{2|\boldsymbol{u}^\bullet - \overline{\boldsymbol{u}}|}{|N'|} \quad , \tag{7.225}$$

where α is a proportionality constant, nearly equal to 0.5 for the box filter. This completes the description of the model.

7.7.7 Multi-level Simulations

This class of simulation relies on the resolution of an evolution equation for $\boldsymbol{u}'_a$ on the auxiliary grid. These simulations can be analyzed within the framework of the multiresolution representation of the data [293, 295, 17, 294], or similar theories such as the Additive Turbulent Decomposition [471, 338, 80].

Let us consider N filters $G_1, ..., G_N$, with associated cutoff lengths $\overline{\Delta}_1 \leq ... \leq \overline{\Delta}_N$. We define the two following sets of velocity fields:

$$\overline{\boldsymbol{u}}^n = G_n \star ... \star G_1 \star u = \mathcal{G}_1^n \star \boldsymbol{u} \quad , \tag{7.226}$$

$$\boldsymbol{v}^n = \overline{\boldsymbol{u}}^n - \overline{\boldsymbol{u}}^{n+1} = (\mathcal{G}_1^n - \mathcal{G}_1^{n+1}) \star \boldsymbol{u} = \mathcal{F}_n \star \boldsymbol{u} \quad . \tag{7.227}$$

The fields $\overline{\boldsymbol{u}}^n$ and $\boldsymbol{v}^n$ are, respectively, the resolved field at the nth level of filtering and the nth level details. We have the decomposition

$$\overline{\boldsymbol{u}}^n = \overline{\boldsymbol{u}}^{n-k} + \sum_{l=1,k} \boldsymbol{v}^{n-l} \quad , \tag{7.228}$$

yielding the following multiresolution representation of the data:

$$\boldsymbol{u} \equiv \{\overline{\boldsymbol{u}}^N, \boldsymbol{v}^1, ..., \boldsymbol{v}^{N-1}\} \quad . \tag{7.229}$$

The multilevel simulations are based on the use of embedded computational grids or a hierarchical polynomial basis to solve the evolution equations associated with each filtering level/details level. The evolution equations are expressed as

$$\frac{\partial \overline{\boldsymbol{u}}^n}{\partial t} + \mathcal{NS}(\overline{\boldsymbol{u}}^n) = -\tau^n = -[\mathcal{G}_1^n \star, \mathcal{NS}](\boldsymbol{u}), \quad n \in [1, N] \quad , \tag{7.230}$$

where $\mathcal{NS}$ is the symbolic Navier–Stokes operator and $[.,.]$ the commutator operator. The equations for the details are

$$\frac{\partial \boldsymbol{v}^n}{\partial t} + \mathcal{NS}(\overline{\boldsymbol{v}}^n) = -\tau^n = -[\mathcal{F}_n \star, \mathcal{NS}](\boldsymbol{u}), \quad n \in [1, N-1] \quad , \tag{7.231}$$

or, equivalently,

$$\frac{\partial \boldsymbol{v}^n}{\partial t} + \mathcal{NS}(\overline{\boldsymbol{u}}^n) - \mathcal{NS}(\overline{\boldsymbol{u}}^{n+1}) = -\tau^n + \tau^{n+1}, \quad n \in [1, N-1] \quad . \tag{7.232}$$

There are three possibilities for reducing the complexity of the simulation with respect to Direct Numerical Simulation:

- The use of a cycling strategy between the different grid levels. Freezing the high-frequency details over some time while integrating the equations for the low-frequency part of the solution results in a reduction of the simulation complexity. This is referred to as the quasistatic approximation for the

high frequencies. The main problem associated with the cycling strategy is the determination of the time over which the high frequencies can be frozen without destroying the quality of the solution. Some examples of such a cycling strategy can be found in the Multimesh method of Voke [736], the Non-Linear Galerkin Method [174, 579, 705, 222, 223, 224, 91], the Incremental Unknowns technique [73, 201, 128, 200], Tziperman's MTS algorithm [722], Liu's multigrid method [452, 453] and the Multilevel algorithm proposed by Terracol et al. [711, 710, 633].
- The use of simplified evolution equations for the details instead of (7.231). A linear model equation is often used, which can be solved more easily than the full nonlinear mathematical model. Some examples among others are the Non-Linear Galerkin method, early versions of the Variational Multiscale approach proposed by Hughes et al. [332, 331, 336, 335, 333, 334], and the dynamic model of Dubrulle et al. [203, 426]. Another possibility is to assume that the nth-level details are periodic within the filtering cell associated with the $(n-1)$th filtering level. Each cell can then be treated separately from the others. An example is the Local Galerkin method of McDonough [468, 466, 467]. It is interesting to note that this last assumption is shared by the Homogenization approach developed by Perrier and Pironneau (see Sect. 7.2.3). Menon and Kemenov [382] use a simplified set of one-dimensional equations along lines. A reminiscent approach was developed by Kerstein on the grounds of the stochastic One-Dimensional Turbulence (ODT) model (presented in Sect. 7.7.3).
- The use of a limited number of filtering levels. In this case, even at the finest description level, subgrid scales exist and have to be parametrized. The gain is effective because it is assumed that simple subgrid models can be used at the finest filtering level, the associated subgrid motion being closer to isotropy and containing much less energy than at the coarser filtering levels. Examples, among others, are the Multilevel algorithm of Terracol [710, 633], the Modified Estimation Procedure of Domaradzki [193, 183], and the Resolvable Subfilter Scales (RSFS) model [806].

Some strategies combining these three possibilities can of course be defined. The efficiency of the method can be further improved by using a local grid refinement [699, 62, 378]. Non-overlapping multidomain techniques can also be used to get a local enrichment of the solution [611, 633]. These methods are presented in Chap. 11.

We now present a few multilevel models for large-eddy simulation. The emphasis is put here on methods based on two grid levels, and which can be interpreted as models, in the sense that they rely on some simplifications and cannot be considered just as multilevel algorithms applied to classical large-eddy simulations. In these methods, the secondary grid level is introduced to compute the fluctuations $\boldsymbol{u}'$, and not for the purpose of reducing the cost of the primary grid computation. This latter class of methods, which escapes the simple closure problem, is presented in Chap. 11. It is worth noting that

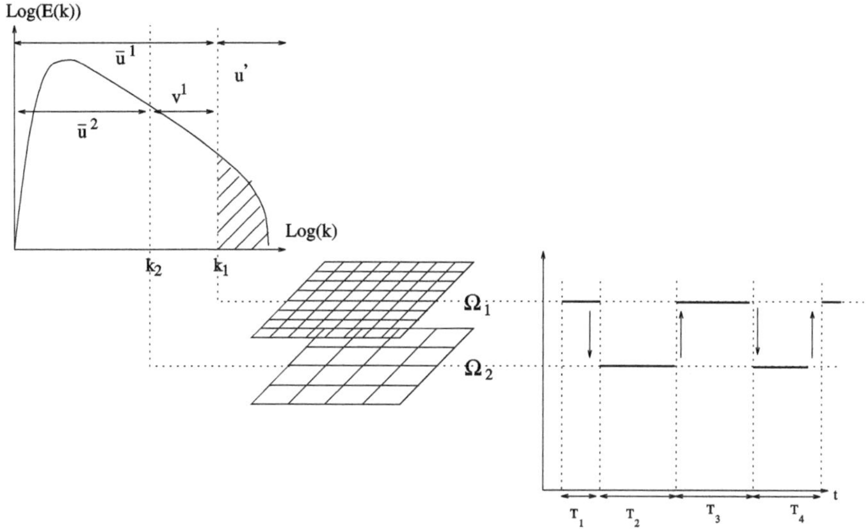

Fig. 7.3. Schematic of the three-level models. *Left*: spectral decomposition; *Middle*: computational grid; *Right*: time cycling.

these two-grid methods correspond to a three-level decomposition of the exact solution: two filters are applied in order to split the exact solution into three spectral bands (see Fig. 7.3).

Using Harten's representation (7.229), this decomposition is expressed as

$$\boldsymbol{u} = \{\overline{\boldsymbol{u}}^2, \boldsymbol{v}^1, \boldsymbol{v}^0\} \quad , \tag{7.233}$$

where $\overline{\boldsymbol{u}}^2$ is the resolved filtered field at the coarsest level, $\overline{\boldsymbol{u}}^1 = \overline{\boldsymbol{u}}^2 + \boldsymbol{v}^1$ is the resolved filtered field at the finest level, and $\boldsymbol{v}^0$ is the unresolved field at the finest level, i.e. the true subgrid velocity field. The detail $\boldsymbol{v}^1$ corresponds to the part of the unresolved field at the coarsest level which is resolved at the finest filtering level.

The coupling between these three spectral bands and the associated closure problem can be understood by looking at the nonlinear term. For the sake of simplicity, but without restricting the generality of the results, we will assume here that the filtering operators perfectly commute with differential operators, and that the domain is unbounded. The remaining coupling comes from the nonlinear convective term. For the exact solution, it is expressed as

$$\mathcal{B}(\boldsymbol{u}, \boldsymbol{u}) = \mathcal{B}(\overline{\boldsymbol{u}}^2 + \boldsymbol{v}^1 + \boldsymbol{v}^0, \, \overline{\boldsymbol{u}}^2 + \boldsymbol{v}^1 + \boldsymbol{v}^0) \quad , \tag{7.234}$$

where $\mathcal{B}$ is the bilinear form defined by relation (3.27).

At the coarsest resolution level, the nonlinear term can be split as follows:

$$\overline{\mathcal{B}(\boldsymbol{u}, \boldsymbol{u})}^2 = \underbrace{\overline{\mathcal{B}(\overline{\boldsymbol{u}}^2, \overline{\boldsymbol{u}}^2)}^2}_{I}$$

$$
+ \underbrace{\overline{\mathcal{B}(\overline{\boldsymbol{u}}^2, \boldsymbol{v}^1)}^2 + \overline{\mathcal{B}(\boldsymbol{v}^1, \overline{\boldsymbol{u}}^2)}^2 + \overline{\mathcal{B}(\boldsymbol{v}^1, \boldsymbol{v}^1)}^2}_{II}
$$

$$
+ \underbrace{\overline{\mathcal{B}(\overline{\boldsymbol{u}}^2 + \boldsymbol{v}^1, \boldsymbol{v}^0)}^2 + \overline{\mathcal{B}(\boldsymbol{v}^0, \overline{\boldsymbol{u}}^2 + \boldsymbol{v}^1 + \boldsymbol{v}^0)}^2}_{III} \quad . \tag{7.235}
$$

Term I can be computed directly at the coarsest grid level. Term II represents the direct coupling between the two levels of resolution, and can be computed exactly during the simulation. Term III represents the direct coupling with the true subgrid modes. It is worth noting that, at least theoretically, the non-local interaction between $\overline{\boldsymbol{u}}^2$ and $\boldsymbol{v}^0$ is not zero and requires the use of an ad hoc subgrid model.

At the finest resolution level, the analogous decomposition yields

$$
\begin{aligned}
\overline{\mathcal{B}(\boldsymbol{u}, \boldsymbol{u})}^1 = & \underbrace{\overline{\mathcal{B}(\overline{\boldsymbol{u}}^2, \overline{\boldsymbol{u}}^2)}^1}_{IV} \\
& + \underbrace{\overline{\mathcal{B}(\overline{\boldsymbol{u}}^2, \boldsymbol{v}^1)}^1 + \overline{\mathcal{B}(\boldsymbol{v}^1, \overline{\boldsymbol{u}}^2)}^1}_{V} \\
& + \underbrace{\overline{\mathcal{B}(\boldsymbol{v}^1, \boldsymbol{v}^1)}^1}_{VI} \qquad\qquad (7.236) \\
& + \underbrace{\overline{\mathcal{B}(\overline{\boldsymbol{u}}^2 + \boldsymbol{v}^1, \boldsymbol{v}^0)}^1 + \overline{\mathcal{B}(\boldsymbol{v}^0, \overline{\boldsymbol{u}}^2 + \boldsymbol{v}^1)}^1 + \overline{\mathcal{B}(\boldsymbol{v}^0, \boldsymbol{v}^0)}^1}_{VII} \ .
\end{aligned}
$$

Terms IV and V represent the coupling with the coarsest resolution level, and can be computed explicitly. Term VI is the nonlinear self-interaction of the detail $\boldsymbol{v}^1$, while term VII is associated with the interaction with subgrid scales $\boldsymbol{v}^0$ and must be modeled.

By looking at relations (7.235) and (7.236), we can see that specific subgrid models are required at each level of resolution. Several questions arise dealing with this closure problem:

1. Is a subgrid model necessary in practice for terms III and VII?
2. What kind of model should be used?
3. Is it possible to use the same model for both terms?

Many researchers have worked on these problems, leading to the definition of different three-level strategies. A few of them are presented below:

1. The Variational Multiscale Method (VMS) proposed by Hughes et al. (p. 267).
2. The Resolvable Subfilter-scale Model (RSFR) of Zhou et al. (p. 269).
3. The Dynamic Subfilter-scale Model (DSF) developed by Dubrulle et al. (p. 270).

4. The Local Galerkin Approximation (LGA), as defined by McDonough et al. (p. 270).
5. The Two-Level-Simulation (TLS) method, proposed by Menon et al. (p. 271).
6. The Modified Subgrid-scale Estimation Procedure (MSEP) of Domaradzki et al. (p. 269).
7. Terracol's multilevel algorithm (TMA) with explicit modeling of term III (p. 271).

The underlying coupling strategies are summarized in Table 7.3.

Table 7.3. Characteristics of multilevel subgrid models. + means that the term is taken into account, and – that it is neglected.

Model	I	II	III	IV	V	VI	VII
VMS	+	+	–	+	+	+	+
RSFR	+	+	–	+	+	+	+
MSEP	+	+	–	+	+	+	–
DSF	+	+	–	+	+	–	–
LGA	+	+	–	–	+	+	–
TLS	+	+	–	–	+	+	–
TMA	+	+	+	+	+	+	+

Variational Multiscale Method. Hughes et al. [332, 331, 336, 335, 333, 334] first introduced the Variational Multiscale Method within the framework of finite element methods, and then generalized it considering a fully general framework.

The coupling term III is neglected (see Table 7.3), at least in the original formulation of the method. The need for a full coupling including term III was advocated by Scott Collis [660].

In practical applications, the coarse resolution cutoff length scale $\overline{\Delta}^2$ is taken equal to twice the fine resolution cutoff length scale $\overline{\Delta}^1$, and the solution is integrated with the same time step at the two levels.

The subgrid term VII is parametrized using a Smagorinsky-like functional model

$$(VII)_{ij} = -2\nu_{\text{sgs}} \left(\frac{\partial v_j^1}{\partial x_i} + \frac{\partial v_i^1}{\partial x_j} \right) \quad , \tag{7.237}$$

with two possible variants:

– The small–small model[13]:

[13] This form of the dissipation is very close to the variational embedded stabilization previously proposed by Hughes (see Sect. 5.3.4). Similar expressions for the dissipation term have been proposed by Layton [429, 428] and Guermond [280].

$$\nu_{\text{sgs}} = (C_{\text{S}}\overline{\Delta}_1)^2 \sqrt{2|S^1|}, \quad S^1_{ij} = \left(\frac{\partial v^1_j}{\partial x_i} + \frac{\partial v^1_i}{\partial x_j} \right) \quad . \tag{7.238}$$

– The large–small model:

$$\nu_{\text{sgs}} = (C_{\text{S}}\overline{\Delta}_1)^2 \sqrt{2|\overline{S}^1|}, \quad \overline{S}^1_{ij} = \left(\frac{\partial \overline{u}^1_j}{\partial x_i} + \frac{\partial \overline{u}^1_i}{\partial x_j} \right) \quad . \tag{7.239}$$

The recommended value of the constant C_{S} is 0.1 for isotropic turbulence and plane channel flow. This value is arbitrary, and numerical experiments show that it could be optimized.

Numerical results show that both variants lead to satisfactory results on academic test cases, including non-equilibrium flow. This can be explained by the fact that the subgrid tensor is evaluated using the S^1 tensor, instead of $\overline{S}^2$ in classical subgrid-viscosity methods. Thus, the subgrid model dependency is more local in Fourier space, the emphasis being put on the highest resolved frequency, yielding more accurate results (see Sect. 5.3.3).

This increased localness in terms of wavenumber with respect to the usual Smagorinsky model is emphasized rewriting the VMS method as a special class of hyperviscosity models. The link between these two approaches is enlightened assuming that the following differential approximation for secondary filter utilized to operate the splitting $\overline{\boldsymbol{u}}^1 = \overline{\boldsymbol{u}}^2 + \boldsymbol{v}^1$ holds[14]:

$$\overline{\boldsymbol{u}}^2 = G_2 \star \overline{\boldsymbol{u}}^1 \simeq (Id + \alpha \overline{\Delta}_2^2 \nabla^2) \overline{\boldsymbol{u}}^1 \tag{7.240}$$

where α is a filter-dependent parameter, yielding

$$\boldsymbol{v}^1 = -\alpha \overline{\Delta}_2^2 \nabla^2 \overline{\boldsymbol{u}}^1 \tag{7.241}$$

Inserting that definition for $\boldsymbol{v}^1$ into previous expressions for both the Small-Small (7.238) and the Large-Small (7.239) models shows that these models are equivalent to fourth-order hyperviscosity models (see p. 121). Higher-order hyperviscosities are recovered using higher-order elliptic filters to operate the splitting.

The splitting of the resolved field into two parts can be interpreted as the definition of a more complex accentuation technique (see Sect. 5.3.3, p. 156). In the parlance of Hughes, the filtered Smagorinsky model corresponds the the Small-Large model (while the classical Smagorinsky model is the Large-Large model).

The accuracy of the results is observed to depend on the spectral properties of the filter used to extract $\boldsymbol{v}^1$ from $\overline{\boldsymbol{u}}^1$. It is observed that kinetic energy pile-up can occur if a spectral sharp cutoff is utilized. The reason why is that in this case the subgrid viscosity acts only on scales encompassed within $\boldsymbol{v}^1$

[14] It is known from results of Sect. 2.1.6 that this approximation is valid for smooth symmetric filters and for most discrete filters.

and non-local energy transfers between largest scales and $\boldsymbol{v}^0$ are neglected. The use of a smooth filter which allows a frequency overlap between $\overline{\boldsymbol{u}}^1$ and $\boldsymbol{v}^1$ alleviates this problem.

The use of the classical Smagorinsky model may also lead to an excessive damping of scales contained in $\boldsymbol{v}^1$. To cure this problem, Holmen et al. [316] proposed to use the Germano-Lilly procedure to evaluate the constant of the Large-Small model.

Another formulation for the Variational Multiscale Method is proposed by Vreman [745] who build some analogous models on the grounds of the test filter G_2. Keeping in mind that one essential feature of the VMS approach is that the action of the subgrid scale is restricted to the small scale field $\boldsymbol{v}^1$, the following possibilities arise

– Model 1: restriction of the usual Smagorinsky model

$$(VII)_{ij} = (Id - G_2) \star \left(-2(C_S \overline{\Delta}^1)^2 \sqrt{2 |\overline{S}^1| \overline{S}^1_{ij}} \right) \quad . \tag{7.242}$$

– Model 2 : Smagorinsky model based on the small scales

$$(VII)_{ij} = -2(C_S \overline{\Delta}^1)^2 \sqrt{2 |S^1| S^1_{ij}} \quad . \tag{7.243}$$

– Model 3 : restriction of the Smagorinsky model based on the small scales

$$(VII)_{ij} = (Id - G_2) \star \left(-2(C_S \overline{\Delta}^1)^2 \sqrt{2 |S^1| S^1_{ij}} \right) \quad . \tag{7.244}$$

Models 2 and 3 are equivalent if the filter G_2 is a sharp cutoff filter, but are different in the general case. If a differential second-order elliptic filter is used, model 3 will be equivalent to a sixth-order hyperviscosity, while model 2 is associated to a fourth-order hyperviscosity.

Resolvable Subfilter-scale Model. Zhou, Brasseur and Juneja [806] developed independently a three-level model which is almost theoretically equivalent to the Variational Multiscale Method of Hughes. The terms taken into account at the two resolution levels are the same (term III is ignored in both cases), and term VII is modeled using the Smagorinsky model. The model used is the small–small model (7.238) following Hughes' terminology. As in the VMS implementation described above, simulations of homogeneous anisotropic turbulence are carried out with $\overline{\Delta}^2 = 2\overline{\Delta}^1$.

Modified Subgrid-scale Estimation Procedure. Domaradzki et al. [782, 193, 183] proposed a modification of the original subgrid-scale estimation procedure (see Sect. 7.7.6) in order to improve its robustness. The key point of this method is to account directly for the production of small scales $\boldsymbol{v}^1$ by the forward energy cascade rather than using the production estimate (7.222).

The governing equations of this method are the same as those of VMS and RSFR, with the exception that the subgrid term VII in the detail equation is neglected. The main difference with VMS and RSFR is the time integration procedure: in the two previous approaches the coarsely and finely resolved fields are advanced at each time step, while Domaradzki and Yee proposed advancing the fine grid solution $\overline{\boldsymbol{u}}^1$ over an evolution time T between 1% and 3% of the large-eddy turnover time.

Dynamic Subfilter-scale Model. The dynamic subfilter-scale model of Dubrulle et al. [203, 426] appears as a linearized version of the three-level approach: the nonlinear term VI in the evolution equation of the details is neglected, and the subgrid term VII is not taken into account. This linearization process renders the detail equation similar to those of the Rapid Distortion Theory. Only a priori tests have been carried out on parallel wall-bounded flows.

Local Galerkin Approximation. The Local Galerkin Approach proposed by McDonough et al. [468, 466, 467, 470, 471] can be seen as a simplification of a typical three-level model. All subgrid terms are neglected, and, due to the fact that the original presentation of the method is not based on a filtering operator, term IV is not taken into account.

The key idea of the method is to make the assumption that the fluctuating field $\boldsymbol{v}^1$ is periodic in space within each cell associated with the coarse level resolution (see Fig. 7.4). Consequently, a spectral simulation is performed within each cell of size $\overline{\Delta}^2$, but the field $\boldsymbol{v}^1$ is not continuous at the interface of each cell. The number of Fourier modes determine the cutoff length scale $\overline{\Delta}^1$.

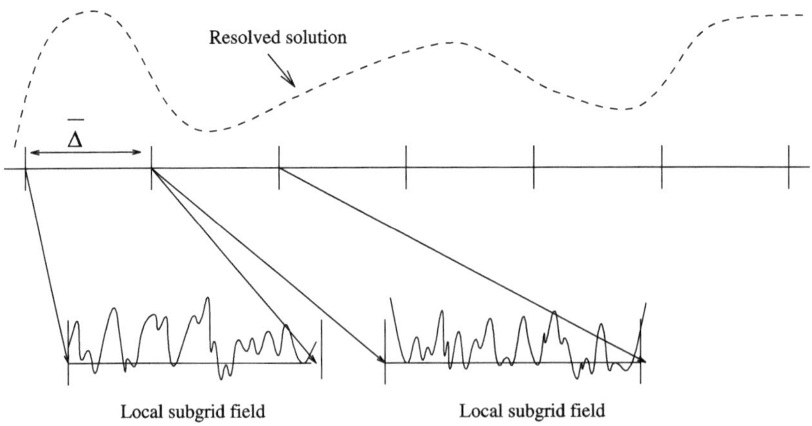

Fig. 7.4. Schematic of the Local Galerkin Approach.

This method can be seen as a dynamic extension of the Kinematic Simulation approach (see Sect. 7.7.4) and is very close, from a practical point of view, to the Perrier–Pironneau homogenization technique (see Sect. 7.2.3).

Menon's Two-Level-Simulation Method. Menon and Kemenov [382] developed a two-level method based on a simplified model for the subgrid scales. Instead of defining a three-dimensional grid to compute the subgrid modes inside each cell of the large-eddy simulation grid, the authors chose to solve simplified one-dimensional equations along lines (one in each direction in practice) inside each cell, leading to a large cost reduction. This feature makes it reminiscent of the Kerstein subgrid closure based on the stochastic ODT model (see Sect. 7.7.3). In Menon's approach, terms III and VII are neglected.

In each cell, the three-dimensional subgrid velocity field is modeled as a family of one-dimensional velocity vector fields defined on the underlying family of lines $\{l_1, l_2, l_3\}$ (which are in practice aligned with the axes of the reference Cartesian frame). Assuming that the derivatives of the modeled subgrid velocity field are such that

$$\frac{\partial v_i^1}{\partial l_1} \sim \frac{\partial v_i^1}{\partial l_2} \sim \frac{\partial v_i^1}{\partial l_3}, \quad i = 1, 2, 3 \quad , \tag{7.245}$$

and that the incompressibility constraint can be expressed as

$$\frac{\partial}{\partial l_j} \left(v_{1,j}^1 + v_{2,j}^1 + v_{3,j}^1 \right) = 0 \quad , \tag{7.246}$$

where $v_{k,j}^1$ refers to the jth component of the subgrid field computed along the line of index k, the following mometum-like equations are found:

$$\frac{\partial v_{i,j}^1}{\partial t} + NL(v_{i,j}^1, \overline{\boldsymbol{u}}, l_j) = -\frac{\partial p^1}{\partial l_j} + 3\nu \frac{\partial^2 v_{i,j}^1}{\partial l_j^2} \quad , \tag{7.247}$$

where the non-linear term $NL(v_{i,j}^1, \overline{\boldsymbol{u}}, l_j)$ contains the surrogates for terms IV, V and VI.

Terracol's Multilevel Algorithm. The last three-level model presented in this section is the multilevel closure proposed by Terracol et al. [710, 633, 711]. It is the only one which considers the full closure problem by taking into account the non-local interaction term III, and can then be considered as the most general one. The original method presented in [711] is able to handle an arbitrary number of filtering levels, but the present presentation will be restricted to the three-level case.

The key points of the method are:

- The use of a specific closure at each level $\overline{\boldsymbol{u}}^2$ and $\overline{\boldsymbol{u}}^1$. The proposed model at the coarse level is an extension of the one-parameter dynamic mixed model (see Sect. 7.4.2, p. 240), where the scale-similarity part is replaced by the explicitly computable term II. Term III is modeled using the Smagorinsky part of the mixed dynamic model, with a dynamically computed constant. At the fine resolution level, term VII is parametrized using a one-parameter dynamic mixed model. Numerical results demonstrate that the use of a specific model for term III is mandatory, since the use of a classical dynamic Smagorinsky model yields poor results.
- The definition of a cycling strategy between the different resolution levels, in order to decrease the computational cost while maintaining the accuracy of the results. The idea is here to freeze the details $\boldsymbol{v}^1$ and to carry out the computation at the coarse level only during a time T.[15] The problem is to find the optimal T in order to maximize the cost reduction while limiting the loss of coherence between $\overline{\boldsymbol{u}}^2$ and $\boldsymbol{v}^1$. A simple solution is to advance the solution for one time step at each level, alternatively, with the same value of the Courant number at each level. Numerical experiments show that this solution leads to good results with a gain of about a factor two.

Results obtained using this method on a plane mixing layer configuration are illustrated in Fig. 7.5.

7.8 Direct Identification of Subgrid Terms

Introduction. This section is dedicated to the presentation of approaches which aim at reconstructing the subgrid terms using direct identification mathematical tools.

Like all other subgrid models, either of functional or structural types, they answer the following question: given a filtered velocity field $\overline{\boldsymbol{u}}$, what is the subgrid acceleration? Subgrid models described previously were all based on some a priori knowledge of the nature of the interactions between resolved and subgrid scales, on a description of the filter, or on a structure of the subgrid scales hypothesized a priori. The models presented in this section do not require any of this information, and do not rely on any assumptions about the internal structure of the subgrid modes.

They are based on mathematical tools which are commonly used within the framework of pattern recognition, and do not really correspond to what is usually called a "model". Using Moser's words, they represent a *radical approach* to large-eddy simulation.

[15] This part is close to the quasistatic approximation for small scales introduced within the context of the nonlinear Galerkin method [174].

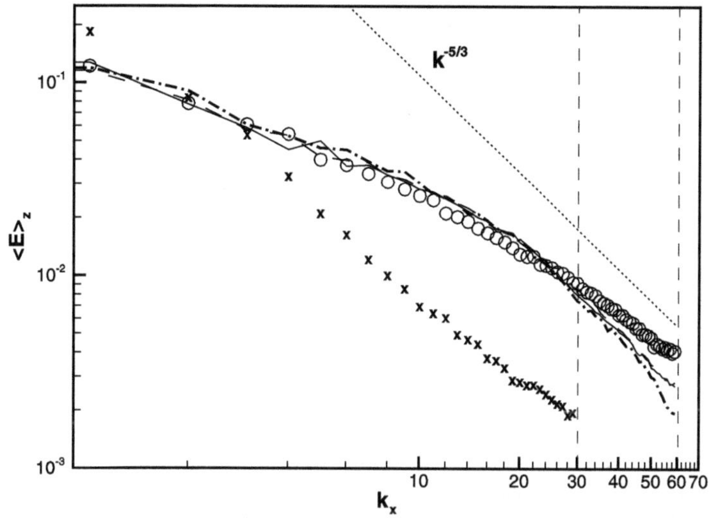

Fig. 7.5. Terracol's three-level method. Plane mixing layer. Streamwise energy spectrum during the self-similar phase. Crosses: large-eddy simulation on the coarse grid only. Other symbols and lines: direct numerical simulation and large-eddy simulation using the multilevel closure. The vertical *dashed lines* denote the cutoff wave numbers of the two grids. Courtesy of M. Terracol, ONERA.

Let us consider a scalar-, vector- or tensor-valued variable ϕ, which is to be estimated, and a set of solutions $(\boldsymbol{u}, \phi)_n, 1 \leq n \leq N$. What we are looking for is an estimation of the value of ϕ for any new arbitrary velocity field. This problem is equivalent to estimating the following functional

$$\phi \longrightarrow \mathcal{M}_\phi(\boldsymbol{u}, \mathcal{K}(\boldsymbol{u})) \quad , \tag{7.248}$$

where $\mathcal{K}(\boldsymbol{u})$ can be any arbitrary function based on the solution (gradient, correlations, ...). The two classes of approaches presented below are:

1. The approach developed by Moser et al., which relies on linear stochastic estimation (Sect. 7.8.1).
2. The proposal of Sarghini et al. dealing with the use of neural networks (Sect. 7.8.2).

Both approaches share the same very difficult practical problem: they require the existence of a set of realizations to achieve the identification process (computation of correlation tensors in the first case, and training phase of the neural network in the second case). In other approaches, this systematic identification process is replaced by the subgrid modeling phase, in which the modeler plays the role of the identification algorithm. As a consequence, the potential success of these identification methods depends on trade-off between the increase of computing power and the capability of researchers to improve typical subgrid models.

7.8.1 Linear-Stochastic-Estimation-Based Model

Moser et al. [419, 549, 548, 165, 296, 741, 797] proposed an identification procedure based on Linear Stochastic Estimation. This approach can be interpreted in several ways. An important point is that it is closely tied to the definition of large-eddy simulation as an optimal control problem, where the subgrid model plays the role of a controller. This interpretation is discussed in Sect. 9.1.4 and will not be repeated here. The linear stochastic estimation approach can also be seen as the best linear approximation for the subgrid term in the least-squares sense. Starting from the general formulation (7.248), the linear estimation is written as

$$\mathcal{M}_\phi(E) = \langle \phi \rangle + L \cdot E^T, \text{ with } \quad E = (\boldsymbol{u}, \mathcal{K}(\boldsymbol{u})) \quad , \tag{7.249}$$

where the tensorial dimension of L depends on those of $\phi, \boldsymbol{u}$ and $\mathcal{K}(\boldsymbol{u})$. The linear stochastic estimation procedure leads to the best values of the coefficient of L in the least-squares sense. Considering vectorial unknowns, we obtain the following spatially non-local estimation

$$\phi_i(\boldsymbol{x}) = \langle \phi_i \rangle + \int L_{ij}(\boldsymbol{x}, \boldsymbol{x}') E_j(\boldsymbol{x}') d\boldsymbol{x}' \quad ; \tag{7.250}$$

the best coefficients L_{ij} are computed by solving the following linear problem:

$$\langle E_i(\boldsymbol{x}') \phi_j(\boldsymbol{x}) \rangle = \int L_{jk}(\boldsymbol{x}, \boldsymbol{x}'') \langle E_j(\boldsymbol{x}) E_k(\boldsymbol{x}'') \rangle d\boldsymbol{x}'' \quad . \tag{7.251}$$

Local estimates can also be defined using one-point correlations instead of two-point correlations to define L.

Practial applications of this approach have been carried out in homogeneous turbulence and plane channel flow [549, 548]. Numerical experiments have shown that both the subgrid stresses and the subgrid energy transfer must be taken into account to obtain stable and accurate numerical simulations. This means that both $\overline{u}_i \tau_{ij}$ and $\overline{S}_{ij} \tau_{ij}$ must be recovered. This is achieved by estimating the subgrid acceleration

$$\phi_i = \frac{\partial}{\partial x_j} \tau_{ij} \quad , \tag{7.252}$$

as a function of the velocity field and its gradients

$$E = (\boldsymbol{u}, \nabla \boldsymbol{u}) \quad . \tag{7.253}$$

The mean value $\langle \phi \rangle$ is computed from the original data set and stored. In practice, some simplifications can be assumed in definitions (7.252) and (7.253) in parallel shear flows.

7.8.2 Neural-Network-Based Model

Sarghini et al. [645] proposed estimating the subgrid terms using a multilayer, feed-forward neural network. Rather than estimating directly the subgrid acceleration in the momentum equation, the authors decided to decrease the complexity of the problem by identifying the value of a subgrid-viscosity coefficient, yielding a new dynamic Smagorinsky model. To this end, a three-layer network is employed. The number of neurons in each layer is 15, 12 and 6, respectively, with a single output (the subgrid viscosity). The input vector is

$$E = (\nabla \overline{\boldsymbol{u}}, \boldsymbol{u}' \otimes \boldsymbol{u}') \in \mathbb{R}^{15} \quad . \tag{7.254}$$

The output is the subgrid model constant: $\phi = C_{\mathrm{S}} \in \mathbb{R}$. The training of the neural network is achieved using data fields originating from a classical large-eddy simulation of the same plane channel flow configuration. The learning rule used to adjust the weights and the biases of the network is chosen so as to minimize the summed-squared-error between the output of the network and the original set of data. Six thousand samples were found necessary for the training and validation steps. The training was performed through a backpropagation with weight decay technique in less than 500 iterations.

A priori and a posteriori tests show that the resulting model leads to stable numerical simulations, whose results are very close to those obtained using typical subgrid viscosity models. An interesting feature of the model is that the predicted subgrid viscosity exhibits the correct asymptotic behavior in the near-wall region (see p. 159).

7.9 Implicit Structural Models

The last class of structural subgrid models discussed in this chapter is the implicit structural model family. These models are structural ones, i.e. they do not rely on any foreknowledge about the nature of the interactions between the resolved scales and the subgrid scales. They can be classified as implicit, because they can be interpreted as improvements of basic numerical methods for solving the filtered Navier–Stokes equations, leading to the definition of higher-order accurate numerical fluxes. We note that, because the modification of the numerical method can be isolated as a new source term in the momentum equation, these models could also be classified as exotic formal expansion models. A major specificity of these models is that they all aim at reproducing directly the subgrid force appearing in the momentum equation, and not the subgrid tensor τ. They differ from the stabilized numerical methods presented in Sect. 5.3.4 within the MILES framework because they are not designed to induce numerical dissipation.

The two models presented in the following are:

1. The Local Average Method of Denaro (p. 276), which consists in a particular reconstruction of the discretized non-linear fluxes associated to the convection term. This approach incorporates a strategy to filter the subgrid-scale by means of an integration over a control volume and to recover the contribution of the subgrid scales with an integral formulation. It can be interpreted as a high-order space-time reconstruction procedure for the convective numerical fluxes based on a defiltering process.
2. The Scale Residual Model of Maurer and Fey (p. 278). As for the Approximate Deconvolution Procedure, the purpose is to evaluate the commutation error which defines the subgrid term. This evaluation is carried out using the residual between the time evolution of the solutions of the Navier–Stokes equations on two different grids (i.e. at two different filtering levels) and assuming some self-similarity properties of this residual. This model can be considered as: (i) a generalization of the previous one, which does not involve the deconvolution process anymore, but requires the use of the second computational grid and (ii) a generalization of the scale-similarity models, the use of a test filter for defining the test field being replaced by the explicit computation (by solving the Navier–Stokes equations) of the field at the test filter level.

Other implicit approaches for large-eddy simulation exist, which make it possible to obtain reliable results without subgrid scale model (in the common sense given to that term), and without explicit addition of numerical diffusion[16]. An example is the Spectro-Consistent Discretization proposed by Verstappen and Veldman [730, 729]. Because these approaches rely on numerical considerations only, they escape the modeling concept and will not be presented here.

7.9.1 Local Average Method

An other approach to the traditional large-eddy simulation technique was proposed by Denaro and his co-workers in a serie of papers [175, 169, 170]. It is based on a space-time high-order accurate reconstruction/deconvolution of the convective fluxes, which account for the subgrid-scale contribution. As a consequence, it can be seen as a particular numerical scheme based on a differential approximation of the filtering process. For sake of simplicity, we will present the method in the case of a dummy variable ϕ advected by a velocity field $\boldsymbol{u}$, whose evolution equation is (only convective terms are retained):

$$\frac{\partial \phi}{\partial t} = -\nabla \cdot (\boldsymbol{u}\phi) = A(\boldsymbol{u}, \phi) \quad . \tag{7.255}$$

[16] Dissipative numerical methods should be classified as Implicit Functional Modeling.

The local average of ϕ in a filtering cell Ω is defined as the mean value of ϕ in this cell[17]:

$$\overline{\phi}(\boldsymbol{x},t) \equiv \frac{1}{V}\int_{\Omega} \phi(\boldsymbol{\xi},t)d\boldsymbol{\xi} = \overline{\phi}(t), \quad \forall x \in \Omega \quad , \tag{7.256}$$

where V is the measure of Ω. We now consider an arbitrary filtering cell. Applying this operator to equation (7.255), and integrating the resulting evolution equation over the time interval $[t, t+\Delta t]$, we get:

$$(\overline{\phi}(t+\Delta t) - \overline{\phi}(t))V = \int_t^{t+\Delta t}\int_{\partial\Omega} \boldsymbol{n}\cdot\boldsymbol{u}\phi(\boldsymbol{\xi},t')d\boldsymbol{\xi}dt' \quad , \tag{7.257}$$

where $\partial\Omega$ is the boundary of Ω, and $\boldsymbol{n}$ the vector normal to it. The right-hand side of this equation, which appears as the application of a time-box filter to the boundary fluxes, can be approximated by means of a differential operator, exactly in the same way as for the space-box filter (see Sect. 7.2.1), yielding:

$$\int_t^{t+\Delta t}\int_{\partial\Omega} \boldsymbol{n}\cdot\boldsymbol{u}\phi(\boldsymbol{\xi},t')d\boldsymbol{\xi}dt' \simeq \Delta t\int_{\partial\Omega} \boldsymbol{n}\cdot\boldsymbol{u}\left(Id + \sum_{l=1,\infty}\frac{\Delta t^{l-1}}{l!}\frac{\partial^l}{\partial t^l}\right)\phi(\boldsymbol{\xi},t)d\boldsymbol{\xi}. \tag{7.258}$$

The time expansion is then writen as a space differential operator using the balance equation (7.255):

$$\left(Id + \sum_{l=1,\infty}\frac{\Delta t^{l-1}}{l!}\frac{\partial^l}{\partial t^l}\right)\phi(\boldsymbol{\xi},t) = \left(Id + \sum_{l=1,\infty}\frac{\Delta t^{l-1}}{l!}A^{l-1}(\boldsymbol{u},\cdot)\right)\phi(\boldsymbol{\xi},t) \quad , \tag{7.259}$$

with

$$A^l(\boldsymbol{u},\phi) \equiv \underbrace{A(\boldsymbol{u},\cdot)\circ A(\boldsymbol{u},\cdot)\circ \ldots \circ A(\boldsymbol{u},\phi)}_{l \text{ times}} \quad .$$

The second step of this method consists in the reconstruction step. At each point x located inside the filtering cell Ω, we have

$$\phi(\boldsymbol{x},t) = \overline{\phi}(t) + \phi'(\boldsymbol{x},t) \quad , \tag{7.260}$$

$$\begin{aligned} \phi(\boldsymbol{x},t+\Delta t) &= \overline{\phi}(t+\Delta t) + \phi'(\boldsymbol{x},t+\Delta t) \\ &= \overline{\phi}(t) + (\overline{\phi}(t+\Delta t) - \overline{\phi}(t)) + \phi'(x,t+\Delta t) \\ &\quad + (\phi'(\boldsymbol{x},t+\Delta t) - \phi'(\boldsymbol{x},t)) \\ &= \phi(\boldsymbol{x},t) + (\overline{\phi}(t+\Delta t) - \overline{\phi}(t)) \\ &\quad + (\phi'(\boldsymbol{x},t+\Delta t) - \phi'(\boldsymbol{x},t)) \quad . \end{aligned} \tag{7.261}$$

[17] This filtering operator corresponds to a modification of the box filter defined in Sect. 2.1.5: the original box filter is defined as a $\mathbb{R} \to \mathbb{R}$ operator, while the local average is a $\mathbb{R} \to \mathbb{N}$ operator. It is worth noting that the local average operator is a projector.

The first term in the left-hand side of relation (7.261) is known. The second one, which corresponds to the contribution of the low frequency part of the solution (i.e. *the local averaged part*), is computed using equation (7.257). The third term remains to be evaluated. This is done using the differential operator (2.52), leading to the final expression:

$$\phi(\boldsymbol{x}, t + \Delta t) = \phi(\boldsymbol{x}, t) + (Id - P_d)\left(\overline{\phi}(t + \Delta t) - \overline{\phi}(t)\right) \quad , \tag{7.262}$$

with

$$P_d = \sum_{l=1,\infty} \frac{1}{l!} \left(\frac{1}{V} \int_{\Omega} \left(\sum_{i=1,d} (\xi_i - x_i^c) \frac{\partial}{\partial x_i} \right)^l \right) d\xi \quad , \tag{7.263}$$

where d is the dimension of space and x_i^c the ith coordinate of the center of the filtering cell. In practice, the serie expansions are truncated to a finite order. The repeated use of equation (7.262) makes it possible to compute the value of the new pointwise value at each time step.

7.9.2 Scale Residual Model

Maurer and Fey [500] propose to evaluate the full subgrid term, still defined as the commutation error between the Navier–Stokes operator and the filter (see Chap. 3 or equation (7.2)), by means of a two-grid level procedure. A deconvolution procedure is no longer needed, but some self-invariance properties of the subgrid term have to be assumed. First we note that a subgrid model, referred to as $m(\overline{\boldsymbol{u}})$, is defined in order to minimize the residual E, with

$$E = [\mathcal{NS}, G\star](\boldsymbol{u}) - m(G \star \boldsymbol{u}) \quad . \tag{7.264}$$

Assuming that the filter G has the two following properties:

- G is a projector,
- G commutes with the Navier–Stokes operator in the sense that

$$\mathcal{NS} \circ (G\star)\boldsymbol{u} = \mathcal{NS} \circ (G\star) \circ (G\star)\boldsymbol{u} = (G\star) \circ \mathcal{NS} \circ (G\star)\boldsymbol{u} \quad ,$$

the residual can be rewritten as

$$E = (G\star) \circ (\mathcal{NS} \circ Id - \mathcal{NS} \circ (G\star))\boldsymbol{u} - m \circ (G\star)\boldsymbol{u} \quad . \tag{7.265}$$

We now introduce a set of filter $G_k, k = 0, N$, whose characteristic lengths $\overline{\Delta}_k$ are such that $0 = \overline{\Delta}_N < \overline{\Delta}_{N-1} < .... < \overline{\Delta}_0$. The residual E_k obtained for the kth level of filtering is easily deduced from relation (7.265):

$$\begin{aligned} E_k &= (G_k\star) \circ (\mathcal{NS} \circ (G_N\star) - \mathcal{NS} \circ (G_k\star))\boldsymbol{u} \\ &\quad - m \circ (G_k\star)\boldsymbol{u} \end{aligned} \tag{7.266}$$

$$= (G_k\star) \circ \sum_{j=0,k-1} (\mathcal{NS} \circ (G_j\star) - \mathcal{NS} \circ (G_{j+1}\star))\boldsymbol{u}$$
$$-m \circ (G_k\star)\boldsymbol{u} \quad . \tag{7.267}$$

To construct the model m, we now make the two following assumptions:

- The interactions between spectral bands are local, in that sense that the influence of each spectral band gets smaller with decreasing values of $j < k$.
- The residuals between two filtering levels have the following self-invariance property:

$$(\mathcal{NS}\circ(G_{j+1}\star)-\mathcal{NS}\circ(G_j\star))\boldsymbol{u} = \alpha(\mathcal{NS}\circ(G_j\star)-\mathcal{NS}\circ(G_{j-1}\star))\boldsymbol{u}, \tag{7.268}$$

where $\alpha < 1$ is a constant parameter. It is important noting that this can only be true if the cutoffs occur in the inertial range of the spectrum (see the discussion about the validity of the dynamic procedure in Sect. 5.3.3). Using these hypotheses, the following model is derived:

$$m \circ (G_k\star)\boldsymbol{u} = G_k \star (\sum_{j=1,k} \alpha^j)(\mathcal{NS} \circ (G_k\star) - \mathcal{NS} \circ (G_{k+1}\star))\boldsymbol{u} \quad , \tag{7.269}$$

where the operator $(G_k\star) \circ \mathcal{NS} \circ (G_{k+1}\star)$ corresponds to a local reconstruction of the evolution of the coarse solution $G_{k+1} \star \boldsymbol{u}$ according to the fluctuations of the fine solution $G_k \star \boldsymbol{u}$. The implementation of the model is carried out as follows: a short history of both the coarse and the fine solutions are computed on two different computational grids, and the model (7.269) is computed and added as a source term into the momentum equations solved on the fine grid. This algorithm can be written in the following symbolic form:

$$\boldsymbol{u}_k^{n+1} = \left(\mathcal{NS}^2_{k,\Delta t} + \omega(\mathcal{NS}^2_{k,\Delta t} - \mathcal{NS}^1_{k+1,2\Delta t})\right) \boldsymbol{u}_k^{n+1} \quad , \tag{7.270}$$

where $\boldsymbol{u}_k^{n+1}$ designates the solution on the fine grid (kth filtering level) at the $(n+1)$th time step, $(\mathcal{NS}^n_{k,\Delta t}$ refers to n applications of the discretized Navier–Stokes operator on the grid associated to the filtering level k with a time step Δt (i.e. the computation of n time steps on that grid without any subgrid model), and ω is a parameter deduced from relation (7.269). The weight α is evaluated analytically through some inertial range consideration, and is assumed to be equal to the ratio of the kinetic energy contained in the two spectral bands (see equation (5.132)). An additional correction factor (lower than 1) can also be introduced to account for the numerical errors.

8. Numerical Solution: Interpretation and Problems

This chapter is devoted to analyzing certain practical aspects of large-eddy simulation.

The first point concerns the differences between the filtering such as it is defined by a convolution product and such as it is imposed on the solution during the computation by the subgrid model. We distinguish here between static and dynamic interpretations of the filtering process. The analysis is developed only for subgrid viscosity models because their mathematical form makes this possible. However, the general ideas resulting from this analysis can in theory be extended to other types of models. The second point has to do with the link between the filter cutoff length and the mesh cell size used in the numerical solution. It is important to note that all of the previous developments proceed in a continuous, non-discrete framework and make no mention of the spatial discretization used for solving the equations of the problem numerically. The third point addressed is the comparative analysis of the numerical error and the subgrid terms. We propose here to compare the amplitude of the subgrid terms and numerical discretization errors to try to establish criteria for the required numerical scheme accuracy so that the errors committed will not overly mar the computed solution.

8.1 Dynamic Interpretation of the Large-Eddy Simulation

8.1.1 Static and Dynamic Interpretations: Effective Filter

The approach that has been followed so far in explaining large-eddy simulation consists in filtering the momentum equations explicitly, decomposing the non-linear terms that appear, and then modeling the unknown terms. If the subgrid model is well designed (in a sense defined in the following chapter), then the energy spectrum of the computed solution, for an exact solution verifying the Kolmogorov spectrum, is of the form

$$E(k) = K_0 \varepsilon^{2/3} k^{-5/3} \widehat{G}^2(k) \quad , \tag{8.1}$$

where $\widehat{G}(k)$ is the transfer function associated with the filter. This is the classical approach corresponding to a static and explicit view of the filtering process.

An alternate approach is proposed by Mason et al. [495, 496, 497], who first point out that the subgrid viscosity models use an intrinsic length scale denoted $\Delta_{\rm f}$, which can be interpreted as the mixing length associated with the subgrid scales. A subgrid viscosity model based on the large scales is written thus (see Sect. 5.3.2):

$$\nu_{\rm sgs} = \Delta_{\rm f}^2 |\overline{S}| \quad . \tag{8.2}$$

The ratio between this mixing length and the filter cutoff length $\overline{\Delta}$ is:

$$\frac{\Delta_{\rm f}}{\overline{\Delta}} = C_{\rm s} \quad . \tag{8.3}$$

Referring to the results explained in the section on subgrid viscosity models, $C_{\rm s}$ can be recognized as the subgrid model constant. Varying this constant is therefore equivalent to modifying the ratio between the filter cutoff length and the length scale included in the model. These two scales can consequently be considered as independent. Also, during the simulation, the subgrid scales are represented only by the subgrid models which, by their effects, impose the filter on the computed solution[1]. But since the subgrid models are not perfect, going from the exact solution to the computed one does not correspond to the application of the desired theoretical filter. This switch is ensured by applying an implicit filter, which is intrinsically contained in each subgrid model. Here we have a dynamic, implicit concept of the filtering process that takes the modeling errors into account. The question then arises of the qualification of the filters associated with the different subgrid models, both for their form and for their cutoff length.

The discrete dynamical system represented by the numerical simulation is therefore subjected to two filtering operations:

- The first is imposed by the choice of a level of representation of the physical system and is represented by application of a filter using the Navier–Stokes equations in the form of a convolution product.
- The second is induced by the existence of an intrinsic cutoff length in the subgrid model to be used.

In order to represent the sum of these two filtering processes, we define the effective filter, which is the filter actually seen by the dynamical system. To qualify this filter, we therefore raise the problem of knowing what is the share of each of the two filtering operations mentioned above.

---

[1] They do so by a dissipation of the resolved kinetic energy $-\tau_{ij}\overline{S}_{ij}$ equal to the flux $\widetilde{\varepsilon}$ through the cutoff located at the desired wave number.

8.1.2 Theoretical Analysis of the Turbulence Generated by Large-Eddy Simulation

We first go into the analysis of the filter associated with a subgrid viscosity model.

This section resumes Muschinsky's [551] analysis of the properties of a homogeneous turbulence simulated by a Smagorinsky model. The analysis proceeds by establishing an analogy between the large-eddy simulation equations incorporating a subgrid viscosity model and those that describe the motions of a non-Newtonian fluid. The properties of the latter are studied in the framework of isotropic homogeneous turbulence, so as to bring out the role of the different subgrid model parameters.

Analogy with Generalized Newtonian Fluids. Smagorinsky Fluid. The constitutive equations of large-eddy simulation for a Newtonian fluid, at least in the case where a subgrid viscosity model is used, can be interpreted differently as being those that describe the dynamics of a non-Newtonian fluid of the generalized Newtonian type, in the framework of direct numerical simulation, for which the constitutive equation is expressed

$$\sigma_{ij} = -p\delta_{ij} + \nu_{\rm sgs} S_{ij} \quad , \tag{8.4}$$

where σ_{ij} is the stress tensor, S the strain rate tensor defined as above, and $\nu_{\rm sgs}$ will be a function of the invariants of S. Effects stemming from the molecular viscosity are ignored because this is a canonical analysis using the idea of an inertial range. It should be noted that the filtering bar symbol no longer appears because we now interpret the simulation as a direct one of a fluid having a non-linear constitutive equation. If the Smagorinsky model is used, i.e.

$$\nu_{\rm sgs} = \Delta_{\rm f}^2 |S| = \left(C_{\rm s}\overline{\Delta}\right)^2 |S| \quad , \tag{8.5}$$

such a fluid will be called a Smagorinsky fluid.

Laws of Similarity of the Smagorinsky Fluid. The first step consists in extending the Kolmogorov similarity hypotheses (recalled in Appendix A):

1. First similarity hypothesis. $E(k)$ depends only on ε, $\Delta_{\rm f}$ and $\overline{\Delta}$.
2. Second similarity hypothesis. $E(k)$ depends only on ε and $\overline{\Delta}$ for wave numbers k much greater than $1/\Delta_{\rm f}$.
3. Third similarity hypothesis. $E(k)$ depends only on ε and $\Delta_{\rm f}$ if $\overline{\Delta} \ll \Delta_{\rm f}$.

The spectrum can then be put in the form:

$$E(k) = \varepsilon^{2/3} k^{-5/3} G_{\rm s}(\Pi_1, \Pi_2) \quad , \tag{8.6}$$

where $G_{\rm s}$ is a dimensionless function whose two arguments are defined as:

$$\Pi_1 = k\Delta_{\rm f}, \qquad \Pi_2 = \frac{\overline{\Delta}}{\Delta_{\rm f}} = \frac{1}{C_{\rm s}} \quad . \tag{8.7}$$

By analogy, the limit in the inertial range of G_s is a quantity equivalent to the Kolmogorov constant for large-eddy simulation, denoted $K_{les}(C_s)$:

$$K_{les}(C_s) = G_s(0, \Pi_2) \quad . \tag{8.8}$$

By introducing the shape function

$$f_{les}(k\Delta_f, C_s) = \frac{G_s(\Pi_1, \Pi_2)}{G_s(0, \Pi_2)} \quad , \tag{8.9}$$

the spectrum is expressed:

$$E(k) = K_{les}(C_s)\varepsilon^{2/3}k^{-5/3}f_{les}(k\Delta_f, C_s) \quad . \tag{8.10}$$

By analogy with Kolmogorov's work, we define the dissipation scale of the non-Newtonian fluid η_{les} as:

$$\eta_{les} = \left(\frac{\nu_{sgs}^3}{\varepsilon}\right)^{1/4} \quad . \tag{8.11}$$

For the Smagorinsky model, by replacing ε with its value, we get:

$$\eta_{les} = \Delta_f = C_s\overline{\Delta} \quad . \tag{8.12}$$

Using this definition and postulating that Kolmogorov's similarity theory for the usual turbulence remains valid, the third similarity hypothesis stated implies, for large values of the constant C_s:

$$E(k) = \left(\lim_{C_s \to \infty} K_{les}(C_s)\right)\varepsilon^{2/3}k^{-5/3}\left(\lim_{C_s \to \infty} f_{les}(k\eta_{les}, C_s)\right) \quad , \tag{8.13}$$

which allows us to presume that the two following relations are valid:

$$\lim_{C_s \to \infty} K_{les}(C_s) = K_0 \quad , \tag{8.14}$$

$$\lim_{C_s \to \infty} f_{les}(x, C_s) = f(x) \quad , \tag{8.15}$$

where $f(x)$ is the damping function including the small scale viscous effects, for which the Heisenberg–Chandresekhar, Kovazsnay, and Pao models have already been discussed in Sect. 5.3.2.

The corresponding normalized spectrum of the dissipation[2] is of the form:

$$g_{les}(x, C_s) = x^{1/3}f_{les}(x, C_s) \quad , \tag{8.16}$$

where x is the reduced variable $x = k\eta_{les}$.

[2] The dissipation spectrum, denoted $D(k)$, associated with the energy spectrum $E(k)$ is defined by the relation:

$$D(k) = k^2E(k) \quad .$$

By comparing the dissipation computed by integrating this spectrum with the one evaluated from the energy spectrum (8.10), the dependency of the Kolmogorov constant as a function of the Smagorinsky constant is formulated as:

$$K_{\rm les}(C_{\rm s}) = \frac{1}{2\int_0^\infty g_{\rm les}(x, C_{\rm s})} \approx \frac{1}{2\int_0^{C_{\rm s}\pi} g_{\rm les}(x, C_{\rm s})} \quad . \tag{8.17}$$

When this expression is computed using the formulas of Heisenberg–Chandrasekhar and Pao, it shows that the function $K_{\rm les}$ does tend asymptotically to the value $K_0 = 1.5$ for large values values of $C_{\rm s}$, as the error is negligible beyond $C_{\rm s} = 0.5$. The variation of the parameter $K_{\rm les}$ as a function of $C_{\rm s}$ for the spectra of Heisenberg–Chandrasekhar and Pao is presented in Fig. 8.1. When $C_{\rm s}$ is less than 0.5, the Kolmogorov constant is over-evaluated, as has actually been observed in the course of numerical experiments [478, 479]. These numerical simulations, carried out by Magnient et al. [479], have shown that:

– The damping function depends on $C_{\rm s}$. A clear bifurcation is observed in the behavior of the models. The theoretical value of $C_{\rm s}$, referred to as $C_{\rm s0}$, obtained by the canonical analysis corresponds to the case where the resolved kinetic energy transfer is equal to the energy transfer across the wave number $\pi/\overline{\Delta}$. In this case, we obtain $f_{\rm les} = 1$ for all subgrid viscosity models. For larger values, the resulting damping function is not equal to a Heaviside function, and depends on the subgrid model. An interesting feature is that scales larger than $\overline{\Delta}$ are progressively damped. This damping

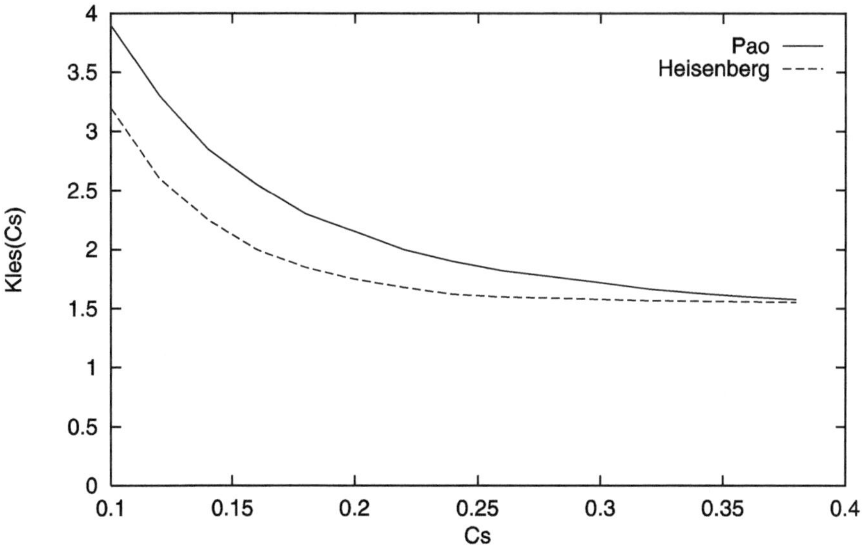

Fig. 8.1. Variation of the Kolmogorov constant as a function of the Smagorinsky constant for the Heisenberg–Chandrasekhar spectrum and the Pao spectrum.

originates from two different phenomena: (i) the energy drain induced by the subgrid-viscosity model, and (ii) the forward energy cascade, which is responsible for a net drain of kinetic energy by the modes located within the spectral band $[(C_s/C_{s0})\pi/\overline{\Delta}, \pi/\overline{\Delta}]$.
- The damping function is subgrid-model dependent: each subgrid model leads to a different equilibrium between the two sources of resolved energy drain, yielding different spectra and associated damping functions.

Interpretation of Simulation Parameters.

Effective Filter. The above results allow us to refine the analysis concerning the effective filter. For large values of the Smagorinsky constant ($C_s \geq 0.5$), the characteristic cutoff length is the mixing length produced by the model. The model then dissipates more energy than if it were actually located at the scale $\overline{\Delta}$ because it ensures the energy flux balance through the cutoff associated with a longer characteristic length. The effective filter is therefore fully determined by the subgrid model. This solution criterion should be compared with the one defined for hot-wire measurements, which recommends that the wire length be less than twice the Kolmogorov scale in developed turbulent flows.

For small values of the constant, it is the cutoff length $\overline{\Delta}$ that plays the role of characteristic length and the effective filter corresponds to the usual analytical filter. It should be noted in this case that the energy drainage induced by the model is less than the transfer of kinetic energy through the cutoff, so the energy balance is no longer maintained. This is reflected in an accumulation of energy in the resolved scales, and the pertinence of the simulation results should be taken with caution.

For intermediate values of the constant, i.e. values close to the theoretical one predicted in Sect. 5.3.2 (i.e. $C_s \approx 0.2$), the effective filter is a combination of the analytical filter and model's implicit filter, which makes it difficult to interpret the dynamics of the smallest resolved scales. The dissipation induced by the model in this case correctly insures the equilibrium of the energy fluxes through the cutoff.

Microstructure Knudsen Number. It has already been seen (relation (8.12)) that the mixing length can be interpreted as playing a role analogous to that of the Kolmogorov scale for the direct numerical simulation. The cutoff length $\overline{\Delta}$, for its part, can be linked to the mean free path for Newtonian fluids. We can use the ratio of these two quantities to define an equivalent of the microstucture Knudsen number K_{nm} for the large-eddy simulation:

$$K_{nm} = \frac{\overline{\Delta}}{\Delta_f} = \frac{1}{C_s} \quad . \tag{8.18}$$

Effective Reynolds Number. Let us also note that the effective Reynolds number of the simulation, denoted Re_{les}, which measures the ratio of the inertia effects to the dissipation effects, is taken in ratio to the Reynolds number Re corresponding to the exact solution by the relation:

$$Re_{\mathrm{les}} = \left(\frac{\eta}{\eta_{\mathrm{les}}}\right)^{4/3} Re \quad , \tag{8.19}$$

where η is the dissipative scale of the full solution. This decrease in the effective Reynolds number in the simulation may pose some problems, if the physical mechanism determining the dynamics of the resolved scales depends explicitly on it. This will, for example, be the case for all flows where critical Reynolds numbers can be defined for which bifurcations in the solution are associated[3].

Subfilter Scale Concept. By analysis of the decoupling between the cutoff length of the analytical filter $\overline{\Delta}$ and the mixing length Δ_{f}, we can define three families of scales [495, 551] instead of the usual two families of resolved and subgrid scales. These three categories, illustrated in Fig. 8.2, are the:

1. Subgrid scales, which are those that are excluded from the solution by the analytical filter.
2. Subfilter scales, which are those of a size less than the effective filter cutoff length, denoted Δ_{eff}, which are scales resolved in the usual sense but whose dynamics is strongly affected by the subgrid model. Such scales exist only if the effective filter is determined by the subgrid viscosity model. There is still the problem of evaluating Δ_{eff}, and depends both on the presumed shape of the spectrum and on the point beyond which we consider to be "strongly affected". For example, by using Pao's spectrum and defining the non-physically resolved modes as those for which the energy level is reduced by a factor $e = 2.7181...$, we get:

$$\Delta_{\mathrm{eff}} = \frac{C_{\mathrm{s}}}{C_{\mathrm{theo}}}\overline{\Delta} \quad , \tag{8.20}$$

where C_{theo} is the theoretical value of the constant that corresponds to the cutoff length $\overline{\Delta}$.
3. Physically resolved scales, which are those of a size greater than the effective filter cutoff length, whose dynamics is perfectly captured by the simulation, as in the case of direct numerical simulations.

Characterization of the Filter Associated with the Subgrid Model. The above discussion is based on a similarity hypothesis between the properties of isotropic homogeneous turbulence and those of the flow simulated

[3] Numerical experiments show that too strong a dissipation induced by the subgrid model in such flows may inhibit the flow driving mechanisms and consequently lead to unreliabable simulations. One known example is the use of a Smagorinsky model to simulate a plane channel flow: the dissipation is strong enough to prevent the transition to turbulence.

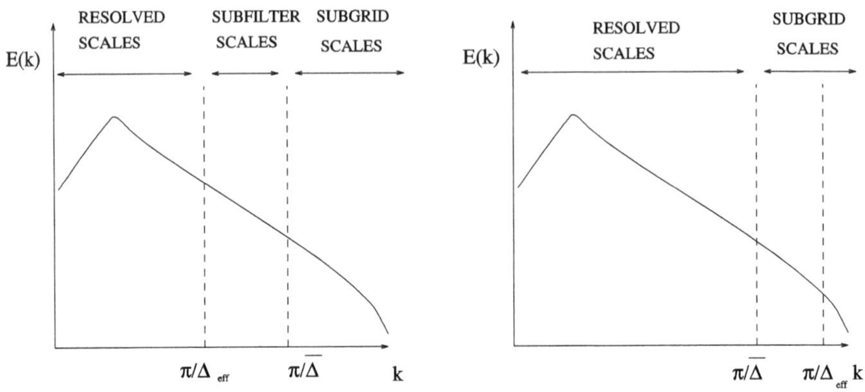

Fig. 8.2. Representation of different scale families in the cases of $\Delta_{\rm eff} < \overline{\Delta}$ (*Right*) and $\Delta_{\rm eff} > \overline{\Delta}$ (*Left*).

using a subgrid viscosity model. This is mainly true of the dissipative effects, which are described using the Pao spectrum or that of Heisenberg–Kovazsnay. So here, we adopt the hypothesis that the subgrid dissipation acts like an ordinary dissipation (which was already partly assumed by using a subgrid viscosity model). The spectrum $E(k)$ of the solution from the simulation can therefore be interpreted as the product of the spectrum of the exact solution $E_{\rm tot}(k)$ by the square of the transfer function associated with the effective filter $\widehat{G}_{\rm eff}(k)$:

$$E(k) = E_{\rm tot}(k)\widehat{G}^2_{\rm eff}(k) \quad . \tag{8.21}$$

Considering that the exact solution corresponds to the Kolmogorov spectrum, and using the form (8.10), we get:

$$\widehat{G}_{\rm eff}(k) = \sqrt{\frac{K_{\rm les}(C_{\rm s})}{K_0} f_{\rm les}(k\Delta_{\rm f}, C_{\rm s})} \quad . \tag{8.22}$$

The filter associated with the Smagorinsky model is therefore a "smooth" filter in the spectral space, which corresponds to a gradual damping, very different from the sharp cutoff filter.

8.2 Ties Between the Filter and Computational Grid. Pre-filtering

The above developments completely ignore the computational grid used for solving the constitutive equations of the large-eddy simulation numerically. If we consider this new element, it introduces another space scale: the spatial discretization step Δx for simulations in the physical space, and the maximum wave number $k_{\rm max}$ for simulations based on spectral methods.

The discretization step has to be small enough to be able to correctly integrate the convolution product that defines the analytical filtering. For filters with fastly-decaying kernel, we have the relation:

$$\Delta x \leq \overline{\Delta} \quad . \tag{8.23}$$

The case where $\Delta x = \overline{\Delta}$ is the optimal case as concerns the number of degrees of freedom needed in the discrete system for performing the simulation. This case is illustrated in Fig. 8.3.

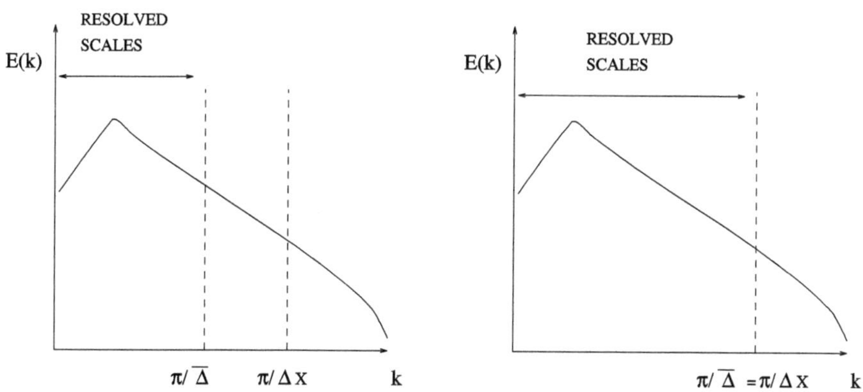

Fig. 8.3. Representation of spectral decompositions associated with pre-filtering (*Left*) and in the optimal case (*Right*).

The numerical errors stemming from the resolution of the discretized system still have to be evaluated. To ensure the quality of the results, the numerical error committed on the physically resolved modes has to be negligible, and therefore committed only on the subfilter scales. The theoretical analysis of this error by Ghosal in the simple case of isotropic homogeneous turbulence is presented in the following.

As the numerical schemes used are consistent, the discretization error cancels out as the space and time steps tend toward zero. One way of minimizing the effect of numerical error is to reduce the grid spacing while maintaining the filter cutoff length, which comes down to increasing the ratio $\overline{\Delta}/\Delta x$ (see Fig. 8.3). This technique, based on the decoupling of the two space scales, is called pre-filtering [14], and aims to ensure the convergence of the solution regardless of the grid[4]. It minimizes the numerical error but induces more computations because it increases the number of degree of freedoms in the numerical solution without increasing the number of degrees of freedom in the physically resolved solution, and requires that the analytical filtering be

[4] A simplified analysis shows that, for an nth-order accurate numerical method, the weight of the numerical error theoretically decreases as $(\overline{\Delta}/\Delta x)^{-n}$. A finer estimate is given in the remainder of this chapter.

performed explicitly [14] [59]. Because of its cost[5], this solution is rarely used in practice.

Another approach is to link the analytical filter to the computed grid. The analytical cutoff length is associated with the space step using the optimal ratio of these quantities and the form of the convolution kernel is associated with the numerical method. Let us point out a problem here that is analogous to that of the effective filter already mentioned: the effective numerical filter and therefore the effective numerical cutoff length, are generally unknown. This method has the advantage of reducing the size of the system as best possible and not requiring the use of an analytical filter, but it allows no explicit control of the effective numerical filter, which makes it difficult to calibrate the subgrid models. This method, because of its simplicity, is used by nearly all the authors.

8.3 Numerical Errors and Subgrid Terms

8.3.1 Ghosal's General Analysis

Ghosal [259] proposes a non-linear analysis of the numerical error in the solution of the Navier–Stokes equations for an isotropic homogeneous turbulent flow whose energy spectrum is approximated by the Von Karman model.

Classification of Different Sources of Error. In order to analyze and estimate the discretization error, we first need a precise definition of it. In all of the following, we consider a uniform Cartesian grid of N^3 points, which are the degrees of freedom of the numerical solution. Periodicity conditions are used on the domain boundaries.

A first source of error stems from the approximation we make of a continuous solution $\overline{\boldsymbol{u}}$ by a making a discrete solution $\boldsymbol{u}_\mathrm{d}$ with a set of N^3 values. This is evaluated as:

$$|\boldsymbol{u}_\mathrm{d} - \mathcal{P}(\overline{\boldsymbol{u}})| \quad , \tag{8.24}$$

where $\mathcal{P}$ is a definite projection operator of the space of continuous solutions to that of the discrete solutions. This error is minimum (in the L^2 sense) if $\mathcal{P}$ is associated with the decomposition of the continuous solution on a finite base of trigonometric polynomials, with the components of $\boldsymbol{u}_\mathrm{d}$ being the associated Fourier coefficients. This error is intrinsic and cannot be canceled. Consequently, it will not enter into the definition of the numerical

---

[5] For a fixed value of $\overline{\Delta}$, increasing the ratio $\overline{\Delta}/\Delta x$ by a factor n leads to an increase in the number of points of the simulation by a factor of n^3 and increases the number of time steps by a factor n in order to maintain the same ratio between the time and space steps. In all, this makes an overall increase in the cost of the simulation by a factor n^4.

error discussed in this present section. The best possible discrete solution is $\boldsymbol{u}_{\rm opt} \equiv \mathcal{P}(\overline{\boldsymbol{u}})$.

The equations of the continuous problem are written in the symbolic form:

$$\frac{\partial \overline{\boldsymbol{u}}}{\partial t} = \mathcal{NS}(\overline{\boldsymbol{u}}) \quad , \tag{8.25}$$

where $\mathcal{NS}$ is the Navier–Stokes operator. The optimal discrete solution $\boldsymbol{u}_{\rm opt}$ is a solution of the problem:

$$\frac{\partial \mathcal{P}\overline{\boldsymbol{u}}}{\partial t} = \mathcal{P} \circ \mathcal{NS}(\overline{\boldsymbol{u}}) \quad , \tag{8.26}$$

where $\mathcal{P} \circ \mathcal{NS}$ is the optimal discrete Navier–Stokes operator which, in the fixed framework, corresponds to the discrete operators obtained by a spectral method.

Also, we note the discrete problem associated with a fixed discrete scheme as:

$$\frac{\partial \boldsymbol{u}_{\rm d}}{\partial t} = \mathcal{NS}_{\rm d}(\boldsymbol{u}_{\rm d}) \quad . \tag{8.27}$$

By taking the difference between (8.26) and (8.27), it appears that the best possible numerical method, denoted $\mathcal{NS}_{\rm opt}$, is the one that verifies the relation:

$$\mathcal{NS}_{\rm opt} \circ \mathcal{P} = \mathcal{P} \circ \mathcal{NS} \quad . \tag{8.28}$$

The numerical error $E_{\rm num}$ associated with the $\mathcal{NS}_{\rm d}$ scheme, and which is analyzed in the following, is defined as:

$$E_{\rm num} \equiv (\mathcal{P} \circ \mathcal{NS} - \mathcal{NS}_{\rm d} \circ \mathcal{P})\,(\overline{\boldsymbol{u}}) \quad . \tag{8.29}$$

This represents the discrepancy between the numerical solution and the optimal discrete one. To simplify the analysis, we consider in the following that the subgrid models are perfect, i.e. that they induce no error with respect to the exact solution of the filtered problem. By assuming this, we can clearly separate the numerical errors from the modeling errors.

The numerical error $E_{\rm num}(k)$ associated with the wave number k is decomposed as the sum of two terms of distinct origins:

- The differentiation error $E_{\rm df}(k)$, which measures the error the discrete operators make in evaluating the derivatives of the wave associated with k. Let us note that this error is null for a spectral method if the cutoff frequency of the numerical approximation is high enough.
- The spectrum aliasing error $E_{\rm rs}(k)$, which is due to the fact that we are computing non-linear terms in the physical space in a discrete space of finite dimension. For example, a quadratic term will bring in higher frequencies than those of each of the arguments in the product. While some of these frequencies are too high to be represented directly on the discrete base,

they do combine with the low frequencies and introduce an error in the representation of them[6].

Estimations of the Error Terms. For a solution whose spectrum is of the form proposed by Von Karman:

$$E(k) = \frac{a\ k^4}{(b + k^2)^{17/6}} \quad , \tag{8.30}$$

with $a = 2.682$ and $b = 0.417$, and using a quasi-normality hypothesis for evaluating certain non-linear terms, Ghosal proposes a quantitative evaluation of the different error terms, the subgrid terms, and the convection term, for various Finite Difference schemes as well as for a spectral scheme. The convection term is written in conservative form and all the schemes in space are centered. The time integration is assumed to be exact.

The exact forms of these terms, available in the original reference work, are not reproduced here. For a cutoff wave number k_c and a sharp cutoff filter, simplified approximate estimates of the average amplitude can be derived for some of these terms.

The amplitude of the subgrid term $\sigma_{sgs}(k_c)$, defined as

$$\sigma_{sgs}(k_c) = \left[\int_0^{k_c} |\tau(k)| dk \right]^{1/2} \quad , \tag{8.31}$$

where $\tau(k)$ is the subgrid term for the wave number k, is bounded by:

$$\sigma_{sgs}(k_c) = \left\{ \begin{array}{ll} 0.36\ k_c^{0.39} & \text{upper limit} \\ 0.62\ k_c^{0.48} & \text{lower limit} \end{array} \right. \quad , \tag{8.32}$$

that of the sum of the convection term and subgrid term by:

$$\sigma_{tot}(k_c) = 1.04\ k_c^{0.97} \quad , \tag{8.33}$$

[6] Let us take the Fourier expansions of two discrete functions u and v represented by N degrees of freedom. At the point of subscript j, the expansions are expressed:

$$u_j = \sum_{n=-N/2}^{N/2-1} \widehat{u}_n e^{(\mathrm{i}(2\pi/N)jn)}, \quad v_j = \sum_{m=-N/2}^{N/2-1} \widehat{v}_m e^{(\mathrm{i}(2\pi/N)jm)} \quad j = 1, N \quad .$$

The Fourier coefficient of the product $w_j = u_j v_j$ (without summing on j) splits into the form:

$$w_k = \sum_{n+m=k} \widehat{u}_n \widehat{v}_m + \sum_{n+m=k \pm N} \widehat{u}_n \widehat{v}_m \quad .$$

The last term in the right-hand side represents the spectrum aliasing error. These are terms of frequencies higher than the Nyquist frequency, associated the sampling, which will generate spurious resolved frequencies.

in which

$$\frac{\sigma_{\rm sgs}(k_{\rm c})}{\sigma_{\rm tot}(k_{\rm c})} \approx k_{\rm c}^{-0.5} \quad . \tag{8.34}$$

The amplitude of the differentiation error $\sigma_{\rm df}(k_{\rm c})$, defined by:

$$\sigma_{\rm df}(k_{\rm c}) = \left[\int_0^{k_{\rm c}} E_{\rm df}(k) dk \right]^{1/2} \quad , \tag{8.35}$$

is evaluated as:

$$\sigma_{\rm df}(k_{\rm c}) = k_{\rm c}^{0.75} \times \left\{ \begin{array}{cc} 1.03 & \text{(second order)} \\ 0.82 & \text{(fourth order)} \\ 0.70 & \text{(sixth order)} \\ 0.5 & \text{(heigth order)} \\ 0 & \text{(spectral)} \end{array} \right. \quad , \tag{8.36}$$

and the spectrum aliasing error $\sigma_{\rm rs}(k_{\rm c})$, which is equal to:

$$\sigma_{\rm rs}(k_{\rm c}) = \left[\int_0^{k_{\rm c}} E_{\rm rs}(k) dk \right]^{1/2} \tag{8.37}$$

is estimated as:

$$\sigma_{\rm rs} = \left\{ \begin{array}{cc} 0.90\ k_{\rm c}^{0.46} & \text{(minimum estimation, spectral, no de-aliasing)} \\ 2.20\ k_{\rm c}^{0.66} & \text{(maximum estimation, spectral, no de-aliasing)} \\ 0.46\ k_{\rm c}^{0.41} & \text{(minimum estimation, second order)} \\ 1.29\ k_{\rm c}^{0.65} & \text{(maximum estimation, second order)} \end{array} \right. \quad . \tag{8.38}$$

The spectrum aliasing error for the spectral method can be reduced to zero by using the 2/3 rule, which consists of not considering the last third of the wave numbers represented by the discrete solution. It should be noted that, in this case, only the first two-thirds of the modes of the solution are correctly represented numerically. The error of the finite difference schemes of higher order is intermediate between that of the second-order accurate scheme and that of the spectral scheme.

From these estimations, we can see that the discretization error dominates the subgrid terms for all the finite difference schemes considered. The same is true for the spectrum aliasing error, including for the finite difference schemes. Finer analysis on the basis of the spectra of the various terms shows that the discretization error is dominant for all wave numbers for the second-order accurate scheme, whereas the subgrid terms are dominant at the low frequencies for the heigth-order accurate scheme. In the former case, the effective numerical filter is dominant and governs the solution dynamics. Its cutoff length can be considered as being of the order of the size of the computational domain. In the latter, its cutoff length, defined as the wavelength of the mode beyond which it becomes dominant with respect to the subgrid terms, is smaller and there exist numerically well-resolved scales.

8.3.2 Pre-filtering Effect

The pre-filtering effect is clearly visible from relations (8.32) to (8.38). By decoupling the analytical from the numerical filter, two different cutoff scales are introduced and thereby two different wave numbers for evaluating the numerical error terms and the subgrid terms: while the cutoff scale $\overline{\Delta}$ associated with the filter remains constant, the scale associated with the numerical error (i.e. Δx) is now variable.

By designating the ratio of the two cutoff lengths by $C_{\mathrm{rap}} = \Delta x/\overline{\Delta} < 1$, we see that the differentiation error $\sigma_{\mathrm{df}}(k_{\mathrm{c}})$ of the finite difference scheme is reduced by a factor $C_{\mathrm{rap}}^{-3/4}$ with respect to the previous case, since it varies as $k_{\mathrm{c}}^{3/4}$. This reduction is much greater than the one obtained by increasing the order of accuracy of the the schemes.

Thus, more detailed analysis shows that, for the second-order accurate scheme, the dominance of the subgrid term on the whole of the solution spectrum is ensured for $C_{\mathrm{rap}} = 1/8$. For a ratio of $1/2$, this dominance is once again found for schemes of order of accuracy of 4 or more.

These theoretical evaluations do not take into account the nonlinear feedback of the computed solution on the numerical error. Numerical experiments were conducted by Chow and Moin [138] in isotropic turbulence to assess Ghosal's results. Their results show that to ensure that the subgrid terms will dominate the numerical error, a filter-grid ratio $\overline{\Delta}/\Delta x$ of at least four is desired for a second-order centered finite difference scheme. This minimum is a decreasing function of the scheme order of accuracy, and a ratio of two is found to be sufficient for a sixth-order centered Padé scheme.

The efficiency of the prefiltering technique was exhaustively checked by Geurts and Fröhlich [256, 257] on the plane mixing layer configuration. The main conclusion of this study is that the best solution for improving the results of a large-eddy simulation in practice is to refine the computational grid, i.e. to lower Δx, while keeping a low value of the ratio $\overline{\Delta}/\Delta x$. The numerical tests conducted by these authors show that $\overline{\Delta}/\Delta x = 1$ is optimal for high-resolution large-eddy simulations, i.e. for simulations which are close to direct numerical simulation. For coarser grid simulation, the optimum was found to be $\overline{\Delta}/\Delta x \approx 2-3$, combined with a fourth-order accurate non-dissipative numerical method. Nevertheless, the recommended *strategy* to improve the results at a fixed computational cost is to refine the grid using $\overline{\Delta}/\Delta x = 1-2$ rather than augmenting the ratio $\overline{\Delta}/\Delta x$ for a given value of $\overline{\Delta}$. This conclusion is confirmed by a large set of results published by Gullbrand and Chow [283], who carried out several simulations of turbulent plane channel flow with second-order and fourth-order accurate finite difference methods. In these simulations, the use of a prefilter with size $2\Delta x$ didn't lead to a clear improvement of the results.

The finest study of the effect of prefiltering was conducted by Meyers, Geurts and Baelmans [515] in large-eddy simulation of isotropic turbulence using the Smagorinsky model. Defining the error $e_{\mathrm{total}}(\overline{\Delta}, \Delta x)$ for a dummy variable ϕ as the difference between the filtered exact solution (obtained by Direct Numerical Simulation) and the solution found using large-eddy simulation with cutoff length $\overline{\Delta}$ on a mesh of size Δx:

$$e_{\text{total}}(\overline{\Delta}, \Delta x) = \overline{\phi_{\text{DNS}}} - \phi_{\text{LES}}(\overline{\Delta}, \Delta x) \quad , \tag{8.39}$$

the authors investigate the relative influence of different sources of error. Subgrid modeling and discretization errors being respectively defined as

$$e_{\text{model}} = \overline{\phi_{\text{DNS}}} - \phi_{\text{LES}}(\overline{\Delta}, 0) \quad , \tag{8.40}$$

$$e_{\text{discr}} = \phi_{\text{LES}}(\overline{\Delta}, 0) - \phi_{\text{LES}}(\overline{\Delta}, \Delta x) \quad , \tag{8.41}$$

exhaustive analyses carried out using a direct numerical simulation database yields the following conclusions:

– The global error behavior is parametrized by the *subgrid activity* parameter s defined like

$$s = \frac{\varepsilon}{\varepsilon + \varepsilon_\nu} \quad , \tag{8.42}$$

where $\varepsilon = -\tau_{ij}\overline{S}_{ij}$ is the subgrid dissipation and $\varepsilon_\nu = \nu S_{ij} S_{ij}$ is the molecular dissipation. Direct numerical simulation corresponds to $s = 0$, while $s = 1$ characterizes large-eddy simulation in the limit of infinite Reynolds number. Tests prove there there exists a threshold value s_c[7]: for $s \leq s_c$ the total error is dominated by discretization error effects, while for $s > s_c$ the modeling error is the most important source of uncertainty.

– For $s \leq s_c$, the relative error δ_{err}:

$$\delta_{\text{err}} = \frac{\int e_{\text{total}}^2 dt}{\int \overline{\phi_{\text{DNS}}}^2 dt} \quad , \tag{8.43}$$

can be either a decreasing or increasing function of s, depending on s and Δx when it is based on the kinetic energy (see Fig. 8.4). No general trend is observed in this regime, in which strong interactions can occur between discretization errors and modeling errors. Partial cancellation sometimes occurs leading to a significant reduction of the total error. For $s > s_c$, the relative error is a monotone exponentially increasing function of s.

– The relative error based on the Taylor length scale has a different behavior in the $s \leq s_c$ regime (see Fig. 8.5): strong interactions between modeling and discretization errors are observed, but the relative error exhibits a monotonic increasing behavior as a function of s_c. Partial error cancellation is less intense than in the case of the turbulent kinetic energy.

– Because of the strong non-linear interactions between modeling and discretization errors, which leads to a non-monotone behavior of the total

[7] Typical values are $s_c = 0.4$ for isotropic turbulence at $Re_\lambda = 50$, $s_c = 0.8$ for isotropic turbulence at $Re_\lambda = 100$ and $s_c = 0.5$ for a plane mixing layer.

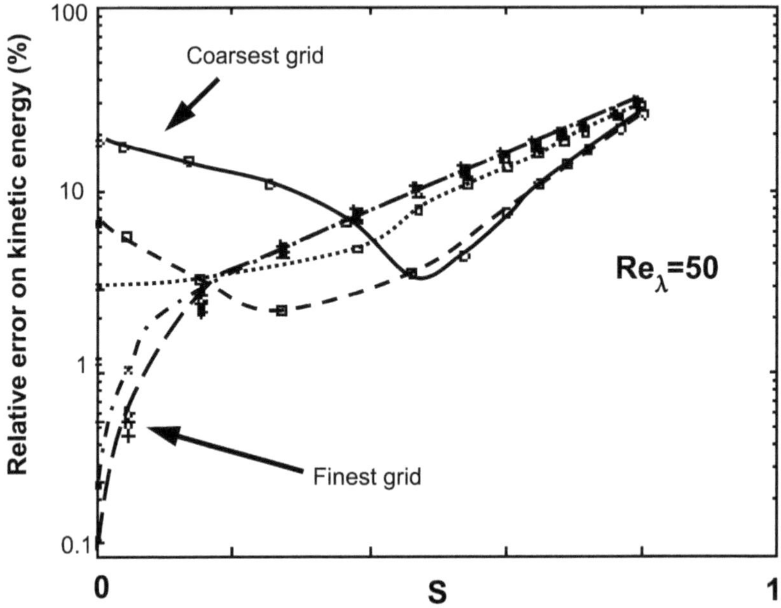

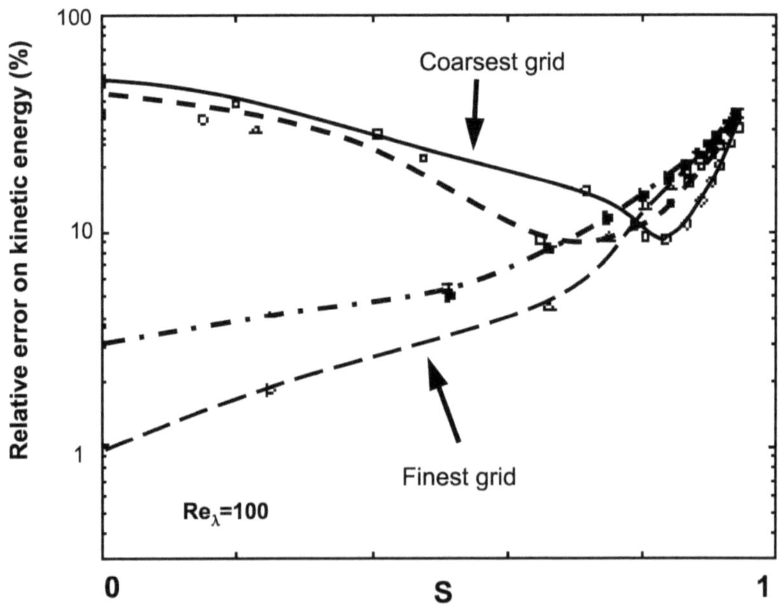

Fig. 8.4. Evolution of the relative error on kinetic energy as a function of the subgrid activity, in isotropic turbulence (Courtesy of J. Meyers and B. Geurts, Univ. Twente). *Top:* $Re_\lambda = 50$, *Bottom:* $Re_\lambda = 100$.

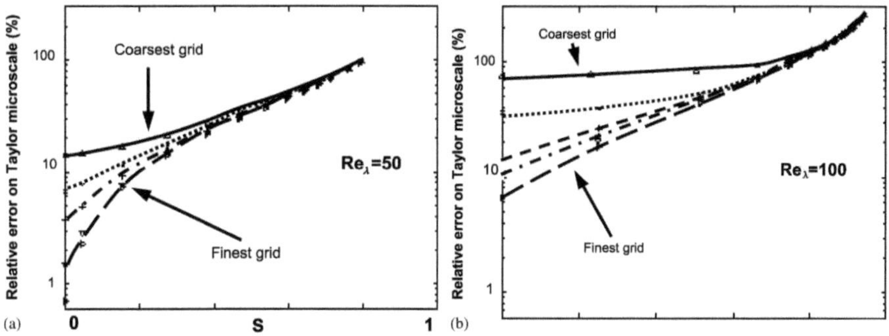

Fig. 8.5. Evolution of the relative error on the Taylor scale as a function of the subgrid activity, in isotropic turbulence (Courtesy of J. Meyers and B. Geurts, Univ. Twente). *Left:* $Re_\lambda = 50$, *Right:* $Re_\lambda = 100$.

error. As a consequence, grid refinement (decrease of Δx at constant $\overline{\Delta}$) must be coupled to a change in the Smagorinsky constant C_S to obtain an optimal error minimization. The ratio $C_S\overline{\Delta}/\Delta x$ which yields the minimum error follows non-trivial trajectories in both the $(\Delta x, C_S\overline{\Delta})$ and the $(\Delta x, s/s_c)$ planes. This behavior is illustrated in Fig. 8.6.

8.3.3 Conclusions

This analysis can be used only for reference, because it is based on very restrictive hypotheses. It nonetheless indicates that the numerical error is not negligible and that it can even be dominant in certain cases over the subgrid terms. The effective numerical filter is then dominant over the scale separation filter.

This error can be reduced either by increasing the order of accuracy of the numerical scheme or by using a pre-filtering technique that decouples the cutoff length of the analytical filter of the discretization step. Ghosal's findings seem to indicate that a combination of these two techniques would be the most effective solution.

These theoretical findings are confirmed by the numerical experiments of Najjar and Tafti [563] and Kravenchko and Moin [409], who observed that the effect of the subgrid models is completely or partially masked by the numerical error when second-order accurate methods are employed. It should be noted here that practical experience leads us to less pessimistic conclusions than the theoretical analyses: large-eddy simulations performed with a scheme accurate to the second order show a dependency with respect to the subgrid model used. The effects of these models are not entirely masked, which justifies using them. However, no precise qualification exists today of the information loss due to the use of a given scheme. These observations are made empirically, case by case [76, 480, 94].

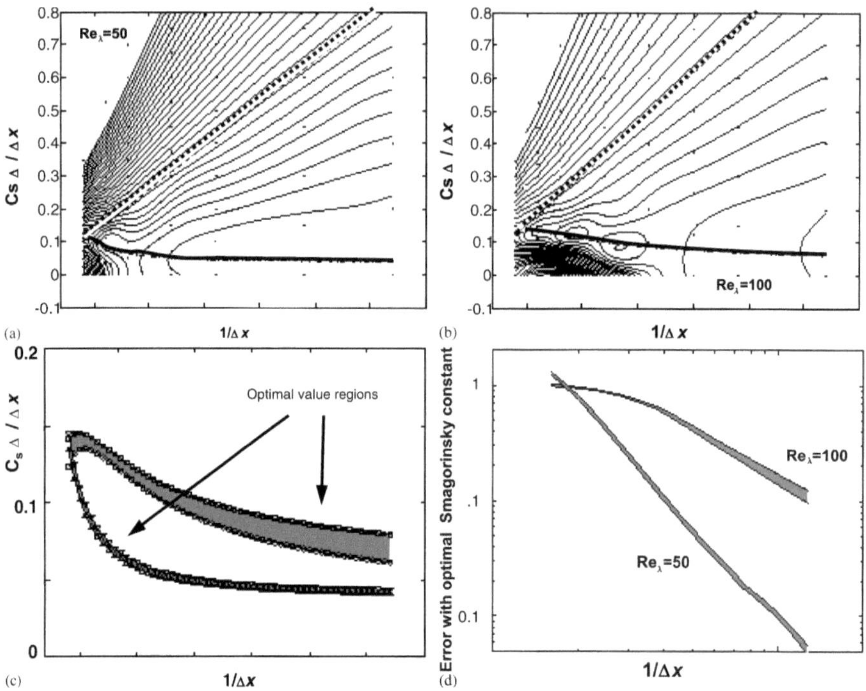

Fig. 8.6. *Top:* Map of the relative error in isotropic turbulence (Courtesy of J. Meyers and B. Geurts, Univ. Twente). *Left:* $Re_\lambda = 50$, *Right:* $Re_\lambda = 100$. The *dashed-dotted line* is related to the optimal trajectory corresponding to the minimal error. The *dashed line* corresponds to the trajectory associated to a fixed value of the subgrid activity parameter. *Bottom:* Optimal refinement strategy for different Reynolds numbers and error definitions, shown in different planes. The *shadded areas* are related to minimal value of the error.

Another effect was emphasized by Geurts et al. [255, 515], who found that partial cancellation of modeling and numerical errors may occur, leading to a significant improvement of the accuracy of the simulation. This cancellation was observed in both plane mixing layer configuration and isotropic turbulence for fourth-order and spectral schemes. An important and counter-intuitive conclusion is that using higher-order accurate schemes or improved subgrid models may lead to worse results with regards to a simulation in which this cancellation occurs. This behavior is observed to be strongly dependent on the value of the subgrid activity parameter s and the considered definition of the error.

An important finding dealing with the prefiltering technique is that refining the the grid at constant filter cutoff (i.e. $\overline{\Delta}/\Delta x \longrightarrow \infty$ at fixed $\overline{\Delta}$) must be coupled to a change in subgrid viscosity model constant to obtain the optimal error reduction.

8.3.4 Remarks on the Use of Artificial Dissipations

Many comments have been made over recent decades on the sensitivity of large-eddy simulation results, for example concerning the formulation of the convection term [319, 409], the discrete form of the test filter [597, 563, 81, 641, 605], and the formulation of the subgrid term [634], but there are far too many, too dispersed, and too far from general to be detailed here. Moreover, countless analyses have been made of the numerical error associated with various schemes, especially as concerns the treatment of the non-linear terms, which will not be resumed here, but we will still take more special note of the findings of Fabignon et al. [211] concerning the characterization of the effective numerical filter of several schemes.

Special attention should still be paid to the discretization of the convective terms. To capture strong gradients without having the numerical solution polluted with spurious high-frequency wiggles, the scheme is very often stabilized by introducing artificial dissipation. This dissipation is added explicitly or implicitly using an upwind scheme for the convection term. Introducing an additional dissipation term for the large-eddy simulation is still controversial [563, 521, 239, 228, 665, 501, 94] because the effective filter is then very similar in form to that which would be imposed by subgrid viscosity model, making for two competing resolved kinetic energy spectrum mechanisms. The similarity between the numerical dissipation and that associated with the energy cascade model is still being investigated, but a few conclusions have already been drawn.

It seems that the total numerical dissipation induced by most upwind schemes is still greater than that of the subgrid viscosity models, if no prefiltering method is used. This is true even for seventh-order accurate upwind schemes [52].

Garnier et al. [239] developed the *generalized Smagorinsky constant* as a tool to compare numerical and physical subgrid dissipations. The generalized Smagorinsky constant is the value that should take the constant of the Smagorinsky model to obtain a total dissipation equal to the numerical dissipation. Numerical tests carried out on decaying isotropic turbulence have shown that all the numerical upwind schemes, up to the fifth order of accuracy, are more dissipative than the usual Smagorinsky model. Typical results are displayed in Fig. 8.7.

It has also been demonstrated [501] that the use of a stabilized sixth-order accurate scheme may lead to the same quality of results as a second-order accurate scheme, because of the very high dissipation applied to the highest resolved frequency, which is responsible for the largest part of the interactions with subgrid modes. These two dissipations are correlated in space (especially in the case of the Smagorinsky model), but have different spectral distributions: a subgrid viscosity model corresponds to a second-order dissipation associated with a spectrum of the form $(k/k_c)^2E(k)$, while an nth-order numerical dissipation is associated with a spectrum of the form $(k/k_c)^nE(k)$.

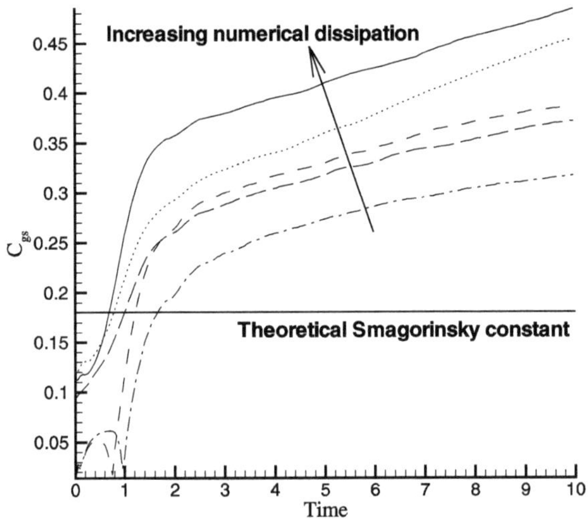

Fig. 8.7. Time history of the generalized Smagorinsky constant for various dissipative schemes (second- to fifth-order of accuracy) in freely decaying isotropic turbulence. Courtesy of E. Garnier, ONERA.

For $n > 2$ (resp. $n < 2$), the numerical (resp. subgrid) dissipation may be dominant for the highest resolved frequencies and the subgrid (resp. numerical) dissipation will govern the dynamics of the low frequencies. This point is illustrated in Figs. 8.8 and 8.9.

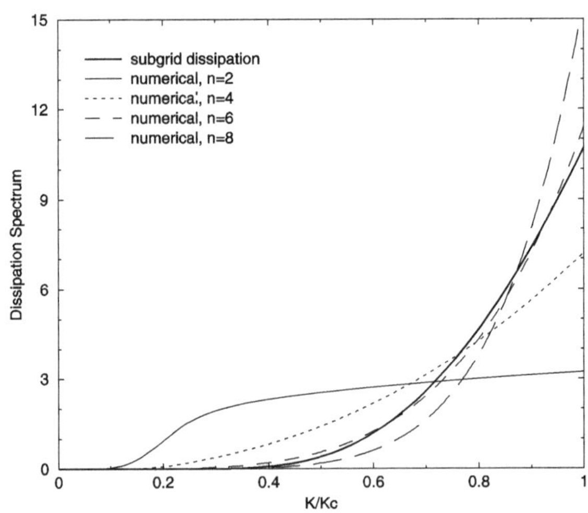

Fig. 8.8. Numerical and subgrid dissipations for a Von Karman spectrum. The peak of the Von Karman spectrum is at $k_C/5$. The dissipation spectra have been normalized so that the total dissipation is the same in all cases. It is worth noting that typical numerical schemes lead to a total dissipation higher than the subgrid dissipation.

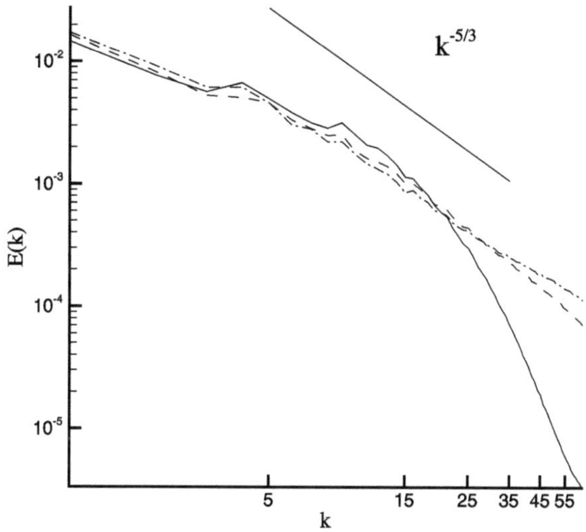

Fig. 8.9. Turbulent kinetic energy spectra (isotropic turbulence) computed by large-eddy simulation. *Solid line*, fifth-order accurate dissipative scheme without subgrid model. *Dashed lines*, second-order accurate non-dissipative scheme with (i) Smagorinsky model and (ii) dynamic Smagorinsky model. Courtesy of E. Garnier, ONERA.

The studies that have been made show a sensitivity of the results to the subgrid model used, which proves that the effects of the model are not entirely masked by the numerical dissipation. The theoretical analysis presented above should therefore be taken relative to this. But consistent with it, Beaudan et al. [52] have observed a reduction in the numerical cutoff length as the order of accuracy of the scheme increases. This type of finding should nonetheless be treated with caution, because the conclusions may be reversed if we bring in an additional parameter, which is the grid refinement. Certain studies have shown that, for coarse grids, i.e. high values of the numerical cutoff length, increasing the order of accuracy of the upwind scheme can lead to a degradation of the results [701]. But some specific numerical stabilization procedures can be defined, which tune the numerical dissipation in such a way that the results remain sensitive to subgrid modeling [94, 15, 95]. These numerical methods allow the use of relatively coarse grids, but no general theory for them exist at present time.

This relative similarity between artificial dissipation and the direct energy cascade model has induced certain authors to perform "no-model" large-eddy simulations, with the filtering based entirely on the numerical method (leading to the Implicit Large-Eddy Simulation technique, see Sect. 5.3.4 for a detailed discussion). Thus many flow simulations have been seen in complex geometries, based on the use of an third-order accurate upwind scheme proposed by Kawamura and Kuwahara [381], yielding interesting results. In the compressible case, this approach has been called the Monotone Integrated

Large-Eddy Simulation (MILES) method (see Sect. 5.3.4 for a description of ILES numerical methods). The "experimental" analysis of some particular numerical methods reveals that there exists some numerical dissipation procedures which mimic very well the theoretical subgrid viscosity [192]. This is illustrated in Fig. 8.10, which compares the computes spectral viscosity associated to the MPDATA scheme with the theoretical spectral viscosity profile. The observed agreement prove that the Implict Large-Eddy Simulation approach can yield very good results if the numerical scheme is carefully chosen.

The use of artificial dissipation therefore raises many questions, but is very common in simulations that are physically very strongly under-resolved in complex configurations, because experience shows that adding subgrid models does not ensure a monotonic solution. To ensure that certain variables remain positive, such as concentrations of pollutants or the temperature, it seems to be necessary to resort to such numerical methods. Alternatives based on local refinement of the solution, i.e. decreasing the effective cutoff length by enriching or adapting the grid, have been studied by certain authors but no final conclusion has been drawn.

A few studies dealing with some particular stabilization techniques such as the Galerkin Least-Square method, reveal that the numerical dissipation may happen to be be too low to prevent a pile-up in the resolved kinetic energy, leading to the growth of bounded wiggles. In such a case, the use of a subgrid model seems to be required to recoved physical results. The main problem arising in this case is that the use of a basic subgrid model will lead to a global overdamping of the computed solution. The use of self-adaptive subgrid models able to tune their induced dissipation so that the sum of

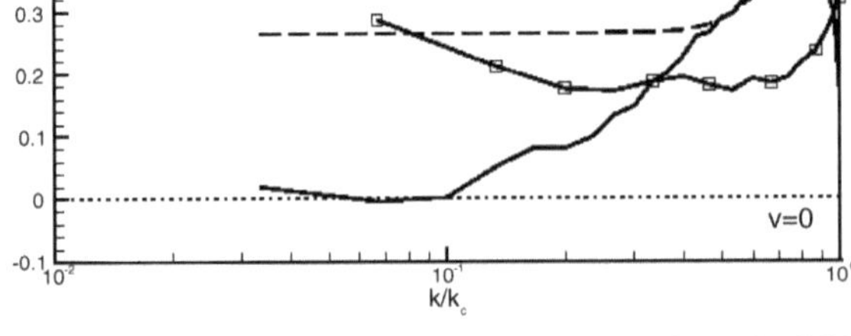

Fig. 8.10. Analysis of the numerical viscosity associated to the MPDATA scheme. *Dashed line:* theoretical spectral subgrid viscosity. *Solid line:* MPDATA equivalent spectralk viscosity. *Solid line with symbol:* spectral viscosity recomputed from kinetic energy transfer of the unfiltered velocity field. From [192].

the numerical and subgrid dissipation will reach the required level to obtain physical results. The identification of subgrid models and/or procedures to define such subgrid models is still a open issue. An a priori requirement is that the subgrid model must be local in terms of wave number, i.e. it must emphasized the highest resolved frequencies. Numerical experiments show that self-adaptive models and some multilevel approaches have this property.

While the Navier–Stokes equations contain energy information, they also contain information concerning the signal phase. Using centered schemes for the convection term therefore raises problems too, because of the dispersive errors they induce in the highest resolved frequencies.

Generally, estimates of the wave number beyond which the modes are considered to be well resolved numerically vary from $2\Delta x$ to $20\Delta x$, depending on the schemes and authors [604].

8.3.5 Remarks Concerning the Time Integration Method

Large-eddy simulation is ordinarily addressed using a spatial filtering, but without explicitly stating the associated time filtering. This is due to the fact that most computations are made for moderate time steps ($\mathrm{CFL} \equiv u\Delta t/\Delta x < 1$) and it is felt that the time filtering effects are masked by those of the space filtering. Choi and Moin [131], however, have shown by direct simulations of a plane channel flow that the time filtering effects can be very large, even for CFLs of the order of 0.5, since the turbulence cannot be maintained numerically if the time step is greater than the characteristic time associated with the Kolmogorov scale. Most authors use second-order accurate integration methods, but no complete study has been published to date to determine what timescales are well resolved numerically and physically. We should also note the results of Beaudan and Moin [52] and Mittal and Moin [521], who showed that the use of artificial viscosity affects the solution of a very large share of the simulated time frequencies (about 75% for the particular case studied).

9. Analysis and Validation of Large-Eddy Simulation Data

9.1 Statement of the Problem

9.1.1 Type of Information Contained in a Large-Eddy Simulation

The solution to the equations that define the large-eddy simulation furnishes explicit information only on the scales that are resolved, i.e. those that are left after reduction of the number of degrees of freedom in the exact solution. We are therefore dealing with information that is truncated in space and time. The time filtering is induced implicitly by the spatial filtering because, as the filtering eliminates certain space scales, it eliminates the corresponding time scales with them (see p. 19).

The information of use for analysis or validation is what is contained in those scales that are both physically and numerically well-resolved. It should be remembered that, since the effective numerical and physical filters are unknown, the usable scales are most often identified empirically.

Adopting the assumption that all the scales represented by the simulation are physically and numerically well-resolved, the statistical average of the usable resolved field is expressed $\langle \overline{\boldsymbol{u}} \rangle$. The statistical fluctuation of the resolved field, denoted $\overline{\boldsymbol{u}}''$, is defined by:

$$\overline{u}_i'' = \overline{u}_i - \langle \overline{u}_i \rangle \quad . \tag{9.1}$$

The difference between the statistical average of the resolved scales and that of the exact solution is defined as:

$$\langle u_i \rangle - \langle \overline{u}_i \rangle = \langle u_i' \rangle \quad , \tag{9.2}$$

which corresponds to the statistical average of the unresolved scales. The Reynolds stresses computed from the resolved scales are equal to $\langle \overline{u}_i'' \overline{u}_j'' \rangle$. The difference from the exact stresses $\langle u_i'^e u_j'^e \rangle$, where the exact fluctuation is defined as $\boldsymbol{u}'^e = \boldsymbol{u} - \langle \boldsymbol{u} \rangle$, is:

$$\begin{aligned}
\langle u_i'^e u_j'^e \rangle &= \langle (u_i - \langle u_i \rangle)(u_j - \langle u_j \rangle) \rangle \\
&= \langle u_i u_j \rangle - \langle u_i \rangle \langle u_j \rangle \\
&= \langle \overline{u}_i \overline{u}_j + \tau_{ij}' \rangle - \langle \overline{u}_i + u_i' \rangle \langle \overline{u}_j + u_j' \rangle
\end{aligned}$$

$$= \langle \overline{u}_i \overline{u}_j \rangle + \langle \tau'_{ij} \rangle - \langle \overline{u}_i \rangle \langle \overline{u}_j \rangle - \langle u'_i \rangle \langle \overline{u}_j \rangle - \langle \overline{u}_i \rangle \langle u'_j \rangle - \langle u'_i \rangle \langle u'_j \rangle$$
$$= \langle \overline{u}''_i \overline{u}''_j \rangle + \langle \tau'_{ij} \rangle - \langle u'_i \rangle \langle \overline{u}_j \rangle - \langle \overline{u}_i \rangle \langle u'_j \rangle - \langle u'_i \rangle \langle u'_j \rangle \quad ,$$

where $\tau'_{ij} = \overline{u}_i u'_j + \overline{u}_j u'_i + u'_i u'_j$ is the defiltered subgrid tensor.

Since the subgrid scales are not known, the terms containing the contribution $\langle \boldsymbol{u}' \rangle$, cannot be computed from the simulation. When the statistical average of the subgrid modes is very small compared with the other terms, we get:

$$\langle u'^e_i u'^e_j \rangle \simeq \langle \overline{u}''_i \overline{u}''_j \rangle + \langle \tau'_{ij} \rangle \quad . \tag{9.3}$$

The two terms on the right-hand side can be evaluated from the numerical simulation, but the quality of the model's representation of the subgrid tensor partly conditions that of the result. We can easily see that subgrid-viscosity models, which only account for the deviatoric part of the subgrid-stress tensor, make it possible only to recover the deviatoric part of the Reynolds stresses [760, 397], at least theoretically.

9.1.2 Validation Methods

The subgrid models and their various underlying hypotheses can be validated in two ways [218]:

- A priori validation. The exact solution, which is known in this case, is filtered analytically, leading to the definition of a fully determined resolved field and subgrid field. The various hypotheses or models can then be tested. The exact solutions are usually generated by direct numerical simulations at moderate or low Reynolds numbers, which limits the field of investigation. A priori tests like this have also been performed using experimental data, making it possible to reach higher Reynolds numbers. This type of validation raises a fundamental problem, though. By comparing the exact subgrid stresses with those predicted by a subgrid model evaluated on the basis of the filtered exact solution, the effects of the modeling errors are neglected and the implicit filter associated with the model is not considered[1]. This means that the results of a priori validations are only relative in value.
- A posteriori validation. Here, we perform a large-eddy simulation computation and validate by comparing its results with a reference solution. This is a dynamic validation that takes all the simulation factors into consideration, while the previous method is static. Experience shows that models yielding poor a priori results can be satisfactory *a posteriori*, and vice

[1] This field could not have been obtained by a large-eddy simulation since it is a solution of the filtered momentum equations in which the exact subgrid tensor appears. In the course of a simulation, the subgrid model is applied to a velocity field that is a solution of the momentum equation where the modeled subgrid tensor appears. These two fields are therefore different in theory. Consequently, in order to be fully representative, an a priori test has to be performed on the basis of a velocity field that can be obtained from the subgrid model studied.

versa [597]. It is more advantageous to validate models *a posteriori* because it corresponds to their use in the simulation; but it is sometimes difficult to draw any conclusions on a precise point because of the multitude of often imperfectly controlled factors at play in a numerical simulation.

9.1.3 Statistical Equivalency Classes of Realizations

The subgrid models are statistical models and it seems pointless to expect them to produce deterministic simulations in which the resolved scales coincide exactly with those of other realizations, for example of the experimental sort. On the other hand, large-eddy simulation should correctly reproduce the statistical behavior of the scales making up the resolved field. Equivalency classes can thus be defined among the realizations [503] by considering that one of the classes consists of realizations that lead to the same values of certain statistical quantities computed from the resolved scales.

Belonging to the same class of equivalency as a reference solution is a validation criterion for the other realizations. If we set aside the numerical errors, we can define the necessary conditions on the subgrid models such that two realizations will be equivalent, by verifying these validity criteria. These conditions will be discussed in the following sections. A subgrid model can thus be considered validated if it can generate realizations that are equivalent to a reference solution, in a sense defined below.

Theoretically, while we overlook the effect of the discretization on the modeling, it can be justifiably thought that a model reproducing the interscale interactions exactly will produce good results, whereas the opposite proposition is not true. That is, the idea of sufficient complexity of a model has to be introduced in order to obtain a type of result on a given configuration with a tolerated margin of error in order to say what a good model is. The idea of a universal or best model might not be rejected outright, but should be taken relatively. The question is thus raised of knowing what statistical properties two subgrid models should share in order for the two resulting solutions to have common properties.

Let $\overline{\boldsymbol{u}}$ and $\overline{\boldsymbol{u}}^*$ be the filtered exact solution and the solution computed with a subgrid model, respectively, for the same filter. The exact (unmodeled) subgrid tensor corresponding to $\overline{\boldsymbol{u}}$ is denoted τ_{ij}, and the modeled subgrid tensor computed from the $\overline{\boldsymbol{u}}^*$ field is denoted $\tau_{ij}^*(\overline{\boldsymbol{u}}^*)$. The two velocity fields are solutions of the following momentum equations:

$$\frac{\partial \overline{\boldsymbol{u}}}{\partial t} + \nabla \cdot (\overline{\boldsymbol{u}} \otimes \overline{\boldsymbol{u}}) = -\nabla \cdot \overline{p} + \nu \nabla^2 \overline{\boldsymbol{u}} - \nabla \cdot \tau \quad , \tag{9.4}$$

$$\frac{\partial \overline{\boldsymbol{u}}^*}{\partial t} + \nabla \cdot (\overline{\boldsymbol{u}}^* \otimes \overline{\boldsymbol{u}}^*) = -\nabla \cdot \overline{p}^* + \nu \nabla^2 \overline{\boldsymbol{u}}^* - \nabla \cdot \tau^*(\overline{\boldsymbol{u}}^*) \quad . \tag{9.5}$$

A simple analysis shows that, if all the statistical moments (at all points of space and time) of τ_{ij} conditioned by the $\overline{\boldsymbol{u}}$ field are equal to those of

$\tau^*_{ij}(\overline{\boldsymbol{u}}^*)$ conditioned by $\overline{\boldsymbol{u}}^*$, then all the statistical moments of $\overline{\boldsymbol{u}}$ and $\overline{\boldsymbol{u}}^*$ will be equal. This is a full statistical equivalency, which implies that the subgrid models fulfill an infinity of conditions. To relax this constraint, we define less restrictive equivalency classes of solutions which are described in the following sections. They are defined in such a way as to bring out the necessary conditions applying to the subgrid models, in order to qualify them [503]. We try to define conditions such that the statistical moments of moderate order[2] (1 and 2) of the field resulting from the large-eddy simulation $\overline{\boldsymbol{u}}^*$ are equal to those of a reference solution $\overline{\boldsymbol{u}}$.

Equivalency of First-Order Moments. The equivalency relation is built on the equality of the first-order statistical moments of the realizations. A velocity and a pressure field are associated with each realization. Let $(\overline{\boldsymbol{u}}, \overline{p})$ and $(\overline{\boldsymbol{u}}^*, \overline{p}^*)$ be the doublets associated with the first and second realizations, respectively. The evolution equations of the first-order statistical moments of the velocity field of these two realizations are expressed:

$$\begin{aligned} \frac{\partial \langle \overline{\boldsymbol{u}} \rangle}{\partial t} + \nabla \cdot (\langle \overline{\boldsymbol{u}} \rangle \otimes \langle \overline{\boldsymbol{u}} \rangle) &= -\nabla \cdot \langle \overline{p} \rangle + \nu \nabla^2 \langle \overline{\boldsymbol{u}} \rangle - \nabla \cdot \langle \tau \rangle \\ &\quad -\nabla \cdot (\langle \overline{\boldsymbol{u}} \otimes \overline{\boldsymbol{u}} \rangle - \langle \overline{\boldsymbol{u}} \rangle \otimes \langle \overline{\boldsymbol{u}} \rangle) \quad , \end{aligned} \tag{9.6}$$

$$\begin{aligned} \frac{\partial \langle \overline{\boldsymbol{u}}^* \rangle}{\partial t} + \nabla \cdot (\langle \overline{\boldsymbol{u}}^* \rangle \otimes \langle \overline{\boldsymbol{u}}^* \rangle) &= -\nabla \cdot \langle \overline{p}^* \rangle + \nu \nabla^2 \langle \overline{\boldsymbol{u}}^* \rangle - \nabla \cdot \langle \tau^*(\overline{\boldsymbol{u}}^*) \rangle \\ &\quad -\nabla \cdot (\langle \overline{\boldsymbol{u}}^* \otimes \overline{\boldsymbol{u}}^* \rangle - \langle \overline{\boldsymbol{u}}^* \rangle \otimes \langle \overline{\boldsymbol{u}}^* \rangle) \quad , \end{aligned} \tag{9.7}$$

where $\langle \ \rangle$ designates an ensemble average performed using independent realizations. The two realizations will be called equivalent if their first- and second-order moments are equivalent, i.e.

$$\langle \overline{u}_i \rangle = \langle \overline{u}^*_i \rangle \quad , \tag{9.8}$$

$$\langle \overline{p} \rangle = \langle \overline{p}^* \rangle \quad , \tag{9.9}$$

$$\langle \overline{u}_i \overline{u}_j \rangle = \langle \overline{u}^*_i \overline{u}^*_j \rangle \quad . \tag{9.10}$$

Analysis of evolution equations (9.6) and (9.7) shows that one necessary condition is that the resolved and subgrid stresses be statistically equivalent. The last condition is expressed:

$$\langle \tau_{ij} \rangle = \langle \tau^*_{ij} \rangle + C_{ij} \quad , \tag{9.11}$$

where C_{ij} is a null-divergence tensor. This condition is not sufficient because a model that leads to a good prediction of the mean stresses can generate an error on the mean field if the mean resolved stresses are not correct. To obtain a sufficient condition, the equivalency of the stresses $\langle \overline{u}_i \overline{u}_j \rangle$ and $\langle \overline{u}^*_i \overline{u}^*_j \rangle$ must be ensured by another relation.

[2] Because these are the quantities sought in practice.

Equivalency of Second-Order Moments. We now base the equivalency relation on the equality of the second-order moments of the resolved scales. Two realizations will be called equivalent if the following conditions are satisfied:

$$\langle \overline{u}_i \rangle = \langle \overline{u}_i^* \rangle \quad , \tag{9.12}$$

$$\langle \overline{u}_i \overline{u}_j \rangle = \langle \overline{u}_i^* \overline{u}_j^* \rangle \quad , \tag{9.13}$$

$$\langle \overline{u}_i \overline{u}_j \overline{u}_k \rangle = \langle \overline{u}_i^* \overline{u}_j^* \overline{u}_k^* \rangle \quad , \tag{9.14}$$

$$\langle \overline{p} \overline{u}_i \rangle = \langle \overline{p}^* \overline{u}_i^* \rangle \quad , \tag{9.15}$$

$$\langle \overline{p} \overline{S}_{ij} \rangle = \langle \overline{p}^* \overline{S}_{ij}^* \rangle \quad , \tag{9.16}$$

$$\left\langle \frac{\partial \overline{u}_i}{\partial x_k} \frac{\partial \overline{u}_i}{\partial x_k} \right\rangle = \left\langle \frac{\partial \overline{u}_i^*}{\partial x_k} \frac{\partial \overline{u}_i^*}{\partial x_k} \right\rangle \quad . \tag{9.17}$$

Analysis of the equation for the second-order moments $\langle \overline{u}_i \overline{u}_j \rangle$ shows that, in order for two realizations to be equivalent, the following necessary condition must be satisfied:

$$\langle \tau_{ik} \overline{S}_{kj} \rangle + \langle \tau_{jk} \overline{S}_{ki} \rangle - \frac{\partial}{\partial x_k} \left(\langle \overline{u}_i \tau_{jk} \rangle + \langle \overline{u}_j \tau_{ik} \rangle \right) =$$

$$\langle \tau_{ik}^* \overline{S}_{kj}^* \rangle + \langle \tau_{jk}^* \overline{S}_{ki}^* \rangle - \frac{\partial}{\partial x_k} \left(\langle \overline{u}_i^* \tau_{jk}^* \rangle + \langle \overline{u}_j^* \tau_{ik}^* \rangle \right) \quad .$$

This condition is not sufficient. To obtain such an realization, the equality of the third-order moments also has to be ensured. It is noted that the non-linear coupling prohibits the definition of sufficient conditions on the subgrid model to ensure the equality of the nth-order moments of the resolved field without adding necessary conditions on the equality of the $(n+1)$th-order moments.

Equivalency of the Probability Density Functions. We now base the definition of the equivalency classes on the probability density function $f_{\text{prob}}(\boldsymbol{V}, \boldsymbol{x}, t)$ of the resolved scales. The field $\boldsymbol{V}$ is the test velocity field from which the conditional average is taken. The function f_{prob} is defined as the statistical average of the one-point probabilities:

$$f_{\text{prob}}(\boldsymbol{V}, \boldsymbol{x}, t) \equiv \langle \delta(\overline{\boldsymbol{u}}(\boldsymbol{x}, t) - \boldsymbol{V}) \rangle \quad , \tag{9.18}$$

and is a solution of the following transport equation:

$$\frac{\partial f_{\text{prob}}}{\partial t} + V_j \frac{\partial f_{\text{prob}}}{\partial x_j} = \frac{\partial}{\partial V_j} \left\{ f_{\text{prob}} \left\langle \frac{\partial p}{\partial x_j} + \frac{\partial \tau_{ij}}{\partial x_j} - \nu \nabla^2 \overline{u}_j | \overline{\boldsymbol{u}} = \boldsymbol{V} \right\rangle \right\} \quad . \tag{9.19}$$

Two realizations can be called equivalent if:

$$f_{\text{prob}}(\boldsymbol{V}, \boldsymbol{x}, t) = f_{\text{prob}}^*(\boldsymbol{V}, \boldsymbol{x}, t) \quad , \tag{9.20}$$

$$\langle \overline{u}_i(\boldsymbol{y}) | \overline{\boldsymbol{u}} = \boldsymbol{V} \rangle = \langle \overline{\boldsymbol{u}}_i^*(\boldsymbol{y}) | \overline{\boldsymbol{u}} = \boldsymbol{V} \rangle \quad , \tag{9.21}$$

$$\langle \overline{u}_i(\boldsymbol{y}) \overline{u}_j(\boldsymbol{y}) | \overline{\boldsymbol{u}} = \boldsymbol{V} \rangle = \langle \overline{\boldsymbol{u}}_i^* \overline{\boldsymbol{u}}_j^*(\boldsymbol{y}) | \overline{\boldsymbol{u}} = \boldsymbol{V} \rangle \quad . \tag{9.22}$$

Once the pressure gradient is expressed as a function of the velocity (by an integral formulation using a Green function) and the conditional average of the strain rate tensor is expressed using gradients of the two-point conditional averages, equation (9.19) can be used to obtain the following necessary condition:

$$-\frac{1}{4\pi}\int \frac{\partial^2}{\partial y_i \partial y_k}\langle \tau_{ik}|\overline{\boldsymbol{u}}(\boldsymbol{x}) = \boldsymbol{V}\rangle \frac{x_j - y_j}{|\boldsymbol{x}-\boldsymbol{y}|} d^3\boldsymbol{y} + \lim_{\boldsymbol{y}\to\boldsymbol{x}} \frac{\partial}{\partial y_i}\langle \tau_{ij}|\overline{\boldsymbol{u}}(\boldsymbol{x}) = \boldsymbol{V}\rangle$$

$$= -\frac{1}{4\pi}\int \frac{\partial^2}{\partial y_i \partial y_k}\langle \tau^*_{ik}|\overline{\boldsymbol{u}}^*(\boldsymbol{x}) = \boldsymbol{V}\rangle \frac{x_j - y_j}{|\boldsymbol{x}-\boldsymbol{y}|} d^3\boldsymbol{y}$$

$$+ \lim_{\boldsymbol{y}\to\boldsymbol{x}} \frac{\partial}{\partial y_i}\langle \tau^*_{ij}|\overline{\boldsymbol{u}}^*(\boldsymbol{x}) = \boldsymbol{V}\rangle + C_j \quad ,$$

in which the divergence of vector C_j is null. It is noted that the condition defined from the one-point probability density uses two-point probabilities. We again find here the problem of non-localness already encountered when the equivalency class is based on statistical moments. A more restrictive condition is:

$$\langle \tau_{ik}|\overline{\boldsymbol{u}}(\boldsymbol{x}) = \boldsymbol{V}\rangle = \langle \tau^*_{ik}|\overline{\boldsymbol{u}}^*(\boldsymbol{x}) = \boldsymbol{V}\rangle \quad . \tag{9.23}$$

9.1.4 Ideal LES and Optimal LES

An abstract subgrid model can be defined, which is in all senses ideal [419]. An LES using this model will exactly reproduce all single-time, multipoint statistics, and at the same time will exhibit minimum possible error in instantaneous dynamics. Such a LES will be referred to as *ideal LES*. Using the same notations as in Sect. 9.1.3, ideal LES is governed by the conditional average

$$\frac{d\overline{\boldsymbol{u}}^*}{dt} = \left\langle \frac{d\boldsymbol{u}}{dt} \middle| \overline{\boldsymbol{u}} = \overline{\boldsymbol{u}}^* \right\rangle \quad , \tag{9.24}$$

where $\overline{\boldsymbol{u}}$ and $\overline{\boldsymbol{u}}^*$ are the solution of the exact LES equation and the LES equation with a subgrid model, respectively. It can be shown that such ideal LES is associated to the minimum mean-square error between the evolution of the LES field $\overline{\boldsymbol{u}}^*(t)$ and the exact solution $\overline{\boldsymbol{u}}(t)$, defined as an instantaneous pointwise measurement on $\partial\overline{\boldsymbol{u}}/\partial t$:

$$e_i(\boldsymbol{x}) = \frac{\partial \overline{u}_i^*}{\partial t} - \frac{\partial \overline{u}_i}{\partial t} \quad . \tag{9.25}$$

Equivalently, this error can be evaluated using the exact and the modeled subgrid forces, referred to as $\boldsymbol{M} = \nabla \cdot \tau$ and $\boldsymbol{m} = \nabla \cdot \tau^*$:

$$\boldsymbol{e}(\boldsymbol{x}) = \boldsymbol{M}(\boldsymbol{x}) - \boldsymbol{m}(\boldsymbol{x}) \quad . \tag{9.26}$$

The ideal subgrid model τ^* is then such that

$$\nabla \cdot \tau^* = \boldsymbol{m} = \langle \boldsymbol{M} | \overline{\boldsymbol{u}} = \overline{\boldsymbol{u}}^* \rangle \quad . \tag{9.27}$$

This model is written as an average over the real turbulent fields whose resolved scales match the current LES field, making it impossible to compute in practical applications. In order to approximate in an optimal sense this ideal model, several authors [419, 54, 6, 137] propose to formally approximate the conditional average by a stochastic estimation. These new models can be referred to as *optimal* or *nearly-optimal models*, leading to *optimal LES*. The estimation of the subgrid force is based on the convolution of an estimation kernel K_{ij} with velocity event data at N points $(\xi_1, ..., \xi_N)$:

$$m_i(\boldsymbol{x}) = \int K_{ij}(\boldsymbol{x}, \xi_1, ..., \xi_N) E_j(\overline{\boldsymbol{u}}^*; \xi_1, ..., \xi_N) d\xi_1 ... d\xi_N \quad , \tag{9.28}$$

where E_j is an event vector. Chosing

$$\boldsymbol{E}(\xi_1, ..., \xi_N) = (1, \overline{u}_i^*(\xi_1), \overline{u}_j^*(\xi_1)\overline{u}_k^*(\xi_2), ...) \quad , \tag{9.29}$$

we recover the expansion

$$\begin{aligned} m_i(\boldsymbol{x}) &= A_i(\boldsymbol{x}) + \int B_{ij}(\boldsymbol{x}, \xi_1)\overline{u}_j^*(\xi_1) d\xi_1 \\ &+ \int C_{ijk}(\boldsymbol{x}, \xi_1, \xi_2)\overline{u}_j^*(\xi_1)\overline{u}_k^*(\xi_2) d\xi_1 d\xi_2 \quad . \end{aligned} \tag{9.30}$$

The random mean square error between M_i and m_i is minimal when

$$\langle e_i(\boldsymbol{x}) E_k(\eta_1, ..., \eta_N) \rangle = 0 \quad , \tag{9.31}$$

yielding the following definition of the optimal kernel K_{ij}

$$\begin{aligned} \langle M_i(\boldsymbol{x}) E_k(\eta_1, ..., \eta_N) \rangle &= \int K_{ij}(\boldsymbol{x}, \xi_1, ..., \xi_N) \\ &\times \langle E_j(\xi_1, ..., \xi_N) E_k(\eta_1, ..., \eta_N) \rangle \, d\xi_1 ... d\xi_N \end{aligned} \tag{9.32}$$

The resulting optimal subgrid models have the property that the correlation of the parametrized subgrid force with any event data is the same as the correlation of the exact subgrid force with the same event data:

$$\langle m_i(\boldsymbol{x}) E_j(\xi_1, ..., \xi_N) \rangle = \langle M_i(\boldsymbol{x}) E_j(\xi_1, ..., \xi_N) \rangle \quad . \tag{9.33}$$

9.1.5 Mathematical Analysis of Sensitivities and Uncertainties in Large-Eddy Simulation

The developments presented above deal with the nature of the information retrieved from large eddy simulation and some statistical equivalency constraints. But a key problem in practical cases is to evaluate the sensitivity of the computed results with respect to the subgrid model (or its inputs), i.e. to obtain an estimate for the uncertainties associated to the fact that all scales are not resolved and that small ones are parameterized. The mathematical framework for such an analysis was proposed by Anitescu and Layton [13]. Starting from the governing equations for large eddy simulation in the case where a subgrid viscosity model is utilized

$$\frac{\partial \overline{\boldsymbol{u}}}{\partial t} + \nabla \cdot (\overline{\boldsymbol{u}} \otimes \overline{\boldsymbol{u}}) = -\nabla \overline{p} + \nu \nabla^2 \overline{\boldsymbol{u}} + \nabla \cdot (\nu_{\text{sgs}}(\overline{\Delta}, \overline{\boldsymbol{u}}) S(\overline{\boldsymbol{u}})) \quad , \tag{9.34}$$

$$\nabla \cdot \overline{\boldsymbol{u}} = 0 \quad , \tag{9.35}$$

where the notation $S(\overline{\boldsymbol{u}})$ was used instead of the usual $\overline{S}$ for the sake of convenience. Let us now introduce the sensitivities of the velocity and the pressure defined as

$$\boldsymbol{v} \equiv \frac{\partial \overline{\boldsymbol{u}}}{\partial \overline{\Delta}}, \quad q \equiv \frac{\partial \overline{p}}{\partial \overline{\Delta}} \quad . \tag{9.36}$$

These sensitivities are solutions of the following equations, which are obtained differentiating (9.34) and (9.35) with respect to $\overline{\Delta}$

$$\begin{aligned} \frac{\partial \boldsymbol{v}}{\partial t} &+ \nabla \cdot (\boldsymbol{v} \otimes \overline{\boldsymbol{u}} + \overline{\boldsymbol{u}} \otimes \boldsymbol{v}) = -\nabla q + \nu \nabla^2 \boldsymbol{v} \\ &+\nabla \cdot \left(\left[\frac{\partial}{\partial \overline{\Delta}} \nu_{\text{sgs}}(\overline{\Delta}, \overline{\boldsymbol{u}}) + \frac{\partial}{\partial \overline{\boldsymbol{u}}} \nu_{\text{sgs}}(\overline{\Delta}, \overline{\boldsymbol{u}}) \cdot \boldsymbol{v} \right] S(\overline{\boldsymbol{u}}) \right) \\ &+\nabla \cdot \left(\nu_{\text{sgs}}(\overline{\Delta}, \overline{\boldsymbol{u}}) S(\boldsymbol{v}) \right) \quad , \end{aligned} \tag{9.37}$$

$$\nabla \cdot \boldsymbol{v} = 0 \quad . \tag{9.38}$$

Thus, once the large-eddy simulation fields $\overline{\boldsymbol{u}}$ and $\overline{p}$ are computed, their sensitivities can be evaluated solving the linear problem given above. Let us illustrate this approach considering the Smagorinsky model (5.90), which corresponds to

$$\nu_{\text{sgs}}(\overline{\Delta}, \overline{\boldsymbol{u}}) = (C_S \overline{\Delta})^2 |S(\overline{\boldsymbol{u}})| \quad . \tag{9.39}$$

Applying the differentiation rules, one obtains the following Jacobian of the subgrid viscosity

$$\left[\frac{\partial}{\partial \overline{\Delta}} \nu_{\text{sgs}} + \frac{\partial}{\partial \overline{\boldsymbol{u}}} \nu_{\text{sgs}} \cdot \boldsymbol{v} \right] = 2 C_S^2 \overline{\Delta} |S(\overline{\boldsymbol{u}})| + (C_S \overline{\Delta})^2 \left(\frac{S(\overline{\boldsymbol{u}})}{|S(\overline{\boldsymbol{u}})|} \right) : S(\boldsymbol{v}) \quad . \tag{9.40}$$

That expression is also linear with respect to the sensitivity vector $\boldsymbol{v}$ and can easily be computed.

Once the sensitivies of the solution are known, one can derive an estimate for the uncertainties on the quantities calculated from it. Let $\mathcal{J}(\overline{\Delta}, \overline{\boldsymbol{u}})$ be any smooth functional estimated using the large-eddy simulation results (e.g. drag and lift of an immersed body). The best value one can sought is the one associated to the exact, unfiltered velocity field $\boldsymbol{u}$, i.e. $\mathcal{J}(0, \boldsymbol{u})$.

The error commited on $\mathcal{J}$ can be expressed as a function of the cutoff length $\overline{\Delta}$ writing the following first-order Taylor series expansion:

$$\mathcal{J}(0, \boldsymbol{u}) = \mathcal{J}(\overline{\Delta}, \overline{\boldsymbol{u}}) - \Delta \mathcal{J}'(\overline{\Delta}, \overline{\boldsymbol{u}}) \cdot \boldsymbol{v} \quad , \tag{9.41}$$

where $\mathcal{J}'$ is the Jacobian of $\mathcal{J}$. The error is given by the second term in the right hand side. An interesting point is that this relation can also be used to estimate the exact value of the functional by correcting the value computed from the large-eddy simulation data.

9.2 Correction Techniques

As relations (9.2) and (9.3) show, the statistical moments computed from the resolved field cannot be equal to those computed from the exact solution. In order to be able to compare these moments for validation purposes, or analyze the large-eddy simulation data, the error term has to be evaluated or eliminated. Several possible techniques are described in the following for doing this.

9.2.1 Filtering the Reference Data

The first solution is to apply the same filtering as was used for the scale separation to the reference solution [533, 7]. Strict comparisons can be made with this technique, but it does not provide access to theoretically usable values, which makes it difficult to use the data generated by large-eddy simulation for predicting physical phenomena, because only filtered data are available. In order for physical analyses to be fully satisfactory, they should be made on complete data. However, analysis is possible when the quantities considered are independent or weakly dependent on the subgrid scales[3].

Moreover, this approach is difficult to apply when the effective filter is not known analytically, because the reference data cannot be filtered consistently. It may also be difficult to apply an analytical to experimental data, because in order to do so, access is needed to the data spectra that are to serve for validation or analysis. We see another source of problems cropping up here [562]: experimentally measured spectra are time spectra in the vast majority of cases, while the large-eddy simulation is based on space filtering. This may introduce essential differences, especially when the flow is highly anisotropic in space, as it is in the regions near a solid wall. Similar remarks can be made concerning the spatial filtering of data from a direct numerical simulation for a priori test purposes: applying a one- or two-dimensional filter can produce observations that are different from those that would be obtained with a three-dimensional filter.

[3] As is generally the case for the mean velocity field. See the examples given in Chap. 14.

9.2.2 Evaluation of Subgrid-Scale Contribution

A second solution is to evaluate the error term and reconstruct from the filtered solution moments that are equal to those obtained from the full field.

Use of a De-filtering Technique. One way is to try to reconstruct the full field from the resolved one, and compute the statistical moments from the reconstructed field. In theory, this makes it possible to obtain exact results if the reconstruction itself is exact. This reconstruction operation can be interpreted as de-filtering, i.e. as an inversion of the scale separation operation. As was seen in Chap. 2, this operation is possible if the filter is an analytical one not belonging to the class of Reynolds operators. In other cases, i.e. when the effective filter is unknown or possesses projector properties, this technique is not strictly applicable and we have to do with an approximate recontruction. We then use a technique based on the differential interpretation of the filter analogous to the one described in Sect. 7.2.1. With this interpretation, we can express the filtered field $\overline{\boldsymbol{u}}$ as:

$$\overline{\boldsymbol{u}} = \left(Id + \sum_{n=1}^{\infty} C_n \overline{\Delta}^{2n} \frac{\partial^{2n}}{\partial x^{2n}} \right) \boldsymbol{u} \quad . \tag{9.42}$$

This relation can be formally inverted writing:

$$\boldsymbol{u} = \left(Id + \sum_{n=1}^{\infty} C_n \overline{\Delta}^{2n} \frac{\partial^{2n}}{\partial x^{2n}} \right)^{-1} \overline{\boldsymbol{u}} \quad , \tag{9.43}$$

and, by interpreting the differential operator as an expansion function of the small parameter $\overline{\Delta}$, we get:

$$\boldsymbol{u} = \left(Id + \sum_{n=1}^{\infty} C'_n \overline{\Delta}^{2n} \frac{\partial^{2n}}{\partial x^{2n}} \right) \overline{\boldsymbol{u}} \quad . \tag{9.44}$$

By truncating the series at some arbitrary order, we thus get a recontruction method that is local in space and easy to use. The difficulty resides in the choice of the coefficients C_n, which describe the effective filter and can only be determined empirically.

Use of a Subgrid Model. Another means that is easier to use is to compute the contribution of the subgrid terms by means of the subgrid stresses representation generated by the model used in the simulation. This technique cannot evaluate all the error terms present in (9.2) and (9.3) and can only reduce the error committed in computing the second-order moments.

It does, however, offer the advantage of not requiring additional computations as in the recontruction technique.

It should be noted here that this technique seems to be appropriate when the models used are structural, representing the subgrid tensor, but that it is no longer justified when functional models are used because these ensure only an energy balance.

It is worth noting that different subgrid models can be used for different purposes: a model can be used during the computation to close the filtered momentum equation, while another model can be used to recover a better reconstruction of subgrid contributions. This is more particulary true when the data extraction is related to multiphysics purposes. As an example, specific subgrid models have been derived for predicition of subgrid acoustic sources by Seror et al. [666, 667], while an usual functional model was used in the momentum equations.

9.2.3 Evaluation of Subgrid-Scale Kinetic Energy

The specific case of the evaluation of the subgrid kinetic energy $q^2_{\mathrm{sgs}} = \overline{u'_i u'_i}/2$ (not to be assimilated to the generalized subgrid kinetic energy, see p. 54 received a lot of attention, since it is required in many applications. Several proposals have been made, which are presented below:

1. Yoshizawa's method (p. 315), which relies on a simple dimensional analysis and requires the same inputs as the Smagorinsky functional model. Its dynamic variants are also discussed.
2. Knaepen's dynamic model (p. 316), which is based on scale similarity concepts.
3. Models based on the integration of a spectrum shape (p. 317). These methods require the foreknowledge of the analytical spectrum shape, whose parameters (dissipation rate, Kolmogorov scale) are computed using simple subgrid models.

This section presents the original version of the models. But it is observed that many elements of each model can be transposed in the other ones, leading to the definition of new models.

Yoshizawa's Model. A simple dimensional analysis yields

$$q^2_{\mathrm{sgs}} = 2C_I \overline{\Delta}^2 |\overline{S}|^2 \quad . \tag{9.45}$$

The value of the constant is taken equal to $C_I = 1/\pi^2$ in [231] and to 0.01 in [538]. A dynamic evaluation of the parameter C_I based on the Germano identity was proposed by Moin et al. [538]. Denoting the test filter level with a *tilde* symbol, one obtains

$$C_I = \frac{1}{2} \frac{\widetilde{\overline{u}_k \overline{u}_k} - \widetilde{\overline{u}}_k \widetilde{\overline{u}}_k}{\widetilde{\overline{\Delta}}^2 |\widetilde{\overline{S}}|^2 - \widetilde{\overline{\Delta}^2 |\overline{S}|^2}} \quad . \tag{9.46}$$

This dynamic model may requires some numerical stabilization (like averaging over homogeneous directions) in practical applications. Another dynamic evaluation procedure is proposed by Wong and Lilly [766], which is based on the so-called Kolmogorov scaling:

$$q_{\text{sgs}}^2 = C_I \overline{\Delta}^{4/3} |\overline{S}| \quad , \tag{9.47}$$

where the *dimensional constant* is evaluated as

$$C_I = \frac{1}{2} \frac{\widetilde{\overline{u}_k \overline{u}_k} - \widetilde{\overline{u}}_k \widetilde{\overline{u}}_k}{\left(\widetilde{\overline{\Delta}}^{4/3} - \overline{\Delta}^{4/3} \right) |\widetilde{\overline{S}}|} \quad . \tag{9.48}$$

This new formulation ensures that the realizability constraint q_{sgs}^2 is fulfilled.

Knaepen's Model. Rewriting the multiplicative Germano identity (3.80) and taking the trace of it, one obtains (the test filter level being noted with the *tilde* symbol)

$$L_{ii} \equiv \widetilde{\overline{u}_i \overline{u}_i} - \widetilde{\overline{u}}_i \widetilde{\overline{u}}_i = \overline{u}_i \overline{u}_i - \widetilde{\overline{u}}_i \widetilde{\overline{u}}_i \quad , \tag{9.49}$$

showing that the trace of the Leonard tensor L is equal to twice the difference of the resolved kinetic energy at the two considered filtering level. Assuming that the two filters are sharp cutoff filters with cutoff wavenumbers k_c and k_c', the following expression is recovered

$$\frac{1}{2} L_{ii} = \int_{k_c'}^{k_c} E(k) dk \quad . \tag{9.50}$$

If the case the spectrum shape is of the form $E(k) = A\ k^{-5/3}$, (9.50) yields

$$A = \frac{1}{3} \frac{L_{ii}}{k_c^{-2/3} - {k_c'}^{-2/3}} \quad . \tag{9.51}$$

Using this last relationship and assuming that the inertial range spectrum shape is valid at all subgrid wave numbers, one obtains the following estimate for the subgrid kinetic energy

$$q_{\text{sgs}}^2 = \int_{k_c}^{\infty} E(k) dk = \frac{1}{2} \frac{L_{ii}}{\left(\frac{\widetilde{\overline{\Delta}}}{\overline{\Delta}} \right)^{2/3} - 1} \quad . \tag{9.52}$$

This model can be modified to account for the existence of the dissipative range of the spectrum, leading to

$$q_{\text{sgs}}^2 = \int_{k_c}^{\infty} E(k) dk = \frac{1}{2} \frac{L_{ii}}{\left(\frac{\widetilde{\overline{\Delta}}}{\overline{\Delta}} \right)^{2/3} - 1} (1 - \gamma^{2/3}) \quad , \tag{9.53}$$

where $\gamma = \eta_K/\overline{\Delta}$ is the ratio of the filter cutoff length and the Kolmogorov scale. The Kolmogorov scale can also be evaluated in the same way. Assuming that the local equilibrium hypothesis applies and that most of the viscous dissipation occurs in the inertial range, one obtains

$$\int_{k_c}^{k_\eta} 2\nu k^2 E(k) = -\tau_{ij}\overline{S}_{ij} \quad , \tag{9.54}$$

where $k_\eta = 2\pi/\eta_K$ and τ is the subgrid tensor (to be evaluated with any subgrid model). As a consequence, one obtains

$$k_\eta = \left(k_c^{4/3} - \frac{2\tau_{ij}\overline{S}_{ij}}{3\nu A} \right)^{3/4} \quad , \tag{9.55}$$

leading to a closed model.

Dissipation-based models. Another way to compute the subgrid kinetic energy is to assume a spectrum shape and to integrate it for all wave numbers larger than the cutoff k_c [517, 735, 608].

A simple spectrum shape is obtained assuming that the Kolmogorov spectrum shape is valid from k_c to Jk_η (with k_η the Kolmogorov wave number and J a cutoff parameter) and that no scale exist at higher wave number, leading to:

$$E(k) = \begin{cases} K_0 \varepsilon^{2/3} k^{-5/3} & k_c \leq k \leq Jk_\eta \\ 0 & \text{otherwise} \end{cases} \quad . \tag{9.56}$$

Integrating this expression, one obtains

$$q^2_{\text{sgs}} = \begin{cases} \frac{3K_0\varepsilon^{2/3}}{2k_c^{2/3}} \left(1 - \left(\frac{k_c}{Jk_\eta} \right)^{2/3} \right) & k_c \leq Jk_\eta \\ 0 & \text{otherwise} \end{cases} \quad . \tag{9.57}$$

The subgrid dissipation rate ε is not known a priori and must be evaluated. Assuming that the subgrid scale are in statistical equilibrium, the relation $\varepsilon = -\tau_{ij}\overline{S}_{ij}$ is valid an can be used to compute ε, the subgrid stresses τ_{ij} being parameterized using an arbitrary subgrid model. A simple subgrid viscosity model is usually sufficient to this end, but more complex models can be used to account for anisotropy, as proposed by Pullin [608] who uses a stretched-vortex structural model. Equation (9.57) is explicit if $J = \infty$. For finite values of J, the factor Jk_η remains to be evaluated. A simple method to do that [517, 735] is to assume a local balance between the total dissipation on one side and the sum of the resolved-scale dissipation and the subgrid dissipation on the other side

$$\varepsilon = \nu|\overline{S}|^2 - \tau_{ij}\overline{S}_{ij} \quad . \tag{9.58}$$

Expressing ε and/or τ_{ij} as a function of the kinetic energy spectrum, one obtains a non-linear equation for Jk_η which can be solved thanks to Newton algorithm.

9.3 Practical Experience

Practice shows that nearly all authors make comparisons with reference data or analyze large-eddy simulation data with no processing of the data. The agreements observed with the reference data can then be explained by the fact that the quantities being compared are essentially related to scale ranges contained in the resolved field. This is generally true of the first-order moments (i.e. the mean velocity field) and, in certain cases, of the second-order moments (the Reynolds stresses). This lack of processing prior to data analysis seems to be due mainly to the uncertainties in the techniques for evaluating the contributions of the subgrid scales and to the difficulty of ad hoc filtering of the reference data. Large-eddy simulation also allows a satisfactory prediction of the time frequency associated with large-scale periodic or quasi-periodic phenomena (such as vortex shedding) and the first harmonics of this frequency for fine mesh.

One has to be careful when trying to recover high-order statistical data from large-eddy simulation. The two main reasons are:

- The properties of the exact filtered solution, i.e. the solution of an ideal large-eddy simulation without numerical and modeling errors referred to as u_{Π} in Chap. 1, may be different from those of the exact unfiltered solution.
- Numerical and modeling errors can corrupt the field, yielding new errors.

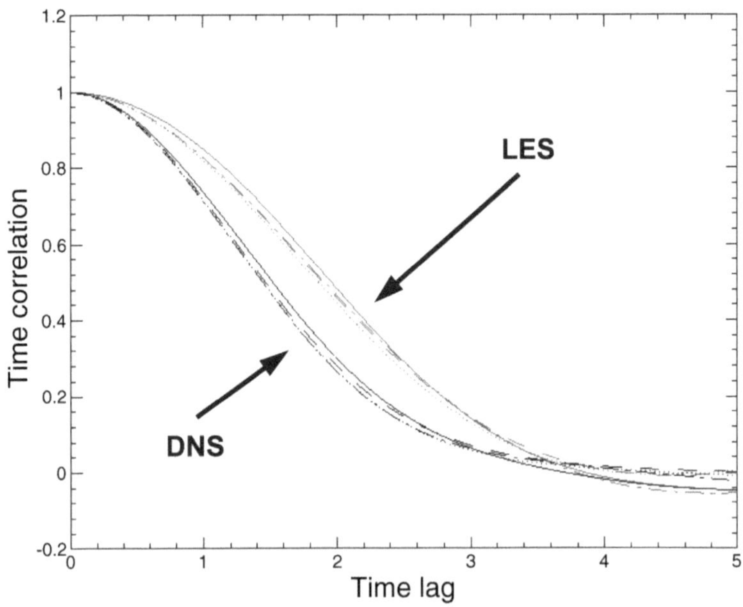

Fig. 9.1. Comparison of time correlation versus a normalized time lag in isotropic turbulence. *Upper curves* are from Large-Eddy Simulation and *lower curves* from Direct Numerical Simulation. Courtesy of S. Rubinstein, NASA.

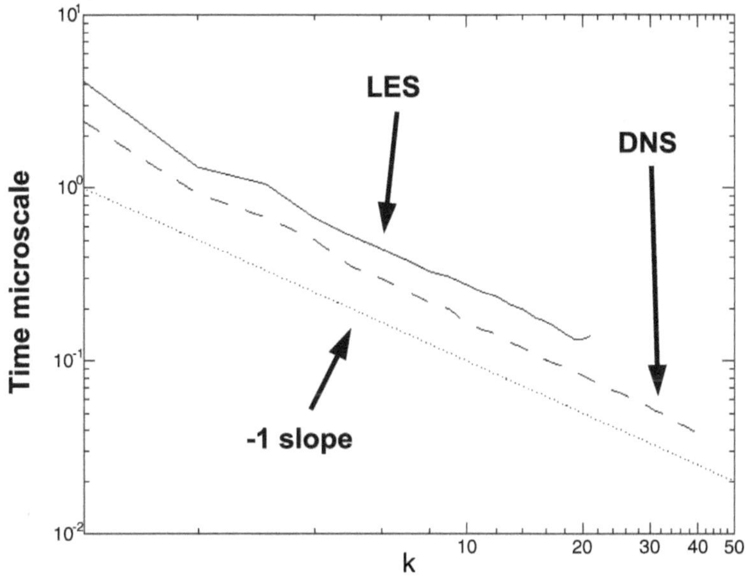

Fig. 9.2. Time microscale versus wave number in isotropic turbulence. *Solid line:* Large-Eddy Simulation; *Dashed line:* Direct Numerical Simulation; *Dotted line:* theoretical -1 slope in the inertial range. Courtesy of S. Rubinstein, NASA.

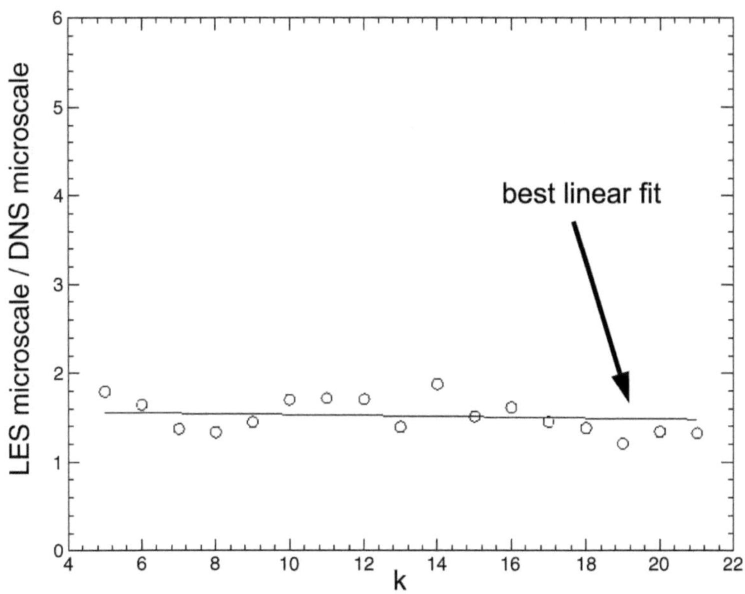

Fig. 9.3. Ratio of LES time microscale with the DNS value (*Symbols*). *Solid line* shows the best linear fit. Courtesy of S. Rubinstein, NASA.

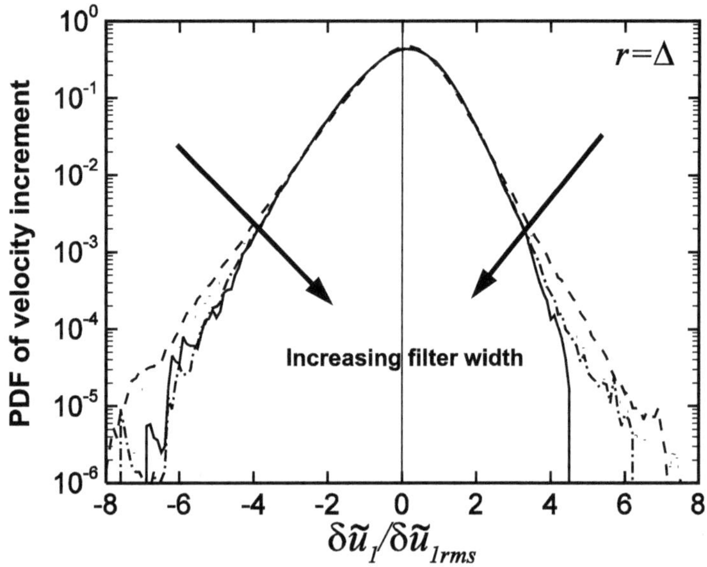

Fig. 9.4. Probability density function of filtered velocity increments (experimental data), for different filter size. Tails of the PDF is observed to be damped with increasing filter size. Courtesy of C. Meneveau, Johns Hopkins Univ.

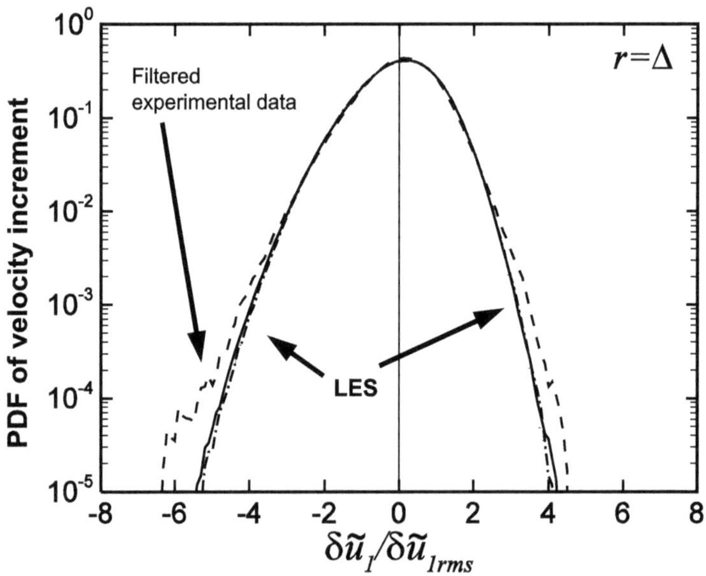

Fig. 9.5. Comparison of probability density functions of filtered velocity increments. *Dashed line:* filtered experimental data. *Others:* LES computations based on different subgrid models (Smagorinsky, dynamic Smagorinsky, dynamic functional-strucrtural models). Courtesy of C. Meneveau, Johns Hopkins Univ.

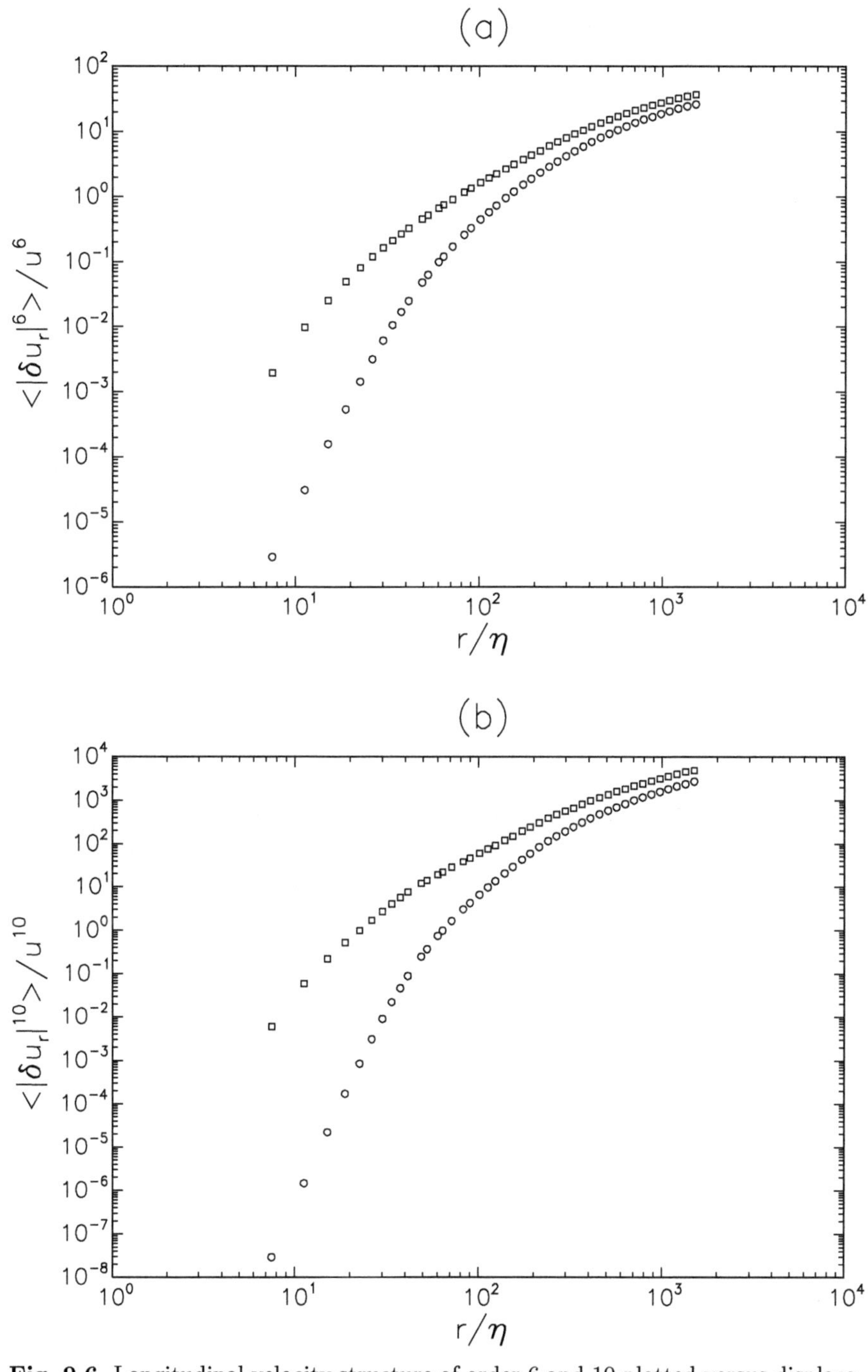

Fig. 9.6. Longitudinal velocity structure of order 6 and 10 plotted versus displacement. *Squares:* unfiltered data; *Circles:* filtered data. From [116].

Very little information is available on these two sources of error. He et al. [300, 301] have shown that spectral viscosity models generate fields which are more correlated than the corresponding unfiltered or perfect large-eddy simulation solutions (see Fig. 9.1). This increased correlation is shown to lead to a modification of Eulerian time correlations. The subgrid closure is observed to have a significant impact on the time microscale at all wave numbers (see Figs. 9.2 – 9.3): the value found in Large-Eddy Simulation is 1.8 times greater than in Direct Numerical Simulation, while theoretical analyses show that the value in filtered Direct Numerical Simulation (or ideal Large-Eddy Simulation) should lie in the range 1–1.15.

From experimental grid and wake turbulence data, Cerutti and Meneveau [116] and Kang et al. [375] examined fundamental differences between filtered and unfiltered velocity fields through probability density functions and the scaling behavior of high-order structure functions (see Figs. 9.4 – 9.6). This comparative study dealing with probability density functions of velocity increments yields the conclusion that the tails of the distributions are affected by the filtering even at scales much larger than the filter scale. Large discrepancies are also observed with respect to the scaling of structure functions. But it is worth noting that using simple shell models of turbulence, Benzi et al. [53] have shown that the use of subgrid viscosity-like models does not preclude internal intermittency. Inertial intermittent exponents were observed to be fairly independent of the way energy is dissipated at small scales for this very simplified dynamical model of turbulence. Other effects, such as the limited size of the computational domain, can also generate some discrepancies with experimental data [7]. Nevertheles, Large-Eddy Simulation can be used to recover some useful informations if the numerical error is controlled and the grid fine enough: Alvelius and Johansson [11] observed a good prediction of two-point pressure-velocity correlations in homogeneous turbulence.

10. Boundary Conditions

Like all the other approaches mentioned in the introduction, large-eddy simulation requires the setting of boundary conditions in order to fully determine the system and obtain a mathematically well-posed problem. This chapter is devoted to questions of determining suitable boundary conditions for large-eddy simulation computations. The first section is a discussion of general order, the second is devoted to the representation of solid walls, and the third discusses methods used for representing an unsteady upstream flow.

10.1 General Problem

10.1.1 Mathematical Aspects

Many theoretical problems dealing with boundary conditions for large-eddy simulation can be identified. Two of the most important ones are the possible interaction between the filter and the exact boundary conditions, and the specification of boundary conditions leading to the definition of a well-posed problem.

From the results of Sects. 2.2 and 3.4, we deduce that the general form of the filtered Navier–Stokes equations on a bounded domain Ω is

$$\frac{\partial \overline{u_i}}{\partial t} + \frac{\partial}{\partial x_j}(\overline{u_i u_j}) + \frac{\partial \overline{p}}{\partial x_i} - \nu \frac{\partial}{\partial x_j}\left(\frac{\partial \overline{u}_i}{\partial x_j} + \frac{\partial \overline{u}_j}{\partial x_i}\right) =$$
$$-\int_{\partial\Omega} G(\boldsymbol{x}-\boldsymbol{\xi})\left[u_i(\boldsymbol{\xi})u_j(\boldsymbol{\xi})+p(\boldsymbol{\xi})-\nu\left(\frac{\partial u_i(\boldsymbol{\xi})}{\partial x_j}+\frac{\partial u_j(\boldsymbol{\xi})}{\partial x_i}\right)\right] n_j(\boldsymbol{\xi})d\boldsymbol{\xi} \quad , \tag{10.1}$$

$$\frac{\partial \overline{u}_i}{\partial x_i} = -\int_{\partial\Omega} G(\boldsymbol{x}-\boldsymbol{\xi})u_i(\boldsymbol{\xi})n_i(\boldsymbol{\xi})d\boldsymbol{\xi} \quad , \tag{10.2}$$

where n is the outward unit normal vector to the boundary of Ω, $\partial\Omega$. It must be noted that only terms induced by the interaction of the filter and the boundaries have been retained, i.e. other commutation errors are neglected. Right-hand side terms in (10.1) and (10.2) are additional source terms, that must be modeled because they involve the non-filtered velocity and pressure fields.

It is seen from these equations that two possibilities arise when filtering the Navier–Stokes equations on bounded domains:

- The first one, which is the most commonly adopted (*classical approach*), consists of considering that the filter width decreases when approaching the boundaries such that the interaction term cancels out (see Fig. 10.1). Thus, the source term can be neglected and the basic filtered equations are left unchanged. The remaining problem is to define classical boundary conditions for the filtered field.
- The second solution (*embedded boundary conditions*), first advocated by Layton et al. [206, 234, 429, 357, 207], consists of filtering through the boundary (see Fig. 10.1). As a consequence, there exists a layer along the boundary, whose width is of the order of the filter cutoff lengthscale, in which the source term cannot be neglected. This source term must then be explicitly computed, i.e. modeled.

The discussions so far clearly show that the constitutive equations of large-eddy simulation can be of a degree different from that of the original Navier–Stokes equations. This is trivially verified by considering the differential interpretation of the filters: the resolved equations are obtained by applying a differential operator of arbitrarily high order to the basic equations. Moreover, it has been seen that certain subgrid models generate high-order derivatives of the velocity field.

This change of degree in the resolved equations raises the problem of determining the associated boundary conditions, because those associated with the equations governing the evolution of the exact solution can no longer be used in theory for obtaining a mathematically well-posed problem [728, 234]. This problem is generally not considered, arguing the fact that the higher-order terms appear only in the form of $O(\overline{\Delta}^p), p \geq 1$ perturbations of the Navier–Stokes equations and the same boundary conditions are used both for the large-eddy simulation and for direct numerical simulation of the Navier–Stokes equations. Moreover, when the effective filter is unknown, it is no longer possible to derive suitable boundary conditions strictly, which also leads to the use of the boundary conditions of the basic problem.

10.1.2 Physical Aspects

The boundary conditions, along with the similarity parameters of the equations, specify the flow, i.e. determine the solution. These conditions represent the whole fluid domain beyond the computational domain. To specify the solution completely, these conditions must apply to all of its scales, i.e. to all the space-time modes it comprises.

So in order to characterize a particular flow, the amount of information in the boundary conditions is a function of the number of degrees of freedom of the boundary condition system. This poses the problem of representing a particular solution, in order to be able to reproduce it numerically. We

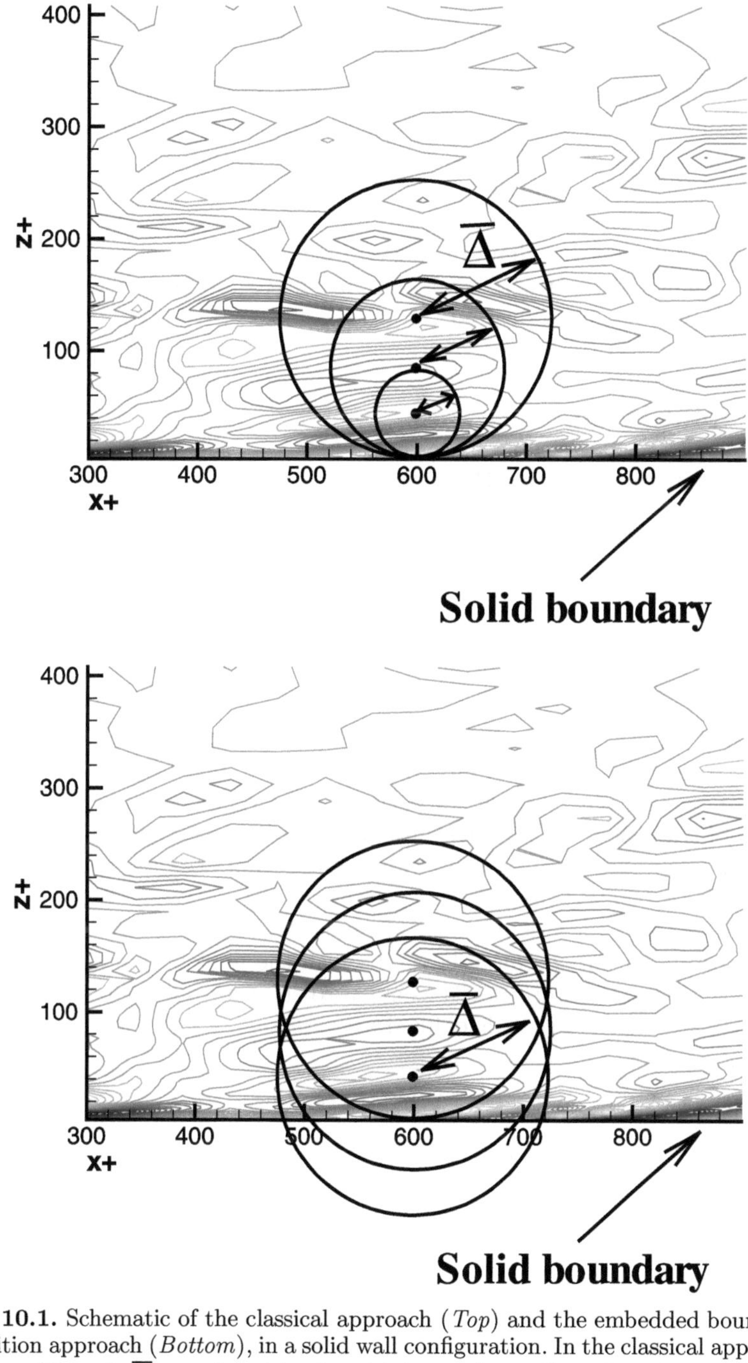

Fig. 10.1. Schematic of the classical approach (*Top*) and the embedded boundary condition approach (*Bottom*), in a solid wall configuration. In the classical approach the cutoff length $\overline{\Delta}$ is reduced in the vicinity of the wall so that the interaction term cancels out. In the second approach, the cutoff length is constant. Courtesy of E. Garnier, ONERA.

have a new modeling problem here, which is that of modeling the physical test configuration.

This difficulty is increased for the large-eddy simulation and direct numerical simulation, because these simulations contain a large number of degrees of freedom and require a precise space-time deterministic representation of the solution at the computational domain boundaries.

Two special cases will be discussed in the following sections: that of representing solid walls and that of representing a turbulent inflow. The problem of the outflow conditions, which is not specific to the large-eddy simulation technique, will not be addressed[1].

10.2 Solid Walls

10.2.1 Statement of the Problem

Specific Features of the Near-Wall Region. The structure of the boundary layer flow has certain characteristics that call for special treatment in the framework of large-eddy simulation. In this section, we describe the elements characteristic of the boundary layer dynamics and kinematics, which shows up the difference with an isotropic homogeneous turbulence. For a detailed description, the reader may refer to [648, 150].

Definitions. Here we adopt the ideal framework of a flat-plate, turbulent boundary layer, without pressure gradient. The external flow is in the (Ox) direction and the (Oz) direction is normal to the wall. The external velocity is denoted $U_{\rm e}$. In the following, the Cartesian coordinate system will be denoted either (x, y, z) or (x_1, x_2, x_3), for convenience. Similarly, the velocity vector is denoted (u, v, w) or (u_1, u_2, u_3).

We first recall a few definitions. The boundary layer thickness δ is defined as the distance from the plate beyond which the fluid becomes irrotational, and thus where the fluid velocity is equal to the external velocity.

The wall shear stress $\tau_{\rm p}$ is defined as:

$$\tau_{\rm p} = \sqrt{\tau_{{\rm p},13}^2 + \tau_{{\rm p},23}^2} \quad , \tag{10.3}$$

in which $\tau_{{\rm p},ij} = \nu \overline{S}_{ij}(x, y, 0)$. The friction velocity u_τ is defined as:

$$u_\tau = \sqrt{\tau_{\rm p}} \quad . \tag{10.4}$$

In the case of the canonical boundary layer, we get:

$$u_\tau = \sqrt{\nu \frac{\partial u_1}{\partial z}(x, y, 0)} \quad . \tag{10.5}$$

[1] See [156] for a specific study of exit boundary conditions for the plane channel flow case.

We define the Reynolds number Re_τ by:

$$Re_\tau = \frac{\delta u_\tau}{\nu} \quad . \tag{10.6}$$

The reduced velocity $\boldsymbol{u}^+$, expressed in wall units, is defined as:

$$\boldsymbol{u}^+ = \boldsymbol{u}/u_\tau \quad . \tag{10.7}$$

The wall coordinates (x^+, y^+, z^+) are obtained by the transformation:

$$(x^+, y^+, z^+) = (x/l_\tau, y/l_\tau, z/l_\tau) \quad , \tag{10.8}$$

where the viscous length l_τ is defined as $l_\tau = \nu/u_\tau$.

Statistical Description of the Canonical Boundary Layer. The boundary layer is divided into two parts: the inner region ($0 \leq z < 0.2\delta$) and the outer region ($0.2\delta \leq z$). This decomposition is illustrated in Fig. 10.2. In the inner region, the dynamics is dominated by the viscous effects. In the outer region, it is controlled by the turbulence. Each of these regions is split into several layers, corresponding to different types of dynamics.

In the case of the canonical boundary layer, we have three layers in the inner region in which the mean longitudinal velocity profile follows special laws. The positions of these layers are referenced in the reduced coordinate system, because the dynamics of the inner region is dominated by the wall effects and l_τ is the pertinent length scale for describing the dynamics. The characteristic velocity scale is the friction velocity. These three layers are the:

– Viscous sublayer: $z^+ \leq 5$, in which

$$\langle u_1^+(z^+) \rangle = z^+ \quad . \tag{10.9}$$

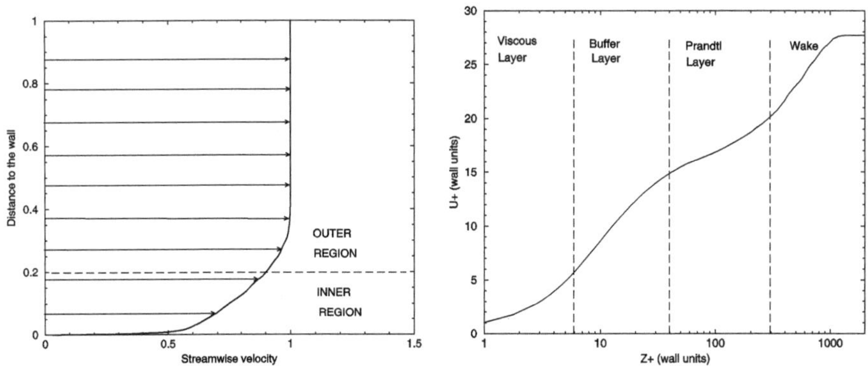

Fig. 10.2. Mean streamwise velocity profile for the canonical turbulent boundary layer, and its decomposition into inner and outer regions. *Left*: mean velocity profile (external units). *Right*: mean velocity profile (wall units).

– Buffer layer: $5 < z^+ \leq 30$, where

$$\langle u_1^+(z^+)\rangle \simeq 5 \ln z^+ - 3.05 \quad . \tag{10.10}$$

– Prandtl or logarithmic inertial layer: $30 < z^+; z/\delta \ll 1$, for which

$$\langle u_1^+(z^+)\rangle \simeq \frac{1}{\kappa} \ln z^+ + 5,5 \pm 0,1, \quad \kappa = 0,4 \quad . \tag{10.11}$$

The outer region includes the end of the logarithmic inertial region and the wake region. In this zone, the characteristic length is no longer l_τ but rather the thickness δ. The characteristic velocity scale remains unchanged, though. The average velocity profiles are described by:

– For the logarithmic inertial region:

$$\frac{\langle u_1(z)\rangle}{u_\tau} = A \ln \frac{z u_\tau}{\nu} + B \quad , \tag{10.12}$$

where A and B are constants;

– For the wake region:

$$\frac{\langle u_1(z)\rangle}{u_\tau} = A \ln \frac{z u_\tau}{\nu} + B + \frac{\Pi}{\kappa} W\left(\frac{z}{\delta}\right) \quad , \tag{10.13}$$

where A, B and Π are constants and W the wake function defined by Clauser as:

$$W(x) = 2 \sin^2(\pi x/2) \quad . \tag{10.14}$$

Concerning the Dynamics of the Canonical Boundary Layer. Experimental and numerical studies have identified dynamic processes within the boundary layer. We will summarize here the main elements of the boundary layer dynamics that originate the turbulence in the near-wall region.

Observations show that the flow is highly agitated very close to the wall, consisting of pockets of fast and slow fluid that organize in ribbons parallel to the outer velocity (streaks, see Fig. 10.3). The low-velocity pockets migrate slowly outward in the boundary layer (ejection) and are subject to an instability that makes them explode near the outer edge of the inner region. This burst is followed by an arrival of fast fluid toward the wall, sweeping the near-wall region almost parallel to it. These highly intermittent events in time and space induce strong variation in the unsteady Reynolds stresses and originate a very large part of the production and dissipation of the turbulent kinetic energy. These variations produce fluctuations in the subgrid dissipation that can reach 300% of the average value and can make it change sign. Analyses of direct numerical simulations [291, 598, 449, 363] indicate that a very intense small scale dissipation in the buffer region is correlated with the presence of sheared layers that form the interfaces between the fluid pockets of different velocities. These mechanisms are highly anisotropic. Their

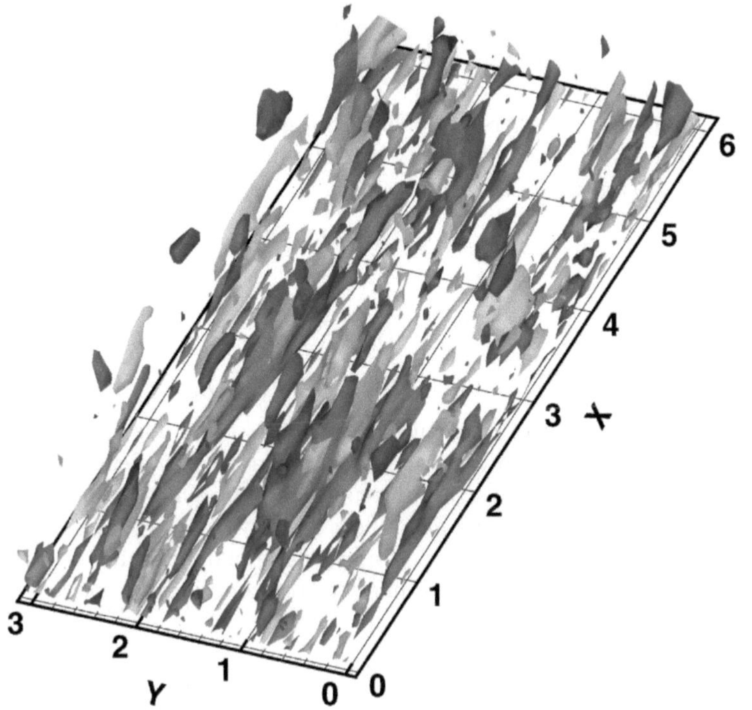

Fig. 10.3. Visualization of the streaks in a plane channel flow ($Re_\tau = 590$). *Dark* and *light gray* denote opposite sign of streamwise vorticity. Courtesy of M. Terracol, ONERA.

characteristic scales in the longitudinal and transverse directions λ_x and λ_y, respectively, are such that $\lambda_x^+ \approx 200 - 1000$ and $\lambda_y^+ \approx 100$. The maximum turbulent energy production is observed at $z^+ \approx 15$. This energy production at small scales gives rise to a high backward energy cascade and associated with the sweeping type events. The forward cascade, for its part, is associated with the ejections.

In the outer regions of the boundary layer where the viscous effects no longer dominate the dynamics, the energy cascade mechanism is predominant. Both cascade mechanisms are associated preferentially with the ejections.

Numerical and theoretical results [353, 354, 355, 750] show that wall-bounded turbulence below $z^+ \approx 80$ is a relatively autonomous system, which is responsible for the generation of a significant part of the turbulent kinetic energy dissipated in the outer part of the boundary layer. This turbulent cycle is often referred to as an autonomous cycle because it is observed to remain active in the absence of the outer flow. This cycle involves the formation of velocity streaks from the advection of the mean velocity profile by streamwise vortices, and the generation of the vortices from the instability of the streaks.

The presence of the wall seems to be only necessary to maintain the mean shear. The way that this inner turbulent cycle and the outer flow interact are still under investigation.

Härtel and his coworkers [291, 290, 289] give a more precise analysis of the subgrid transfer in the boundary layer by introducing a new splitting[2] of the subgrid dissipation ε

$$\varepsilon = -\tau_{ij}\overline{S}_{ij} = \varepsilon^{\mathrm{MS}} + \varepsilon^{\mathrm{FS}} \quad , \tag{10.15}$$

with

$$\varepsilon^{\mathrm{MS}} = -\langle\tau_{ij}\rangle\langle\overline{S}_{ij}\rangle \quad , \tag{10.16}$$

$$\varepsilon^{\mathrm{FS}} = -\langle(\tau_{ij} - \langle\tau_{ij}\rangle)(\overline{S}_{ij} - \langle\overline{S}_{ij}\rangle)\rangle \quad . \tag{10.17}$$

The $\varepsilon^{\mathrm{MS}}$ is related to the mean strain, and accounts for an enhancement of subgrid kinetic energy in the presence of mean-flow gradients. The second term, which is linked to the strain fluctuations, represents the redistribution of energy without affecting the mean flow directly.

A priori tests [291, 290, 289] perfomed using plane channel flow and circular pipe data reveal that the net effect of the coupling is a forward energy transfer, and:

- The mean strain part is always associated to a net forward kinetic energy cascade.
- The fluctuating strain part results in a net backward kinetic cascade in a zone located in the buffer layer, with a maximun near $z^+ = 15$. This net backward cascade is correlated to the presence of coherent events associated to turbulence production.

Typical distributions of $\varepsilon^{\mathrm{MS}}$ and $\varepsilon^{\mathrm{FS}}$ are shown in Fig. 10.4.

Kinematics of the Turbulent Boundary Layer. The processes described above are associated with existence of coherent structures [617].

The buffer layer is dominated by isolated quasi-longitudinal structures that form an average angle with the wall of 5° at $z^+ = 15$ and 15° at $z^+ = 30$. Their mean diameter increases with their distance from the wall[3].

The logarithmic inertial region belongs both to the inner and outer regions, and thus contains characteristic space scales, which is compatible with the existence of two different types of structures. The dynamics is governed by quasi-longitudinal and arch structures. The quasi-longitudinal structures can be connected to transverse structures and form an angle with the surface that varies from 15° to 30°. The span of the arch structures is of the order of the width of the slow-fluid pockets at the bottom of the layer, and

[2] It differs from the splitting proposed by Shao (see Sect. 7.4.1).

[3] It should be noted that contradictory observations can be found. Lamballais [421] observes that the most probable angle of the vorticity (projected on a plane perpendicular to the wall) is close to 90° for $5 < z^+ < 25$, which goes against the model of longitudinal vortices at the wall.

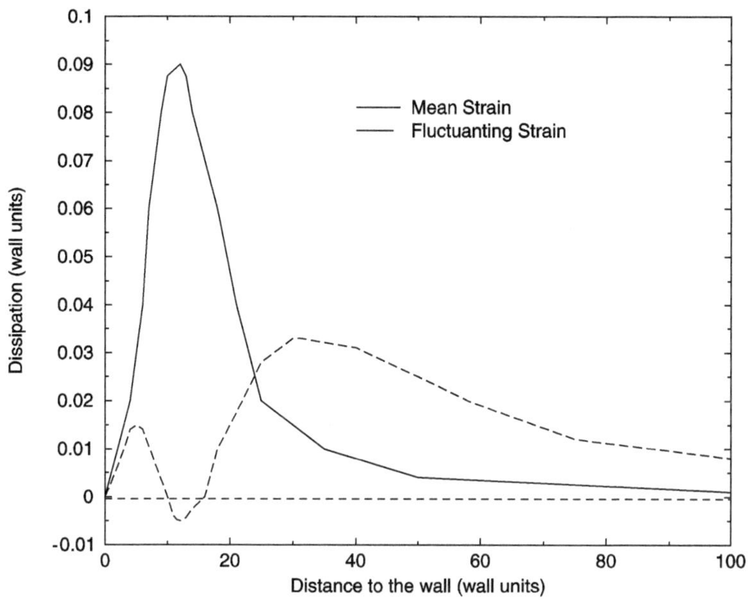

Fig. 10.4. Distribution of mean strain ($\varepsilon^{\mathrm{MS}}$) and fluctuating strain ($\varepsilon^{\mathrm{FS}}$) dissipations in a plane channel flow. Fluctuating strain is observed to yield dominant backscatter near $z^+ \sim 15$, but the total dissipation remains positive.

increases linearly with the distance from the wall. The relative number of quasi-longitudinal structures decreases with the distance from the wall, until it cancels out at the beginning of the wake region.

The wake region is populated with arch structures forming an angle of 45° with the wall. Their x and y spacing is of the order of δ.

Resolving or Modeling. The description we have just made of the boundary layer flow structure clearly shows the problem of applying the large-eddy simulation technique in this case. Firstly, the mechanisms originating the turbulence, i.e. the flow driving mechanisms, are associated with fixed characteristic length scales on the average. Also, this turbulence production is associated with a backward energy cascade, which is largely dominant over the cascade mechanism in certain regions of the boundary layer. These two factors make it so that the subgrid models presented in the previous chapters become inoperative because they no longer permit a reduction of the number of degrees of freedom while ensuring at the same time a fine representation of the flow driving mechanisms. There are then two possible approaches [536]:

– Resolving the near-wall dynamics directly. Since the production mechanisms escape the usual subgrid modeling, if we want to take them into account, we have to use a sufficiently fine resolution to capture them. The solid wall is then represented by a no-slip condition: the fluid velocity is set equal to that of the solid wall. This equality implicitly relies on the hypothesis that the mean free path of the molecules is small compared with the characteristic scales of the motion, and that these scales are large compared with the distance of the first grid point from the wall. In practice, this is done by placing the first point in the zone ($0 \leq z^+ \leq 1$). To represent the turbulence production mechanisms completely, Schumann [655] recommends a spatial resolution such that $\overline{\Delta}_1^+ < 10, \overline{\Delta}_2^+ < 5$ and $\overline{\Delta}_3^+ < 2$. Also, Zang [798] indicates that the minimum resolution for capturing the existence of these mechanisms is $\overline{\Delta}_1^+ < 80, \overline{\Delta}_2^+ < 30$ and that three grid points should be located in the $z^+ \leq 10$ zone. Zahrai et al. [795] indicate that $\overline{\Delta}_1^+ \simeq 100, \overline{\Delta}_2^+ = 12$ should be used as an upper limit if a second-order accurate numerical method is used. These values are given here only for reference, since larger values can also be found in the literature. For example, Piomelli [591] uses $\overline{\Delta}_1^+ = 244$ for a plane channel flow. Chapman [119] estimates that representing the dynamics of the inner region, which contributes about one percent to the thickness of the full boundary layer, requires $O(Re^{1.8})$ degrees of freedom, while only $O(Re^{0.4})$ are needed to represent the outer zone. This corresponds to $\overline{\Delta}_1^+ \simeq 100, \overline{\Delta}_2^+ \simeq 20$ and $\overline{\Delta}_3^+ < 2$. Considering that non-isotropic modes must be directly resolved, Bagget et al. [32] show that the number of degrees of freedom of the solution (in space) scales as Re_τ^2.

- Modeling the near-wall dynamics. To reduce the number degrees of freedom and especially avoid having to represent the inner region, we use a model for representing the dynamics of that zone. This is a special subgrid model called the wall model. Since the distance from the first grid point to the wall is greater than the characteristic scales of the modes existing in the modeled region, the no-slip condition can no longer be used. The boundary condition will apply to the values of the velocity components and/or their gradients, which will be provided by the wall model. This approach makes it possible to place the first point in the logarithmic layer (in practice, $20 \leq z^+ \leq 200$). The main advantage of this approach is that the number of degrees of freedom in the simulation can be reduced greatly; but since a part of the dynamics is modeled, it constitutes an additional source of error.

10.2.2 A Few Wall Models

In the following, we present the most popular wall models for large-eddy simulation. These models all represent an impermeable wall, and most of them have been implemented using a staggered grid (see Fig. 10.5). The discussion will be restricted to wall models defined on Cartesian body-fitted computational grids. Details dealing with implementation on curvilinear body-fitted grids or on Cartesian grids using the immersed boundary technique [35, 713, 271, 270] will not be discussed, since they would require an

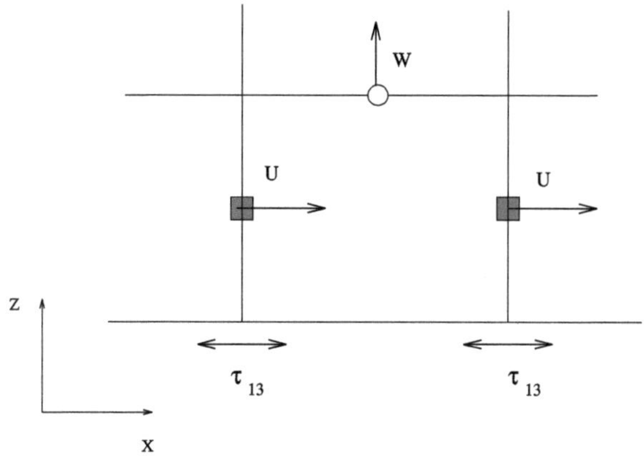

Fig. 10.5. Illustration of a staggered grid system in the streamwise/wall-normal plane.

extensive discussion about numerical methods which is far beyond the scope of this book.

Existing strategies for the definition of wall models for large-eddy simulation following the classical approach can be grouped in several classes:

- *Higher-order boundary conditions*: velocity gradients at the wall are controlled by enforcing boundary conditions on second-order derivatives. This class is represented by Deardorff's model (p. 337).
- *Wall stress models*: the first mesh at the wall is chosen to be very large, so that it is not able to respresent correctly the dynamics of the inner layer. Typical dimensions are 100 wall units in the wall-normal and spanwise directions and 500 wall units in the streamwise direction. This is illustrated in Fig. 10.7, where it is clearly observed that the near-wall events are averaged over the first grid cell. The wall model should provide the value of wall stresses, which cannot be accurately directly computed on the grid because of its coarseness (see Fig. 10.6), and the value of the wall-normal velocity component.
  The basic form of these stress models is

$$\tau_{i,3} = \mathcal{F}(\overline{u}_1, z_2) \quad , \quad i = 1, 2 \quad , \tag{10.18}$$

  where z_2 is the height of the first cell. If the flow is bidimensional and laminar, a trivial relation is

$$\tau_{1,3} = -\nu \frac{\overline{u}_1}{z_2/2} \quad . \tag{10.19}$$

  A first group of wall-stress models is based on the extrapolation of this linear, laminar law, the molecular viscosity being replaced by an effective turbulent viscosity, ν_{eff}. Recalling that the total mean shear stress is almost

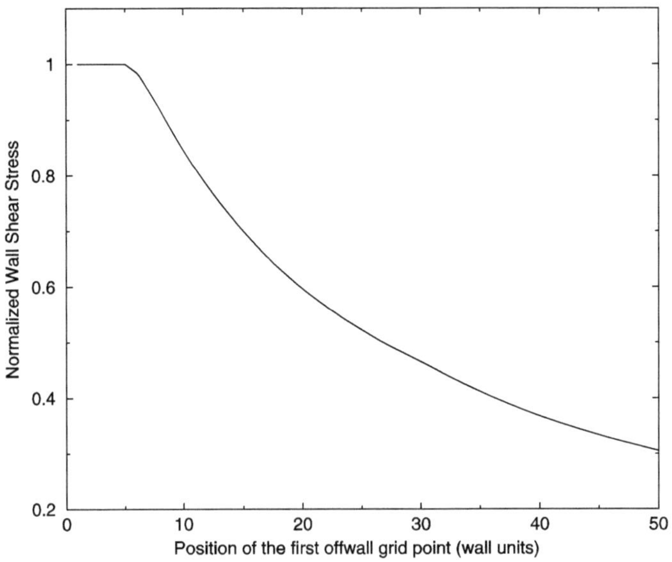

Fig. 10.6. Computed value of the mean wall shear stress (with reference to its exact value), expressed as a function of the distance of the first off-wall grid point.

constant across the inner part of the boundary layer, we have:

$$\langle \tau_{13}(z) \rangle = -\nu_{\rm tot}(z) \frac{\partial \langle \overline{u}_1(z) \rangle}{\partial z} \simeq \langle \tau_{{\rm p},13} \rangle \quad , \tag{10.20}$$

where $\nu_{\rm tot}(z)$ is the sum of the molecular and the turbulent viscosity. By integrating (10.20) in the wall-normal direction, we obtain:

$$\int_0^z \frac{\langle \tau_{13}(z) \rangle}{\nu_{\rm tot}(z)} dz = -\int_0^z \frac{\partial \langle \overline{u}_1(z) \rangle}{\partial z} dz \quad , \tag{10.21}$$

leading to

$$\langle \tau_{{\rm p},13} \rangle \int_0^z \frac{1}{\nu_{\rm tot}(z)} dz = -\langle \overline{u}_1(z) \rangle \quad , \tag{10.22}$$

and, finally,

$$\langle \tau_{{\rm p},13} \rangle = -\underbrace{\left(\frac{1}{z} \int_0^z \frac{1}{\nu_{\rm tot}(z)} dz \right)^{-1}}_{\nu_{\rm eff}} \frac{\langle \overline{u}_1(z) \rangle}{z} \quad . \tag{10.23}$$

The wall-stress models discussed below are:

1. The Schumann model (p. 339), which relies on a linear relation between the wall stress and the velocity component at the first off-wall grid point. The skin friction is an entry parameter for the model.

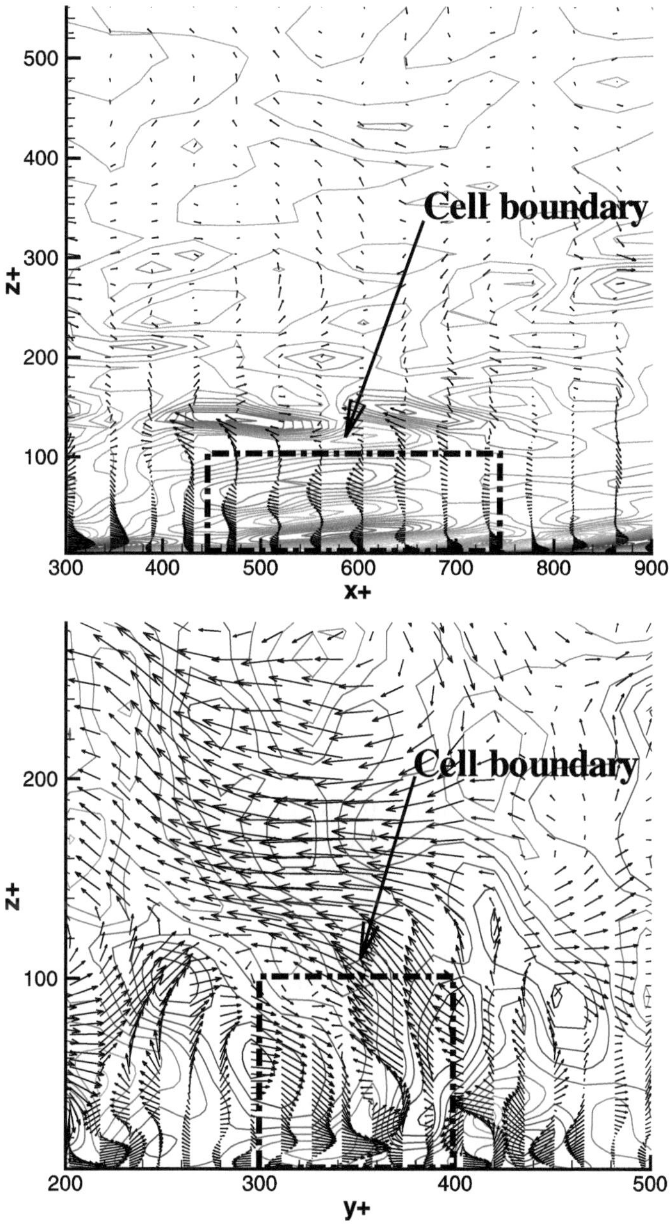

Fig. 10.7. Representation of the wall-stress approach. The contours of a typical cell are superposed onto a boundary-layer instantaneous flow obtained with a wall-resolving mesh (instantaneous velocity vectors and isolevels of streamwise velocity fluctuation). Length scales are expressed in wall units. The dimensions of the cell are 500 wall units in the streamwise direction and 100 wall units in the spanwise and wall-normal directions. *Top*: view in an (x, z) plane. *Bottom*: view in an (y, z) plane. Courtesy of E. Tromeur and E. Garnier, ONERA.

2. The Grötzbach model (p. 339), which is an extension of the Schumann model. The skin friction is now computed assuming that the flow corresponds to the canonical flat-plate boundary layer. The skin friction is computed by inverting the logarithmic law profile for the streamwise velocity.
3. The shifted correlations model (p. 340), which extends the Grötzbach model by taking into account explicitly the fact that fluctuations are governed by coherent structures with given time- and lengthscales.
4. The ejection model (p. 341), which is another extension of the model of Grötzbach. It takes into account the effects of sweep and ejection events on the wall shear stress.
5. The optimized ejection model (p. 341), which is based on experimental correlation data and yields better correlation coefficient in a priori tests.
6. The model of Werner and Wengle (p. 344), which can be seen as a variant of the Grötzbach model based on power-law profiles for the streamwise velocity instead of the logarithmic law. The main advantage is that the power law can be inverted explicitly.
7. The modified Werner–Wengle model (p. 345), which accounts for the existence of ejection, as the model proposed by Piomelli.
8. The model of Murakami et al. (p. 343), which can be interpreted as a simplified version of the preceding model.
9. The model of Mason and Callen for rough walls (p. 340).
10. The suboptimal-control-based wall models (p. 345). Numerical experiments show that the previous wall stress models are not robust, i.e. they lead to disappointing results on very coarse mesh for high Reynolds numbers. This is mainly due to the fact that they can not account for large numerical and physical errors occuring on such coarse grids. These new models, developed within the framework of suboptimal control, aim at producing the best possible results.

The second group of wall-stress models relies on the use of an internal layer near the wall, leading to the definition of two-layer simulations. Another set of governing equations is solved on an auxiliary grid located inside the first cell of the large-eddy simulation grid. An effective gain is obtained if the auxiliary simulation can be run on a much coarser grid than large-eddy simulation while guaranteeing the accuracy of the results, or if the auxiliary equations are much simpler than the Navier–Stokes equations. Models belonging to this group are:

1. The thin-boundary-layer equation model developed by Balaras et al. (p. 342). The governing equations solved in the inner layer are the boundary layer equations derived from the Navier–Stokes equations. The gain comes from the fact that the pressure is assumed to be constant across the boundary layer, and the pressure is not computed.
2. The wall model based on Kerstein's ODT approach (p. 350). This model relies on the reconstruction of the solution in the inner layer via

a simplified one-dimensional stochastic system. It can be interpreted as a surrogate of the model based on the full boundary layer equations.
3. Hybrid RANS/LES approaches, in which a RANS-like simulation is performed in the near-wall region, while the core of the flow is treated by large-eddy simulation. All these models and techniques are discussed in Chap. 12, with emphasis in Sect. 12.2.

– *Off-wall boundary conditions*: the mesh is built so that the first point is located in the fluid region, and not on the wall. The first grid line parallel to the wall must be located in the inner region of the boundary layer. This approach is illustrated in Fig. 10.8. An important point is that the grid must still be able to represent details of the flow, and thus the mesh should be the same as those used for wall-resolving simulation away from the wall. Several types of boundary conditions can be used following this approach, dealing either with the velocity components or their derivatives [30, 568, 356]. Poor results have been obtained using that approach, which requires very accurate structural information on the fluctuations to yield accurate results. Jimenez and Vasco [356] showed that the flow is very sensitive to the prescribed wall-normal velocity component (*transpiration velocity*), which must satisfy the continuity constraint.[4] The need for an instantaneous non-zero transpiration velocity can easily be understood by looking at the velocity field displayed in Fig. 10.8.
The usual failure of these models leads to the appearance of a strong, spurious boundary layer above the artificial boundary.
– *Deterministic minimal boundary-layer unit simulation.* Pascarelli et al. [580] proposed performing a wall-resolved temporal large-eddy simulation on the smallest domain allowing the existence of the near-wall autonomous cycle, and duplicating it. This approach leads to the definition of a crystal of elementary chaotic dynamical systems. The associated numerical technique corresponds to a multiblock approach. Nonlinear interactions are expected to scramble the data in the outer part of the boundary layer and to break possible periodicity.

The wall-model developed by Das and Moser within the framework of embedded boundary conditions is presented on p. 349.

Deardorff's Model. In the framework of a plane channel simulation with infinite Reynolds number, Deardorff [172] proposes using the following conditions for representing the solid walls:

$$\frac{\partial^2 \overline{u}_1}{\partial z^2} = -\frac{1}{\kappa (z_2/2)^2} + \frac{\partial^2 \overline{u}_1}{\partial y^2} \quad , \tag{10.24}$$

[4] This conclusion must be considered together with the fact that the best results obtained with wall-stress models are with non-zero transpiration velocity (see suboptimal models, p. 345). A general conclusion is that even for impermeable walls a non-zero wall-normal velocity must be prescribed to accurately describe the near-wall dynamics.

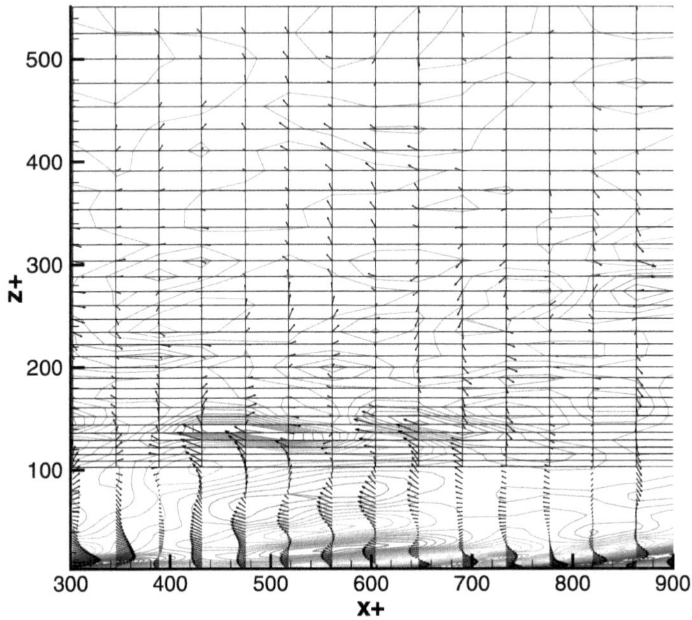

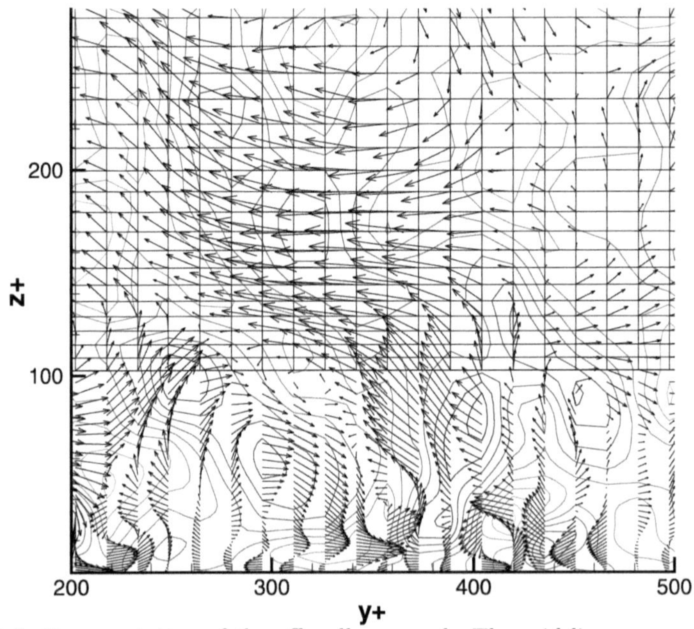

Fig. 10.8. Representation of the off-wall approach. The grid lines are superposed to a boundary-layer instantaneous flow obtained with a wall-resolving mesh (instantaneous velocity vectors and isolevels of streamwise velocity fluctuation). Length scales are expressed in wall units. The first grid point is located at 100 wall units. *Top*: view in an (x, z) plane. *Bottom*: view in an (y, z) plane.

$$\overline{u}_3 = 0 \quad , \tag{10.25}$$

$$\frac{\partial^2 \overline{u}_2}{\partial z^2} = \frac{\partial^2 \overline{u}_2}{\partial x^2} \quad , \tag{10.26}$$

where z_2 is the distance from the first point to the wall and $\kappa = 0.4$ the Von Karman constant. The first condition assumes that the average velocity profile verifies the logarithmic law and that the second derivatives of the fluctuation $\boldsymbol{u}'' = \overline{\boldsymbol{u}} - \langle \boldsymbol{u} \rangle$ in the y and z directions are equal. The impermeability condition (10.25) implies that the resolved stresses $\overline{u}_1\overline{u}_3$, $\overline{u}_3\overline{u}_3$ and $\overline{u}_2\overline{u}_3$ are zero at the wall. This model suffers from a number of defects. Namely, it shows no dependency as a function of the Reynolds number, and assumes that the shear-stress near the wall is entirely due to the subgrid scales.

Schumann Model. Schumann [653] has developed a wall model for performing a plane channel flow simulation at a finite Reynolds number. It is based on the extended turbulent relation (10.23). Using dimensional analysis, the effective viscosity can be evaluated using

$$\nu_{\text{eff}} = \frac{z_2}{2} \frac{\langle \tau_{\text{p}} \rangle}{\langle \overline{u}_1(x, y, z_2) \rangle} \quad . \tag{10.27}$$

The resulting boundary conditions are:

$$\tau_{\text{p},13}(x, y) = \left(\frac{\overline{u}_1(x, y, z_2)}{\langle \overline{u}_1(x, y, z_2) \rangle} \right) \langle \tau_{\text{p}} \rangle \quad , \tag{10.28}$$

$$\overline{u}_3 = 0 \quad , \tag{10.29}$$

$$\tau_{\text{p},23}(x, y) = \frac{2}{Re_\tau} \left(\frac{\overline{u}_3(x, y, z_2)}{z_2} \right) \quad , \tag{10.30}$$

where $\langle\rangle$ designates a statistical average (associated here with a time average), and z_2 the distance of the first point to the wall. The condition (10.28) is equivalent to adopting the hypothesis that the longitudinal velocity component at position z_2 is in phase with the instantaneous wall shear stress. The mean velocity profile can be obtained by the logarithmic law, and the mean wall shear stress $\langle \tau_{\text{p}} \rangle$ is, for a plane channel flow, equal to the driving pressure gradient. This wall model therefore implies that the mean velocity field verifies the logarithmic law and can be applied only to plane channel flows for which the value of the driving pressure gradient is known a priori. The second condition is the impermeability condition, and the third corresponds to a no-slip condition for the transverse velocity component $\overline{u}_2$.

Grötzbach Model. Grötzbach [278] proposes extending the Schumann model to avoid having to know the mean wall shear stress *a priori*. To do this, the statistical average $\langle\rangle$ is now associated with a mean on the plane parallel to the solid wall located at $z = z_2$. Knowing $\langle \overline{u}_1(z_2) \rangle$, the mean wall shear stress $\langle \tau_\mathrm{p} \rangle$ is computed from the logarithmic law. The friction velocity is computed from (10.11), i.e.:

$$u_1^+(z_2) = \langle \overline{u}_1(z_2) \rangle / u_\tau = \frac{1}{\kappa} \log(z_2 u_\tau / \nu) + 5.5 \pm 0.1 \quad , \tag{10.31}$$

then $\langle \tau_\mathrm{p} \rangle$, by relation (10.4). This model is more general than Schumann's, but it still requires that the mean velocity profile verify the logarithmic law. Another advantage of Grötzbach's modification is that it allows variations of the total mass flux through the channel.

Shifted correlations Model. Another modification of Schumann's model can be made on the basis of the experimental works of Rajagopalan and Antonia [614]. These two authors observed that the correlation between the wall shear stress and the velocity increases when we consider a relaxation time between these two evaluations. This phenomenon can be explained by the existence of coherent inclined structures that are responsible for the velocity fluctuations and the wall shear stress. The modified model is expressed [595]:

$$\tau_{\mathrm{p},13}(x,y) = \left(\frac{\overline{u}_1(x + \Delta_\mathrm{s}, y, z_2)}{\langle \overline{u}_1(x,y,z_2) \rangle} \right) \langle \tau_\mathrm{p} \rangle \quad , \tag{10.32}$$

$$\overline{u}_3 = 0 \quad , \tag{10.33}$$

$$\tau_{\mathrm{p},23}(x,y) = \left(\frac{\overline{u}_2(x + \Delta_\mathrm{s}, y, z_2)}{\langle \overline{u}_1(x,y,z_2) \rangle} \right) \langle \tau_\mathrm{p} \rangle \quad , \tag{10.34}$$

where the value of the length Δ_s is given by the approximate relation:

$$\Delta_\mathrm{s} = \begin{cases} (1 - z_2) \cot(8^\circ) & \text{for } 30 \leq z_2^+ \leq 50\text{–}60 \\ (1 - z_2) \cot(13^\circ) & \text{for } z_2^+ \geq 60 \end{cases} \quad . \tag{10.35}$$

Rough Wall Model. Mason and Callen [497] propose a wall model including the roughness effects. The three velocity components are specified at the first computation point by the relations:

$$\overline{u}_1(x,y,z_2) = \cos\theta \left(\frac{u_\tau(x,y)}{\kappa} \right) \ln(1 + z_2/z_0) \quad , \tag{10.36}$$

$$\overline{u}_2(x,y,z_2) = \sin\theta \left(\frac{u_\tau(x,y)}{\kappa} \right) \ln(1 + z_2/z_0) \quad , \tag{10.37}$$

$$\overline{u}_3(x,y,z_2) = 0 \quad , \tag{10.38}$$

where z_0 is the roughness thickness of the wall and angle θ is given by the relation $\theta = \arctan(\overline{u}_2(z_2)/\overline{u}_1(z_2))$. These equations can be used to compute the friction velocity u_τ as a function of the instantaneous velocity components

$\overline{u}_1$ and $\overline{u}_2$. The instantaneous surface friction vector $\boldsymbol{u}_\tau^2$ is then evaluated as:

$$\boldsymbol{u}_\tau^2 = \frac{1}{M}|\boldsymbol{u}_\parallel|\boldsymbol{u}_\parallel \quad , \tag{10.39}$$

where $\boldsymbol{u}_\parallel$ is the vector $(\overline{u}_1(x,y,z_2), \overline{u}_2(x,y,z_2), 0)$ and

$$\frac{1}{M} = \frac{1}{\kappa^2}\ln^2(1 + z_2/z_0) \quad .$$

The instantaneous wall shear stresses in the x and y directions are then evaluated respectively as $|u_\tau^2|\cos\theta$ and $|u_\tau^2|\sin\theta$. This model is based on the hypothesis that the logarithmic distribution is verified locally and instantaneously by the velocity field. This becomes even truer as the grid is coarsened, and the large scale velocity approaches the mean velocity.

Ejection Model. Another wall model is proposed by Piomelli, Ferziger, Moin, and Kim [595] in consideration of the fact that the fast fluid motions toward or away from the wall greatly modify the wall shear stress. The impact of fast fluid pockets on the wall causes the longitudinal and lateral vortex lines to stretch out, increasing the velocity fluctuations near the wall. The ejection of fast fluid masses induces the inverse effect, i.e. reduces the wall shear stress. To represent the correlation between the wall shear stress and the velocity fluctuations, the authors propose the following conditions:

$$\tau_{\mathrm{p},13}(x,y) = \langle \tau_\mathrm{p} \rangle - C u_\tau \overline{u}_3(x + \Delta_\mathrm{s}, y, z_2) \quad , \tag{10.40}$$

$$\tau_{\mathrm{p},23}(x,y) = \left(\frac{\langle \tau_\mathrm{p} \rangle}{\langle \overline{u}_1(z_2) \rangle} \right) \overline{u}_2(x + \Delta_\mathrm{s}, y, z_2) \quad , \tag{10.41}$$

$$\overline{u}_3(x,y) = 0 \quad , \tag{10.42}$$

where C is a constant of the order of unity, $\langle \tau_\mathrm{p} \rangle$ is computed from the logarithmic law as it is for the Grötzbach model, and Δ_s is computed by the relation (10.35).

Marusic's Optimized Ejection Model. Piomelli's ejection wall model was futher improved by Marusic et al. [492] on the grounds of very accurate wind tunnel experiments. The proposed generalization for relation (10.40) based on the experimental correlations is

$$\tau_{\mathrm{p},13}(x,y) = \langle \tau_\mathrm{p} \rangle - \alpha u_\tau \left(\overline{u}_1(x + \Delta_\mathrm{s}, y, z_2) - \langle \overline{u}_1(x,y,z_2) \rangle \right) \quad , \tag{10.43}$$

where α is a parameter taken equal to 0.1 for zero pressure gradient flows. This new ejection model, which is based on the streamwise velocity component instead of the wall-normal velocity in Piomelli's original model, is found to yield better results on priori tests carried out using experimental data: the computed peak correlation coefficient is in the range 0.34–0.53 for the new model while it is between 0.19 and 0.24 for the original model.

Thin Boundary Layer Models. Balaras et al. [37] and Cabot [86, 87] propose more sophisticated models based on a system of simplified equations derived from the boundary layer equations. A secondary grid is embedded within the first cell at the wall (see Fig. 10.9), on which the following system is resolved:

$$\frac{\partial \overline{u}_i}{\partial t} + \frac{\partial}{\partial x}(\overline{u}_1\overline{u}_i) + \frac{\partial}{\partial z}(\overline{u}_3\overline{u}_i) = -\frac{\partial \overline{p}}{\partial x_i} + \frac{\partial}{\partial z}\left((\nu + \nu_{\mathrm{sgs}})\frac{\partial \overline{u}_i}{\partial z}\right), \quad i = 1,2 \quad , \tag{10.44}$$

where z is the direction normal to the wall. Equation (10.44) can be recast as an equation for the shear stresses $\tilde{\tau}_{i3} = \partial \overline{u}_i/\partial z, i = 1,2$:

$$\frac{\partial}{\partial z}((\nu + \nu_{\mathrm{sgs}})\tilde{\tau}_{i3}) = \frac{\partial \overline{u}_i}{\partial t} + \frac{\partial}{\partial x}(\overline{u}_1\overline{u}_i) + \frac{\partial}{\partial z}(\overline{u}_3\overline{u}_i) + \frac{\partial \overline{p}}{\partial x_i} \quad i = 1,2 \quad . \tag{10.45}$$

Simplified models can be derived by neglecting some source terms in the right-hand side of (10.45) or by approximating them using values from the outer flow [90].

This approach is equivalent to assuming that the inner zone of the boundary layer behaves like a Stokes layer forced by the outer flow. Balaras et al. propose computing the viscosity ν_{sgs} by the simplified mixing length model:

$$\nu_{\mathrm{sgs}} = (\kappa z)^2 D_{\mathrm{b}}(z)|\overline{S}| \quad , \tag{10.46}$$

where z is the distance to the wall, κ the Von Karman constant, and $D_{\mathrm{b}}(z)$ the damping function:

$$D_{\mathrm{b}}(z) = \left(1 - \exp(-(z^+/A^+)^3)\right) \quad , \tag{10.47}$$

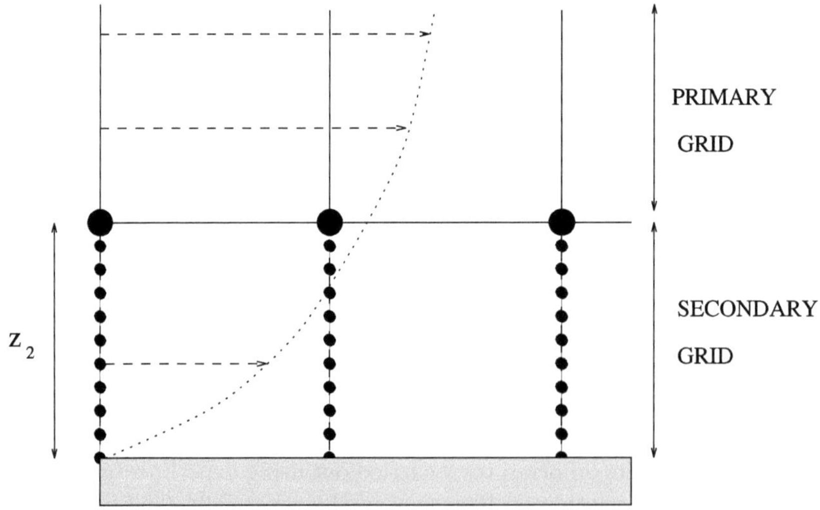

Fig. 10.9. Representation of the primary and secondary grids.

with $A^+ = 25$. Cabot proposes the alternate definition:

$$\nu_{\text{sgs}} = \kappa u_{\text{s}} z D_C^2(z) \quad , \tag{10.48}$$

in which

$$D_C(z) = (1 - \exp(-z u_{\text{d}}/A\nu)) \quad , \tag{10.49}$$

where u_{s} and u_{d} are velocity scales to be determined, and $A = 19$. The simplest choice is $u_{\text{s}} = u_{\text{d}} = u_\tau$.

Cabot and Moin [90] observed that the constant of the mixing length model must be lowered in regard to its usual value in RANS computations in order to account for resolved stresses. A dynamic evaluation of this constant was achieved by Wang [751] to deal with flows with strong favorable/adverse pressure gradient and incipient separation. The dynamic adjustment is performed by imposing that the mixing-length viscosity and the subgrid viscosity are equal at the interface of the two simulations.

When this system is solved, it generates longitudinal and transverse velocity component distributions at each time step, so that the value of the wall shear stress can be calculated for solving the filtered Navier–Stokes equations on the main grid. The pressure gradient appears as a source term, because this is obtained using the relation $\partial \overline{p}/\partial x_n = 0$.

The vertical velocity component is obtained from the continuity equation:

$$\overline{u}_3(x, y, z_2) = - \int_0^{z_2} \left(\frac{\partial \overline{u}_1}{\partial x}(x, y, \xi) + \frac{\partial \overline{u}_2}{\partial y}(x, y, \xi) \right) d\xi \quad . \tag{10.50}$$

The boundary conditions applied to the secondary system are:

- On the solid wall: no-slip condition;
- On the upper boundary: Dirichlet condition obtained from the value of the velocity field computed on the first cell of the main grid.

Model of Murakami et al. Murakami, Mochida, and Hibi [556] developed a wall model for dealing with the case of the separated flow around a cube mounted on a flat plate. This model is based on power-law solutions for the mean longitudinal velocity profile of the form:

$$\frac{\langle \overline{u}_1(z) \rangle}{U_{\text{e}}} \simeq \left(\frac{z}{\delta} \right)^n \quad . \tag{10.51}$$

The authors recommend using $n = 1/4$ on the flat plate and $n = 1/2$ on the cube surface. When the first grid point is located close enough to the wall, the following boundary conditions are used:

$$\overline{u}_i(x, y) = \left(\frac{z_2}{z_2 + \Delta z} \right)^n \overline{u}_i(x, y, z_2 + \Delta z), \quad i = 1, 2 \quad , \tag{10.52}$$

$$\overline{u}_3(x, y) = 0 \quad , \tag{10.53}$$

where Δz is the size of the first cell. The first equation is obtained by assuming that the instantaneous profile also verifies the law (10.51). When the distance of the first point from the wall is too large for the convection effects to be neglected, the relation (10.53) is replaced by:

$$\frac{\partial \overline{u}_3}{\partial z} = 0 \quad . \tag{10.54}$$

Werner–Wengle Model. In order to be able to compute the same flow as Murakami et al., Werner and Wengle [757] propose a wall model based on the following hypotheses:

- The instantaneous tangential velocity components at the wall $u_2(x, y, z_2)$ and $u_3(x, y, z_2)$ are in phase with the associated instantaneous wall shear stresses.
- The instantaneous velocity profile follows the law:

$$u^+(z) = \begin{cases} z^+ & \text{if } z^+ \leq 11.81 \\ A(z^+)^B & \text{otherwise} \end{cases} \quad , \tag{10.55}$$

in which $A = 8.3$ and $B = 1/7$.

The values of the tangential velocity components can be related to the corresponding values of the wall shear stress components by integrating the velocity profile (10.55) over the distance separating the first cell from the wall. This allows a direct analytical evaluation of the wall shear stress components from the velocity field:

- If $|\overline{u}_i(x, y, z_2)| \leq \frac{\nu}{2z_m} A^{2/(1-B)}$, then:

$$\tau_{\mathrm{p},i3}(x, y) = \frac{2\nu \overline{u}_i(x, y, z_2)}{z_2} \quad , \tag{10.56}$$

- and otherwise:

$$\begin{aligned} \tau_{\mathrm{p},i3}(x, y) &= \frac{\overline{u}_i(x, y, z_2)}{|\overline{u}_i(x, y, z_2)|} \left[\frac{1-B}{2} A^{\frac{1+B}{1-B}} \left(\frac{\nu}{z_2} \right)^{1+B} \right. \\ &\quad \left. + \frac{1+B}{A} \left(\frac{\nu}{z_2} \right)^B |\overline{u}_i(x, y, z_2)| \right]^{\frac{2}{1+B}} \quad , \end{aligned} \tag{10.57}$$

where z_m is the distance to the wall that corresponds to $z^+ = 11.81$. This model has the advantage of not using average statistical values of the velocity and/or wall shear stresses, which makes it easier to use for inhomogeneous configurations. An impermeability condition is used to specify the value of the velocity component normal to the wall:

$$\overline{u}_3 = 0 \quad . \tag{10.58}$$

Werner–Wengle-Type Ejection Model. A version of the Werner–Wengle model which accounts for the shift that exists in the correlation between the wall friction and the instantaneous velocity is proposed by Hassan and Barsamian [298]. The authors recommend to account for this shift as in Piomelli's shifted model, leading to the following modifications for relation (10.57)

$$\begin{aligned}\tau_{\mathrm{p},i3}(x,y) &= \frac{\overline{u}_i(x+\Delta_{\mathrm{s}},y,z_2)}{|\overline{u}_i(x+\Delta_{\mathrm{s}},y,z_2)|}\left[\frac{1-B}{2}A^{\frac{1+B}{1-B}}\left(\frac{\nu}{z_2}\right)^{1+B}\right.\\ &\left. + \frac{1+B}{A}\left(\frac{\nu}{z_2}\right)^{B}|\overline{u}_i(x+\Delta_{\mathrm{s}},y,z_2)|\right]^{\frac{2}{1+B}}\\ &-Cu_\tau\overline{u}_3(x+\Delta_{\mathrm{s}},y,z_2) \quad , \end{aligned} \tag{10.59}$$

where C is a constant of the order of the unity and Δ_{s} is computed using the relation (10.35). Equation (10.56) is kept unchanged, since the shift is assumed to be negligible in the viscous sublayer.

Suboptimal-Control-Based Wall Models. The goal of this approach is to provide numerical boundary conditions so that the overall error (to be defined) is minimum in a given norm. The boundary conditions (wall stresses and wall-normal velocity component) are used as a control to minimize a cost function at each time step. Many variants of this approach can be defined [566, 33], considering different degrees of freedom at the boundary (i.e. different controllers), different cost functions and different ways to evaluate the gradient of the cost function with respect to the controller.

Nicoud et al. [566] and Bagget et al. [33] considered a control vector ϕ on the boundary whose components are the usual output of a wall-stress model:

$$\phi = (\tau_{\mathrm{p},13}, \overline{u}_3, \tau_{\mathrm{p},23}) \quad . \tag{10.60}$$

The control can then be exerted by modifying both the stress at the wall and the wall-normal transpiration velocity.

The general form of the cost function $\mathcal{J}(\overline{u};\phi)$ is defined as

$$\mathcal{J}(\overline{u};\phi) = \sum_{i=1}^{3}\mathcal{J}_{\mathrm{mean},i}(\overline{u};\phi) + \sum_{i=1}^{3}\mathcal{J}_{\mathrm{rms},i}(\overline{u};\phi) + \sum_{i=1}^{3}\mathcal{J}_{\mathrm{penalty}}(\phi) \quad , \tag{10.61}$$

where the terms appearing on the right-hand side are, from left to right: the part of the cost based on the mean flow, the part of the cost based on the rms velocity fluctuations, and a penalty term representing the cost of the control.

The mean-flow part of the cost function is typically a measure of the difference between the computed mean flow and a target mean flow $\boldsymbol{u}_{\mathrm{ref}}$. For the plane channel flow, it can be expressed as

$$\mathcal{J}_{\mathrm{mean},i}(\overline{u};\phi) = \alpha_i\frac{1}{2h}\int_{-h}^{h} e_{u_i}(z)^2 dz \quad , \tag{10.62}$$

with

$$e_{u_i}(z) = \frac{1}{A} \int (u_i(x,y,z) - U_{\mathrm{ref},i}(z)) dxdy \quad , \tag{10.63}$$

where A is the surface of computational planes parallel to the walls and α_i is an arbitrary weighting factor. The target mean flow can be prescribed using experimental data, RANS simulations or theory. In a similar way, the rms-based cost function is defined as

$$\mathcal{J}_{\mathrm{rms},i}(\overline{u};\phi) = \beta_i \frac{1}{2h} \int_{-h}^{h} e_{u_i'}(z)^2 dz \quad , \tag{10.64}$$

with

$$e_{u_i'}(z) = \frac{1}{A} \int \left((u_i(x,y,z) - \langle u_i \rangle (z))^2 - u'^2_{\mathrm{ref},i}(z) \right) dxdy \quad , \tag{10.65}$$

where $\langle u_i \rangle$ is the average of the computed flow over homogeneous directions, β_i is an arbitrary parameter, and $u'^2_{\mathrm{ref},i}(z)$ are prescribed rms velocity profiles.

Bagget et al. proposed the following form of the penalty term:

$$\mathcal{J}_{\mathrm{penalty}}(\phi) = \frac{\gamma_i}{A} \int_{z=\pm h} \phi_{u_i}^2 dxdy + \frac{\lambda}{A} \int_{z=\pm h} \delta_{i3} \phi_{u_3}^4 dxdy \quad , \tag{10.66}$$

where γ_i and λ are arbitrary constants. The last term of the penalty term prevents the transpiration velocity from becoming too high.

Several tests have been carried out, dealing with different weights of the three parts of the cost function and the possibility of having a non-zero transpiration velocity. The main results are:

- Suboptimal-control-based models lead to better results than usual wall-stress models (see Fig. 10.10). An interesting feature of these models is that they are able to break the spurious linear dependence of the predicted wall stresses with respect to the instantaneous velocity at the first off-wall grid point. This is observed by looking at Figs. 10.11 and 10.12.
- The use of a non-zero transpiration velocity makes it possible to improve significantly the mean flow profile compared to usual wall-stress models, if the rms part of the cost function is not considered ($\beta_i = 0$). But, in that case, rms velocity profiles are not improved and prediction can even be worse.
- Rms velocity profiles can be improved if both $\mathcal{J}_{\mathrm{mean},i}$ and $\mathcal{J}_{\mathrm{rms},i}$ are taken into account, but the improvement of the prediction of the mean velocity profile is less important than in the previous case. The use of the non-zero transpiration velocity is also observed to be beneficial.

This class of wall model based on the suboptimal control theory can be considered as the best achievable wall-stress model, and it seems difficult to get much better results controlling the same parameters (wall stresses and transpiration velocity). Then, an interesting conclusion is that getting the

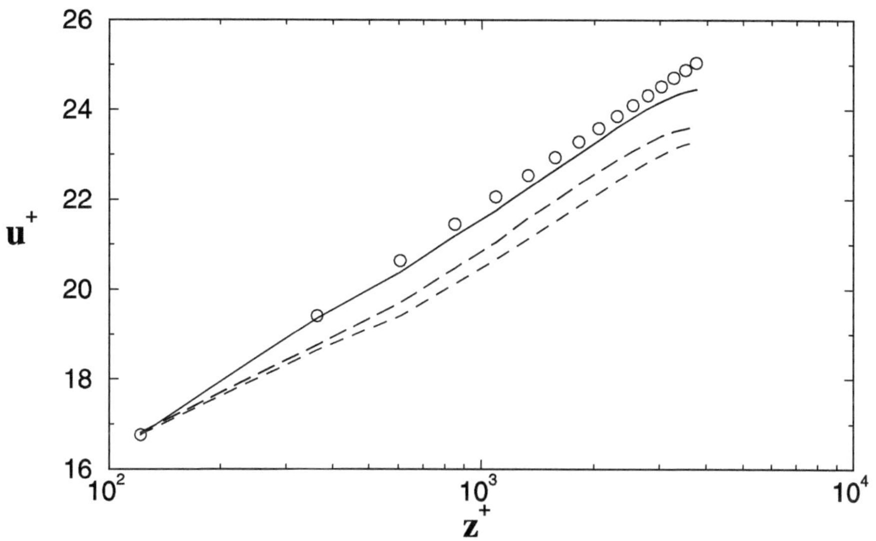

Fig. 10.10. Large-eddy simulations of plane channel flow ($Re_\tau = 4000$) on a uniform $32 \times 32 \times 32$ grid. Mean velocity profile: *Symbols:* $2.41 \log(z^+) + 5.2$; *Solid line*: suboptimal wall stress model, without transpiration velocity; *Upper dashed line*: suboptimal estimation of τ_{13}, with shifted wall-stress model for τ_{23}; *Lower dashed line*: shifted wall-stress model. Courtesy of F. Nicoud, University of Montpellier.

correct mean velocity profile and the correct rms velocities may be competing objectives.

The preceding models are based on suboptimal control theory, and necessitate computing the gradient of the cost function. This involve a large computational effort, whatever solution is adopted to compute the gradient (finite differences or solving the adjoint problem).[5] The use of the incomplete gradient approach of Mohammadi and Pironneau [535] was observed to yield poor results for this problem by Templeton et al. [707]. Following the pioneering works of Bagwell et al. [34], Nicoud et al. [566] proposed a more practical wall model based on linear stochastic estimation. This model is the best possible least-square estimate of the suboptimal wall stresses as explicit functions of the local velocity field. It can be expressed as the conditional average of the wall stress given the local velocity field:

$$\langle \tau_{i,3} | E \rangle \quad , \tag{10.67}$$

where E is a vector of events containing the local instantaneous velocity. This formal expression does not lead to a tractable wall model, and it is approximated via a polynomial expansion. Restricting this expansion to the

[5] The reported cost of a large-eddy simulation based on these models is 20 times greater than that of a simulation on the same grid with explicit wall-stress models.

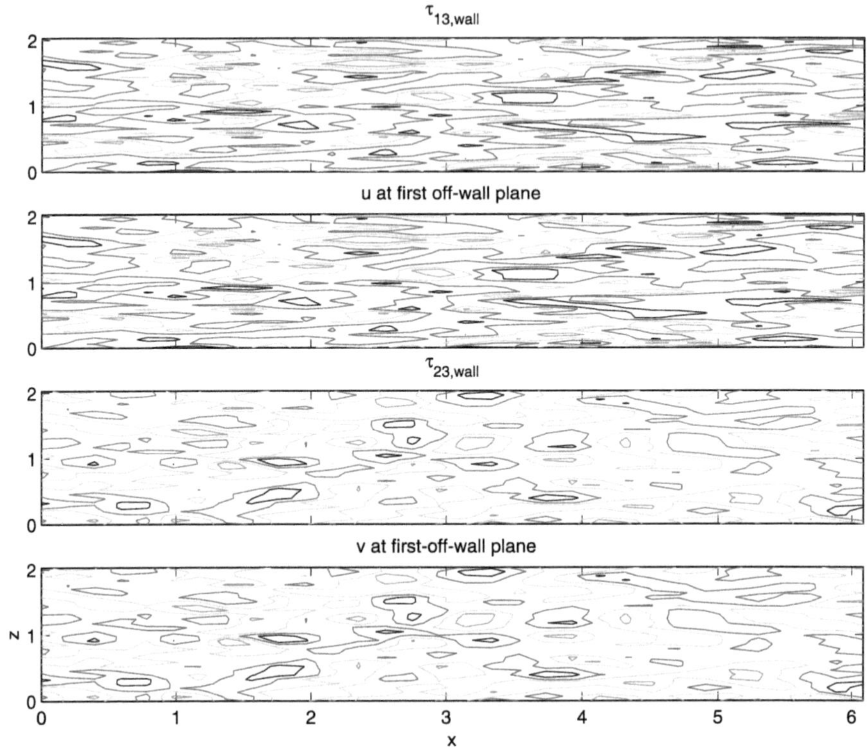

Fig. 10.11. Large-eddy simulations of plane channel flow ($Re_\tau = 4000$) on a uniform $32 \times 32 \times 32$ grid. Shifted correlation model. Instantaneous isocontours of predicted stresses and velocity components at the first off-wall grid point. Courtesy of F. Nicoud, University of Montpellier.

first term, one obtains the following linear stochastic estimate for the wall stresses:

$$\langle \tau_{i,3} | E \rangle \approx L_{ij} E_j \quad , \tag{10.68}$$

where the estimation coefficients L_{ij} are governed by

$$\langle \tau_{i,3} E_k \rangle = L_{ij} \langle E_j E_k \rangle \quad . \tag{10.69}$$

In practice, these coefficients are computed using reference data (direct numerical simulation data or suboptimal-control-based prediction of the wall stresses). The resulting model is an explicit model, whose cost is of the same order as those of the other explicit wall-stress models, which are able to reproduce accurately the results of the suboptimally controlled simulations. Unfortunately, numerical experiments have shown this model to be very sensitive to numerical and modeling errors, indicating that it might be impossible to find an accurate and robust[6] linear wall stress model.

[6] A model whose accuracy will be the same whatever numerical method and subgrid model are employed.

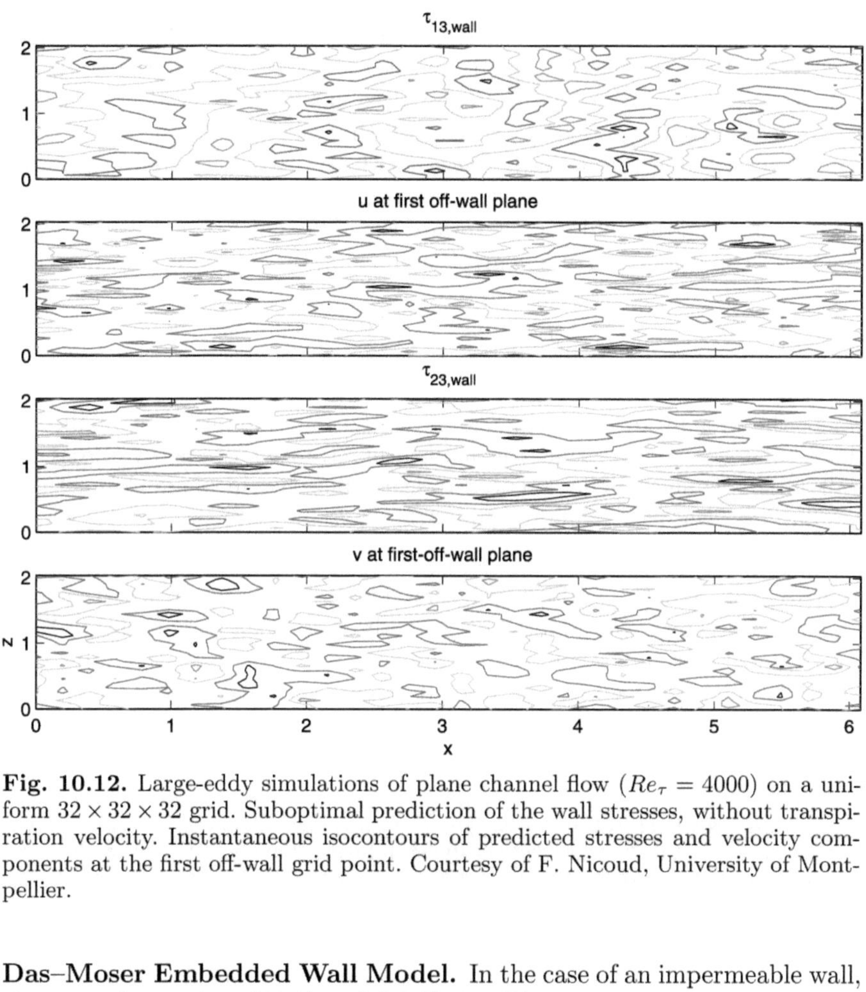

Fig. 10.12. Large-eddy simulations of plane channel flow ($Re_\tau = 4000$) on a uniform $32 \times 32 \times 32$ grid. Suboptimal prediction of the wall stresses, without transpiration velocity. Instantaneous isocontours of predicted stresses and velocity components at the first off-wall grid point. Courtesy of F. Nicoud, University of Montpellier.

Das–Moser Embedded Wall Model. In the case of an impermeable wall, the general equations (10.1) and (10.2) simplify to

$$\frac{\partial \overline{u_i}}{\partial t} + \frac{\partial}{\partial x_j}(\overline{u_i u_j}) + \frac{\partial \overline{p}}{\partial x_i} - \nu \frac{\partial}{\partial x_j}\left(\frac{\partial \overline{u}_i}{\partial x_j} + \frac{\partial \overline{u}_j}{\partial x_i}\right) =$$

$$-\int_{\partial \Omega} G(\boldsymbol{x} - \boldsymbol{\xi}) \left[p(\boldsymbol{\xi}) - \nu \left(\frac{\partial u_i}{\partial x_j}(\boldsymbol{\xi}) + \frac{\partial u_j}{\partial x_i}(\boldsymbol{\xi}) \right) \right] n_j(\boldsymbol{\xi}) d\boldsymbol{\xi} \quad , \quad (10.70)$$

$$\frac{\partial \overline{u}_i}{\partial x_i} = 0 \quad . \quad (10.71)$$

The source term in the continuity equation cancels out, while the momentum source term is associated with the total force exerted on the wall. When filtering through the wall, it is necessary to prolong the velocity field inside the solid wall. This is done by assuming a zero velocity field inside the wall in a buffer zone Ω_*.

In the present formulation, the problem is to compute the source term in (10.70). Because the unfiltered velocity is set to zero in the wall, the wall stress is the surface forcing required to ensure that momentum is not transfered to a buffer domain located inside the wall, i.e. that the velocity will remain zero. Using this remark, Das and Moser [164] suggested computing the source term at each time step in order to minimize the transport of momentum from the fluid domain toward the buffer zone. The embedded wall boundary condition is then a control minimizing the following cost function:

$$\mathcal{J} = \int_{\Omega_*} \left(|\overline{\boldsymbol{u}}|^2 + \alpha \left| \frac{\partial \overline{\boldsymbol{u}}}{\partial t} \right|^2 \right) d\boldsymbol{x} \quad , \tag{10.72}$$

where the first term forces the energy in the buffer zone to be small, and the second one ensures that the momentum transfer of energy across the wall surface is small. The constant α scales like Δt^2 in plane channel flow computation, with Δt being the time step.

ODT-Based Wall Model. Structural models presented in Sect. 7.7 can be used to reconstruct subgrid fluctuations inside the first cell near the solid walls. This strategy was adopted by Schmidt et al. [650], who used the ODT-based model (see p. 257) to predict the near wall dynamics, the outer part of the boundary layer being computed using a classical large-eddy simulation method. It is worth noting that this approach is reminiscent of the one proposed by Balaras based on thin boundary layer equations within the first grid cell.

The proposed coupling strategy, based on the existence of an overlap region between ODT and large-eddy simulation (see p. 257 for a detailed description of the ODT-based model), is the following. This zonal decomposition is illustrated in Fig. 10.13. An ODT-type simulation is performed within the first grid cell of the large-eddy simulation grid. The large-eddy simulation provides the ODT with boundary conditions at the top of the ODT domain, and also appears in the evaluation of the advection speed in the ODT equation. The ODT simulation provides the large-eddy simulation with subgrid fluxes in two different regions: (i) in the *inner region* (i.e. the first grid cell off the wall), the full fluxes appearing in the large-eddy simulation equations is computed using the ODT fluctuations and (ii) in the *outer region*, which has a thickness L (L being the maximum size for eddy events allowed in the ODT simulation within the inner region). In the outer region, ODT equations are not solved, but the influence of large events belonging to the ODT solution computed in the inner region is taken into account. This is achieved by adding the contribution of ODT eddy events of sufficient size to reach the considered position in the large-eddy simulation grid to the conventional fluxes in that region.

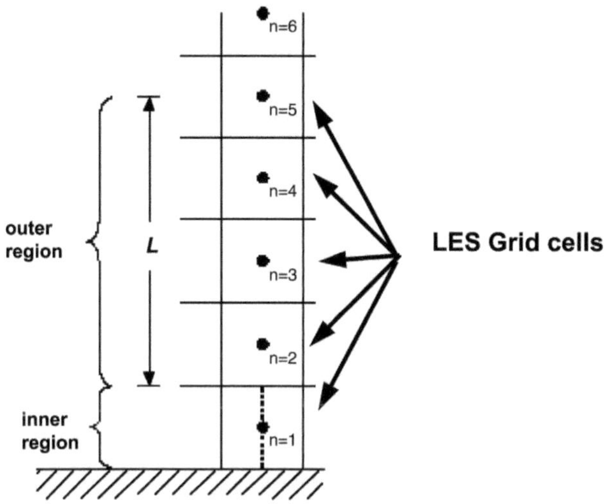

Fig. 10.13. Schematic of the ODT-based wall model.

10.2.3 Wall Models: Achievements and Problems

Most of the wall models presented above have been developed on the grounds of the dynamics of the zero-pressure gradient, equilibrium flate plate boundary layer. In practical cases, most of them exhibit the same behavior:

- In attached, equilibrium flows, satisfactory results (skin friction predicted within a 20% error) on the mean flow are recovered on medium grids, i.e. on grids such that the first cell off the wall has the following dimensions:

$$\Delta x^+ \leq 500, \quad \Delta y^+ \leq 200-300, \quad \Delta z^+ = 50-150 \quad .$$

  On such a grid, spurious bumps are observed in the turbulence intensities just near the first grid point (see Fig. 10.14). These unphysical overshoots are associated to the existence of large spurious streaky structures, whose size can be governed by either the mesh size or the numerical and subgrid model dissipation. The growth of these structures might be explained by the same mechanisms as the physical streaky vortices: they might be parts of an autonomous self-sustaining cycle feeded by the mean shear involving low-/high-speed streaks and streamwise vortices. Another possible cause is that they could arise from the splatting of turbulent eddies coming from the outer part of the boundary layer on the boundary condition (almost all models account for the impermeability constraint enforcing a vanishing wall normal velocity).
  No general cure for this problem is known. These bumps can be damped by scrambling the spurious streamwise vortices by adding a random noise in the region where they are detected [592]. Partial error cancellation when using a very dissipative subgrid model has also been observed: the spurious

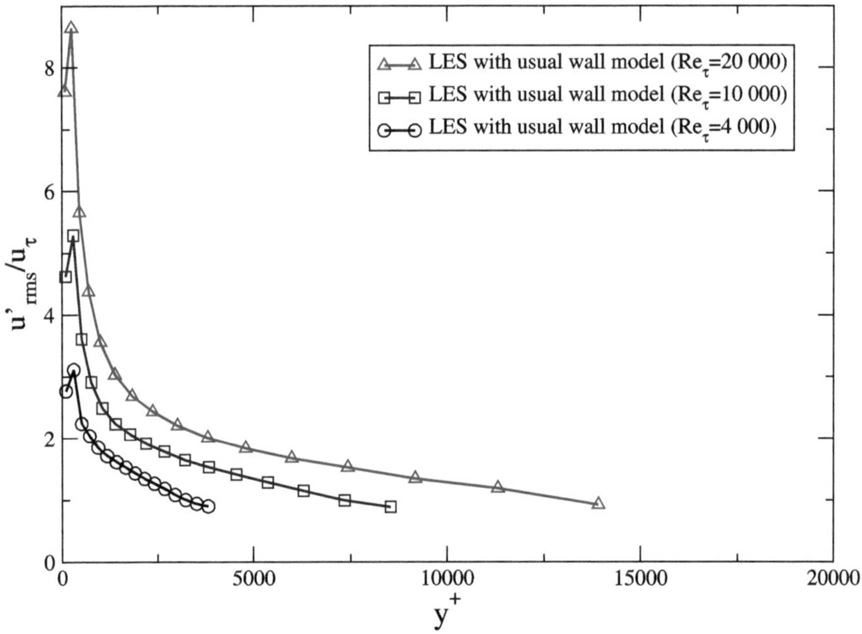

Fig. 10.14. Resolved Reynolds stresses in a plane channel flow with Grötzbach wall stress model. Courtesy of Y. Benarafa and F. Ducros, CEA.

streaks are then damped. It was also shown by Nicoud that some control on the turbulence intensities can be achieved playing of the wall stresses and the wall normal velocity, but with a higher error level on the mean flow profile. It is worth noticing that the problem of the reduction of this error is close to the one of the active control of the boundary layer dynamics.

- On very coarse grids, the models are no longer ables to yield an accurate prediction of the mean flow profile: the correct logarithmic slope is not recovered in the logarithmic layer, the skin friction is poorly predicted (see Fig. 10.15), ... It must be observed that on very coarse grids the problem is much more complicated: (i) the near-wall layer dynamics is not resolved, (ii) large subgrid modeling errors occur in the core of the flow since a large part of the turbulent kinetic energy is contained in subgrid scales (and most subgrid scale models are not good at taking into account a large part of the full turbulent kinetic energy), and (iii) since the mesh is coarse, numerical errors may become dominant.
- Most models are not very efficient at predicting separation, since they are based on very stringent assumptions. This was observed, among other studies, by Temmerman et al. [706] in a wavy channel configuration where the location of the separation point (and therefore of the reattachment point) is seen to be sensitive to the wall model. A noticeable exception is the Thin Boundary Layer Model, which has been proved to yield a satisfac-

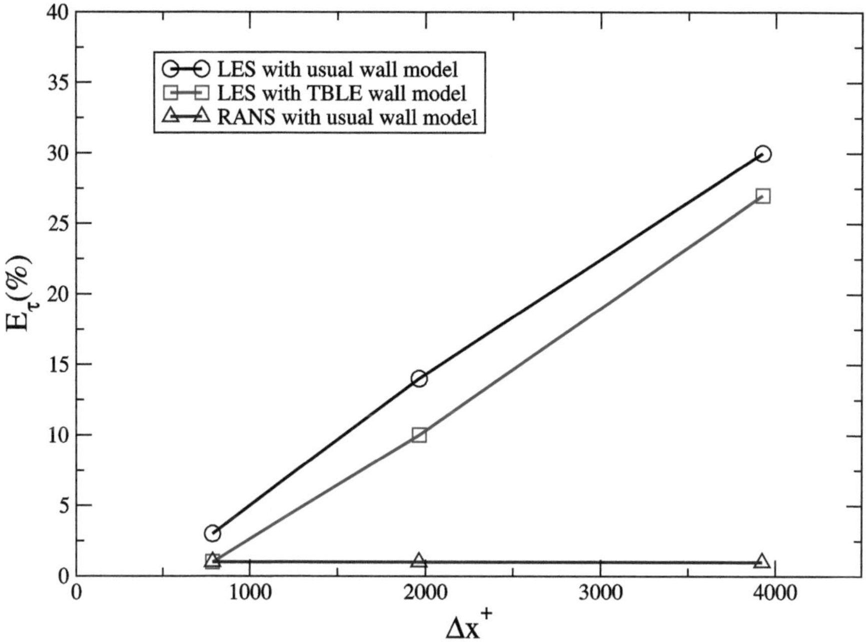

Fig. 10.15. Error on the computed skin friction versus the size of the mesh in the streamwise direction (in wall units) in plane channel flow computation, with different wall models. Steady RANS solution is shown for comparison. Courtesy of Y. Benarafa and F. Ducros, CEA.

tory prediction of an incipient separated region on smooth airfoil trailing edge [754] and on a circular cylinder at high Reynolds number [113]. In configurations where the separation point is imposed like in the backward facing step case investigated by Cabot [87], main features of the flow are relatively insensitive to the way the solid wall is treated inside the separated region. But skin friction in the seprated region is of course subject to consequent errors. First trials for the definition of wall models devoted to separated regions have been done [706, 370], but satisfactory results have not yet been obtained. A main difficulty is the lack of universal scaling law for the mean velocity profile in separated flows.

- Some wall-stress models require values related to the mean flow as inputs. This need lead to a severe limitation of the models, since the mean flow must be known, or some homogeneous directions must exists to enable the evaluation of some statistiscal moments of the instantaneous fields at each time step. Possible solutions to overcome this problem are: (i) to run several statistically equivalent simulations in parallel and to perform true statistical average or (ii) to use a local spatial averaging instead of a statistical average. The latter solution is used by Hassan and Barsamian [298] to obtain a localized version of the Grötzbach model.

10.3 Case of the Inflow Conditions

10.3.1 Required Conditions

Representing the flow upstream of the computational domain also raises difficulties when this flow is not fully known deterministically, because the lack of information introduces sources of error. This situation is encountered for transitional or turbulent flows that generally contain a very large number of space–time modes [631]. Several boundary condition generation techniques are used for furnishing information about all the modes contained to the large-eddy simulation computation.

Apart the purely mathematical problem of defining well-posed inflow boundary conditions, the Large-Eddy Simulation and the Direct Numerical Simulation techniques raise the problem of reconstructing the turbulent fluctuations at the inlet plane in an accurate way. The exact definition of accurate turbulent inflow conditions is still an open question, but the accumlulated experience proves that both kinetic energy and coherence of the inlet fluctuations must be taken into account to minimize the size of the buffer region that exists downstream the inlet plane, in which turbulent fluctuations consistent with the Navier–Stokes dynamics are reconstructed by the non-linear effects. This need is illustrated in Fig. 10.16, which displays results obtained in Direct Numerical Simulation of a two-dimensional mixing layer.

10.3.2 Inflow Condition Generation Techniques

Stochastic Recontruction from a Statistical One-Point Description. When the freestream flow is described statistically (usually the mean velocity field and the one-point second-order moments), the deterministic information is definitively lost. The solution is then to generate instantaneous realizations that are statistically equivalent to the freestream flow, i.e. that have the same statistical moments.

In practice, this is done by superimposing random noises having the same statistical moments as the velocity fluctuations, on the mean statistical profile. This is expressed as

$$\boldsymbol{u}(x_0, t) = U(x_0) + \boldsymbol{u}'(x_0, t) \quad , \tag{10.73}$$

where the mean field U is given by experiment, theory or steady computations, and where the fluctuation $\boldsymbol{u}'$ is generated from random numbers. This technique makes it possible to remain in keeping with the energy level of the fluctuations as well as the one-point correlations (Reynolds stresses) in the directions of statistical homogeneity of the solution, but does not reproduce the two-point (and two-time) space–time correlations [432, 540, 493]. The information concerning the phase is lost, which can have very harmful consequences when the consistency of the fluctuations is important, as is the case

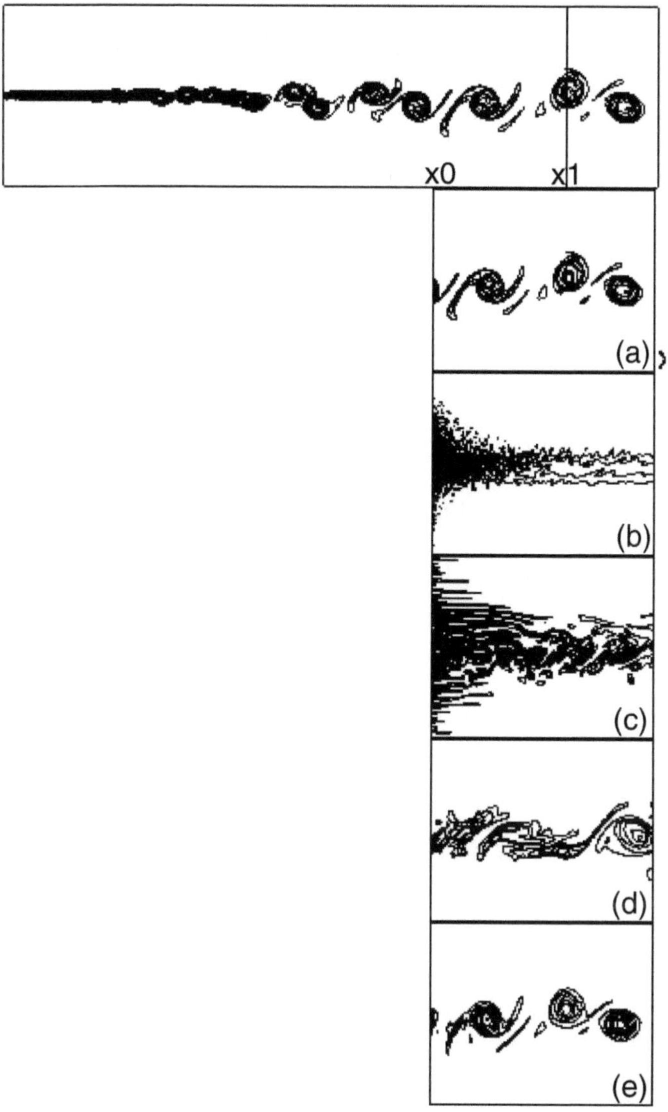

Fig. 10.16. Illustration of the influence of the turbulent inlet boundary condition (DNS of a 2D mixing layer). Iso-contours of instantaneous vorticity are shown. *Top:* reference 2D simulation. *Below:* Truncated simulation using as inflow conditions: a) exact instantaneous velocity field stored at the x0 section; b) random velocity fluctuations spatially and temporally uncorrelated (white noise) having the same Reynolds stress tensor components as in case (a); c) instantaneous velocity field preserving temporal two point correlation tensor of case (a); d) instantaneous velocity field preserving spatial two point correlation tensor of case (a); e) reconstructed velocity field with the aid of Linear Stochastic Estimation procedure from the knowledge of exact instantaneous velocity field at 3 reference locations (center of the mixing layer and $\pm\delta_\omega/2$ where δ_ω is the local vorticity thickness). Courtesy of Ph. Druault and J.P. Bonnet, LEA.

for shear flows (mixing layer, jet, boundary layer, and so forth). That is, the computations performed show the existence of a region in the computational domain in which the solution regenerates the space–time consistency specific to the Navier–Stokes equations [139]. The solution is not usable in this region, which can cover a large part[7] of the computational domain, and this entails an excess cost for the simulation. Also, it appears that this technique prevents the precise control of the dynamics of the solution, in the sense that it is very difficult to reproduce a particular solution for a given geometry.

A few ways to generate the random part of the inlet flow are now presented:

1. The Lee–Lele–Moin procedure (p. 356).
2. The Smirnov–Shi–Celik procedure and Batten's simplified version (p. 356).
3. The Li–Wang procedure (p. 358).
4. The Weighted Amplitude Wave Superposition procedure (p. 358).
5. The digital filter based method (p. 359).
6. The Arad procedure (p. 360).
7. The Yao–Sandham model (p. 361).

The LLM Procedure. The first method was proposed by Lee, Lele and Moin [432] for a flow evolving in the direction x and homogeneous in the two other directions, and statistically stationary in time. Assuming that the energy spectrum of a flow variable ϕ, $E_{\phi\phi}$, is prescribed in terms of frequency and two transverse wave numbers, the Fourier coefficients of the fluctuating part of ϕ are prescribed as

$$\hat{\phi}(k_y, k_z, \omega, t) = \sqrt{E_{\phi\phi}(k_y, k_z, \omega)} \quad \exp\left[\imath\psi_r(k_y, k_z, \omega, t)\right] \quad , \tag{10.74}$$

where ψ_r is the phase factor and $\imath = \sqrt{-1}$. The dependence of this phase factor on time and transverse wave numbers is necessary so that the signal generated is not periodic. The authors propose changing ψ_r only once in a given time interval T_r at a random instance by a random bounded amount $\Delta\psi_r$. The resulting signal is not continuous, and the frequency spectrum of the generated turbulence is not equal to $E_{\phi\phi}$.

The authors get satisfactory results for decaying isotropic turbulence by applying this procedure to each fluctuating velocity component u_i', but its application to advanced transitional flows or turbulent flows yields the occurrence of large non-physical transition regions.

The SSC Procedure. Another procedure was proposed by Smirnov, Shi and Celik [677] with application to wall-bounded flows. It involves scaling and orthogonal transformation operations applied to a continuous field generated as a superposition of harmonic functions.

[7] Numerical experiments show that this region can cover more than 50% of the total number of simulation points.

Let $R_{ij} = \langle u_i' u_j' \rangle$ be the (anisotropic) velocity correlation tensor at the inlet plane (see Appendix A for a precise definition). The first step of the SSC procedure consists of finding an orthogonal transformation tensor A_{ij} that would diagonalize R_{ij} (without summation over Greek indices)

$$A_{\alpha i} A_{\beta j} R_{ij} = \delta_{\alpha\beta} \lambda_\beta^2 \quad , \tag{10.75}$$

$$A_{ik} A_{kj} = \delta_{ij} \quad , \tag{10.76}$$

where the coefficients λ_1, λ_2 and λ_3 play the role of turbulent fluctuating velocities u_1', u_2' and u_3' in the new coordinate system. It is worth noting that both the transformation matrix and new coefficients are functions of space.

The second step consists of generating a transient flow-field $\boldsymbol{V}$ in a three-dimensional domain and rescaling it. This field is computed using the modified Kraichnan's method:

$$V_i(\boldsymbol{x}, t) = \sqrt{\frac{2}{N}} \sum_{n=1}^{N} \left[a_i^n \cos(\tilde{k}_j^n \tilde{x}_j + \omega_n \tilde{t}) + b_i^n \sin(\tilde{k}_j^n \tilde{x}_j + \omega_n \tilde{t}) \right] \quad , \tag{10.77}$$

with

$$\tilde{x}_j = \frac{x_j}{L_0}, \quad \tilde{t} = \frac{t}{T_0}, \quad c = \frac{L_0}{T_0}, \quad \tilde{k}_j^n = k_j^n \frac{c}{\lambda_j} \quad , \tag{10.78}$$

$$a_i^n = \epsilon_{ijm} \zeta_j^n k_m^n, \quad b_i^n = \epsilon_{ijk} \xi_j^n k_m^n \quad , \tag{10.79}$$

$$\zeta_j^n, \xi_j^n, \omega_n \in N(0,1), \quad k_i^n \in N(0, 1/2) \quad , \tag{10.80}$$

where L_0 and T_0 are the length- and timescales of turbulence, ϵ_{ijm} is the permutation tensor, and $N(M, \sigma)$ is a normal distribution with mean M and standard deviation σ. Quantities ω_n and k_j^n represent a sample of n frequencies and wave number vectors of the turbulent spectrum. In practice, the authors use the following model spectrum:

$$E(k) = 16 \sqrt{\frac{2}{\pi}} k^4 \exp(-2k^2) \quad . \tag{10.81}$$

The last step consists of applying scaling and orthogonal transformations to $\boldsymbol{V}$ to recover the synthetic fluctuating field in physical space:

$$W_\alpha = \lambda_\alpha V_\alpha, \quad u_i' = A_{ik} W_k \quad . \tag{10.82}$$

The resulting fluctuating field is nearly divergence-free, and has correlation scales L_0 and T_0 with the correlation tensor R_{ij}. This method was successfully applied to boundary-layer flows.

A simplified formulation is proposed by Batten et al. [49], which does not require the use of the orthogonal transformation. This simplification is achieved redefining the parameters λ_j as

$$\lambda_j = \sqrt{\frac{3 R_{lm} k_l^j k_m^j}{2 k_n^j k_n^j}} \quad , \tag{10.83}$$

and using the relation (10.77) to prescribe directly the velocity fluctuations $\boldsymbol{u}'$ instead of $\boldsymbol{V}$.

The Li–Wang Procedure. Another random generation technique for fluctuations in a boundary-layer flow was proposed by Li and Wang [446]. Fluctuations are reconstructed using the following equation:

$$\begin{aligned} u'(x,y,z,0) &= \sqrt{2}\sum_{l=1}^{N_1}\sum_{m=1}^{N_2}\sum_{n=1}^{N_3}\sqrt{E_u(\omega_{xl},\omega_{ym},\omega_{zn})\Delta\omega_x\Delta\omega_y\Delta\omega_z} \\ &\times \cos(\omega_{xl}' x + \omega_{ym}' y + \omega_{zn} z + \phi_{lmn}) \quad , \end{aligned} \tag{10.84}$$

where E_u is the target spectrum, ϕ_{lmn} a random phase with uniform distribution, and $\omega_{xl} = (l-1)\Delta\omega_x$ the angular frequency in the x direction. The periodicity is eliminated by defining

$$\omega_{xl}' = \omega_{xl} + \delta\omega_x \quad , \tag{10.85}$$

where $\delta\omega_x$ is a small random frequency. The time-evolving fluctuating field at the inlet plane is then reconstructed using Taylor's frozen turbulence hypothesis:

$$u'(0,y,z,t) = u'(x',y,z,0), \quad x' = U_c t \quad , \tag{10.86}$$

where U_c is an advecting velocity.

Weighted Amplitude Wave Superposition (WAWS) spectral representation method. Another procedure relying on modified random time series to generate velocity fluctuations was proposed by Glaze and Frankel [265]. This method, referred to as the WAWS method, is capable of simulating both spatial and temporal correlation. It is based on the regeneration of the fluctuating signal from its cross-spectral density at the inlet plane. As a result, both spatial correlation across the inlet plane and the power spectrum of each velocity component can be enforced.

Let M and N be the number of grid points at the inlet plane and the number of frequencies to be prescribed, respectively. Each velocity component is synthetized at the ith grid point using the following relation

$$u_i(t) = \sqrt{2}\sum_{m=1}^{i}\sum_{n=1}^{N} |H_{im}(\omega_n)|\sqrt{2\Delta_\omega}\cos(\omega_n' t + \theta_{im}(\omega_n) + \phi_{mn}) \quad , \tag{10.87}$$

with

$$\omega_n = \left(n - \frac{1}{2}\right)\Delta_\omega, \quad \omega_n' = \omega_n + \delta\omega_n \quad , \tag{10.88}$$

where $\Delta_\omega = \omega_u/N$, ω_u and $\delta\omega_n \in [-\Delta_\omega/20, +\Delta_\omega/20]$ are the frequency resolution, the maximum frequency and a small random perturbation, respectively. The parameters ϕ_{mn} are random phases perturbations uniformly distributed between 0 and 2π. The key parameters are the components of the transfer function matrix, $H_{im}(\omega_n)$, which are linked to the cross spectral density matrix components $S_{ij}(\omega_n)$ through the relation

$$S_{ij}(\omega_n) = H_{ik}(\omega_n)H^*_{kj}(\omega_n) \tag{10.89}$$

where H^* is the Hermitian transpose of H. Since the H matrix is not unique for a given S, a method should be chosen to compute it. Glaze and Frankel used a Cholesky decomposition to obtain a lower-triangular matrix, leading to simple calculations. The $\theta_{im}(\omega_n)$ parameter in (10.87) are defined writing the transfer function matrix components in polar form:

$$H_{im}(\omega_n) = |H_{im}(\omega_n)|e^{\imath\theta_{im}(\omega_n)} \quad . \tag{10.90}$$

Since the foreknowledge of the cross-spectral density matrix S for each velocity component is not a realistic requirement in practice, the next step consists in modeling it from available data. Noticing that it can be expressed as a function of one-point spectral density and the complex coherence function γ_{im}:

$$S_{im}(\omega_n) = \gamma_{im}\sqrt{S_{ii}(\omega_n)S_{mm}(\omega_n)} \quad , \tag{10.91}$$

the problem is equivalent to prescribing the power spectrum at each grid point and the coherence function. This is achieved using informations available on each flow.

The case of the near-field of a turbulent jet studied in [265] is given below as an illustration. The coherence function is estimated like follows

$$\gamma_{im} = \exp\left(-\frac{A(r_{im} + Br^2_{im})}{U_{im}}\right), \quad A = 1, \quad B = 4 \quad , \tag{10.92}$$

where r_{im} is the distance between points i and m and U_{im} is the average mean velocity between these points. The power spectrum is obtained using the von Karman model:

$$S_{ii}(f) = \frac{4\tilde{f}(u')^2}{f(1+70.8\tilde{f}^2)^{5/6}} \quad , \tag{10.93}$$

where u' is the rms turbulence intensity, f the frequency (in hertz), $\tilde{f} = L_u f/U$ the associated Strouhal number, with U the mean velocity and L_u the turbulent integral scale.

Digital Filter Based Method. Klein, Sadiki and Janicka [396] introduced a new approach based on signal modeling through the use of linear non-recursive filters. The general form of the discrete time series for the u velocity component at any grid point of the inflow plane is

$$u'_m = \sum_{n=-N}^{N} b_n r_{m+n} \quad , \tag{10.94}$$

where r_m is a series of random data with zero mean and such that $\langle r_m r_n \rangle = \delta_{mn}$ and b_n are the digital filter coefficients. The autocorrelation of the synthetized signal is

$$\frac{\langle u'_m u'_{m+k} \rangle}{\langle u'_m u'_m \rangle} = \frac{\sum_{j=-N+k}^{N} b_j b_{j+k}}{\sum_{j=-N}^{N} b_j^2} \quad . \tag{10.95}$$

The problem consists in inverting this relation to compute the coefficients of the digital filter associated to a given autocorrelation tensor. The authors propose to use a model autocorrelation function to obtain a simple form of the filter coefficients. They use the following form, which is valid for fully developed homogeneous turbulence (autocorrelation of the u' component in the direction associated to it):

$$R_{u'u'}(r, 0, 0) = \exp\left(\frac{\pi r^2}{4L^2}\right) \quad , \tag{10.96}$$

where L is a prescribed integral length scale. Combining this model autocorrelation function with relation (10.95) and setting $L = n\Delta x$, one obtains

$$\frac{\sum_{j=-N+k}^{N} b_j b_{j+k}}{\sum_{j=-N}^{N} b_j^2} = \exp\left(-\frac{\pi(k\Delta x)^2}{4(n\Delta x)^2}\right) = \exp\left(-\frac{\pi k^2}{4n^2}\right) \quad , \tag{10.97}$$

whose accurate approximate explicit solution is

$$b_k = \frac{a_k}{\sqrt{\sum_{j=-N}^{N} a_k^2}}, \quad a_k = \exp\left(-\frac{\pi k^2}{4n^2}\right) \quad . \tag{10.98}$$

This method yields the generation of a set of values with the targeted autocorrelation. Cross-correlations between velocity component can be enforced using the same change of variable as in the SSC procedure. Extension to the three-dimensional case straightforward, applying the procedure in sequentially in the three directions. This method was shown to give satifactory results in plane jet simulations.

Arad's Procedure. Arad [16] proposed a reconstruction technique for an initial condition based on physical observations related to the turbulence production process in a boundary layer. Assuming a general form of the perturbation corresponding to linearly unstable modes of the mean profile,

$$u'_i(x, y, z, t) = \hat{u}_i(z) \exp(\imath(\alpha x + \beta y - \omega t)), \quad \hat{u}_i(z) = A_i \phi \exp(-\gamma z^2) \quad , \tag{10.99}$$

where A_i is the amplitude of the mode and ϕ a random number. This idea of Arad is to design the fluctuations so that their growth rate will be maximized, resulting in a short unphysical transient region near the inlet plane.

Remarking that in a boundary layer the turbulence production term has the following form

$$P = \frac{\langle u'w' \rangle}{u_\tau^2} \frac{dU^+}{dz^+} \quad , \tag{10.100}$$

and taking into account the fact that quadrant Q2 (ejection: $u' < 0, w' > 0$) and Q4 (sweep: $u' > 0, w' < 0$) events have a dominant contribution to the shear stress $\langle u'w' \rangle$, Arad proposed introducing a phase shift between u' and w':

$$u'(x, y, z, t) = \hat{u}(z) \cos(\alpha x + \beta y - \omega t) \quad , \tag{10.101}$$

$$w'(x, y, z, t) = \hat{u}(z) \cos(\alpha x + \beta y - \omega t + \pi) \quad . \tag{10.102}$$

The spanwise component fluctuation, v', is assumed to be in phase with u' in Arad's work. The resulting fluctuations correspond to sweep and ejection events.

The Yao–Sandham Procedure. A more sophisticated procedure was proposed by Yao and Sandham [781, 644], which relies on the observations that fluctuations in the inner and outer parts of the boundary layer have different characteristic scales. As a consequence, specific disturbances are introduced in each part of the boundary layer. The inner-part fluctuations, $\boldsymbol{u}'^{\mathrm{inner}}$ are designed to represent lifted streaks with an energy maximum at $z_{p,j}^+$:

$$u_i'^{\mathrm{inner}}(y, z, t) = c_{ij} \exp(-z^+/z_{p,j}^+) \sin(\omega_j t) \cos(k_{y,j} y + \phi_j) \quad . \tag{10.103}$$

The outer-part fluctuation is assumed to be of the following form (with a peak at $z_{p,j}$)

$$u_i'^{\mathrm{outer}}(y, z, t) = c_{ij} \frac{z}{z_{p,j}} \exp(-z/z_{p,j}) \sin(\omega_j t) \cos(k_{y,j} y + \phi_j) \quad , \tag{10.104}$$

where subscripts $i = 1, 2$ and j are related to the velocity component and to the mode indices, respectively, and c_{ij} are constants. The $+$ superscript refers to inner coordinates (wall units). The ϕ_j are phase shifts, ω_j are forcing frequencies, and $k_{y,j}$ are spanwise wave numbers. These parameters are to be adjusted using information on the boundary-layer dynamics. The spanwise velocity component is deduced from the continuity constraint.

In the inner region of the boundary layer, it is assumed that the disturbances travel downstream for a distance of 1000 wall units at a convective velocity $U_c \approx 10\, u_\tau$, where u_τ is the friction velocity within a time period. The wave numbers $k_{y,j}$ are chosen such that there will be four streaks with a typical characteristic length of 100 wall units.

In the outer region, the downstream travelling distance is taken equal to 16 and the convection velocity is $U_c \approx 0.75\, U_\infty$, where U_∞ is the external velocity. The spanwise wave number is chosen to be of the order of the spanwise extent of the computational domain.

Yao and Sandham applied this procedure to a turbulent boundary layer, taking one mode in the inner region and three in the outer region. Corresponding parameters are given in Table 10.1. They also add a random noise with a maximum amplitude of 4% of the external velocity to prevent possible spurious symmetries.

Table 10.1. Coefficients of the four-mode Yao–Sandham model of fluctuations for boundary-layers.

	j	c_{1j}	c_{2j}	ω_j	$k_{y,j}$	ϕ_j	$z^+_{p,j}$	$z_{p,j}$
inner region	0	0.1	−0.0016	0.1	π	0.	12	–
outer region	1	0.3	−0.06	0.25	0.75π	0.	–	1.
outer region	2	0.3	−0.06	0.125	0.5π	0.1	–	1.5
outer region	3	0.3	−0.06	0.0625	0.25π	0.15	–	2.0

Deterministic Computation.

Precursor Simulation. One way of minimizing the errors is to perform a simulation of the upstream flow [740, 226, 649], called a precursor simulation, with a degree of resolution equivalent to that desired for the final simulation (see Fig. 10.17). This technique almost completely eliminates the errors encountered before, and offers very good results. On the other hand, it is hardly practical in the general case because it requires reproducing the entire history

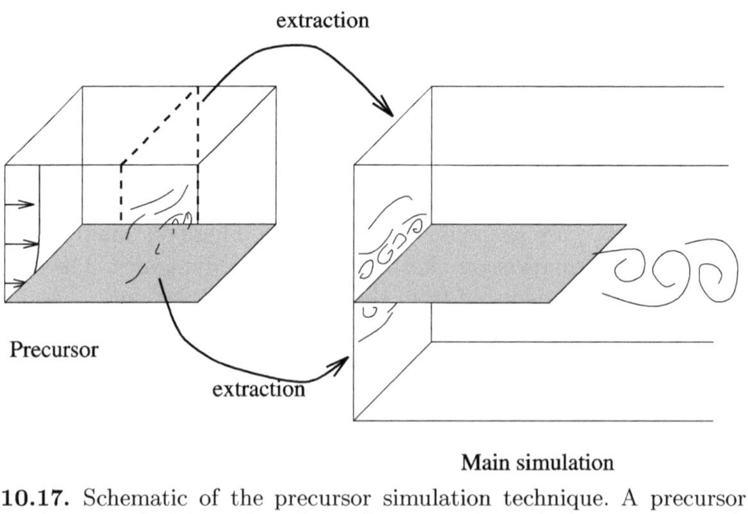

Fig. 10.17. Schematic of the precursor simulation technique. A precursor simulation of an attached boundary layer flow is performed. An extraction plane is defined, whose data are used as an inlet boundary condition for a simulation of the flow past a trailing edge.

of the flow which, for complex configurations, implies very high computation costs. Another problem stemming from this approach is that of causality: since the precursor is computed separately, no feedback of information from the second simulation is possible. This is a one-way coupling between two simulations that can become problematic when a signal (acoustic wave, for example) is emitted by the second.

Li et al. [445] reduced the cost of the precursor technique by storing the results of the precursor simulation over a (relatively) short time, and cycling in time over these data to generate the flow at the inlet plane of the main computation. In practice, the precursor results are stored over a time of the order of the integral timescale of the flow, and windowed to get a periodic signal. This technique has been applied to a plane mixing layer flow. Numerical results show that non-linear interactions quickly eliminate the spurious periodicity imposed at the inlet plane (about 25% of the total computational domain is contaminated). The efficiency of the method for flows with lower scrambling effects, such as wall-bounded flows, remains to be investigated.

Lund's Extraction/Rescaling Technique. Lund et al. [463] developed a variant of the precursor approach for boundary layers, in which the information at the inlet plane is produced from that contained in the computation. There is no longer any need for a precursor. The heart of the method is a means of estimating the velocity at the inlet plane, based on the velocity field extracted from the simulation on a plane downstream, as illustrated in Fig. 10.18.

The main difficulty arises from the fact that the mean flow is not parallel, i.e. the boundary layer thickness increases, and the flow at the extraction plane must first be rescaled before being used at the inlet plane.

The first step consists of decomposing the extracted flow, $\boldsymbol{u}^{\mathrm{e}}(\boldsymbol{x}, t)$, as the sum of a mean and a fluctuating part:

$$u_i^{\mathrm{e}}(x,y,z,t) = u_i^{\mathrm{e}\prime}(x,y,z,t) + U_i^{\mathrm{e}}(y,z) \quad . \tag{10.105}$$

The second step consists of rescaling the mean flow part using classical scalings related to the mean velocity profile of the turbulent boundary layer (see Sect. 10.2.1). In practice, the rescaling is carried out according to the law of the wall in the inner region and the defect law in the outer region of the boundary layer, leading to the following relations for the streamwise component:

$$U^{\mathrm{inner}} = u_\tau(x) f_1(z^+) \quad , \tag{10.106}$$
$$U_\infty - U^{\mathrm{outer}} = u_\tau(x) f_2(\eta) \quad , \tag{10.107}$$

where x is assumed to be the streamwise direction, z the wall-normal direction, U_∞ the external velocity, u_τ the friction velocity, $\eta = z/\delta$ the outer coordinate, δ the boundary-layer thickness, and f_1 and f_2 two universal functions to be determined. These two scaling laws dictate that the extracted mean

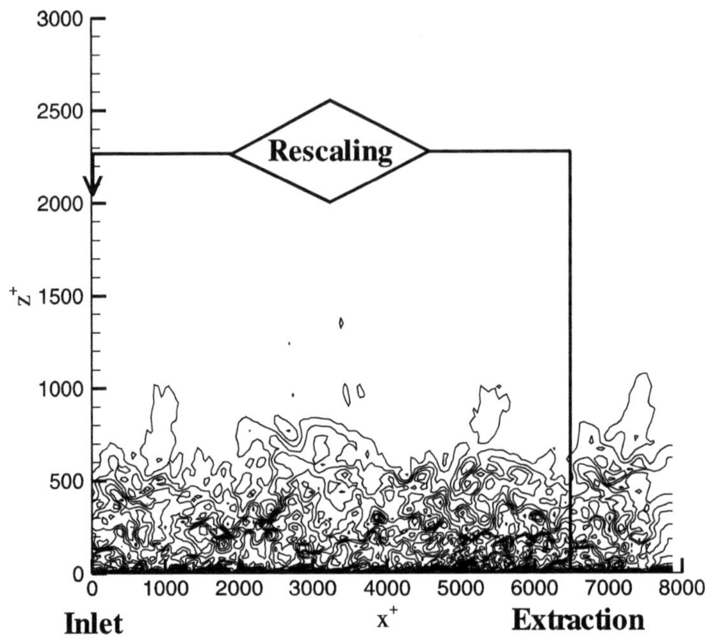

Fig. 10.18. Schematic of Lund's extraction/rescaling technique. Instantaneous isolevels of streamwise velocity in a boundary layer are shown. Courtesy of E. Tromeur and E. Garnier, ONERA.

velocity U^{e} and the rescaled mean velocity at the inflow, U^{r}, are related in the inner and outer regions via

$$U^{\mathrm{r,inner}} = \gamma U^{\mathrm{e}}(z^{+,\mathrm{r}}) \quad , \tag{10.108}$$

$$U^{\mathrm{r,outer}} = \gamma U^{\mathrm{e}}(\eta^{\mathrm{r}}) + (1-\gamma)U_{\infty} \quad , \tag{10.109}$$

with

$$\gamma = \frac{u_\tau^{\mathrm{r}}}{u_\tau^{\mathrm{e}}} \quad , \tag{10.110}$$

and where u_τ^{r} and u_τ^{e} are skin friction at the inlet plane and extraction plane, respectively, $z^{+,\mathrm{r}}$ is the inner coordinate computed at the inlet plane, and η^{r} is the external coordinate computed at the inlet plane. A linear interpolation is used between the grid points of the planes.

A similar technique is used to rescale the mean wall-normal component, yielding:

$$W^{\mathrm{r,inner}} = W^{\mathrm{e}}(z^{+,\mathrm{r}}) \quad , \tag{10.111}$$

$$W^{\mathrm{r,outer}} = W^{\mathrm{e}}(\eta^{\mathrm{r}}) \quad . \tag{10.112}$$

The third step consists of rescaling the fluctuating part of the instantaneous field:

$$u_i^{\prime,\mathrm{r,inner}} = \gamma u^{\prime,\mathrm{e}}(y, z^{+,\mathrm{r}}, t) \quad , \tag{10.113}$$

$$u_i^{\prime,\mathrm{r,outer}} = \gamma u^{\prime,\mathrm{e}}(y, \eta^{\mathrm{r}}, t) \quad . \tag{10.114}$$

The last step consists of writing a composite profile for the full instantaneous velocity at the inlet plane, $\boldsymbol{u}^{\mathrm{r}}$, that is approximately valid over the entire boundary layer. It is defined as a weighted average of the inner and outer profiles

$$u_i^{\mathrm{r}} = \left[U_i^{\mathrm{r,inner}} + u_i^{\prime,\mathrm{r,inner}}\right](1 - \beta(\eta^{\mathrm{r}})) + \left[U_i^{\mathrm{r,outer}} + u_i^{\prime,\mathrm{r,outer}}\right]\beta(\eta^{\mathrm{r}}) \quad , \tag{10.115}$$

with

$$\beta(\eta) = \frac{1}{2}\left[1 + \tanh\left(\frac{\alpha(\eta - b)}{(1 - 2b)\eta + b}\right)/\tanh(\alpha)\right] \quad , \tag{10.116}$$

where $\alpha = 4$ and $\beta = 0.2$.

This extraction/rescaling technique is observed to be efficient in practice, but must be used with care. The first point is that the extraction plane must be located far enough from the inlet to prevent spurious couplings in the computed solution. This constraint is satisfied by taking a distance between the two planes larger than the correlation length of the fluctuations in the streamwise direction. The second point is that it is valid for fully turbulent self-similar boundary layers only, and that the scaling laws must hold to obtain a relevant procedure.

Spille-Kohoff–Kaltenbach Method. The extraction/rescaling technique presented above suffers some lack of generality, because it relies on self-similarity assumptions and can introduce some spurious couplings inside the computational domain. A more general method was proposed by Spille-Kohoff and Kaltenbach [686], with application to a boundary-layer.

The core of the method is the definition of a buffer region, referred to as the control region, near the inlet plane, where a body force is adjusted in order to recover targeted profiles of turbulent fluctuations at a position located downstream of this buffer region. The body force is defined within the closed-loop control theory, and makes it possible, at least theoretically, to control both rms profiles and integral properties of the boundary layer. This procedure is illustrated in Fig. 10.19.

A random fluctuation is first specified at the inlet plane, together with a mean velocity profile. The body force is applied to the wall-normal velocity component only. The rationale for this is the observation that $-\langle w'w'\rangle dU/dz$

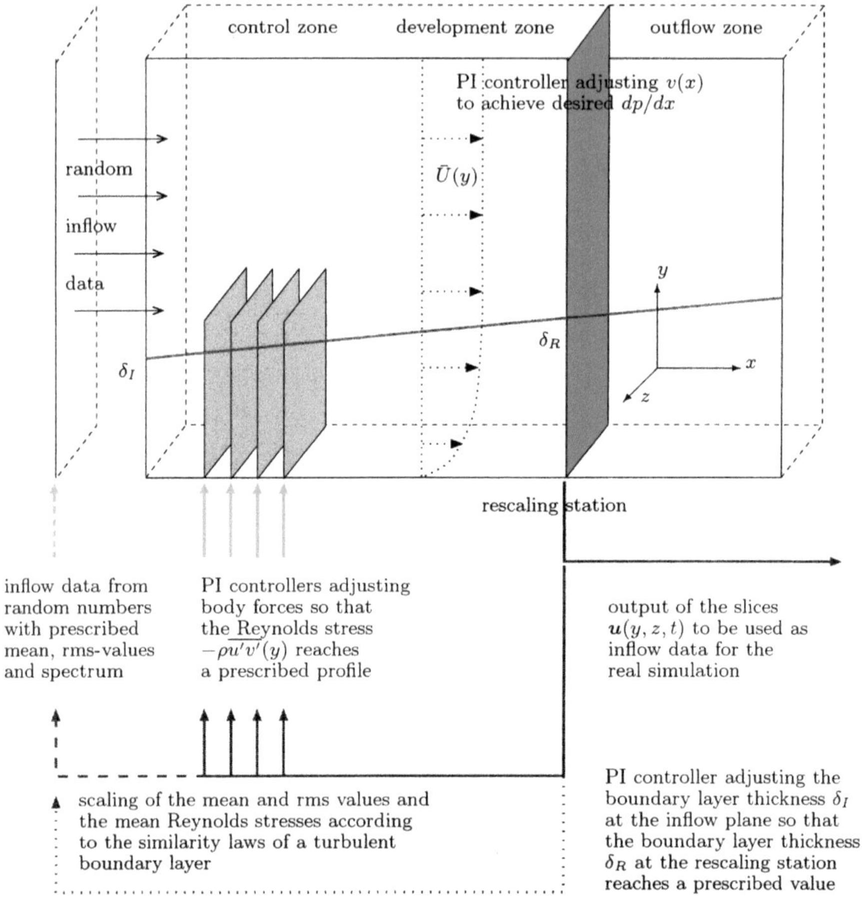

Fig. 10.19. Schematic of the Spille-Kohoff–Kaltenbach method. Courtesy of H. Kaltenbach, University of Berlin.

is the dominant production term in the balance equation for the shear stress $-\langle u'w' \rangle$. The amplitude of the body force at a streamwise position x_0 is adjusted via a PI controller in order to achieve a prescribed shear–stress profile $-\langle u'w' \rangle_{\text{target}}$. This target can be defined using experimental data or RANS simulations. Another way to define the targeted profile is to extract and rescale it from a position x_R downstream of the control region.[8]

The instantaneous body force in the plane $x = x_0$ is computed as follows

$$f(x_0, y, z, t) = A(z, t) \left[u(x_0, y, z, t) - \langle u \rangle_{y,t}(x_0, z) \right] \quad , \tag{10.117}$$

[8] It is important to remark that only mean profiles are extracted, and not instantaneous fields as in Lund's approach. This prevents the occurrence of spurious feedback.

where the amplitude $A(z,t)$ is given by

$$A(z,t) = \alpha e(z,t) + \beta \int_0^t e(z,t')dt' \quad , \tag{10.118}$$

where α and β are two arbitrary parameters. The averaging operator $\langle\,\rangle_{y,t}$ is associated with the average in the spanwise (homogeneous) direction and in time over a sliding window of width equal to $O(10)\delta/U_\infty$. The error term $e(z,t)$ is a measure of the difference between the computed and the prescribed shear stress:

$$e(z,t) = \langle u'w' \rangle_{\text{target}}(x_0, z) - \langle u'w' \rangle_{y,t}(x_0, z, t) \quad . \tag{10.119}$$

In order to prevent unphysically large values of the shear stress, the body force is applied only at grid points satisfying the following four instantaneous constraints:

$$|u'| < 0.6\, U_\infty, \quad |w'| < 0.4\, U_\infty, \quad u'w' < 0, \quad |u'w'| > 0.0015\, U_\infty^2 \quad . \tag{10.120}$$

If the extraction technique is used to specify target values inside the control region, another closed-loop controller is used to control the boundary-layer thickness at the inlet plane until the target value is obtained at the extraction plane. Similarly, the wall-normal velocity component at the top of the computational domain is controlled in order to obtain the desired streamwise pressure gradient. In this case, the authors observed that the adjustment time for reaching statistically steady values is of the order of $100\ \delta/U_\infty$.

Semideterministic Recontruction. Bonnet et al. [64, 199] propose an intermediate approach between the two previous ones, to recover the two-point correlations of the inflow with no preliminary computations. The signal at the inflow plane is decomposed in the form

$$\boldsymbol{u}(x_0, t) = U(x_0) + U_\text{c}(x_0, t) + \boldsymbol{u}'(x_0, t) \quad , \tag{10.121}$$

where $U(x_0)$ is the mean field, $U_\text{c}(x_0,t)$ the coherent part of turbulent fluctuations, and $\boldsymbol{u}'(x_0,t)$ the random part of these fluctuations. In practice, this last part is generated by means of random variables and the coherent part is provided by a dynamical system with a low number of degrees of freedom (like the POD, as seen in the Introduction), or by linear stochastic estimation, which gives access to the two-point correlations.

11. Coupling Large-Eddy Simulation with Multiresolution/Multidomain Techniques

11.1 Statement of the Problem

This chapter is devoted to the presentation of the coupling of large-eddy simulation with multiresolution and/or multidomain approaches. The main purpose of these couplings is to decrease the computational cost of the large-eddy simulations by clustering the degrees of freedom in regions of interest. The key idea is to adapt locally the cutoff length scale of the simulation, i.e. to refine the computational grid.[1] This grid refinement is associated with the definition of different subdomains with varying resolution.[2]

The methods proposed by various research groups can be classified as follows:

- Methods relying on fully overlapping subdomains (Sect. 11.2), as shown in Fig. 11.1. The term *fully overlapping* means here that the subdomain with the finest resolution is totally embedded within the coarsest resolution subdomain.

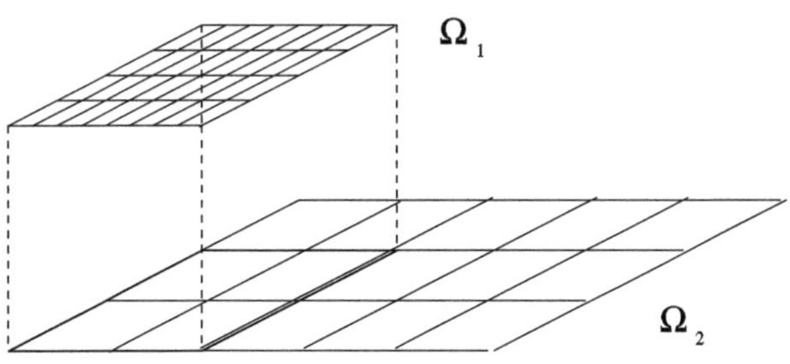

Fig. 11.1. Multiresolution decomposition with full overlap.

---

[1] This corresponds to h-adaptivity within the framework of h–p methods. We will focus on h-adaptivity only, because these are the most employed methods, even within the finite-element framework.

[2] The problem of mesh adaptation on unstructured grids for large-eddy simulation will not be discussed here, because it has not yet been treated.

The full overlap feature makes it possible to define two different strategies:
- Cycling between the different grid resolutions, which will be considered as a time-consistent extension of the multigrid acceleration technique for steady simulations. The emphasis is put here on the fact that the cycling strategies discussed below are not associated with convergence acceleration for implicit methods for unsteady simulations but are based on consistent time-integration at each grid level.
- Global resolution methods, in which time integration on all the subdomains is carried out at each time step.

- Methods with partial or no overlap between the subdomains (Sect. 11.3), as illustrated in Fig. 11.2.

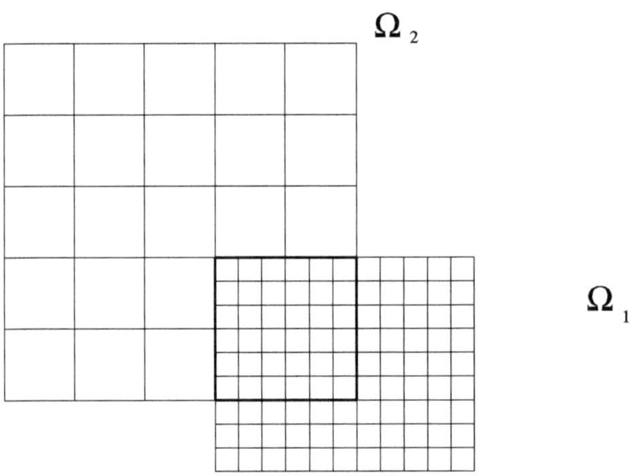

Fig. 11.2. Partial-overlap configuration.

From mathematical and physical points of view, the underlying problem can be interpreted as coupling two solutions obtained with different filter kernels and different cutoff length scales. An idealized problem with two domains is discussed below, which shows the full complexity of the problem.

Let us note G_1 and G_2, the filter kernels associated with the fine and coarse resolution levels, respectively. The associated cutoff length scales are $\overline{\Delta}_1$ and $\overline{\Delta}_2$. The domains with fine and coarse resolutions are referred to as Ω_1 and Ω_2, respectively.

The solution is decomposed as

$$\boldsymbol{u} = \overline{\boldsymbol{u}}^1 + \boldsymbol{u}_1' \text{ in } \Omega_1, \quad \boldsymbol{u} = \overline{\boldsymbol{u}}^2 + \boldsymbol{u}_2' \text{ in } \Omega_2 \quad . \tag{11.1}$$

The key problem is the transfer of information between Ω_1 and Ω_2. Let $\boldsymbol{v}_{12}$ be the complementary field defined as

$$\boldsymbol{v}_{12} = \overline{\boldsymbol{u}}^1 - \overline{\boldsymbol{u}}^2 = G_1 \star \boldsymbol{u} - G_2 \star \boldsymbol{u} = (G_1 - G_2) \star \boldsymbol{u} = G_{12} \star \boldsymbol{u} \quad , \tag{11.2}$$

where G_{12} is the restriction operator governing the transfer from the fine to the coarse resolution level.

The problem of interfacing the two domains is thus the following:

- Transfer from Ω_1 to Ω_2: restrict $\overline{\boldsymbol{u}}^1$ on Ω_2, or equivalently substract $\boldsymbol{v}_{12}$ from $\overline{\boldsymbol{u}}^1$. This corresponds to the definition of the new filtering operator G_1^2 such that
$$G_2 \star \boldsymbol{u} = G_1^2 \star G_1 \star \boldsymbol{u} \quad . \tag{11.3}$$
- Transfer from Ω_2 to Ω_1: defilter $\overline{\boldsymbol{u}}^2$ at the $\overline{\Delta}_1$ level, or equivalently add $\boldsymbol{v}_{12}$ to $\overline{\boldsymbol{u}}^2$.

It is worth noting that most methods presented in this chapter can be defined as static methods, because the number of levels of resolution is arbitrarily fixed before the computation. Results dealing with dynamic methods, in which the number of levels is not fixed but automatically adjusted, are very rare. Most advanced results dealing with the coupling of large-eddy simulation with Adaptive Mesh Refinement (AMR) are given in Sect. 11.4.

11.2 Methods with Full Overlap

According to the notation used in the previous section, the full overlap corresponds to $\Omega_1 \cap \Omega_2 = \Omega_1$.

The methods described in this section are:

- Methods with separate time-integration at each level. These methods are true multidomain methods, in the sense that the solution is integrated separately on each subdomain. The solutions are coupled via information transfer from time to time. The methods presented below are:
  1. The one-way coupling procedure of Khanna and Brasseur (Sect. 11.2.1). This method is the simplest one, and represents the minimal degree of coupling between the two domains, the coarse level solution being independent of the fine level solution.
  2. The two-way coupling procedure of Sullivan et al. (Sect. 11.2.2). This method can be seen as an extension of the previous one, because the information transfer is now taken into account in both ways: from fine to coarse level, and from coarse to fine level.
  3. The multilevel of Terracol et al. (Sect. 11.2.3), which is the most complete one. It relies on a two-way coupling, and also incorporates a dynamic cycling strategy. It is also the only one to incorporate a specific subgrid modeling for multilevel computations.[3]
- Methods with a single time-integration step. These methods are not true multidomain techniques, but rather multiblock methods. Time-integration

[3] This multilevel method is an extension of the multilevel closure presented in Sect. 7.7.7 (p. 271).

is carried out at the same time at each level, without any distinction between the different resolution levels. The example given below is the method proposed by Kravchenko et al. (Sect. 11.2.4), which is based on a Galerkin method with overlapping trial functions.

11.2.1 One-Way Coupling Algorithm

Khanna and Brasseur [365] developed a one-way coupling embedded grid technique for large-eddy simulation of atmospheric boundary layers. The fine resolution (i.e. fine grid) domain Ω_1 is located in the near-wall region, in order to permit a more accurate capture of details of the flow in that zone. The coarse resolution domain is noted Ω_2.

The solution is integrated in each domain independently for an arbitrary time. The coupling is enforced by imposing boundary conditions on Ω_1 using data coming from the coarse solution Ω_2.

The proposed boundary conditions are the following:

– Dirichlet conditions on the velocity components:

$$\overline{\boldsymbol{u}}^1\big|_{\partial\Omega_1} = \overline{\boldsymbol{u}}^2\big|_{\partial\Omega_1} + \boldsymbol{v}_{12} \quad , \tag{11.4}$$

where the coarse resolution field $\overline{\boldsymbol{u}}^2$ is obtained on the boundary of Ω_1, $\partial\Omega_1$ by a simple linear interpolation procedure. The complementary field $\boldsymbol{v}_{12}$ defined by relation (11.2) is here approximated by adding random noise perturbation following a $k^{-5/3}$ law within the spectral band $k \in [\pi/\overline{\Delta}_2, \pi/\overline{\Delta}_1]$.

– Rescaling of the subgrid viscosity. The second boundary condition is applied to the subgrid viscosity, which is rescaled in order to take into account the grid refinement. Assuming that both cutoffs are within the inertial subrange, (5.36) leads to the following rescaling law for the subgrid viscosity at the boundary

$$\nu_{\mathrm{sgs}}^1\big|_{\partial\Omega_1} = \left(\frac{\overline{\Delta}^1}{\overline{\Delta}^2}\right)^{4/3} \nu_{\mathrm{sgs}}^2\big|_{\partial\Omega_1} \quad , \tag{11.5}$$

where ν_{sgs}^1 and ν_{sgs}^2 are the values of the subgrid viscosity in Ω_1 and Ω_2, respectively.

In practice, the coupling is operated at the end of each time step.

11.2.2 Two-Way Coupling Algorithm

A two-way coupling procedure was proposed by Sullivan, McWilliams and Moeng [698] (also used in [62]). These authors consider a set of nested grids with increasing resolution for simulating the planetary boundary layer. The presentation of the method is restricted to a two-grid case for the sake of simplicity. The extension to an arbitrary number of grids is straightforward.

The coupling is achieved at each time step in the following way:

- From the coarse resolution level to the fine resolution level: boundary conditions on $\partial\Omega_1$ are obtained by interpolating the low-resolution field $\overline{\boldsymbol{u}}^2$ (see (11.4)). The difference between this and the one-way coupling algorithm presented above is that the complementary field $\boldsymbol{v}_{12}$ is now neglected.
- From the fine resolution level to the coarse resolution level: the numerical fluxes at the coarse resolution level are computed by filtering the fluxes computed at the fine resolution level on the overlap region. This can be written as follows:

$$\frac{\partial \overline{\boldsymbol{u}}^2}{\partial t} + G_1^2 \star \mathcal{NS}(\overline{\boldsymbol{u}}^1) = 0, \text{ in } \Omega_1 \cap \Omega_2 \quad , \tag{11.6}$$

where $\mathcal{NS}$ denotes the Navier–Stokes operator. Another possibility, as proposed by Manhart and Friedrich [483, 482] for direct numerical simulation, is to replace the coarse velocity field with the restricted finely resolved velocity field:

$$\overline{\boldsymbol{u}}^2 = G_1^2 \star \overline{\boldsymbol{u}}^1 \text{ in } \Omega_1 \cap \Omega_2 \quad . \tag{11.7}$$

The data transfer is done at each time step.

11.2.3 FAS-like Multilevel Method

We now discuss the most general approaches, which can be seen as generalizations of the Full Approximation Scheme (FAS) classical multigrid acceleration technique. The method proposed by Terracol et al. [633, 711, 710, 712] appears as the most general one. It deals with the use of an arbitrary number of nested resolution levels, and the gain in computational time is optimized by considering a self-adaptive time-cycling strategy between the different levels.

Other FAS-type methods have been proposed by Voke [736], Tziperman et al. [722] and Liu et al. [452, 453].

Using the multilevel framework developed in Sect. 7.7.7 (p. 271), the governing equations at each resolution level can be expressed as

$$\frac{\partial \overline{\boldsymbol{u}}^n}{\partial t} + \mathcal{NS}(\overline{\boldsymbol{u}}^n) = -\tau^n = -[G_n\star, \mathcal{NS}](\boldsymbol{u}), \quad n = 1, 2 \quad . \tag{11.8}$$

Equation (11.8) shows that the coupling between the different levels is theoretically achieved through the generalized subgrid term which arises on the right-hand side. This is seen by decomposing the commutation error at the coarse resolution level as

$$\begin{aligned} \tau^2 &= [G_2\star, \mathcal{NS}](\boldsymbol{u}) \\ &= G_1^2 \star G_1 \star \mathcal{NS}(\boldsymbol{u}) - \mathcal{NS}(G_1^2 \star G_1 \star \boldsymbol{u}) \\ &= G_1^2 \star (\mathcal{NS}(G_1 \star \boldsymbol{u}) + [G_1\star, \mathcal{NS}](\boldsymbol{u})) - \mathcal{NS}(G_1^2 \star \overline{\boldsymbol{u}}^1) \\ &= G_1^2 \star \mathcal{NS}(\overline{\boldsymbol{u}}^1) + G_1^2 \star \tau^1 - \mathcal{NS}(\overline{\boldsymbol{u}}^2) \quad . \end{aligned} \tag{11.9}$$

The first and last terms on the right-hand side of (11.9) correspond to the direct coupling term between the two levels of resolution. It is worth noting that it is equivalent to the so-called forcing function in FAS multigrid methods. The remaining term represents the coupling between unresolved scales ("true" subgrid scales) and the coarsest level of resolution. It is the only term which requires a physical modeling, all the other terms being directly computable.

On the grounds of the preceding developments, Terracol proposed a multilevel algorithm whose main elements are:

- The subgrid term at level $\overline{\Delta_2}$ is computed according to relation (11.9). The subgrid term at the finest resolution level can be computed with any subgrid model. In Liu's multigrid approach, the subgrid term is replaced by a numerical stabilization term, which is essentialy equivalent to low-pass filtering.
- Time cycling is optimized by freezing the velocity field at the finest level during a time T. During this period, time integration is carried out at the coarse resolution level only. The coupling from the fine to the coarse level is carried out when evaluating τ^2. The coupling from the coarse to the fine level is performed by refreshing the low-frequency part of $\overline{\boldsymbol{u}}^1$ at the end of the period T using the new value of $\overline{\boldsymbol{u}}^2$:

$$\overline{\boldsymbol{u}}^1(t+T) = \overline{\boldsymbol{u}}^2(t+T) + \boldsymbol{v}_{12}(t) \quad . \tag{11.10}$$

The time scale T is evaluated in order to satisfy the following constraint:

$$T \left| \frac{\partial |\boldsymbol{v}_{12}|^2}{\partial t} \right|_2 \leq \epsilon_{\max} \left| \overline{\boldsymbol{u}}^1 \cdot \overline{\boldsymbol{u}}^1 \right|_2 \quad , \tag{11.11}$$

where $\epsilon_{\max}$ is a prescribed error tolerance. Typical values for this parameter range from 10^{-4} to 10^{-3}. The resulting self-adaptive cycling obtained in a plane mixing layer configuration is illustrated in Fig. 11.3. These bounds are experimentally observed to allow a decrease of the cost by a factor N for an N-grid simulation for both wall-bounded and free shear flows. Voke's and Liu's methods are based on static cycling strategies, with empirically determined integration times at each level. Typical results are illustrated in Fig. 11.4, which displays the computed time evolution of the momentum thickness of the mixing layer obtained with different subgrid closures at the finest resolution level.

11.2.4 Kravchenko et al. Method

A zonal embedded grid technique for wall bounded flows was proposed by Kravchenko et al. [411]. Contrary to the other methods with full overlap presented in this section, it does not rely on a separated time integration

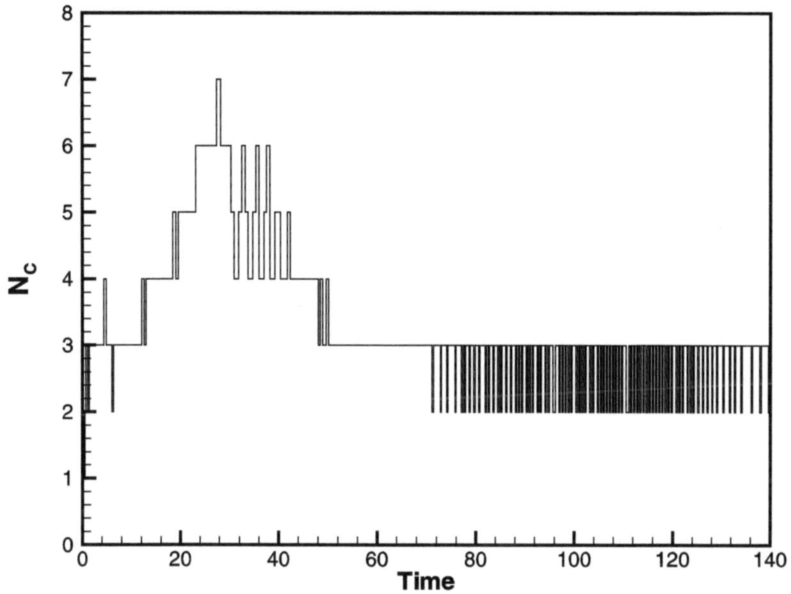

Fig. 11.3. Terracol's self-adaptive time-cycling algorithm. Two-grid simulation of a plane mixing layer flow. Time history of the number of consecutive time-steps on the coarse grid. Courtesy of M. Terracol, ONERA.

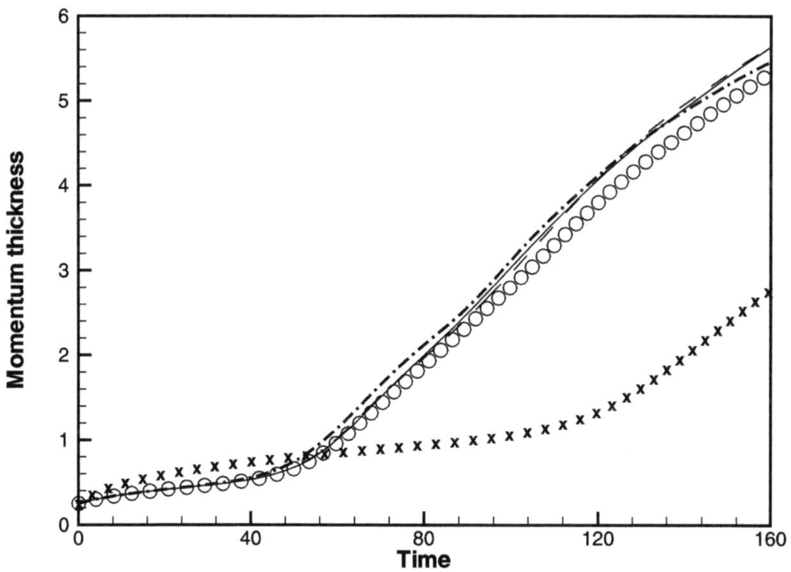

Fig. 11.4. Large-eddy simulation of a plane mixing layer using Terracol's self-adaptive time-cycling algorithm with two resolution levels. Time evolution of the momentum thickness for different subgrid closures at the finest resolution level. Crosses correspond to classical large-eddy simulation on the coarse level. Courtesy of M. Terracol, ONERA.

on each grid. The main element of the method is the use of a Galerkin-type numerical method with wide-stencil trial and weighting functions (fully non-local, such as Fourier basis, or B-splines). The transfer of information is intrinsically achieved thanks to the fact that the basis functions are not orthogonal.

In fact, this method can be interpreted as a particular case of the implementation of Galerkin methods on unstructured grids.

11.3 Methods Without Full Overlap

We now describe the methods without full overlap, i.e. methods designed for the case where the fine-resolution domain Ω_1 is not totally embedded into the coarse-resolution domain Ω_2: $\Omega_1 \cap \Omega_2 \neq \Omega_1$. The partial overlap precludes the use of multigrid-type algorithms like those presented in the previous section, and time integration must be carried synchronously in each domain. The data transfer is done at each time step (or substep for multistep schemes) in order to provide refreshed boundary conditions to the subdomains.

Quéméré et al. [612] carried out an extensive analysis of the interfacing problem between subdomains with different resolutions within the large-eddy simulation framework. The main results of their analysis are the following:

- From fine to coarse resolution subdomain: boundary conditions are obtained by filtering the data (point values or fluxes):

$$\boldsymbol{u}^2\big|_\Gamma = G_1^2\, \boldsymbol{u}^1\big|_\Gamma \,, \text{ and/or } G_2 \star \mathcal{NS}(\boldsymbol{u})|_\Gamma = G_1^2 \star G_1 \star \mathcal{NS}(\boldsymbol{u})|_\Gamma \quad , \tag{11.12}$$

  with

$$\Gamma = \partial\Omega_1 \cap (\Omega_1 \cap \Omega_2) \quad . \tag{11.13}$$

- From coarse to fine resolution subdomain: the definition of boundary conditions for $\overline{\boldsymbol{u}}^1$ requires two successive steps:
  1. Interpolation of $\overline{\boldsymbol{u}}^2$ at the fine resolution level $\overline{\Delta}_1$. This step is easy to implement, and does not induce specific theoretical problems.
  2. Reconstruction of the complementary field $\boldsymbol{v}_{12}$. If Γ corresponds to an exit boundary for the subdomain Ω_1, numerical experiments show that the enrichment step is not mandatory and $\boldsymbol{v}_{12} = 0$ can be used. On the contrary, if Γ is an inflow boundary condition for Ω_1, the reconstruction of $\boldsymbol{v}_{12}$ is necessary, and the enrichment step is theoretically equivalent to the definition of turbulent inflow conditions (see Sect. 10.3). All the methods proposed for generating inflow turbulence can be used to predict $\boldsymbol{v}_{12}$. A simple extrapolation procedure from the interior of Ω_1 can also be used, if the elementary convection length $U_c \Delta t$ (with U_c the characteristic advection velocity across Γ and Δt the time step) is much smaller than the local correlation length scale of turbulent fluctuations.

Quéméré et al. emphasized the fact that a weighted extrapolation can be used to account for the spatial variation of the mean Reynolds stress profiles.

3. Rescaling of the subgrid models. The coupling can also be completed by enforcing some compatibility conditions on the subgrid terms at the interface. Inertial range arguments make it possible to derive proportionality factors. The following technique for rescaling subgrid viscosity models was proposed in [612]. Starting with the classical relations

$$\nu_{\rm sgs}^1 \propto \overline{\Delta}_1 \sqrt{\boldsymbol{u}_1' \cdot \boldsymbol{u}_1'}, \quad \nu_{\rm sgs}^2 \propto \overline{\Delta}_2 \sqrt{\boldsymbol{u}_2' \cdot \boldsymbol{u}_2'} \quad , \tag{11.14}$$

and assuming the orthogonality property

$$\boldsymbol{u}_2' \cdot \boldsymbol{u}_2' = (\boldsymbol{u}_1' + \boldsymbol{v}_{12}) \cdot (\boldsymbol{u}_1' + \boldsymbol{v}_{12}) \approx \boldsymbol{u}_1' \cdot \boldsymbol{u}_1' + \boldsymbol{v}_{12} \cdot \boldsymbol{v}_{12} \quad , \tag{11.15}$$

we obtain the following relation

$$\nu_{\rm sgs}^2 \propto \left(\frac{\overline{\Delta}_2}{\overline{\Delta}_1} \right) \overline{\Delta}_1 \sqrt{\boldsymbol{u}_1' \cdot \boldsymbol{u}_1' + \boldsymbol{v}_{12} \cdot \boldsymbol{v}_{12}} \quad . \tag{11.16}$$

From this last relation, we deduce a scaling law between the two viscosities

$$\left(\frac{\nu_{\rm sgs}^2}{\nu_{\rm sgs}^1} \right)^2 \propto \left(\frac{\overline{\Delta}_2}{\overline{\Delta}_1} \right)^2 \left(1 + \frac{\boldsymbol{v}_{12} \cdot \boldsymbol{v}_{12}}{\boldsymbol{u}_1' \cdot \boldsymbol{u}_1'} \right) \quad , \tag{11.17}$$

which can be used at the interface.

11.4 Coupling Large-Eddy Simulation with Adaptive Mesh Refinement

11.4.1 Statement of the Problem

The theoretical advantage of coupling with Adaptive Mesh Refinement (AMR) is twofold:

- Decrease in the mesh size Δx results in a reduction of the discretization error, yielding an improved numerical accuracy.
- If the cutoff length $\overline{\Delta}$ is tied to Δx, mesh refinement will also induce a decrease in $\overline{\Delta}$ and make the subgrid model less influential on the results (since a larger part of the exact solution is directly captured).

The philosophy of the AMR approach is that the grid refinement must be performed automatically during the computation, leading to the definition of several open problems:

- Finding an error estimate (the global error or each error source individually) and defining a criterion (usually a threshold value) that will trigger the refinement.
- Finding a bound for the algorithm: since projection error, discretization error and modeling error are present, an accurate error sensor should detect all of them, and a consistent unbounded AMR algorithm will converge toward a Direct Numerical Simulation, leading to very high computational cost. The method must then be bounded in the sense that the final solution should correspond to a large-eddy simulation with potentially large projection error. This bound can be imposed in several ways: by fixing a maximum number of degrees of freedom, by giving a minimum mesh size or playing on the threshold level in the evaluation of the error.

11.4.2 Error Estimation

The criterion used to refine the grid is based on the definition of an error estimate. Therefore, it relies on an a priori choice dealing with the quantities whose accuracy is the most important for the considered application. This is an arbitrary, user-dependent decision.

Since it should account for both the numerical and modeling errors, the error estimate is closely tied to the numerical method and the subgrid model used in the simulation. The optimal choice is therefore case dependent, and reveals a certain degree of empiricism.

The most striking achievements have been obtained by Hoffman [314, 310, 312, 313] using an a posteriori error estimate within a finite element framework. Less developed methods [149, 518, 268] will not be described here. The definition of an error estimate brings in many questions dealing with mathematical analysis that will not be mentioned here. The interested reader can refer to the original publications and the references given therein for more details. The presentation given below will be restricted to the main features of the method.

The purpose is to improve the accuracy of the simulation by error control of the quantity

$$err(\boldsymbol{u}_{\Pi} - \boldsymbol{u}_{\mathrm{d}}) = \int_Q (\boldsymbol{u}_{\Pi} - \boldsymbol{u}_{\mathrm{d}}) \cdot \psi dt d\boldsymbol{x} \quad , \tag{11.18}$$

where ψ is a vectorial test function and the integration domain Q is defined as $Q = \Omega \times I$, where Ω is the volume of the computational domain and I a time interval. The two velocity fields u_{Π} and u_{d} are related to the ideal solution in which only the projection error is present and the computed solution including modeling and discretization errors, respectively. The aim of an AMR procedure based on (11.18) is therefore to eliminate modeling and discretization errors. This definition of the error is the most general one, and allows the definition of optimal control methods.

Hoffman introduces the following linearized dual problem to define an optimal error control method:

$$-\frac{\partial \boldsymbol{u}_\phi}{\partial t} - \boldsymbol{u}_\Pi \cdot \nabla \boldsymbol{u}_\phi + \nabla \boldsymbol{u}_{\mathrm{d}} \cdot \boldsymbol{u}_\phi + \nabla p_\phi - \nu \nabla^2 \boldsymbol{u}_\phi = \psi \quad , \tag{11.19}$$

$$\nabla \cdot \boldsymbol{u}_\phi = 0 \quad , \tag{11.20}$$

where $\boldsymbol{u}_\phi$ and p_ϕ are the dual variables. The dual system is supplemented with adequate boundary and initial conditions which account for the definition of the physical quantity of interest (drag/lift of a body, vortex shedding frequency, ...). The exact projected field $\boldsymbol{u}_\Pi$ being unknown, it is replaced in practice by $\boldsymbol{u}_{\mathrm{d}}$. Using these new variables, the error can be expanded as

$$\begin{aligned} err(\boldsymbol{u}_\Pi - \boldsymbol{u}_{\mathrm{d}}) \quad = \quad & \underbrace{\int_Q (\tau(\boldsymbol{u}_{\mathrm{d}}) - \nu \nabla \boldsymbol{u}_{\mathrm{d}}) : \nabla \boldsymbol{u}_\phi d\boldsymbol{x} dt}_{\text{discretization error}} \\ & \underbrace{+ \int_Q \left(-\frac{\partial \boldsymbol{u}_{\mathrm{d}}}{\partial t} + \boldsymbol{u}_{\mathrm{d}} \cdot \nabla \boldsymbol{u}_{\mathrm{d}} \right) \cdot \boldsymbol{u}_\phi d\boldsymbol{x} dt}_{\text{discretization error}} \\ & \underbrace{+ \int_Q (\nabla \cdot \boldsymbol{u}_{\mathrm{d}}) p_\phi + p_{\mathrm{d}} (\nabla \cdot \boldsymbol{u}_\pi) d\boldsymbol{x} dt}_{\text{discretization error}} \\ & \underbrace{+ \int_Q (\Pi(\tau(u)) - \tau(\boldsymbol{u}_{\mathrm{d}})) : \nabla \boldsymbol{u}_\phi d\boldsymbol{x} dt}_{\text{modeling error}} \quad , \end{aligned} \tag{11.21}$$

where $\tau(\boldsymbol{u}_{\mathrm{d}})$ denotes the modeled subgrid tensor and $\Pi(\tau(u))$ the projection of the exact subgrid tensor onto the considered basis of degrees of freedom. This expression enlights the definition of both the modeling and the numerical errors. The discretization error is a closed quantity which can be used in a straightforward way, to the contrary of the modeling error, which requires the knowledge of the projection of the exact subgrid tensor and therefore appears as an unclosed quantity. To close this equation, Hoffman proposes to use a structural subgrid model which exhibits a high degree of correlation with the exact subgrid tensor, like the Bardina model or approximate deconvolution models. More accurate formula including interpolation errors on the dual variables can be derived, which are not presented here for the sake of simplicity.

It is worth noting that the error estimate is based on a volume integral, even in the case where the physical quantities can be expressed as surface integrals (e.g. drag, lift). This volumic formulation makes it possible to refine the grid in regions where the error originates.

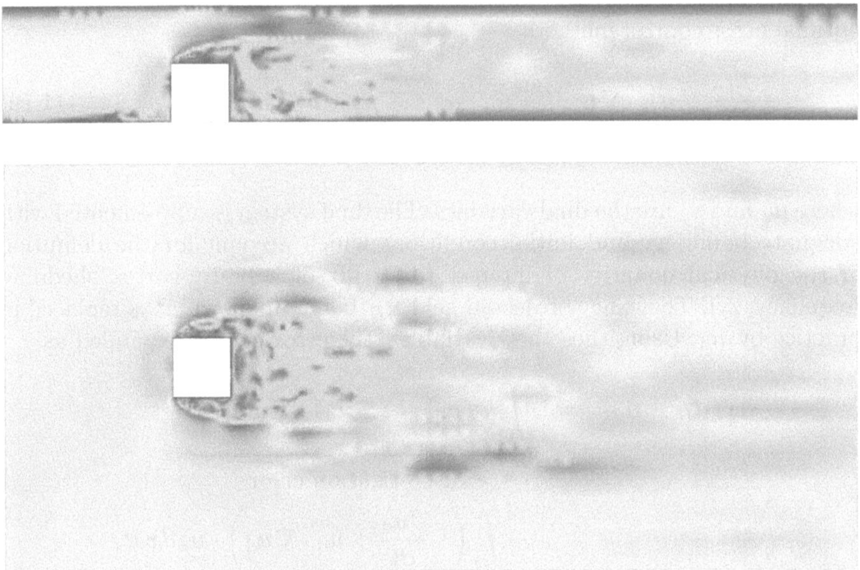

Fig. 11.5. LES-AMR simulation of the flow around a surface-mounted cube. View of the instantaneous flow *Top:* (x–y) plane: *Bottom:* (x–z) plane. Courtesy of J. Hoffman, Courant Institute.

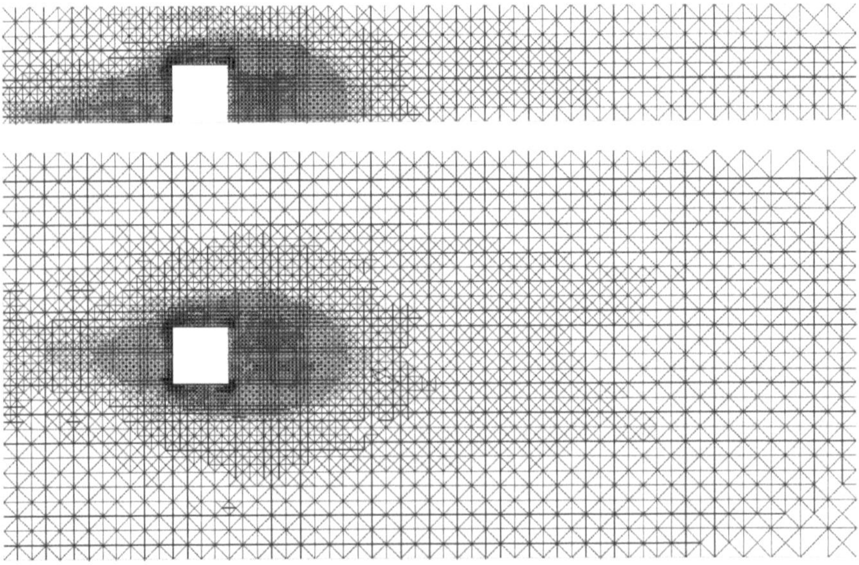

Fig. 11.6. LES-AMR simulation of the flow around a surface-mounted cube. View of the grid after 9 refinement steps. Courtesy of J. Hoffman, Courant Institute .

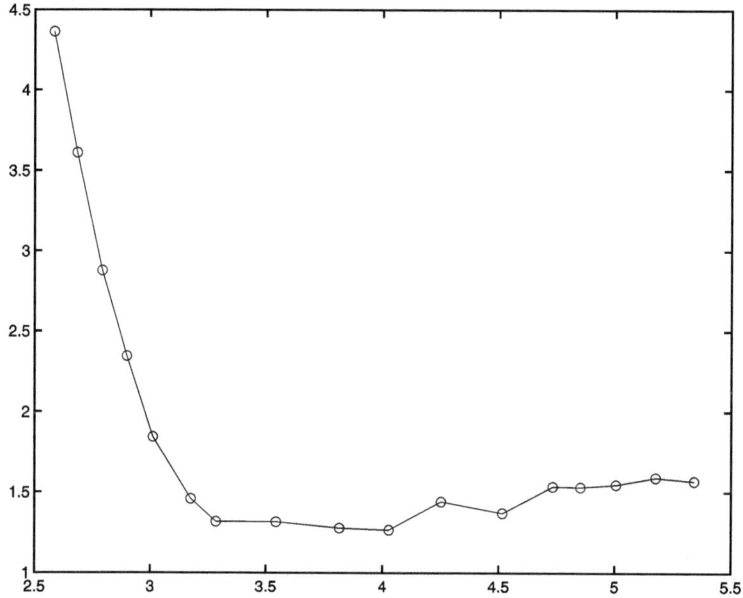

Fig. 11.7. LES-AMR simulation of the flow around a surface-mounted cube. Computed cube drag versus the number of grid points (in arbitrary units). Courtesy of J. Hoffman, Courant Institute .

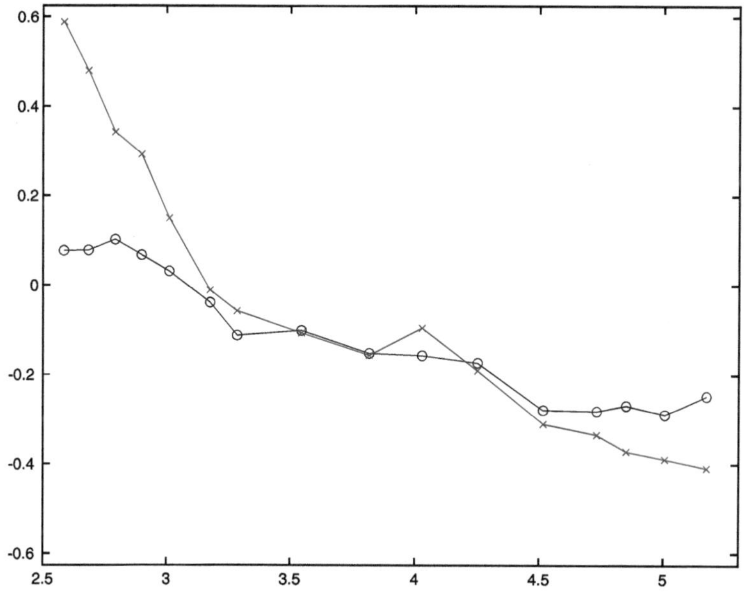

Fig. 11.8. LES-AMR simulation of the flow around a surface-mounted cube. Discretization (*Circle*) and modelling (*Cross*) errors (as defined in 11.21) versus the number of grid points (in arbitrary units). Courtesy of J. Hoffman, Courant Institute.

The resulting algorithm is

1. Compute the solution $(\boldsymbol{u}_{\mathrm{d}}, p_{\mathrm{d}})$ solving the large-eddy simulation governing equation over a time interval I.
2. Solve the dual problem (11.19) – (11.20) to obtain $(\boldsymbol{u}_{\phi}, p_{\phi})$ over I and compute the discretization and modeling error according to (11.21).
3. Refine the grid in zones where one or both error sources are higher than a fixed threshold, taking into account possible bounds.
4. Loop until the targeted level of accuracy is reached, i.e. until $err(\boldsymbol{u}_{\Pi} - \boldsymbol{u}_{\mathrm{d}})$ is lower than a required value.

Good results were obtained dealing with the prediction of aerodynamic forces on bluff bodies, using a finite-element method on unstructured meshes. Typical results obtained dealing with the prediction of the drag of a surface-mounted cube are displayed in Figs. 11.5 – 11.8.

12. Hybrid RANS/LES Approaches

12.1 Motivations and Presentation

This chapter is devoted to the presentation of hybrid RANS/LES methods. The main motivation for hybridizing the two methods is to decrease the cost of the traditional large-eddy simulation method, which is large because of:

- the requirement to directly capture all the scales of motion responsible for turbulence production;
- the observed inability of most subgrid models to correctly account for anisotropy and disequilibrium.

These two weaknesses lead to the use of very-fine-resolution meshes, which can be a real problem if the characteristic lengthscale of turbulence-production events is a decreasing function of the Reynolds number. A famous problematic example is the inner region of boundary layers, whose intrinsic scale is the wall unit (see Sect. 10.2).

In order to alleviate this problem, a possible solution is to blend large-eddy simulation with another technique which must be able to provide relevant lower-frequency solutions at a much lower cost. A natural proposition is to use the RANS approach, which relies on a statistical average of the exact solution and leads to a very large reduction of the number of degrees of freedom in comparison with large-eddy simulation.

The presently existing techniques can be classified as follows:

- *Zonal decomposition* (Sect. 12.2): the global computational domain is divided into subdomains, some of them being treated with the RANS method, the other ones being computed using large-eddy simulation. The gain comes from the fact that the grid resolution can be coarser in the RANS subdomains. Further cost reduction can be obtained when one or two spatial directions can be suppressed in the RANS subdomains thanks to statistical homogeneity of the flow in these directions.
- *Nonlinear Disturbance Equations* (Sect. 12.3): the idea here is to compute the mean flowfield or the low-frequency part of the solution using a RANS or unsteady RANS simulation, and to reconstruct the missing part of the fluctuating field using a large-eddy-type simulation. This approach can be interpreted as a zonal multidomain approach in the frequency domain.

– *Universal modeling* (Sect. 12.4): the subgrid model is replaced by a new model, which appears as a generalized turbulence model defined as a combination of a RANS model and a typical subgrid model. The hope here is to obtain robust subgrid models, which are able to deal with very coarse grids similar to those used in unsteady RANS computation. The merging with a RANS model is expected to introduce more physics into the subgrid model, rendering it efficient if the cutoff is located in the low-frequency part of the spectrum, outside or at the very beginning of the inertial range.

12.2 Zonal Decomposition

12.2.1 Statement of the Problem

The zonal decomposition approach can be recast as a generalized multidomain/multiresolution problem. The main difference with the multidomain methods presented in Chap. 11 is that the coupling is now performed between subdomains where the scale separation is performed using operators of a different nature: filtering in large-eddy simulation and statistical averaging in RANS.

Equations presented in Sect. 11.1 can be reused in the present framework, keeping in mind that now G_2 is related to statistical averaging:

$$\overline{\phi}^2 \equiv G_2(\phi) = \frac{1}{N} \sum_{i=1,N} \phi_i = \langle \phi \rangle \quad , \tag{12.1}$$

where N is the number of samples chosen to compute the statistical average, while G_1 is still defined as a convolution filter.

All the theoretical analyses presented in Sect. 11.1 can be directly extended to the RANS/LES coupling method. The main practical difference is that now the problem of the reconstruction of the complementary field $\boldsymbol{v}_{12}$ is strictly equivalent to that of the definition of turbulent inflow conditions for large-eddy simulation described in Sect. 10.3. The mean flow is now predicted using the RANS computation.

Another pratical difference is the definition of the restriction operator G_1^2, which is defined so that:

$$G_1^2(G_1 \star \boldsymbol{u}) = G_2(\boldsymbol{u}) \quad , \tag{12.2}$$

or, equivalently, and using the usual notation:

$$G_1^2(\overline{\boldsymbol{u}}^1) = \langle \boldsymbol{u} \rangle \quad . \tag{12.3}$$

From a purely theoretical point of view, we see that G_1^2 must be defined as the sequential application of a deconvolution operator and statistical averaging. It can be simplified as a simple statistical average when the mean velocity profiles of the exact and filtered solutions are equal.[1]

[1] This condition is satisfied if the filter is applied in homogeneous directions only.

Two main approaches are identified:

- Sharp transition (Sect. 12.2.2): the reconstruction of the complementary field is carried out explicitly at the interface between RANS and LES subdomains, yielding a sharp transition over one grid mesh from one solution to the other.
- Smooth transition (Sect. 12.2.3): the reconstruction of $\boldsymbol{v}_{12}$ is not performed at the interface, and the high-frequency part of the fluctuating field must be regenerated by instabilities of the mean flow and the forward energy cascade. This approach leads to the existence of a transition region near the interface, whose width is case-dependent.

12.2.2 Sharp Transition

A method based on the sharp transition approach was proposed by Quéméré et al. [610], and is an extension of the multidomain method proposed by the same authors (see Sect. 11.3). It relies on a strict analogy between the reconstruction of the complementary field $\boldsymbol{v}_{12}$ and the definition of the turbulent inflow condition.

The G_1^2 restriction operator is simply computed as a statistical average, without incorporating a defiltering operator. Several techniques for reconstructing the fluctuations at the interface have been implemented (listed in decreasing order of efficiency): use of a predictor simulation, Lund's extraction/rescaling technique, and random perturbation. Numerical experiments show that the smaller the transition region near the interface the more realistic the fluctuating field option becomes. When a predictor simulation is implemented, it is reduced to one grid cell.

The method was applied with varying success to plane channel flow and flow around a blunt trailing edge. The RANS model was the Jones–Launder k–ε model. The boundary conditions for the turbulent quantities at the interface were derived from a simple extrapolation procedure. Typical results for a plane channel flow configuration are presented in Figs. 12.1 and 12.2.

A simplified version of Quéméré's approach was implemented by Georgiadis et al. [241], who operated the switch from RANS to LES while neglecting the reconstruction of the complementary field. The purpose was the simulation of a mixing layer flow, in which the primary instabilities have a very large growth rate, resulting in a small influence of the exact definition of the complementary field.

Applications of this approach to the definition of sophisticated two-layer wall models for large-eddy simulation have also been proposed by Davidson et al. [167] and Diurno et al. [180]. In both cases, the complementary field is not reconstructed. The resolved field is assumed to be continuous at the interface in Davidson's method, which is based on a $k-\omega$ RANS model, while the coupling is achieved by transmiting wall stresses at the interface in Diurno's simulations relying on the Spalart–Allmaras model. Additional boundary conditions must be provided for turbulent quantities at the RANS/LES

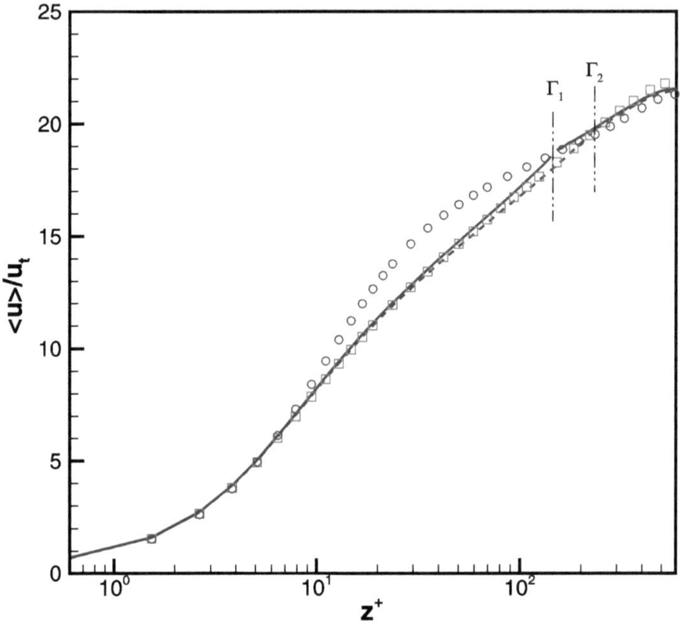

Fig. 12.1. Large-eddy simulation of plane channel flow with zonal RANS/LES coupling. RANS subdomains are used in the near-wall region, while the core of the channel is simulated using LES. The computed mean velocity profile is compared with reference data, for two position of the interface Γ. *Solid* and *dashed lines*: hybrid RANS/LES computations. *Squares*: classical RANS simulation. *Crosses*: classical LES simulation. Courtesy of P. Quéméré, ONERA.

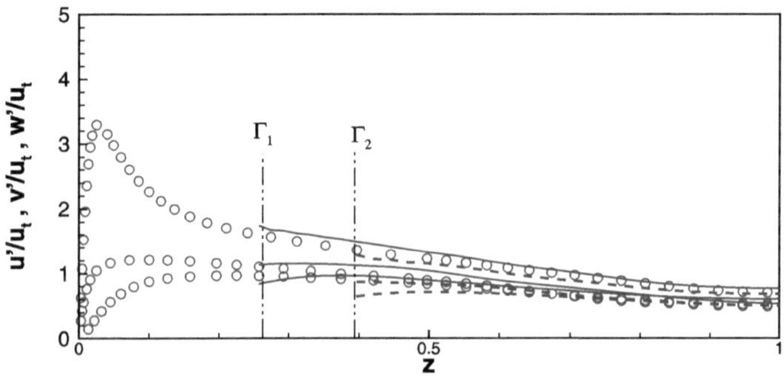

Fig. 12.2. Large-eddy simulation of plane channel flow with zonal RANS/LES coupling. RANS subdomains are used in the near-wall region, while the core of the channel is simulated using LES. The computed rms velocity profiles are compared with reference data obtained via typical LES simulation, for two positions of the interface Γ. *Solid* and *dashed lines*: hybrid RANS/LES computations. *Symbols*: classical LES simulation. Courtesy of P. Quéméré, ONERA.

interface. Davidson assumed no coupling at the interface for variables related to models and imposed arbitrary conditions, while explicit coupling is enforced in Diurno's method. The latter assumes that the total stress is equal on both sides of the interface.

12.2.3 Smooth Transition

Hybrid RANS/LES strategies presenting a smooth transition have all been derived to alleviate the classical grid resolution requirements of classical large-eddy simulation in the near-wall region. Consequently, they can be interpreted as hybrid wall-models for large-eddy simulation. Combinations between the two basic techniques are derived in two different ways:

- By performing an explicit blending of the two basic models. Bagget [31] classified the existing approaches in two groups. The first one corresponds to the linear combination of the predicted turbulent/subgrid stresses:

$$\tau_{ij} - \frac{1}{3}\tau_{kk}\delta_{ij} = -\nu_{\text{sgs}}\left[\overline{S}_{ij} - (1-\beta(z))\langle\overline{S}_{ij}\rangle\right] - \beta(z)\nu_{\text{rans}}\langle\overline{S}_{ij}\rangle \quad , \tag{12.4}$$

  where ν_{sgs} and ν_{rans} are the subgrid and turbulent viscosities, respectively. The arbitrary weight $\beta(z)$ is expressed as a function of the distance to the wall, and should be prescribed. This form of hybrid RANS/LES model is an extension of the splitting technique proposed by Schumann (see Sect. 6.3.3, p. 200) to account for anisotropy in the near-wall region. Methods belonging to the second group are based on the blending of the modeled viscosities:

$$\tau_{ij} - \frac{1}{3}\tau_{kk}\delta_{ij} = -\left[(1-\beta(z))\nu_{\text{sgs}} + \beta(z)\nu_{\text{rans}}\right]\overline{S}_{ij} \quad . \tag{12.5}$$

  In both cases $\beta = 0$ corresponds to classical large-eddy simulations and $\beta = 1$ to classical RANS simulations.
- Replacing the characteristic scales which explicitly appear in the RANS model equations by new scales corresponding to the LES filter. The important point is that the turbulent viscosity or the turbulent stresses are not directly rescaled, and consequently these quantities exhibit a continuous behaviour, yielding smooth transition. The most famous example belonging to this category is the Detached Eddy Simulation proposed by Spalart et al. [680, 569]. This model is based on a modification of the Spalart–Allmaras RANS model, in which the distance to the wall d is replaced by $\tilde{d}$:

$$\tilde{d} = \min(d, C_{\text{des}}\overline{\Delta}) \quad , \tag{12.6}$$

  in the destruction term of the transport equation. The constant C_{des} is calibrated in order to recover a Smagorinsky-like behaviour in isotropic turbulence at equilibrium, yielding $C_{\text{des}} = 0.65$. The Detached Eddy Simulation approach has been extented by Strelets [16] to a large family of

RANS two-equation models. For a k–ω model, the characteristic length $l_{k-\omega} \propto \sqrt{k}/\omega$ is replaced in the dissipative term of the k-transport equation by

$$\tilde{l} = \min(l_{k-\omega}, C_{\rm des}\overline{\Delta}), \quad C_{\rm des} = 0.78 \quad . \tag{12.7}$$

A similar approach was developed by Tucker and Davidson [720] which relies on two-equation k–l models. In the RANS domain, a usual k–l model is used, while the Yoshizawa k–l subgrid model is used in the LES subdomain. The equations of these two models are formally equivalent, but with different values of the constants appearing in them. The switch is achieved by changing the length scale in the transport equations: the usual RANS length scale is used in the RANS subdomain, while the subgrid lengthscale $\overline{\Delta}$ is considered in the LES subdomain. This switch results in a discontinuous value of the length scale at the interface. Tucker and Davidson obtain a regular distribution at the interface by applying a 3-point test filter across the interface to the lengthscale.
Another smooth transition method was proposed by Hamba [284]. In this method, the RANS and LES subdomains are overlapping, and the length scale in the subgrid model is assumed to vary linearly from the RANS length scale to the usual subgrid length scale tied to the computational mesh. Hamba applied this technique to a channel flow, hybridizing a k–ε model and a Smagorinsky model. This method yields some error in the mean velocity profile, and was further improved in [285]. In the improved method, the dissipation ε is also linearly interpolated between the RANS value and the one predicted using the subgrid model. But the most important improvement factor is that the convective fluxes on each side at the interface are estimated using an ad hoc interpolated velocity field which is more compatible with the discrete mass preservation constraint. This blending of the two solutions at the interface can also be interpreted as a surrogate for the complementary field $\boldsymbol{v}_{12}$. The improved method is observed to yield much satisfactory results on a plane channel configuration.

12.2.4 Zonal RANS/LES Approach as Wall Model

Both sharp- and smooth-transition hybrid methods have been used to derive new robust wall models for large-eddy simulation. Results are generally disappointing for methods which neglect the reconstruction of the complementary field at the interface. An artificial turbulent boundary layer develops, which is composed of overly large streamwise streaks and vortices. This turbulent process is observed to be self-sustaining. This can be explained by theoretical and numerical results dealing with the existence of a near-wall autonomous cycle [353, 354, 355, 750], which is created by the mean shear.

For sharp-transition methods, the problem is identical to that of wall models for large-eddy simulation, and it seems that zonal modeling cannot perform better than the suboptimal-control-theory-based wall stress models (see Sect. 10.2).

For smooth-transition methods, the interface is replaced by a transition region. An explanation [31] for the existence of a spurious near-wall cycle is that the no-slip boundary condition produces a near-wall viscous region in which the mean pressure gradient is balanced only by the mean viscous stress. The flow is fully turbulent outside the transition region, so that the mean shear scales as $1/z$, z being the wall-normal coordinate. In the transition region, the mean shear has to be reduced by wall-normal streamwise momentum transport, i.e. mean Reynolds stress, to couple the core flow to the wall. If the subgrid model does not carry a large enough amount of Reynolds stress, the resolved motion will adjust in order to enforce the necessary balance. Some authors tried to cure this problem by artificially enhancing the dissipation in the transition region, but this method did not prove to be efficient and general.

For high-Reynolds-number boundary layers, the use of too coarse a grid also yields significant errors in the mean velocity profiles [569]. This demonstrates that, even if the RANS region is able to provide the large-eddy simulation with correct boundary conditions, the grid resolution in the LES region must be the same as for classical wall-resolving large-eddy simulations.

Another point is that Quéméré's results seem to indicate that accounting for the complementary field $\boldsymbol{v}_{12}$ at the interface improves the results and can prevent the appearance of the spurious cycle.[2]

Applications to massively separated flows have also been perfomed [180, 167, 146, 680, 695] which demonstrate that these hybrid approaches can yield much more satisfactory results than for attached flows. The main reason is that massively separated flows are mostly governed by inviscid, large-scale instabilities of the mean flow, which occur in the LES region. The quality of the results is then equivalent to that of a typical large-eddy simulation, but with an effective cost reduction. Strelets [695] remarked that the numerical scheme must be adjusted to the selected approach in each region to obtain reliable results: a low-order accurate dissipative method can be used in the RANS region, but not in the LES region. As a consequence, the zonal approach for physical modeling can also imply the definition of a zonal approach for the numerical method.

The concept of hybridizing RANS and large-eddy simulation in the near-wall region is fully general and does not depends, from a purely formal standpoint, on the modeling approaches used in both parts. The RANS simulation is almost always carried out using models relying on zero to three additional equations, while subgrid models involving zero to two equations are very often met. A noticeable exception is provided by Tucker [719], who couples a one-equation RANS model with a MILES approach in the large-eddy simulation domain.

[2] This is consistent with the finding that the use of a non-zero transpiration velocity for wall stress models yields improved results.

12.3 Nonlinear Disturbance Equations

Another hybrid RANS/LES method was proposed by Morris et al. [545, 140, 288, 547], which is referred to as the Nonlinear Disturbance Equations (NLDE) method. The underlying idea is to split the field into a low-frequency or steady part on the one hand, and a high-frequency fluctuating part. The former can be computed using steady or unsteady RANS simulations, or theoretical laws, while the latter is computed via large-eddy simulation. The resulting set of governing equations appears as a generalized form of the Navier–Stokes equations in perturbation form:

$$\frac{\partial \overline{\boldsymbol{u}}^2}{\partial t} + \mathcal{NS}(\overline{\boldsymbol{u}}^2) = \tau^2 \quad , \tag{12.8}$$

$$\frac{\partial \boldsymbol{v}_{12}}{\partial t} + \mathcal{NS}(\overline{\boldsymbol{u}}^1) - \mathcal{NS}(\overline{\boldsymbol{u}}^2) = \tau^1 - \tau^2 \quad . \tag{12.9}$$

An interesting difference with other methods presented above is that the equation for the detail $\boldsymbol{v}_{12}$ is solved, rather than the equation for the filtered field $\overline{\boldsymbol{u}}^1$. If the carrying field $\overline{\boldsymbol{u}}^2$ is steady, the time derivative in (12.8) cancels out, leading to

$$\frac{\partial \boldsymbol{v}_{12}}{\partial t} = \frac{\partial \overline{\boldsymbol{u}}^1}{\partial t} \quad . \tag{12.10}$$

If it is laminar, then $\tau^2 = 0$.

Using the bilinear form (3.27), the nonlinear convective term appearing in (12.9) can be recast as

$$\mathcal{B}(\overline{\boldsymbol{u}}^1, \overline{\boldsymbol{u}}^1) - \mathcal{B}(\overline{\boldsymbol{u}}^2, \overline{\boldsymbol{u}}^2) = \underbrace{\mathcal{B}(\overline{\boldsymbol{u}}^2, \boldsymbol{v}_{12}) + \mathcal{B}(\boldsymbol{v}_{12}, \overline{\boldsymbol{u}}^2)}_{I} + \underbrace{\mathcal{B}(\boldsymbol{v}_{12}, \boldsymbol{v}_{12})}_{II} \quad . \tag{12.11}$$

Terms I and II are related to the coupling between the two levels of resolution and to the nonlinear interactions between the fluctuations, respectively. Another coupling is achieved through the source term in the right-hand side of (12.9), which is defined as the difference between the subgrid force and the Reynolds forcing term.

This approach can be interpreted as a generalized multilevel simulation (see Sects. 7.7.7 and 11.2), in which the coarsest level is defined using a statistical-average operator and not a convolution filter.

This technique was first developed by Morris et al. to evaluate acoustic sources from a steady jet computation, on the basis of a nonlinear inviscid model equation for the fluctuations. Their second main approximation is that mean-flow source terms in the nonlinear disturbance equations can be neglected. More recently, these authors added a subgrid model to the disturbance equations, in order to take into account the effect of unresolved modes, leading to a true hybrid RANS/LES approach. They also reintroduced a part of the mean flow source term into the nonlinear disturbances equation. The

NLDE approach was used by Hansen et al. [288] to simulate the unsteady, two-dimensional laminar flow around a circular cylinder, corresponding to the first coupling between NLDE and an unsteady mean flow. The first attempt to use it for wall-bounded flow is due to Chyczewski et al. [140], with fairly good results due to the use of very coarse grids.

A fully general expression of the governing equations for nonlinear disturbance equations for compressible flows was derived by Labourasse and Sagaut [416] who also proposed using this method to locally reconstruct the unsteady turbulent motion in subdomains embedded within the full flow configuration. This corresponds to a coupling between the nonlinear disturbance equations approach and the multidomain method presented above. The expected gain with respect to the classical large-eddy simulation approach is twofold:

- Firstly, the mean field being prescribed, the errors commited on turbulent fluctuations will not pollute it, and we can expect this method to be more robust than classical large-eddy simulation.
- Secondly, if the turbulence-producing events are localized in a small region, it will be possible to restrict the LES-type computation to a small subdomain included in the global domain, while classical large-eddy simulation would require us to consider the full domain.

Numerical experiments show that this method is more robust than the usual large-eddy simulation if the mean flow is correctly prescribed, i.e. coarser grids can be employed without loss of accuracy. Errors on the mean flow field $\overline{\boldsymbol{u}}^2$ can be corrected using the hybrid approach on a fine grid: the fluctuating field will adjust such that $(\overline{\boldsymbol{u}}^2 + G_1^2 \star \boldsymbol{v}_{12})$ will be correct.

12.4 Universal Modeling

The last class of hybrid RANS/LES methods presented in this chapter is that of the universal models. The underlying idea here is to design new models which can asymptotically recover typical RANS or typical LES capabilities. The existing models are based on the rescaling of the RANS models thanks to inertial range arguments. They all aim at decreasing the resolved kinetic energy dissipation induced by the model for the unresolved scales. This decrease in the induced dissipation leads to a weaker damping of high frequencies/high wavenumbers which can be sustained on the selected computational grid, yielding the recovery more irregular "LES-like" flowfields. Despite this strategy is clear and does not suffer any flaw, it is worth noting that the different methods used to achieve this reduction in the small scale damping[3] are mostly empirical. The most common strategies found in the literature

[3] The most common method is to reduce the amplitude of the turbulent viscosity of the RANS model.

deal with the modification of eddy-viscosity-type RANS models, and can be grouped in two classes:

1. The methods based on a rescaling of the eddy viscosity provided by the RANS model.
2. The methods based on the modification of some terms in the transport equations of the original RANS model, so as to reduce the amplitude of the eddy viscosity. In the case of most RANS models which include an evolution equation for the turbulent kinetic energy, k, this is achieved by increasing the destruction of k and/or modifying the second variable (if any).

A few hybrid models are presented below:

1. Germano's mixed modeling (Sect. 12.4.1).
2. Speziale's general rescaling method for turbulent stresses (Sect. 12.4.2).
3. Arunajatesan's modified two-equation model (Sect 12.4.4).
4. Bush–Mani limiters (Sect. 12.4.5).

12.4.1 Germano's Hybrid Model

A general approach for the definition of hybrid RANS/LES models was proposed by Germano [249]. This approach relies on two basic hypotheses:

- The statistical value of a filtered quantity is equal to the statistical value of the unfiltered quantity:

$$\langle \overline{\phi} \rangle = \langle \phi \rangle \quad . \tag{12.12}$$

- The filtered solution is characterized by the associated mean production of subgrid kinetic energy:

$$\langle \tau_{\mathrm{les},ij} \rangle \langle \overline{S}_{ij} \rangle = C_{\mathrm{f}} \, \tau_{\mathrm{rans},ij} \langle S_{ij} \rangle \quad , \tag{12.13}$$

  where $\tau_{\mathrm{les}}(\boldsymbol{u})$ and $\tau_{\mathrm{rans}}(\boldsymbol{u})$ are the subgrid and Reynolds stress tensor, respectively, and C_{f} is a constant.

The first hypothesis makes it possible to decompose the modeled Reynolds stress tensor as follows:

$$\tau_{\mathrm{rans}}(\boldsymbol{u}) = \langle \tau_{\mathrm{les}}(\boldsymbol{u}) \rangle + \tau_{\mathrm{rans}}(\overline{\boldsymbol{u}}) \quad , \tag{12.14}$$

while the second hypothesis leads to

$$\langle \tau_{\mathrm{les},ij} \rangle \langle S_{ij} \rangle = \frac{C_{\mathrm{f}}}{1 - C_{\mathrm{f}}} \tau_{\mathrm{rans},ij}(\overline{\boldsymbol{u}}) \langle S_{ij} \rangle \quad . \tag{12.15}$$

This equation establishes a bridge between the modeled tensors at both filtered and statistically averaged levels. For subgrid viscosity models, we obtain the following definition:

$$\nu_{\rm les} = \frac{C_{\rm f}\,|\overline{S}|}{2(1-C_{\rm f}\,)}\frac{\tau_{{\rm trans},ij}(\overline{\boldsymbol{u}})\langle S_{ij}\rangle}{\langle|\overline{S}|\overline{S}_{ij}\rangle\langle S_{ij}\rangle} \quad . \tag{12.16}$$

The parameter $C_{\rm f} \in [0,1]$ determines the respective weights of the filtered and the Reynolds-averaged parts. The RANS solution is recovered as $C_{\rm f} \longrightarrow 1$. The convergence of the method depends on the capability of the RANS model for $\tau_{{\rm trans},ij}$, once applied to the filtered field $\overline{\boldsymbol{u}}$, to alleviate the singular behavior of (12.16).

12.4.2 Speziale's Rescaling Method and Related Approaches

Models belonging to this category can be expressed as

$$\tau_{ij} = \mathcal{F}_{\mathcal{R}}\tau_{ij}^{\rm rans} \quad , \tag{12.17}$$

where $\mathcal{F}_{\mathcal{R}}$ is the rescaling function to be defined, and $\tau_{ij}^{\rm rans}$ the modeled turbulent stress tensor predicted using any RANS model.

Speziale [685, 684] proposed a general rescaling function

$$\mathcal{F}_{\mathcal{R}} = \left[1 - \exp(-\beta\overline{\Delta}/\eta_K)\right]^n \quad , \tag{12.18}$$

where β is a constant, n an arbitrary power, and η_K the Kolmogorov length scale computed from the variables of the RANS model. The direct numerical simulation regime is recovered as the limit of a very fine resolution, $\overline{\Delta}/\eta_K \longrightarrow 0$. As the resolution becomes very coarse, $\overline{\Delta}/\eta_K \longrightarrow \infty$, the RANS model is recovered. As noted by Magnient [477], a consistency problem of the Speziale's transition function is that large-eddy simulation is not recovered as a limit for very high Reynolds number.[4] In practice, Speziale recommended applying this scaling law to complex RANS models able to deal with anisotropy and disequilibrium, such as Reynolds Stress Models or Explicit Algebraic Reynolds Stress Models. Proposed values in [685, 684] are $\beta = 0.001$ and $n = 1$, but the model is expected to be very sensitive to these parameters.

Fasel [212] later used $\beta = 0.004$, $\overline{\Delta} = (\overline{\Delta}_1\overline{\Delta}_2\overline{\Delta}_3)^{1/3}$ and $\eta_K = Re^{-3/4}\varepsilon^{-1/4}$ together with an Explicit Algebraic Resynolds Stress Model to compute a flat plate boundary layer (ε being evaluated from the outputs of the RANS model). The same authors used a slightly modified version to compute a wall jet flow. Parameters for this configuration are

$$\mathcal{F}_{\mathcal{R}} = \left[1 - \exp(-\max(0, 5\overline{\Delta} - 10\eta_K)/N\eta_K)\right], \quad \overline{\Delta} = \frac{1}{3}\sqrt{\overline{\Delta}_1^2\overline{\Delta}_2^2\overline{\Delta}_3^2} \quad , \tag{12.19}$$

[4] A well known example is isotropic turbulence in the limit of an infinite Reynolds number.

where N is an adjustable parameter that governs the cutoff, which taken in the range 1000–2000 in the wall-jet application.

A much simpler rescaling for computing subgrid viscosity was proposed by Peltier et al. [796, 584]. Using inertial range argument and numerical tests, these authors proposed computing the subgrid viscosity $\nu_{\rm sgs}$ from the RANS eddy viscosity $\nu_{\rm rans}$ as

$$\nu_{\rm sgs} = \left(\frac{\overline{\Delta}}{l}\right)^2 \nu_{\rm rans} \quad , \tag{12.20}$$

where l is the turbulent length computed using the RANS model outputs. Magnient [477] proposed a similar rescaling law for the RANS viscosity with the power 4/3 instead of 2.

The rescaling approach was further developed by Batten and his co-workers [46, 47, 48, 49] as the *Limited-Numerical-Scales method*, who introduced the following rescaling factor:

$$\mathcal{F}_{\mathcal{R}} = \frac{\min\left[(L_{\rm sgs}U_{\rm sgs}), (L_{\rm rans}U_{\rm rans})\right]}{(L_{\rm rans}U_{\rm rans})} \quad , \tag{12.21}$$

where $L_{\rm sgs}, U_{\rm sgs}, L_{\rm rans}, U_{\rm rans}$ are the characteristic subgrid length scale, subgrid velocity scale, turbulent length scale and turbulent velocity scale, respectively. In the simple case where eddy viscosity type models are used in both cases, the products LU are equal to the eddy viscosities. The original implementation by Batten for several flows rely on the nonlinear k–ε model of Goldberg and the Smagorinsky model. It is important noticing that the subgrid closure is used here only to evaluate the rescaling factor $\mathcal{F}_{\mathcal{R}}$ through (12.21). This method, coined as the Limited Numerical Scale approach, was mainly used through a zonal implementation. To improve the results, Batten strongly recommend to add explicit fluctuations at the interface, i.e. to synthetize explicitly the field $\boldsymbol{v}_{12}$.

12.4.3 Baurle's Blending Strategy

A more complex method, which can be interpreted as an evolution of Batten's proposal using the same strategy as the one underlying Menter's hybrid RANS model, is presented by Baurle and his co-workers [50]. The underlying idea is still to hybridize a classical RANS model with a standard subgrid scale model. A requirement is that both RANS and LES models include an equation on the unresolved turbulent kinetic energy (TKE). The hybrid model is therefore defined by a linear combination of each model equation set:

$$\begin{aligned} \text{Hybrid RANS/LES} \qquad & \text{TKE equation} = \\ & \mathcal{F} \times [\text{RANS TKE equation}] \\ & +(1-\mathcal{F}) \times [\text{LES TKE equation}] \quad , \end{aligned} \tag{12.22}$$

and

$$\nu_{\text{hybrid}} = \mathcal{F}\nu_{\text{rans}} + (1 - \mathcal{F})\nu_{\text{sgs}} \quad . \tag{12.23}$$

where $\mathcal{F}$ is a weighting function to be discussed below. As remarked by Baurle, an algebraic subgrid viscosity model or even an implicit model can also be considered in this approach by constructing a production/destruction balance equation for the subgrid kinetic energy that recovers the desired subgrid viscosity. The key parameter in this approach is the weighting function $\mathcal{F}$. In the case $\mathcal{F}$ depends only on geometric factors such as the distance to solid walls and on grid topology, the blending strategy results in a purely zonal approach reminiscent to those described in Sect. 12.2. The blending function proposed in Ref. [50] is more general and includes some information tied to the level of resolution of the flow. It is defined as

$$\mathcal{F} = \max(\tanh(\xi^4), F_1) \quad , \tag{12.24}$$

whith

$$\xi = \max\left(\frac{L_{\text{rans}}}{d}, 500\nu\frac{C_d L_{\text{rans}}}{\sqrt{k_{\text{rans}}}d^2}\right) \quad , \tag{12.25}$$

where L_{rans}, d and k_{rans} are the turbulent length scale computed from the RANS model outputs, the distance to the nearest wall and the turbulent kinetic energy provided by the RANS model, respectively. The constant C_d is taken equal to 0.01. The function F_1 is defined on the grounds of Batten's blending function (12.21):

$$F_1 = \begin{cases} 1 & \text{if } \nu_{\text{sgs}} < \nu_{\text{rans}} \\ 0 & \text{otherwise} \end{cases} \quad . \tag{12.26}$$

The resulting weigthing function is such that the RANS treatment will be invoked near solid surfaces, while the LES treatment will govern the simulation in separated and free shear regions. The introduction of the F_1 parameter allows to maintain the RANS treatment when sudden grid refinements are encountered and that no resolved fluctuations are present.

A simplified version of the weighting function was developed in [772], which is expressed as

$$\mathcal{F} = \tanh\left(\frac{\nu_{\text{sgs}}}{\nu_{\text{rans}}}\right)^2 \quad . \tag{12.27}$$

This approach is now illustrated considering the case of the two-equations $k - \zeta$ model (ζ being the turbulent enstrophy) treated in [772]. The RANS equation for the turbulent kinetic energy k is:

$$\begin{aligned} \frac{\partial \overline{\rho} k}{\partial t} + \frac{\partial \overline{\rho}\overline{u}_j k}{\partial x_j} &= \frac{\partial}{\partial x_j}\left(\left[\frac{\mu}{3} + \frac{\mu_{\text{rans}}}{\sigma_k}\right]\frac{\partial k}{\partial x_j}\right) + \tau_{ij}\frac{\partial \overline{u}_i}{\partial x_j} \\ &\quad - \frac{1}{C_k}\frac{\mu_{\text{rans}}}{\overline{\rho}^2}\frac{\partial \overline{\rho}}{\partial x_i}\frac{\partial \overline{p}}{\partial x_i} - C_1\frac{\overline{\rho} k}{\tau_p} - \mu\zeta \quad , \end{aligned} \tag{12.28}$$

where

$$\nu_{\text{rans}} = \frac{\mu_{\text{rans}}}{\overline{\rho}} = C_\mu \frac{k^2}{\nu\zeta} \quad , \tag{12.29}$$

with $C_\mu = 0.09$, C_k, C_1 and σ_k model-dependent parameters. The subgrid viscosity being evaluated as

$$\nu_{\text{sgs}} = C_s \overline{\Delta}\sqrt{k}, \quad C_s = 0.01 \quad , \tag{12.30}$$

the hybrid viscosity is defined according to (12.23) while the original turbulent kinetic energy equation (12.28) is transformed into

$$\begin{aligned}
\frac{\partial \overline{\rho} k}{\partial t} + \frac{\partial \overline{\rho}\,\overline{u}_j k}{\partial x_j} &= \frac{\partial}{\partial x_j}\left(\left[\frac{\mu}{3} + \frac{\mu_{\text{hybrid}}}{\sigma_k}\right]\frac{\partial k}{\partial x_j}\right) + \tau_{ij}\frac{\partial \overline{u}_i}{\partial x_j} \\
&\quad -(1-\mathcal{F})\left(\frac{1}{C_k}\frac{\mu_{\text{hybrid}}}{\overline{\rho}^2}\frac{\partial \overline{\rho}}{\partial x_i}\frac{\partial \overline{p}}{\partial x_i} + C_1 \frac{\overline{\rho} k}{\tau_p} + \mu\zeta\right) \\
&\quad -\mathcal{F} C_d \overline{\rho}\frac{k^{3/2}}{\overline{\Delta}} \quad ,
\end{aligned} \tag{12.31}$$

where $C_d = 0.01$ is a constant. The RANS turbulent length scale appearing in the computation of the blending function is

$$L_{\text{rans}} = \frac{k^{3/2}}{\nu\zeta} \quad . \tag{12.32}$$

12.4.4 Arunajatesan's Modified Two-Equation Model

A modified two-equation k–ε model was proposed by Arunajatesan et al. [20, 21, 22] as a basis for hybrid RANS/LES simulations. The key idea is similar to that of the Detached Eddy Simulation, i.e. some characteristic scales appearing in the classical RANS model are replaced by new ones associated with the filtering operator. Two transport equations are solved along with the Navier–Stokes equations: one for the subgrid kinetic energy q_{sgs} and one for the total dissipation rate ε. Based on the values of these quantities and the filter length $\overline{\Delta}$, a RANS and a subgrid viscosity referred to as ν_{rans} and ν_{sgs}, respectively, are computed. The subgrid viscosity is used in the equations for momentum and subgrid kinetic energy, while the RANS viscosity is used in the equation for the turbulent dissipation rate.

The proposed transport equations for the subgrid kinetic energy and the dissipation rate are

$$\frac{\partial q_{\text{sgs}}}{\partial t} + \frac{\partial}{\partial x_i}\left(\overline{u}_i q_{\text{sgs}} - (\nu + \nu_{\text{sgs}}/\sigma_q)\frac{\partial q_{\text{sgs}}}{\partial x_i}\right) = P_q - \varepsilon \quad , \tag{12.33}$$

$$\frac{\partial \varepsilon}{\partial t} + \frac{\partial}{\partial x_i}\left(\overline{u}_i \varepsilon - (\nu + \nu_{\text{rans}}/\sigma_\varepsilon)\frac{\partial \varepsilon}{\partial x_i}\right) = P_\varepsilon - D_\varepsilon \quad , \tag{12.34}$$

where σ_k and σ_ε are modeling constants equal to their usual RANS counterparts. Production and dissipation terms appearing on the right-hand sides of (12.33) and (12.34) are identical to those of the original RANS model and will not be detailed here (see [424]).

Once these two quantities are known, it is assumed that the local turbulent energy spectrum can be represented everywhere using the following hybrid Von Karman/Pao form:

$$\breve{E}(\breve{k}) = C_1 \breve{k}_*^{-5/3} \left(\frac{\breve{k}}{\breve{k}_*}\right)^4 \left(1 + \left(\frac{\breve{k}}{\breve{k}_*}\right)^2\right)^{-17/6} \exp\left(\frac{-9}{4}\breve{k}^{4/3}\right) \quad , \tag{12.35}$$

where $\breve{k} = k\eta_K$ is the normalized wave number, η_K the Kolmogorov length scale, k_* the energy-containing wave number, $\breve{E} = E/(\nu^5\varepsilon)^{1/4}$ the normalized spectrum, and C_1 a constant to be determined.

The Kolmogorov scale is computed as $\eta_K = (\nu^3/\varepsilon)^{1/4}$. The constant C_1 is chosen so that the integral of the dissipation range spectrum is equal to the local turbulence dissipation rate.

The energy-containing wave number k_* is computed in order to enforce the following relation:

$$q_{\text{sgs}} = \int_{\pi/\overline{\Delta}}^{\infty} E(k)dk \quad . \tag{12.36}$$

Once the local turbulence spectrum is completely determined, the authors propose computing the value of the subgrid viscosity for any cutoff wave number using the constant spectral subgrid viscosity model (5.19), yielding

$$\nu_{\text{sgs}} = 0.28\sqrt{\frac{E(\pi/\overline{\Delta})}{\pi\overline{\Delta}}} \quad , \tag{12.37}$$

$$\nu_{\text{rans}} = 0.28\sqrt{\frac{E(k_*)}{k_*}} \quad . \tag{12.38}$$

12.4.5 Bush–Mani Limiters

Another possibility proposed by Bush and Mani [792] for deriving a hybrid model is to generalize the basic idea underlying Spalart's Detached Eddy Simulation by applying it to all the turbulent variables. Considering the two-equation RANS model of the general form $k - \phi$, the proposed solution is summarized in Table 12.1. In order to account for both space- and time-filterings, the authors proposed redefining the filter length in the dissipation term as

$$\widetilde{\Delta} = \max(\overline{\Delta}, |\overline{\boldsymbol{u}}|\Delta t, \sqrt{k}\Delta t) \quad , \tag{12.39}$$

Table 12.1. Bush–Mani limiters for hybrid RANS/LES models. C_i are constants, and $\widetilde{\Delta}$ is the cutoff length.

ϕ	limiter
l	$\min(l, C_1\widetilde{\Delta})$
ε	$\max(\varepsilon, C_2 k^{3/2}/\widetilde{\Delta})$
ω	$\max(\omega, C_3 k^{1/2}/\widetilde{\Delta})$

where $\overline{\Delta}$ is the usual cutoff length used in large-eddy simulation, and Δt the time step of the computation.

The same approach was followed by Allen and Mendonça [10], who tested extended versions of Spalart's Detached Eddy Simulation based on several two-equations models. The values of the constants appearing in Table 12.1 given by these authors are $C_2 = 0.73$ and $C_3 = 0.61$. But numerical experiments reported by several authors indicate that the optimal value is case-dependent.

12.4.6 Magagnato's Two-Equation Model

A modified two-equation model approach is proposed by Magnagato and Gabi [476], which includes an explicit random backscatter term. The original formulation of the model is based on either a non-linear $k-\varepsilon$ or a non-linear $k-\tau$ RANS model, in which the characteristic length scale L is evaluated as

$$L = \max\left(\overline{\Delta}, |\overline{\boldsymbol{u}}|\Delta t\right) \quad . \tag{12.40}$$

Here, quantities computed solving the equations of the RANS model with the modified length scale are assumed to be related to the unresolved turbulent fluctuations. The dissipative part of the model is then written as

$$\tau_{ij} = -C_\mu k\tau\overline{S}_{ij} + \frac{2}{3}k\delta_{ij} \quad , \tag{12.41}$$

for the $k-\tau$ model, and

$$\tau_{ij} = -C_\mu \frac{k^2}{\varepsilon}\overline{S}_{ij} + \frac{2}{3}k\delta_{ij} \quad , \tag{12.42}$$

for the $k-\varepsilon$ model. In both cases, the value $C_\mu = 0.09$ is retained. The authors thus add a non-dissipative term to the expressions given above for the subgrid tensor to account for the backscatter phenomenon. The generic form of the backscatter model is

$$\tau_{ij} = v_i v_j - \frac{2}{3}k\delta_{ij} \tag{12.43}$$

where the random velocities v_i are calculated at each time step using a Langevin-type equation. Introducing ζ_i an independent random vector ranging from -1 to 1, the value of v_i at the nth time step is defined as

$$v_i^n = v_i^{n-1}\left(1 - \frac{\Delta t}{\tau}\right) + \sqrt{\frac{\Delta t}{\tau}\left(2 - \frac{\Delta t}{\tau}\right)\frac{2}{3}k\zeta_i^n} \quad , \tag{12.44}$$

where the subgrid time scale τ is directly provided by the $k-\tau$ model and is evaluated as

$$\tau = \frac{L}{\sqrt{k_\infty}} \quad , \tag{12.45}$$

for the $k-\varepsilon$ model, where the length scale L is given by relation (12.40) and k_∞ is the sum of the resolved and the unresolved kinetic energy. The full model formulation is obtained by adding the backscatter term to the selected dissipative part.

12.5 Toward a Theoretical Status for Hybrid RANS/LES Approaches

Since they are based on the hybridization of the classical Large-Eddy Simulation and Reynolds-Averaged Numerical Simulation methods, most of the hybrid approaches presented above escape the usual theoretical framework developed to present them. As a matter of fact, the resulting flow field can neither be interpreted in terms of statistical average of the exact solution nor as the result of a conventional filtering operation. Therefore, the question arises of finding a relevant paradigm to analyze and understand these hybrid approaches.

A powerful framework to understand the properties of these methods is the *unresolved-scale-model induced effective filter* paradigm developed by Muchinsky (see Sect. 8.1.1). This analysis is based on the observation that, during the computation, the only term which carries some information about the unresolved scales is the turbulence/subgrid model[5]. Therefore, it is the model for the unresolved scales which determines the amount of damping (the numerical errors are assumed to be negligible here for the sake of clarity) of the resolved scales and governs the effective filtering of the exact solution.

This is illustrated writing the resolved discrete problem as

$$\frac{\delta u_\mathrm{d}}{\delta t} + F_\mathrm{d}(u_\mathrm{d}, u_\mathrm{d}) = S_\mathrm{d} \quad , \tag{12.46}$$

where u_d, $\delta/\delta t$ and $F_\mathrm{d}(\cdot,\cdot)$ are the discrete approximations of the exact terms u, $\partial/\partial t$ and $F(\cdot,\cdot)$ on the computational grid, respectively. Here, $F(\cdot,\cdot)$ is

[5] In the case of the Implicit Large-Eddy Simulation approach, this information is contained in the numerical errors.

related to the fluxes in the exact Navier–Stokes equations. The source term S_d stands for the model for the unresolved scales.

The discrete equation for the resolved kinetic energy is

$$\frac{1}{2}\frac{\delta u_\mathrm{d}^2}{\delta t} + u_\mathrm{d} \cdot F_\mathrm{d}(u_\mathrm{d}, u_\mathrm{d}) = u_\mathrm{d} \cdot S_\mathrm{d} \quad . \tag{12.47}$$

The term in the right hand side of this new equation shows how the model for unresolved scales governs the dissipation process. A decrease in this dissipation will results in a larger amount of resolved kinetic energy, and therefore to the existence of smaller scales (the limit being fixed by the grid resolution). All hybrid RANS/LES methods based on the modification or the rescaling of a RANS model aim at decreasing this term.

Following Muchinsky's analysis, which relies on the idea that the RANS or LES computations can be seen as direct numerical simulations of a non-newtonian fluid, the observed effects of the modification of the original RANS models can be easily interpreted. The steady RANS solution corresponds to a steady laminar non-newtonian flow. Reduction in the dissipation results in a higher Reynolds number. Unsteady RANS solutions are therefore similar to direct numerical simulations of unsteady non-newtonian flow. The bifurcations sometimes observed in unsteady RANS simulations when the total dissipation is further reduced can be interpreted as analogous to bifurcations of a laminar newtonian flow at low-Reynolds number (transtition to three-dimensional modes, growth of small scales, ...). The further dissipation reduction achieved by the definition of hybrid models leads to higher Reynolds numbers, and to the possiblity for smaller scales to be sustained.

13. Implementation

This chapter is devoted to the practical details of implementing the large-eddy simulation technique. The following are described:

- Cutoff length computation procedures for an arbitrary grid;
- Discrete test filters used for computing the subgrid models or in a pre-filtering technique;
- Computing the Structure Function model on an arbitrary grid.

This part of the implementation of large-eddy simulation is more and more recognized as one of the keys of the success of a simulation. Most of the theoretical developments rely on an abstract filter, which is characterized by its cutoff frequency. But practical experience show that computational results can be very sensitive to the effective properties (transfer function, cutoff length) of the fitering operators used during the simulation.

This is especially true of subgrid models relying on the use of a test filter, such as dynamic models and scale-similarity models. Thus, the problem of the consistency of the filter with the subgrid model must be taken into account [597, 563, 81, 641, 605].

13.1 Filter Identification. Computing the Cutoff Length

The theoretical developments of the previous chapters have identified several filters of different origins:

1. Analytical filter, represented by a convolution product. This is the filter used for expressing the filtered Navier–Stokes equations.
2. Filter associated with a given computational grid. No frequency higher than the Nyquist frequency associated with this grid can be represented in the simulation.
3. Filter induced by the numerical scheme. The error committed by approximating the partial derivative operators by discrete operators modifies the computed solution mainly the high-frequency modes.
4. Filter associated with the subgrid model, which acts like a control process on the computed solution.

The computed solution is the result of these four filtering processes constituting the simulation effective filter. When performing a computation, then, the question arises as to what the effective filter is, that governs the dynamics of the numerical solution, in order to determine the characteristic cutoff length. This length is needed for several reasons.

- In order to be able to determine the physically and numerically well-resolved scale beyond which we will be able to start using the results for analysis.
- In order to be able to use the subgrid models like the subgrid viscosity models that use this cutoff length explicitly.

While the filters mentioned above are definable theoretically, they are almost never quantifiable in practice. This is particularly true of the filter associated with the numerical schemes used. In face of this uncertainty, practitioners have one of two positions they can adopt:

1. Arrange it so that one of the four filters becomes predominant over the others and is controllable. The effective filter is then known. This is done in practice by using a pre-filtering technique.
   Normally, this is done by ensuring the dominance of the analytical filter, which allows us strict control of the form of the filter and of its cutoff length, so that we can get the most out of the theoretical analyses and thereby minimize the relative uncertainty concerning the nature of the computed solution. In the numerical solution, an analytical filter is then applied here to each computed term. In order for this filter to be dominant, its cutoff length must be large compared with the other three. Theoretically, this analytical filter should be a convolution filter which, to keep the computation cost within acceptable limits, can only be applied for simulations performed in the spectral space[1]. For the simulations performed in the physical space, discrete filters are used, based on weighted averages with compact support. These operators enter into the category of explicit discrete filters, which are discussed in the following section.
   We may point out here that the methods based on implicit diffusion with no physical subgrid model can be re-interpreted as a pre-filtering method, in which case it is the numerical filter that is dominant. We can see the major problem of this approach looming here: the filter associated with a numerical method is often unknown and is highly dependent on the simulation parameters (grid, boundary conditions, regularity of the solution, and so forth). This approach is therefore an empirical one that offers little in the way of an a priori guarantee of the quality of the results. It does, however, have the advantage of minimizing the computation costs because we are then limited to solving the Navier–Stokes equations without implanting any subgrid model or explicit discrete filter.

[1] The convolution product is then reduced to a simple product of two arrays.

2. Considering that the effective filter is associated with the computational grid. This position, which can be qualified as minimalist on the theoretical level, is based on the intuitive idea that the frequency cutoff associated with a fixed computational grid is unavoidable and that this filter is therefore always present. The problem then consists in determining the cutoff length associated with the grid at each point, in order to be able to use the subgrid models.
In the case of a Cartesian grid, we take the filtering cell itself as Cartesian. The cutoff length $\overline{\Delta}$ is evaluated locally as follows:
   - For uniform grid, the characteristic filtering length in each direction is taken equal to the mesh size in this same direction:

$$\overline{\Delta}_i = \Delta x_i \quad . \tag{13.1}$$

     The cutoff length is then evaluated by means of one of the formulas presented in Chap. 6.
   - For a variable mesh size grid, the cutoff length in the ith direction of the grid point of index l is computed as:

$$\overline{\Delta}_i|_l = (x_i|_{l+1} - x_i|_{l-1})/2 \quad . \tag{13.2}$$

     The cutoff length is then computed locally according to the results of Chap. 6.

   In the case of a curvilinear structured grid, two options are possible depending on the way the partial derivative operators are constructed:
   - If the method is of the finite volume type in the sense of Vinokur [733], i.e. if the control volumes are defined directly on the grid in the physical space and their topologies are described by the volume of the control cells, by the area and the unit normal vector to each of their facets, the filter cutoff length can be computed at each point either by taking it equal to the cube root of the control volume to which the point considered belongs, or by using what Bardina et al. propose (see Sect. 6.2.3).
   - If the method is of the finite differences type in the Vinokur sense [733], i.e. if the partial derivative operators are computed on a uniform Cartesian grid after a change of variables whose Jacobian is denoted J, then the cutoff length can be evaluated at the point of index (l,m,n) either by Bardina's method or by the relation:

$$\overline{\Delta}_{l,m,n} = \left(J_{l,m,n} \Delta\xi \Delta\eta \Delta\zeta\right)^{1/3} \quad , \tag{13.3}$$

     where $\Delta\xi$, $\Delta\eta$ and $\Delta\zeta$ are the grid steps in the reference space.

   In the case of an unstructured grid, we use the same evaluations as for a structured curvilinear grid with a finite volume type method, in the sense given above.

13.2 Explicit Discrete Filters

Several techniques and subgrid models described in the previous chapters use a test filter. For reference, these are the:

- Pre-filtering technique;
- Soft deconvolution models and scale similarity models;
- Mixed Scale Model;
- Dynamic constant adjustment procedures;
- Models incorporating a structural sensor;
- Accentuation procedure.

The corresponding theoretical developments all assume that we are able to apply an analytical filter in the simulation. This operation comes down to a product of two arrays in the spectral space, which is a simple operation of little cost, and all the analytical filters whose transfer function is known explicitly can be used. The problem is very different, though, when we consider the simulations performed in the physical space on bounded domains: applying a convolution filter becomes very costly and non-local filters cannot be employed. In order to be able to use the models and techniques mentioned above, we have to use discrete filters with compact support in the physical space. These are described in the rest of this section. These discrete filters are defined as linear combinations of the values at points neighboring the one where the filtered quantity is computed [632, 728, 563, 461].

The weighting coefficients of these linear combinations can be computed in several ways, which are described in the following. We first present the one-dimensional case and then that of the Cartesian grids of more than one dimension, and lastly extend this to arbitrary grids.

The discrete approximation of the convolution filters is then discussed.

13.2.1 Uniform One-Dimensional Grid Case

We restrict ourselves here to the case of a uniform one-dimensional grid of mesh size Δx. The abscissa of the grid point of index i is denoted x_i, such that we can say $x_{i+1} - x_i = \Delta x$. The filtered value of the variable ϕ at the grid point of index i is defined by the relation:

$$\overline{\phi}_i \equiv \sum_{l=-N}^{N} a_l \phi_{i+l} \quad , \tag{13.4}$$

where N is the radius of the discrete filter stencil. The filter is said to be symmetrical if $a_l = a_{-l} \forall l$ and anti-symmetrical if $a_0 = 0$ and $a_l = -a_{-l} \forall l \neq 0$. The constant preservation property is represented by the following relation:

$$\sum_{l=-N}^{N} a_l = 1 \quad . \tag{13.5}$$

A discrete filter defined by the relation (13.4) is associated with the continuous convolution kernel:

$$G(x-y)=\sum_{l=-N}^{N} a_l \delta(x-y+l\Delta x) \quad , \tag{13.6}$$

where δ is a Dirac function. Simple computations show that the associated transfer function $\widehat{G}(k)$ is of the form:

$$\widehat{G}(k)=\sum_{l=-N}^{N} a_l e^{\imath k l \Delta x} \quad . \tag{13.7}$$

The real and imaginary parts of this transfer function are:

$$\begin{aligned}\Re(\widehat{G}(k)) &= a_0+\sum_{l=1}^{N}(a_l+a_{-l})\cos(kl\Delta x) \quad , \\ \Im(\widehat{G}(k)) &= \sum_{l=1}^{N}(a_l-a_{-l})\sin(kl\Delta x) \quad .\end{aligned}$$

The continuous differential operator can be associated with the discrete filter (13.4). To do this, we introduce the Taylor expansion of the variable ϕ about the point i:

$$\phi_{i\pm n}=\sum_{l=0}^{\infty}\frac{(\pm n\Delta x)^l}{l!}\left(\frac{\partial^l \phi}{\partial x^l}\right)_i \quad . \tag{13.8}$$

By substituting in relation (13.4), we get:

$$\overline{\phi}_i=\left(1+\sum_{l=1}^{\infty} a_l^* \Delta x^l \frac{\partial^l}{\partial x^l}\right)\phi_i \quad , \tag{13.9}$$

in which

$$a_l^*=\frac{1}{l!}\sum_{n=-N}^{N} a_n n^l \quad .$$

We note that these filters belong to the class of elliptic filters as defined in Sect. 2.1.3. In practice, the filters most used are the two following three-point symmetrical filters:

$$a_0=\frac{1}{2}, a_{-1}=a_1=\frac{1}{4} \quad , \tag{13.10}$$

$$a_0=\frac{2}{3}, a_{-1}=a_1=\frac{1}{6} \quad . \tag{13.11}$$

Table 13.1. Coefficients of discrete nonsymmetrical filters. N is the number of vanishing moments

N	a_{-2}	a_{-1}	a_0	a_1	a_2	a_3	a_4
1		1/4	1/2	1/4			
2			7/8	3/8	−3/8	1/8	
2		1/8	5/8	3/8	−1/8		
3			15/16	1/4	−3/8	1/4	−1/16
3		1/16	3/4	3/8	−1/4	1/16	
3	−1/16	1/4	5/8	1/4	−1/16		

Vasilyev et al. [728] have defined nonsymmetric filters, which have a large number of vanishing moments[2]. These filters are presented in Table 13.1.

Linearly constrained filters can also be defined, which satisfy additional constraints.

Optimized filters, whose coefficients are computed to minimize the functional

$$\int_0^{\pi/\Delta x} (\Re\{\widehat{G}(k) - \widehat{G}_t(k)\})^2 dk + \int_0^{\pi/\Delta x} (\Im\{\widehat{G}(k) - \widehat{G}_t(k)\})^2 dk \quad , \tag{13.12}$$

where $\widehat{G}_t(k)$ is the targeted transfer function, have been proposed [632, 728]. These filters ensure a better spectral response of the filter, resulting in a better localization of the information in spectral space.

For certain uses, such as in the Germano-Lilly dynamic procedure, the characteristic length of the discrete filter, denoted $\Delta_{\rm d}$, has to be known. For a definite positive filter, one measure of this length is obtained by computing the standard deviation of the associated convolution filter [563, 461]:

$$\Delta_{\rm d} = \sqrt{12 \int_{-\infty}^{+\infty} \xi^2 G(\xi) d\xi} \quad . \tag{13.13}$$

The characteristic lengths of the two three-point filters mentioned above are $2\Delta x$ for the $(1/6, 2/3, 1/6)$ filter and $\sqrt{6}\Delta x$ for the $(1/4, 1/2, 1/4)$ filter. This method of evaluating the characteristic lengths of the discrete filters is inefficient for filters whose second-order moment is zero. One alternative is work directly with the associated transfer function and define the wave number associated with the discrete filter, as for the one for which the transfer function takes the value 1/2. Let $k_{\rm d}$ be this wave number. The discrete filter cutoff length is now evaluated as:

$$\Delta_{\rm d} = \frac{\pi}{k_{\rm d}} \quad . \tag{13.14}$$

[2] These filters are necessary to obtain high-order commuting discrete filters (see Sect. 2.2.2).

Implementation of test filters for the dynamic procedure within the spectral element framework is discussed by Blackburn and Schmidt [60]. The general unstructured case is discussed in [297]. The case of the finite element method is addressed by Kollman et al. [400].

13.2.2 Extension to the Multi-Dimensional Case

For Cartesian grids, we extend to the multidimensional case by applying a one-dimensional filter in each direction of space. This application can be performed simultaneously or sequentially. When simultaneously, the multidimensional filter is written symbolically as a summation:

$$G^n = \frac{1}{n}\sum_{i=1}^{n} G_i \quad , \tag{13.15}$$

where n is the dimension of the space and G_i the one-dimensional filter in the ith direction of space. If applied sequentially, the resulting filter takes the form of a product:

$$G^n = \prod_{i=1}^{n} G_i \quad . \tag{13.16}$$

The multidimensional filters constructed by these two techniques from the same one-dimensional filter are not the same in the sense that their transfer functions and equivalent differential operators are not the same. In practice, it is the product construction that is most often used, for two reasons:

- This approach makes it possible to call the easily implemented one-dimensional filtering routines sequentially.
- Such filters are more sensitive to the cross modes than are the filters constructed by summation, and allow a better analysis of the three-dimensional aspect of the field.

13.2.3 Extension to the General Case. Convolution Filters

For structured curvilinear grids (or Cartesian grids with variable mesh size), one method is to employ the filters defined in the uniform Cartesian grid and take no account of the variations of the metric coefficients. This method, which is equivalent to applying the filter in a reference space, is very easy to implement but allows no control of the discrete filter transfer function or its equivalent differential operator. So it should be used only for grids whose metric coefficients vary slowly.

Another method that is completely general and applicable to unstructured grids consists in defining the discrete filter by discretizing a chosen differential operator. The weighting coefficients of the neighboring nodes are then the coefficients of the discrete scheme associated with this differential operator. In practice, this method is most often used by discretizing second-order elliptic operators:

$$\overline{\phi} = (Id + \alpha\overline{\Delta}^2\nabla^2)\phi \quad , \tag{13.17}$$

where α is a positive constant and $\overline{\Delta}$ the desired cutoff length. Limiting the operator to the second order yields filters with compact support using only the immediate neighbors of each node. This has the advantages of:

- Making it possible to define operators that cost little to implement;
- Making a multiblock and/or multidomain technique easier to use, and the boundary conditions easier to process.

The fast-decay convolution filters (box or Gaussian) can thus be approximated by discretizing the differential operators associated with them. These operators are described in Sect. 7.2.1. The sharp cutoff filter, which is not of compact support, is used only when fast Fourier transforms are usable, which implies that the grid step is constant and the data periodic.

Another possibility for deriving discrete filters on general meshes is to compute the weight of neighbouring points by solving a linear system based on Taylor series expansions [632, 490].

13.2.4 High-Order Elliptic Filters

Convolution filters are non-local, and may sometimes be difficult to use together with complex numerical algorithms (multidomain topology, unstructured grid, ...). An alternative, that can be implemented with all numerical methods, consists in high-order elliptic filters [553].

The filtered variable is computed as being the solution of the general elliptic equation:

$$[-(\nabla^2)^m + \alpha Id]\overline{\phi} = \alpha\phi, \quad m \geq 1 \quad . \tag{13.18}$$

High values of m make it possible to obtain very sharp filters in the spectral space. Mullen and Fischer show that the solution of equation (13.18) can be approximated through numerical solution of a much simpler problem, namely the Poisson equation

$$-\nabla^2\psi = \phi \quad . \tag{13.19}$$

13.3 Implementation of the Structure Function Models

In order to use the subgrid viscosity model based on the second-order structure function or the third-order structure function of the velocity (see p. 124 and p. 126), we have to establish a discrete approximation of the operator:

$$D_{\rm LL}(\boldsymbol{x}, r, t) = \int_{|\boldsymbol{x}'|=r} [\boldsymbol{u}(\boldsymbol{x}, t) - \boldsymbol{u}(\boldsymbol{x} + \boldsymbol{x}', t)]^2 \, d^3\boldsymbol{x}' \quad . \tag{13.20}$$

In practice, this integration is approximated as a sum of the contributions of the neighboring points. In the case of uniform Cartesian grid with $\Delta x = r$, the structure function is evaluated at the index point (i,j,k) by the relation:

$$\begin{aligned} D_{\mathrm{LL}}(\Delta x,t)_{i,j,k} &= \frac{1}{6}\left(\left|\boldsymbol{u}_{i,j,k}-\boldsymbol{u}_{i+1,j,k}\right|^2+\left|\boldsymbol{u}_{i,j,k}-\boldsymbol{u}_{i-1,j,k}\right|^2\right. \\ &+ \left|\boldsymbol{u}_{i,j,k}-\boldsymbol{u}_{i,j+1,k}\right|^2+\left|\boldsymbol{u}_{i,j,k}-\boldsymbol{u}_{i,j-1,k}\right|^2 \\ &+ \left.\left|\boldsymbol{u}_{i,j,k}-\boldsymbol{u}_{i,j,k+1}\right|^2+\left|\boldsymbol{u}_{i,j,k}-\boldsymbol{u}_{i,j,k-1}\right|^2\right) . \end{aligned} \qquad (13.21)$$

When the grid is non-uniform or when $\Delta x \neq r$, an interpolation technique has to be used to compute the integral. Rather than use a linear interpolation, it is recommended that the interpolation method be based on physical knowledge. So in the isotropic homogeneous turbulence case, when we see that we have:

$$\begin{aligned} D_{\mathrm{LL}}(\boldsymbol{x},r,t) &= 4.82K_0(\varepsilon r)^{2/3} \quad , \\ D_{\mathrm{LL}}(\boldsymbol{x},r',t) &= 4.82K_0(\varepsilon r')^{2/3} \quad , \end{aligned}$$

we deduce the proportionality relation:

$$D_{\mathrm{LL}}(\boldsymbol{x},r,t) = D_{\mathrm{LL}}(\boldsymbol{x},r',t)\left(\frac{r}{r'}\right)^{2/3} \quad . \qquad (13.22)$$

Relation (13.21) is thus generalized to the form:

$$D_{\mathrm{LL}}(\boldsymbol{x},r,t) = \frac{1}{n}\sum_{i=1}^{n}|\boldsymbol{u}(\boldsymbol{x})-\boldsymbol{u}(\boldsymbol{x}+\Delta_i)|^2\left(\frac{r}{\Delta_i}\right)^{2/3} \quad , \qquad (13.23)$$

where n is the number of neighboring points retained for computing the structure function and Δ_i the distance of the ith point to the point where this function is evaluated.

It has already been said that the second-order Structure Function model in its original form exhibits defects similar to those of the Smagorinsky model because of the uncertainty relation that prevents any good frequency localization of the information. One way of at least partly remedying this problem is to look for the structure function evaluation information only in the directions of statistical homogeneity of the solution. This is done by evaluating the structure function only from points located in the directions of periodicity of the solution. This way, the mean gradient of the solution is not taken into account in the evaluation of the subgrid viscosity. We again find here an idea similar to the one on which the splitting technique is based, in Sect. 6.3.3.

14. Examples of Applications

This chapter gives a few examples of large-eddy simulation applications that are representative of their accomplishments in the sense that they correspond either to flows that are very frequently treated or to configurations that stretch the technique of today to its limits.

14.1 Homogeneous Turbulence

14.1.1 Isotropic Homogeneous Turbulence

Problem Description. Isotropic homogeneous turbulence is the simplest turbulent flow on which subgrid models can be validated. The physical description of this flow is precisely the one on which the very great majority of these models are constructed. Moreover, the flow's statistical homogeneity makes it possible to use periodicity conditions for the computation, and high-accuracy numerical methods: pseudo-spectral methods can be used, optimally reducing the effect of the numerical error on the solution.

Because of the great simplicity of this flow, most subgrid models yield very satisfactory results in terms of the statistical moments of the velocity field and the integral scales, which reduces the discriminatory range of this test case. It is nonetheless widely used for fundamental type studies of turbulence and modeling.

Two types of such flow are considered:

- Freely decaying isotropic homogeneous turbulence in which the energy is initially distributed in a narrow spectral band and then, as the energy cascade sets in, is directed toward the small scales and finally dissipated at the cutoff by the subgrid model. During the time the cascade is setting in, the kinetic energy remains constant, and later declines. The computation can be validated by comparison with decay laws developed by analytical theories (see [439]) or by comparison with experimental data.
- Sustained isotropic homogeneous turbulence, in which total dissipation of the kinetic energy is prevented by injecting energy at each time step, for example by maintaining a constant energy level in the wave vectors of a given norm. After a transitory phase, an equilibrium solution is established including an inertial range. The computation is validated by comparison

with theoretical or experimental data concerning the inertial region, and quantities associated with the large scales.

A few Realizations. The first large-eddy simulations of the free-decaying type were performed at the end of the seventies and early eighties [136] with resolutions of the order of 16^3 and 32^3. Self-similar solutions could not be obtained with these resolutions because the integral scale becomes larger than the computational domain. However, the comparison with filtered experimental data turns out to be satisfactory [40]. More recent simulations (for example [441, 514]) performed with different subgrid models on grids of 128^3 points have yielded data in agreement analytical theories for the kinetic energy decay. Higher-resolution simulations have been performed.

In the sustained case, Chasnov [120] is an example of achieving self-similar solutions in agreement with theory for resolutions of 64^3 and 128^3, though with an over-evaluation of the Kolmogorov constant. More recently, Fureby et al. [231] have tested six subgrid models and a case of implicit numerical diffusion on a 32^3 grid. The conclusions of this work are that the different realizations, including the one based on artificial dissipation, are nearly indiscernable in terms of the quantities linked to the resolved field, and are in good agreement with data yielded by a direct numerical simulation.

Though isotropic homogeneous turbulence is statistically the simplest case of turbulent flow, it possesses a complex dynamics resulting from the interactions of very many elongated vortex structures called "worms". These structures are illustrated in Fig. 14.1, which comes from a large-eddy simulation of freely decaying isotropic homogeneous turbulence on a 128^3 grid. Obtaining good results therefore implies that the simulation is capable of reflecting the dynamics of these structures correctly. We clearly see here the difference with the RANS approach (see Chap. 1), for which isotropic homogeneous turbulence is a zero-dimension problem: for the large-eddy simulation, this problem is fully three-dimensional and reveals all the aspects of this technique (modeling errors, filter competition, and so forth).

14.1.2 Anisotropic Homogeneous Turbulence

Anisotropic homogeneous turbulence allows a better analysis of the subgrid models because the dynamics is more complex, while optimal numerical methods are retained. So it can be expected that this type of flow offers more discriminatory test cases for the subgrid models than do isotropic flows.

Bardina et al. [40] performed a set of simulations corresponding to the following three cases in the early eighties:

- Homogeneous turbulence subjected to a solid-body rotation. Good agreement is measured with experimental data using a de-filtering technique, on a 32^3 grid with a Smagorinsky model (5.90). The effects of rotation on the turbulence are confirmed, i.e. a reduction in the dissipation of the kinetic energy.

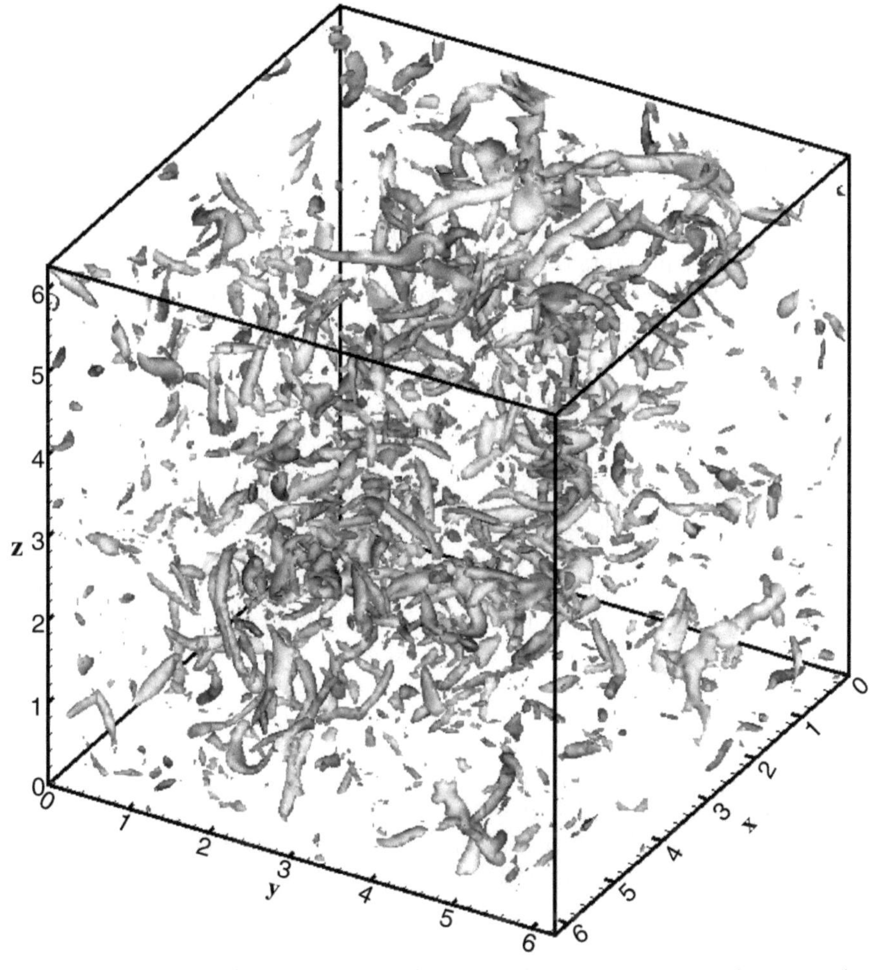

Fig. 14.1. Isotropic homogeneous turbulence. Instantaneous view of vortices (illustrated by an iso-value surface of the vorticity). Courtesy of E. Garnier, ONERA.

- Homogeneous turbulence subjected to pure strain: still a 32^3 grid, with results in good agreement with experimental data concerning the turbulent intensity, using the Smagorinsky model and mixed Smagorinsky–Bardina model (7.125). The best results are obtained with the latter.
- Homogeneous turbulence subjected to a deformation and rotation: simulations are performed on a 32^3 grid with the two previously mentioned models. No validation is presented, for lack of reference data.

Simulations of homogeneous turbulence subjected to sequential shearing have also been performed by Dang [162] on a 16^3 grid with several effective viscosity models, yielding good results concerning the prediction of the anisotropy of the resolved scales. Similar computations have also been performed by Aupoix [23].

14.2 Flows Possessing a Direction of Inhomogeneity

These flows represent the next level of complexity. The presence of a direction of inhomogeneity prompts the use of lower-order numerical methods, at least for this inhomogeneity, and boundary conditions. Also, more complex physical mechanisms are at play that can exceed the possibilities of the subgrid models.

14.2.1 Time-Evolving Plane Channel

Problem Description. The time-evolving plane channel flow is a flow between two infinite parallel flat plates having the same velocity. The time character is due to the fact that we consider the velocity field as being periodic in both directions parallel to the plates. Since the pressure is not periodic, a forcing term corresponding to the mean pressure gradient is added in the form of a source term in the momentum equations. The flow is characterized by the fluid viscosity, the distance between the plates, and the fluid velocity. This academic configuration is used for investigating the properties of a turbulent flow in the presence of solid walls, and is a widely used test case. Turbulence is generated within the boundary layers that develop along each solid wall (see Sect. 10.2.1). It is the driving mechanism here, which must imperatively be simulated with accuracy to obtain reliable results. To do so, the grid has to be refined near the surfaces, which raises numerical problems with respect to the homogeneous turbulence. Moreover, the subgrid models must be able to preserve these driving mechanisms.

The flow topology is illustrated in the iso-value surface plot of the streamwise velocity in Fig. 14.2.

A Few Realizations. There are dozens of numerical realizations of plane channel flows. The first are from Deardorff [172] in 1970. The first landmark results obtained by solving the dynamics of the near-wall region are due to Moin and Kim [537] in 1982. The characteristics of the computations presented in the four reference works [537, 591, 653, 411] are reported in Table 14.1. These computations are representative of the various techniques employed by most authors. The Table summarizes the following information:

- The Reynolds number Re_c referenced to the channel mid-height and mean velocity at the center of the channel.
- The dimensions of the computational domain expressed as a function of the channel mid-height. The domain dimensions must be greater than those of the driving mechanisms in the near-wall region (see Sect. 10.2.1).

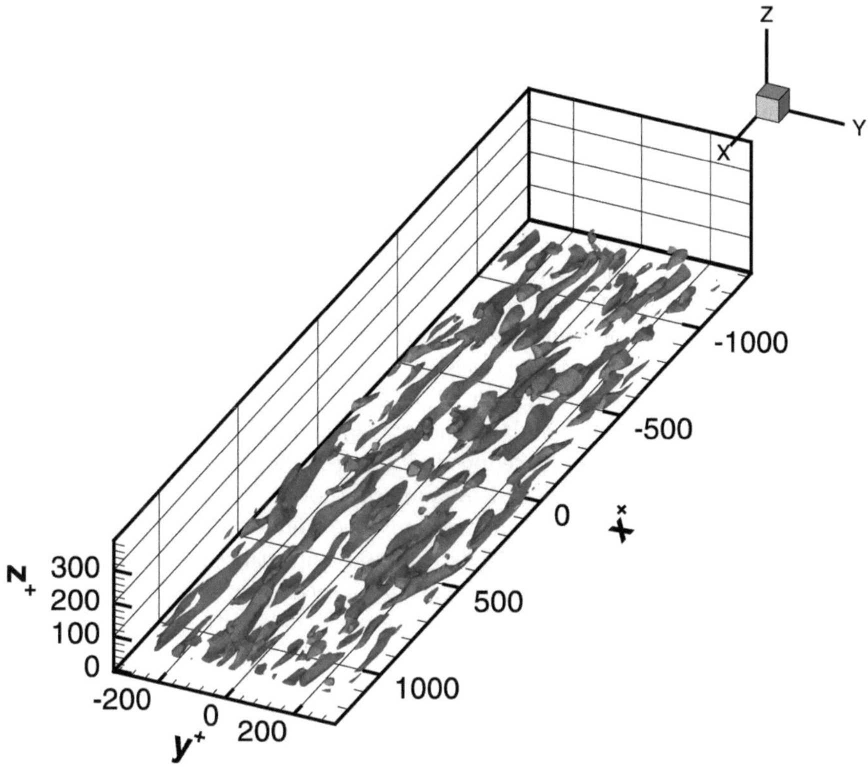

Fig. 14.2. Plane channel flow. Iso-surface of instantaneous streamwise velocity fluctuations. Courtesy of E. Montreuil, ONERA.

- The number of grid points. Simulations generally include few points because the solution is bi-periodical. The computations at high Reynolds number without wall model presented [411] use a hierarchic grid technique with nine grid levels (symbol "+H").
- The subgrid model used ("Sc" is the Schumann subgrid viscosity model (6.59) and "Dyn" the dynamic Smagorinsky model (5.149)). Only two

Table 14.1. Characteristics of time-evolving plane channel flow computations.

Ref.	[537]	[591]	[653]	[411]
Re_c	13800	47100	$\approx 1,5.10^5$	$1,09.10^5$
$Lx \times Ly \times Lz$	$2\pi \times \pi \times 2$	$5\pi/2 \times \pi/2 \times 2$	$4 \times 2 \times 1$	$2\pi \times \pi/2 \times 2$
$Nx \times Ny \times Nz$	$64 \times 64 \times 128$	$64 \times 81 \times 80$	$64 \times 32 \times 32$	$\simeq 2.10^6$+H
SGS Model	Sc	Dyn	Sc	Dyn
Wall	–	–	MSc	–
$O(\Delta x^\alpha)$	S/2	S/T	2	S/Gsp
$O(\Delta t^\beta)$	2	3	2	3

models are used in the computations presented, but most existing models have been applied to this configuration.

- The treatment of the solid walls ("–" is the no-slip condition, "MSc" the Schumann wall model (10.28) to (10.30)). A single computation based on a wall model is presented, knowing that nearly all the models mentioned in Chap. 10 have been used for dealing with this flow.
- The accuracy of the space discretization schemes. Since the directions of statistical homogeneity are linked to directions of periodicity in the solution, pseudo-spectral methods are often used for processing them. This is true of all the computations presented, identified by an "S", except for reference [653], which presents a second-order accurate finite volume method. In the normal direction, three cases are presented here: use of second-order accurate schemes (identified by a "2"), of a Chebyshev method ("T"), and a Galerkin method based on B-splines ("Gsp"). The effect of the numerical error on the solution can be reduced by using higher-order methods, which are consequently recommended by many authors.
- The accuracy of the time integration. The convection term is usually treated explicitly (Runge-Kutta or Adams-Bashforth scheme) and the diffusion terms implicitly (Crank-Nicolson or second-order backward Euler scheme). Nearly all the computations are performed with second- or third-order accuracy.

The results obtained on this configuration are usually in good agreement with experimental data, and especially as concerns the first-order (mean field) and second-order (Reynolds stresses) statistical moments. Examples of data for these quantities are shown in Figs. 14.3 and 14.4. The mean longitudinal

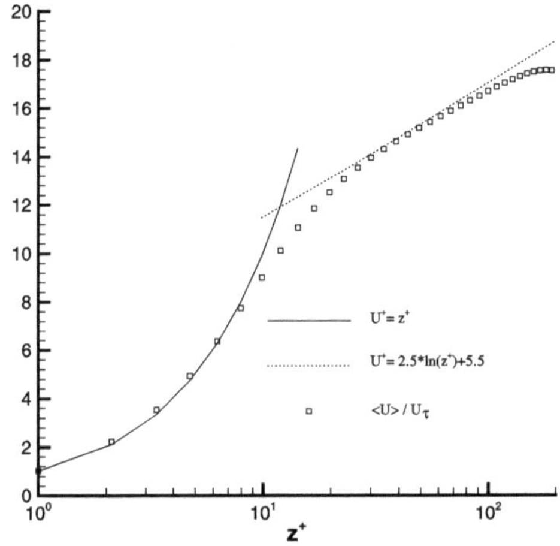

Fig. 14.3. Plane channel flow. Mean longitudinal velocity profile referenced to the friction velocity, compared with a theoretical turbulent boundary layer profile. *Small circle symbols:* LES computation. *Lines:* theoretical profile. Courtesy of E. Montreuil, ONERA.

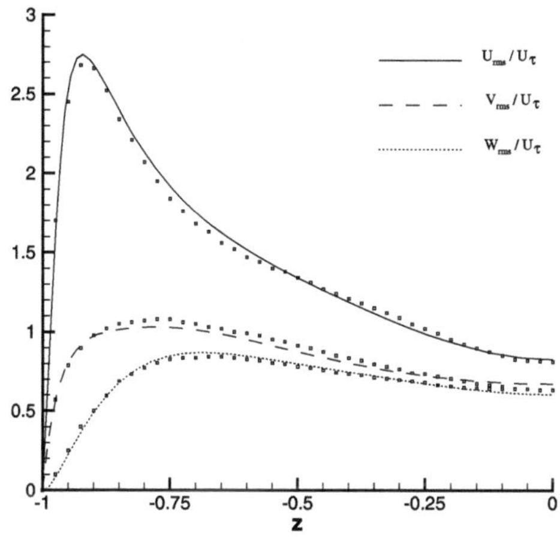

Fig. 14.4. Plane channel flow. Profiles of solved Reynolds stresses with respect to the friction velocity, compared with data from a direct numerical simulation computation. *Dot symbols:* direct numerical simulation. *Lines:* LES computation. Courtesy of E. Montreuil, ONERA.

velocity profile is compared here with a theoretical turbulent boundary layer solution, and very good agreement with it is observed. It should be noted that the logarithmic region is relatively small, which is due to the fact that the Reynolds number for the computation is low ($Re_\tau = 180$). The profiles of the three main Reynolds stresses are compared with those obtained by direct numerical simulation on a grid including about twenty times more degrees of freedom. Although these stresses are calculated only from the resolved field, such that the contribution of the subgrid scales is not included, we observe that the agreement with the reference solution is very satisfactory. This illustrates the fact that data obtained by large-eddy simulation can be used directly in practice without recourse to a de-filtering technique. In the present case, the very good quality of the results can be explained by the fact that a large part of the kinetic energy of the exact solution is contained in the resolved scales.

The quality of the results is essentially due to the resolution of the dynamics in the near-wall region ($z^+ < 100$). This implies that, if a wall model is not used, the computational grid is fine enough to represent the dynamics of the vortex structures present, and that the subgrid models employed do not alter this dynamics. Because of the necessary volumes of the grids, this resolution constraint limits the possible Reynolds number. The largest friction Reynolds number achieved to date, using a hierarchic grid generation method, is $Re_\tau = 4000$ [411]. More results dealing with high Reynolds number simulations can be found in [36]. The standard subgrid viscosity models (Smagorinsky, Structure Function, and so forth) are generally too dissipative and have to be used with caution (modification of the value of the model constant, wall damping function, and so forth) [635]. Results concerning the transition to turbulence in this configuration are available in [600, 599]. Lastly, the results obtained for this flow have been found to be very sensitive to numerical errors induced either by the discrete numerical scheme or by the continuous form of the convection term [409, 563].

14.2.2 Other Flows

Other examples of shear flows treated in the framework of the time-evolution approximation can be found for:

- plane mixing layer, in [674, 38];
- rotating boundary layer, in [152];
- free-surface flow, in [672];
- boundary layers, in [401, 496, 498, 533];
- round jet, in [214];
- plane wake, in [263];
- rotating plane channel, in [523, 524, 702, 422, 596, 421];
- plane jet, in [431].

As in the case of the plane channel flows described above, periodicity conditions are used in the directions of statistical homogeneity. The numerical methods are generally dedicated to the particular configuration being treated (with spectral methods used in certain directions) and are therefore optimal. A forcing term is added in the momentum equations to take the driving pressure gradient into account or avoid diffusion of the base profile.

Transitional flows are more sensitive to the subgrid model and to the numerical errors, as an inhibition of the transition or re-laminarization of the flow are possible. This is more especially true of flows (for example boundary layers) for which there exists a critical Reynolds number: the effective Reynolds number of the simulation must remain above the threshold within which the flow is laminar.

It should be noted that the boundary conditions in the inhomogeneous direction raises little difficulty for the flow configurations mentioned above. These are either solid walls that are easily included numerically (except for the procedure of including the dynamics), or outflow conditions in regions where the flow is potential. In the latter case, the computation domain boundary is generally pushed back as far as possible from the region being studied, which reduces any spurious effects.

The types of results obtained, and their quality, are comparable to what has already been presented for the plane channel flow.

14.3 Flows Having at Most One Direction of Homogeneity

This type of flow introduces several additional difficulties compared with the previous cases. The limited number or total absence of directions of homogeneity makes it necessary in practice to use numerical methods of moderate order of accuracy (generally two, rarely more than four), and highly anisotropic grids. The effect of the numerical error will therefore be high. Moreover, most of these flows are in spatial expansion and the problems related to the definition of the inflow and outflow conditions then appear. Lastly, the flow dynamics becomes very complex, which accentuates the modeling problems.

14.3.1 Round Jet

Problem Description. The example of the round jet flow in spatial expansion is representative of the category of free shear flows in spatial expansion. The case is restricted here to an isothermal, isochoric round jet flow piped into a uniform, steady outer flow in a direction parallel to that of the jet. Two main regions can be identified:

- First, we find a region at the pipe exit where the flow consists of a laminar core called a potential cone, which is surrounded by an annular mixing layer. The mixing layer is created by the inflectional instability associated

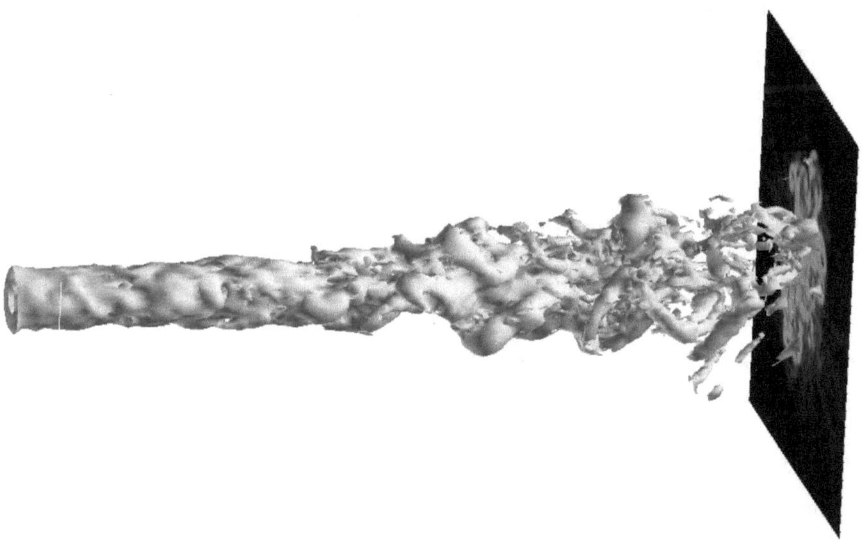

Fig. 14.5. Round jet. Iso-surface of instantaneous vorticity (LES computation). Exit plane in black. Courtesy of P. Comte, LEGI.

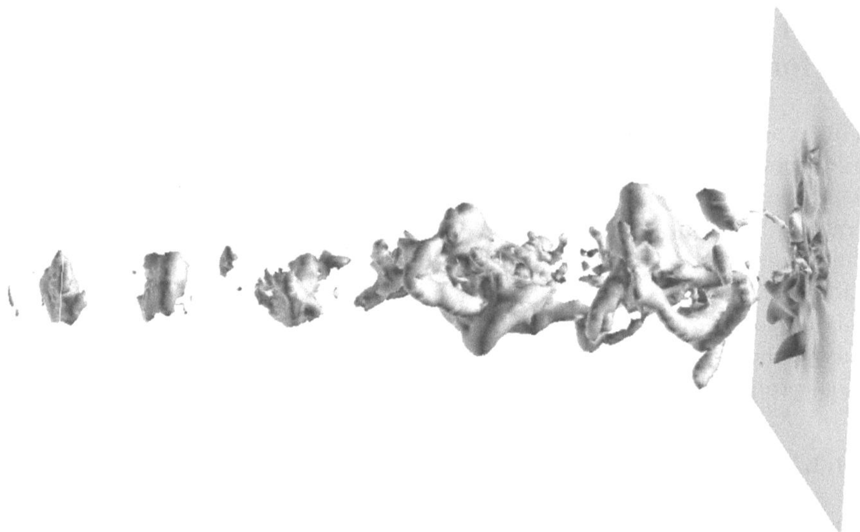

Fig. 14.6. Round jet. Iso-surface of instantaneous vorticity (LES computation). Courtesy of P. Comte, LEGI.

with the deficit velocity profile of the boundary layer on the wall of the circular pipe. As the mixing layer thickens while moving away from the pipe exit section, it reduces the diameter of the potential cone and also induces an increase in the jet diameter.

- After the potential cone disappears, we have a "pure jet" region where the flow gradually reaches a regime corresponding to a similarity solution.

The first region can be decomposed into two: a "transition" region where the mixing layer has not yet reached its self-similar regime, and the similarity region where it has. This organization is illustrated in Figs. 14.5 and 14.6, representing respectively the iso-surfaces of vorticity and pressure obtained from large-eddy simulation results. The vorticity field very clearly shows the transition of the annular mixing layer. The topology of the pressure field shows the existence of coherent structures.

Experimental and numerical analyses have shown that this flow is strongly dependent on many parameters, which makes it highly discriminatory.

A Few Realizations. There are far fewer round jet simulations in the literature than there are plane channel flows. This is mainly due to the increased difficulty. Four of these realizations are described in the following, with their characteristics listed in Table 14.2, which gives:

- the Reynolds number Re_D referenced to the initial jet diameter D and the maximum of the mean initial velocity profile;
- the computational domain dimensions referenced to the length D;

Table 14.2. Characteristics of the round jet computations.

Ref.	[634]	[573]	[83]
Re_D	21000	50.10^4	21000
$Lx \times Ly \times Lz$	$10 \times 11 \times 11$	$12 \times 8 \times 8$	$10 \times 11 \times 11$
$Nx \times Ny \times Nz$	$101 \times 121 \times 121$	≈ 270000+H	$101 \times 288 \times 288$
SGS Model	MSM	Dyn	FSF
Inflow	U+b	U+b	U+b
$O(\Delta x^{\alpha})$	2+up3	3+up3	S/6
$O(\Delta t^{\beta})$	2	2	3

- the number of grid points. All the grids used by the authors mentioned are Cartesian. The symbol H designates the use of embedded grids (four grid levels for [573]).
- the subgrid model ("MSM" standing for the Mixed Scale Model (5.127); "Dyn" the dynamic model (5.149); "FSF" the filtered Structure Function model (5.264)). It should be noted that, for all the known realizations of this flow, only the subgrid viscosity models have been used.
- the freestream condition generation mode. The symbol $U + b$ indicates that the non-steady inflow condition was generated by superimposing an average steady profile and a random noise, as indicated in Sect. 10.3.2.
- the overall order of accuracy of accuracy in space of the numerical method. The symbol $+up3$ indicates that a third-order accurate upwind scheme is used for the convection term to ensure computation stability. The computations presented in [83] rely on spectral schemes in the directions normal to that of the jet.
- the time accuracy of the method employed.

Examples of results obtained on this configuration are compared with experimental data in Figs. 14.7 to 14.11. The axial evolution of the location of the point where the mean velocity is equal to half the maximum velocity is represented in Fig. 14.7. This quantity, which gives some indication concerning the development of the annular mixing layer, remains constant during the first phases of evolution of the jet, which confirms the existence of a potential cone. After the cone disappears, this quantity increases, which indicates the beginning of the pure jet region. It is observed that the length of the potential cone predicted by the computation is less than is observed experimentally. Similar conclusions can be drawn from the axial evolution of the average longitudinal velocity, which is presented in Fig. 14.8. The too-rapid expansion of the pure jet region is accompanied by a strong decay of the mean velocity[1]. These symptoms are observed on all known large-eddy simulation computations on this configuration and still have no precise explanation. Several hypotheses have been formulated concerning the dependency

[1] This results from the conservation of the mass.

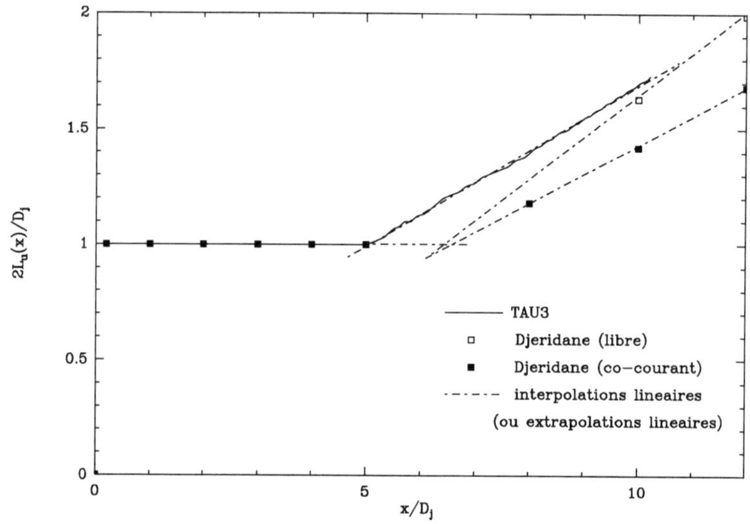

Fig. 14.7. Round jet. Axial evolution of the radial position of the point where the mean velocity is half the maximum velocity. *Dots:* experimental data. *Dot-dashed lines:* extrapolation of this data. *Solid line:* LES computation. Courtesy of P. Comte, LEGI.

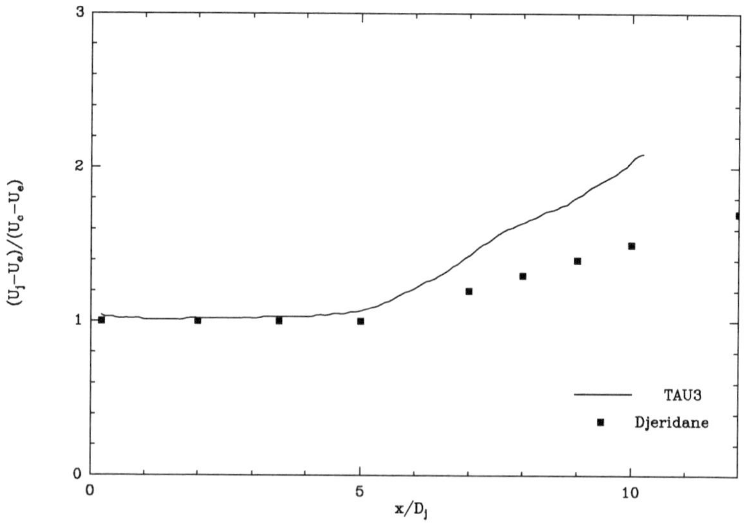

Fig. 14.8. Round jet. Axial evolution of the mean longitudinal velocity. *Dots:* experimental data. *Line:* LES computation. Courtesy of P. Comte, LEGI.

on the initial perturbation, on the boundary conditions, or on the computational grid. The axial profiles of two main Reynolds stresses are presented in Figs. 14.9 and 14.10. These results are qualitatively correct. The Reynolds stresses increase along the axis and exhibit a maximum in a region close to the tip of the potential cone, which is in agreement with the experimental observations. It is noted that the level of the longitudinal stress predicted by the computation is higher than the experimental level in the pure jet region. The peak observed on the downstream boundary of the computational domain is a spurious effect that is no doubt related to the outflow condition. Generally, it is noticed that the quality of the results is not as good as in the case of the plane channel flow, which illustrates the fact that this flow is a more complicated case for large-eddy simulation.

Lastly, the velocity spectra generated from the computation are presented in Fig. 14.11. Over a decade, the computations recover a slope close to the $-5/3$ predicted by theory, and which is the foundation of the theoretical analyses presented in the previous chapters. This indicates that the resolved turbulent scales have "physical" behavior.

More generally, the conclusions given by the various authors are the following.

- The dynamics of the numerical solution is consistent, i.e. the values produced are located within the bounds fixed by the collected set of experimental measurements and the topology of the simulated flow exhibits the expected characteristics (potential cone, annular mixing layer, and so forth).
- While the dynamics is consistent, it is nonetheless very difficult to reproduce a particular realization (for example with fixed potential cone length and maximum turbulent intensity).
- The numerical solution exhibits a strong dependency on many parameters, among which we find:
  - the subgrid model, which allows a more or less rapid transition of the annular mixing layer are consequently influences the length of the potential cone and the turbulent intensity by modifying the effective Reynolds number in the simulation. More dissipative models delay the development of the mixing layer, inducing the existence of a very long potential cone.
  - the inflow condition: the mixing layer transition is also strongly dependent on the amplitude and shape of the perturbations.
  - the numerical error, which can affect the turbulent of the annular mixing layer and of the developed jet, especially during the transition phases. Here it is a matter of an error controlled by the computational grid and the numerical method. A dispersive error will have a tendency to accelerate the transition and thereby shorten the potential cone. A dissipative error will have the inverse effect. With too coarse a grid, the annular mixing layer cannot be represented correctly, which can induce

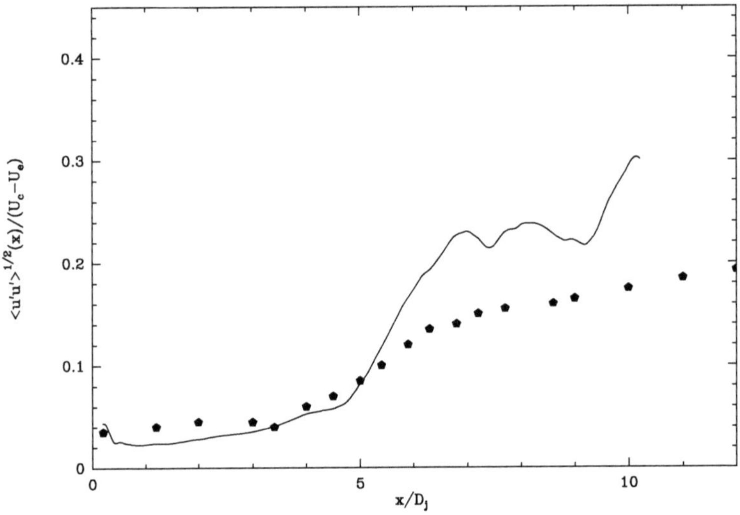

Fig. 14.9. Round jet. Axial evolution of the normalized longitudinal turbulent intensity. *Dots:* experimental data. *Line:* LES computation. Courtesy of P. Comte, LEGI.

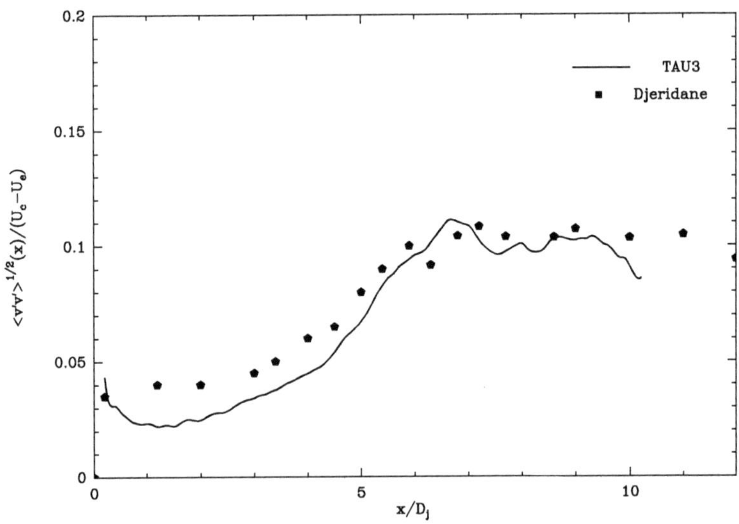

Fig. 14.10. Round jet. Axial evolution of the normal turbulent intensity. *Dots:* experimental data. *Line:* LES computation. Courtesy of P. Comte, LEGI.

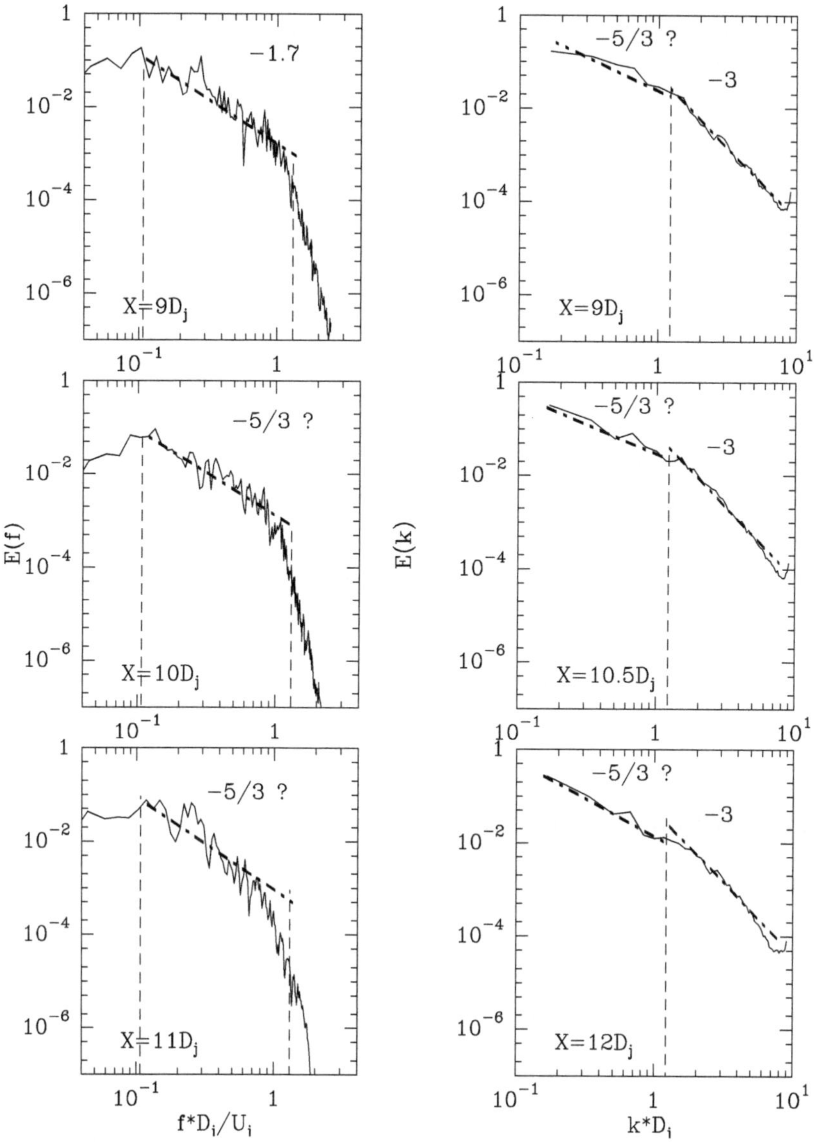

Fig. 14.11. Round jet. Time (*Left*) and space (*Right*) spectra of solved turbulent kinetic energy at different positions along the axis. Courtesy of P. Comte, LEGI.

it to thicken too quickly and thereby decrease the length of the potential cone.

– the size of the computational domain. The computation is sensitive to the size of the computational domain, which modulates the effect of the boundary conditions, especially the outflow condition.

– All the computations predict the dominant time frequency of the jet correctly, which is therefore not a pertinent parameter for analyzing the models finely.
– The quality of the simulation is not a global character. Certain parameters can be correctly predicted while others are not. This diversity in the robustness of the results with respect to the simulation parameters sometimes makes it difficult to define discriminatory parameters.

Other Examples of Free Shear Flows. Other examples of free shear flows in spatial expansion have been simulated:

– plane mixing layer (see [674, 38]);
– planar jet (see [159]);
– rectangular jet (see [272]);
– swirling round jet (see [472]);
– controlled round jet (see [157]);
– plane wake (see [269, 740, 237]).

The conclusions drawn from the analysis of these different cases corroborate those explained previously for the round jet as concerns the quality of the results and their dependency as a function of the computation parameters (subgrid model, grid, inflow condition, computational domain, and so forth). These conclusions are therefore valid for all free shear flows in spatial expansion.

14.3.2 Backward Facing Step

Problem description. The flow over a backward facing step of infinite span is a generic example for understanding separated internal flows. It involves most of the physical mechanisms encountered in this type of flow and is doubtless the best documented, both experimentally and numerically, of the flows in this category. Its dynamics can be decomposed as follows. The boundary layer that develops upstream of the step separates at the step corner, becoming a free shear layer. This layer expands in the recirculation region, thereby entraining turbulent fluid volumes. This entrainment phenomenon may influence the development of the shear layer, which curves inward toward the wall in the reattachment region and impacts with it. After the reattachment, the boundary layer re-develops while relaxing toward a profile in equilibrium. The topology of this flow is illustrated in Fig. 14.12, which is developed from large-eddy simulation data. We observe first the transition of the separated shear layer, the formation of vortex structures in the impact area, and then of hairpin structures in the boundary layer after the reattachment.

This flow brings out difficulties in addition to those of the round jet, because it adds the dynamics both of the free shear layers and of the near wall region.

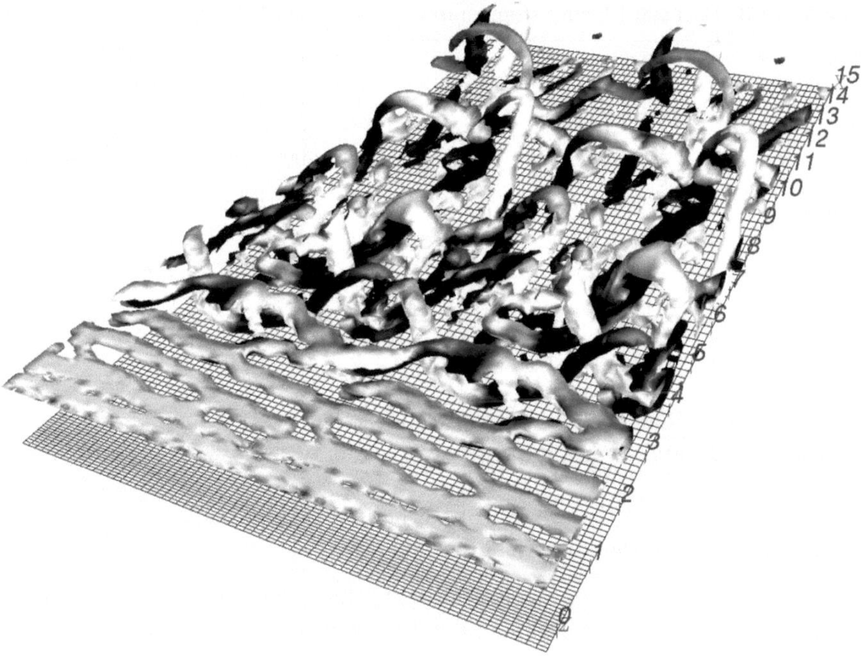

Fig. 14.12. Backward Facing Step. Iso-surface of instantaneous vorticity. Courtesy of F. Delcayre, LEGI.

A Few Realizations. The methods used for simulating this flow are illustrated by four computations presented in Table 14.3. The parameters indicated are.

- Reynolds number Re_H, referenced to the step height H and the inflow velocity profile;
- the dimensions of the computational domain, referenced to the length H;
- the number of grid points used;
- subgrid model used ("Sc" means the Schumann model (6.59); "MSM" the Mixed Scale Model (5.127); "SF" the Structure Function model (5.102); "DynLoc" the constrained localized dynamic model (5.207)). As before for the round jet, only subgrid viscosity models have been used in the configuration.
- inflow condition generation mode: $U + b$ means the same thing as before, while P designates the use of a precursor, which in this case is a large-eddy simulation of a plane channel flow in [226]; Ca indicates the use of an inflow channel to allow the development of a "realistic" turbulence upstream of the separation. Depending on the author, the length of this channel is between four and ten H.
- treatment of the solid walls: "–" is the no-slip condition; MSc the Schumann wall model, (10.28) to (10.30); MGz the Grötzbach wall model (10.31). It

Table 14.3. Backward facing step computation characteristics.

Ref.	[226]	[637]	[673]	[261]
Re_H	$1,65.10^5$	11200	38000	28 000
$Lx \times Ly \times Lz$	$16 \times 4 \times 2$	$20 \times 4 \times 2,5$	$30 \times 5 \times 2,5$	$30 \times 3 \times 4$
$Nx \times Ny \times Nz$	$128 \times 32 \times 32$	$201 \times 31 \times 51$	$200 \times 30 \times 30$	$244 \times 96 \times 96$
SGS Model	Sc	MSM	FS	DynLoc
Inflow	P	U+b	U+b	U+b,Ca
Wall	MSc	–	MLog	–
$O(\Delta x^\alpha)$	2	2+up3	2+up3	2
$O(\Delta t^\beta)$	2	2	2	3

can be seen that the use of wall models reduces the number of points considerably and makes it possible to simulate flows with high Reynolds numbers.
- spatial accuracy of the numerical method;
- time accuracy of the numerical method.

The results the various authors have obtained are generally in good qualitative agreement with experimental data: the flow topology is recovered and the realizations show the existence of coherent structures similar to those observed in the laboratory. On the other hand, there is much more difficulty obtaining quantitative agreement whenever this is possible at all (only reference [261] produces results in satisfactory agreement on the mean velocity field and turbulent intensity). This is due to the very high sensitivity of the result to the computation parameters. For example, variations of the order of 100% of the average length of the recirculation region have been recorded when the inflow boundary condition or subgrid model are manipulated. This sensitivity stems from the fact that the flow dynamics is governed by that of the separated shear layer, so the problems mentioned before concerning free shear flows crop up here. We also note a tendency to under-estimate the value of the mean velocity in the recirculation area. However, as in the case of the round jet, the simulated physics does correspond to that of a backward facing step flow. This is illustrated by the mean velocity profiles and resolved Reynolds stresses in Fig. 14.13, and the pressure spectra in Fig. 14.14. The good agreement with experimental data in the prediction of the mean field and Reynolds stresses proves the theoretical consistency of the computation. This agreement is even clearer if we analyze the spectra. Near the step, the mixing layer dynamics is dominated by frequencies associated with the Kelvin-Helmholtz instability. The predicted value of the dominant frequency is in very good agreement with experimental observations. The double peak at the second measurement point shows that the simulation is capable of reflecting the low-frequency flapping mechanism of the separated region, and still remain in good agreement with experimental observations.

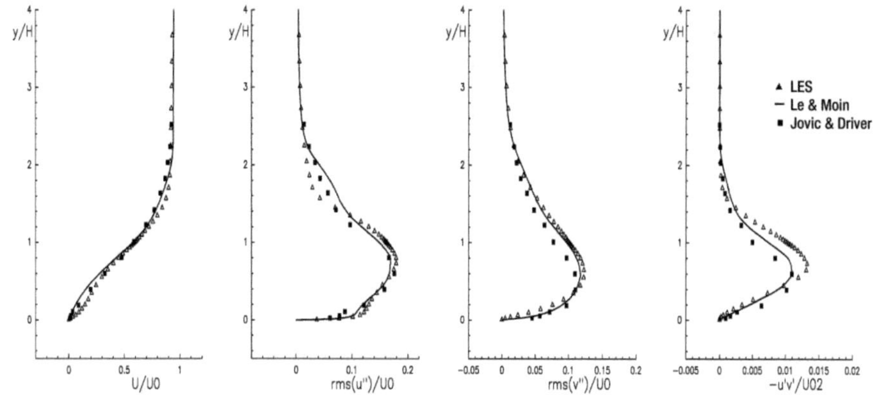

Fig. 14.13. Backward Facing Step. Mean velocity and Reynolds stresses profiles at the reattachment point. *Triangles* and *solid line:* LES computations. *Squares:* experimental data. Courtesy of F. Delcayre, LEGI.

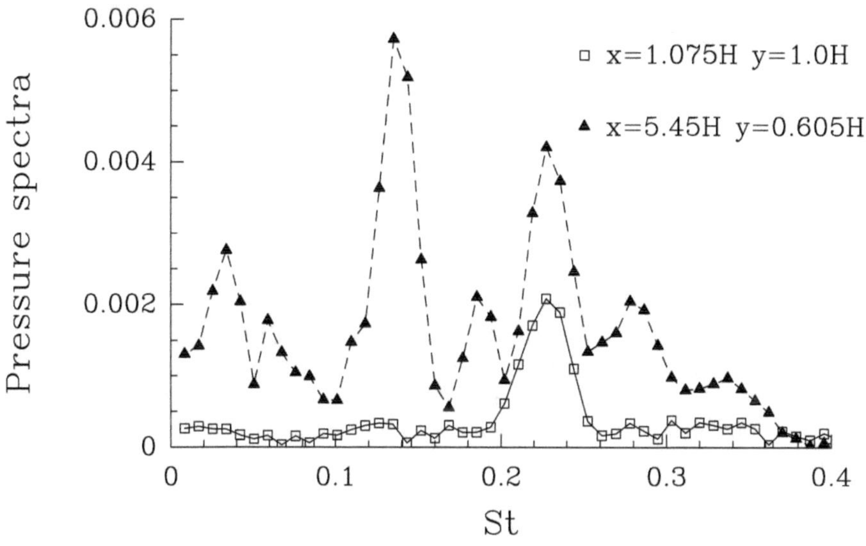

Fig. 14.14. Backward Facing Step. Pressure spectra. *Squares:* in the free shear layer near the step corner. *Triangles:* in the separated region near the reattachment point. Courtesy of F. Delcayre, LEGI.

Also, it seems that the use of wall models does not affect the dynamics of this shear layer greatly. It becomes possible to deal with higher Reynolds numbers, but at the price of losing some of the quality of the results as concerns the wall surface terms (friction, pressure coefficient) in the recirculation zone [87]. Solutions for this configuration turn out to be highly dependent on the subgrid model: a too dissipative model will delay the development of the separated shear layer, pushing away the position of the reattachment point.

The results are also found to be dependent on the size of the domain and the fineness of the mesh in the spanwise direction, because these parameters affect the development of the mixing layer emanating from the step corner. A spanwise domain width of 4 to 6 H is considered to be a minimum in order to be able to capture the three-dimensional mechanisms at low frequencies. Lastly, the time frequencies associated with the separated zone dynamics are robust parameters in the sense that they are often predicted with precision.

14.3.3 Square-Section Cylinder

Problem description. The square-section infinite-span cylinder is a good example of separated external flows around bluff bodies. This type of configuration involves phenomena as diversified as the impact of the flow on a body, its separation (and possible reattachment) on the body surface, the formation of a near-wake region, and alternating vortex street, and the development of the wake up to a self-similar solution. Each of these phenomena poses its own particular numerical and modeling problems.

Realizations. This flow was chosen as an international test case for large-eddy simulation, and has consequently served as a basis for many computations, which are mostly summarized in [618] (see also [555, 729, 730, 678, 233, 94, 652] for a discussion of this test case). The test case parameters are: a Reynolds number Re_D, referenced to the length D of the cylinder edge and the freestream velocity, equal to 22,000, and a computational domain of $20D \times 4D \times 14D$. The span is assumed to be infinite and a periodicity condition is used in this direction.

None of the sixteen computations collected in [618] produces an overall good agreement with experimental data, i.e. is capable of predicting all of the following parameters with an error of less than 30%: average lift and drag; drag and lift variances; main vortex shedding frequency; and average length of the separated region behind the cylinder. Average lift and drag, as well as the vortex shedding frequency, are very often predicted very satisfactorily. This is due to the fact that these quantities do not depend on the small scale turbulence and are governed by Von Karman structures, which are very large in size. The length of the recirculation region behind the cylinder is very often under-estimated, as is the amplitude of the mean velocity in this region. Also, the mean velocity in the wake is only very rarely in agreement the experimental data.

The numerical methods used are of moderate order of accuracy (at most second-order in space and third-order in time), and third-order upwind schemes are often used for the convection term. Only subgrid viscosity models have been used (Smagorinsky model, various dynamic models, Mixed Scale Model). Certain authors use wall models at the cylinder surface.

The lack of agreement with experimental data is explained by the very high sensitivity of the different driving mechanisms to the numerical errors and to the diffusion introduced by the models. So we again find here the problems mentioned above for the case of the backward facing step, but now they are amplified by the fact that, in order to master the impact phenomenon numerically, the numerical diffusion introduced is much stronger than in the former case. Also, as most of the grids used are Cartesian and monodomain, the resolution near the cylinder is too weak to allow a satisfactory representation of the boundary layers.

14.3.4 Other Examples

Many other flows have been examined by large-eddy simulation.

Among wall-bounded flows without separation, we may mention: flat plate boundary layer [237, 205, 463, 457, 585]; boundary layer on a curved surface in the presence of Görtler vortices [460]; flow in a circular-section toric pipe [653]; three-dimensional boundary layer [767, 366]; juncture flow [687]; flow in a rotating pipe [777, 215]; flow in a rotating square duct [576]; flow in a square/rectangular annular duct [774, 65].

Examples of recirculating flows are: confined coaxial round jets [9]; flow around a wing section of infinite span at incidence [347, 371, 348, 349, 425, 78, 502, 494] (see Fig. 14.15); flow in a planar asymmetric diffuser [367, 213, 372, 369]; flow around a cube mounted on a flat plate [618, 556, 757, 493, 652, 407]; flow around a circular-section cylinder [52, 519, 520, 362, 75, 410, 715, 77, 360, 76]; flow in tube bundles [620]; flow in a lid-driven cavity [799, 179]; flow in a ribbed channel/duct [779, 142, 778, 151, 361]; jet

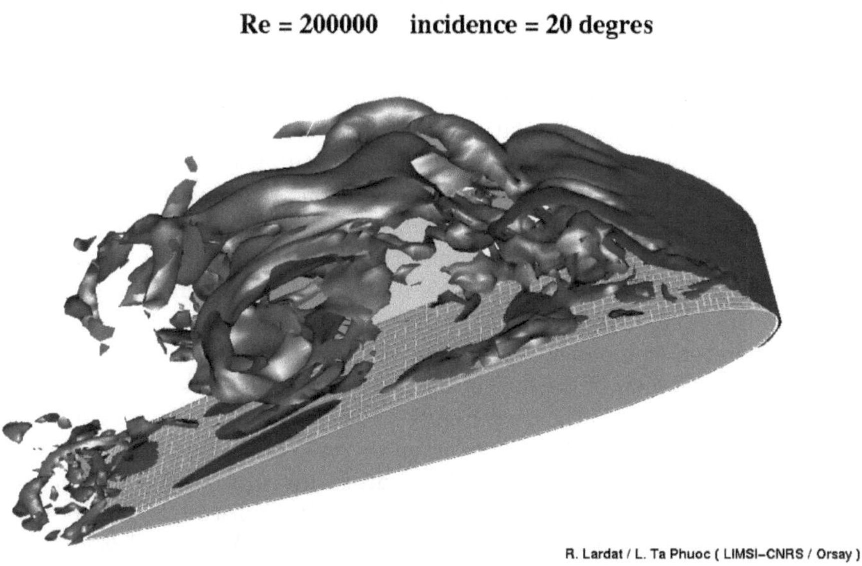

Fig. 14.15. Flow around a wing at high incidence: isosurface of instantaneous vorticity. Courtesy of R. Lardat and L. Ta Phuoc, LIMSI.

impacting a flat plate [739, 616, 574, 51, 718]; boundary layer on a wavy surface or a bump [412, 197, 770, 302, 768, 769, 642, 522]; flow over a swept wedge [330]; flow past a blunt trailing edge [484, 754, 564, 571]; separated boundary layer [88, 758]; aircraft wake vortices [286]; axisymmetric piston-cylinder flow [732]; flow around an oscillating cylinder [721]; flow around a road vehicle [731, 408]; flow around a 3D wing [329]; flow around a square cylinder [399, 393]; flow around a forward-backward facing step [5]; flow in reversing systems [61]; flow in a blade passage [147, 516]; flow pas a swept fence [178]; flow over a sphere [146].

14.4 Industrial Applications

14.4.1 Large-Eddy Simulation for Nuclear Power Plants

The increasing power of supercomputers and of numerical description of unsteady flows allows the simulation of complex flows relevant to industrial configurations in accident scenarios obviously not available through experiments. Simulation is used to capture large features of the flow and to focus on major events, at least concerning qualitative behavior. The present example deals with induced rupture of the primary circuit in a pressurized water reactor. The simulation domain includes the reactor vessel, the steam generator and their connection through the *hot leg* (see Fig. 14.16): the dimensions of the real domain cover several meters. For the calculation, the steam generator tubes have been grouped into nine, thus decreasing the number of control volumes (a model has been developed to take into account this change in the geometry). Furthermore, the tube's length has been reduced in order to limit the computer time consumption.

In the case of the total loss of the heat sink (very hypothetical severe accident scenario) the flow circulation driven by the primary pump is stopped, the coolant is pure vapour ($T > 1000\,^{\circ}$C) and natural convection develops in the circuit. The question here is to investigate the temperature distribution in the wall of the hot leg and in the tubes of the steam generator, in order to analyze their mechanical constraints. For that it is crucial to get the best discretization of the related zones. This is achieved by concentrating the grid points in this area. Figure 14.16 shows the mesh of each tube of a "simplified steam generator" composed of 400 tubes (3300 in reality). Figure 14.17 (top) provides a global view of the flow in the hot leg, showing a clear stratification and a return of flow into the core. Figure 14.17 (bottom) displays a cut in the hot leg at different locations, showing an increase of turbulence and mixing as the section is closer to the steam generator: this mixing leads to homogenization of the temperature and reduced mechanical constraints. The present simulation was performed using the TRIO-U/PRICELES code developed at CEA Grenoble (in the DTP/SMTH/LDTA laboratory): it required around 10^6 grid points using a tetrahedral discretization. The underlying numerical

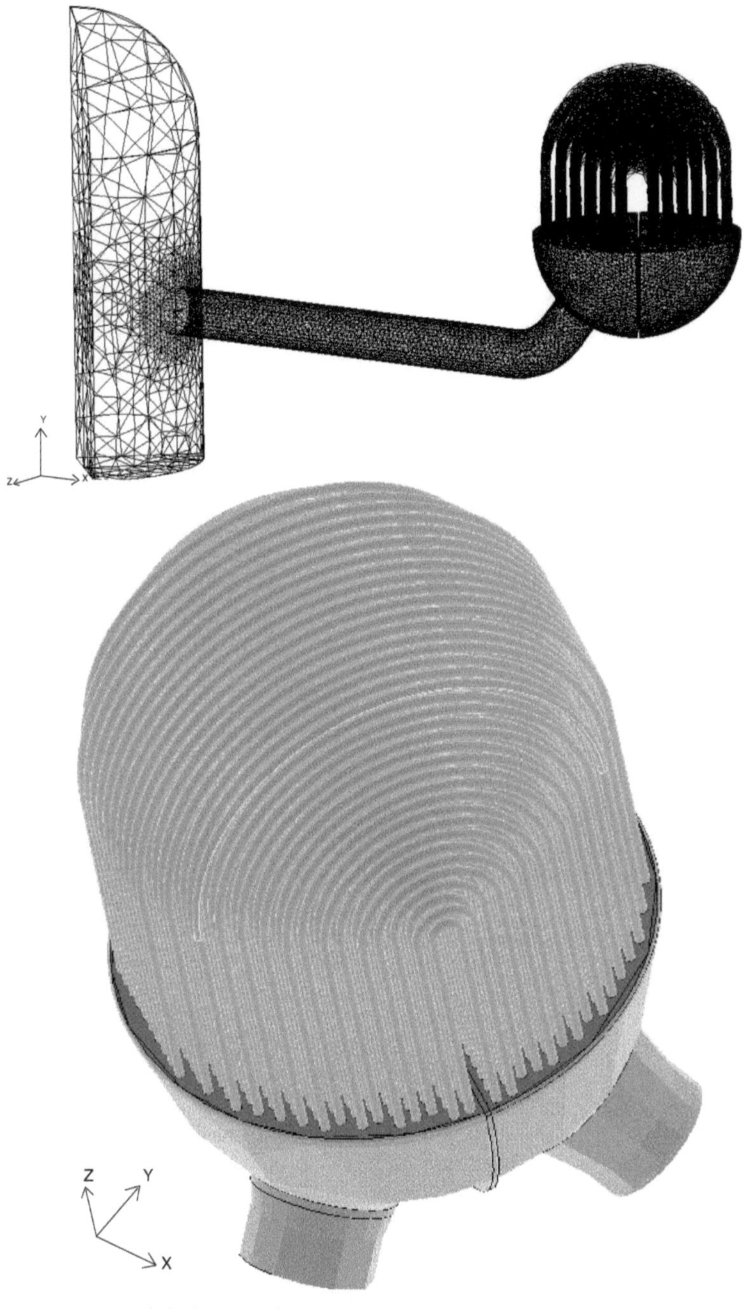

Fig. 14.16. *Top*: global view of the unstructured mesh. A coarse discretization is chosen for the vessel, taking into account here a *realistic boundary condition*. The resolution is finer in a region of interest, say the steam generator region (tubes and plenum). *Bottom*: local view of the mesh of the steam generator: 400 individual tubes are independently meshed. Courtesy of IRSN.

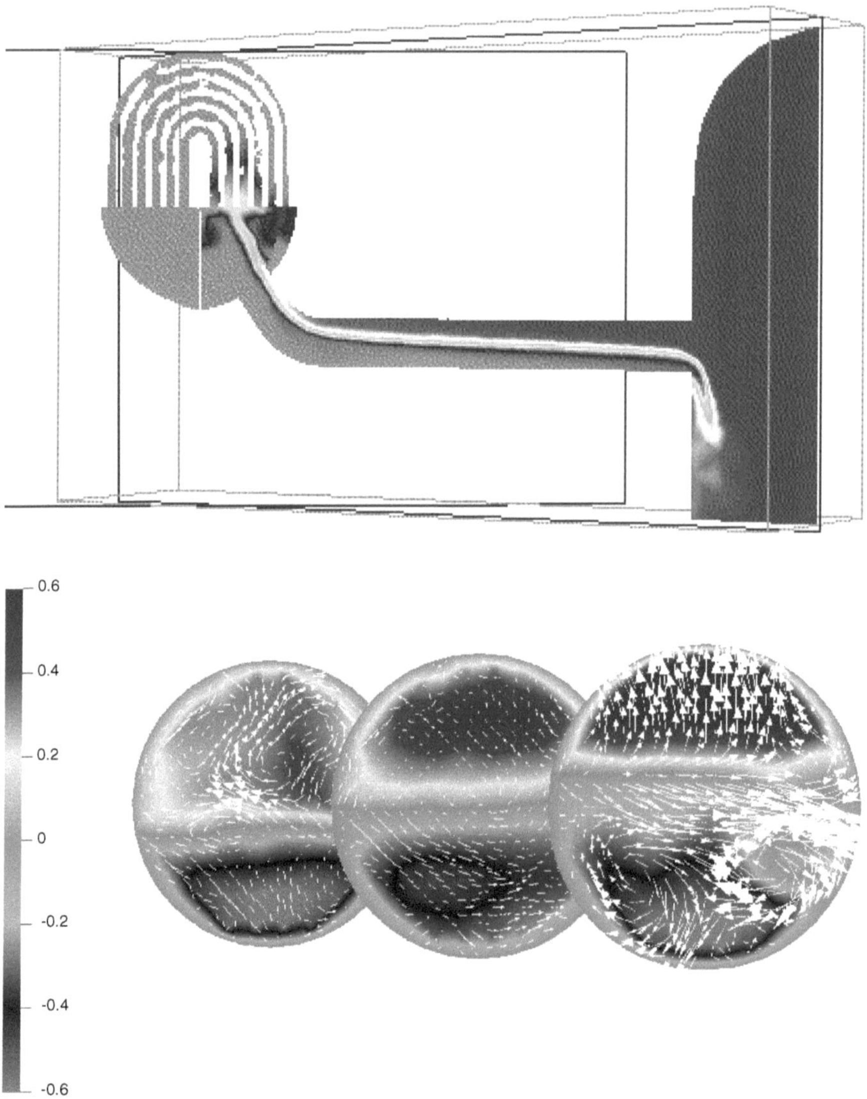

Fig. 14.17. *Top*: a cut in the instantaneous temperature field in the whole system, showing a stratification in the "hot leg" that propagates in the direction of the steam generator tubes; a back flow develops in the vessel. *Bottom*: cuts of the same field near the steam generator, plus velocity vectors. Stratification is perturbated by the turbulent flow. Courtesy of IRSN.

method is a centered second-order accurate Finite Element based discretization, used together with a standard Smagorinsky model. All simulations were performed by U. Bieder (CEA) and H. Mutelle (IRSN) for the French Nuclear Safety Institute (IRSN).

For other situations where experimental data are available, qualitative as well as quantitative features are looked at, as in more traditional disciplines.

14.4.2 Flow in a Mixed-Flow Pump

This example of the use of large-eddy simulation for turbomachinery flows is due to C. Kato et al. [380]. These authors computed the internal flow in a mixed-flow pump stage with a high designed specific speed.

A view of the computational domain is shown in Fig. 14.18. The computational domain includes: the upstream inlet pipe, the two regulation plates, a four-blade impeller and the diffuser, leading to a very complex configuration, both from the geometrical and physical points of view.

In order to take the rotation into account, the authors make use of a Chimera-type technique: a moving boundary interface in the flow field is treated with overset grids from multiple frames of reference. A computational grid that rotates along with the impeller is used to compute the flow within the impeller, and stationary grids are used for stationary parts of the pump. Each grid includes appropriate margins of overlap with its neighboring grids in order to allow appropriate interpolations, and coordinate transformation to take into account the different frames of reference.

Two mesh resolutions have been investigated: a coarse grid with a total of 1.7×10^6 points and a fine grid with 5×10^6 points. A partial view of the surface mesh on the impeller is shown in Fig. 14.19. The boundary conditions are the following: no-slip boundary conditions are used on solid walls, and the turbulent pipe flow at the inlet is obtained by performing an auxiliary large-eddy simulation of an infinite pipe flow.

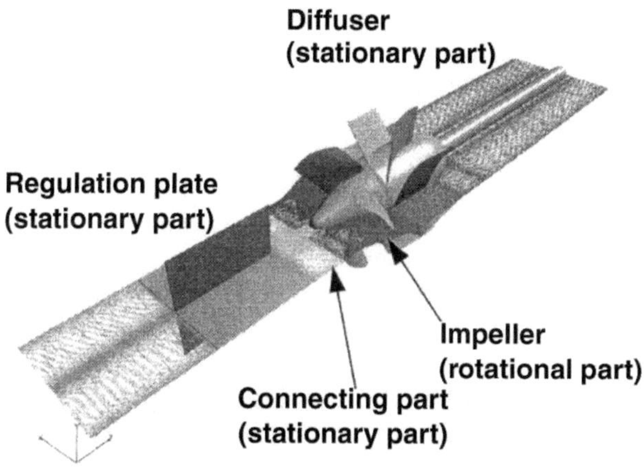

Fig. 14.18. Flow in a mixed-flow pump. View of the global computational domain. Courtesy of C. Kato, University of Tokyo.

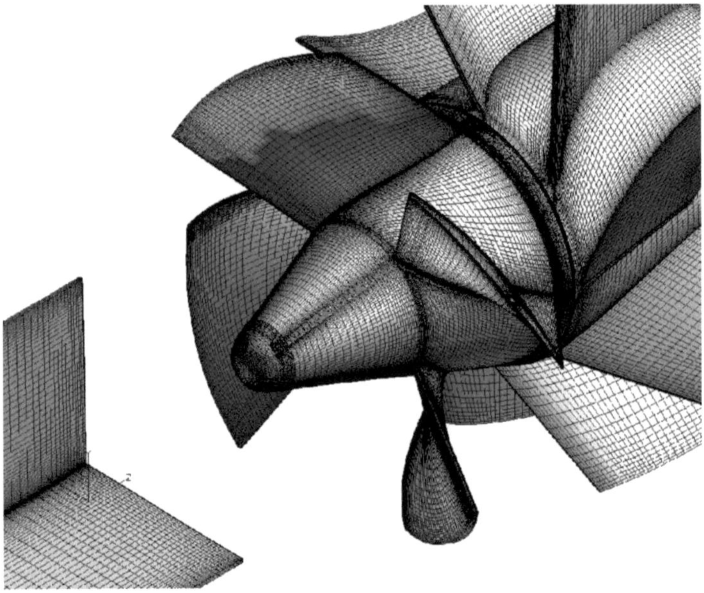

Fig. 14.19. Partial view of the surface mesh. Courtesy of C. Kato, University of Tokyo.

A classical Smagorinsky model is employed, with a Van Driest damping function enforcing a vanishing subgrid viscosity in the near-wall region. A streamwise-upwind finite-element method is used to discretize the Navier–Stokes equations. Time integration is based on the explicit Euler scheme, but shifts the spatial residuals of the governing equations in the upstream direction of the local flow. The magnitude of this shift is one half of the time increment multiplied by the magnitude of the local flow velocity, yielding an exact cancellation of the first-order error terms. The resulting method is second-order accurate in space and time.

Several operating conditions of the pump have been investigated using large-eddy simulations. All of them correspond to partial load conditions, i.e. off-design operating conditions.

Comparison of computed and measured head-flow characteristics is shown in Fig. 14.20, showing a very good agreement between fine-grid simulations and experiments. Coarse grid simulation is seen to yield larger discrepancies when design operating conditions are approached, because the mesh is not fine enough to capture accurately the attached boundary-layer dynamics. The normalized head-flow characteristic is defined here as the relation between the flow-rate of the pump and its total-pressure rise. After stall onset ($Q/Q_{\rm d} \leq 0.55$, where Q is the mass flowrate), both grids yield very similar descriptions of the flow. As in the previous case, $Q_{\rm d}$ is the mass flow-rate related to the design point.

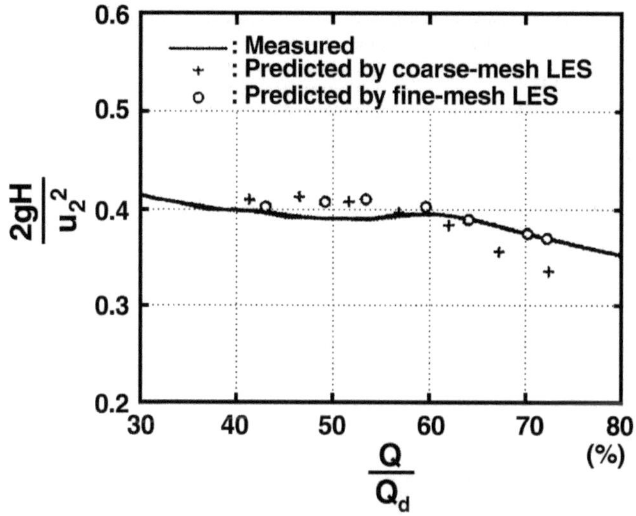

Fig. 14.20. Comparison of measured and predicted head-flow characteristics as a function of the mass flow rate Q with reference to the design flow rate Q_d. Courtesy of C. Kato, University of Tokyo.

An example of comparisons of computed and experimental data is shown in Fig. 14.21.

14.4.3 Flow Around a Landing Gear Configuration

The flow around a realistic landing gear configuration was studied by Souliez et al. [679]. The main purpose of this work was the prediction of the far-field noise generated by the turbulent fluctuations.

The most complex geometry includes four wheels and two lateral struts. An unstructured mesh with 135 000 triangles on the surface of the landing gear and 1.2 million tetrahedral cells was used.

The simulation is run without an explicit subgrid model, and then belongs to the MILES-like group of large-eddy simulations. The PUMA code used for this simulation is based on the finite volume approach. Explicit time-integration is carried out using a Runge–Kutta scheme.

Typical results dealing with the mean surface pressure and the topology of the instantaneous flow are shown in Figs. 14.22 and 14.23.

14.4.4 Flow Around a Full-Scale Car

Large-eddy simulations is also used in the automotive industry during the design of new cars. The main purposes are the prediction of traditional aerodynamic parameters (drag, lift, etc.) and aeroacoustics.

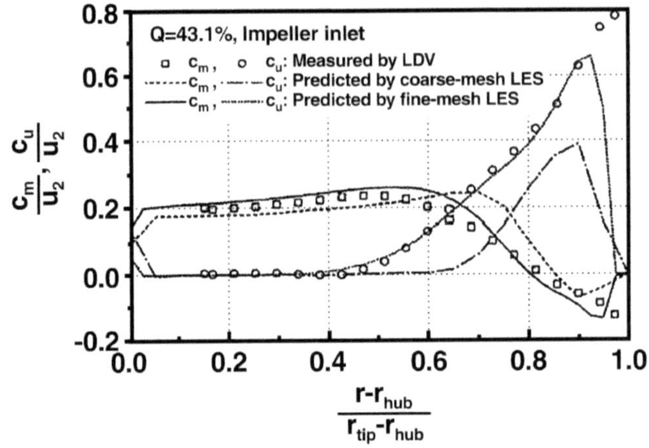

Fig. 14.21. Predicted and measured phase-averaged velocity profiles at the impeller's inlet cross-section for $Q/Q_{\mathrm{d}} = 0.43$ (post-stall operating condition). Courtesy of C. Kato, University of Tokyo.

The example presented here deals with the large-eddy simulations of the flow around a full-scale car model. The wheels and the floor are fixed, as in the wind tunnel experiments. The Reynolds number is 2 684 563 per meter.

The numerical simulation is carried out using the Powerflow code, which is based on the Lattice Boltzmann approach (see [127] for an introduction). A Cartesian grid is used, which is composed of 20×10^6 cells. The size of the smallest mesh is 7 mm. Subgrid scales are taken into account using a two-equation $k - \tau$ subgrid model. A wall model is used at solid boundaries.

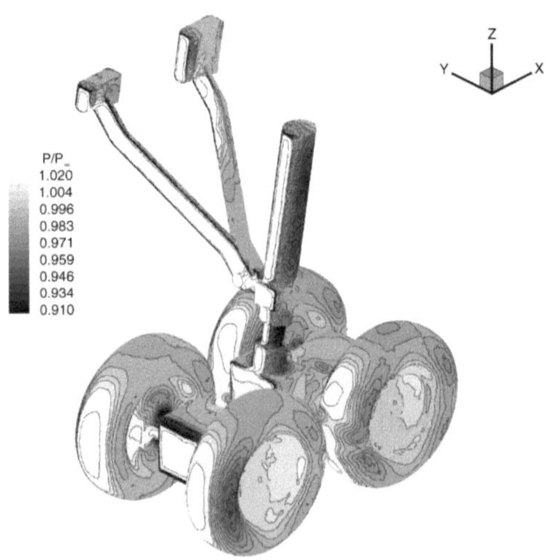

Fig. 14.22. Predicted mean surface pressure on the landing gear. Courtesy of F. Souliez and L. Long, Pennsylvania State University.

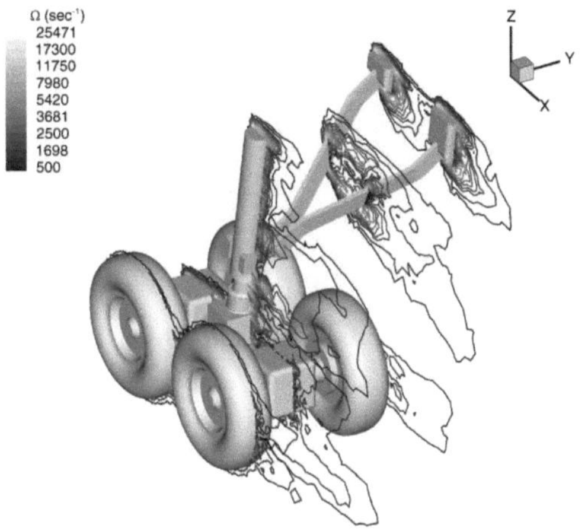

Fig. 14.23. Instantaneous vorticity filaments. Courtesy of F. Souliez and L. Long, Pennsylvania State University.

Typical results are displayed in Figs. 14.24–14.27. The very complex topology of the flow is recovered, exhibiting very intense vortical structures and pressure variations. The drag is predicted to be within 1% error in comparison with experimental data.

14.5 Lessons

14.5.1 General Lessons

We can draw the following lessons concerning the large-eddy simulation technique from the computations mentioned above:

- When the technique is used for dealing with the ideal case in which it was derived (homogeneous turbulence, optimal numerical method), it yields very good results. The vast majority of subgrid models produce results that are indiscernable from reality, which removes any discriminatory character from this type of test case, which in fact can only be used to assess the consistency of the method.
- Extending the technique to inhomogeneous cases brings up many other problems, concerning both the physical modeling (subgrid models) and the numerical method. The latter point becomes crucial because the use of numerical methods of moderate order of accuracy (generally two) greatly increases the effect of the numerical error. This is accentuated by the use of artificial dissipations for stabilizing the simulation in "stiff" cases (strong under-resolution, strong gradients). This error seems to be reducable by refining the computational grid, which is done more and more by using adaptive grids (local adaptation or enrichment).

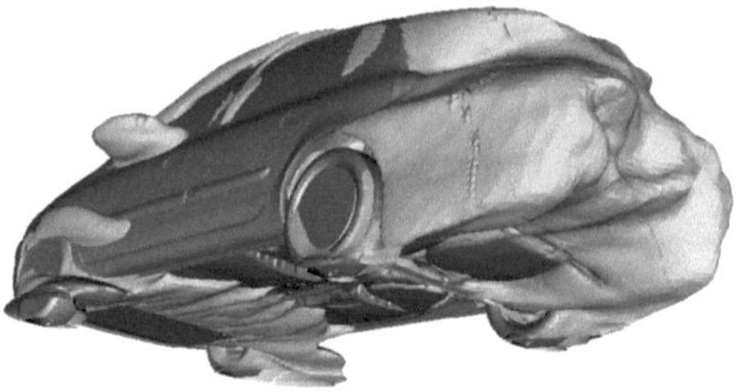

Fig. 14.24. Isocontours of total pressure. Courtesy of Renault-Aerodynamic Department.

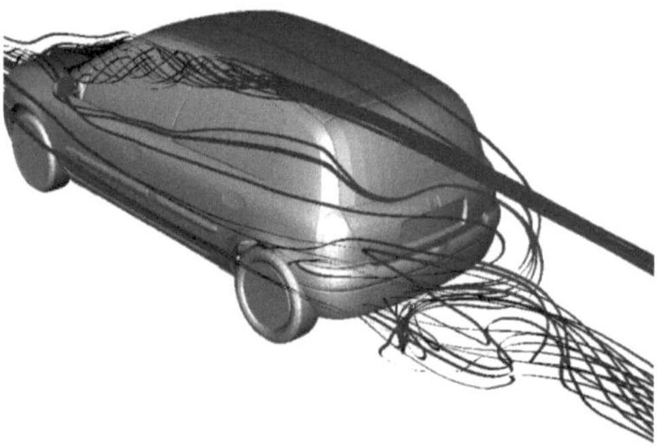

Fig. 14.25. Streamlines. Courtesy of Renault-Aerodynamic Department.

Fig. 14.26. Streamlines. Courtesy of Renault-Aerodynamic Department.

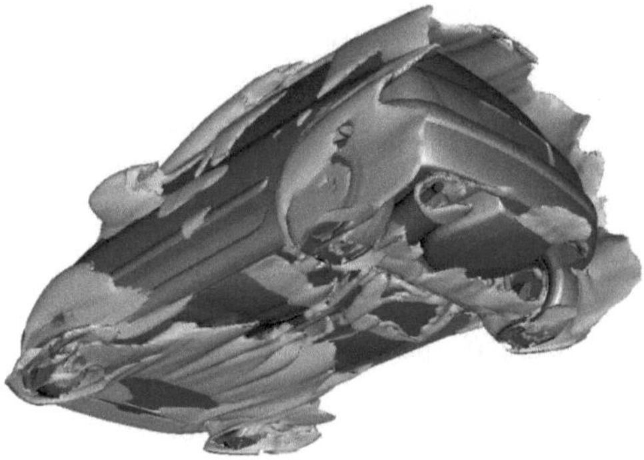

Fig. 14.27. Streamlines. Courtesy of Renault-Aerodynamic Department.

- Shear flows show themselves to be very strongly dependent on the inflow condition when this is unsteady. Generating these conditions is still an open problem.
- The quality of the results is variable but, for each configuration, robust, correctly predicted parameters exist. The physics simulated is often consistent in that it exhibits the generic features that are observed experimentally but does not necessarily correspond to a desired target realization. This is due to the dependency on the many simulation parameters.
- The quality of the results is subordinate to the correct representation of the flow driving mechanisms (transition, near-wall dynamics, and so forth). Low numerical error and consistent modeling are therefore mandatory in those regions where these mechanisms occur. The other regions of the flow where the energy cascade is the dominant mechanism are of lesser importance.
- When the flow dynamics becomes complex, subgrid viscosity models are often used. This is because they provide a clear kinetic energy dissipation and therefore stabilize the simulation. This stabilizing character seems to become predominant compared with the physical quality of the modeling insofar as the numerical difficulties increase (with the presence of strong shear and highly heterogeneous grids, and so forth).
- There is a consensus today that the numerical method used must be accurate to at least the second order in space and time. First-order accurate numerical dissipations are totally proscribed. Third-order accurate methods in time are rarely used. As concerns the spatial accuracy, satisfactory results are obtained by certain authors with second-order accurate methods, but higher-order accurate schemes are often used. Numerical stabilization methods (upwind scheme, artificial dissipation, smoothing, and so forth) should be used only when absolutely necessary.
- Large-eddy simulation is presently a powerful tool for investigating massively separated industrial flows, where the large scales and the turbulence production is not driven by fine details of the dynamics of the boundary

layers. For other flows, such as fully attached flows, large-eddy simulation is still too expensive to be used on full-scale configurations for realistic Reynolds numbers because of the huge number of grid points required to obtain an accurate resolution of the boundary layers. Existing wall models have not been validated in realistic cases up to now, and the question of their validity for such flows remains an open question.

14.5.2 Subgrid Model Efficiency

Here we will try to draw some conclusions concerning the efficiency of the subgrid models for processing a few generic flows. These conclusions should be taken with caution. As we have seen all through this book, very many factors (numerical method, grid, subgrid model, and others) are involved and are almost indissociable, so it is very difficult to try to isolate the effect of a model in a simulation. The conclusions presented are statistical in the sense that they are the fruit of an analysis of simulations performed on similar (at least geometrically) flow configurations with different methods. A "deterministic" analysis could lead to contradictory conclusions, depending on the other. Also, there is no question of ranking the models, as the available information is too lacunary to draw up a reliable list. Lastly, very many factors like the discretization of the subgrid models still remain to be studied.

We may, however, sketch out the following conclusions.

1. To simulate a homogeneous flow:
   a) All subgrid models including a subgrid viscosity yield similar results. The efficiency of functional models for the forward energy cascade in isotropic turbulence, despite their lack of accuracy in representing the subgrid tensor eigenvectors, was analyzed by Jimenez [351]. It is explained by the existence of a feedback between the resolved scales and the net energy drain provided by the subgrid model. Errors in subgrid models are localized at the highest resolved frequency, and do not contaminate the low frequency which is responsible for the turbulence production. The error accumulation at high frequencies leads to an adjustment of the subgrid model, which is expressed as a function of the resolved scales. This is easily seen by writing the induced subgrid dissipation: $\varepsilon = -\tau_{ij}\overline{S}_{ij} = 2\nu_{\rm sgs}|\overline{S}|^2$. A classical example is the Smagorinsky model, which leads to $\varepsilon = C_{\rm S}|\overline{S}|^3$. A local underestimation of $C_{\rm S}$ will result in an energy accumulation at the cutoff, leading to an increase of the resolved shear $|\overline{S}|$ and an increase of the net drain of kinetic energy. This drain mostly affects the highest resolved frequencies, leading to a decrease of $|\overline{S}|$. The global effect on the flow is small, because the highest resolved frequencies contain a small part of the total resolved kinetic energy. The efficiency of this adjustment depends of course on the way the subgrid viscosity is computed.

For anisotropic flows the loss of efficiency of the basic subgrid viscosity models is a well recognized fact. This is explained by the fact that the shear magnitude $|\overline{S}|$ is not governed by the highest resolved frequency any more, but by very large scales. Thus, the dynamic feedback loop described above is not sufficient any more to yield good results. Self-adaptive models (dynamic, filtered, selective, etc.) must be used to ensure the quality of the results. The available experimental data suggest that the correlation obtained using the subgrid-viscosity model is of the order of 20%. This implies that it is impossible to get both the right energy spectrum and the right stresses from a subgrid-viscosity model if its effect on the resolved field is not negligible. Grid refinement is known to yield improved results for two reasons: the error is commited on a smaller fraction of the total kinetic energy, and the Kolmogorov theory predicts that the small scales are more isotropic than the large ones, rendering the structural error less important. For a weak shear S, the Kolmogorov theory leads to the following scaling laws for the normal-stresses spectrum E_{11} and off-diagonal stress cospectrum E_{12}:

$$E_{11} \propto \varepsilon^{2/3} k^{-5/3}, \quad E_{12} \propto S \varepsilon^{1/3} k^{-7/3} \quad ,$$

and

$$\tau_{12} \propto (\pi L_S/\overline{\Delta})^{-4/3}, \quad \tau_{12}/\tau_{11} \propto (\pi L_S/\overline{\Delta})^{-2/3} \quad ,$$

where the shear length $L_S = \sqrt{\varepsilon/S^3}$ is proportional to the integral scale L_ε in equilibrium flows. These scaling laws show that refining the resolution, i.e. decreasing $\overline{\Delta}$, makes it possible to reduce quickly the anisotropy of the subgrid scales. Jimenez estimates that 1% error is obtained for $L_\varepsilon/\overline{\Delta} = 10$–$20$.

b) Scale similarity or soft-deconvolution-type models do not yield good results if used alone. This is also true for all the other types of flows. A possible explanation for the improvement observed when using mixed models or explicit random models for the backward energy cascade is that these models weaken the spurious correlation between the resolved strain rate tensor and the modeled subgrid tensor. This improvement is also expected from theoretical results dealing with the full deconvolution problem or the rapid/slow decomposition introduced by Shao et al. It is important to note that on very coarse grids mixed models may not result in improved results.

c) These results hold locally for all other flows.

2. To simulate a free shear flow (mixing layer, jet, wake):
   a) Subgrid viscosity models based on large scales can delay the transition. This problem can be remedied by using a dynamic procedure, a selection function, or an accentuation technique. The other subgrid viscosity models seem to allow the transition without any harmful effects.

   b) Using a mixed structural/functional model improves the results obtained with a subgrid viscosity model based on the large scales.
3. To simulate a boundary layer or plane channel flow:
   a) Subgrid viscosity models based on the resolved scales may inhibit the driving mechanisms and relaminarize the flow. As before, this problem is resolved by using a dynamic procedure, selection function, or accentuation technique. The other subgrid viscosity models do not seem to exhibit this defect.
   b) Using a mixed functional/structural model can improve the results by better taking the driving mechanisms into account.
   c) Using a model for the backward cascade can also improve the results.
4. For separated flows (backward facing step, for example), use a model that can yield good data on a free shear flow (to capture the dynamics of the recirculating area) and on a boundary layer (to represent the dynamics after the reattachment point).
5. For transitional flows:
   a) Subgrid viscosity models based on the gradients of the resolved scales generally yield poor results because they are too dissipative and damp the phenomena. This problem can be remedied by using a dynamic procedure, a selection function, or the accentuation technique.
   b) Anisotropic tensorial models can inhibit the growth of certain three-dimensional modes and lead to unexpected scenarios of transition to turbulence.
6. For fully developed turbulent flows, the problems with subgrid viscosity models based on the large scales are less pronounced than in the previous cases. Because these flows have a marked dissipative character, they produce results that are sometimes better than the other models because they ensure numerical stability properties in the simulation.

14.5.3 Wall Model Efficiency

Numerical experiments show that wall stress models based on a linear relationship between wall stresses and the instantaneous velocity component yield satisfactory results for well-resolved large-eddy simulation of attached flows.

For separated flows, this class of wall models is inadequate, leading to a poor prediction of the skin friction inside the separation zone.

For very coarse grids, i.e. when the cutoff is not located inside the inertial range of the spectrum, large errors are observed and the mean velocity profile is not recovered [363, 806, 89]. Several reasons for this can be identified:

- On coarse grids, the numerical error is large and pollutes the solution, yielding erroneous input for the wall model.
- On coarse grids, most subgrid models induce large errors on the resolved scales. This is especially true for all subgrid viscosity models, which are

not able to account for the strong flow anisotropy in the near-wall region. It has been shown [363, 806] that this error is mainly due to the fact that the subgrid acceleration term in the filtered momentum equations is not properly predicted on very coarse grids.[2] This bad prediction can lead to the occurrence of spurious coupling with linear wall models, yielding very bad results. The use of dynamic models may result in worse results if the test filter is applied outside the inertial range [89].

A very important conclusion of studies dealing with suboptimal-based wall models is that predicting the mean velocity profile and recovering the best possible rms velocity profile seem to be competing objectives.

Wall models based on auxiliary simulations performed on secondary embedded grids (RANS, thin boundary layer equations) may be an alternative, which remains to be assessed in critical situations. The use of more complex subgrid models, at least in the resolved near-wall region, which are able to predict the subgrid acceleration is also expected to improve the response of wall stress models. Multilevel subgrid models have demonstrated a clear superiority in academic test cases.

14.5.4 Mesh Generation for *Building Blocks* Flows

We discuss here the basic rules for mesh generation for two classical *building blocks* of complex flows, namely the attached equilibrium boundary layer and the plane mixing layer.

The two basic rules are:

- The key idea is that the driving mechanisms, i.e. the events responsible for turbulence production and mean profile instabilities, must be correctly captured by the simulation to recover reliable results.
- The size of the computational domain must be larger than the correlation length of the fluctuations, in each direction of space. Too small a domain size will yield spurious coupling between the dynamics of the flow and the boundary conditions.

Exact values of mesh size, number of grid points and their repartition, and domain size for 'plug and play' simulations are not available: these parameters depend on many parameters, including numerical methods. What are presented below are the commonly accepted ideas underlying the design of a large number of published works.

Equilibrium Boundary Layer. As discussed in Sect. 10.2, the boundary layer exhibits two different scalings.

- In the inner layer the viscous length l_τ is relevant to describe the dynamics. The typical correlation lenghts are 1000 wall units in the streamwise

[2] It is recalled that subgrid viscosity models are designed to yield the correct amount of dissipation, not the proper subgrid acceleration.

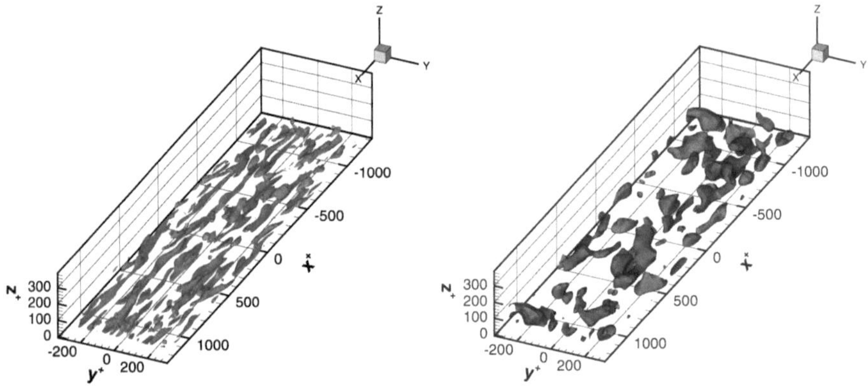

Fig. 14.28. Streaks in the near-wall region of a plane channel flow ($Re_\tau = 180$). *Left*: well-resolved large-eddy simulation. *Right*: Coarse-grid large-eddy simulation. Courtesy of E. Montreuil, ONERA.

(x) direction, and 100 wall units in the spanwise direction (y). Based on a study of the anisotropy of the flow, Bagget et al. [32] derived the following general criterion: $\overline{\Delta}/L \simeq 0.1$, where the local inertial length $L = k^{3/2}/\varepsilon$ is associated with the peak of the turbulent energy spectrum (with k the turbulent kinetic energy and ε the dissipation rate). Typical criteria dealing with the mesh size for a wall-resolving simulation are:

$$\Delta x^+ \simeq 50, \quad \Delta y^+ \leq 15 \quad . \tag{14.1}$$

The first grid point must be located at one wall unit from the wall, with three grid points in the viscous region $1 \leq z^+ \leq 10$. Thirty to fifty grid points across the boundary layer are generally enough to get acceptable results. Grids which are too coarse to allow a good resolution of the near-wall events usually yield to the occurrence of an overshoot in the streamwise rms velocity profile. This peak is associated with fat streaky structures (see Fig. 14.28). In order to prevent spurious correlations the minimum size of the computational domain in the streamwise direction should be larger than 2000 wall units, with a spanwise extent of 400 wall units.

– In the outer layer, the visual boundary layer thickness δ_{99} is the relevant scale. The large scales have a correlation length which scales with δ_{99}, and are advected in the streamwise direction at a speed roughly equal to 0.8 U_∞. Consequently, a minimum domain size of $3\,\delta_{99}$ to $5\,\delta_{99}$ is required in the streamwise direction, with a spanwise extent of $2\,\delta_{99}$ to $3\,\delta_{99}$.

Plane Mixing Layer. The choice of the extent of the computational domain and the mesh size for the plane mixing layer is often guided by results from the linear instability theory. For a hyperbolic tangent type mean velocity profile, it is known that the two-dimensional linearly most unstable mode has a streamwise wavelength λ_x nearly equal to $7\delta_\omega$, where the vorticity thickness is defined as $\delta_\omega = \Delta U / \max(d\langle U \rangle / dz)$, with ΔU the velocity jump across the shear layer. The associated spanwise length scale is $\lambda_y \simeq 2\lambda_x/3$.

Classical analysis of numerical errors shows that at least 5 to 20 grid points per wavelength are required to get a reliable description of physical phenomena. Thus, in the streamwise direction, large scales associated with Kelvin–Helmholtz instability should be captured with $\Delta x \simeq \delta_\omega/2$. Square-like meshes in the (x, y) plane are recommended, yielding $\Delta x = \Delta y$. Numerical experiments show that a minimum of 20 grid points is required across the shear layer, yielding $\Delta z \simeq \delta_\omega/20$.

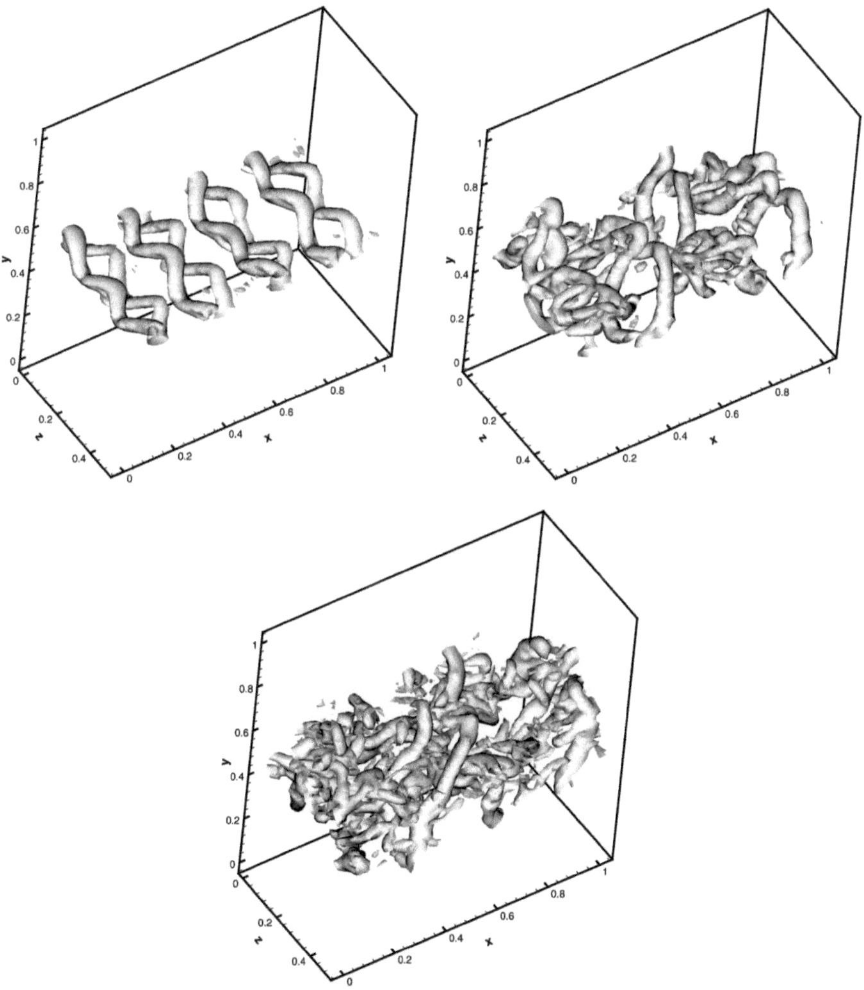

Fig. 14.29. Large-eddy simulation of a time-developing plane mixing layer. Instantaneous view of the coherent structures at three different times. *Left*: early transition stage. *Right*: advanced transition stage. *Bottom*: fully developed flow. Courtesy of M. Terracol, ONERA.

The size of the domain is dictated by the purpose of the simulation. For unforced mixing layers, the turbulent self-similar state is reached after the second pairing. For temporal simulations, this indicates that the size of the domain in the streamwise direction must be greater than or equal to $8\lambda_x = 56\,\delta_\omega$. The need for a third pairing originates from the fact that, after the last pairing permitted by the size of the computational domain, fluctuations are correlated over the domain, leading to corrupted results. The self-similar state can be observed between the second and the last pairing.

Typical results for a time-developing plane mixing layer are displayed in Fig. 14.29.

15. Coupling with Passive/Active Scalar

15.1 Scope of this Chapter

This chapter is devoted to the extension of the previous results dealing with Large-Eddy Simulation of incompressible flows to the case where a scalar quantity is added to the velocity and the pressure to describe the physical system. Depending on the application, the scalar can be related to the temperature (or temperature increment with respect to a given value), the density, the concentration of a pollutant, ...

The chapter details two different cases:

1. The case of the *passive scalar* (Sect. 15.2), where there is no feedback of the scalar equation in the momentum equation. In this regime, which represents a one-way coupling between the velocity field and the scalar field, the scalar dynamics is enslaved to the velocity field, while the dynamics of the later is not affected. The new closure problem is therefore restricted to the scalar equation, the treatment of the momentum equation being exactly the same as in the previous chapters.
2. The case of the *active scalar* (Sect. 15.3), which corresponds to physical systems in which a two-way coupling exist between the velocity and the scalar. The emphasis will be put on simple (destabilizing) buoyancy and stabilizing stratification effects. In the first case the turbulent production is enhanced by the coupling with the scalar dynamics, while in the second case a new turbulence damping mechanism is involved. These two simple cases are used to illustrate the new problems dealing with subgrid modeling induced by the active scalar model: the filtered scalar equation need to be closed, but the subgrid closure in the momentum equation must a priori be changed to account for the fact that the dynamics is now more complex than the simple kinetic energy cascade process.

For the sake of simplicity, the chapter is not made exhaustive: in many cases, strategies developed to close the scalar equation are nothing but direct extrapolations of models/methods developed to close the mometum equations. In these cases, the models for the scalar equations will not be extensively described, and the way the original model is extended is indicated. A detailed description will be given only when the closure strategy or the model is not a straightforward extension of a former proposal. It is worth noticing here that many models developed for the momentum equation have not yet been extended to the scalar problem, despite it may be done very easily.

This chapter is restricted to presentation of the classical subgrid closure problem, and the issues dealing with the development of wall models and turbulent inlet conditions for the scalar (and the velocity field in the active scalar case) will not be discussed.

15.2 The Passive Scalar Case

15.2.1 Physical Model

Definitions and Filtered Equations in Physical Space. The following passive scalar equation is used as a starting point:

$$\frac{\partial \theta}{\partial t} + \nabla \cdot (\boldsymbol{u}\theta) = \kappa \nabla^2 \theta \quad , \tag{15.1}$$

where θ is the scalar quantity (temperature, pollutant concentration, ...) and κ the associated molecular diffusivity. Since there is no feedback in the momentum equation, the later will not be considered here. Applying a filter[1] to (15.1), one obtains

$$\frac{\partial \overline{\theta}}{\partial t} + \nabla \cdot (\overline{\boldsymbol{u}\theta}) = \kappa \nabla^2 \overline{\theta} \quad . \tag{15.2}$$

Splitting the filtered non-linear term $\overline{\boldsymbol{u}\theta}$ into a resolved and a subgrid part, one obtains

$$\frac{\partial \overline{\theta}}{\partial t} + \nabla \cdot (\overline{\boldsymbol{u}}\,\overline{\theta}) = \kappa \nabla^2 \overline{\theta} - \nabla \cdot \tau_\theta \quad , \tag{15.3}$$

where the subgrid scalar flux τ_θ is defined as

$$\tau_\theta \equiv (\overline{\boldsymbol{u}\theta} - \overline{\boldsymbol{u}}\,\overline{\theta}) \quad . \tag{15.4}$$

The subgrid scalar flux can be further decomposed in exactly the same way as the subgrid tensor in the momentum equation (see Sect. 3.3), yielding

$$\tau_\theta = \underbrace{\overline{\overline{\boldsymbol{u}}\,\overline{\theta}} - \overline{\boldsymbol{u}}\,\overline{\theta}}_{L_\theta} + \underbrace{\overline{\boldsymbol{u}'\overline{\theta}} + \overline{\overline{\boldsymbol{u}}\theta'}}_{C_\theta} + \underbrace{\overline{\boldsymbol{u}'\theta'}}_{R_\theta} \quad . \tag{15.5}$$

[1] As for the momentum equation, the mathematical model used here to represent the true Large-Eddy Simulation problem is the convolution filter paradigm. But other mathematical models presented in Chap. 4 can also be used to this end.

The subgrid fluxes L_θ, C_θ and R_θ are analogous to the Leonard stress tensor, the Cross stress tensor and the subgrid Reynolds stress tensor, respectively. The subgrid scalar flux can also be decomposed using Germano's consistent decomposition approach (Sect. 3.3.3):

$$\tau_\theta = \tau_{\mathrm{G}}(\boldsymbol{u},\theta) = \underbrace{\tau_{\mathrm{G}}(\overline{\boldsymbol{u}},\overline{\theta})}_{\mathcal{L}_\theta} + \underbrace{\tau_{\mathrm{G}}(\overline{\boldsymbol{u}},\theta') + \tau_{\mathrm{G}}(\boldsymbol{u}',\overline{\theta})}_{\mathcal{C}_\theta} + \underbrace{\tau_{\mathrm{G}}(\boldsymbol{u}',\theta')}_{\mathcal{R}_\theta} \quad , \tag{15.6}$$

where $\tau_{\mathrm{G}}(\phi,\psi) \equiv \overline{\phi\psi} - \overline{\phi}\,\overline{\psi}$ is the generalized central moment of ψ and ϕ associated to the filter kernel G.

Two quantities of interest to characterize the scalar dynamics are the subgrid scalar flux τ_θ (which is equal to $\overline{\boldsymbol{u}'\theta'}$ if the filter belongs to the class of the Reynolds operators) and the scalar subgrid variance $\theta^2_{\mathrm{sgs}} \equiv \overline{\theta'\theta'}$. The former is related to the mixing/stirring transport process of the scalar field at small scales, while the latter is tied to the existence of fluctuations of the scalar field at the subgrid level and therefore is a measure of its *unmixedness*. It is worth noting that these two quantities are generalized within the Large-Eddy Simulation framework when arbitrary convolution filters are used in the same way that the subgrid kinetic energy $\overline{u'_i u'_i}/2$ is extended considering the trace of the subgrid tensor. In the scalar case, $\overline{\boldsymbol{u}'\theta'}$ and the subgrid scalar variance τ_θ are extended as $\tau_{\mathrm{G}}(\boldsymbol{u},\theta)$ and $\tau_{\mathrm{G}}(\theta,\theta)$, respectively.

Transport equations for these two quantities are easily derived using the scalar equation and the momentum equation. The most general expressions based on the generalized central moments are

$$\begin{aligned}
\frac{\partial \tau_{\mathrm{G}}(u_i,\theta)}{\partial t} + \overline{u}_k \frac{\partial \tau_{\mathrm{G}}(u_i,\theta)}{\partial x_k} \quad = \quad & \underbrace{-\,\tau_{\mathrm{G}}(u_k,u_i)\frac{\partial \overline{\theta}}{\partial x_k}}_{I} \underbrace{-\,\tau_{\mathrm{G}}(u_k,\theta)\frac{\partial \overline{u}_i}{\partial x_k}}_{II} \\
& \underbrace{+\frac{\partial}{\partial x_k}\left(\nu\tau_{\mathrm{G}}\left(\frac{\partial u_i}{\partial x_k},\theta\right) + \kappa\tau_{\mathrm{G}}\left(\frac{\partial \theta}{\partial x_k},u_i\right)\right)}_{III} \\
& \underbrace{-\,(\nu+\kappa)\tau_{\mathrm{G}}\left(\frac{\partial u_i}{\partial x_k},\frac{\partial \theta}{\partial x_k}\right)}_{IV} \underbrace{-\,\tau_{\mathrm{G}}\left(\theta,\frac{\partial p}{\partial x_i}\right)}_{V} \\
& \underbrace{-\,\frac{\partial \tau_{\mathrm{G}}(u_k,u_i,\theta)}{\partial x_k}}_{VI} \quad ,
\end{aligned} \tag{15.7}$$

for the generalized subgrid scalar flux and

$$\frac{\partial \tau_{\mathrm{G}}(\theta,\theta)}{\partial t} + \overline{u}_k \frac{\partial \tau_{\mathrm{G}}(\theta,\theta)}{\partial x_k} = \underbrace{-\,2\tau_{\mathrm{G}}(u_k,\theta)\frac{\partial \overline{\theta}}{\partial x_k}}_{VII}$$

$$+ \underbrace{\kappa \frac{\partial^2 \tau_G(\theta,\theta)}{\partial x_k \partial x_k}}_{VIII} \underbrace{- 2\kappa\tau_G\left(\frac{\partial \theta}{\partial x_k}, \frac{\partial \theta}{\partial x_k}\right)}_{IX}$$
$$\underbrace{- \frac{\partial \tau_G(u_k,\theta,\theta)}{\partial x_k}}_{X} \quad , \qquad (15.8)$$

for the generalized subgrid scalar variance, where it is recalled that

$$\tau_G(a,b,c) = \overline{abc} - \overline{a}\tau_G(b,c) - \overline{b}\tau_G(a,c) - \overline{c}\tau_G(a,b) - \overline{a}\,\overline{b}\,\overline{c} \quad .$$

The physical meaning of the terms appearing in the preceding equations are

- I: Production by interaction between the subgrid stresses and the resolved scalar gradient
- II: Production by interaction between the subgrid scalar fluxes and the resolved velocity gradient
- III: Viscous diffusion
- IV: Viscous dissipation
- V: Scalar-pressure subgrid flux
- VI: Diffusion by subgrid motion
- VII: Subgrid scalar variance production by interaction with the resolved scalar gradient
- VIII: Viscous diffusion
- IX: Subgrid scalar variance diffusion, referred to as ε_θ
- X: Diffusion by subgrid motion.

Definitions and Filtered Equations in Spectral Space. Physical quantities and corresponding equations can be rewritten in the spectral space performing a Fourier transform. Denoting $\widehat{\theta}(\boldsymbol{k})$ the Fourier transform of $\theta(\boldsymbol{x})$, one has the following relation dealing with the two-point correlation

$$\begin{aligned}\left\langle \widehat{\theta}(\boldsymbol{k}')\widehat{\theta}(\boldsymbol{k})\right\rangle &= \left(\frac{1}{2\pi}\right)^6 \int e^{-\imath(\boldsymbol{k}'\cdot\boldsymbol{x}+\boldsymbol{k}\cdot\boldsymbol{k}')} \left\langle \theta(\boldsymbol{x})\theta(\boldsymbol{x}')\right\rangle d\boldsymbol{x}d\boldsymbol{x}' \\ &= \frac{E_\theta(k)}{2\pi k^2}\delta(\boldsymbol{k}+\boldsymbol{k}') \quad ,\end{aligned} \qquad (15.9)$$

where $E_\theta(k)$ is the scalar spectrum (defined here as an average over shell $k =$ cste.). The scalar variance is equal to

$$\frac{1}{2}\left\langle \theta(\boldsymbol{x})\theta(\boldsymbol{x})\right\rangle = \int_0^{+\infty} E_\theta(k)dk \quad . \qquad (15.10)$$

The spectral analogue of (15.1) is

$$\left(\frac{\partial}{\partial t} + \kappa k^2\right)\widehat{\theta}(\boldsymbol{k}) = T^\theta(\boldsymbol{k}) \quad , \qquad (15.11)$$

in which the non-linear term $T^\theta(\boldsymbol{k})$ is equal to

$$T^\theta(\boldsymbol{k}) = -\imath k_j \int\int\int \widehat{u}_j(\boldsymbol{k}-\boldsymbol{p})\widehat{\theta}(\boldsymbol{p})d^3\boldsymbol{p} \quad . \tag{15.12}$$

The corresponding evolution equation for the non-stationary spectrum of scalar variance is

$$\left(\frac{\partial}{\partial t} + 2\kappa k^2\right) E_\theta(\boldsymbol{k}) = T^{\theta\theta}(\boldsymbol{k}) \quad , \tag{15.13}$$

where the scalar spectrum transfer is expressed as

$$T^{\theta\theta}(\boldsymbol{k}) = -8\imath k^2 \int k_m \left\langle \widehat{\theta}(\boldsymbol{k}-\boldsymbol{p})\widehat{u}_m(\boldsymbol{p})\widehat{\theta}(-\boldsymbol{k})\right\rangle d^3\boldsymbol{p} \quad . \tag{15.14}$$

The conservation property for the scalar variance takes the following form:

$$\int T^{\theta\theta}(\boldsymbol{k})d^3\boldsymbol{k} = 0 \quad . \tag{15.15}$$

The filtered equation associated to the filtered variable $\widehat{\overline{\theta}}(\boldsymbol{k}) \equiv \widehat{G}(\boldsymbol{k})\widehat{\theta}(\boldsymbol{k})$, where $G(\boldsymbol{k})$ denotes the transfer function of the selected filter, is

$$\left(\frac{\partial}{\partial t} + \kappa k^2\right) \widehat{\overline{\theta}}(\boldsymbol{k}) = \widehat{G}(\boldsymbol{k})T^\theta(\boldsymbol{k}) \tag{15.16}$$

$$= T^\theta_{\mathrm{r}}(\boldsymbol{k}) + T^\theta_{\mathrm{sgs}}(\boldsymbol{k}) \quad , \tag{15.17}$$

where $T^\theta_{\mathrm{r}}(\boldsymbol{k})$ and $T^\theta_{\mathrm{sgs}}(\boldsymbol{k})$ are the resolved and subgrid spectral scalar fluxes, respectively. The former involves resolved quantities $\widehat{\overline{\theta}}$ and $\widehat{\overline{\boldsymbol{u}}}$ only, while the latter contains all terms involving at least one subgrid component among $(1-\widehat{G}(\boldsymbol{k}))\widehat{\theta}(\boldsymbol{k})$ and $(1-\widehat{G}(\boldsymbol{k}))\widehat{\boldsymbol{u}}(\boldsymbol{k})$. The closure problem in the spectral space consists in finding an expression for $T^\theta_{\mathrm{sgs}}(\boldsymbol{k})$ which involves only known quantities.

15.2.2 Dynamics of the Passive Scalar

This section is devoted to a brief survey of the dynamics of the passive scalar in isotropic turbulence. The main purpose here is to enlight the fact that the modeling task is far from being a trivial one, even within this simplified framework. The very reason for this is that several physical regimes exist, which are associated to different values of the molecular Prandtl number (or Schmidt number, or Peclet number depending on the physical significance of the scalar field) defined as

$$Pr \equiv \frac{\nu}{\kappa} \quad . \tag{15.18}$$

The spectral properties of the three ideal regimes $Pr \ll 1, Pr \sim 1$ and $Pr \gg 1$ are surveyed in Sect. 15.2.2. The interested reader can refer to specialized books [714, 439, 464, 708] for a detailed discussion. Results obtained via Direct Numerical Simulation and EDQNM analysis dealing with the spectral dynamics of the passive scalar are displayed first (see p. 456). The concept of subgrid Prandtl number is introduced and discussed in a second step (p. 459).

Different Regimes and Associated Dynamics. Three regimes for the passive scalar in isotropic turbulence are identified, each one being associated with a range of values for the Prandtl number. The existence of these three archetypal[2] regimes originates in the difference between the viscous cutoff scales for the scalar and the velocity field. Each regime is associated to a specific scalar spectrum shape (the kinetic energy spectrum does not vary, since the scalar is strictly passive).

To analyze the spectral characteristic features of these dynamical regimes, we introduce the scalar diffusion cutoff length, referred to as the Obukhov–Corrsin scale, which is the analogue of the Kolmogorov scale η_K for the velocity:

$$\eta_\theta = \left(\frac{\varepsilon}{\kappa^3}\right)^{1/4} = \left(\frac{1}{Pr}\right)^{3/4} \eta_K \quad . \tag{15.19}$$

The ratio of the wave numbers associated to these cutoffs is therefore estimated as

$$\frac{k_\theta}{k_\eta} = Pr^{3/4} \quad . \tag{15.20}$$

The definition of Obukhov–Corrsin cutoff scale is not valid if $Pr \gg 1$, since it is based on Kolmogorov-type hypotheses on the nature of the fluctuations which are no longer adequate. In this case, Batchelor derived the following diffusive cutoff wave number (referred to as the Batchelor wave number):

$$k_{\mathrm{B}} = \left(\frac{\varepsilon}{\nu\kappa^2}\right)^{1/4} \quad . \tag{15.21}$$

These three regimes are

1. $Pr \ll 1$: the molecular diffusivity is much larger than the molecular viscosity. Looking at relations (15.19) and (15.20), it is seen that k_θ is much smaller than k_η, meaning that the the scalar diffusive cutoff occurs within the inertial range of the Kolmogorov spectrum for the kinetic energy. One can therefore infer that two inertial ranges will be observed in the scalar spectrum (see Fig. 15.1):
   a) The *inertial-convective range*, which corresponds to wave numbers $k \ll k_\theta \ll k_\eta$. These scales are not subject to viscous and diffusive effects accordingly to Kolmogorov's picture. Scalar fluctuations are

[2] It is important noticing that the results presented here deal with asymptotic cases, and that real-life flows are more complex.

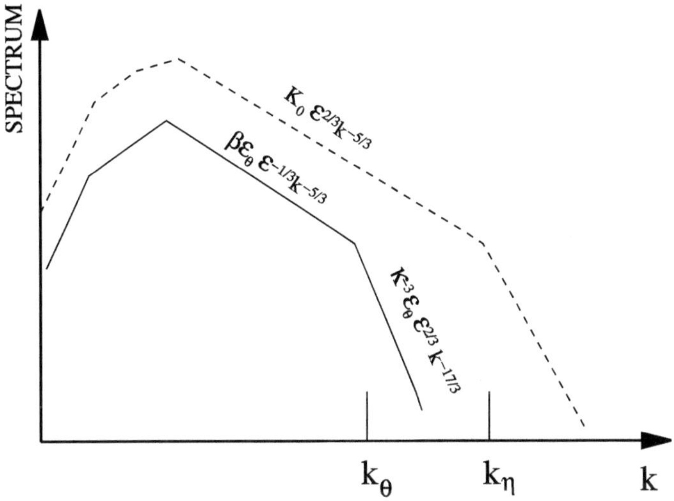

Fig. 15.1. Schematic of the kinetic energy spectrum and the scalar variance spectrum in the very-low Prandtl number case.

driven by the stirring induced by velocity fluctuations. The corresponding form of the scalar spectrum is

$$E_\theta(k) = \beta \varepsilon_\theta \varepsilon^{-1/3} k^{-5/3} \quad , \tag{15.22}$$

where Obukhov–Corrsin constant β is in the range $0.68 - 0.83$.

b) The *inertial-diffusive range*, which is observed for scales which belong to the Kolmogorov intertial range, meaning that the velocity fluctuations are not directly sensitive to the molecular viscosity, but at which scalar fluctuations experience a strong action of the molecular diffusivity. The corresponding wave number band is $k_\theta \ll k \ll k_\eta$. The dynamics of the scalar at these scales is governed by a balance between the turbulent advection and the molecular diffusion. The scalar spectrum shape is

$$E_\theta(k) = \frac{K_0}{3} \varepsilon_\theta \varepsilon^{2/3} \kappa^{-3} k^{-17/3} \quad . \tag{15.23}$$

Smaller scales ($k \geq k_c$) are governed by molecular viscosity and diffusivity effects, and exhibit an exponentially decaying behavior.

2. $Pr \simeq 1$. The two cutoff scales are very close, leading to the existence of a unique *inertial-convective* inertial range. The scalar spectrum shape is the same as in the previous case (15.22).
3. $Pr \gg 1$. This last case corresponds to configurations in which the diffusive cutoff scale is much smaller than the viscous cutoff scale for the velocity fluctuations. Two inertial ranges exist (see Fig. 15.2):

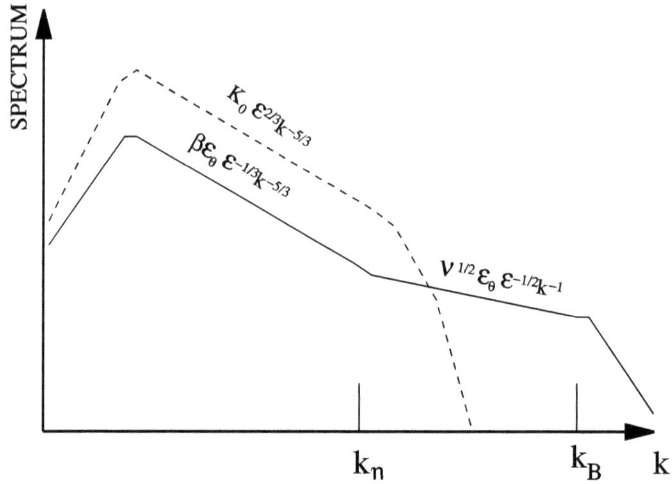

Fig. 15.2. Schematic of the kinetic energy spectrum and the scalar variance spectrum in the very-high Prandtl number case.

a) An *inertial-convective range*, for wave numbers such that $k \ll k_\theta \ll k_B$, which is similar to those already described above.
b) A *viscous-convective range*, which is associated to scales ($k_\theta \ll k \ll k_B$) where the velocity fluctuations are severely damped by viscous effects but scalar fluctuations are not affected by molecular diffusion. The associated spectrum is

$$E_\theta(k) = \frac{1}{2}\sqrt{\frac{\nu}{\varepsilon}}\varepsilon_\theta k^{-1} \quad . \tag{15.24}$$

This brief survey obviously shows that the spectral scalar transfers across the large-eddy simulation cutoff may strongly depend on the type of inertial range it is located in. As a consequence, the development of a closure based on the functional modeling approach may be much more difficult than in the previous case of the momentum equations, since the latter exhibits only one type of inertial range dynamics.

Interscale Scalar Transfers. We now present some results dealing with the analyis of passive scalar fluxes in the spectral space. These results are the corner stone of the functional modeling approach. The emphasis is put on theoretical results coming from the EDQNM analysis of the passive scalar dynamics in isotropic turbulence (see B.3 for a short presentation).

The main results of the EDQNM analysis deal with the non-local interactions in the scalar variance equation. Considering a non-local triad $(\boldsymbol{k}, \boldsymbol{p}, \boldsymbol{q})$ satisfying the constraint $\boldsymbol{k} + \boldsymbol{p} + \boldsymbol{q} = 0$, two cases are identified

– The considered scale, associated with $\boldsymbol{k}$, is interacting with one much larger scale : $p \ll k \simeq q$. The EDQNM form of the associated non-local scalar variance flux is

$$
\begin{aligned}
T_{+}^{\theta\theta}(k) &= \frac{2}{15}\int_0^{ak} \Theta^\theta_{kkq} q^2 E(q) dq \left(2kE_\theta(k) - k^2 \frac{\partial E_\theta}{\partial k}(k) \right) \\
&+ \frac{1}{4}\int_0^{ak} \Theta^\theta_{kkq} q^3 E_\theta(q) dq E(k) \\
&- \frac{1}{4}\int_0^{ak} \Theta^\theta_{kkq} q^5 dq \frac{E(k)E_\theta(k)}{k^2} \quad .
\end{aligned}
\tag{15.25}
$$

The parameter a defines the local/non-local triads (see p. 93). The first term in the right hand side corresponds to interactions which are responsible for the existence of the viscous-convective range.

– The scale $\boldsymbol{k}$ is involved in a non-local triadic interaction with much smaller scales: $k \ll p \simeq q$. The asymptotic expression for the scalar variance flux is

$$
\begin{aligned}
T_{-}^{\theta\theta}(k) &= -\frac{4}{3}\int_0^{k} q^2 E_\theta(q) dq \int_{\sup(k,q/a)}^{+\infty} \Theta^\theta_{qpp} E(p) dp \\
&+ \frac{4}{3}\int_0^{k} q^4 dq \int_{\sup(k,q/a)}^{+\infty} \Theta^\theta_{qpp} \frac{E(p)E_\theta(p)}{p^2} dp \quad .
\end{aligned}
\tag{15.26}
$$

A simplified expression, which is valid at low wave numbers, is

$$
\begin{aligned}
T_{-}^{\theta\theta}(k) &= -\frac{4}{3}k^2 E_\theta(k) \int_{ak}^{+\infty} \Theta^\theta_{0pp} E(p) dp \\
&+ \frac{4}{3}k^4 \int_{ak}^{+\infty} \Theta^\theta_{0pp} \frac{E(p)E_\theta(p)}{p^2} dp \quad .
\end{aligned}
\tag{15.27}
$$

The first term in the right hand side is associated with a forward scalar variance cascade, whose intensity is governed by both the kinetic energy and scalar variance spectral distribution. The second term represents a backward scalar cascade.

The analysis of Direct Numerical Simulation results [783] reveals that, in the inertial-convective range, the overall spectral transfer is forward cascading. It is dominated by the energy-containing scales in the velocity field. More precisely, the dominant physical process at high wave numbers belonging to this inertial range is a local transfer associated with a non-local triadic interaction that links two high wave number scalar modes and one low wave number velocity mode. The other non-local interactions (one high wave number scalar mode, one low wave number scalar mode and one high wave number velocity mode) and local interactions are much weaker. Therefore, the suggested dominant physical process is the breaking of "blobs" of scalar fluctuations into smaller-scale fragments by large energetic eddies.

The large-eddy simulation closure problem is now directly addressed by writing the equation associated to the resolved scalar variance density, denoted $\overline{E}_\theta(k) = \widehat{G}^2(k)E_\theta(k)$:

$$\left(\frac{\partial}{\partial t} + 2\kappa k^2\right)\overline{E}_\theta(\boldsymbol{k}) = T_{\rm r}^{\theta\theta}(\boldsymbol{k}) + T_{\rm sgs}^{\theta\theta}(\boldsymbol{k}) \quad , \tag{15.28}$$

where $T_{\rm r}^{\theta\theta}(\boldsymbol{k})$ and $T_{\rm sgs}^{\theta\theta}(\boldsymbol{k})$ are the resolved and the subgrid spectral scalar variance transfer, respectively. In a way similar to the one used to characterize subgrid kinetic energy transfers in Sect. 5.1.2, the intensity of these transfers across a cutoff wave number $k_{\rm c}$ can be parameterized defining an effective spectral diffusivity $\kappa_{\rm e}(k|k_{\rm c})$ such that

$$T_{\rm sgs}^{\theta\theta}(\boldsymbol{k}) = -2\kappa_{\rm e}(k|k_{\rm c})E_\theta(k) \quad . \tag{15.29}$$

In the case the cutoff is located within the inertial-conductive range, and assuming that all the basic hypotheses of the canonical analysis (inertial ranges extending to infinity, sharp cutoff filter, see Sect. 5.1.2), both the EDQNM analysis conducted by Chollet and Zhou's RNG theory [811] show that the spectral effective diffusivity shape is strictly similar to the one of the effective spectral viscosity $\nu_{\rm e}(k|k_{\rm c})$:

- for $k \ll k_{\rm c}$, it is independent of the wave number. The value of the plateau deduced from the non-local forward cascade term in (15.27) is

$$\kappa_{\rm e}(k|k_{\rm c}) = \kappa_{\rm e}^\infty \equiv \frac{2}{3}\int_{k_c}^{+\infty} \Theta_{0pp}^\theta E(p)dp \tag{15.30}$$

  In practice, this plateau is observed for wave numbers in the range $0 \leq k \leq k_{\rm c}/3$. Theoretical analysis carried out by Chollet and Lesieur using the EDQNM analysis in the case of a very wide inertial-convective range also shows that the value of the plateau depends on both the kinetic energy at the cutoff and the slope of the spectrum. A general expression is [514]:

$$\kappa_{\rm e}^\infty = \left\{ \begin{array}{ll} \dfrac{4}{3a}\dfrac{\sqrt{3-m}}{m+1}\sqrt{\dfrac{E(k_{\rm c})}{k_{\rm c}}} & m < 3 \\ \dfrac{1}{3}\dfrac{1}{m-1}\dfrac{1}{\sqrt{D_{\rm r}}}\dfrac{E(k_{\rm c})}{k_{\rm c}} & m > 3 \end{array} \right. \quad , \tag{15.31}$$

  where $-m$ is the slope of the kinetic energy spectrum, $a = 0.218K_0$ a structural EDQNM parameter and $D_{\rm r} = \int_0^{k_{\rm c}} k^2E(k)dk$ a norm of the resolved velocity gradient. Considering a Kolmogorov spectrum ($m = 5/3, K_0 = 1.4$) one obtains $\kappa_{\rm e}^\infty = 0.3$.
- Near the cutoff, i.e. at wave numbers $k_{\rm c}/3 \leq k \leq k_{\rm c}$, the effective diffusivity exhibits a cusp, showing that the spectral transfers are more intense. In this region, local interactions are not negligible. It is important noting that this cusp is as sensitive as the one in the effective viscosity: it is

not observed with smooth filters, and can also disappear if the cutoff is located at the very beginning of the inertial range. In the canonical case of an infinite inertial-convective range, the maximum value of the subgrid diffusivity found using EDQNM analysis is $\kappa_e(k_c|k_c) = 0.6$.

Detailed investigations of the spectral transfers across a cutoff located within the viscous-convective or the inertial-diffusive range are missing. This lack in the theory may be not very important for most practical applications, since putting the cutoff within one of these inertial range requires the use of very fine computational grids.

Numerical experiments carried out by several authors [717] also show that the effective diffusivity spectral shape is very sensitive to both the scalar variance spectrum shape and the kinetic energy spectrum: it has been observed to be either an increasing or a decreasing function of the wave number. Effective viscosities computed from simulated isotropic turbulence by Métais and Lesieur are displayed in Figs. 15.3 and 15.4.

The Subgrid Prandtl Number Paradigm. The molecular Prandtl number (15.18) is a useful tool to compare turbulent scales which characterize dissipative/diffusive cutoff scales of the velocity and scalar fields. This is why the idea of introducing a *subgrid Prandtl number*, Pr_{sgs}, is attractive. But preceding remarks dealing with the sensitivity of the effective spectral diffu-

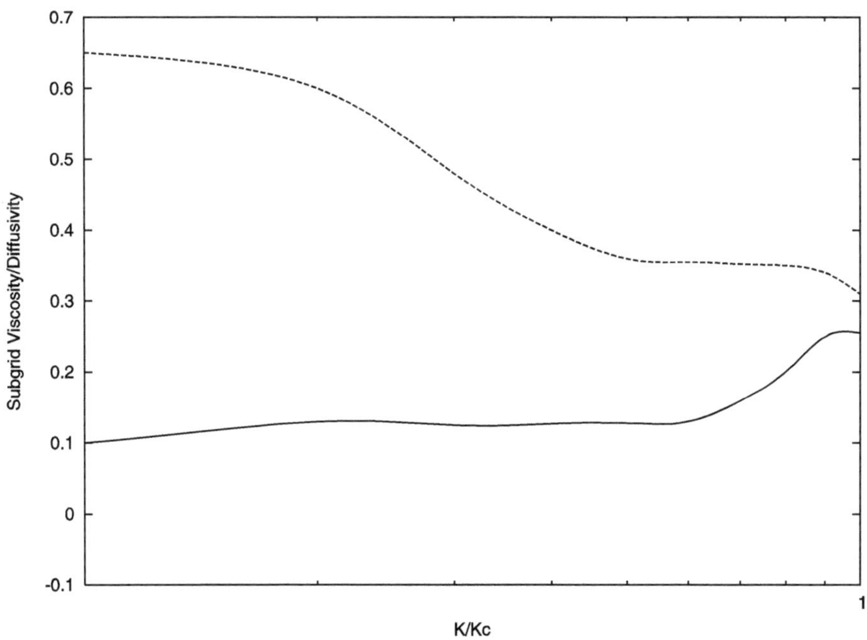

Fig. 15.3. Effective subgrid viscosity (*Solid line*) and subgrid diffusivity (*Dashed line*) normalized by $\sqrt{E(k_c)/k_c}$ in the Lesieur–Métais simulation (Case 1).

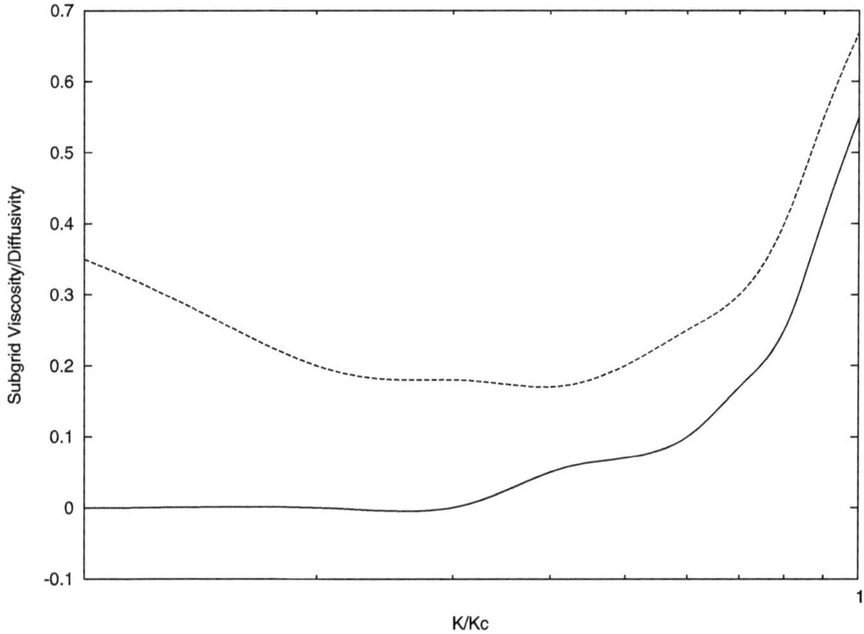

Fig. 15.4. Effective subgrid viscosity (*Solid line*) and subgrid diffusivity (*Dashed line*) normalized by $\sqrt{E(k_c)/k_c}$ in the Lesieur–Métais simulation (Case 2).

sivity indicate that a universal distribution for the spectral effective subgrid Prandtl number

$$Pr^{\mathrm{e}}_{\mathrm{sgs}}(k|k_{\mathrm{c}}) = \frac{\nu_{\mathrm{e}}(k|k_{\mathrm{c}})}{\kappa_{\mathrm{e}}(k|k_{\mathrm{c}})} \quad , \tag{15.32}$$

cannot exist. As a consequence, subgrid closure strategies based on the concept of subgrid Prandtl number necessarily introduce some errors, since they do not account for subtle discrepancies that exist between the velocity and the scalar dynamics.

In the very simple case of a quasi-infinite inertial-convective range, the value found for wave numbers located within the plateau ($k \leq 0.3k_{\mathrm{c}}$) via the EDQNM analysis is [514]

$$Pr^{\mathrm{e}}_{\mathrm{sgs}}(k|k_{\mathrm{c}}) \simeq \frac{5-m}{20} \quad . \tag{15.33}$$

But such a simple expression should not mask the fact that the subgrid Prandtl number is fully case dependent, since it characterizes the differences that may exist between kinetic energy and scalar spectral transfers. As an example, in the very simple case of passive scalar in isotropic turbulence, Lesieur and Rogallo [441] observed two very different spectral subgrid Prandtl distributions (see Fig 15.5). In the first case, the scalar field exhibits a k^{-1}

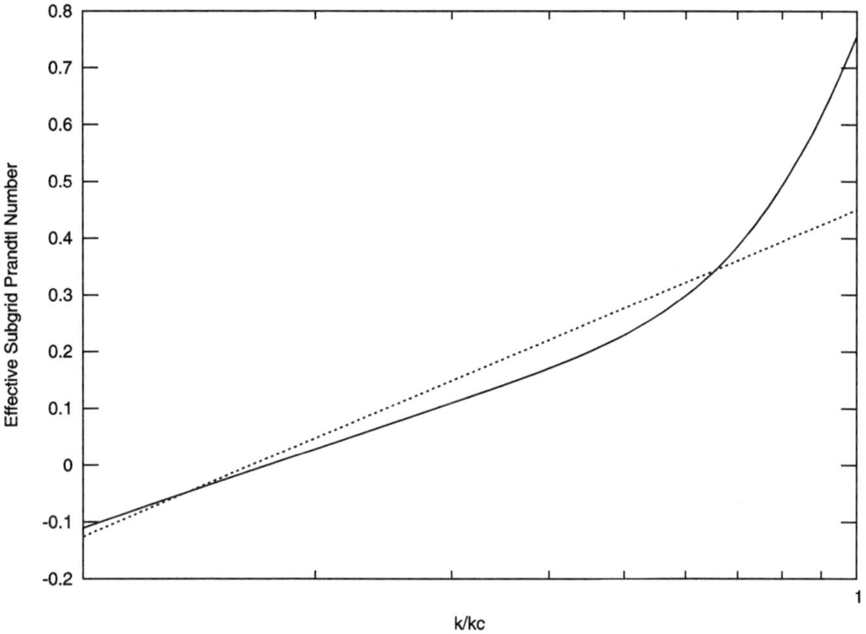

Fig. 15.5. Effective Prandtl number versus the wavenumber in Lesieur–Rogallo simulations of isotropic turbulence.

inertial range (this anomalous exponent is supposed to be due to shearing by large scales), and the subgrid Prandtl number approximately follows the law

$$Pr^{\mathrm{e}}_{\mathrm{sgs}}(k|k_{\mathrm{c}}) = 0.35 + 0.2\log(k/k_{\mathrm{c}}) + 9e^{-3.099k_{\mathrm{c}}/k} \quad , \tag{15.34}$$

while in the second set of simulations the following distribution was observed

$$Pr^{\mathrm{e}}_{\mathrm{sgs}}(k|k_{\mathrm{c}}) = 0.45 + 0.25\log(k/k_{\mathrm{c}}) \quad . \tag{15.35}$$

15.2.3 Extensions of Functional Models

We now discuss the extension of functional models in the physical space. The construction of functional models in the Fourier space is a straightforward utilization of spectral subgrid difusivity laws given in the preceding section, and will not be further detailed.

All explicit functional models for the passive scalar equations are based on the subgrid diffusivity paradigm, yielding the following general closure relation:

$$\tau_\theta = -\kappa_{\mathrm{sgs}}\nabla\overline{\theta} \quad . \tag{15.36}$$

As a consequence, the emphasis is put on the dominant mechanism observed in isotropic turbulence, namely the forward scalar cascade. This approach is strictly equivalent to the use of a subgrid viscosity to close the filtered momentum equations. Almost all functional subgrid models and/or closure strategies proposed for the momentum equations and presented in Chaps. 5 and 6 can be very easily extended to the scalar equation. An exhaustive description of all possibilities is not of interest, and it is worth noting that many straightforward extensions have not been published yet[3]. The implicit Large-Eddy Simulation approach can also be used, if an adequate numerical scheme for the scalar equation is utilized.

General Expression. Scalar Subgrid Diffusivity. As in the case of subgrid viscosity, a fully general expression of the subgrid diffusivity κ_{sgs} as a function of a set of selected basic quantities can be used to obtain a unified view of most published models. A possible expression is

$$\kappa_{\text{sgs}} = \kappa_{\text{sgs}}(Pr, \overline{\Delta}, |\overline{S}|, q^2_{\text{sgs}}, \varepsilon, |\nabla\overline{\theta}|, \theta^2_{\text{sgs}}, \varepsilon_\theta) \quad , \tag{15.37}$$

where quantities directly tied to subgrid quantities (i.e. $q^2_{\text{sgs}}, \varepsilon, \theta^2_{\text{sgs}}, \varepsilon_\theta$) can be evaluated solving corresponding prognostic equations, or using some local equilibrium and/or scale similarity assumptions to recover expressions involving resolved scales only. The Prandtl number is a priori included as an input of this general expression, since it can be used to distinguish the different inertial range regimes. Weighting coefficients in the constitutive equation (15.37) are found performing a simple dimensional analysis.

An example is the model proposed by Schmidt and Schumann [651]:

$$\kappa_{\text{sgs}} = C_\kappa \overline{\Delta} \sqrt{q^2_{\text{sgs}}} \quad , \tag{15.38}$$

where interial-conductive range considerations lead to

$$C_\kappa = \sqrt{\frac{2}{3K_0}\frac{1}{3\beta\pi}} \quad . \tag{15.39}$$

Yoshizawa [789] proposes a more complex model:

$$\kappa_{\text{sgs}} = C\frac{(\theta^2_{\text{sgs}})^2\varepsilon}{\varepsilon^2_\theta}, \quad C = 0.446 \quad . \tag{15.40}$$

A subgrid diffusivity model which accounts for molecular Prandtl number effects was developed by Grötzbach [277, 279] to describe the temperature field in liquid metals, which have very small Prandtl number ($Pr = 0.007$ for liquid sodium under nuclear reactor conditions). We reproduce here the method developed by Grötzbach to account for small Prandtl number effects rather than its model, since it can be applied to any subgrid viscosity model with one arbitrary constant. Let us consider a generic subgrid diffusivity

[3] The interested reader will quickly find material for a few dozens of papers dealing with “new improved models”!

model under the form

$$\kappa_{\text{sgs}} = C\mathcal{K} \quad , \tag{15.41}$$

where C is the constant to be adjusted and $\mathcal{K}$ a constant-free parameter which has the dimension of a diffusivity. The method proposed by Grötzbach relies on the local equilibrium assumption (i.e. production = dissipation) for the scalar variance:

$$\tau_\theta \cdot \nabla\overline{\theta} = \varepsilon_\theta - \kappa\nabla\overline{\theta} \cdot \nabla\overline{\theta} \quad , \tag{15.42}$$

where ε_θ is the full turbulent scalar variance dissipation. Now introducing the model in the left hand side of this relation, one obtains the following relation for the constant C:

$$C = \frac{\varepsilon_\theta - \kappa\nabla\overline{\theta} \cdot \nabla\overline{\theta}}{-\mathcal{K}\nabla\overline{\theta} \cdot \nabla\overline{\theta}} \quad . \tag{15.43}$$

The last parameter to be evaluated is ε_θ. Assuming that the cutoff occurs in a very wide inertial-convective range, an analytic expression for the scalar variance dissipation as a function of the kinetic energy dissipation can be found, yielding

$$C \simeq \frac{1}{\beta\sqrt{K_0}}\left(1 - \beta\kappa\varepsilon^{-1/3}\overline{\Delta}^{-4/3}\right) \quad , \tag{15.44}$$

where the subgrid kinetic energy dissipation rate ε is retrieved from the work done to close the momentum equation, leading to a fully determined definition of the subgrid diffusivity.

The Subgrid Prandtl Number Approach. A very common closure approach is to use a subgrid Prandtl number, leading to

$$\kappa_{\text{sgs}} = \frac{\nu_{\text{sgs}}}{Pr_{\text{sgs}}} \quad , \tag{15.45}$$

where the subgrid viscosity ν_{sgs} can be evaluated using any model described in the preceding chapters devoted to the momentum equation closure. Despite it is flawed from a purely theoretical point of view, this approach is very often used in simulations in the physical space and the subgrid Prandtl number appears as an adjustable parameter which can be tuned in an ad hoc way to obtain the best fit with the reference data. Values found in the literature range from 0.1 to 1, the most common one being 0.6. It is important noting that this approach is not valid in cases where the velocity field if fully resolved, yielding $\nu_{\text{sgs}} = 0$, while some subgrid scalar fluctuations exist. A typical example would be to put the resolution cutoff within the viscous-convective range.

The subgrid Prandtl number can be made more accurate, but still not adequate to represent the cases mentioned above, rendering it self-adaptive, meaning that the value of the Prandtl number will be made space and time dependent. Such a modification is expected to make the subgrid Prandtl number based models adequate to treat the case where all scalar fluctuations are resolved while subgrid kinetic energy transfers exist. A common example is to use the Germano identity to define a dynamic procedure, as proposed by Moin et al. [538]. The Germano identity for the scalar equation is

$$\underbrace{\left(\widetilde{\overline{\boldsymbol{u}}\overline{\theta}}-\widetilde{\overline{\boldsymbol{u}}}\widetilde{\overline{\theta}}\right)}_{L_\theta}=\underbrace{\left(\widetilde{\overline{\boldsymbol{u}\theta}}-\widetilde{\overline{\boldsymbol{u}}}\widetilde{\overline{\theta}}\right)}_{T_\theta}-\underbrace{\left(\widetilde{\overline{\boldsymbol{u}\theta}-\overline{\boldsymbol{u}}\overline{\theta}}\right)}_{\widetilde{\tau}_\theta} \quad , \tag{15.46}$$

where the *tilde* symbol refers to the test filter level. Vectors T_θ and τ_θ are the subgrid scalar fluxes at the grid and test filter levels, respectively. As in the analogous relation for the momentum equation, the left hand side of (15.46) can be directly computed, while replacing subgrid fluxes in the right hand side by the corresponding subgrid models makes it possible the find the best value of the model constant in the least-square sense.

Denoting $\nu_{\text{sgs}}(\overline{\Delta})$ and $\nu_{\text{sgs}}(\widetilde{\overline{\Delta}})$ the subgrid viscosity values computed at the grid and test filter levels, respectively, the residual associated to the Germano identity (15.46) is equal to

$$\mathcal{E}_\theta = L_\theta + \frac{\nu_{\text{sgs}}(\widetilde{\overline{\Delta}})}{Pr_{\text{sgs}}}\nabla\widetilde{\overline{\theta}} - \frac{\widetilde{\nu_{\text{sgs}}(\overline{\Delta})}}{Pr_{\text{sgs}}}\nabla\overline{\theta} \quad . \tag{15.47}$$

Assuming that the subgrid Prandtl number is the same at the two levels and that it does not vay significantly over distances of the order of $\widetilde{\overline{\Delta}}$, the least-square minimization of $\mathcal{E}_\theta$ leads to

$$Pr_{\text{sgs}} = -\frac{m_\theta \cdot m_\theta}{m_\theta \cdot L_\theta} \quad , \tag{15.48}$$

where

$$m_\theta \equiv \nu_{\text{sgs}}(\widetilde{\overline{\Delta}})\nabla\widetilde{\overline{\theta}} - \widetilde{\nu_{\text{sgs}}(\overline{\Delta})\nabla\overline{\theta}} \quad . \tag{15.49}$$

Other expressions can be derived. As an example, Moin et al. [538] do not use the least-square minimization but chose to find the value of Pr_{sgs} which corresponds to a zero of $\mathcal{E}_\theta \cdot \nabla\widetilde{\overline{\theta}}$:

$$Pr_{\text{sgs}} = -\frac{m_\theta \cdot \nabla\widetilde{\overline{\theta}}}{\nabla\widetilde{\overline{\theta}} \cdot L_\theta} \quad . \tag{15.50}$$

Wong and Lilly [766] define a dynamic procedure based on dimensional parameters, yielding

$$Pr_{\text{sgs}} = \frac{(L : \widetilde{\overline{S}})}{|\widetilde{\overline{S}}|^2}\frac{(L_\theta \cdot \nabla\widetilde{\overline{\theta}})}{\nabla\widetilde{\overline{\theta}} \cdot \nabla\widetilde{\overline{\theta}}} \quad . \tag{15.51}$$

All these dynamic subgrid Prandtl number definitions suffer some numerical stability problems. The methods used to cure this problem are the same as for the momentum equation (see Sect. 5.3.3): clipping, time/space averaging, ... All variants of the dynamic procedure described in preceding chapters can be applied to compute the dynamic subgrid Prandtl number.

Anisotropic Subgrid Diffusivity. All models mentioned above are based on inertial range considerations, which are characteristic features of homogeneous isotropic turbulence. Practical applications involve much more complex configurations, in which the scalar field can be non homogeneous while the velocity field remains isotropic, or even more complex flows were both the velocity and the scalar fields are not homogeneous. Therefore, the question arises of accounting for the velocity field anisotropy in the scalar diffusion. This task is expected to be much more complex than the equivalent one for the velocity field, since the scalar variance spectrum exhibits much less universal features than the kinetic energy spectrum. Experiments conducted by Kang et al. [376] also show that the trend of decreasing anisotropy at small scales observed on velocity fluctuations is not recovered on scalar fluctuations. It is also known that, in the presence of a mean scalar gradient, structure functions and the derivative skewness of the scalar field do not follow predictions from isotropy at inertial and even dissipative scales. This local isotropy breakdown is tied to the direct action of large velocity structures on small scalar scales (this is consistent with the EDQNM analyis results dealing with dominant triadic interactions).

Taking anisotropy into account can be an important issue when the cutoff is such that the subgrid scalar fluctuations are governed by velocity fluctuations with high local Reynolds number (the localness is understood here in terms of wave number). In such cases, anisotropy in the velocity field will have an effect on the scalar diffusion process, and the most simple idea consists in defining a tensorial subgrid diffusivity.

Yoshizawa [789] proposes the following anisotropic diffusivity, which is derived using a two-scale expansion:

$$\left(\kappa_{\text{sgs}}\right)_{ij} = -C \left(\frac{\theta^2_{\text{sgs}}}{\varepsilon_\theta} \right)^3 \varepsilon \overline{S}_{ij} \quad , \tag{15.52}$$

where $C = 0.366$. The subgrid scalar dissipation rate and the subgrid scalar variance remain to be evaluated. Models for these quantities are presented in Sect. 15.2.6. The use of the resolved strain tensor $\overline{S}$ is coherent with the simple physical picture that scalar “blobs” are stretched by the velocity fluctuations, and will therefore be elongated in the principal shear direction, leading to the development of smaller scalar scales in transverse directions. A similar model is proposed by Peng and Davidson [587]. Rotational effects can also be included by taking into account the skewsymmetric part of the resolved velocity gradient, $\overline{\Omega}_{ij}$:

$$\left(\kappa_{\text{sgs}}\right)_{ij} = -2\varepsilon \left(\frac{\theta^2_{\text{sgs}}}{\varepsilon_\theta} \right)^3 \left[C_1 \overline{S}_{ij} + C_2 \overline{\Omega}_{ij} \right] \quad , \tag{15.53}$$

with $C_1 = 0.29$ and $C_2 = 0.08$.

Another tensorial subgrid diffusivity model was proposed by Pullin [608], who extended the stretched-vortex approach (see Sect. 7.6) to the passive scalar case. In the extended model, it is assumed that the subgrid mixing is restricted to the plane normal to the vortex axis. A secondary property of this model is that the subgrid scalar gradient induced by the transport is orthogonal to the subgrid vorticity, the latter being represented by the stretched vortex. This is consistent with the observations that the probability density of the alignement between the vorticity and the scalar gradient is maximum when these vectors are orthogonal. The corresponding formulation is

$$(\kappa_{\rm sgs})_{ij} = \frac{\gamma\pi}{2k_c}\sqrt{q_{\rm sgs}^2}\left(\delta_{ij} - e_i^v e_j^v\right) \quad , \tag{15.54}$$

where the vector e^v is the same as in Sect. 7.6. Assuming that the cutoff occurs within an infinite inertial-convective range, the structural parameter γ is given by

$$\gamma = \frac{2}{\pi}\sqrt{\frac{2}{3K_0}\frac{1}{\beta}} \quad . \tag{15.55}$$

15.2.4 Extensions of Structural Models

Many structural models have been applied to the subgrid scalar flux vector. The most illustrative ones are presented below.

Approximate Deconvolution and Scale Similarity Models. Scale similarity, or, in an equivalent way, soft approximate deconvolution models are defined writing the following approximation:

$$\tau_\theta \equiv \overline{\boldsymbol{u}\theta} - \overline{\boldsymbol{u}}\overline{\theta} \approx \overline{\boldsymbol{u}^\bullet\theta^\bullet} - \overline{\boldsymbol{u}}\overline{\theta} \quad , \tag{15.56}$$

where the approximate defiltered fields are expressed as

$$\boldsymbol{u}^\bullet = G_l^{-1} \star \overline{\boldsymbol{u}}, \quad \theta^\bullet = G_l^{-1} \star \overline{\theta} \quad , \tag{15.57}$$

with G_l^{-1} an approximate inverse of the filter (see Sect. 7.2.1 for a detailed presentation). A large number of structural models can be generated using this simple form, as for the momentum equation: models based on iterative deconvolution, models based on Taylor series expansions and tensor diffusivity models. Because they are all approximations of the exact soft deconvolution solution, these models do not take into account scalar transfer towards subgrid scales and they must be supplemented with a dissipative term, which can be either a numerical regularization or a functional subgrid model. That leads to the definition of linear combination models for the subgrid scalar fluxes.

The Bardina-type model obtained using the lowest-order deconvolution is

$$\tau_\theta = \overline{\overline{\boldsymbol{u}}\overline{\theta}} - \overline{\overline{\boldsymbol{u}}}\,\overline{\overline{\theta}} \quad . \tag{15.58}$$

The associated gradient-type model is (for a Gaussian or Box filter)

$$\tau_\theta = \frac{\overline{\Delta}^2}{24} \nabla^2(\overline{\boldsymbol{u}}\overline{\theta}) \quad . \tag{15.59}$$

Differential Scalar Fluxes Model. Another solution consists in solving a differential equation for each component of the subgrid scalar flux vector. This approach requires to close (15.7), in which terms V (pressure-scalar correlations) and VI (diffusion by subgrid fluctuations) are not directly computable.

Deardorff [173] proposes to close term V as

$$\tau_{\mathrm{G}}\left(\theta, \frac{\partial p}{\partial x_i}\right) = -C_1 \frac{\sqrt{q_{\mathrm{sgs}}^2}}{\overline{\Delta}} \tau_{\mathrm{G}}(u_i, \theta) \quad , \tag{15.60}$$

and to express the third-order term as

$$\tau_{\mathrm{G}}(u_k, u_i, \theta) = -C_2 \overline{\Delta} \sqrt{q_{\mathrm{sgs}}^2} \left(\frac{\partial}{\partial x_i} \tau_{\mathrm{G}}(u_k, \theta) + \frac{\partial}{\partial x_k} \tau_{\mathrm{G}}(u_i, \theta) \right) \quad , \tag{15.61}$$

with $C_1 = 4.8$ and $C_2 = 0.2$.

Other closures are derived by Sheikhi et al. [671], who use the velocity-scalar filtered density function approach. This method is strictly equivalent to the one developed by Gicquel et al. for the momentum equation (see Sect. 7.5.3): a stochastic field whose probability density function is the solution of the evolution equation of the filtered density probability function is computed via a Lagrangian-Monte-Carlo method. In the present case, the method is utilized to reconstruct both $\boldsymbol{u}'$ and θ'. These reconstructed stochastic variables can be used to close the filtered momentum and scalar equations directly, or to close the equations for the subgrid scalar fluxes and the subgrid scalar variance. The latter approach is considered here, since the former belongs to the family of the methods based on a direct evaluation of the subgrid scales (to be discussed below). This approach yields the following closure relation for the sum of the pressure-scalar correlations and the dissipation:

$$2\nu\tau_{\mathrm{G}}\left(\frac{\partial u_i}{\partial x_k}, \frac{\partial \theta}{\partial x_k}\right) + \tau_{\mathrm{G}}\left(\theta, \frac{\partial p}{\partial x_i}\right) = \omega\left(\frac{3}{2} + \frac{3}{4}C_0\right) \tau_{\mathrm{G}}(u_i, \theta) \quad , \tag{15.62}$$

where $C_0 = 2.1$ and the time scale ω is equal to

$$\omega = \frac{\sqrt{q_{\mathrm{sgs}}^2}}{\overline{\Delta}} \quad . \tag{15.63}$$

The triple correlation term is directly computed using the stochastic subgrid field.

Explicit Evaluation of Subgrid Scales, Multilevel Simulations and Others. The technique consisting in regenerating the subgrid scales of the scalar field on a finer mesh using a low-cost stochastic model, a simplified determinitic model or multilevel simulation has also been extended to the passive scalar case. These extensions being very reminiscent of their counterparts for the momentum equation, they will not be detailed here.

15.2.5 Generalized Subgrid Modeling for Arbitrary Non-linear Functions of an Advected Scalar

The preceding sections addressed the problem of finding subgrid closures for the convection term, the other ones being assumed to be linear. But in many flows, additional physical processes associated to scalar source/sink are present, which appear as non-linear functions of the scalar field θ. Common examples are chemical reactions, fluids with temperature-dependent molecular viscosity/diffusivity, or coupling of the scalar field (density, temperature, concentration) with micro-physics such as radiative transfer. Let us consider a general form for this additional source term:

$$\frac{\partial \theta}{\partial t} + \nabla \cdot (\boldsymbol{u}\theta) = \kappa \nabla^2 \theta + \underbrace{f(\theta)}_{\text{source term}} \quad . \tag{15.64}$$

The corresponding source term in the filtered scalar equation is $\overline{f(\theta)}$, which is decomposed as

$$\overline{f(\theta)} = \underbrace{f(\overline{\theta})}_{\text{computable}} + \underbrace{\left(\overline{f(\theta)} - f(\overline{\theta})\right)}_{\tau_f\text{:subgrid term}} \quad . \tag{15.65}$$

The new closure problem consists in finding a surrogate for τ_f. Since we are addressing a fully general problem which can cover a very wide range a physical mechanisms, the definition of a general functional model is hopeless. Specific functional models can be found for individual process, as it is done for the convection term.

The most general approach being the structural modeling, Pantano and Sarkar [578] propose to use an approximate deconvolution approach to model general subgrid source terms (in practice, they applied this strategy to a chemical reaction term) . The key idea developed by these authors is to approximate θ by a synthetic field $\theta^\bullet$ which is optimized so that it will minimize an error functional, leading to

$$\overline{f(\theta)} - f(\overline{\theta}) = \overline{f(\theta^\bullet)} - f(\overline{\theta^\bullet}) \quad . \tag{15.66}$$

Since physical processes may be very sensitive to small errors and are associated to high-order non-linearities in terms of θ (e.g. the radiative transfer flux behaves as θ^4, where θ is the temperature), it is important to enforce some accuracy requirements when defining $\theta^\bullet$.

The simple Taylor series expansion

$$\overline{f(\theta)} - f(\overline{\theta}) = \frac{1}{2} f''(\overline{\theta})(\overline{\theta^2} - \overline{\theta}^2) + ... \quad , \tag{15.67}$$

illustrates the fact that a good control of the error committed on the subgrid scalar variance is necessary to ensure a satisfactory representation of the subgrid source/sink term. As a consequence, Pantano and Sarkar propose to modify the usual soft deconvolution procedure, leading to

$$\theta^\bullet = \overline{\theta} + C_0(\overline{\theta} - \overline{\overline{\theta}}) + C_1(\overline{\theta} - 2\overline{\overline{\theta}} + \overline{\overline{\overline{\theta}}}) + ... \quad , \tag{15.68}$$

were the coefficients C_i are allowed to vary, instead of being fixed to be unity as in the original approximate deconvolution procedure based on the Van Cittert iterative method (7.8). These coefficients are chosen in order to make the statistical mean filtered moments that appear in the Taylor series expansion of the modeled field equal to their counterparts defined using the exact field. In practice, it is aimed to enforce the following global constraints

$$\int_\Omega \overline{\theta^{\bullet k}} - \overline{\theta^\bullet}^k d\boldsymbol{x} = \int_\Omega \overline{\theta^k} - \overline{\theta}^k d\boldsymbol{x}, \quad k = 2, ..., N \quad , \tag{15.69}$$

where N is an arbitrary parameter and Ω is the fluid domain under consideration. Practical subgrid models are derived truncating expansion (15.68) at an arbitrary order and inserting it in (15.69). To get a closed set of non-linear polynomial equations for the coefficients C_i, it is also required to select an analytic expression of the scalar spectrum and the filter transfer function.

15.2.6 Models for Subgrid Scalar Variance and Scalar Subgrid Mixing Rate

This section is devoted to models aiming at evaluating the two following subgrid quantities, which characterize the dynamics of the subgrid fluctuations of the scalar field θ:

- The subgrid scalar variance, θ^2_{sgs}, also referred to as the scalar *unmixedness*, since it measures the degree of local non homogeneity of θ within the volume of characteristic diameter $\overline{\Delta}$: a uniform distribution is associated with a zero subgrid variance.
- The scalar variance dissipation rate, ε_θ, also referred to as the scalar variance destruction rate, which is related to the stirring and mixing process at subgrid scales. A high destruction rate means that diffusion is quickly homogenezing the scalar field.

A large number of subgrid models for these two quantities have been proposed, since they are important inputs of subgrid models for combustion related terms. Since the specific problem of reactive flows is beyond the scope of this chapter, this section illustrates the different modeling ways and does not provide the reader with an exhaustive description.

Models for the Subgrid Scalar Variance. It is worth noting that the issue of predicting $\theta^2_{\rm sgs}$ is reminiscent of the one dealing with the evaluation of the subgrid kinetic energy, $q^2_{\rm sgs}$. Therefore, all methods and models defined to compute the latter (see Sect. 9.2.3) can be modified to evaluate the former. This work is straightforward and is left to the interested reader. Only general approaches will be discussed in this section.

The subgrid scalar variance being defined as

$$\theta^2_{\rm sgs} \equiv \overline{\theta^2} - \overline{\theta}^2 \quad , \tag{15.70}$$

a natural way to model it is to use a deconvolution-type approach (or its optimized version proposed by Pantano and Sarkar for arbitrary non-linear functions of θ, see Sect. 15.2.5). The usual deconvolution procedure yields

$$\theta^2_{\rm sgs} \simeq \overline{\theta^{\bullet 2}} - \overline{\theta^{\bullet}}^2 \quad , \tag{15.71}$$

where $\theta^{\bullet} = G_l^{-1} \star \overline{\theta}$ is the approximate defiltered field. Using the zeroth-order expansion of the deconvolution operator, one obtains a scale-similarity type model:

$$\theta^2_{\rm sgs} \simeq C(\overline{\overline{\theta}^2} - \overline{\overline{\theta}}^2) \quad , \tag{15.72}$$

where the constant C is equal to one if the Van Cittert method is used. Other values can be found using a dynamic procedure based on a Germano-type identity. A gradient type model is recovered replacing the convolution filters by their equivalent differential expansion, and then truncating these expansion at an arbitrary order. For box and Gaussian filters, the first order term is

$$\theta^2_{\rm sgs} \simeq C' \overline{\Delta}^2 |\nabla \overline{\theta}|^2 \quad , \tag{15.73}$$

where the parameter C' is filter dependent and can be adjusted using a dynamic procedure.

The subgrid scalar variance can also be easily retrieved if a structural model based on an explicit reconstruction of the subgrid scales is used to close the filtered scalar equation. In this case, it is directly computed from the synthetic subgrid scalar field. It is also straightforwardly extracted if a differential model for the scalar fluxes is used.

Another possiblity consists in solving a prognostic equation for $\theta^2_{\rm sgs}$, in a way similar to what is done for the subgrid kinetic energy. Using (15.8) as a starting point, a closed equation is obtained. The resulting model is presented in (15.135) for the sake of brevity. Prognostic equations with dynamic coefficients can also be derived using techniques presented in Sect. 5.4 (p. 173) for the transport equation of the subgrid kinetic energy.

A simple expression is obtained in the simplified case where the inertial-convective range spectrum (15.22) is assumed to be valid at all subgrid wave numbers:

$$\theta_{\text{sgs}}^2 = \int_{\pi/\overline{\Delta}}^{+\infty} E_\theta(k)dk = \frac{3\beta}{2\pi^{2/3}}\varepsilon_\theta \varepsilon^{-1/3}\overline{\Delta}^{2/3} \quad . \tag{15.74}$$

Models for the Subgrid Scalar Dissipation Rate. A first simple evaluation of ε_θ is obtained assuming that the cutoff is located within an infinite inertial-convective range:

$$\varepsilon_\theta = C\frac{\theta_{\text{sgs}}^2\sqrt{q_{\text{sgs}}^2}}{\overline{\Delta}} \quad , \tag{15.75}$$

where the constant C can by either computed analytically or adjusted using a dynamic procedure. Considering an infinite inertial-convective range, one obtains

$$C = \frac{2\pi}{3\beta}\sqrt{\frac{2}{3K_0}} \quad . \tag{15.76}$$

The other way to evaluate the subgrid scalar variance dissipation rate is to assume that the local equilibrium hypothesis holds: in this case, it is equal to the local subgrid scalar variance production rate, leading to

$$\varepsilon_\theta = -\tau_\theta \cdot \nabla\overline{\theta} \quad , \tag{15.77}$$

where the subgrid scalar flux τ_θ can be evaluated using any adequate subgrid model. If a subgrid diffusity model is utilized, one obtains

$$\varepsilon_\theta = 2\kappa_{\text{sgs}}|\nabla\overline{\theta}|^2 \quad . \tag{15.78}$$

This expression can be modified to account for the dissipation at the resolved scales, yielding

$$\varepsilon_\theta = 2(\kappa + \kappa_{\text{sgs}})|\nabla\overline{\theta}|^2 \quad . \tag{15.79}$$

In order to obtain a more general expression which does not rely on the local equilibrium assumption, Jimenez et al. [352] assumed that the characteristic subgrid mixing time is proportional to the subgrid turbulent time:

$$\frac{\varepsilon_\theta}{\theta_{\text{sgs}}^2} \propto \frac{\varepsilon}{q_{\text{sgs}}^2} \quad , \tag{15.80}$$

yielding the following model

$$\varepsilon_\theta = C\frac{\theta_{\text{sgs}}^2\varepsilon}{q_{\text{sgs}}^2} \quad , \tag{15.81}$$

where C is a parameter. Tests show that $C = 1/Pr$ leads to satisfactory results. All quantities which appear in the right hand side of (15.81) can be evaluated using ad hoc models.

15.2.7 A Few Applications

- Heat transfer at free surface [93]
- Heat transfer in plane channel flow [92, 755, 196, 753]
- Isotropic turbulence [121, 441, 538, 122, 220, 608, 344, 345, 752]
- Complex cavities and ducts [279, 590, 639]
- Homogeneous turbulence [344, 345]
- Time-developing shear layer [344, 345]
- Flow past a backward facing step [27]
- Jet impinging on a plate [738, 155, 126, 236]
- Flow in a corrugated passage [141]
- Mixing Layer [168]
- Flow in rotating/steady duct with/without rib [557, 558, 559, 560]
- Flow in S-shaped duct [561]
- Wall-mounted cube matrix [565]

15.3 The Active Scalar Case: Stratification and Buoyancy Effects

We now turn to the case of the coupling with an *active* scalar, i.e. with a field which has a feedback effect on the velocity field, leading to a two-way coupling between the Navier–Stokes and the scalar equations. Since there are a very huge number of possible interactions, it is chosen to put the emphasis on buoyancy and stable stratification effects. Other physical models, such as Eulerian–Eulerian models for two-phase flows, will not be considered.

15.3.1 Physical Model

Buoyancy and stable stratification effects originate in the force to which a blob of fluid is submitted when it is immersed in a fluid with different density. Therefore, a natural physical parameter to describe these effects is the density. When very weak compressiblity effects are taken into account, the density of "usual" fluids is assumed to decrease when the temperature is increased. Thus, the temperature can also be used to represent the dynamics of these flows. Both solutions are found in the literature, depending on the authors and specific purposes of the studies. In the following, the scalar field θ will be related to the temperature field.

Using the Boussinesq approximation (see reference books for a detailed discussion of the range of validity of this model, e.g. [714, 708, 439]), the basic set of unfiltered equations consists in the Navier–Stokes equations with a gravitational source term supplemented by a scalar equation identical to (15.1):

$$\frac{\partial \boldsymbol{u}}{\partial t} + \nabla(\boldsymbol{u} \otimes \boldsymbol{u}) = -\nabla p + \nu \nabla^2 \boldsymbol{u} + \underbrace{\frac{\theta - \Theta}{\Theta_0} \boldsymbol{g}}_{\text{feedback term}} \quad , \tag{15.82}$$

$$\nabla \cdot \boldsymbol{u} = 0 \quad , \tag{15.83}$$

$$\frac{\partial \theta}{\partial t} + \nabla \cdot (\boldsymbol{u}\theta) = \kappa \nabla^2 \theta \quad , \tag{15.84}$$

where θ is *potential temperature*, defined as the thermodynamic temperature increased by the normalized gravitational potential, $\boldsymbol{g} = (0, 0, -g)$ the gravity vector and Θ is the potential temperature field associated to the hydrostatic equilibrium. The vector $\boldsymbol{g}$ is related to the gravitational acceleration. Θ_0 is related to a reference value, which is assumed to be unique for the whole flow domain. The molecular viscosity and the molecular diffusivity are assumed to be constant and uniform.

The corresponding set of filetered equations is very similar to those used in previous chapters, since the original set of equations differs by only one source term:

$$\frac{\partial \overline{\boldsymbol{u}}}{\partial t} + \nabla(\overline{\boldsymbol{u}} \otimes \overline{\boldsymbol{u}} + \tau) = -\nabla \overline{p} + \nu \nabla^2 \overline{\boldsymbol{u}} + \frac{\overline{\theta} - \Theta}{\Theta_0} \boldsymbol{g} \quad , \tag{15.85}$$

$$\nabla \cdot \overline{\boldsymbol{u}} = 0 \quad , \tag{15.86}$$

$$\frac{\partial \overline{\theta}}{\partial t} + \nabla \cdot (\overline{\boldsymbol{u}}\overline{\theta} + \tau_\theta) = \kappa \nabla^2 \overline{\theta} \quad , \tag{15.87}$$

where it has been assumed that the scale separation operator perfectly commutes with all partial derivatives operator and that the mean field Θ is varying slowly enough in space to have $\overline{\Theta} = \Theta$. The subgrid fluxes τ and τ_θ have exactly the same expressions as in the passive scalar case.

A deeper insight into the two-way coupling is gained rewriting the evolution equations for the subgrid quantities. Assuming that the vertical direction is associated to x_3, the equations for the subgrid mometum fluxes are:

$$\begin{aligned} \frac{\partial \tau_{\mathrm{G}}(u_i, u_j)}{\partial t} + \overline{u}_k \frac{\partial \tau_{\mathrm{G}}(u_i, u_j)}{\partial x_k} = & -\tau_{\mathrm{G}}(u_k, u_i) \frac{\partial \overline{u}_j}{\partial x_k} - \tau_{\mathrm{G}}(u_k, u_j) \frac{\partial \overline{u}_i}{\partial x_k} \\ & + \nu \frac{\partial^2 \tau_{\mathrm{G}}(u_i, u_j)}{\partial x_k \partial x_k} - 2\nu \tau_{\mathrm{G}} \left(\frac{\partial u_i}{\partial x_k}, \frac{\partial u_j}{\partial x_k} \right) \\ & - \tau_{\mathrm{G}} \left(u_j, \frac{\partial p}{\partial x_i} \right) - \frac{\partial \tau_{\mathrm{G}}(u_k, u_i, u_j)}{\partial x_k} \\ & + \underbrace{\frac{g}{\Theta_0} (\delta_{i3} \tau_{\mathrm{G}}(u_j, \theta) + \delta_{j3} \tau_{\mathrm{G}}(u_i, \theta))}_{\text{coupling term}} \quad , \end{aligned} \tag{15.88}$$

and the corresponding equations for the generalized subgrid kinetic energy is:

$$\begin{aligned} \frac{\partial \tau_{\mathrm{G}}(u_i,u_i)}{\partial t} &= \frac{\partial}{\partial x_j}\left(\frac{1}{2}\tau_{\mathrm{G}}(u_i,u_i,u_j)+\tau_{\mathrm{G}}(p,u_j)-\nu\frac{\partial \tau_{\mathrm{G}}(u_i,u_i)}{\partial x_j}\right) \\ &- \nu\tau_{\mathrm{G}}\left(\frac{\partial u_i}{\partial x_j},\frac{\partial u_i}{\partial x_j}\right)-\tau_{\mathrm{G}}(u_i,u_j)\frac{\partial \overline{u}_i}{\partial x_j} \\ &+ \underbrace{\delta_{i3}\frac{g}{\Theta_0}\tau_{\mathrm{G}}(\theta,u_i)}_{\text{coupling term}} \quad . \end{aligned} \tag{15.89}$$

The subgrid scalar fluxes are solutions of

$$\begin{aligned} \frac{\partial \tau_{\mathrm{G}}(u_i,\theta)}{\partial t}+\overline{u}_k\frac{\partial \tau_{\mathrm{G}}(u_i,\theta)}{\partial x_k} &= -\tau_{\mathrm{G}}(u_k,u_i)\frac{\partial \overline{\theta}}{\partial x_k}-\tau_{\mathrm{G}}(u_k,\theta)\frac{\partial \overline{u}_i}{\partial x_k} \\ &+\frac{\partial}{\partial x_k}\left(\nu\tau_{\mathrm{G}}\left(\frac{\partial u_i}{\partial x_k},\theta\right)+\kappa\tau_{\mathrm{G}}\left(\frac{\partial \theta}{\partial x_k},u_i\right)\right) \\ &-(\nu+\kappa)\tau_{\mathrm{G}}\left(\frac{\partial u_i}{\partial x_k},\frac{\partial \theta}{\partial x_k}\right)-\tau_{\mathrm{G}}\left(\theta,\frac{\partial p}{\partial x_i}\right) \\ &-\frac{\partial \tau_{\mathrm{G}}(u_k,u_i,\theta)}{\partial x_k}+\underbrace{\delta_{i3}\frac{g}{\Theta_0}\tau_{\mathrm{G}}(\theta,\theta)}_{\text{coupling term}} \quad , \end{aligned} \tag{15.90}$$

while the generalized subgrid scalar variance equation is kept unchanged, since there is no gravitational source term in the scalar equation. Therefore, the subgrid variance is not directly affected by the feedback effect on the velocity field, but it is sensitive to it via changes in the latter.

15.3.2 Some Insights into the Active Scalar Dynamics

Flows governed by (15.82)–(15.84) exhibit a very wide and complex range of physical mechanisms, which originate in the coupling that may exists between convection, diffusion, stratification and other features such as mean shear, rotation and boundary conditions. It is meaningless to try to give an exhaustive description of all these possibilities. The present section will focus on very simple cases and the emphasis will be put on results dealing with interscale transfers, which are of primary interest for subgrid modeling.

The influence of the stratification effects on the subgrid kinetic energy is seen looking at the last term of (15.89), which can lead to an increase associated to buoyancy effect or a decrease due to stable stratification damping, depending on its sign. Its relative importance with respect to the production term is measured by the flux Richardson number:

$$Ri_f=\frac{\frac{g}{\Theta_0}\tau_{\mathrm{G}}(u_3,\theta)}{\tau_{\mathrm{G}}(u_i,u_k)\frac{\partial \overline{u}_i}{\partial x_k}} \quad , \tag{15.91}$$

High values of Ri_f are associated to flows in which stratification effects are dominant, while they can be neglected when $Ri_f \ll 1$. Neglecting energy transport across the flow, one obtains

$$Ri_f \simeq 1 - \frac{\varepsilon}{-\tau_G(u_i, u_k)\frac{\partial \overline{u}_i}{\partial x_k}} \quad , \tag{15.92}$$

showing that destabilizing effects are present if $Ri_f > 1$ (the denominator of the last term being assumed to be positive).

Two basic cases are discussed in the present section:

- The case of stable stratification (p. 475), in which stratification effects tend to damp the subgrid kinetic energy. An important feature of flows with stable stratification is the existence of dispersive internal gravity waves. These waves are important in many applications dealing with meteorology and oceanology. They induce irrotational large-scale horizontal motions, whose interactions with small scale turbulence will be one of the most important issue discussed below.
- The case of unstable stratification (p. 480), where buoyancy effects generate turbulence, a well known example being thermal plumes.

Another relevant parameter is the group

$$\mathcal{N} = \frac{g}{\Theta_0}\frac{\partial \overline{\theta}}{\partial z} \quad . \tag{15.93}$$

Under the assumption that the subgrid heat flux $\tau_G(\theta, \boldsymbol{u})$ obeys a Fickian-like law, i.e. $\tau_G(\theta, \boldsymbol{u}) \simeq -C\nabla\overline{\theta}$, where C is a positive constant, it is seen from (15.88), (15.89) and (15.90) that the local stablizing/destabilizing effect of stratification depends on the sign of $\mathcal{N}$, i.e. on the local sign of the vertical resolved gradient $\partial\overline{\theta}/\partial z$:

1. If $\partial\overline{\theta}/\partial z > 0$, the effect is stabilizing, since it appears as a sink term in the subgrid kinetic energy equation. In this regime, gravity waves are stable and their frequency is characterized by the Brunt–Väisälä frequency, $N \equiv \sqrt{\mathcal{N}}$.
2. If $\partial\overline{\theta}/\partial z < 0$, gravity waves are unstable and break up into turbulence: the buoyancy effects are seen to increase the production of subgrid kinetic energy. The buoyancy time scale is estimated as $T_b \equiv 1/\sqrt{-\mathcal{N}}$.

Energy Transfers in Stably Stratified Flows. Energy transfers in stably stratified flows have been addressed by many authors, and are observed to be case dependent. Therefore, no general description of the transfer across a cutoff similar to what exists for the momentum equation is available, the main explanation for that being that such a general description is nearly impossible. The difficult points which preclude such an analysis are the following

- These flows are strongly anisotropic, and it has already been seen that kinetic energy transfers within homogeneous anisotropic flows cannot be described in a simple and general way, since they are case-dependent.
- The description in terms of interscale transfers is not sufficient to obtain an accurate picture of the governing physical processes. It appears that the concept of modes must be introduced to achieve a meaningful analysis: the energy must be split into the *vortex kinetic energy* and the *total wave energy*, the latter being further decomposed into the *potential energy* and the *wave kinetic energy* [266]. The analysis of triadic interactions using this scheme leads to a global description which is too complex to yield results that can be straightforwardly utilized for subgrid parameterization. This modal decomposition results from a local decomposition of the velocity and temperature field in the Fourier space:

$$\widehat{\boldsymbol{u}}(\boldsymbol{k}) = \widehat{\boldsymbol{u}}_1(\boldsymbol{k}) + \widehat{\boldsymbol{u}}_2(\boldsymbol{k}) \quad , \tag{15.94}$$

where $\widehat{\boldsymbol{u}}_1(\boldsymbol{k})$ and $\widehat{\boldsymbol{u}}_2(\boldsymbol{k})$ are related to the vortex and internal gravity wave components, respectively, with

$$\widehat{\boldsymbol{u}}_1(\boldsymbol{k}) = \widehat{\phi}_1(\boldsymbol{k})\boldsymbol{e}_1(\boldsymbol{k}), \quad \widehat{\boldsymbol{u}}_2(\boldsymbol{k}) = \widehat{\phi}_2(\boldsymbol{k})\boldsymbol{e}_2(\boldsymbol{k}) \quad , \tag{15.95}$$

where

$$\boldsymbol{e}_1(\boldsymbol{k}) = \frac{\boldsymbol{k} \times \boldsymbol{g}}{|\boldsymbol{k} \times \boldsymbol{g}|}, \quad \boldsymbol{e}_2(\boldsymbol{k}) = \frac{\boldsymbol{k} \times \boldsymbol{e}_1(\boldsymbol{k})}{|\boldsymbol{k} \times \boldsymbol{e}_1(\boldsymbol{k})|} \quad . \tag{15.96}$$

The two vectors $(\boldsymbol{e}_1(\boldsymbol{k}), \boldsymbol{e}_2(\boldsymbol{k}))$ generate an orthonormal basis for the plane perpendicular to $\boldsymbol{k}$. The vortex kinetic energy spectrum, $\Phi_1(\boldsymbol{k})$, the wave kinetic energy spectrum, $\Phi_2(\boldsymbol{k})$, and the potential energy spectrum $P(k)$ are defined as follows:

$$\Phi_1(\boldsymbol{k}) = \frac{1}{2}\left\langle \widehat{\phi}_1(\boldsymbol{k})\widehat{\phi}_1(-\boldsymbol{k})\right\rangle, \quad \Phi_2(\boldsymbol{k}) = \frac{1}{2}\left\langle \widehat{\phi}_2(\boldsymbol{k})\widehat{\phi}_2(-\boldsymbol{k})\right\rangle \quad ,$$

$$P(k)(\boldsymbol{k}) = \frac{1}{2N^2}\left\langle \widehat{\theta}(\boldsymbol{k})\widehat{\theta}(-\boldsymbol{k})\right\rangle \quad ,$$

where N^2 is equal to the square of the Brunt-Väisälä frequency built on the mean temperature gradient:

$$N^2 = \frac{g}{\Theta_0}\frac{d\Theta}{dz} \quad . \tag{15.97}$$

The total kinetic energy spectrum is recovered summing the vortex and wave contributions: $E(\boldsymbol{k}) = \Phi_1(\boldsymbol{k}) + \Phi_2(\boldsymbol{k})$.

As a consequence, this section will be devoted to the description of the main features of very simple cases which have been very accurately analyzed using both theoretical tools and direct numerical simulation.

The most simple case is initially isotropic turbulence submitted the strong stabilizing effects. The main observed phenomenon is the breakdown of isotropy, associated to a drastic restriction of motion along the mean stratification direction. After some times, the flow is organized in thin layers with a strong variability along the stratification direction (*pancake-shaped* large-scale vortices), leading to the occurance of a velocity field which is almost *two-component*[4] Detailed investigations of energy transfers using Direct Numerical Simulation and EDQNM models have been carried out [235, 304, 513, 266, 689, 267]. The main results are the following

- Several phases are observed, the number of which being dependent on the inital condition and the value of the Richardson number. The first phase (if initial potential energy is large enough) corresponds to a flow whose large scales are governed by stratification effects and inertial gravity waves and exhibit a N^2k^{-3} kinetic energy spectrum, while small scales are not affected by stratification effects and are governed by the usual isotropic turbulent dynamics. In the last phase, all scales are anisotropic and stratification effects drive the whole flow.
- During the first phase (pre-collapse phase), the flow dynamics is controlled by the potential energy being transferred from large to small scales at a higher rate than kinetic energy, leading to a higher total energy decay as in unstratified isotropic turbulence. The kinetic energy along the stratification direction is converted into potential energy, feeding the irreversible forward cascade of potential energy. A consequence of the potential energy cascade is the existence of a persisting counter-gradient scalar flux. Such a counter-gradient scalar flux should be associated to a *negative* subgrid diffusivity at small scales, making all previous models irrelevant. But this counter-gradient flux seems to play no major dynamical role on the smallest scales during the first phase if the molecular diffusivity is high enough.
- The final phase (post-collapse phase) is associated to a scalar flux collapse at large scales. The ratio of the kinetic energy along the the stratification direction and the potential energy reach a constant value, while large-scale energetic motion in directions perpendicular to the mean stratification direction arise from turbulence, forming the vortex part of the flow. The small scale dynamics is dominated by stratification effects. An explanation is that, because of the collapse of large scales which reduces the rate of the forward energy cascade, the smallest scales are made sensitive to stratification.
- A detailed analysis of the transfers reveals that the pure vortex-vortex interactions (i.e. triadic interactions involving vortex modes only) are initially important among all triadic interactions, the resonant interactions involving at least one wavy mode being less important. The former are

[4] But not *two-dimensional*, since its dynamics is very different from the one of two-dimensional turbulence. In particular, no strong backward energy cascade is observed.

responsible for the isotropy breakdown and the blocking of a possible backward energy cascade. A very important feature, which limits the accuracy of the modeling of triadic transfers in the physical space, is that the energy transfers are strongly dependent on both the modulus and the direction of the three wave numbers. Oscillatory exchanges between wave kinetic energy and potential energy are also observed (their sum exhibiting a non-oscillatory behavior), associated to irreversible anisotropy creation, leading to angular variations (in the spectral space) of the vortex kinetic energy and the total wave energy.
- The counter-gradient scalar flux at small scales can inhibit mixing if the molecular diffusivity is too low, leading to a complex behavior of the scalar variance dissipation rate.

These results show that the triadic transfer patterns are much more complicated than in unstratified flows. The physical picture presented above is to be further complexified to account for additional physical mechanisms, such as turbulence generation by a forcing term, interaction with a mean shear, coupling with dynamics, etc.

The main information retrieved from the previous analysis is that the forward kinetic energy cascade is decreased by stratification effects in the post-collapse phase. The analysis of the transfers across a cutoff wave number during the pre- and post-collapse phase was carried out by Métais and Lesieur [514] using numerical simulations. Using the orthogonality of the local reference frame in the Fourier space, the authors extend the analysis presented in Sect. 5.1.2 by decomposing the total kinetic energy transfer into a vortex and wave part, and introduce a spectral subgrid viscosity representation for each part:

$$\nu_{\mathrm{e}}^{i}(\boldsymbol{k}|k_{\mathrm{c}}) = -\frac{T_{\mathrm{sgs}}^{\mathrm{e,i}}(\boldsymbol{k}|k_{\mathrm{c}})}{2k^2\Phi_i(\boldsymbol{k})} \quad , \tag{15.98}$$

where $i = 1, 2$ refers to the subgrid transfer associated with the corresponding mode in the modal decomposition. As in the unstratified case, these subgrid effective viscosties can be normalized using the kinetic energy at the cutoff, yielding

$$\nu_{\mathrm{e}}^{i}(\boldsymbol{k}|k_{\mathrm{c}}) = \nu_{\mathrm{e}}^{i+}(\boldsymbol{k}|k_{\mathrm{c}})\sqrt{\frac{E(k_{\mathrm{c}})}{k_{\mathrm{c}}}} \quad . \tag{15.99}$$

The total effective subgrid viscosity, $\nu_{\mathrm{e}}(\boldsymbol{k}|k_{\mathrm{c}})$, is obtained using the following relationship

$$\nu_{\mathrm{e}}(\boldsymbol{k}|k_{\mathrm{c}})E(k) = \nu_{\mathrm{e}}^{1}(\boldsymbol{k}|k_{\mathrm{c}})\Phi_1(\boldsymbol{k}) + \nu_{\mathrm{e}}^{2}(\boldsymbol{k}|k_{\mathrm{c}})\Phi_2(\boldsymbol{k}) \quad . \tag{15.100}$$

This last expression illustrates the high difficulty which arises in the modeling of the subgrid transfers via a single subgrid viscosity, since its value depends on both the energy cascade rate of the vortex and wave components and the repartition of the kinetic energy among these two modes. This double dependency shows that the total effective viscosity must be sensitive to the initial conditions, which seems to preclude any general accurate definition. The subgrid potential temperature transfer is formally defined as in the passive scalar case, since the evolution equations are the same.

It is observed in the numerical simulations carried out by Métais and Lesieur that

- In the pre-collapse phase (see Fig. 15.6), $\nu_e^{1+}(\boldsymbol{k}|k_c)$ and $\nu_e^{2+}(\boldsymbol{k}|k_c)$ are almost identical and do not differ from the effective subgrid viscosity observed in unstratified isotropic turbulence: the same plateau (at low wave numbers) and cusp (near the cutoff) behaviors are observed. The effective subgrid diffusivity exhibits a plateau at low wave numbers but do not show any cusp at high wave numbers.
- In the post-collapse phase (see Fig. 15.7), the cusp in the two subgrid viscosities is higher than in the pre-collapse phase because of the shifting of the spectrum maxima towards low wave numbers. The vortex-related viscosity, $\nu_e^{1+}(\boldsymbol{k}|k_c)$, exhibits a plateau with constant value 0.09 at low wave numbers, while the wave mode related viscosity exhibits the same plateau value at intermediate scales but increases with decreasing wave number at low wave numbers. The cusp is also increased in the subgrid diffusivity. Its is important noting that despite this increase in the relative intensity of the transfer at small scales, the absolute level of the effective

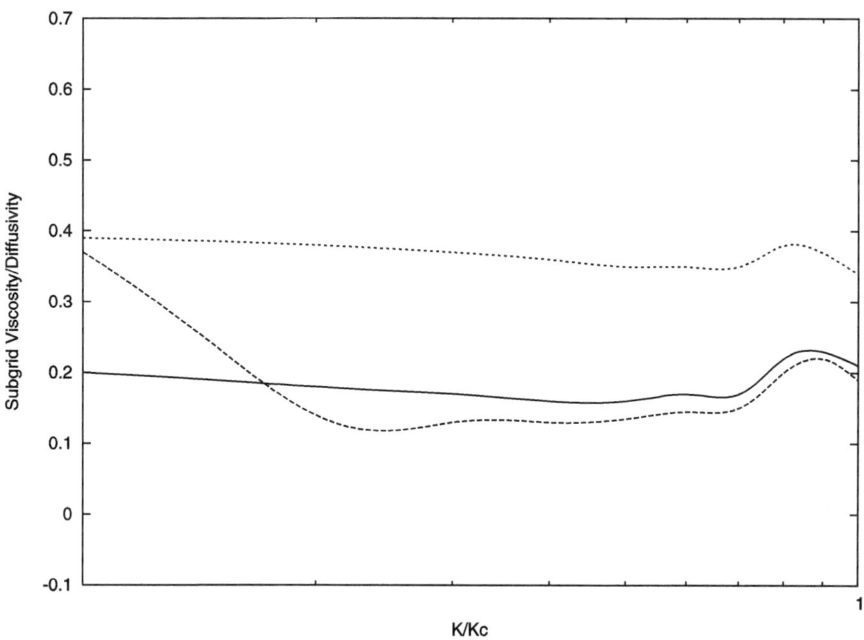

Fig. 15.6. Effective subgrid viscosities $\nu_e^{1+}(\boldsymbol{k}|k_c)$ (*Solid line*) and $\nu_e^{2+}(\boldsymbol{k}|k_c)$ (*Dashed line*) and subgrid diffusivity (*Dotted line*) in the Lesieur–Métais stably-stratified case (pre-collapse phase).

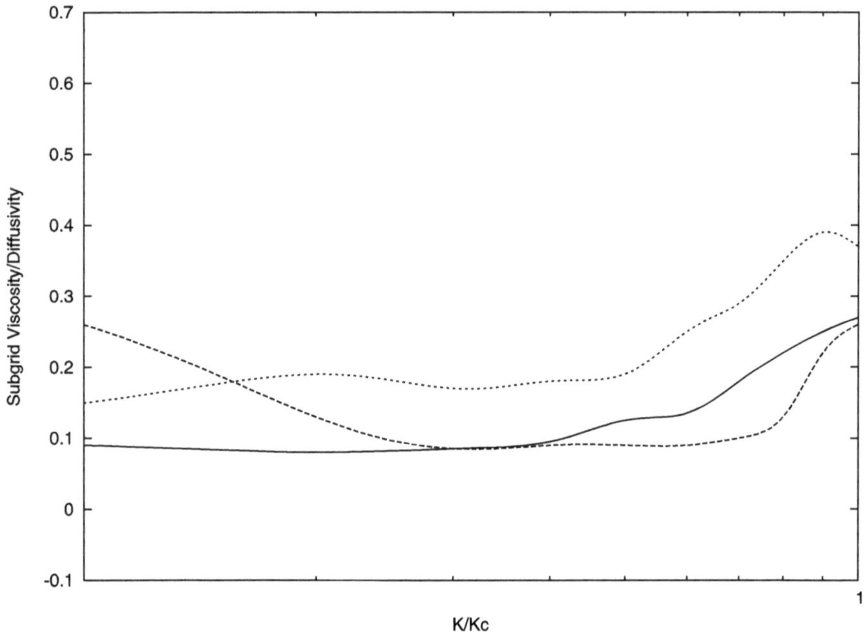

Fig. 15.7. Effective subgrid viscosities $\nu_e^{1+}(\boldsymbol{k}|k_c)$ (*Solid line*) and $\nu_e^{2+}(\boldsymbol{k}|k_c)$ (*Dashed line*) and subgrid diffusivity (*Dotted line*) and subgrid diffusivity (*Dotted line*) in the Lesieur–Métais stably-stratified case (post-collapse phase).

subgrid viscosities, which scales like $\sqrt{E(k_c)}$, is decreased in the post-collapse phase, since stable stratification inhibits the forward kinetic energy cascade, leaving less energy in the small scales.

These results show that, depending on the considered flow regime and initial conditions [287], both the intensity and the spectral shape of the global transfer operator vary, rendering its accurate modeling in the physical space nearly impossible.

Energy Transfers in Buoyancy-Driven Flows. We now turn to the case were the buoyancy force acts as a turbulence generator, as in *natural convection* flows. In these flows, buoyancy-generated instabilities create unsteady motion which will lead to turbulence. Similarly to the case of stable stratification effects, buoyancy-driven flows exhibit a wide range of physical mechanisms which prevent an exhaustive detailed analysis.

As an example, let us consider the analysis of energy transfers in thermal plumes carried out by Baastians et al. [29]. Using an experimental database, these authors observe a time mean backscatter in the lower part of the plume and instantaneous backscatter in the total up-flowing area. The energy transfer rate across the cutoff exhibits large fluctuations, instantaneous values being about ten times larger than the mean value. Forward and backward energy cascades are equally important in the mean in the plume region. These results show that a reliable subgrid model should be able to account for backscatter, and to distinguish between different regions of the flow, rendering the modeling task even more difficult as in the stably stratified flow case.

But it is important to remark that, in the basic philosophy which underlies most of the developments in the field of Large-Eddy Simulation, it is assumed that driving mechanisms must be directly captured. In the case of buoyancy-driven flows, buoyancy effects should therefore be directly simulated if they are the only source of turbulence production (free convection). In the case of mixed convection or forced convection, the requirement might be less stringent since other physical mechanisms are at play. This is why some authors use unmodified passive scalar models (and usual closures for the momentume equations) to treat Rayleigh–Bénard convection at medium Prandtl number.

15.3.3 Extensions of Functional Models

We now discuss the most illustrative functional models in the physical space for the active scalar case. Since a two-way coupling exists, subgrid models for both the momentum and the scalar equations must be revisited to account for stratification/buoyancy effects. Modeling strategies based on the definition of scalar subgrid viscosity/diffusivity parameters are first considered (p. 481). Dynamic models based on the Germano identity are presented in the second part of the section (p. 486).

Scalar Subgrid Viscosity/Diffusivity Models. The brief review of results dealing with interscale energy transfers in flows with active scalar given in the previous section obviously shows that they are strongly affected in both stable and unstable stratification cases. This is why most scalar models are derived considering a simplified kinetic energy balance equation which includes buoyancy effects. The most popular simplified energy balance expression is obtained by neglecting all diffusive and convective effects in (15.89), yielding an extended local equilibrium (production = dissipation) assumption:

$$\varepsilon = -\tau : \overline{S} - \boldsymbol{b} \cdot \tau_\theta, \quad \boldsymbol{b} \equiv \frac{1}{\Theta_0}\boldsymbol{g} \quad , \tag{15.101}$$

where it is recalled that ε, τ and τ_θ are the subgrid kinetic energy dissipation rate, the momentum subgrid tensor and the potential temperature subgrid flux vector, respectively. Now specializing Eq. (15.101) by inserting scalar functional models of the form

$$\tau = -2\nu_{\mathrm{sgs}}\overline{S}, \quad \tau_\theta = -\kappa_{\mathrm{sgs}}\nabla\overline{\theta} \quad , \tag{15.102}$$

one obtains

$$\nu_{\mathrm{sgs}}|\overline{S}|^2 - \kappa_{\mathrm{sgs}}N_c^2 = \varepsilon \quad , \tag{15.103}$$

where the local parameter N_c^2 is defined as

$$N_c^2 \equiv -\boldsymbol{b} \cdot \nabla \overline{\theta} = \frac{g}{\Theta_0} \frac{\partial \theta}{\partial z} \quad , \tag{15.104}$$

and can be either positive or negative. Negative (resp. positive) values are associated with a decrease (resp. increase) of the subgrid kinetic energy dissipation rate by the buoyancy effects.

Equation (15.103) shows that the two-way coupling observed in the active scalar case precludes any decoupled definitions for $\nu_{\rm sgs}$ and $\kappa_{\rm sgs}$.

Now assuming that the cutoff is located within an inertial-convective-like range, dimensional analysis yields

$$\nu_{\rm sgs} = C_\nu |\varepsilon|^{1/3} \overline{\Delta}^{4/3}, \quad \kappa_{\rm sgs} = C_\kappa |\varepsilon|^{1/3} \overline{\Delta}^{4/3} \quad , \tag{15.105}$$

where C_ν and C_κ are positive parameters. The combination of (15.103) and (15.105) leads to the following expression for the dissipation rate:

$$\varepsilon = |\varepsilon|^{1/3} \overline{\Delta}^{4/3} \left(C_\nu |\overline{S}|^2 - C_\kappa N_c^2 \right) . \tag{15.106}$$

This expression leads to the following definition for the subgrid viscosity and the subgrid diffusivity:

$$\nu_{\rm sgs} = C_\nu \overline{\Delta}^2 \left| C_\nu |\overline{S}|^2 - C_\kappa N_c^2 \right|^{1/2} = C'_\nu \overline{\Delta}^2 \underbrace{\left| |\overline{S}|^2 - \frac{N_c^2}{Pr_{\rm sgs}} \right|^{1/2}}_{\mathcal{T}} \quad , \tag{15.107}$$

$$\kappa_{\rm sgs} = C_\kappa \overline{\Delta}^2 \left| C_\nu |\overline{S}|^2 - C_\kappa N_c^2 \right|^{1/2} = C'_\kappa \overline{\Delta}^2 \underbrace{\left| |\overline{S}|^2 - \frac{N_c^2}{Pr_{\rm sgs}} \right|^{1/2}}_{\mathcal{T}} \quad , \tag{15.108}$$

where the subgrid Prandtl number is defined as $Pr_{\rm sgs} \equiv C_\nu / C_\kappa = C'_\nu / C'_\kappa$. The main modification with respect to the passive scalar lies in the evaluation of the characteristic time scale $\mathcal{T}^{-1}$: it now combines the velocity gradient and the buoyancy effects. Model (15.107) can be seen as an extension of the classical Smagorinsky subgrid viscosity model (5.90). It is seen that stable stratification, which is associated with positive values of N_c^2, corresponds to a decrease of the subgrid viscosity and diffusivity. This is in agreement with results dealing with the evolution of subgrid transfers discussed above. Unstable stratification corresponds to an increase of subgrid viscosity and diffusivity, in agreement with the physical picture that the forward kinetic energy cascade is enhanced since the subgrid kinetic energy is increased.

The subgrid time scale in (15.107) and (15.108) is defined as an absolute value and leads to the definition of positive subgrid parameters. The reason is that the rough estimates used to obtain the expressions given above do not ensure that realizability is enforced, i.e. that the energy destroyed by stabilizing stratification effects is smaller than the available subgrid kinetic energy. But when the backscatter is dominant, the definition is no longer relevant to account for subgrid transfers if the subgrid Prandtl number is kept positive. In this case, Eidson [209] proposes to clip negative values of N_c^2 so that $\mathcal{T}$ remains well-posed. A simple solution is to consider $N_c^{2+} = \max(0, N_c^2)$. The resulting subgrid viscosity model can be rewritten as a function of the original Smagorinsky model (5.90):

$$\nu_{\rm sgs} = \underbrace{C_\nu' \overline{\Delta}^2 |\overline{S}|}_{\rm Smagorinsky} \sqrt{1 - \frac{1}{Pr_{\rm sgs}} \frac{N_c^{2+}}{|\overline{S}|^2}} \quad . \tag{15.109}$$

This modified expression can be recast in a form which emphasizes the role of the Richardson number [690]:

$$\nu_{\rm sgs} = \underbrace{C_\nu' \overline{\Delta}^2 |\overline{S}|}_{\rm Smagorinsky} \sqrt{1 - \frac{Ri}{Ri_{\rm ref}}} \quad , \tag{15.110}$$

where the local subgrid Richardson number Ri is defined as $Ri = N_c^{2+}/|\overline{S}|^2$ and $Ri_{\rm ref} \equiv Pr_{\rm sgs}$ is a reference value. This new expression also illustrates the physical meaning of the subgrid Prandtl number in stably stratified flows.

A modified evaluation of the time scale is proposed by Peng and Davidson [587], which is always well-posed from a mathematical standpoint:

$$\mathcal{T} = \left(|\overline{S}| - \frac{N_c^2}{|\overline{S}| Pr_{\rm sgs}} \right) \quad . \tag{15.111}$$

A more general scalar model is proposed by Schumann, which contains an anisotropic residual heat flux associated to buoyancy effects. It is derived considering simplified evolution equation for the subgrid stresses and subgrid heat fluxes and assuming that the local equilibrium applies, yielding

$$\tau = -2 C_\nu \overline{\Delta}^2 \mathcal{T} \overline{S}, \quad \tau_\theta = -C_\kappa \overline{\Delta}^2 \mathcal{T} (\boldsymbol{b}' \nabla \overline{\theta}) \quad , \tag{15.112}$$

where the time scale $\mathcal{T}^{-1}$ and the tensor $\boldsymbol{b}'$ are defined as

$$\boldsymbol{b}' = Id + \frac{C_2}{\mathcal{T}^2 + C_2 N_c^2} \boldsymbol{b} \nabla \overline{\theta} \quad , \tag{15.113}$$

$$\mathcal{T}^2 = \frac{1}{2} \left(|\overline{S}|^2 - (C_1 + C_2) N_c^2 \right) + \sqrt{\frac{1}{4} \left(|\overline{S}|^2 - (C_1 - C_2) N_c^2 \right)^2 + C_1 C_2 N_m^4} \quad , \tag{15.114}$$

where N_c^2 is given by (15.104) and

$$N_m^2 = \sqrt{(\boldsymbol{b} \cdot \boldsymbol{b})(\nabla \overline{\theta} \cdot \nabla \overline{\theta})} \quad . \tag{15.115}$$

Table 15.1. Values of the constants in buoyancy-affected subgrid models.

Model	C_1	C_2
Passive scalar	0	0
Buoyant	$1/Pr_{\text{sgs}}$	0
Eidson	2.5	0.
Schumann	2.5	3.

As noted by Cabot [85], this general model encompasses the previous ones. Corresponding values of the constants C_1 and C_2 are summarized in Table 15.1.

The Schumann model appears to be the most complex one, and includes more information related to anisotropy. It is therefore expected that it should yield better results in buoyancy-driven flows with high Prandtl numbers.

More complex models relying on the definition of tensorial subgrid viscosity/diffusivity to account for anisotropy can also be defined. Such models, very similar to the one proposed by Yoshizawa for the passive scalar case (Sect. 15.2.3, p. 465) were tested by Peng and Davidson [587] for buoyancy-driven flows. These authors use the following expression for the subgrid heat flux

$$\tau_{\text{G}}(u_i, \theta) = -CT_{\text{sgs}}\tau_{\text{G}}(u_i, u_k)\frac{\partial \overline{\theta}}{\partial x_k} \quad , \tag{15.116}$$

where C is a constant, T_{sgs} and appropriate time scale and the subgrid momentum flux is modeled by an ad hoc model. In the case a subgrid-viscosity type model is considered for $\tau_{\text{G}}(\boldsymbol{u}, \boldsymbol{u})$ and the cascade time scale is evaluated as $T_{\text{sgs}} = \overline{\Delta}^2/\nu_{\text{sgs}}$, the resulting model is

$$\tau_{\text{G}}(u_i, \theta) = -C\overline{\Delta}^2\overline{S}_{ik}\frac{\partial \overline{\theta}}{\partial x_k} \quad . \tag{15.117}$$

The value of the subgrid viscosity can be modified accounting for this expression in the following way. Starting from the local equilibrium hypothesis (15.101) and approximating the subgrid dissipation rate as $\varepsilon \propto \nu_{\text{sgs}}^3/\overline{\Delta}^4$, one obtains

$$\nu_{\text{sgs}} = C\overline{\Delta}^2\sqrt{|\overline{S}|^2 - \frac{1}{\nu_{\text{sgs}}}\frac{g_i}{\Theta_0}\tau_{\text{G}}(u_i, \theta)} \quad . \tag{15.118}$$

Now using (15.116), one obtains the following evaluation of the subgrid viscosity

$$\nu_{\text{sgs}} = C\overline{\Delta}^2\sqrt{|\overline{S}|^2 - \frac{\gamma}{|\overline{S}|}\frac{g_i}{\Theta_0}\overline{S}_{ik}\frac{\partial \overline{\theta}}{\partial x_k}} \quad , \tag{15.119}$$

where C and γ are adjustable parameters, which can be computed using inertial-range considerations or a dynamic procedure.

All functional models presented above are built on the same bases, and are based on the resolved strain. They can be interpreted as extensions of the Smagorinsky model. Another class of models, which is very popular among researchers working in the field of meteorology, is the one of models based on the subgrid kinetic energy:

$$\nu_{\rm sgs} = C_\nu l \sqrt{q^2_{\rm sgs}}, \quad \kappa_{\rm sgs} = C_\kappa l \sqrt{q^2_{\rm sgs}} \quad , \tag{15.120}$$

where C_ν and C_κ are constants and l a characteristic length scale. The subgrid kinetic energy must be evaluated using a prognostic equation or a simplified approximation (see Sect. 15.3.5). Since the time scale $\mathcal{T}$ does not appear explicitely in these expressions, the stratification effect must be taken into account in the definition of the length scale l. The subgrid Prandtl number associated with this corrected length scale is

$$Pr_{\rm sgs} = \frac{\overline{\Delta}}{\overline{\Delta} + 2l} \quad . \tag{15.121}$$

In the case of unstable stratification, the length scale is taken equal to its usual value tied to the filter cutoff, i.e. $l = \overline{\Delta}$. In the case of stably stratified flows, a first simple model proposed by Schumann for meteorology-related purpose is

$$l = \min\left(\overline{\Delta}, L_N\right), \quad L_N = 0.76\sqrt{\frac{q^2_{\rm sgs}}{N_c^2}} \quad . \tag{15.122}$$

A more general corrected expression, which accounts for both stable stratification and shear effects is proposed by Canuto and Cheng [99]. It is written as follows:

$$l = \overline{\Delta} f(|\overline{S}|, N_c) \quad , \tag{15.123}$$

with

$$f(|\overline{S}|, N_c) = \left(\int_0^1 [1 - \mathcal{X}\log(1 + a\mathcal{Q}^2)]^2 d\mathcal{Q}\right)^{3/2} \quad , \tag{15.124}$$

$$\mathcal{X} = \frac{\sqrt{3}}{16} K_0^{3/2} \left(\frac{Pr_{\rm sgs} S_h^2}{Fr_i^2} - 1\right), \quad a = \frac{2}{\pi^2} Fr_i^2 f(|\overline{S}|, N_c)^{2/3},$$
$$\mathcal{Q} = (k\overline{\Delta}/\pi)^{2/3} \quad , \tag{15.125}$$

where the non-dimensional shear number and the inverse Froude number are defined as

$$S_h = \frac{\overline{\Delta}|\overline{S}|}{\sqrt{q^2_{\rm sgs}}}, \quad Fr_i = \frac{\overline{\Delta} N_c}{\sqrt{q^2_{\rm sgs}}} \quad . \tag{15.126}$$

Dynamic Models. A weakness shared by all the models presented above is the existence of preset constants. They can be evaluated using inertial range considerations, or they can be automatically adjusted using a dynamic procedure. The use of a dynamic procedure based on the Germano identity yields much more difficult problems than in other cases, because of the intrinsic coupling between the momentum and the temperature subgrid fluxes. This problem is clearly seen considering the following generic closures

$$\tau_{ij} = C_\nu f_{ij}(\overline{\Delta}, \overline{\boldsymbol{u}}, \overline{\theta}, Pr_{\rm sgs}), \quad \tau_{\theta i} = \frac{C_\nu}{Pr_{\rm sgs}} h_i(\overline{\Delta}, \overline{\boldsymbol{u}}, \overline{\theta}, Pr_{\rm sgs}) \quad . \qquad (15.127)$$

The two unknown parameters are C_ν and $Pr_{\rm sgs}$. Inserting these expressions in the Germano identities for the momentum and temperature fluxes and using the classical assumptions, one defines the two following residuals:

$$\mathcal{E}_{ij} = L_{ij} - C_\nu \left(f_{ij}(\widetilde{\overline{\Delta}}, \widetilde{\overline{\boldsymbol{u}}}, \widetilde{\overline{\theta}}, Pr_{\rm sgs}) - \widetilde{f_{ij}}(\overline{\Delta}, \overline{u}, \overline{\theta}, Pr_{\rm sgs}) \right) \quad , \qquad (15.128)$$

$$\mathcal{E}_{\theta i} = h_i - \frac{C_\nu}{Pr_{\rm sgs}} \left(h_i(\widetilde{\overline{\Delta}}, \widetilde{\overline{\boldsymbol{u}}}, \widetilde{\overline{\theta}}, Pr_{\rm sgs}) - \widetilde{h_i}(\overline{\Delta}, \overline{u}, \overline{\theta}, Pr_{\rm sgs}) \right) \quad . \qquad (15.129)$$

The two unknown parameters are chosen so as to minimize the global error in the least square sense, leading to the definition of four constraints:

$$\frac{\partial \mathcal{E} : \mathcal{E}}{\partial C_\nu} = 2\mathcal{E} : \frac{\partial \mathcal{E}}{\partial C_\nu} = 0, \quad \frac{\partial \mathcal{E} : \mathcal{E}}{\partial Pr_{\rm sgs}} = 2\mathcal{E} : \frac{\partial \mathcal{E}}{\partial Pr_{\rm sgs}} = 0 \quad , \qquad (15.130)$$

$$\frac{\partial \mathcal{E}_\theta : \mathcal{E}_\theta}{\partial C_\nu} = 2\mathcal{E}_\theta : \frac{\partial \mathcal{E}_\theta}{\partial C_\nu} = 0, \quad \frac{\partial \mathcal{E}_\theta : \mathcal{E}_\theta}{\partial Pr_{\rm sgs}} = 2\mathcal{E}_\theta : \frac{\partial \mathcal{E}_\theta}{\partial Pr_{\rm sgs}} = 0 \quad . \qquad (15.131)$$

The complexity of this problem is enlightened looking at the definition of the subgrid Prandtl number: since both f_{ij} and h_i are non linear functions of $Pr_{\rm sgs}$, the use of either (15.130) or (15.131) leads to the definition of a system of non-linear equations for it, which cannot be solved explicitely as in previous cases. Thus, the dynamic evaluation of the constants now require to solve a non linear system, leading to a subsequent increase in the computational complexity. It also raises the problem of finding the best among the multiple possible roots of the system.

This is why many authors use the same dynamic models as in the passive scalar case (i.e. with basic models which are linear with respect to the subgrid Prandtl number), with the idea that, if the grid is fine enough, most of the stratification effects will be captured. This is especially true of the stable stratification effects, where the decrease in the subgrid energy transfer may be partially captured using a dynamic procedure, since resolved scales are already affected. The case of buoyancy driven flows is much more difficult, since turbulence production exist at small scales which cannot be inferred from the resolved scale dynamics. Therefore, the use of purposely modified models for buoyancy-driven flows is more important than in the case of stable stratification.

15.3.4 Extensions of Structural Models

We now consider structural models. As in the passive scalar case, models relying on the explicit evaluation of subgrid scales (soft deconvolution models, scale similarity models, linear combination models [19], explicit stochastic or deterministic models for subgrid fluctuations, multilevel simulations) can be used in a straightforward way, since they do not rely on any assumptions dealing with the nature of subgrid transfers. Consequently, these models will not be explicitly detailed in this section, since they don't need to be modified to account for the active scalar dynamics. Another comment is that many structural models have not yet been tested in stably/unstably stratified flows.

In this section, we will put the emphasis on models which contain explicit modifications and whose accuracy has already been assessed in real Large-Eddy Simulations:

- Deardorff's differential stress model (p. 487), which requires to solve ten additional transport equations.
- Schumann's algebraic stress model (p. 488), wich can be seen as a simplification of the previous model and necessitates to solve only one additional prognostic equation to compute the subgrid kinetic energy.

Deardorff Differential Model. The differential model proposed by Deardorff [173] relies on solving closed expressions of (15.88) and (15.90).

In the equations for the subgrid momentum fluxes (15.88), the only new terms with respect to the case treated in Sect. 7.5.1 is the coupling term. Since the closures proposed by Deardorff are kept unmodified in the active scalar case, all the dynamic coupling effects are contained in the coupling term. Since this term is directly proportional to the subgrid heat fluxes which are explicitely computed, no further closure assumptions is needed for these equations.

The equations for the subgrid scalar fluxes and the associated closures have already been discussed in Sect. 15.2.4. Among already closed terms in the passive scalar case, only the pressure-temperature term is modified to account for stratification effects: Equation (15.60) is changed into

$$\tau_{\mathrm{G}}\left(\theta, \frac{\partial p}{\partial x_i}\right) = -C_1 \frac{\sqrt{q^2_{\mathrm{sgs}}}}{\overline{\Delta}} \tau_{\mathrm{G}}(u_i, \theta) - \frac{1}{3} b_i \tau_{\mathrm{G}}(\theta, \theta) \quad , \tag{15.132}$$

which necessitates an evaluation of the generalized subgrid scalar variance, $\tau_{\mathrm{G}}(\theta, \theta)$. The new term appearing in (15.90) is the coupling term, which is also proportional to the generalized subgrid scalar variance.

Therefore, an additional prognostic equation for $\tau_{\mathrm{G}}(\theta, \theta)$ deduced from (15.8) is solved. Term VII is directly computed, since the subgrid scalar fluxes are explicitly computed. Molecular diffusion IX, is neglected, while the molecular dissipation IX is evaluated as

$$2\kappa\tau_{\mathrm{G}}\left(\frac{\partial\theta}{\partial x_k},\frac{\partial\theta}{\partial x_k}\right)=C_{IX}\frac{\sqrt{q_{\mathrm{sgs}}^2}}{\overline{\Delta}}\tau_{\mathrm{G}}(\theta,\theta),\quad C_{IX}=0.42\quad, \tag{15.133}$$

while the subgrid diffusion cubic term X is closed as

$$\tau_{\mathrm{G}}(u_k,\theta,\theta)=-C_X\overline{\Delta}\sqrt{q_{\mathrm{sgs}}^2}\frac{\partial}{\partial x_k}\tau_{\mathrm{G}}(\theta,\theta),\quad C_X=0.2\quad. \tag{15.134}$$

The resulting solvable equation for the subgrid temperature variance is

$$\begin{aligned}\frac{\partial\tau_{\mathrm{G}}(\theta,\theta)}{\partial t}+\overline{u}_k\frac{\partial\tau_{\mathrm{G}}(\theta,\theta)}{\partial x_k} &= -2\tau_{\mathrm{G}}(u_k,\theta)\frac{\partial\overline{\theta}}{\partial x_k}-C_{IX}\frac{\sqrt{q_{\mathrm{sgs}}^2}}{\overline{\Delta}}\tau_{\mathrm{G}}(\theta,\theta)\\ &\quad +C_X\frac{\partial}{\partial x_k}\left(\overline{\Delta}\sqrt{q_{\mathrm{sgs}}^2}\frac{\partial}{\partial x_k}\tau_{\mathrm{G}}(\theta,\theta)\right)\end{aligned} \tag{15.135}$$

achieving the description of the model.

Schumann Algebraic Stress Model. A much simpler model, which does not require to solve a large amount of additional transport equations, was developed by Schmidt and Schumann [651] to analyze the convective boundary layer dynamics. This model can be interpreted as a simplification of the Deardorff model, in which several contributions will be neglected.

Starting from (15.88) and making the following assumptions:

1. The subgrid fluxes respond instantaneously to large-scale forcing, i.e. the time derivative term is negligible.
2. Diffusive terms are small, and can be neglected.
3. Importance of the deviatoric part of the subgrid stress tensor in the production terms is small, with respect to the one of the isotropic part.
4. The pressure-velocity term can be modeled as

$$\begin{aligned}\tau_{\mathrm{G}}\left(p,\frac{\partial u_i}{\partial x_k}\right) &= C_1 q_{\mathrm{sgs}}^2\overline{S}_{ik}-C_2\frac{\sqrt{q_{\mathrm{sgs}}^2}}{\overline{\Delta}}\tau_{\mathrm{G}}^*(u_i,u_j)+C_3\frac{g}{\Theta_0}\\ &\quad\times\left(\tau_{\mathrm{G}}(u_i,\theta)\delta_{k3}+\tau_{\mathrm{G}}(u_k,\theta)\delta_{i3}-\frac{2}{3}\tau_{\mathrm{G}}(u_3,\theta)\delta_{ij}\right)\quad,\end{aligned} \tag{15.136}$$

   where C_1, C_2 and C_3 are model parameters. The *star* superscript denotes the deviatoric part of the tensor.
5. The subgrid dissipation term is approximated as

$$2\nu\tau_{\mathrm{G}}\left(\frac{\partial u_j}{\partial x_k},\frac{\partial u_i}{\partial x_k}\right)=\frac{2}{3}q_{\mathrm{sgs}}^2\overline{S}_{ij}\quad. \tag{15.137}$$

one obtains the following linear relationship for the subgrid stresses:

$$\begin{aligned} C_{Rm}\frac{\sqrt{q^2_{\rm sgs}}}{\overline{\Delta}}\tau^*_{\rm G}(u_i,u_j) &= (1-C_{Bm})\frac{g}{\Theta_0}\times\left(\tau_{\rm G}(u_i,\theta)\delta_{k3}+\tau_{\rm G}(u_k,\theta)\delta_{i3}\right.\\ &\left.-\frac{2}{3}\tau_{\rm G}(u_3,\theta)\delta_{ij}\right)-(1-C_{Gm})\frac{2}{3}q^2_{\rm sgs}\overline{S}_{ij} \quad , \end{aligned} \tag{15.138}$$

where $C_{Rm}=3.5, C_{Bm}=0.55$ and $C_{Gm}=0.55$. The subgrid heat fluxes are known since they are also outputs of the model, and the subgrid kinetic energy is obtained solving a prognostic equation derived from (15.89). Since the exact equation differs from the one associated to the unstratified case only because of the coupling term, and that this coupling term is directly proportional to the subgrid heat flux which is an output of the model, a simple solution consists in using the same closed equation as in the unstratified case (see Sect. 5.4.2, p. 173) with the additional coupling term. More complex variants of the prognostic equation for the subgrid kinetic energy are presented in a dedicated section (Sect. 15.3.5).

The subgrid heat flux equation (15.90) is simplified using similar assumptions:

1. The time derivative is negligible.
2. Convective and diffusive fluxes are negligible.
3. The dissipation term is negligible.
4. Contribution of the deviatoric part of the subgrid stress tensor in the production term is very small compared to the one of its isotropic part.
5. The pressure-temperature subgrid flux can be modeled as

$$\begin{aligned} \nu\tau_{\rm G}\left(p,\frac{\partial\theta}{\partial x_i}\right) &= 2a_1q^2_{\rm sgs}\frac{\partial\overline{\theta}}{\partial x_i}-a_2\sqrt{2}\frac{\sqrt{q^2_{\rm sgs}}}{\overline{\Delta}}\tau_{\rm G}(\theta,u_i)\\ &+2a_3\frac{g}{\Theta_0}\tau_{\rm G}(\theta,\theta)\delta_{i3} \quad , \end{aligned} \tag{15.139}$$

with a_i are adjustable parameters.

The resulting algebraic equation is

$$\begin{aligned} \left(C_{R\theta}\frac{\sqrt{q^2_{\rm sgs}}}{\overline{\Delta}}\delta_{ik}+\frac{\partial\overline{u}_i}{\partial x_k}\right)\tau_{\rm G}(\theta,u_k) &= -\frac{2}{3}(1-C_{G\theta})q^2_{\rm sgs}\frac{\partial\overline{\theta}}{\partial x_i}\\ &+(1+C_{B\theta})\frac{g}{\Theta_0}\tau_{\rm G}(\theta,\theta)\delta_{i3} \quad . \end{aligned} \tag{15.140}$$

To obtain a fully explicit model for the subgrid heat flux, it is necessary to eliminate the term related to the resolved velocity gradient in the left hand side, i.e. to neglect production effects associated with the interaction of the subgrid heat flux with the resolved velocity gradient, yielding

$$C_{R\theta}\frac{\sqrt{q_{\rm sgs}^2}}{\overline{\Delta}}\tau_{\rm G}(\theta,u_i) = -\frac{2}{3}(1-C_{G\theta})q_{\rm sgs}^2\frac{\partial\overline{\theta}}{\partial x_i} + 2(1-C_{B\theta})\frac{g}{\Theta_0}\tau_{\rm G}(\theta,\theta)\delta_{i3} \quad . \tag{15.141}$$

The constants are set equal to: $C_{R\theta} = 1.63, C_{G\theta} = 0.5, C_{B\theta} = 0.5$. The subgrid scalar variance equation is simplified using a local equilibrium hypothesis, leading to the following explicit expression

$$C_{\varepsilon\theta}\frac{\sqrt{q_{\rm sgs}^2}}{\overline{\Delta}}\tau_{\rm G}(\theta,\theta) = 2\tau_{\rm G}(\theta,u_i)\frac{\partial\overline{\theta}}{\partial x_i} \quad , \tag{15.142}$$

with $C_{\varepsilon\theta} = 2.02$. The explicit model is obtained by solving (15.138), (15.141) and (15.142) supplemented with a prognostic equation for the subgrid kinetic energy. Therefore, it appears much more simple than a differential stress model. Another interesting feature of this model is that it do not rely on the eddy diffusivity/viscosity paradigm and is fully anisotropic. Nevertheless, since it is based on strong assumptions, its domain of validity is expected to be narrower than the one of the full differential model.

In the absence of buoyancy, the model simplifies as a simple subgrid viscosity/subgrid diffusivity model:

$$\nu_{\rm sgs} = C_\nu\overline{\Delta}\sqrt{q_{\rm sgs}^2}, \quad \kappa_{\rm sgs} = C_\kappa\overline{\Delta}\sqrt{q_{\rm sgs}^2} \quad , \tag{15.143}$$

with

$$C_\nu = \frac{2}{3}\frac{1-C_{Gm}}{C_{Rm}}, \quad C_\kappa = \frac{2}{3}\frac{1-C_{G\theta}}{C_{R\theta}} \quad . \tag{15.144}$$

The corresponding value of the subgrid Prandtl number is

$$Pr_{\rm sgs} = \frac{\beta}{2K_0} \quad . \tag{15.145}$$

15.3.5 Subgrid Kinetic Energy Estimates

The modeled prognostic equation for the subgrid kinetic energy $q_{\rm sgs}^2 \equiv \frac{1}{2}\tau_{\rm G}(u_i,u_i)$ can be modified to account for buoyancy/stratification effects [402]. Rewritting the transport equation for this quantity as

$$\frac{\partial q_{\rm sgs}^2}{\partial t} + \overline{u}_j\frac{\partial q_{\rm sgs}^2}{\partial x_j} = P - D - \varepsilon + \frac{g}{\Theta_0}\tau_{\rm G}(u_3,\theta) \quad , \tag{15.146}$$

where the production P and the diffusion D are defined as

$$P = -\tau : \overline{S}, \quad D = \nabla \cdot (\overline{\boldsymbol{u}} q_{\mathrm{sgs}}^2 + pId) \quad , \tag{15.147}$$

and where ε is the dissipation rate. The model for the production term and the diffusion term are not modified to account for active scalar dynamics

$$P = C_e \overline{\Delta} \sqrt{q_{\mathrm{sgs}}^2} |\overline{S}|^2 \quad , \tag{15.148}$$

$$D = -2C_e \frac{\partial}{\partial x_k} \left(\overline{\Delta} \sqrt{q_{\mathrm{sgs}}^2} \frac{\partial q_{\mathrm{sgs}}^2}{\partial x_k} \right) \quad , \tag{15.149}$$

where C_e is a constant. The subgrid kinetic energy dissipation rate is modified to account for stabilizing stratification effects by writing it as

$$\varepsilon = C_e \frac{\left(q_{\mathrm{sgs}}^2\right)^{3/2}}{L_\varepsilon} \quad , \tag{15.150}$$

where the characteristic dissipative length scale L_ε is defined as

$$L_\varepsilon = \left(\frac{1}{\overline{\Delta}^2} + \frac{1}{L_N^2} + \frac{1}{L_S^2} \right)^{-1/2} \quad . \tag{15.151}$$

The subgrid buoyancy lenghtscale, L_N, and the subgrid shear length scale, L_S, are defined as

$$L_N = 0.76 \sqrt{\frac{q_{\mathrm{sgs}}^2}{N_c^2}}, \quad L_S = 2.76 \sqrt{\frac{q_{\mathrm{sgs}}^2}{|\overline{S}|^2}} \quad , \tag{15.152}$$

where N_c^2 is defined as in (15.104). This expression was derived to account for the dynamics of a shear-driven stable atmospheric boundary layer. The predicted value of the dissipation length scale is larger or equal to $\overline{\Delta}$, yielding a decrease of the predicted dissipation, in agreement with the observed diminution in the forward kinetic energy cascade rate.

An algebraic expression which takes into account the correction in the definition of the dissipation length scale is obtained by assuming that the dissipation is locally balanced by conversion of the mean flow energy into subgrid kinetic energy (production = dissipation):

$$\varepsilon = P - D \quad . \tag{15.153}$$

Using (15.150) and (15.103), one obtains

$$q_{\mathrm{sgs}}^2 = \left[\frac{L_\varepsilon}{C_e} \left(\nu_{\mathrm{sgs}} |\overline{S}|^2 - \kappa_{\mathrm{sgs}} N_c^2 \right) \right]^{2/3} \quad . \tag{15.154}$$

This equation is fully non-linear if models based on the subgrid kinetic energy are used for the subgrid viscosity and the subgrid diffusivity, and/or if the corrected model for L_ε is utilized. In the absence of stability correction for the dissipation length scale, and using the following subgrid kinetic energy-based models:

$$\nu_{\rm sgs} = C_\nu \overline{\Delta} \sqrt{q_{\rm sgs}^2}, \quad \kappa_{\rm sgs} = C_\kappa \overline{\Delta} \sqrt{q_{\rm sgs}^2} \quad , \tag{15.155}$$

one obtains the following explicit algebraic expression [690]:

$$q_{\rm sgs}^2 = \frac{\overline{\Delta}^2}{C_e} \left(C_\nu |\overline{S}|^2 - C_\kappa N_c^2 \right) \quad . \tag{15.156}$$

More expressions for the subgrid kinetic energy can be found using other models for the subgrid viscosity and diffusivity.

15.3.6 More Complex Physical Models

The active scalar model discussed above can be further complexified to account for more complex physics. This is currently done in studies dealing with the atmospheric boundary layer, in which detailed microphysical models (icing, moisture) and infrared radiative cooling are taken into account. The related subgrid models are made more complex to account for new physical mechanisms. They are nor presented here, since they are very specific to the field of application. The interested reader can refer to the original publications.

15.3.7 A Few Applications

Stably stratified flows:

- Stably stratified channel flow [19, 195, 238]
- Forced homogeneous stably stratified turbulence [110]
- Decaying homogeneous stably stratified flows [514, 373]
- Turbulent penetrative convection [124]
- Wake in a weakly stably stratified fluid [194]

Buoyancy-driven flows:

- Thermal plume [29, 28]
- Turbulent penetrative convection [124]
- Buoyancy-generated homogeneous turbulence [123]
- Rayleigh–Bénard convection [209, 394, 766, 85, 575]
- Rotating Rayleigh–Bénard flow [152]
- Forced and mixed convection in rotating and non-rotating square duct [577]
- Natural convection in a cavity [586, 800]

- Turbulent convection driven by free-surface cooling [814]
- Buoyant jet [808]
- Buoyant wake [539]
- Buoyant pipe flow [433]

Meteorology-related applications: [601, 488, 525, 444, 698, 651, 130, 656, 173, 402, 277, 690, 337, 615, 109, 154, 111, 527, 534, 528, 317, 657, 529, 658, 530, 450, 699, 697, 690, 638, 526, 532, 531]

A. Statistical and Spectral Analysis of Turbulence

A.1 Turbulence Properties

Flows qualified as "turbulent" can be found in most fields that make use of fluid mechanics. These flows posses a very complex dynamics whose intimate mechanisms and repercussions on some of their characteristics of interest to the engineer should be understood in order to be able to control them. The criteria for defining a turbulent flow are varied and nebulous because there is no true definition of turbulence. Among the criteria most often retained, we may mention [150]:

- the random character of the spatial and time fluctuations of the velocities, which reflect the existence of finite characteristic scales of statistical correlation (in space and time);
- the velocity field is three-dimensional and rotational;
- the various modes are strongly coupled, which is reflected in the non-linearity of the mathematical model retained (Navier–Stokes equations);
- the large mixing capacity due to the agitation induced by the various scales;
- the chaotic character of the solution, which exhibits a very strong dependency on the initial condition and boundary conditions.

A.2 Foundations of the Statistical Analysis of Turbulence

A.2.1 Motivations

The very great dynamical complexity of turbulent flows makes for a very lengthy deterministic description of them. To analyze and model them, we usually refer to a statistical representation of the fluctuations. This reduces the description to that of the various statistical moments in the solution, which sharply reduces the volume of information. Moreover, the random character of the fluctuations make this approach natural.

A.2.2 Statistical Average: Definition and Properties

We use $\langle\phi\rangle$ to denote the stochastic mean (or statistical average, or mathematical expectation, or ensemble average) of a random variable ϕ calculated from n independent realizations of the same phenomenon $\{\phi_l\}$:

$$\langle\phi\rangle = \lim_{n\to\infty} \frac{1}{n}\sum_{l=1}^{n} \phi_l \quad . \tag{A.1}$$

The turbulent fluctuation ϕ'_l associated with the realization ϕ_l is defined as its deviation from the mathematical expectation:

$$\phi'_l = \phi_l - \langle\phi\rangle \quad . \tag{A.2}$$

By construction, we have the property:

$$\langle\phi'\rangle \equiv 0 \quad . \tag{A.3}$$

On the other hand, fluctuation moments of second or higher order are not necessarily zero. The standard deviation σ can be defined as:

$$\sigma^2 = \langle{\phi'}^2\rangle \quad . \tag{A.4}$$

We define the *turbulence intensity* as $\sigma/\langle\phi\rangle$.

The correlation at two points in space and two times, $(\boldsymbol{x}, \boldsymbol{x}')$ and (t, t') of the two random variables ϕ and ψ, denoted $R_{\phi\psi}(\boldsymbol{x}, \boldsymbol{x}', t, t')$ is:

$$R_{\phi\psi}(\boldsymbol{x}, \boldsymbol{x}', t, t') = \langle\phi(\boldsymbol{x}, t)\psi(\boldsymbol{x}', t')\rangle \quad . \tag{A.5}$$

A.2.3 Ergodicity Principle

When ϕ is a random steady function in time (i.e. its probability density function is independent of time), we can apply the ergodicity principle according to which it is equivalent, statistically speaking, to consider indefinitely repeated experiments with a single drawing or a single experiment with an infinite number of drawings. We will therefore admit that a single experiment of infinite duration can be considered as representative of all possible scenarios.

The theorem of ergodicity says that the quadratic mean of the random function $\phi_T(t)$ defined by:

$$\phi_T(t) = \frac{1}{T}\int_t^{t+T} \phi(t')dt' \quad , \tag{A.6}$$

converges to a non-random limit equal to the stochastic mean $\langle\phi\rangle$ as $T \to \infty$ only on the condition that:

$$\lim_{T\to\infty} \frac{1}{T}\int_0^T R_{\phi'\phi'}(t)dt = 0 \quad , \tag{A.7}$$

where $R_{\phi'\phi'}(t)$ is the time autocorrelation (or covariance) of the fluctuations of ϕ over time interval t:

$$R_{\phi'\phi'}(t) = \langle(\phi(t') - \langle\phi\rangle)(\phi(t'+t) - \langle\phi\rangle)\rangle \quad . \tag{A.8}$$

For turbulent fluctuations, the random character reflects the fact that $R_{\phi'\phi'}(t) \to 0$ as $t \to \infty$. So if we define the mean in time $\overline{\phi}$ as the limit of ϕ_T as $T \to \infty$, i.e.:

$$\overline{\phi} = \lim_{T\to\infty} \frac{1}{T}\int_0^T \phi(t)dt \quad , \tag{A.9}$$

we get the equality:

$$\overline{\phi} = \langle\phi\rangle \quad . \tag{A.10}$$

We establish that the standard error varies as $1/\sqrt{T}$ for sufficiently large T. Another way of estimating $\langle\phi\rangle$ is to construct the "experimental" average ϕ_n defined as the arithmetic mean from experiments:

$$\phi_n(t) = \frac{1}{n}\sum_{i=1}^{n} \phi_i(t) \quad , \tag{A.11}$$

where the time t is arbitrary since the flow is assumed to be statistically steady. We show that the standard error decreases as $1/\sqrt{n}$ if the experiments ϕ_l are independent.

Let ϕ and ψ be two random variables. The operator $\langle\,\rangle$ thus defined verifies the following properties, sometimes called Reynolds rules:

$$\langle\phi + \psi\rangle = \langle\phi\rangle + \langle\psi\rangle \quad , \tag{A.12}$$

$$\langle a\phi\rangle = a\langle\phi\rangle \quad a = \text{const.} \quad , \tag{A.13}$$

$$\langle\langle\phi\rangle\psi\rangle = \langle\phi\rangle\langle\psi\rangle \quad , \tag{A.14}$$

$$\left\langle\frac{\partial\phi}{\partial s}\right\rangle = \frac{\partial\langle\phi\rangle}{\partial s} \qquad s = \boldsymbol{x}, t \quad , \tag{A.15}$$

$$\left\langle\int \phi(\boldsymbol{x},t)d^3\boldsymbol{x}dt\right\rangle = \int\langle\phi(\boldsymbol{x},t)\rangle d^3\boldsymbol{x}dt \quad . \tag{A.16}$$

Any operator that verifies these properties is called a Reynolds operator. We deduce from these relations the properties:

$$\langle\langle\phi\rangle\rangle = \langle\phi\rangle \quad , \tag{A.17}$$

$$\langle\phi'\rangle = 0 \quad . \tag{A.18}$$

A.2.4 Decomposition of a Turbulent Field

Decomposition Principle. One technique very commonly used for describing a turbulent field is statistical representation. The velocity field at time t and position $\boldsymbol{x}$ splits into:

$$\boldsymbol{u}(\boldsymbol{x},t) = \langle \boldsymbol{u}(\boldsymbol{x},t)\rangle + \boldsymbol{u}'(\boldsymbol{x},t) \quad . \tag{A.19}$$

Using this decomposition and the stochastic mean, we define an evolution equation for the quantity $\langle \boldsymbol{u}(\boldsymbol{x},t)\rangle$. To recover all the information contained in the $\boldsymbol{u}(\boldsymbol{x},t)$ field, we have to handle an infinite set of equations for the statistical moments of it. The quadratic non-linearity of the Navier–Stokes equations induces an intrinsic coupling among the various moments of the solution: the evolution equation of the moment of order n in the solution uses the moment of order $(n+1)$. To recover all the information in the exact solution, it is thus necessary to solve an infinite hierarchy of coupled equations. As this is impossible in practice, this hierarchy is truncated at an arbitrarily chosen level so as to obtain a finite number of equations. This truncation brings out an unknown term that will be modeled using closure hypotheses. If the degree of precision of the information obtained theoretically increases with the number of equations retained, the consequences of the truncation and of the hypotheses used are difficult to predict.

Equations of the Stochastic Moments. The evolution equations of the mean field are obtained by applying the averaging operator to the Navier–Stokes equations. By applying the rules of commutation with the derivation in the case of an incompressible Newtonian fluid and with no external forces, we get

$$\frac{\partial \langle u_i\rangle}{\partial t} + \frac{\partial}{\partial x_j}\langle u_i u_j\rangle = -\frac{\partial \langle p\rangle}{\partial x_i} + \nu \frac{\partial^2 \langle u\rangle}{\partial x_j \partial x_j} \quad , \tag{A.20}$$

$$\frac{\partial \langle u_i\rangle}{\partial x_i} = 0 \quad , \tag{A.21}$$

where ν is the kinematic viscosity. The non-linear term $\langle u_i u_j\rangle$ is unknown and has to be decomposed as a function of $\langle \boldsymbol{u}\rangle$ and $\boldsymbol{u}'$. By introducing relation (A.19) and considering the properties (A.12) to (A.18), we get:

$$\langle u_i u_j\rangle = \langle u_i\rangle\langle u_j\rangle + \langle u_i' u_j'\rangle \quad . \tag{A.22}$$

The last term of the right-hand side, called the Reynolds tensor, is unknown and has to be evaluated. It represents the coupling between the fluctuations and the mean field. This evaluation can be made by solving the corresponding evolution equation, either by employing a model, called closure or turbulence model.

A.2.5 Isotropic Homogeneous Turbulence

Definitions. A field is said to be statistically homogeneous along the parameter x, or imprecisely just "homogeneous", if its statistical moments are independent of the value of x where the measurements are made. This is expressed:

$$\frac{\partial}{\partial x}\langle \phi_1....\phi_n \rangle = 0 \quad . \tag{A.23}$$

A homogeneous field is said to be statistically isotropic (in the Taylor sense), or more simply "isotropic", if all statistical moments relative to a set of points $(x_1, ..., x_n)$ at times $(t_1, ..., t_n)$ remains invariant when the set of n points and the coordinate axis are rotated, and if there is statistical invariance for symmetry about an arbitrary plane.

We may note that there exists an idea of quasi-isotropy introduced by Moffat, which does not require the invariance by symmetry.

A Few Properties. A turbulent field is said to be homogeneous (resp. homogeneous isotropic) if its velocity fluctuation $\boldsymbol{u}'$ is homogeneous (resp. homogeneous isotropic). One necessary condition for achieving homogeneity is that the mean velocity gradient be constant in space:

$$\frac{\partial \langle u_i \rangle}{\partial x_j} = \text{const.} \tag{A.24}$$

Isotropy requires that the mean field $\langle \boldsymbol{u} \rangle$ be zero. When the turbulence is isotropic, only the diagonal elements of the Reynolds tensor are non-zero. Moreover, these are mutually equal:

$$\langle u_i' u_j' \rangle = \frac{2}{3} K \delta_{ij} \quad , \tag{A.25}$$

where K is the turbulent kinetic energy.

A.3 Introduction to Spectral Analysis of the Isotropic Turbulent Fields

A.3.1 Definitions

The tensor of correlations at two points $R_{\alpha\beta}(\boldsymbol{r})$ of a statistically homogeneous vector field $\boldsymbol{u}$ defined as:

$$R_{\alpha\beta}(\boldsymbol{r}) = \langle u_\alpha(\boldsymbol{x} + \boldsymbol{r}) u_\beta(\boldsymbol{x}) \rangle \tag{A.26}$$

can be related to a spectral tensor $\Phi_{\alpha\beta}(\boldsymbol{k})$ by the following two relations:

$$R_{\alpha\beta}(\boldsymbol{r}) = \int \Phi_{\alpha\beta}(\boldsymbol{k}) e^{\mathrm{i}k_j r_j} d^3\boldsymbol{k} \quad , \tag{A.27}$$

$$\Phi_{\alpha\beta}(\boldsymbol{k}) = \frac{1}{(2\pi)^3} \int R_{\alpha\beta}(\boldsymbol{r}) e^{-\mathrm{i}k_j r_j} d^3\boldsymbol{k} \quad , \tag{A.28}$$

where $i^2 = -1$. The tensor at the origin, $R_{\alpha\beta}(0)$, is the Reynolds tensor. In the case of an isotropic field, the general form of the correlation tensor becomes:

$$R_{\alpha\beta}(r) = K\left([f(r) - g(r)]\frac{r_\alpha r_\beta}{r^2} + g(r)\delta_{\alpha\beta}\right) \quad , \tag{A.29}$$

where $f(r)$ and $g(r)$ are two real scalar functions. When the velocity field is solenoidal, these two functions are related by:

$$g(r) = f(r) + \frac{r}{2}\frac{\partial f(r)}{\partial r} \quad . \tag{A.30}$$

The incompressibility constraint also allows us to establish the following relation for the tensor $\Phi_{\alpha\beta}(\boldsymbol{k})$:

$$\Phi_{\alpha\beta}(\boldsymbol{k}) = \frac{E(\boldsymbol{k})}{4\pi k^2}\left(\delta_{\alpha\beta} - \frac{k_\alpha k_\beta}{k^2}\right) \quad , \tag{A.31}$$

where the scalar function $E(\boldsymbol{k})$ is called a three-dimensional spectrum. It represents the contribution of the wave vectors of k to the turbulent kinetic energy, i.e. wave vectors whose tips are included in the region located between two spheres of radius k and $k + dk$. The spectral energy density, denoted $A(\boldsymbol{k})$, is therefore equal to $E(\boldsymbol{k})/4\pi k^2$. The three-dimensional spectrum is computed from the spectral tensor by integration over the sphere of radius k:

$$E(k) = \frac{1}{2}\int \Phi_{ii}(\boldsymbol{k})dS(\boldsymbol{k}) \quad , \tag{A.32}$$

where $dS(\boldsymbol{k})$ is the integration element on the sphere of radius k. This quantity can also be related to the function $f(r)$ by the relation:

$$E(k) = \frac{K}{\pi}\int_0^\infty kr\left(\sin(kr) - kr\cos(kr)\right)f(r)dr \quad . \tag{A.33}$$

The turbulent kinetic energy, K, is found by summation over the entire spectrum:

$$K \equiv \frac{\langle u_i' u_i' \rangle}{2} = \int_0^\infty E(\boldsymbol{k})d^3\boldsymbol{k} \quad . \tag{A.34}$$

By construction, the spectral tensor has the property:

$$\Phi_{ij}(-\boldsymbol{k}) = \Phi_{ij}^*(\boldsymbol{k}) \quad , \tag{A.35}$$

where the asterisk indicates the complex conjugate number. The homogeneity property of the turbulent field implies:

$$\Phi_{ij}(\boldsymbol{k}) = \Phi_{ji}^*(\boldsymbol{k}) \quad . \tag{A.36}$$

The spectral tensor can also be related to the velocity fluctuation $\boldsymbol{u}'$ and to its Fourier transform $\widehat{\boldsymbol{u}}'$ defined as:

$$\widehat{\boldsymbol{u}}_i'(\boldsymbol{k}) = \frac{1}{(2\pi)^3} \int \boldsymbol{u}'(\boldsymbol{x}) e^{-\imath k_j x_j} d^3\boldsymbol{x} \quad . \tag{A.37}$$

Simple expansions lead to the equality:

$$\langle \widehat{u}_i'(\boldsymbol{k}') \widehat{u}_j'(\boldsymbol{k}) \rangle = \delta(\boldsymbol{k} + \boldsymbol{k}') \Phi_{ij}(\boldsymbol{k}) \quad . \tag{A.38}$$

So we see that the two modes are correlated statistically only if $\boldsymbol{k}+\boldsymbol{k}' = 0$. An equivalent definition of the spectral tensor is:

$$\Phi_{ij}(\boldsymbol{k}) = \int \langle \widehat{u}_i'^*(\boldsymbol{k}) \widehat{u}_j'(\boldsymbol{k}') \rangle d^3\boldsymbol{k}' \quad . \tag{A.39}$$

A.3.2 Modal Interactions

The nature of the interactions among the various modes can be brought out by analyzing the non-linear term that appears in the evolution equation associated with them. This equation, for the mode associated with the wave vector $\boldsymbol{k}$ (the dependency on $\boldsymbol{k}$ is not expressed, for the sake of simplicity) is:

$$\frac{\partial \widehat{u}_i}{\partial t} + \imath k_j \widehat{a}_{ij} = -\imath k_i \widehat{p} - \nu k^2 \widehat{u}_i \quad . \tag{A.40}$$

The two quantities $\widehat{a}_{ij}$ and $\widehat{p}$ are related to $u_i u_j$ and the pressure p by the relations:

$$u_i(\boldsymbol{x}) u_j(\boldsymbol{x}) = \int \widehat{a}_{ij}(\boldsymbol{k}) e^{\imath k_l x_l} d^3\boldsymbol{k} \quad , \tag{A.41}$$

$$\frac{1}{\rho} p(\boldsymbol{x}) = \int \widehat{p}(\boldsymbol{k}) e^{\imath k_l x_l} d^3\boldsymbol{k} \quad . \tag{A.42}$$

By introducing the spectral decompositions:

$$u_i(\boldsymbol{x}) = \int \widehat{u}_i(\boldsymbol{k}') e^{\imath k_l' x_l} d^3\boldsymbol{k}' \quad , \tag{A.43}$$

$$u_j(\boldsymbol{x}) = \int \widehat{u}_j(\boldsymbol{k}'') e^{\imath k_l'' x_l} d^3\boldsymbol{k}'' \quad , \tag{A.44}$$

the non-linear term becomes:

$$u_i(\boldsymbol{x}) u_j(\boldsymbol{x}) = \int \underbrace{\int \widehat{u}_i(\boldsymbol{k}') \widehat{u}_j(\boldsymbol{k} - \boldsymbol{k}') d^3\boldsymbol{k}'}_{\widehat{a}_{ij}(\boldsymbol{k})} e^{\imath k_l x_l} d^3\boldsymbol{k} \quad , \tag{A.45}$$

where we have performed the variable change $\boldsymbol{k} = \boldsymbol{k}' + \boldsymbol{k}''$. The pressure term is computed by the Poisson equation:

$$\frac{1}{\rho} \frac{\partial^2 p}{\partial x_i \partial x_i} = -\frac{\partial^2 u_i u_j}{\partial x_i \partial x_j} \quad , \tag{A.46}$$

or, in the spectral space:

$$k^2\widehat{p} = -k_l k_m \widehat{a}_{lm} \quad . \tag{A.47}$$

The momentum equation therefore takes the form:

$$\left[\frac{\partial}{\partial t} + \nu k^2\right] \widehat{u}_i(\boldsymbol{k}) = M_{ijm}(\boldsymbol{k}) \int \widehat{u}_m(\boldsymbol{k}')\widehat{u}_j(\boldsymbol{k}-\boldsymbol{k}')d^3\boldsymbol{k}' \quad , \tag{A.48}$$

in which

$$M_{ijm}(\boldsymbol{k}) = -\frac{\imath}{2}\left(k_m P_{ij}(\boldsymbol{k}) + k_j P_{im}(\boldsymbol{k})\right) \quad , \tag{A.49}$$

where $P_{ij}(\boldsymbol{k})$ is the projection operator on the plane orthogonal to the vector $\boldsymbol{k}$. This operator is expressed:

$$P_{ij}(\boldsymbol{k}) = \left(\delta_{ij} - \frac{k_i k_j}{k^2}\right) \quad . \tag{A.50}$$

The linear terms are grouped into the left-hand side and the non-linear terms in the right. The first linear term represents the time dependency and the second the viscous effects. The non-linear term represents the effect of convection and pressure. We can see that the mode $\boldsymbol{k}$ interacts with the modes $\boldsymbol{p} = \boldsymbol{k}'$ and $\boldsymbol{q} = (\boldsymbol{k} - \boldsymbol{k}')$ such that $\boldsymbol{k} + \boldsymbol{p} = \boldsymbol{q}$. This triadic nature of the non-linear interactions is intrinsically related to the mathematical structure of the Navier–Stokes equations.

A.3.3 Spectral Equations

The equations for the spectral tensor components Φ_{ij} are obtained by applying an inverse Fourier transform to the transport equations of the two-point double correlations. After computation, we get:

$$\begin{aligned}\frac{\partial \Phi_{ij}}{\partial t} - \lambda_{lm} k_l \frac{\partial \Phi_{ij}}{\partial k_m} \quad + \quad & \lambda_{il}\Phi_{lj} + \lambda_{jl}\Phi_{il} + 2\nu k^2 \Phi_{ij} = \\ & \left(k_l \Theta_{ilj} + k_l \Theta^*_{jli}\right) + \left(k_i \Sigma_j + k_j \Sigma^*_j\right) \quad ,\end{aligned} \tag{A.51}$$

where:

$$\Theta_{ilj} = \frac{\imath}{(2\pi)^3} \int \langle u'_i(\boldsymbol{x}) u'_l(\boldsymbol{x}) u'_j(\boldsymbol{x}+\boldsymbol{r})\rangle e^{-\imath k_n r_n} d^3\boldsymbol{r} \quad , \tag{A.52}$$

$$\Sigma_j = \frac{\imath}{(2\pi)^3} \int \frac{1}{\rho} \langle p'(\boldsymbol{x}) u'_j(\boldsymbol{x}+\boldsymbol{r})\rangle e^{-\imath k_n r_n} d^3\boldsymbol{r} \tag{A.53}$$

$$= 2\lambda_{lm} \frac{k_l}{k^2} \Phi_{mj} - \frac{k_l k_m}{k^2} \Theta_{mlj} \quad , \tag{A.54}$$

$$\lambda_{ij} = \frac{\partial \langle u \rangle_i}{\partial x_j} \quad . \tag{A.55}$$

By expanding the terms (A.52) and (A.54), equation (A.51) takes the form:

$$\begin{aligned}\left(\frac{\partial}{\partial t}+2\nu k^2\right)\Phi_{ij}(\boldsymbol{k}) &+ \frac{\partial\langle u_i\rangle}{\partial x_l}\Phi_{jl}(\boldsymbol{k})+\frac{\partial\langle u_j\rangle}{\partial x_l}\Phi_{il}(\boldsymbol{k}) \\ &- 2\frac{\partial\langle u_l\rangle}{\partial x_m}\left(k_i\Phi_{jm}(\boldsymbol{k})+k_j\Phi_{mi}(\boldsymbol{k})\right) \\ &- \frac{\partial\langle u_l\rangle}{\partial x_m}\frac{\partial}{\partial k_m}\left(k_l\Phi_{ij}(\boldsymbol{k})\right) \\ &= P_{il}(\boldsymbol{k})T_{lj}(\boldsymbol{k})+P_{jl}(\boldsymbol{k})T_{li}^*(\boldsymbol{k}) \quad , \qquad \text{(A.56)}\end{aligned}$$

where

$$T_{ij}(\boldsymbol{k}) = k_l \int\int\int \langle u_i(\boldsymbol{k})u_l(\mathbf{p})u_j(-\boldsymbol{k}-\mathbf{p})\rangle d^3\mathbf{p} \quad . \qquad \text{(A.57)}$$

The evolution equation for the energy spectrum $E(k)$, derived from (A.51) by integration over the sphere of radius k, is:

$$\frac{\partial E(k)}{\partial t} = P(k)+T(k)+D(k) \quad , \qquad \text{(A.58)}$$

where the kinetic energy production term $P(k)$ by interaction with the mean field, the transfer term $T(k)$ and the dissipation term $D(k)$ are given by:

$$P(k) = -\lambda_{ij}\phi_{ij}(k) \quad , \qquad \text{(A.59)}$$

$$T(k) = \frac{1}{2}\int\left(k_l(\Theta_{ili}+\Theta_{ili}^*)+\lambda_{lm}\frac{\partial(k_l\phi_{ii})}{\partial k_m}\right)dS(\boldsymbol{k}) \quad , \qquad \text{(A.60)}$$

$$D(k) = -2\nu k^2 E(k) \quad , \qquad \text{(A.61)}$$

where the tensor $\phi_{ij}(k)$ is defined as the integral of $\Phi_{ij}(\boldsymbol{k})$ over the sphere of radius k:

$$\phi_{ij}(k) = \int \Phi_{ij}(\boldsymbol{k})dS(\boldsymbol{k}) \quad . \qquad \text{(A.62)}$$

The kinetic energy conservation property for ideal fluid is expressed by:

$$\int_0^\infty T(k)dk = 0 \quad . \qquad \text{(A.63)}$$

We come up with the kinetic energy evolution equation in the physical space by integrating (A.58) over the entire spectrum:

$$\frac{\partial K}{\partial t} = \int_0^\infty \frac{\partial E(k)}{\partial t}dk = \int_0^\infty P(k)dk + \int_0^\infty T(k)dk + \int_0^\infty D(k)dk \quad . \qquad \text{(A.64)}$$

In the isotropic homogeneous case, production is zero and we get:

$$\frac{\partial K}{\partial t} = -\varepsilon \quad , \qquad \text{(A.65)}$$

where the kinetic energy dissipation rate ε is given by:

$$\varepsilon = \int_0^\infty 2\nu k^2 E(k)dk \quad . \qquad \text{(A.66)}$$

A.4 Characteristic Scales of Turbulence

Several characteristic scales of turbulence can be defined. We define the integral scale L^l_{ij} as:

$$L^l_{ij} = \int_{-\infty}^{+\infty} R_{ij}(r) dr \quad . \tag{A.67}$$

This scale is representative of the length over which the turbulent fluctuations are mutually correlated, so it is directly related to the size of the structures that form the turbulent field. Another scale, called the Taylor microscale, is denoted λ_T and is defined as:

$$\frac{\langle u'^2 \rangle}{\lambda_T^2} = \left\langle \left(\frac{\partial u'}{\partial x} \right)^2 \right\rangle \quad . \tag{A.68}$$

While the first scale is associated with all the turbulent structures, this latter scale is related directly to the small scales of the turbulence. Considering that the dissipation , can be written as:

$$\varepsilon = 2\nu \left\langle \left(\frac{\partial u'}{\partial x} \right)^2 \right\rangle \quad , \tag{A.69}$$

we get the relation:

$$\varepsilon = 2\nu \frac{\langle u'^2 \rangle}{\lambda_T^2} \quad . \tag{A.70}$$

The Taylor microscale thus appears as characteristic of the dissipative phenomena.

A.5 Spectral Dynamics of Isotropic Homogeneous Turbulence

A.5.1 Energy Cascade and Local Isotropy

Analyses on the basis of (A.58) show that the dynamical mechanism associated with the term $T(k)$ is a kinetic energy transfer from the small wave numbers to the large. This process is called the energy cascade. It is relatively local in frequency: the transfers are negligible among wave numbers separated by more than two decades. It repeats itself until such time as the structures are so small that the viscous mechanisms, represented by the $D(k)$ term, become preponderant.

The local isotropy hypothesis formalized by Kolmogorov assumes statistical homogeneity and isotropy in a small space–time domain and not throughout the flow. This is equivalent to the hypothesis that the flow scales, while sufficiently small, are governed by a dynamics similar to that of isotropic homogeneous turbulence. They are thus independent of the large scales and their statistical structure acquires a universal character.

Kolmogorov's first hypothesis is that the statistical moments of the scales located in such a domain depend only on the separation distance r, the total dissipation by viscosity per unit mass ε, and the viscosity ν.

The second hypothesis is that the statistical moments for large separation distances become independent of the viscosity and are no longer a function of r and ε.

A.5.2 Equilibrium Spectrum

With the hypothesis of local isotropy, we find three distinct regions in the energy spectrum $E(k)$:

- the large scale region where the turbulence associated with the $P(k)$ term is produced. These scales are coupled with the mean field and are affected by the boundary conditions, so they possess no universal character. However, following arguments related to the finite character of the energy spectral density $A(k)$, we can say that:

$$E(k) \simeq k^4 \text{ or } E(k) \simeq k^2 \text{ for } k \ll 1 \quad \text{(A.71)}$$

- the inertial range, associated with the intermediate scales, in which the energy is transferred by non-linear interaction with no action by viscosity or production. The energy spectrum depends only on k and ε. Since the energy is transferred without loss, ε remains constant. Assuming that there exists a self-similar form of the power-law spectrum, by dimensional arguments we get:

$$E(k) = K_0 \varepsilon^{2/3} k^{-5/3} \quad , \quad \text{(A.72)}$$

  where the constant K_0, called the Kolmogorov constant, is close to 1.5.
- The dissipation region, which comprises the smallest scales where the kinetic energy is dissipated by the viscous effects. In this area, the relaxation time τ_d associated with the viscous effects is at least equal to that of the non-linear transfers, denoted τ_c. For a length scale l, these two times are evaluated as:

$$\tau_d \approx l^2/\nu, \qquad \tau_c \approx \nu^2/\varepsilon \quad . \quad \text{(A.73)}$$

  The dissipation region is characterized by the relation:

$$\tau_d \leq \tau_c \Longrightarrow l \leq \sqrt{\nu^3/\varepsilon} \quad . \quad \text{(A.74)}$$

  We call the Kolmogorov scale, which is denoted η_K, the scale for which these two times are equal and which marks the beginning of the dissipation region:

$$\eta_K = \sqrt{\nu^3/\varepsilon} \quad . \quad \text{(A.75)}$$

The characteristic velocity associated with this scale is:

$$v_{\mathrm{K}} = (\nu\varepsilon)^{1/4} \quad . \tag{A.76}$$

The energy spectrum depends explicitly only on k, ν and ε, or equivalently on k, the Kolmogorov scales η_{K}, and v_{K}. The dimensional arguments do not lead to a unique form $E(k)$, and several solutions have been proposed. Arguments concerning the regularity of the velocity field and these gradients suggest an exponential decay of $E(k)$ in this region.

B. EDQNM Modeling

The EDQNM model is briefly described here in its isotropic and anisotropic versions. For more details concerning the isotropic version, the reader may refer to the work of Lesieur [439].

B.1 Isotropic EDQNM Model

Starting with the Navier–Stokes equations written in symbolic form:

$$\left(\frac{\partial}{\partial t} + \nu k^2 \right) u = uu \quad , \tag{B.1}$$

we derive an infinite hierarchical set of evolution equations as usual for the statistical moments of the velocity u:

$$\left(\frac{\partial}{\partial t} + \nu k^2 \right) \langle uu \rangle = \langle uuu \rangle \quad , \tag{B.2}$$

$$\left(\frac{\partial}{\partial t} + \nu (k^2 + p^2 + q^2) \right) \langle uuu \rangle = \langle uuuu \rangle \quad , \tag{B.3}$$

$$\ldots = \ldots \quad ,$$

where the symbol $\langle \rangle$ designates a statistical average. We then adopt the following hypothesis.

Hypothesis B.1 (Quasi-normality Hypothesis) *The velocity distribution law is close to the Gaussian bell curve and its fourth-order cumulant, denoted* $\langle uuuu \rangle_c$ *is zero.*

The evolution equation of the triple correlations then become:

$$\left(\frac{\partial}{\partial t} + \nu (k^2 + p^2 + q^2) \right) \langle uuu \rangle = \sum \langle uu \rangle \langle uu \rangle \quad . \tag{B.4}$$

The quasi-normal approximation does not provide the realizability condition, i.e. the spectrum $E(k)$ can take negative values. To recover this property, Orszag proposes introducing a triple-correlation damping term and then gets:

$$\left(\frac{\partial}{\partial t} + \nu (k^2 + p^2 + q^2) \right) \langle uuu \rangle = \sum \langle uu \rangle \langle uu \rangle - (\eta_k + \eta_p + \eta_q) \langle uuu \rangle \quad . \tag{B.5}$$

The solution to this equation is:

$$\langle uuu\rangle(t) = \int_0^t \sum \langle uu\rangle\langle uu\rangle e^{-(\mu_k+\mu_p+\mu_q)t} dt \quad , \tag{B.6}$$

with

$$\mu_k = \eta_k + \nu k^2 \quad . \tag{B.7}$$

To get a solution that is easier to calculate, we adopt the following hypothesis:

Hypothesis B.2 *The relaxation time of the triple correlations is small compared with the relaxation time of the double correlations (which is also that of the energy spectrum).*

With this hypothesis, we can "Markovianize" equation (B.6), which leads to:

$$\begin{aligned} \langle uuu\rangle(t) &= \sum \langle uu\rangle\langle uu\rangle \int_0^t e^{-(\mu_k+\mu_p+\mu_q)t} dt \\ &= \sum \langle uu\rangle\langle uu\rangle \frac{1 - e^{-(\mu_k+\mu_p+\mu_q)t}}{\mu_k+\mu_p+\mu_q} \quad . \end{aligned} \tag{B.8}$$

This relation closes the derivation equation of the second-order moments. This closure is equivalent to replacing the solution of the Navier–Stokes equations with that of the following Langevin-type stochastic model:

$$\left(\frac{\partial}{\partial t} + (\nu + \eta(k,t))k^2\right) u = f(k,t) \quad , \tag{B.9}$$

in which

$$\eta(k,t) = \frac{1}{2} \int\int \Theta_{kpq}(t) \frac{p}{kq} b_{kpq} E(q,t) dp dq \quad . \tag{B.10}$$

The forcing term f is such that:

$$\begin{aligned} F(k,t) &\equiv 4\pi k^2 \int_0^t \langle f(k,t) f(k,s)\rangle_{|k|=cste} ds \\ &= \int\int \Theta_{kpq}(t) \frac{k^3}{pq} a_{kpq} E(p,t) E(q,t) dp dq \quad , \end{aligned} \tag{B.11}$$

where a_{kpq} and b_{kpq} are coefficients linked to the geometry of the triad $(\boldsymbol{k}, \boldsymbol{p}, \boldsymbol{q})$, defined as:

$$a_{kpq} = \frac{1}{2}(1 - xyz - 2y^2z^2), \; b_{kpq} = \frac{p}{k}(xy + z^3) \quad , \tag{B.12}$$

where x, y and z are the cosines of the angles of the triangle formed by the wave vectors $(\boldsymbol{k}, \boldsymbol{p}, \boldsymbol{q})$, opposed respectively to $\boldsymbol{k}$, $\boldsymbol{p}$ and $\boldsymbol{q}$. The relaxation time $\Theta_{kpq}(t)$ is evaluated as:

$$\Theta_{kpq}(t) = \frac{1 - e^{-(\mu_k + \mu_p + \mu_q)t}}{\mu_k + \mu_p + \mu_q} \quad , \tag{B.13}$$

where the damping factor μ_k is chosen as follows:

$$\mu_k = \nu k^2 + 0.36\sqrt{\int_0^k p^2 E(p) dp} \quad . \tag{B.14}$$

B.2 Cambon's Anisotropic EDQNM Model

To study anisotropic homogeneous flows, we define the following two spectral tensors [96] (see Appendix A):

$$\Phi_{ij}(\boldsymbol{k}) \simeq \langle \widehat{u}_i'^*(\boldsymbol{k}) \widehat{u}_j'(\boldsymbol{k}) \rangle \quad , \tag{B.15}$$

$$\phi_{ij}(k) = \int \Phi_{ij}(\boldsymbol{k}) dA(\boldsymbol{k}) \quad . \tag{B.16}$$

The evolution equation for the quantities $\phi_{ij}(k)$ in the presence of mean velocity gradients is:

$$\begin{aligned} \left(\frac{\partial}{\partial t} + 2\nu k^2\right) \phi_{ij}(k) &= -\frac{\partial \langle u \rangle_i}{\partial x_k} \phi_{jl}(k) - \frac{\partial \langle u \rangle_j}{\partial x_l} \phi_{il}(k) \\ &+ P_{ij}^{\mathrm{l}}(k) + S_{ij}^{\mathrm{l}}(k) \\ &+ P_{ij}^{\mathrm{nl}}(k) + S_{ij}^{\mathrm{nl}}(k) \quad , \end{aligned} \tag{B.17}$$

where

$$P_{ij}^{\mathrm{l}}(k) = 2\frac{\partial \langle u \rangle_l}{\partial x_m} \int \frac{k_l}{k^2} \left[k_i \Phi_{mj}(\boldsymbol{k}) + k_j \Phi_{mi}(\boldsymbol{k})\right] dA(\boldsymbol{k}) \quad , \tag{B.18}$$

$$S_{ij}^{\mathrm{l}}(k) = \frac{\partial \langle u \rangle_l}{\partial x_m} \int \frac{\partial}{\partial k_m} \left(k_l \Phi_{ij}(\boldsymbol{k})\right) dA(\boldsymbol{k}) \quad , \tag{B.19}$$

$$P_{ij}^{\mathrm{nl}}(k) = -\int \frac{k_l}{k^2} \left[k_i T_{lj}(\boldsymbol{k}) + k_j T_{li}^*(\boldsymbol{k})\right] dA(\boldsymbol{k}) \quad , \tag{B.20}$$

$$S_{ij}^{\mathrm{nl}}(k) = \int \left[T_{ij}(\boldsymbol{k}) + T_{ij}^*(\boldsymbol{k})\right] dA(\boldsymbol{k}) \quad . \tag{B.21}$$

Equation (B.17) is closed by replacing $\Phi(\boldsymbol{k})$ with a modeled form as a function of $\phi(k)$, where the direction of k is controlled in the integrals (B.18) to (B.21):

$$\begin{aligned}P^{\mathrm{l}}_{ij}(k) &= 2E(k)\Bigg[\frac{2}{5}\langle S\rangle_{ij} \\ &- 3D\left(\langle S\rangle_{lj}H_{li}(k)+\langle S\rangle_{li}H_{lj}(k)-\frac{2}{3}\delta_{ij}\langle S\rangle_{lm}H_{lm}(k)\right) \\ &+ \frac{14}{3}\left(D+\frac{4}{7}\right)\left(\langle\Omega\rangle_{il}H_{lj}(k)+\langle\Omega\rangle_{jl}H_{li}(k)\right)\Bigg] \quad , \end{aligned} \tag{B.22}$$

$$\begin{aligned}S^{\mathrm{l}}_{ij}(k) &= -\frac{2}{15}\langle S\rangle_{ij}\frac{\partial}{\partial k}\left(kE(k)\right)+2\langle S\rangle_{il}\frac{\partial}{\partial k}\left(kDE(k)H_{jl}(k)\right) \\ &+ 2\langle S\rangle_{jl}\frac{\partial}{\partial k}\left(kDE(k)H_{il}(k)\right) \\ &- \frac{1}{3}\delta_{ij}\langle S\rangle_{lm}\frac{\partial}{\partial k}\left(\left[2+11D\right]kE(k)H_{lm}(k)\right) \quad , \end{aligned} \tag{B.23}$$

$$\begin{aligned}P^{\mathrm{nl}}_{ij}(k) &= \int\!\!\int \Theta_{kpq}\frac{2}{pq}(x+yz)H_{ij}(q)\Big[k^1pE(p)E(q)\Big(y(z^2-y^2)(a(q)+3) \\ &+ (y+xz)\frac{a(q)}{5}\Big)-p^3E(k)E(q)y(z^2-x^2)(a(q)+3)\Big]dpdq\,, \end{aligned} \tag{B.24}$$

$$\begin{aligned}S^{\mathrm{nl}}_{ij}(k) &= \int\!\!\int \Theta_{kpq}\frac{2}{pq}\Bigg[(xy+z^3)\Bigg(k^2pE(p)E(q)\left\{\frac{1}{3}\delta_{ij}+H_{ij}(p)\right. \\ &\left.+H_{ij}(q)\right\}-p^3E(k)E(q)\left\{\frac{1}{3}\delta_{ij}+H_{ij}(k)+H_{ij}(q)\right\}\Bigg) \\ &+ H_{ij}(q)\,\left(k^2pE(p)E(q)c_{kpq}-p^3E(k)E(q)c_{pkq}\right)\Bigg]\,dpdq\,, \end{aligned} \tag{B.25}$$

where

$$\langle S\rangle=\frac{1}{2}\left(\nabla\langle\boldsymbol{u}\rangle+\nabla^T\langle\boldsymbol{u}\rangle\right),\quad \langle\Omega\rangle=\frac{1}{2}\left(\nabla\langle\boldsymbol{u}\rangle-\nabla^T\langle\boldsymbol{u}\rangle\right) \quad ,$$

and x, y and z are the cosines of the interior angles opposite the wave vectors $\boldsymbol{k}$, $\boldsymbol{p}$ and $\boldsymbol{q}$, respectively, in the triangle formed by these vectors. The anisotropy parameter $a(k)$ is optimized by the Rapid Distortion Theory. The factor D is defined as:

$$D=\frac{2}{7}\left(1+\frac{4}{5}a\right) \quad . \tag{B.26}$$

The energy and anisotropy spectra, denoted respectively $E(k)$ and $H_{ij}(k)$, are given by the relations:

$$E(k) = \frac{1}{2}\phi_{ll}(k) \quad , \tag{B.27}$$

$$H_{ij}(k) = \frac{\phi_{ij}(k)}{2E(k)} - \frac{1}{3}\delta_{ij} \quad . \tag{B.28}$$

The geometric factor c_{kpq} is defined as:

$$c_{kpq} = \frac{1}{2}(xy + z)\left[(y^2 - z^2)(a(q) + 3) + \frac{2}{5}a(q)(1 + z^2)\right] \quad . \tag{B.29}$$

The relaxation time $\Theta_{kpq}(t)$ is evaluated as:

$$\Theta_{kpq}(t) = \frac{1 - e^{-(\mu_k + \mu_p + \mu_q)t}}{\mu_k + \mu_p + \mu_q} \quad , \tag{B.30}$$

where the damping term μ_k is:

$$\mu_k = \nu k^2 + 0,36\left(\int_0^k p^2 E(p)dp + \langle \Omega \rangle_{ij}\langle \Omega \rangle_{ij}\right)^{1/2} \quad . \tag{B.31}$$

It should be noted that fully anisotropic versions, which call for no angular parameter setting, have been proposed and compared with simulations for the case of pure rotation and stable stratification [97].

B.3 EDQNM Model for Isotropic Passive Scalar

The transport equation for the passive scalar (15.1) can be rewritten in the Fourier space under the following symbolic form:

$$\left(\frac{\partial}{\partial t} + \kappa k^2\right)\theta = u\theta \quad , \tag{B.32}$$

leading to a hierarchy of symbolic equtions for the statistical moments:

$$\left(\frac{\partial}{\partial t} + \kappa(k^2 + p^2)\right)\langle u\theta \rangle = \langle \theta\theta u \rangle \quad , \tag{B.33}$$

$$\left(\frac{\partial}{\partial t} + [\kappa(k^2 + p^2) + \nu q^2]\right)\langle \theta\theta u \rangle = \langle \theta\theta uu \rangle \quad , \tag{B.34}$$

$$\ldots = \ldots \tag{B.35}$$

The EDQNM procedure can be applied to this set of equations to obtain a simple, closed equation for the temperature spectrum $E_\theta(k)$. Details will not be given here, since the process that yields the closed EDQNM model involves the same steps as in the isotropic model for the velocity moments discussed above. The reader can refer to Lesieur's book [439] for a detailed presentation. The model is expressed as

$$\left(\frac{\partial}{\partial t} + 2\kappa k^2\right) E_\theta(k) = \int\int \Theta^\theta_{kpq} \frac{k}{pq}(1-y^2)E(q)[k^2 E_\theta(p) - p^2 E_\theta(k)] \times \delta(\boldsymbol{k}-\boldsymbol{p}-\boldsymbol{q})dpdq \quad , \tag{B.36}$$

where the triad geometrical parameter is the same as in above sections, and the characteristic time is defined as

$$\Theta^\theta_{kpq} = \frac{1 - e^{-[\kappa(k^2+p^2)+\nu q^2+\mu_1(k)+\mu_1(p)+\mu_2(q)]t}}{\kappa(k^2+p^2)+\nu q^2+\mu_1(k)+\mu_1(p)+\mu_2(q)} \quad , \tag{B.37}$$

where

$$\mu_i(k) = a_i\sqrt{\int_0^k p^2 E(p)dp}, \quad i = 1,2 \quad . \tag{B.38}$$

The two constants a_1 and a_2 are not uniquely defined, and several choices are possible. Setting $a_1 = 0$ and $a_2 = 1.3$ enables the solve analytically the system and yields a satisfactory prediction from a physical point of view. Another possibility is to take $a_1 = a_2 = 0.218K_0^{3/2}$.

Anisotropic passive scalar models [303] and models for stably stratified flows [266, 689, 267] are very complex and will not be discussed here. This is also the case of EDQNM models for mixed scalars [723].

References

1. Abba, A., Bucci, R., Cercignani, C., Valdettaro, L. (1996): A new approach to the dynamic subgrid scale model. Unpublished
2. Adams, N. (2001): The role of deconvolution and numerical discretization in subgrid-scale modeling. (Direct and large-eddy simulation IV, Geurts, Friedrich and Métais eds.), Kluwer, 311–320
3. Adams, N.A., Leonard, A. (1999): Deconvolution of subgrid-scales for the simulation of shock-turbulence interaction. (Direct and Large Eddy Simulation III, Voke, Sandham and Kleiser eds.) Kluwer, 201–212
4. Adams, N.A., Stolz, S. (2001): Deconvolution methods for subgrid-scale approximation in LES. (Modern simulation strategies for turbulent flow, Geurts ed.), Edwards, 21–44
5. Addad, Y., Laurence, D., Talotte, C., Jacob, M.C. (2003): Large eddy simulation of a forward facing step for acoustic source identification. Int. J. Heat Fluid Flow **24**, 562–571
6. Adrian, J. (1990): Stochastic estimation of subgrid-scale motions. Appl. Mech. Rev. **43**(5), 214–218
7. Agee, E., Gluhovsky, A. (1999): LES model sensitivities to domains, grids, and large-eddy timescales. J. Atmos. Sci. **56**, 599–604
8. Akhavan, R., Ansari, A., Kang, S., Mangiavacchi, N. (2000): Subgrid-scale interactions in a numerically simulated planar turbulent jet an implications for modelling. J. Fluid Mech. **408**, 83–120
9. Akselvoll, K., Moin, P. (1996): Large-eddy simulation of turbulent confined coannular jets. J. Fluid Mech. **315**, 387–411
10. Allen, R., Mendonça, F. (2004): DES validations of cavity acoustics over the subsonic to supersonic range. AIAA Paper 2004-2862
11. Alvelius, K., Johansson, A. (2000): LES computations and comparison with Kolmogorov theory for two-point pressure-velocity correlations and structure functions for globally anisotropic turbulence. J. Fluid Mech. **403**, 23–36
12. Anderson, R., Meneveau, C. (1999): Effects of the similarity model in finite-difference LES of isotropic turbulence using a Lagrangian dynamic mixed model. Flow, Turbulence and Combustion **62**, 201–225
13. Anitescu, M., Layton, W. (2002): Uncertainties in large-eddy simulation and improved estimates of turbulent flow functionals. Unpublished
14. Antonopoulos-Domis, M. (1981): Aspects of large-eddy simulation of homogeneous isotropic turbulence. Int. J. Numer. Meth. Fluids **1**, 273–290
15. Aprovitola, A., Denaro, F. (2004): On the application of congruent upwind discretizations for large eddy simulations. J. Comput. Phys. **194**, 329–343
16. Arad, E. (2001): Analysis of boundary layer separation over a bump using large-eddy simulation. AIAA Paper 2001-2558

17. Arandiga, F., Donat, R., Harten, A. (1999): Multiresolution based on weighted averages of the hat function II: nonlinear reconstruction techniques. SIAM J. Sci. Comput. **20**(3), 1053–1093
18. Armenio, V., Piomelli, U. (2000): A Lagrangian mixed subgrid-scale model in generalized coordinates. Flow, Turbulence and Combustion **65**, 51–81
19. Armenio, V., Sarkar, S. (2002): An investigation of stably stratified turbulent channel flow using large-eddy simulation. J. Fluid Mech. **459**, 1–42
20. Arunajatesan, S., Sinha, N. (2000): Towards hybrid LES-RANS computations of cavity flowfileds. AIAA Paper 2000-0401
21. Arunajatesan, S., Sinha, N. (2001): Unified unsteady RANS-LES simulations of cavity flowfields. AIAA Paper 2001-0516
22. Arunajatesan, S., Sinha, N., Ukeiley, L. (2001): On the application of hybrid RANS-LES and Proper Orthogonal Decomposition techniques to control of cavity flows. (DNS/LES Progress and Challenges, Liu, Sakell and Beutner eds.), Greyden Press, 673–688
23. Aupoix, B. (1984): Eddy viscosity subgrid models for homogeneous turbulence. (Proceedings of "Macroscopic modelling of turbulent flows", Lecture notes in physics, Vol. 230), Springer-Verlag, 45–64
24. Aupoix, B. (1985): Subgrid scale models for homogeneous anisotropic turbulence. (Proceedings of the Euromech Colloquium 199, "Direct and large eddy simulation of turbulence", Notes on numerical fluid mechanics, Vol. 15) Vieweg, 36–66
25. Aupoix, B. (1989): Application of two-point closures to subgrid scale modelling for homogeneous 3D turbulence. (Turbulent shear flows, von Karman Institute for Fluid Dynamics, Lecture Series 1989-03)
26. Aupoix, B., Cousteix, J. (1982): Modèles simples de tensions de sous-maille en turbulence homogène isotrope (in french). Rech. Aéro. **4**, 273–283
27. Avancha, R., Pletcher, R. (2002): Large eddy simulation of the flow past a backward facing step with heat transfer and property variations. Int. J. Heat Fluid Flow **23**, 601–614
28. Baastians, R., Rindt, C., Nieuwstadt, F., van Steenhoven, A. (2000): Direct and large-eddy simulation of the transition of two- and three-dimensional plane plumes in a confined enclosure. Int. J. Heat Mass Transfer **43**, 2375–2393
29. Baastians, R., Rindt, C., van Steenhoven, A. (1998): Experimental analysis of a confined transitional plume with respect to subgrid-scale modeling. Int. J. Heat Mass Transfer **41**, 3989–4007
30. Bagget, J.S. (1997): Some modeling requirements for wall models in large-eddy simulation. Annual Research Briefs – Center for Turbulence Research, 123–134
31. Bagget, J.S. (1998): On the feasability of merging LES with RANS for the near-wall region of attached turbulent flows. Annual Research Briefs – Center for Turbulence Research, 267–277
32. Bagget, J.S., Jimenez, J., Kravchenko, A.G. (1997): Resolution requirements in large-eddy simulations of shear flows. Annual Research Briefs – Center for Turbulence Research, 51–66
33. Bagget, J.S., Nicoud, F., Mohammadi, B., Bewley, T., Gullbrand, J., Botella, O. (2000): Sub-optimal control based wall models for LES including transpiration velocity. Proceedings of the Summer Program – Center for Turbulence Research, 331–342
34. Bagwell, T.G., Adrian, R.J., Moser, R.D., Kim, J. (1993): Improved approximation of wall shear stress boundary conditions for large-eddy simulation. (Near-wall turbulent flows, So, Speziale and Launder eds.), Elsevier

35. Balaras, E. (2003): Modeling complex boundaries using an external force field on fixed Cartesian grids in large-eddy simulations. Computers and Fluids **33**, 375–404
36. Balaras, E., Benocci, C., Piomelli, U. (1995): Finite-difference computations of high Reynolds number flows using the dynamic subgrid-scale model. Theoret. Comput. Fluid Dynamics **7**, 207–216
37. Balaras, E., Benocci, C., Piomelli, U. (1996): Two-layer approximate boundary conditions for large-eddy simulations. AIAA Journal **34**(6), 1111–1119
38. Balaras, E., Piomelli, U., Wallace, J.M. (2001): Self-similar states in turbulent mixing layers. J. Fluid Mech. **446**, 1–24
39. Bardina, J., Ferziger, J.H., Reynolds, W.C. (1980): Improved subgrid scale models for large eddy simulation. AIAA Paper 80-1357
40. Bardina, J., Ferziger, J.H., Reynolds, W.C. (1983): Improved turbulence models based on large eddy simulation of homogeneous, incompressible, turbulent flows. Report TF-19, Thermosciences Division, Dept. Mechanical Engineering, Stanford University
41. Barenblatt, G.I. (1996): Scaling, self-similarity, and intermediate asymptotics. Cambridge University Press, UK
42. Basdevant, C., Lesieur, M., Sadourny, R. (1978): Subgrid-scale modeling of enstrophy transfer in two-dimensional turbulence. J. Atmos. Sci. **35**, 1028–1042
43. Basdevant, C., Sadourny, R. (1983): Modélisation des échelles virtuelles dans la modélisation numérique des écoulements turbulents bidimensionnels (in french). J. Mech. Théor. Appl., numéro spécial, 243–269
44. Bastin, F., Lafon, P., Candel, S. (1997): Computation of jet mixing noise due to coherent structures: the plane jet case. J. Fluid Mech. **335**, 261–304
45. Batchelor, G.K. (1953): The theory of homogeneous turbulence. Cambridge University Press
46. Batten, P., Goldberg, U., Chakravarthy, S. (2000): Sub-grid turbulence moeling for unsteady flow with acoustic resonance. AIAA Paper 2000-0473
47. Batten, P., Goldberg, U., Chakravarthy, S. (2002): Reconstructed subgrid methods for acoustics predictions at all Reynolds numbers. AIAA Paper 2002-2511
48. Batten, P., Goldberg, U., Chakravarthy, S. (2003): Using synthetic turbulence to interface RANS and LES. AIAA Paper 2003-0081
49. Batten, P., Goldberg, U., Chakravarthy, S. (2004): Interfacing statistical turbulence closures with large-eddy simulation. AIAA Journal **42**(3), 485–492
50. Baurle, R.A., Tam, C.J., Edwards, J.R., Hassan, H.A. (2003): Hybrid simulation approach for cavity flows: blending, algorithm, and boundary treatment issues. AIAA Journal **41**(8), 1463–1480
51. Beaubert, F., Viazzo, S. (2003): Large eddy simulations of plane turbulent impinging jets at moderate Reynolds numbers. Int. J. Heat Fluid Flow **24**, 512–519
52. Beaudan, P., Moin, P. (1994): Numerical experiments on the flow past a circular cylinder at subcritical Reynolds number. Report TF-62, Thermosciences Division, Dept. Mechanical Engineering, Stanford University
53. Benzi, R., Biferale, L., Succi, S., Toschi, F. (1999): Intermittency and eddy viscosities in dynamical models of turbulence. Phys. Fluids **11**(5), 1221–1228
54. Berkooz, G. (1994): An observation on probability density equations, or, when do simulations reproduce statistics? Nonlinearity **7**, 313–328
55. Berkooz, G., Holmes, P., Lumley, J.L. (1993): The proper othogonal decomposition in the analysis of turbulent flows. Ann. Rev. Fluid Mech. **25**, 539–575

56. Bertoglio, J.P. (1984): A stochastic subgrid model for sheared turbulence. (Proceedings of Macroscopic modelling of turbulent flows, Lecture notes in physics, Vol. 230) Springer-Verlag, 100–119
57. Bertoglio, J.P., Mathieu, J. (1984): A stochastic subgrid model for large-eddy simulation: general formulation. C. R. Acad. Sc. Paris **299**, série II, (12), 751–754
58. Bertoglio, J.P., Mathieu, J. (1984): A stochastic subgrid model for large-eddy simulation: generation of a stochastic process. C. R. Acad. Sc. Paris **299**, série II, (13), 835–838
59. Biringen, S., Reynolds, W.C. (1981): Large-eddy simulation of the shear-free turbulent boundary-layer. J. Fluid Mech. **103**, 53–63
60. Blackburn, H., Schmidt, S. (2003): Spectral element filtering techniques for large eddy simulation with dynamic estimation. J. Comput. Phys. **186**, 610–629
61. Blin, L., Hadjadj, A., Vervisch, L. (2003): Large eddy simulation of turbulent flows in reversing systems. JOT, 1–19
62. Boersma, B.J., Kooper, M.N., Nieuwstadt, F.T.M., Wesseling, P. (1997): Local grid refinement in large-eddy simulation. J. Engng. Math.**32**, 161–175
63. Bogey, C., Bailly, C. (2004): A family of low dispersive and low dissipative explicit schemes for flow and noise computations. J. Comput. Phys. **194**, 194–214
64. Bonnet, J.P., Delville, J., Druault, P., Sagaut, P., Grohens, R. (1997): Linear stochastic estimation of LES inflow conditions. (Advances in DNS/LES, C. Liu, Z. Liu (eds.)) Greyden Press, 341–348
65. Booij, R. (2003): Measurements and large eddy simulations of the flows in some curved flumes. JOT **4**, 1–17
66. Boris, J.P., Book, D.L. (1973): Flux-Corrected Transport. I. SHASTA, a fluid transport algorithm that works. J. Comput. Phys. **11**, 38–69
67. Boris, J.P., Grinstein, F.F., Oran, E.S., Kolbe, R.L. (1992): New insights into large-eddy simulation. Fluid Dyn. Res. **10**, 199–228
68. Borue, V., Orszag, S.A. (1995): Forced three-dimensional homogeneous turbulence with hyperviscosity. Europhys. Lett. **29**, 687–692
69. Borue, V., Orszag, S.A. (1995): Self-similar decay of three-dimensional turbulence with hyperviscosity. Phys. Rev. E. **52**, R856–859
70. Borue, V., Orszag, S.A. (1996): Numerical study of Kolmogorov flow at high Reynolds numbers. J. Fluid Mech. **306**, 293–323
71. Borue, V., Orszag, S.A. (1996): Kolmogorov's refined similarity hypothesis for hyperviscous turbulence. Phys. Rev. E. **366**, R21–24
72. Borue, V., Orszag, S.A. (1998): Local energy flux and subgrid-scale statistics in three-dimensional turbulence. J. Fluid Mech. **366**, 1–31
73. Bouchon, F., Dubois, T. (1999): Incremental Unknowns: a tool for large-eddy simulations ?. (Direct and Large Eddy Simulation III, Voke, Sandham and Kleiser eds) Kluwer, 275–286
74. Brasseur, J.G., Wei, C.H. (1994): Interscale dynamics and local isotropy in high Reynolds number turbulence within triadic interactions. Phys. Fluids A **6**(2), 842–870
75. Breuer, M. (1998): Large eddy simulation of the subcritical flow past a circular cylinder: numerical and modeling aspects. Int. J. Numer. Meth. Fluids **28**, 1281–1302
76. Breuer, M. (1998): Numerical and modeling influences on large eddy simulations for the flow past a circular cylinder. Int. J. Heat Fluid Flow **19**, 512–521
77. Breuer, M. (2000): A challenging test case for large eddy simulation: high Reynolds number circular cylinder flow. Int. J. Heat Fluid Flow **21**, 648–654

78. Breuer, M., Jovicic, N., Mazaev, K. (2003): Comparison of DES, RANS and LES for the separated flow around a flat plate at high incidence. Int. J. Numer. Meth. Fluids **41**, 357–388
79. Brezzi, F., Houston, P., Marini, D., Süli, E. (2000): Modeling subgrid viscosity for advection-diffusion problems. Comput. Methods Appl. Mech. Engrg. **190**, 1601–1610
80. Brown, R.M., Perry, P., Shen, Z. (1998): The additive turbulent decomposition for the two-dimensional incompressible Navier–Stokes equations: convergence theorems and error estimates. SIAM J. Appl. Math. **59**(1), 139–155
81. Brun, C., Friedrich, R. (2001): Modeling the test SGS tensor T_{ij}: an issue in the dynamic approach. Phys. Fluids **13**(8), 2373–2385
82. Brun, C., Friedrich, R. (2001): The spatial velocity increment as a tool for SGS modeling. (Modern simulation strategies for turbulent flow, Geurts (ed.)), Edwards, 57–84
83. Brun, C., Kessler, P., Comte, P., Lesieur, M. (1997): Simulation des grandes échelles de jets ronds (in french). Rapport de synthèse, Contrat DGA/DRET 95-2557 A
84. Bush, R.H., Mani, M. (2001): A two-equation large eddy stress model for high subgrid shear. AIAA Paper 2001-2561
85. Cabot, W. (1992): Large eddy simulations of time-dependent and buoyancy-driven channel flows. Annual Research Briefs – Center for Turbulence Research, 45–60
86. Cabot, W. (1995): Large-eddy simulations with wall models. Annual Research Briefs – Center for Turbulence Research, 41–49
87. Cabot, W. (1996): Near-wall models in large-eddy simulations of flow behind a backward-facing step. Annual Research Briefs – Center for Turbulence Research, 199–210
88. Cabot, W. (1998): Large-eddy simulation of a separated boundary layer. Annual Research Briefs – Center for Turbulence Research, 279–288
89. Cabot, W., Jimenez, J., Bagget, J.S. (1999): On wakes and near-wall behavior in coarse large-eddy simulation of channel flow with wall models and second-order finite-difference methods. Annual Research Briefs – Center for Turbulence Research, 343–354
90. Cabot, W., Moin, P. (1999): Approximate wall boundary conditions in the large-eddy simulation of high Reynolds number flow. Flow, Turbulence and Combustion **63**, 269–291
91. Calgaro, C., Debussche, A., Laminie, J. (1998): On a multilevel approach for the two-dimensional Navier–Stokes equations with finite elements. Int. J. Numer. Meth. Fluids **27**, 241–258
92. Calmet, I., Magnaudet, J. (1997): Large-eddy simulation of high-Schmidt number mass transfer in a turbulent channel flow. Phys. Fluids **9**(2), 438–455
93. Calmet, I., Magnaudet, J. (1998): High-Schmidt number mass transfer through turbulent gas-liquid interfaces. Int. J. Heat Fluid Flow **19**, 522–532
94. Camarri, S., Salvetti, M., Koobus, B., Dervieux, A. (2002): Large-eddy simulation of a bluff-body flow on unstructurd grids. Int. J. Numer. Meth. Fluids **40**, 1431–1460
95. Camarri, S., Salvetti, M., Koobus, B., Dervieux, A. (2004): A low-diffusion MUSCL scheme for LES on unstructured grids. Computers and Fluids **33**, 1101–1129
96. Cambon, C., Jeandel, D., Mathieu, J. (1981): Spectral modelling of homogeneous non-isotropic turbulence. J. Fluid Mech. **104**, 247–262
97. Cambon, C., Mansour, N.N., Godeferd, F.S. (1997): Energy transfer in rotating turbulence. J. Fluid Mech. **337**, 303–332

98. Cantekin, M.E., Westerink, J.J., Luettich, R.A. (1994): Low and moderate Reynolds number transient flow simulations using space filtered Navier–Stokes equations. Numer. Meth. Partial Diff. Eq. **10**, 491–524
99. Canuto, V.M., Cheng, Y. (1997): Determination of the Smagorinsky-Lilly constant C_S. Phys. Fluids **9**(5), 1368–1378
100. Carati, D., Cabot, W. (1996): Anisotropic eddy viscosity models. Proceedings of the Summer Program – Center for Turbulence Research, 249–259
101. Carati, D., Ghosal, S., Moin, P. (1995): On the representation of backscatter in dynamic localization models. Phys. Fluids **7**(3), 606–616
102. Carati, D., Rogers, M.M. (1998): Ensemble-averaged LES of time-evolving plane wake. Proceedings of the Summer Program – Center for Turbulence Research, 325–336
103. Carati, D., Rogers, M., Wray, A. (2002): Statistical ensemble of large-eddy simulations. J. Fluid Mech. **455**, 195–212
104. Carati, D., Vanden Eijnden, E. (1997): On the self-similarity assumption in dynamic models for large-eddy simulations. Phys. Fluids **9** (7), 2165–2167
105. Carati, D., Winckelmans, G.S., Jeanmart, H. (1999): Exact expansions for filtered-scales modelling with a wide class of LES filters. (Direct and Large Eddy Simulation III, Voke, Sandham and Kleiser eds) Kluwer, 213–224
106. Carati, D., Winckelmans, G., Jeanmart, H. (2001): On the modelling of the subgrid-scale and filtered-scale stress tensors in large-eddy simulation. J. Fluid Mech. **441**, 119–138
107. Carati, D., Wray, A. (2000): Time filtering in large-eddy simulations. Proceedings of the Summer Program – Center for Turbulence Research, 263–270
108. Carati, D., Wray, A., Cabot, W. (1996): Ensemble-averaged dynamic modeling. Proceedings of the Summer Program – Center for Turbulence Research, 237–248
109. Carlotti, P. (2002): Two-point properties of atmospheric turbulence very close to the ground: comparison of a high resolution LES with theoretical models. Boundary-Layer Meteorol. **104**, 381–410
110. Carnevale, G., Briscolini, M., Orlandi, P. (2001): Buoyancy- to inertial-range transition in forced stratified turbulence. J. Fluid Mech. **427**, 205–239
111. Carruthers, D.J., Moeng, C.H. (1987): Waves in the overlying inversion of the convective boundary layer. J. Atmos. Sci. **44**(4), 1801–1808
112. Casalino, D., Boudet, J., Jacob, M.: A shear flow subgrid scale model for large eddy simulations. Submitted
113. Catalano, P., Wang, M., Iaccarino, G., Moin, P. (2003): Numerical simulation of the flow around a circular cylinder at high Reynolds numbers. Int. J. Heat Fluid Flow **24**, 463–469
114. Cerutti, S., Meneveau, C. (1998): Intermittency and relative scaling of subgrid-scale energy dissipation in isotropic turbulence. Phys. Fluids **10** (4), 928–937
115. Cerutti, S., Meneveau, C., Knio, O.M. (2000): Spectral and hyper eddy viscosity in high-Reynolds-number turbulence. J. Fluid Mech. **421**, 307–338
116. Cerutti, S., Meneveau, C. (2000): Statistics of filtered velocity in grid and wake turbulence. Phys. Fluids **12**(5), 1143–1165
117. Chakravarthy, V.K., Menon, S. (2001): Large-eddy simulation of turbulent premixed flames in the flamelet regime. Combust. Sci. and Tech. **162**, 175–222
118. Chakravarthy, V.K., Menon, S. (2001): Subgrid modeling of turbulent premixed flames in the flamelet regime. Flow, Turbulence and Combustion **65**, 133–161
119. Chapman, D.R. (1979): Computational aerodynamics development and outlook. AIAA Journal **17**(12), 1293–1313

120. Chasnov, J.R. (1990): Simulation of the Kolmogorov inertial subrange using an improved subgrid model. Phys. Fluids A **3**(1), 188–200
121. Chasnov, J. (1991): Simulation of the inertial-conductive subrange. Phys. Fluids A **3**(5), 1164–1168
122. Chasnov, J. (1994): Similarity states of passive scalar transport in isotropic turbulence. Phys. Fluids **6**(2), 1036–1051
123. Chasnov, J. (1995): Similarity states of passive scalar transport in buoyancy-generated turbulence. Phys. Fluids **7**(6), 1499–1506
124. Chasnov, J., Tse, K.L. (2001): Turbulent penetrative convection with an internal heat source. Fluid Dyn. Res. **28**, 397–421
125. Chatelain, A., Ducros, F., Métais, O. (2004): LES of turbulent heat transfer: proper convection numerical schemes for temperature transport. Int. J. Numer. Meth. Fluids **44**, 1017–1044
126. Chattopadhyay, H., Saha, S. (2003): Turbulent flow and heat transfer from a slot jet impinging on a moving plate. Int. J. Heat Fluid Flow **24**, 685–697
127. Chen, S., Doolen, G.D. (1998): Lattice Boltzmann method for fluid flows. Ann. Rev. Fluid Mech. **30**, 329–364
128. Chen, M., Temam, R. (1991): Incremental unknowns for solving partial differential equations. Numer. Math. **59**, 255–271
129. Chester, S., Charlette, F., Meneveau, C. (2001): Dynamic model for LES without test filtering: quantifying the accuracy of Taylor series approximations. Theoret. Comput. Fluid Dynamics **15**, 165–181
130. Chlond, A. (1998): Large-eddy simulation of contrails. J. Atmos. Sci. **55**, 796–819
131. Choi, H., Moin, P. (1994): Effects of the computational time step on numerical solutions of turbulent flow. J. Comput. Phys. **113**, 1–4
132. Chollet, J.P. (1983): Two-point closures as a subgrid modelling for large-eddy simulations. (Fourth Symposium on Turbulent Shear Flows, Karlsruhe, Allemagne)
133. Chollet, J.P. (1984): Spectral closures to derive a subgrid scale modelling for large eddy simulation. (Proceedings of Macroscopic modelling of turbulent flows, Lecture notes in physics, Vol. 230), Springer-Verlag, 161–176
134. Chollet, J.P. (1984): Turbulence tridimensionnelle isotrope: modélisation statistique des petites échelles et simulation numérique des grandes échelles (in french). Thèse de Doctorat es-sciences, Grenoble, France
135. Chollet, J.P. (1992): LES and subgrid models for reactive flows and combustion. ERCOFTAC Summer School/Workshop on modelling turbulent flows, Lyon, France
136. Chollet, J.P., Lesieur, M. (1981): Parametrization of small scales of three-dimensional isotropic turbulence utilizing spectral closures. J. Atmos. Sci. **38**, 2747–2757
137. Chorin, A., Kast, A., Kupferman, R. (1998): Optimal prediction of under-resolved dynamics. Proc. Natl. Acad. Sci. USA, Applied Mathematics **95**, 4094–4098
138. Chow, F.K., Moin, P. (2003): A further study of numerical errors in large-eddy simulations. J. Comput. Phys. **184**, 366–380
139. Chung, Y.M., Sung, H.J. (1997): Comparative study of inflow conditions for spatially evolving simulation. AIAA Journal **35**(2), 269–274
140. Chyczewski, T., Morris, P., Long, L. (2000): Large-eddy simulation of wall bounded flow using the nonlinear disturbance equations. AIAA Paper 2000-2007
141. Ciofalo, M. (1996): Investigation of flow and heat transfer in corrugated passages–II. Numerical simulations. Int. J. Heat Mass Transfer **39**(1), 165–192

142. Ciofalo, M., Collins, M.W. (1992): Large-eddy simulation of a turbulent flow and heat transfer in plane and rib-roughened channels. Int. J. Numer. Meth. Fluids **15**, 453–489
143. Clark, R.A., Ferziger, J.H., Reynolds, W.C. (1979): Evaluation of subgrid-scale models using an accurately simulated turbulent flow. J. Fluid Mech. **91**(1), 1–16
144. Codina, R. (2000): Stabilization of incompressibility and convection through orthogonal sub-scales in finite elements methods. Comput. Methods Appl. Mech. Engrg. **190**, 1579–1599
145. Colella, P., Woodward, P.R. (1984): The Piecewise Parabolic Method (PPM) for gas-dynamical simulations. J. Comput. Phys. **54**, 174–201
146. Constantinescu, G.S., Squires, K.D. (2000): LES and DES investigations of turbulent flow over a sphere. *AIAA Paper* 2000-0540
147. Conway, S., Caraeni, D., Fuchs, L. (2000): Large eddy simulation of the flow through the blade of a swirl generator. Int. J. Heat Fluid Flow **21**, 664–673
148. Cook, A.W. (1997): Determination of the constant coefficient in the scale similarity models of turbulence. Phys. Fluids **9**(5), 1485–1487
149. Cook, A. (2001): A consistent approach to large-eddy simulation using adaptive mesh refinement. J. Comput. Phys. **154**, 117–133
150. Cousteix, J. (1989): Turbulence et couche limite (in french). CEPADUES – Editions, France
151. Cui, J., Patel, V., Lin, C.L. (2003): Large eddy simulation of turbulent flow in a channel with rib roughness. Int. J. Heat Fluid Flow **24**, 372–388
152. Cui, A., Street, R.L. (2001): Large-eddy simulation of turbulent rotating convective flow development. J. Fluid Mech. **447**, 53–84
153. Cui, G., Zhou, H., Zhang, Z., Shao, L. (2004): A new dynamic subgrid eddy viscosity model with application to turbulent channel flow. Phys. Fluids **16**(8), 2835–2842
154. Cuxart, J., Bougeault, P., Redelsperger, J.L. (2000): A turbulence scheme allowing for mesoscale and large-eddy simulations. Q.J.R. Meteorol. Soc. **126**, 1–30
155. Cziesla, T., Biswas, G., Chattopadhyay, H., Mitra, N. (2001): Large eddy simulation of flow and heat transfer in an impinging slot jet. Int. J. Heat Fluid Flow **22**, 500-508
156. Cziesla, T., Biswas, G., Mitra, N.K. (1999): Large eddy simulation in a turbulent channel flow using exit boundary conditions. Int. J. Numer. Meth. Fluids **30**, 763–773
157. da Silva, C., Métais, O. (2002): Vortex control of bifurcating jets: a numerical study. Phys. Fluids **14**(11), 3798–3818
158. da Silva, C., Métais, O. (2002): On the influence of coherent structures upon interscale interactions in turbulent plane jets. J. Fluid Mech. **473**, 103–145
159. Dai, Y., Kobayashi, T., Taniguchi, N. (1994): Large eddy simulation of plane turbulent jet flow using a new outflow velocity boundary condition. JSME International Journal Series B **37**(2), 242–253
160. Dakhoul, Y.M., Bedford, K.W. (1986): Improved averaging method for turbulent flow simulation. Part 1: theoretical development and application to Burger's transport equation. Int. J. Numer. Meth. Fluids **6**, 49–64
161. Dakhoul, Y.M., Bedford, K.W. (1986): Improved averaging method for turbulent flow simulation. Part 2: calculations and verification. Int. J. Numer. Meth. Fluids **6**, 65–82
162. Dang, K.T. (1985): Evaluation of simple subgrid-scale models for the numerical simulation of homogeneous turbulence. AIAA Journal **23**(2), 221–227

163. Dantinne, G., Jeanmart, H., Winckelmans, G.S., Legat, V. (1998): Hyperviscosity and vorticity-based models for subgrid scale modeling. Applied Scientific Research **59**, 409–420
164. Das, A., Moser, R.D. (2001): Filtering boundary conditions for LES and embedded boundary simulations. (DNS/LES Progress and Challenges, Liu, Sakell and Beutner eds.), Greyden Press, 389–396
165. Das, A., Moser, R. (2002): Optimal large-eddy simulation of forced burgers equation. Phys. Fluids **14**(12), 4344–4351
166. David, E. (1993): Modélisation des écoulements compressibles et hypersoniques (in french). Thèse de Doctorat de l'INPG, Grenoble, France
167. Davidson, L. (2001): Hybrid LES-RANS: a combination of a one-equation SGS model and a K–ω model for predicting recirculating flows. (Proceedings of ECCOMAS CFD conference)
168. de Bruyn Kops, S., Riley, J. (2000): Re-examining the thermal mixing layer with numerical simulations. Phys. Fluids. **12**(1), 185–192
169. De Stefano, G., Denaro, F.M., Riccardi, G. (1998): Analysis of 3D backward-facing step incompressible flows via a local averaged-based numerical procedure. Int. J. Numer. Meth. Fluids **28**, 1073–1091
170. De Stefano, G., Denaro, F.M., Riccardi, G. (2001): High-order filtering for control volumes flow simulation. Int. J. Numer. Meth. Fluids **37**(7), 797–835
171. De Stefano, G., Vasilyev, O.V. (2002): Sharp cutoff versus smooth filtering in large eddy simulation. Phys. Fluids **14**(1), 362–369
172. Deardorff, J.W. (1970): A numerical study of three-dimensional turbulent channel flow at large Reynolds numbers. J. Fluid Mech. **41**, 453–465
173. Deardorff, J.W. (1973): The use of subgrid transport equations in a three-dimensional model of atmospheric turbulence. ASME J. Fluids Engng., 429–438
174. Debussche, A., Dubois, T., Temam, R. (1995): The nonlinear Galerkin method: a multiscale method applied to simulation of homogeneous turbulent flows. Theoret. Comp. Fluid Dynamics **7**, 279–315
175. Denaro, F.M. (1996): Towards a new model-free simulation of high-Reynolds-number flows: local average direct numerical simulation. Int. J. Numer. Methods Fluids **23**, 125–142
176. Deschamps, V. (1988): Simulation numérique de la turbulence homogène incompressible dans un écoulement de canal plan (in french). ONERA, Note technique 1988-5
177. Desjardin, P.E., Frankel, S.H. (1996): Assessment of turbulent combustion submodels using the linear eddy model. Combust. Flame **104**, 343–357
178. di Mare, L., Jones, W.P. (2003): LES of turbulent flow past a swept fence. Int. J. Heat Fluid Flow **24**, 606–615
179. Ding, X., Tsang, T.T.H. (2001): Large eddy simulation of turbulent flows by a least-squares finite element method. Int. J. Numer. Meth. Fluids **37**, 297–319
180. Diurno, G.V., Balaras, E., Piomelli, U. (2001): Wall-layer models for LES of separated flows. (Modern simulation strategies for turbulent flow, Geurts ed.), Edwards, 207–222
181. Domaradzki, J.A., Adams, N.A. (2002): Direct modelling of subgrid scales of turbulence in large eddy simulations. JOT **3**(024), 1–19
182. Domaradzki, J.A., Holm, D.D. (2001): The Navier–Stokes-alpha model: LES equations with nonlinear dispersion. (Modern simulation strategies for turbulent flow, Geurts ed.), Edwards, 107–122
183. Domaradzki, J.A., Horiuti, K. (2001): Similarity modeling on an expanded mesh applied to rotating turbulence. Phys. Fluids **13**(11), 3510–3512

184. Domaradzki, J.A., Liu, W. (1995): Approximation of subgrid-scale energy transfer based on the dynamics of the resolved scales of turbulence. Phys. Fluids **7**(8), 2025–2035
185. Domaradzki, J.A., Liu, W., Brachet, M.E. (1993): An analysis of subgrid-scale interactions in numerically simulated isotropic turbulence. Phys. Fluids **5**(7), 1747–1759
186. Domaradzki, J.A., Liu, W., Härtel, C., Kleiser, L. (1994): Energy transfer in numerically simulated wall-bounded turbulent flows. Phys. Fluids **6**(4), 1583–1599
187. Domaradzki, J.A., Loh, K.C. (1999): The subgrid-scale estimation model in the physical space representation. Phys. Fluids **11** (8), 2330–2342
188. Domaradzki, J.A., Loh, K.C. (2002): Large eddy simulations using the subgrid-scale estimation model and truncated Navier–Stokes dynamics. Theoret. Comput. Fluid Dynamics **15**, 421–450
189. Domaradzki, J.A., Metcalfe, R.W., Rogallo, R.S., Riley, J.J. (1987): Analysis of subgrid-scale eddy viscosity with the use of results from direct numerical simulations. Phys. Rev. Letter **58**(6), 546–550
190. Domaradzki, J.A., Saiki, E.M. (1997): A subgrid-scale model based on the estimation of unresolved scales of turbulence. Phys. Fluids **9** (7), 2148–2164
191. Domaradzki, J.A., Saiki, E. (1997): Backscatter models for large-eddy simulations. Theoret. Comput. Fluid Dynamics **9**, 75–83
192. Domaradzki, J.A., Xiao, Z., Smolarkiewicz, P. (2003): Effective eddy viscosities in implicit large eddy simulations of turbulent flows. Phys. Fluids **15**(12), 3890–3893
193. Domaradzki, J.A., Yee, P.P. (2000): The subgrid-scale estimation model for high Reynolds number turbulence. Phys. Fluids **12**(1), 193–196
194. Dommermuth, D., Rottman, J., Innis, G., Novikov, E. (2002): Numerical simulation of the wake of a towed sphere in a weakly stratified fluid. J. Fluid Mech. **473**, 83–101
195. Dong, Y.H., Lu, X.Y. (2004): Large eddy simulation of a thermally stratified turbulent channel flow with temperature oscillation on the wall. Int. J. Heat Mass Transfer **47**, 2109–2122
196. Dong, Y.H., Lu, X.Y., Zhuang, L.X. (2003): Large eddy simulation of turbulent channel flow with mass transfer at high Schmidt numbers. Int. J. Heat Mass Transfer **46**, 1529–1539
197. Dornbrack, A., Schumann, U. (1993): Numerical simulation of turbulent convective flow over a wavy terrain. Boundary-Layer Meteorol. **65**, 323–355
198. Dreeben, T., Kerstein, A. (2000): Simulation of vertical slot convection using one-dimensional turbulence. Int. J. Heat Mass TRansfer **43**, 3823–3834
199. Druault, P., Lardeau, S., Bonnet, J.P., Coiffet, F., Delville, J., Lamballais, E., Largeau, J.F., Perret, L. (2004): Generation of three-dimensional turbulent inlet conditions for large-eddy simulation. AIAA Journal **42**(3), 447–456
200. Dubois, T., Bouchon, F. (1998): Subgrid-scale models based on incremental unknowns for large eddy simulations.. Annual Research Briefs – Center for Turbulence Research, 221–235
201. Dubois, T., Jauberteau, F., Temam, R. (1999): Dynamic multilevel methods and the numerical simulation of turbulence. Cambridge University Press
202. Dubois, T., Jauberteau, F., Zhou, Y. (1997): Influences of subgrid scale dynamics on resolvable statistics in large eddy simulations. Physica D **100**, 390–406
203. Dubrulle, B., Laval, J.P., Nazarenko, S., Kevlahan, N.K.R. (2001): A dynamic subfilter-scale model for plane parallel flows. Phys. Fluids **13**(7), 2045–2064

204. Ducros, F. (1995): Simulations numériques directes et des grandes échelles de couches limites compressibles (in french). Thèse de Doctorat de l'INPG, Grenoble, France
205. Ducros, F., Comte, P., Lesieur, M. (1996): Large-eddy simulation of transition to turbulence in a boundary layer developing spatially over a flat plate. J. Fluid Mech. **326**, 1–36
206. Dunca, A., John, V., Layton, W.J. (2002): The commutation error of the spaced averaged Navier–Stokes equations on a bounded domain. Unpublished
207. Dunca, A., John, V., Layton, W.J., Sahin, N. (2001): Numerical analysis of large-eddy simulation. (DNS/LES Progress and Challenges, Liu, Sakell and Beutner eds.), Greyden Press, 359–364
208. Echekki, T., Kerstein, A., Dreeben, T., Chen, J.Y. (2001): One dimensional turbulence simulation of turbulent jet diffusion flames: model formulation and illustrative applications. Combust. Flame **125**, 1083–1105
209. Eidson, T. (1985): Numerical simulation of the turbulent Rayleigh–Bénard problem using subgrid model. J. Fluid Mech. **158**, 245–268
210. Engquist, B., Lötstedt, P., Sjögreen, B. (1989): Nonlinear filters for efficient shock computation. Math. Comput. **52**(186), 509–537
211. Fabignon, Y., Beddini, R.A., Lee, Y. (1995): Analytic evaluation of finite difference methods for compressible direct and large eddy simulations. ONERA, TP 1995-128
212. Fasel, H.F., Seidel, J., Wernz, S. (2002): A methodology for simulations of complex turbulent flows. J. Fluid Engng. **124**, 933–942
213. Fatica, M., Mittal, R. (1996): Progress in the large-eddy simulation of flow through an asymmetric plane diffuser. Annual Research Briefs – Center for Turbulence Research, 249–257
214. Fatica, M., Orlandi, P., Verzicco, R. (1994): Direct and large-eddy simulations of round jets. (Direct and Large Eddy Simulation I, Voke, C-hollet & Kleiser eds.) Kluwer, 49–61
215. Feiz, A.A., Ould-Rouis, M., Lauriat, G. (2003): Large eddy simulation of turbulent flow in a rotating pipe. Int. J. Heat Fluid Flow **24**, 412–420
216. Ferziger, J.H. (1977): Large eddy simulations of turbulent flows. AIAA Journal **15**(9), 1261–1267
217. Ferziger, J.H. (1997): Large eddy simulation: an introduction and perspective. (New tools in turbulence modelling, O. Métais and J. Ferziger eds.) Les éditions de physique, Springer, 29–48
218. Ferziger, J.H., Leslie, D.C. (1979): Large eddy simulation: A predictive approach to turbulent flow computation. (AIAA Comput. Fluid Conf., Williamsburg, USA)
219. Fischer, P., Iliescu, T. (2001): A 3D channel flow simulation at $Re_\tau = 180$ using a rational LES model. (DNS/LES Progress and Challenges, Liu, Sakell and Beutner eds.), Greyden Press, 283–290
220. Flohr, P., Vassilicos, J.C. (2000): A scalar subgrid model with flow structure for large-eddy simulations of scalar variances. J. Fluid Mech. **407**, 315–349
221. Foias, C., Holm, D., Titi, E. (2001): The Navier–Stokes-alpha model of fluid turbulence. Physica D **152-153**, 505–519
222. Foias, C., Manley, O., Temam, R. (1987): Sur l'interaction des petits et grands tourbillons dans des écoulements turbulents (in french). C.R. Acad. Sci. Paris t. **305**, série I, 497–500
223. Foias, C., Manley, O., Temam, R. (1988): Modelling of the interaction of small and large eddies in two dimensional turbulent flows. MMAN **22**(1), 93–114
224. Foias, C., Manley, O., Temam, R. (1991): Approximate inertial manifolds and effective viscosity in turbulent flows. Phys. Fluids A **3**(5), 898–911

225. Franke, J., Frank, W. (2001): Temporal commuation errors in large-eddy simulation. Z. Angew. Math. Mech. **81** S3, S467–S468
226. Friedrich, R., Arnal, M. (1990): Analysing turbulent backward-facing step flow with the lowpass-filtered Navier–Stokes equations. J. Wind Eng. Ind. Aerodyn. **35**, 101–128
227. Fureby, C., Alin, N., Wikström, N., Menon, S., Svanstedt, N., Persson, L. (2004): Large-eddy simulation of high-Reynolds-number wall-bounded flows. AIAA Journal **42**(3), 457–468
228. Fureby, C., Grinstein, F.F. (1999): Monotonically Integrated large eddy simulation of free shear flows. AIAA Journal **37**(5), 544–556
229. Fureby, C., Grinstein, F.F. (2002): Large eddy simulation of high-Reynolds-number free and wall-bounded flows. J. Comput. Phys. **181**, 68–97
230. Fureby, C., Tabor, G. (1997): Mathematical and physical constraints on large-eddy simulations. Theoret. Comput. Fluid Dynamics **9**, 85–102
231. Fureby, C., Tabor, G., Weller, H.G., Gosman, A.D. (1997): A comparative study of subgrid scale models in homogeneous isotropic turbulence. Phys. Fluids **9**(5), 1416–1429
232. Fureby, C., Tabor, G., Weller, H.G., Gosman, A.D. (1997): Differential subgrid stress models in large eddy simulations. Phys. Fluids **9** (11), 3578–3580
233. Fureby, C., Tabor, G., Weller, H.G., Gosman, A.D. (2000): Large-eddy simulations of the flow around a square prism. AIAA Journal **38**(3), 442–452
234. Galdi, G.P., Layton, W.J. (2000): Approximating the larger eddies in fluid motion II: a model for space-filtered flow. Math. Models and Meth. in Appl. Sciences **10**(3), 343–350
235. Galmiche, M., Thual, O., Bonneton, P. (2002): Direct numerical simulation of turbulence-mean field interactions in a stably stratified fluid. J. Fluid Mech. **455**, 213–242
236. Gao, S., Voke, P. (1995): Large eddy simulation of turbulent heat transport in enclosed impinging jets. Int. J. heat Fluid Flow **16**, 349–356
237. Gao S., Voke, P., Gough, T. (1997): Turbulent simulation of flat plate boundary layer and near wake. (Direct and Large Eddy Simulation II, Chollet, Voke and Kleiser eds) Kluwer, 115–124
238. Garg, R., Ferziger, J., Monismith, S., Koseff, J. (2000): Stably stratified turbulent channel flows. I. Stratification regimes and turbulence suppression mechanism. Phys. Fluids **12**(10), 2569–2593
239. Garnier, E., Mossi, M., Sagaut, P., Deville, M., Comte, P. (1999): On the use of shock-capturing schemes for large-eddy simulation. J. Comput. Phys. **153**, 273–311
240. Gatski, T.B., Liu, J.T.C. (1980): On the interactions between large-scale structure and fine-grained turbulence in a free shear flow. III. A numerical solution. Phil. Trans. Royal Soc. Lond. A **293**, 473–509
241. Georgiadis, N.J., Iwan, J., Alexander, D., Reshotko, E. (2001): Development of a hybrid RANS/LES method for compressible mixing layer simulations. AIAA Paper 2001-0289
242. Germano, M. (1986): Differential filters for the large eddy numerical simulation of turbulent flows. Phys. Fluids **29**(6), 1755–1757
243. Germano, M. (1986): Differential filters of elliptic type. Phys. Fluids **29**(6), 1757–1758
244. Germano, M. (1986): A proposal for a redefinition of the turbulent stresses in the filtered Navier–Stokes equations. Phys. Fluids **29**(7), 2323–2324
245. Germano, M. (1987): On the non-Reynolds averages in turbulence. AIAA Paper 87-1297

246. Germano, M. (1992): Turbulence: The filtering approach. J. Fluid Mech. **238**, 325–336
247. Germano, M. (1996): A statistical formulation of the dynamic model. Phys. Fluids **8** (2), 565–570
248. Germano, M. (1998): Fundamentals of large-eddy simulation. (Advanced Turbulent Flows Computations, Peyret and Krause eds.), CISM Courses and Lectures 395, Springer, 81–130
249. Germano, M. (1999): From RANS to DNS: towards a bridging model. (Direct and Large Eddy Simulation III, Voke, Sandham and Kleiser eds.) Kluwer, 225–236
250. Germano, M. (2001): LES overview. (DNS/LES Progress and Challenges, Liu, Sakell and Beutner eds.), Greyden Press, 1–12
251. Germano, M. (2001): Ten years of the dynamic model. (Modern simulation strategies for turbulent flow, Geurts ed.), Edwards, 173–190
252. Germano, M. (2002): On a possible direct effect of the eddy viscosity gradient in turbulence modeling. Phys. Fluids **14**(10), 3745–3747
253. Germano, M., Piomelli, U., Moin, P., Cabot, W.H. (1991): A dynamic subgrid-scale eddy viscosity model. Phys. Fluids A **3**(7), 1760–1765
254. Geurts, B. (1997): Inverse modeling for large-eddy simulation. Phys. Fluids **9** (12), 3585–3587
255. Geurts, B.J. (1999): Balancing errors in LES. (Direct and Large Eddy Simulation III, Voke, Sandham and Kleiser eds) Kluwer, 1–12
256. Geurts, B.J., Froehlich, J. (2001): Numerical effects contaminating LES; a mixed story. (Modern simulation strategies for turbulent flow, Geurts ed.), Edwards, 309–327
257. Geurts, B.J., Froehlich, J. (2002): A framework for predicting accuracy limitations in large-eddy simulation. Phys. Fluids **14**(6), L41–L44
258. Geurts, B.J., Holm, D. (2003): Regularization modeling for large-eddy simulation. Phys. Fluids **15**(1), L13–L16
259. Ghosal, S. (1996): An analysis of numerical errors in large-eddy simulations of turbulence. J. Comput. Phys. **125**, 187–206
260. Ghosal, S. (1999): Mathematical and physical constraints on large-eddy simulation of turbulence. AIAA J. **37** (4), 425–433
261. Ghosal, S., Lund, T.S., Moin, P., Akselvoll, K. (1995): A dynamic localization model for large-eddy simulation of turbulent flows. J. Fluid Mech. **286**, 229–255
262. Ghosal, S., Moin, P. (1995): The basic equations for the large-eddy simulation of turbulent flows in complex geometry. J. Comput. Phys. **118**, 24–37
263. Ghosal, S., Rogers, M. (1997): A numerical study of self-similarity in a turbulent plane wake using large-eddy simulation. Phys. Fluids **9**(6), 1729–1739
264. Gicquel, L., Givi, P., Jaberi, F., Pope, S. (2002): Velocity filtered density function for large eddy simulation of turbulent flows. Phys. Fluids **14**(3), 1196–1213
265. Glaze, D., Frankel, S. (2003): Stochastic inlet conditions for large-eddy simulation of a fully turbulent jet. AIAA Journal **41**(6), 1064–1073
266. Godeferd, F., Cambon, C. (1994): Detailed investigation of energy transfers in homogeneous stratified turbulence. Phys. Fluids **6**(6), 2084–2100
267. Godeferd, F., Staquet, C. (2003): Statistical modelling and direct numerical simulations of decaying stably stratified turbulence. Part 2. Large-scale and small-scale anisotropy. J. Fluid Mech. **486**, 115–159
268. Goldstein, D., Vasilyev, O. (2004): Stochastic coherent adaptive large eddy simulation method. Phys Fluids **16**(7), 2497–2513

269. Gonze, M.A. (1993): Simulation numérique des sillages en transition à la turbulence (in french). Thèse de Doctorat de l'INPG, Grenoble, France
270. Grigoriadis, D., Bartzis, J., Goulas, A. (2003): LES of the flow past a rectangular cylinder using the immersed boundary concept. Int. J. Numer. Meth. Fluids **41**, 615–632
271. Grigoriadis, D., Bartzis, J., Goulas, A. (2004): Efficient treatment of complex geometries for large eddy simulations of turbulent flows. Computers and Fluids **33**, 201–222
272. Grinstein, F.F. (2001): Vortex dynamics and entrainement in rectangular free jets. J. Fluid Mech. **437**, 69–101
273. Grinstein, F., Fureby, C. (2002): Recent progress on MILES for high Reynolds-number flows. AIAA Paper 2002-0134
274. Grinstein, F., Fureby, C. (2002): Recent progress on MILES for high Reynolds Number flows. J. Fluid Engng. **124**, 848–861
275. Grinstein, F.F., Guirguis, R.H. (1992): Effective viscosity in the simulation of spatially evolving shear flows with monotonic FCT models. J. Comput. Phys. **101**, 165–175
276. Grinstein, F., Margolin, L., and Rider, W. (eds.): Implicit Large Eddy Simulation: Computing Turbulent Fluid Dynamics. Cambridge University Press
277. Grötzbach, G. (1983): Spatial resolution requirements for direct numerical simulation of the Rayleigh–Bénard convection. J. Comput. Phys. **49**, 241–264
278. Grötzbach, G. (1987): in Encyclopedia of Fluid Mechanics, N.P. Chereminisoff editor (Gulf, West Orange, NJ), Vol. 6
279. Grötzbach, G., Wörner, M. (1999): Direct and large eddy simulations in nuclear applications. Int. J. Heat Fluid Flow **20**, 222–240
280. Guermond, J.L. (1999): Stabilization of Galerkin approximations of transport equations by subgrid modeling. M2AN **33**(6), 1293–1316
281. Guermond, J.L., Oden, T.J., Prudhomme, S. (2003): An interpretation of the Navier–Stokes-alpha model as a frame-indifferent Leray regularization. Physica D **177**, 23–30
282. Guermond, J.L., Oden, J.T., Prudhomme, S. (2004): Some mathematical issues concerning large-eddy simulation models for turbulent flows. J. Meth. Fluid Mech. **6**(2), 194–248
283. Gullbrand, J., Chow, F.K. (2003): The effect of numerical errors and turbulence models in large-eddy simulations of channel flow, with and without explicit filtering. J. Fluid Mech. **495**, 323–341
284. Hamba, F. (2001): An attempt to combine large eddy simulation with the k-epsilon model in a channel flow calculation. Theoret. Comput. Fluid Dynamics **14**, 323–336
285. Hamba, F. (2003): A hybrid RANS/LES simulation of turbulent channel flow. Theoret. Comput. Fluid Dynamics **16**(5), 387–403
286. Han, J., Lin, Y.L., Schowalter, D.G., Pal Arya, S., Proctor, F.H. (2000): Large-eddy simulation of aircraft wake vortices within homogeneous turbulence: Crow instability. AIAA Journal **38**(2), 292–300
287. Hanazaki, H. (2003): Effects of initial conditions on the passive and active scalar fluxes in unsteady stably stratified turbulence. Phys. Fluids **15**(4), 841–848
288. Hansen, R.P., Long, L.N., Morris, P.J. (2000) Unsteady, laminar flow simulations using the nonlinear disturbance equations. AIAA Paper 2000-1981
289. Härtel, C., Kleiser, L. (1997): Galilean invariance and filtering dependence of near-wall grid-scale/subgrid-scale interactions in large-eddy simulation. Phys. Fluids **9**(2), 473–475

290. Härtel, C., Kleiser, L. (1998): Analysis and modelling of subgrid-scale motions in near-wall turbulence. J. Fluid Mech. **356**, 327–352
291. Härtel, C., Kleiser, L., Unger, F., Friedrich, R. (1994): Subgrid-scale energy transfer in the near-wall region of turbulent flows. Phys. Fluids **6**(9), 3130–3143
292. Harten, A. (1984): On a class of high resolution total-variation-stable finite-difference schemes. SIAM J. Numer. Anal. **21**, 1–23
293. Harten, A. (1993): Discrete multi-resolution analysis and generalized wavelets. Applied Numerical Mathematics **12**, 153–192
294. Harten, A. (1995): Multiresolution algorithms for the numerical solution of hyperbolic conservation laws. Communication on Pure and Applied Mathematics **XLVIII**, 1305–1342
295. Harten, A. (1996): Multiresolution representation of data: a general framework. SIAM J. Numer. Anal. **33**(3), 1205–1256
296. Haselbacher, A., Moser, R., Constantinescu, G., Mahesh, K. (2002): Toward optimal LES on unstructured meshes. Proceedings of the Summer Program, Center for Turbulence Research, 129–140
297. Haselbacher, A., Vasilyev, O. (2003): Commutative discrete filtering on unstructured grids based on least-squares techniques. J. Comput. Phys. **187**, 197–211
298. Hassan, Y., Barsamian, H. (2001): New-wall modeling for complex flows using the large eddy simulation technique in curvilinear coordinates. Int. J. Heat Mass Trans. **44**, 4009–4026
299. Hauke, G., Garcia-Olivares, A. (2001): Variational subgrid scale formulations for the advection-diffusion-reaction equation. Comput. Methods Appl. Mech. Engrg. **190**, 6847–6865
300. He, G.W., Rubinstein, R., Wang, L.P. (2001): Effects of subgrid-scale modeling on time correlations in large-eddy simulation. ICASE Report No. 2001-10
301. He, G.W., Rubinstein, R., Wang, L.P. (2002): Effects of subgrid-scale modeling on time correlations in large-eddy simulation. Phys. Fluids **14**(7), 2186–2193
302. Henn, D.S., Sykes, R.I. (1999): Large-eddy simulation of flow over wavy surfaces. J. Fluid Mech. **383**, 75–112
303. Herr, S., Wang, L.P., Collins, L. (1996): EDQNM model of a passive scalar with uniform mean gradient. Phys. Fluids **8**(6), 1588–1608
304. Herring, J., Métais, O. (1989): Numerical experiments in forced stably stratified turbulence. J. Fluid Mech. **202**, 97–115
305. Hewson, J.C., Kerstein, A. (2002): Local extinction and reignition in non-premixed turbulent CH/H2/N2 jet flames. Combust. Sci. and Tech. **174**, 35–66
306. Hewson, J.C., Kerstein, A. (2001): Stochastic simulation of transport and chemical kinetics in turbulent CH/H2/N2 flames. Combust. Theory Modelling **5**, 669–697
307. Hirsch, C. (1987): Numerical computation of internal and external flows. John Wiley & Son
308. Hoffman, J. (2001): Dynamic subgrid modeling for scalar convection-diffusion-reaction equations with fractal coefficients. Multiscale and multiresolution methods: theory and applications, Lecture Notes in Computational Science and Engineering, vol. 20. Springer, Heidelberg
309. Hoffman, J. (2002): Dynamic subgrid modelling for time dependent convection-diffusion-reaction equations with fractal solutions. Int. J. Numer. Meth. Fluids **40**, 583–592
310. Hoffman, J. (2003): Adaptive DNS/LES: a new agenda in CFD. Chalmers Finite Element Center preprint 2003-23. Chalmers University of Technology

311. Hoffman, J. (2003): Subgrid modeling for convection-diffusion-reaction in 2 space dimensions using a Haar multiresolution analysis. Math. Models & Meth. Appl. Sci. **13**(10), 1515–1536
312. Hoffman, J. (2004): Duality based a posteriori error estimation in various norms and linear functionals for LES. SIAM J. Sci. Comput. **26**(1), 178–195
313. Hoffman, J.: Adaptive finite element methods for LES: computation of the mean drag coefficient in a turbulent flow around a surface mounted cube using adpative mesh refinement. Submitted
314. Hoffman, J., Johnson, C. (2004): Computability and adaptivity in CFD. Encyclopedia of Computational Mechanics, Stein, E., de Horz, R. and Hughes, T.J.R. eds, John Wiley & Sons
315. Hoffman, J., Johnson, C., Bertoluzza, S. (2005): Subgrid modeling for convection-diffusion-reaction in 1 space dimension using a Haar multiresolution analysis. Comupt. Meth. Appl. Mech. Engng. **194**(1), 19–44
316. Holmen, J., Hughes, T.J.R., Oberai, A., Wells, G. (2004): Sensitivity of the scale partition for variational multiscale large-eddy simulation of channel flow. Phys. Fluids **16**(3), 824–827
317. Holtslag, A., Moeng, C.H. (1991): Eddy diffusivity and countergradient transport in the convective atmospheric boundary layer. J. Atmos. Sci. **48**(14), 1690–1698
318. Horiuti, K. (1985): Large eddy simulation of turbulent channel flow by one-equation modeling. J. Phys. Soc. Japan **54**(8), 2855–2865
319. Horiuti, K. (1987): Comparison of conservative and rotational forms in large eddy simulation of turbulent channel flow. J. Comput. Phys. **71**, 343–370
320. Horiuti, K. (1989): The role of the Bardina model in large eddy simulation of turbulent channel flow. Phys. Fluids A **1**(2), 426–428
321. Horiuti, K. (1990): Higher-order terms in the anisotropic representation of Reynolds stresses. Phys. Fluids A **2**(10), 1708–1710
322. Horiuti, K. (1993): A proper velocity scale for modeling subgrid-scale eddy viscosities in large eddy simulation. Phys. Fluids A **5**(1), 146–157
323. Horiuti, K. (1997): Backward scatter of subgrid-scale energy in wall-bounded turbulence and free shear flow. J. Phys. Soc. Japan **66**(1), 91–107
324. Horiuti, K. (1997): A new dynamic two-parameter mixed model for large-eddy simulation. Phys. Fluids **9**(11), 3443–3464
325. Horiuti, K. (2001): Alignment of eigenvectors for strain rate and subgrid-scale stress tensors. (Direct and large-eddy simulation IV, Geurts, Friedrich and Métais eds.), Kluwer, 67–72
326. Horiuti, K. (2001): Rotational transformation and geometrical correlation of SGS models. (Modern simulation strategies for turbulent flow, Geurts ed.), Edwards, 123–140
327. Horiuti, K. (2003): Roles of non-aligned eigenvectors of strain-rate and subgrid-scale stress tensors in turbulence generation. J. Fluid Mech. **491**, 65–100
328. Hossain, M. (1991): Non-diffusive subgrid-scale modelling of turbulence. Phys. Letters A **161**, 277–282
329. Hsiao, C.T., Pauley, L.L. (1999): Direct numerical simulation of unsteady finite-span hydrofoil flow. AIAA Journal **37**(5), 529–536
330. Huai, X., Joslin, R.D., Piomelli, U. (1999): Large-eddy simulation of boundary-layer on a swept wedge. J. Fluid Mech. **381**, 357–380
331. Hughes, T.J.R. (1995): Multiscale phenomena: Green's function, the Dirichlet-to-Neumann formulation, subgrid scale models, bubbles and the origin of stabilized methods. Comput. Methods Appl. Mech. Engrg. **127**, 387–401

332. Hughes, T.J.R., Feijoo, G.R., Mazzei, L., Quincy, J.B. (1998): The variational multiscale method – a paradigm for computational mechanics. Comput. Methods Appl. Mech. Engrg. **166**, 2–24
333. Hughes, T.J.R., Mazzei, L., Jansen, K.E. (2000): Large eddy simulation and the variational multiscale method. Comput. Visual Sci. **3**, 47–59
334. Hughes, T.J.R., Mazzel, L., Oberai, A.A., Wray, A.A. (2001): The multiscale formulation of large-eddy simulation: decay of homogeneous isotropic turbulence. Phys. Fluids **13**(2), 505–512
335. Hughes, T.J.R., Oberai, A.A., Mazzei, L. (2001): Large-eddy simulation of turbulent channel flows by the variational multiscale method. Phys. Fluids **13**(6), 1784–1799
336. Hughes, T.J.R., Stewart, J. (1996): A space-time formulation for multiscale phenomena. J. Comput. Appl. Math. **74**, 217–229
337. Hunt, J.C.R., Carlotti, P. (2001): Statistical structure at the wall of the high Reynolds number turbulent boundary layer. Flow, Turbulence and Combustion **66**, 453–475
338. Hylin, E.C., McDonough, J.M.(1996): Theoretical development of a stochastic model for small-scale turbulence in an additive decomposition of the Navier-Stokes equations. ME Report No CFD-02-96, Dept. of Mechanical Engineering, University of Kentucky
339. Iliescu, T., Fischer, P. (2003): Large eddy simulation of turbulent channel flows by the rational large eddy simulation model. Phys. Fluids **15**(10), 3036–3047
340. Iliescu, T., Fischer, P. (2004): Backscatter in the rational LES model. Computers and Fluids **33**, 783–790
341. Iliescu, T., John, V., Layton, W.J., Matthies, G., Tobiska, L. (2000): An assessment of models in large eddy simulation. Unpublished
342. Iliescu, T., Layton, W.J. (1998): Approximating the larger eddies in fluid motion III: the Boussinesq model for turbulent fluctuations. An. St. Univ. "Al. I. Cuza" **44**, 245–261
343. Iovenio, M., Tordella, D. (2003): Variable scale filtered Navier–Stokes equations: a new procedure to deal with the associated commutation error. Phys. Fluids **15**(7), 1926–1936
344. Jaberi, F., Colucci, P. (2003): Large eddy simulation of heat and mass transport in turbulent flows. Part 1: velocity field. Int. J. Heat Mass Transfer **46**, 1811–1825
345. Jaberi, F., Colucci, P. (2003): Large eddy simulation of heat and mass transport in turbulent flows. Part 2: scalar field. Int. J. Heat Mass Transfer **46**, 1827–1840
346. Jameson, A., Schmidt, W., Turkel, E. (1981): Numerical solutions of the Euler equations by finite volume methods using Runge-Kutta time stepping schemes. AIAA Paper 81-1259
347. Jansen, K. (1994): Unstructured-grid large-eddy simulation of flow over an airfoil. Annual Research Briefs – Center for Turbulence Research, 161–175
348. Jansen, K. (1995): Preliminary large-eddy simulations of flow around a NACA 4412 airfoil using unstructured meshes. Annual Research Briefs – Center for Turbulence Research, 61–73
349. Jansen, K. (1996): Large-eddy simulations of flow around a NACA 4412 airfoil using unstructured grids. Annual Research Briefs – Center for Turbulence Research, 225–233
350. Jiang, G.S., Shu, C.W. (1996): Efficient implementation of weighted ENO schemes. J. Comput. Phys. **126**, 202–228
351. Jimenez, J. (1999): On eddy-viscosity sub-grid models. (Direct and Large Eddy Simulation III, Voke, Sandham and Kleiser eds.) Kluwer, 75–86

352. Jimenez, C., Ducros, F., Cuenot, B., Bédat, B. (2001): Subgrid scale variance and dissipation of a scalar field in large eddy simulations. Phys. Fluids **13**(6), 1748–1754
353. Jimenez, J., Moin, P. (1991) The minimal flow unit in near-wall turbulence. J. Fluid Mech. **225**, 213–240
354. Jimenez, J., Pinelli, A. (1999) The autonomous cycle of near-wall turbulence. J. Fluid Mech. **389**, 335–359
355. Jimenez, J., Simens, M.P. (2001) Low-dimensional dynamics of a turbulent wall flow. J. Fluid Mech. **435**, 81–91
356. Jimenez, J., Vasco, C. (1998): Approximate lateral boundary conditions for turbulent simulations. Proceedings of the Summer Program – Center for Turbulence Research, 399–412
357. John, V., Layton, W.J. (2001): Analysis of numerical errors in large eddy simulation. Unpublished
358. Jordan, S.A. (1999): A large-eddy simulation methodology in generalized curvilinear coordinates. J. Comput. Phys. **148**, 322–340
359. Jordan, S.A. (2001): Dynamic subgrid-scale modeling for large-eddy simulations in complex topologies. J. Fluids Engng. **123**, 619–627
360. Jordan, S.A. (2002): Investigation of the cylinder separated shear-layer physics by large-eddy simulation. Int. J. Heat Fluid Flow **23**, 1–12
361. Jordan, S.A. (2003): The turbulent character and pressure loss produced by periodic symmetric ribs in a circular duct. Int. J. Heat Fluid Flow **24**, 795–806
362. Jordan, S.A., Ragab, S.A. (1998): A large eddy simulation of the near wake of a circular cylinder. Journal of Fluids Engineering **120**, 243–252
363. Juneja, A., Brasseur, J.G. (1999): Characterictics of subgrid-resolved-scale dynamics in anisotropic turbulence, with application to rough-wall boundary layers. Phys Fluids **11** (10), 3054–3068
364. Juneja, A., Lathrop, D.P., Sreenivasan, K.R., Stolovitzky, G. (1994): Synthetic turbulence. Phys. Rev. E. **49**(6), 5179–5194
365. Khanna, S., Brasseur, J.G. (1997): Analysis of Monin-Obukhov similarity from large-eddy simulation. J. Fluid Mech. **435**, 251–286
366. Kannepalli, C, Piomelli, U. (2000): Large-eddy simulation of a three-dimensional shear-driven turbulent boundary layer. J. Fluid Mech. **423**, 175–203
367. Kaltenbach, H.J. (1994): Large-eddy simulation of flow through a plane, asymmetric diffuser. Annual Research Briefs – Center for Turbulence Research, 175–185
368. Kaltenbach, H.J. (1997): Cell aspect ratio dependence of anisotropy measures for resolved and subgrid scale stresses. J. Comput. Phys. **136**, 399–410
369. Kaltenbach, H.J. (1998): Towards a near-wall model for LES of a separated diffuser flow. Annual Research Briefs – Center for Turbulence Research, 255–265
370. Kaltenbach, H.J. (2003): A priori testing of wall models for separated flows. Phys. Fluids **15**(10), 3048–3064
371. Kaltenbach, H.J., Choi, H. (1995): Large-eddy simulation of flow around an airfoil on structured mesh. Annual Research Briefs – Center for Turbulence Research, 51–61
372. Kaltenbach, H.J., Fatica, M., Mittal, R., Lund, T.S., Moin, P. (1999): Study of flow in a planar asymmetric diffuser using large-eddy simulation. J. Fluid Mech. **390**, 151–185
373. Kaltenbach, H., Gerz, T., Schumann, U. (1994): Large-eddy simulation of homogeneous turbulence and diffusion in stably stratified shear flows. J. Fluid Mech. **280**, 1–40

374. Kaneda, Y., Leslie, D.C. (1983): Tests of subgrid models in the near-wall region using represented velocity fields. J. Fluid Mech. **132**, 349–373
375. Kang, H.S., Chester, S., Meneveau, C. (2003): Decaying turbulence in an active-grid-generated flow and comparisons with large-eddy simulation. J. Fluid Mech. **480**, 129–160
376. Kang, H.S., Meneveau, C. (2001): Passive scalar anisotropy in a heated turbulent wake: new observations and implications for large-eddy simulations. J. Fluid Mech. **442**, 161–170
377. Kaniel, S. (1970): On the initial value problem for an incompressible fluid with nonlinear viscosity. J. Math. Mech. **19**(8), 681–706
378. Kanna, S., Brasseur, J.G. (1997): Analysis of Monin-Obukhov similarity from large-eddy simulation. J. Fluid Mech. **345**, 251–286
379. Karamanos, G.S., Karniadakis, G.E. (2000): A spectral vanishing viscosity method for large-eddy simulation. J. Comput. Phys. **163**, 22–50
380. Kato, C., Kaiho, M., Manabe, A. (2001): Industrial applications of LES in mechanical engineering. (DNS/LES Progress and Challenges, Liu, Sakell and Beutner eds.), Greyden Press, 47–58
381. Kawamura, T., Kuwahara, K. (1984): Computation of high Reynolds number flow around a circular cylinder with surface roughness. AIAA Paper 84-0340
382. Kemenov, K., Menon, S. (2003): Two level simulation of high-Reynolds number non-homogeneous turbulent flows. AIAA Paper 2003-0084
383. Kerr, R.M., Domaradzki, J.A., Barbier, G. (1996): Small-scale properties of nonlinear interactions and subgrid-scale energy transfer in isotropic turbulence. Phys. Fluids **8**(1), 197–208
384. Kerstein, A. (1991): Linear-eddy modelling of turbulent transport. Part 6. Microstructure of diffusive scalar mixing fields. J. Fluid Mech. **231**, 361–394
385. Kerstein, A. (1992): Linear-eddy modelling of turbulent transport. Part 7. Finite-rate chemistry and multi-stream mixing. J. Fluid Mech. **240**, 289–313
386. Kerstein, A. (1999): One-dimensional turbulence Part 2. Staircases in double-diffusive convection. Dynamics of Atmospheres and Oceans **30**, 25–46
387. Kerstein, A.R. (1999): One-dimensional turbulence: model formulation and application to homogeneous turbulence, shear flows, and buoyant stratified flows. J. Fluid Mech. **392**, 277–334
388. Kerstein, A. (2002): One dimensional turbulence: a new approach to high-fidelity subgrid closure of turbulent flow simulations. Comput. Phys. Comm. **148**, 1–16
389. Kerstein, A.R., Ashurst, W.T., Wunsch, S., Nilsen, V. (2001): One-dimensional turbulence: vector formulation and application to free shear flows. J. Fluid Mech. **447**, 85–3109
390. Kerstein, A.R., Dreeben, T.D. (2000): Prediction of turbulent free shear flow statistics using a simple stochastic model. Phys. Fluids **12**(2), 418–424
391. Kim, W.W., Menon, S. (1999): An unsteady incompressible Navier–Stokes solver for large-eddy simulation of turbulent flows. Int. J. Numer. Meth. Fluids **31**, 983–1017
392. Kim, W.W., Menon, S., Mongia, H.C. (1999): Large eddy simulation of a gas turbine combustor flow. Combust. Sci. Tech. **143**, 25–62
393. Kim, D.H., Yang, K.S., Send, M. (2004): Large eddy simulation of turbulent flow past a square cylinder confined in a channel. Computers and Fluids **33**, 81–96
394. Kimmel, S.J., Domaradzki, A.J. (2000): Large-eddy simulations of Rayleigh–Bénard convection using subgrid scale estimation model. Phys. Fluids **12**(1), 169–183

395. Kirby, R., Karniadakis, G. (2002): Coarse resolution turbulence simulations with spectral vanishing viscosity–large eddy simulations. J. Fluid Engng. **124**, 886–891
396. Klein, M., Sadiki, A., Janicka, J. (2003): A digital filter based generation of inflow data for spatially developing direct numerical or large eddy simulation. J. Comput. Phys. **186**, 652–665
397. Knaepen, B., Debliquy, O., Carati, D. (2002): Subgrid-scale energy and pseudo pressure in large-eddy simulation. Phys. Fluids **14**(12), 4235–4241
398. Kobayashi, H., Shimomura, Y. (2001): The performance of dynamic subgrid-scale models in the large-eddy simulation of rotating homogeneous turbulence. Phys. Fluids **13**(8), 2350–2360
399. Kogaki, T., Kobayashi, T., Taniguchi, N. (1997): Large eddy simulation of flow around a rectangular cylinder. Fluid Dyn. res. **20**, 11–24
400. Kollman, W., McCallen, R.C., Leone Jr., J.M. (2002): An examination of LES filtering within the finite element method. Commun. Numer. Meth. Engng. **18**, 513–528
401. Kosovic, B. (1997): Subgrid-scale modelling for the large-eddy simulation of high-Reynolds number boundary layers. J. Fluid Mech. **336**, 151–182
402. Kosovic, B., Curry, J. (2000): A large eddy simulation study of a quasi-steady, stably stratified atmospheric boundary layer. J. Atmos. Sci. **57**, 1052–1068
403. Kraichnan, R.H. (1967): Inertial ranges in two-dimensional turbulence. J. Fluid Mech. **10**(7), 1417–1423
404. Kraichnan, R.H. (1971): Inertial-range transfer in two- and three-dimensional turbulence. J. Fluid Mech. **47**(3), 525–535
405. Kraichnan, R.H. (1976): Eddy viscosity in two and three dimensions. J. Atmos. Sci. **33**, 1521–1536
406. Krajnovic, S., Davidson, L. (2002): A mixed one-equation subgrid model for large-eddy simulation. Int. J. Heat Fluid Flow **23**, 413–425
407. Krajnovic, S., Davidson, L. (2002): Large-eddy simulation of the flow around a bluff body. AIAA Journal **40**(5), 927–936
408. Krajnovic, S., Davidson, L. (2003): Numerical study of the flow around a bus-shaped body. J. Fluids Engng. **125**, 500–509
409. Kravchenko, A.G., Moin, P. (1997): On the effect of numerical errors in large-eddy simulations of turbulent flows. J. Comput. Phys. **131**, 310–322
410. Kravchenko, A.G., Moin, P. (2000): Numerical studies of flow over a circular cylinder at $Re_D = 3900$. Phys. Fluids **12**(2), 403–417
411. Kravchenko, A.G., Moin, P., Moser, R. (1996): Zonal embedded grids for numerical simulations of wall-bounded turbulent flows. J. Comput. Phys. **127**, 412–423
412. Krettenauer, K., Schumann, U. (1992): Numerical simulation of turbulent convection over a wavy terrain. J. Fluid Mech. **237**, 261–299
413. Krueger, S. (1993): Linear eddy modeling of entrainment and mixing in stratus clouds. J. Atmos. Sci. **50**(18), 3078–3090
414. Krueger, S., Su, C.W., McMurty, P. (1997): Modeling entrainment and finescale mixing in cumulus clouds. J. Atmos. Sci. **54**(23), 2697–2712
415. Kuerten, J.M.G., Geurts, B.J., Vreman, A.W., Germano, M. (1999): Dynamic inverse modeling and its testing in large-eddy simulations of the mixing layer. Phys. Fluids **11** (12), 3778–3785
416. Labourasse, E., Sagaut, P. (2002) Reconstruction of turbulent fluctuations using a hybrid RANS/LES approach. J. Comput. Phys. **182**, 301–336
417. Ladyzenskaja, O.A. (1970): Modification of the Navier–Stokes equations for large velocity gradients. (Seminar in Mathematics V.A. Stheklov Mathematical Institute, Vol. 7, Boundary value problems of mathematical physics and

related problems of function theory, Part II, New York, London 1970, Consultant Bureau)
418. Ladyzenskaja, O.A. (1970): New equations for description of motion of viscous incompressible fluids and solvability in the large of boundary value problems for them. (Seminar in Mathematics V.A. Stheklov Mathematical Institute, Vol. 102, Boundary value problems of mathematical physics, Part V, Providence, Rhode Island, AMS)
419. Langford, J., Moser, R. (2000): Optimal LES formulations for isotropic turbulence. J. Fluid Mech. **398**, 321–346
420. Langford, J.A., Moser, R.D. (2001): Breakdown of continuity in large-eddy simulation. Phys. Fluids **13**(5), 1524–1527
421. Lamballais, E. (1996): Simulations numériques de la turbulence dans un canal plan tournant (in french). Thèse de Doctorat de l'INPG, Grenoble, France
422. Lamballais, E., Métais, O., Lesieur, M. (1998): Spectral-dynamic model for large-eddy simulations of turbulent rotating channel flow. Theoret. Comput. Fluid Dynamics **12**, 149–177
423. Landau, L., Lifchitz, E. (1967): Fluid Mechanics. MIR, Moscou
424. Launder, B.E., Spalding, D.B. (1972): Mathematical models of turbulence. Academic Press, Londres
425. Lardat, R. (1997): Simulation numériques d'écoulements externes instationnaires décollés autour d'une aile avec des modèles de sous-maille (in french). Notes et documents du LIMSI **97-12**
426. Laval, J.P., Dubrulle, B., McWilliams, J.C. (2003): Langevin models of turbulence: renormalization group, distant interaction algorithms or rapide distortion theory ? Phys. Fluids **15**(5), 1327–1339
427. Layton, W.J. (1996): A nonlinear, subgridscale model for incompressible viscous flow problems. SIAM J. Sci. Comput. **17** (2), 347–357
428. Layton, W.J. (2000): Approximating the larger eddies in fluid motion V: kinetic energy balance of scale similarity models. Mathematical and Computer Modelling **31**, 1–7
429. Layton, W.J. (2000): Analysis of a scale-similarity model of the motion of large eddies in turbulent flows. Unpublished
430. Layton, W., Lewandowski, R. (2002): Analysis of an eddy viscosity model for large eddy simulation of turbulent flows. J. Math. Fluid Mech. **4**, 374–399
431. Le Ribault, C., Sarkar, S., Stanley, S.A. (1999): Large eddy simulation of a plane jet. Phys. Fluids **11**(10), 3069–3083
432. Lee, S., Lele, S.K., Moin, P. (1992): Simulation of spatially evolving turbulence and the application of Taylor's hypothesis in compressible flow. Phys. Fluids A **4**(7), 1521–1530
433. Lee, J.S., Xu, X., Pletcher, R. (2004): Large eddy simulation of heated vertical annular pipe flow in fully developed turbulent mixed convection. Int. J. Heat Fluid Flow **47**, 437–446
434. Leith, C.E. (1990): Stochastic backscatter in a subgrid-scale model: Plane shear mixing layer. Phys. Fluids A **2**(3), 297–299
435. Lele, S.K. (1994): Compressibility effects on turbulence. Ann. Rev. Fluid. Mech. **26**, 211–254
436. Leonard, A. (1974): Energy cascade in large-eddy simulations of turbulent fluid flows. Adv. in Geophys. A **18**, 237–248
437. Leonard, B.P. (1979): A stable and accurate convective modelling procedure based on quadratic upstream interpolation. Comp. Meth. Appl. Mech. Eng. **19**, 59–98
438. Lesieur, M. (1983): Introduction à la turbulence bidimensionnelle (in french). J. Méc. Théor. Appl., numéro spécial, 5–20

439. Lesieur, M. (1997): Turbulence in fluids, 3rd edition. Kluwer
440. Lesieur, M., Métais, O. (1996): New trends in large-eddy simulations of turbulence. Ann. Rev. Fluid Mech. **28**, 45–82
441. Lesieur, M., Rogallo, R.S. (1989): Large-eddy simulation of passive scalar diffusion in isotropic turbulence. Phys. Fluids A **1**(4), 718–722
442. Lesieur, M., Schertzer, D. (1978): Amortissement autosimilaire d'une turbulence à grand nombre de Reynolds (in french). Journal de Mécanique **17**(4), 609–646
443. Leslie, D.C., Quarini, G.L. (1979): The application of turbulence theory to the formulation of subgrid modelling procedures. J. Fluid Mech. **91**(1), 65–91
444. Lewellen, D., Lewellen, W., Yoh, S. (1996): Influence of Bowen ratio on boundary-layer cloud structure. J. Atmos. Sci **53**(1), 175–187
445. Li, N., Balaras, E., Piomelli, U. (2000): Inflow conditions for large-eddy simulations of mixing layers. Phys. Fluids **12**(4), 935–938
446. Li, C.W., Wang, J.H. (2000): Large-eddy simulation of free surface shallow-water flow. Int. J. Numer. Meth. Fluids **34**, 31–46
447. Lilly, D.K. (1967): The representation of small-scale turbulence in numerical simulation experiments. (Proceedings of the IBM Scientific Computing Symposium on Environmental Sciences, Yorktown Heights, USA)
448. Lilly, D.K. (1992): A proposed modification of the Germano subgrid-scale closure method. Phys. Fluids A **4**(3), 633–635
449. Lin, C.C. (1999): Near-grid-scale energy transfer and coherent structures in the convective planetary boundary layer. Phys. Fluids **11** (11), 3482–3494
450. Lin, C.L., McWilliams, J.C., Moeng, C.H., Sullivan, P. (1996): Coherent structures and dynamics in a neutrally stratified planetary boundary layer flow. Phys. Fluids **8**(10), 2626–2639
451. Liu, J.T.C. (1988): Contributions to the understanding of large-scale coherent structures in a developing free turbulent shear flows. Advances in Applied Mechanics **20**, 183–307
452. Liu, C., Liu, Z. (1994): Fourth order finite difference and multigrid methods for modeling instabilities in flat plate boundary layers - 2D and 3D approaches. Computers Fluids **23**(7), 955–982
453. Liu, C., Liu, Z. (1995): Multigrid mapping in box relaxation for simulation of the whole process of flow transition in 3D boundary layers. J. Comput. Phys. **119**, 325–341
454. Liu, S., Katz, J., Meneveau, C. (1999): Evolution and modelling of subgrid scales during rapid straining of turbulence. J. Fluid Mech. **387**, 281–320
455. Liu, S., Meneveau, C., Katz, J. (1994): On the properties of similarity subgrid-scale models as deduced from measurements in a turbulent jet. J. Fluid Mech. **275**, 83–119
456. Liu, J.T.C., Merkine, L. (1976): On the interactions betzeen large-scale structure and fine-grained turbulence in a free shear flow. I. The development of temporal interactions in the mean. Proc. R. Soc. Lond. A **352**, 213–247
457. Lo, S.H., Voke, P.R., Rockliff, N.J. (2000): Eddy structures in a simulated low Reynolds number turbulent boundary layer. Flow, Turbulence and Combustion **64**, 1–28
458. Loh, K.C., Domaradzki, J.A. (1999): The subgrid-scale estimation model on nonuniform grids. Phys. Fluids **11** (12), 3786–3792
459. Love, M.D. (1980): Subgrid modelling studies with Burgers equation. J. Fluid Mech. **100**(1), 87–110
460. Lund, T.S. (1994): Large-eddy simulation of a boundary layer with concave streamwise curvature. Annual Research Briefs – Center for Turbulence Research, 185–197

461. Lund, T.S. (1997): On the use of discrete filters for large eddy simulation. Annual Research Briefs – Center for Turbulence Research, 83–95
462. Lund, T.S., Novikov, E.A. (1992): Parametrization of subgrid-scale stress by the velocity gradient tensor. Annual Research Briefs – Center for Turbulence Research, 27–43
463. Lund, T.S., Wu, X., Squires, K.D. (1998): On the generation of turbulent inflow conditions for boundary-layer simulations. J. Comput. Phys. **140**, 233–258
464. McComb, W.D. (1990): The physics of fluid turbulence. Clarendon Press, Oxford
465. McComb, W.D., Hunter, A., Johnston, C. (2001): Conditional mode-elimination and the subgrid-modeling problem for isotropic turbulence. Phys. Fluids **13**(7), 2030–2044
466. McDonough, J.M., Bywater, R.J. (1985): Effects of local large-scale parameters on the small-scale chaotic solutions to Burgers' equation. AIAA Paper 85-1653
467. McDonough, J.M., Bywater, R.J. (1986): Large-scale effects on local small-scale chaotic solutions to Burgers' equation. AIAA J. **24** (12), 1924–1930
468. McDonough, J.M., Bywater, R.J., Buell, J.C. (1984): An investigation of strange attractor theory and small-scale turbulence. AIAA Paper 84-1674
469. McDonough, J.M., Mukerji, S., Chung, S. (1998): A data-fitting procedure for chaotic time series. Appl. Math. and Comput. **95**, 219–243
470. McDonough, J.M., Saito, K. (1994): Local, small-scale interaction of turbulence with chemical reactions in H2-O2 combustion. Fire Science and Technology **14** (1), 1–18
471. McDonough, J.M., Wang, D. (1995): Additive turbulent decomposition: a highly parallelizable turbulence simulation technique. (New algorithms ans applications, Satofuka, Periaux, Ecer eds.) Elsevier, 129–136
472. Mcllwain, S., Pollard, A. (2002): Large eddy simulation of the effects of mild swirl on the near field of a round free jet. Phys. Fluids **14**(2), 653–661
473. McMurty, P., Menon, S., Kerstein, A. (1993): Linear eddy modeling of turbulent combustion. Energy and Fuels **7**, 817–826
474. McRae, G.J., Goodin, W.R., Seinfeld, J. (1982): Numerical solution of the atmospheric diffusion equation for chemically reacting flows. J. Comput. Phys. **45**, 1–42
475. Maday, Y., Ould Kaber, M., Tadmor, E. (1993): Legendre pseudospectral viscosity method for nonlinear conservation laws. SIAM J. Numer. Anal. **30**(2), 321–342
476. Magnagato, F., Gabi, M. (2002): A new adaptive turbulence model for unstedy flow fields in rotating machinery. Int. J. Rotating Machinery **8**(3), 175–183
477. Magnient, J.C. (2001) Simulation des grandes échelles d'écoulements de fluide quasi-incompressibles (in french). Thèse de Doctorat, Université Paris 11, Orsay, France
478. Magnient, J.C., Sagaut, P., Deville, M. (1999): Analysis of mesh-independent subfilter-scale models for turbulent flows. (Direct and Large Eddy Simulation III, Voke, Sandham and Kleiser eds.) Kluwer, 263–274
479. Magnient, J.C., Sagaut, P., Deville, M. (2001): A study of built-in filter for some eddy viscosity models in large-eddy simulation. Phys. Fluids **13**(5), 1440–1449
480. Majander, P., Siikonen, T. (2002): Evaluation of Smagorinsky-based subgrid scale models in a finite-volume computation. Int. J. Numer. Meth. Fluids **40**, 735–774

481. Maltrud, M.E., Vallis, G.K. (1993): Energy and enstrophy transfer in numerical simulations of two-dimensional turbulence. Phys. Fluids A **5**, 1760–1775
482. Manhart, M. (2004): A zonal grid algorithm for DNS of turbulent boundary layers. Computers and Fluids **33**, 435–461
483. Manhart, M., Friedrich, R. (1999): Towards DNS of separated turbulent boundary layers. (Direct and Large Eddy Simulation III, Voke, Sandham and Kleiser eds.) Kluwer, 429–440
484. Manoha, E., Troff, B., Sagaut, P. (2000): Trailing edge noise prediction using large-eddy simulation and acoustic analogy. AIAA Journal **38**(8), 1340–1350
485. Mansfield, J.R., Knio, O.M., Meneveau, C. (1998): A dynamic LES scheme for the vorticity transport equation: formulation and a priori tests. J. Comput. Phys. **145**, 693–730
486. Mansfield, J.R., Knio, O.M., Meneveau, C. (1999): Dynamic LES of colliding vortex rings using a 3D vortex method. J. Comput. Phys. **152**, 305–345
487. Mansour, N.N., Moin, P., Reynolds, W.C., Ferziger, J.H. (1977): Improved methods for large-eddy simulations of turbulence. (Symposium on Turbulent Shear Flow, Penn State, USA)
488. Margolin, L., Smolarkiewicz, P., Sorbjan, Z. (1999): Large-eddy simulations of convective boundary layers using nonoscillatory schemes. Physica D **133**, 390–397
489. Margolin, L., Smolarkiewicz, P., Wyszogrodzki, A. (2002): Implicit turbulence modeling for high Reynolds number flows. J. Fluid Engng. **124**, 862–867
490. Marsden, A.L., Vasilyev, O.V., Moin, P. (2000): Construction of commutative filters for LES on unstructured meshes. Annual Research Briefs – Center for Turbulence Research, 179–192
491. Marshall, J.S., Beninati, M.L. (2003): Analysis of subgrid scale torque for large-eddy simulation of turbulence. AIAA Journal **41**(10), 1875–1881
492. Marusic, I., Kunkel, G., Porté-Agel, F. (2001): Experimental study of wall boundary conditions fro large-eddy simulation. J. Fluid Mech. **446**, 309–320
493. Maruyama, T., Rodi, W., Maruyama, Y., Hiroaka, H. (1999): Large eddy simulation of the turbulent boundary layer behind roughness elements using an artificially generated inflow. J. Wind Eng. Ind. Aerodyn. **83**, 381–392
494. Mary, I., Sagaut, P. (2003): Large-eddy simulation of the flow around an airfoil near stall. AIAA Journal **40**(6), 1139–1145
495. Mason, P.J. (1994): Large-eddy simulation: A critical review of the technique. Q. J. R. Meteorol. Soc. **120**, 1–26
496. Mason, P.J., Brown, A.R. (1994): The sensitivity of large-eddy simulations of turbulent shear flow to subgrid models. Boundary Layer Meteorol. **70**, 133–150
497. Mason, P.J., Callen, N.S. (1986): On the magnitude of the subgrid-scale eddy coefficient in large-eddy simulations of turbulent channel flow. J. Fluid Mech. **162**, 439–462
498. Mason, P.J., Thomson, D.J. (1992): Stochastic backscatter in large-eddy simulations of boundary layers. J. Fluid Mech. **242**, 51–78
499. Mathew, J., Lechner, R., Foysi, H., Sesterhenn, J., Friedrich, R. (2003): An explicit filtering method for large-eddy simulation of compressible flows. Phys. Fluids **15**(8), 2279–2288
500. Maurer, J., Fey, M. (1999): A scale-residual model for large-eddy simulation. (Direct and Large Eddy Simulation III, Voke, Sandham and Kleiser eds.) Kluwer, 237–248
501. Meinke, M., Schröder, W., Krause, E., Rister, Th. (2002): A comparison of second- and sixth-order methods for large-eddy simulations. Computers and Fluids, in press

502. Mellen, C., Fröhlich, J., Rodi, W. (2003): Lessons from LESFOIL Project on large-eddy simulation of flow around an airfoil. AIAA Journal **41**(4), 573–581
503. Meneveau, C. (1994): Statistics of turbulence subgrid-scale stresses: Necessary conditions and experimental tests. Phys. Fluids **6**(2), 815–833
504. Meneveau, C., Katz, J. (1999): Conditional subgrid force and dissipation in locally isotropic and rapidly strained turbulence. Phys Fluids **11** (8), 2317–2329
505. Meneveau, C., Katz, J. (1999): Dynamic testing of subgrid models in large-eddy simulation based on the Germano identity. Phys. Fluids **11** (2), 245–247
506. Meneveau, C., Katz, J. (2000): Scale-invariance and turbulence models for large-eddy simulation. Ann. Rev. Fluid Mech. **32**, 1–32
507. Meneveau, C., Lund, T.S. (1997): The dynamic Smagorinsky model and scale-dependent coefficients in the viscous range of turbulence. Phys. Fluids **9**(12), 3932–3934
508. Meneveau, C., Lund, T.S., Cabot, W.H. (1996): A Lagrangian dynamic subgrid-scale model of turbulence. J. Fluid Mech. **319**, 353–385
509. Meneveau, C., Lund, T.S., Moin, P. (1992): Search for subgrid scale parametrization by projection pursuit regression. Proceedings of the Summer Program – Center for Turbulence Research, 61–81
510. Meneveau, C., O'Neil, J. (1994): Scaling laws of the dissipation rate of turbulent subgrid-scale kinetic energy. Phys. Rev. E. **49**(4), 2866–2874
511. Menon, S., Yeung, P.K., Kim, W.W. (1996): Effect of subgrid models on the computed interscale energy transfer in isotropic turbulence. Computer and Fluids **25**(2), 165–180
512. Mestayer, P. (1982): Local isotropy and anisotropy in high-Reynolds-number turbulent boundary layer. J. Fluid Mech. **125**, 475–503
513. Métais, O., Herring, J. (1989): Numerical simulations of freely evolving turbulence in stably stratified fluids. J. Fluid Mech. **202**, 117–148
514. Métais, O., Lesieur, M. (1992): Spectral large-eddy simulation of isotropic and stably stratified turbulence. J. Fluid Mech. **256**, 157–194
515. Meyers, J., Geurts, B., Baelmans, M. (2003): Database analysis of errors in large-eddy simulation. Phys. Fluids **15**(9), 2740–2755
516. Michelassi, V., Wissink, J.G., Fröhlich, J., Rodi, W. (2003): Large eddy simulation of flow around low-pressure turbine blade with incoming wakes. AIAA Journal **41**(11), 2143–2156
517. Misra, A., Pullin, D.I. (1997): A vortex-based subgrid stress model for large-eddy simulation. Phys. Fluids **9**(8), 2443–2454
518. Mitran, S. (2001): A comparison of adaptive mesh refinement approaches for large-eddy simulation. (DNS/LES Progress and Challenges, Liu, Sakell and Beutner eds.), Greyden Press, 397–408
519. Mittal, R. (1995): Large-eddy simulation of flow past a circular cylinder. Annual Research Briefs – Center for Turbulence Research, 107–117
520. Mittal, R. (1996): Progress on LES of flow past a circular cylinder. Annual Research Briefs – Center for Turbulence Research, 233–243
521. Mittal, R., Moin, P. (1997): Suitability of upwind-biased finite-difference schemes for large-eddy simulation of turbulent flows. AIAA Journal **35**(8), 1415–1417
522. Mittal, R., Simmons, S.P., Najjar, F. (2003): Numerical study of pulsatile flow in a constricted channel. J. Fluid Mech. **485**, 337–378
523. Miyake, Y., Kajishima, T. (1986): Numerical simulation of the effects of Coriolis force on the stucture of turbulence. Global effects. Bull. JSME **29**, 3341–3346

524. Miyake, Y., Kajishima, T. (1986): Numerical simulation of the effects of Coriolis force on the stucture of turbulence. Structure of turbulence. Bull. JSME **29**, 3347–3351
525. Moeng, C.H. (1984): A large-eddy simulation model for the study of planetary boundary-layer turbulence. J. Atmos. Sci. **41**(13), 2052–2062
526. Moeng, C.H. (1987): Large-eddy simulation of a stratus-topped boundary layer. Part I: structure and budgets. J. Atmos. Sci. **43**(23), 2886–2900
527. Moeng, C.H. (1987): Large-eddy simulation of a stratus-topped boundary layer. Part II: implications for a mixed-layer modeling. J. Atmos. Sci. **44**(12), 1605–1614
528. Moeng, C.H., Rotunno, R. (1990): Vertical-velocity skewness in the buoyancy-driven boundary layer. J. Atmos. Sci. **47**(9), 1149–1162
529. Moeng, C.H., Schumann, U. (1991): Composite structure of plumes in stratus-topped boundary layers. J. Atmos. Sci. **48**(20), 2280–2291
530. Moeng, C.H., Shen, S., Randall, D. (1992): Physical processes within the nocturnal stratus-topped boundary-layer. J. Atmos. Sci. **49**(24), 2384–2401
531. Moeng, C.H., Wyngaard, J.C. (1984): Statistics of conservative scalars in the convective boundary layer. J. Atmos. Sci. **41**(21), 3161–3169
532. Moeng, C.H., Wyngaard, J.C. (1986): An analysis of closures for pressure-scalar covariances in the convective boundary layer. J. Atmos. Sci. **43**(21), 2499–2513
533. Moeng, C.H., Wyngaard, J.C. (1988): Spectral analysis of large-eddy simulations of the convective boundary layer. J. Atmos. Sci. **45**(23), 3573–3587
534. Moeng, C.H., Wyngaard, J.C. (1989): Evaluation of turbulent transport and dissipation closures in second-order modeling. J. Atmos. Sci. **46**(14), 2311–2330
535. Mohammadi, B., Pironneau, O. (2000): Applied shape optimization for fluids. Oxford University Press, UK
536. Moin, P., Jimenez, J. (1993): Large eddy simulation of complex flows. (24th AIAA Fluid Dynamics Conference, Orlando, USA)
537. Moin, P., Kim, J. (1982): Numerical investigation of turbulent channel flow. J. Fluid Mech. **118**, 341–377
538. Moin, P., Squires, K., Cabot, W., Lee, S. (1991): A dynamic subgrid-scale model for compressible turbulence and scalar transport. Phys. Fluids A **3**(11), 2746–2757
539. Morales, R., Balparda, A., Silveira-Neto, A. (1999): Large-eddy simulation of the combined convection around a heated rotating cylinder. Int. J. Heat Mass Transfer **42**, 941–949
540. Morikawa, H., Maruyama, T. (1999): Theory of conditional random field and its application to wind engineering. Kyoto University Press
541. Morinishi, Y., Kobayashi, T. (1990): in (Engineering turbulence modelling and experiments, Rodi and Ganic eds.) Elsevier, New York, 279
542. Morinishi, Y., Vasilyev, O. (1998): Subgrid scale modeling taking the numerical error into consideration. Annual Research Briefs – Center for Turbulence Research, 237–253
543. Morinishi, Y., Vasilyev, O.V. (2001): A recommended modification to the dynamic two-parameter mixed subgrid scale model for large eddy simulation of turbulent flow. Phys. Fluids, submitted
544. Morinishi, Y., Vasilyev, O. (2002): Vector level identity for dynamic subgrid scale modeling in large eddy simulation. Phys. Fluids **14**(10), 3616–3623
545. Morris, P.J., Long, L.N., Bangalore, A., Wang, Q. (1997): A parallel three-dimensional computational aeroacoustics method using nonlinear disturbance equations. J. Comput. Phys., **133**, 56–74

546. Morris, P.J., Long, L.N., Wang, Q., Lockard, D.P. (1997): Numerical prediction of high-speed jet noise. AIAA Paper 97-1598
547. Morris, P.J., Long, L.N., Wang, Q., Pilon, A.R. (1998): High-speed jet noise simulations. AIAA Paper 98-2290
548. Moser, R.D., Langford, J.A., Volker, S (2001): A radical approach to large eddy simulation. AIAA Paper 2001-2835
549. Moser, R.D., Volker, S., Venugopal, P. (2001):Characterization of optimal LES in turbulent channel flow. (DNS/LES Progress and Challenges, Liu, Sakell and Beutner eds.), Greyden Press, 255–262
550. Mossi, M. (1999): Simulation of benchmark and industrial unsteady compressible turbulent fluid flows. PhD Thesis No 1958, EPFL, Lausanne, Switzerland
551. Muchinsky, A. (1996): A similarity theory of locally homogeneous anisotropic turbulence generated by a Smagorinsky-type LES. J. Fluid Mech. **325**, 239–260
552. Mukerji, S., McDonough, J.M., Menguc, M.P., Manickavasagam, S., Chung, S. (1998): Chaotic map models of soot fluctuations in turbulent diffusion flames. Int. J. Heat Mass Transfer **41**, 4095–4112
553. Mullen, J.S., Fischer, P.F. (1999): Filtering technique for complex geometries fluid flows. Commun. Numer. Meth. Engng. **15**, 9–18
554. Murakami, S. (1993): Comparison of various turbulence models applied to a bluff body. J. Wind Eng. Ind. Aerodyn. **46** & **47**, 21–36
555. Murakami, S., Iizuka, S., Ooka, R. (1999): CFD analysis of turbulent flow past square cylinder using dynamic LES. Journal of Fluids and Structures **13**, 1097–1112
556. Murakami, S., Mochida, A., Hibi, K. (1987): Three-dimensional numerical simulation of air flow around a cubic model by means of large-eddy simulation. J. Wind Eng. Ind. Aerodyn. **25**, 291–305
557. Murata, A., Mochizuki, S. (1999): Effect of cross-sectional aspect ratio on turbulent heat transfer in an orthogonally rotating rectangular smooth duct. Int. J. Heat Mass Transfer **42**, 3803–3814
558. Murata, A., Mochizuki, S. (2000): Large eddy simulation with a dynamic subgrid-scale model of turbulent heat transfer in an orthogonally rotating rectangular duct with transverse rib turbulators. Int. J. Heat Mass Transfer **43**, 1243–1259
559. Murata, A., Mochizuki, S. (2001): Effect of centrifugal buoyancy on turbulent heat transfer in an orthogonally rotating square duct with transverse or angled rib turbulators. Int. J. Heat Mass Transfer **44**, 2739–2750
560. Murata, A., Mochizuki, S. (2003): Effect of cross-sectional aspect ratio on turbulent heat transfer in an orthogonally rotating rectangular duct with angled rib turbulators. Int. J. Heat Mass Transfer **46**, 3119–3133
561. Murata, A., Mochizuki, S. (2004): Large eddy simulation of turbulent heat transfer in a rotating two-pass smooth square channel with sharp 180° turns. Int. J. Heat Mass Transfer **47**, 683–698
562. Murray, J.A., Piomelli, U., Wallace, J.M. (1996): Spatial and temporal filtering of experimental data for a priori studies of subgrid-scale stresses. Phys. Fluids **8**(7), 1978–1980
563. Najjar, F.M., Tafti, D.K. (1996): Study of discrete test filters and finite difference approximations for the dynamic subgrid-scale stress model. Phys. Fluids **8**(4), 1076–1088
564. Nakayama, A., Noda, H. (2000): Large-eddy simulation of flow around a bluff body fitted with splitter plate. J. Wind Eng. Ind. Aerodyn. **85**, 85–96

565. Niceno, B., Dronkers, A., Hanjalic, K. (2002): Turbulent heat transfer from a multi-layered wall-mounted cube matrix: large eddy simulation. Int. J. Heat Fluid Flow **23**, 173–185
566. Nicoud, F., Bagget, J.S., Moin, P., Cabot, W. (2001): Large-eddy simulation wall-modeling based on sub-optimal control theory and linear stochastic estimation. Phys. Fluids **13**(10), 2968–2984
567. Nicoud, F., Ducros, F. (1999): Subgrid stress modeling based on the square of the velocity gradient tensor. Flow, Turbulence and Combustion **62**(3), 183–200
568. Nicoud, F., Winckelmans, G., Carati, D., Bagget, J., Cabot, W. (1998): Boundary conditions for LES away from the wall. Proceedings of the Summer Program – Center for Turbulence Research, 413–422
569. Nikitin, N.V., Nicoud, F., Washito, B., Squires, K.D., Spalart, P.R. (2000): An approach to wall modeling in large-eddy simulations. Phys. Fluids **12**(7), 1629–1632
570. O'Neil, J., Meneveau, C. (1997): Subgrid-scale stresses and their modelling in a turbulent plane wake. J. Fluid Mech. **349**, 253–293
571. Oberai, A., Roknaldin, F., Hughes, T.J.R. (2002): Computation of trailing-edge noise due to turbulent flow over an airfoil. AIAA Journal **40**(11), 2206–2216
572. Oberlack, M. (1997): Invariant modeling in large-eddy simulation of turbulence. Annual Research Briefs – Center for Turbulence Research, 3–22
573. Olsson, M., Fuchs, L. (1996): Large eddy simulation of the proximal region of a spatially developing circular jet. Phys. Fluids **8**(8), 2125–2137
574. Olsson, M., Fuchs, L. (1998): Large eddy simulations of a forced semiconfined circular impinging jet. Phys. Fluids **10**(2), 476–486
575. Pallares, J., Cuesta, I., Grau, F. (2002): Laminar and turbulent Rayleigh–Bénard convection in a perfectly conducting cubical cavity. Int. J. Heat Fluid Flow **23**, 346–358
576. Pallares, J., Davidson, L. (2000): Large-eddy simulation of turbulent flow in a rotating square duct. Phys. Fluids **12**(11), 2878–2894
577. Pallares, J., Davidson, L. (2002): Large-eddy simulations of turbulent heat transfer in stationary and rotating square ducts. Phys. Fluids **14**(8), 2804–2816
578. Pantano, C., Sarkar, S. (2001): A subgrid model for nonlinear functions of a scalar. Phys. Fluids **13**(12), 3803–3819
579. Pascal, F., Basdevant, C. (1992):. Nonlinear Galerkin method and subgrid-scale model for two-dimensional turbulent flows. Theoret. Comp. Fluid Dynamics **3**, 267–284
580. Pascarelli, A., Piomelli, U., Candler, G.V. (2000): Multi-block large-eddy simulations of turbulent boundary layers. J. Comput. Phys. **157**(7), 256–279
581. Pasquetti, R.: High-order LES modeling of turbulent incompressible flow, submitted
582. Pasquetti, R., Xu, C. (2002): High-order algorithms for large-eddy simulation of incompressible flows. J. Sci. Comput. **17**(1-4), 273–284
583. Patel, N., Stone, C., Menon, S. (2003): Large eddy simulation of turbulent flow over an axisymmetric hill. AIAA Paper 2003-0967
584. Peltier, L.J., Zajaczkowski, F.J. (2001): Maintenace of the near-wall cycle of turbulence for hybrid RANS/LES of fully developed channel flow. (DNS/LES Progress and Challenges, Liu, Sakell and Beutner eds.), Greyden Press, 829–834
585. Péneau, F., Boisson, H., Djilali, N. (2000): Large eddy simulation of the influence of high free-stream turbulence on a spatially evolving boundary layer. Int. J. Heat Fluid Flow **21**, 640–647

586. Peng, S.H., Davidson, L. (2001): Large eddy simulation for turbulent buoyant flow in a confined cavity. Int. J. Heat Fluid Flow **22**, 323–331
587. Peng, S.H., Davidson, L. (2002): On a subgrid-scale heat flux model for large eddy simulation of turbulent thermal flow. Int. J. Heat Mass Transfer **45**, 1393–1405
588. Perrier, P., Pironneau, O. (1981): Subgrid turbulence modelling by homogeneization. Mathematical Modelling **2**, 295–317
589. Persson, L., Fureby, C., Svanstedt, N. (2002): On homogenization-based methods for large-eddy simulation. J. Fluid Engng. **124**, 892–903
590. Pierce, C., Moin, P. (1998): A dynamic model for subgrid-scale variance and dissipation rate of a conserved scalar. Phys. Fluids **10**(12), 3041–3044
591. Piomelli, U. (1993): High Reynolds number calculations using the dynamic subgrid-scale stress model. Phys. Fluids A **5**(6), 1484–1490
592. Piomelli, U., Balaras, E., Pasinato, H., Squires, K., Spalart, P. (2003): The inner-outer layer interface in large-eddy simulations with wall-layer models. Int. J. heat Fluid Flow **24**, 538–550
593. Piomelli, U., Cabot, W.H., Moin, P., Lee, S. (1990): Subgrid-scale backscatter in transitional and turbulent flows. Proceedings of the Summer Program – Center for Turbulence Research, 19–30
594. Piomelli, U., Coleman, G.N., Kim, J. (1997): On the effects of nonequilibrium on the subgrid-scale stresses. Phys. Fluids **9**(9), 2740–2748
595. Piomelli, U., Ferziger, J.H., Moin, P., Kim, J. (1989): New approximate boundary conditions for large eddy simulations of wall-bounded flows. Phys. Fluids A **1**(6), 1061–1068
596. Piomelli, U., Liu, J. (1995): Large-eddy simulation of rotating channel flows using a localized dynamic model. Phys. Fluids **7** (4), 839–848
597. Piomelli, U., Moin, P., Ferziger, J.H. (1988): Model consistency in large eddy simulation of turbulent channel flows. Phys. Fluids **31**(7), 1884–1891
598. Piomelli, U., Yunfang, X., Adrian, R.J. (1996): Subgrid-scale energy transfer and near-wall turbulence structure. Phys. Fluids **8**(1), 215–224
599. Piomelli, U., Zang, T.A. (1991): Large-eddy simulation of transitional channel flow. Computer Physics Communications **65**, 224–230
600. Piomelli, U., Zang, T.A., Speziale, C.G., Hussaini, M.Y. (1990): On the large-eddy simulation of transitional wall-bounded flows. Phys. Fluids A **2**(2), 257–265
601. Porté-Agel, F., Meneveau, C., Parlange, M.B. (2000): A scale-dependent dynamic model for large-eddy simulation: application to a neutral atmospheric boundary layer. J. Fluid Mech. **415**, 261–284
602. Porter, D.H., Pouquet, A., Woodward, P.R. (1994): Kolmogorov-like spectra in decaying three-dimensional supersonic flows. Phys. Fluids **6**(6), 2133–2142
603. Pruett, C.D. (2000): Eulerian time-domain filtering for spatial large-eddy simulation. AIAA Journal **38**(9), 1634–1642
604. Pruett, C.D. (2001): Toward the de-mystification of LES. (DNS/LES Progress and Challenges, Liu, Sakell and Beutner eds.), Greyden Press, 231–238
605. Pruett, C.D., Adams, N.A. (2000): A priori analyses of three subgrid-scale models for one-parameter families of filters. Phys. Fluids **12**(5), 1133–1142
606. Pruett, C.D., Gatski, T.B., Grosch, C.E., Thacker, W.D. (2003): The temporally filtered Navier–Stokes equations: properties of the residual stress. Phys. Fluids **15**(8), 2127–2140
607. Pruett, C.D., Sochacki, J.S., Adams, N.A. (2001): On Taylor-series expansions of residual stresses. Phys. Fluids **13**(9), 2578–2589
608. Pullin, D.I. (2000): A vortex-based model for the subgrid flux of a passive scalar. Phys. Fluids **12**(9), 2311–2319

609. Pullin, D.I., Saffman, P.G. (1994): Reynolds stresses and one-dimensional spectra for a vortex model of homogeneous anisotropic turbulence. Phys. Fluids **6**(5), 1787–1796
610. Quéméré, P., Sagaut, P.: Zonal multidomain RANS/LES simulations of turbulent flows. Int. J. Numer. Methods Fluids, submitted
611. Quéméré, P., Sagaut, P., Couaillier, V. (2000): Une méthode multidomaine/multirésolution avec application à la simulation des grandes échelles. C. R. Acad. Sci. Paris, Série II.b **328**, 87–90
612. Quéméré, P., Sagaut, P., Couaillier, V. (2001): A new multidomain/ multiresolution technique for large-eddy simulation. Int. J. Numer. Methods Fluids **36**, 391–416
613. Quirk, J.J. (1991): An adaptative grid algorithm for computational shock hydrodynamics. PhD Thesis, College of Aeronautics
614. Rajagopalan, S., Antonia, R.A. (1979): Some properties of the large structure in a fully developed turbulent duct flow. Phys. Fluids **22**(4), 614–622
615. Redelsperger, J.L., Mahé, F., Carlotti, P. (2001): A simple and general subgrid model suitable both for surface layer and free-stream turbulence. Boundary-Layer Meteorol. **101**, 375–408
616. Rizk, M.H., Menon, S. (1988): Large-eddy simulations of axisymmetric excitation effects on a row of impinging jets. Phys. Fluids **31**(7), 1892–1903
617. Robinson, S.K. (1991): The kinematics of turbulent boundary layer structure. NASA, Tech. Memo. TM 103859
618. Rodi, W., Ferziger, J.H., Breuer, M., Pourquié, M. (1997): Status of large-eddy simulation: results of a Workshop. ASME J. Fluid Engng. **119**(2), 248–262
619. Rogallo, R.S., Moin, P. (1984): Numerical simulation of turbulent flows. Ann. Rev. Fluid Mech. **16**, 99–137
620. Rollet-Miet, P., Laurence, D., Ferziger, J. (1999): LES and RANS of turbulent flow in tube bundles. Int. J. Heat Fluid Flow **20**, 241–254
621. Ronchi, C., Ypma, M., Canuto, V.M. (1992): On the application of the Germano identity to subgrid-scale modeling. Phys. Fluids A **4**(12), 2927–2929
622. Rose, H.A. (1977): Eddy diffusivity. eddy noise and subgrid-scale modelling. J. Fluid Mech. **81**(4), 719–734
623. Roy, S., Baker, A.J. (1997): Nonlinear, subgrid embedded finite-element basis for accurate, monotone, steady CFD solutions. Numerical Heat Transfer, Part B **31**, 135–175
624. Sadourny, R., Basdevant, C. (1981): Une classe d'opérateurs adaptés à la modélisation de la diffusion turbulente en dimension deux (in french). C. R. Acad. Sc. Paris **292**, 1061–1064
625. Sadourny, R., Basdevant, C. (1985): Parametrization of subgrid-scale barotropic and baroclinic eddies in quasi-geostrophic models: anticipated potential vorticity method. J. Atmos. Sci. **42**, 1353–1363
626. Sagaut, P. (1996): Numerical simulations of separated flows with subgrid models. Rech. Aéro. **1**, 51–63
627. Sagaut, P. (1998): Introduction à la simulation des grandes échelles pour les écoulements de fluide incompressible (in french). Springer-Verlag, Berlin
628. Sagaut, P., Comte, P., Ducros, F. (2000): Filtered subgrid-scale models. Phys. Fluids **12**(1), 233–236
629. Sagaut, P., Garnier, E., Séror, C. (1999): Généralisation de l'identité de Germano et application à la modélisation sous-maille (in french). C. R. Acad. Sci. Paris, Série II.b, t. **327**, 463–466
630. Sagaut, P., Garnier, E., Terracol, M. (2000): A general algebraic formulation for multi-parameter dynamic subgrid-scale modeling. International Journal of Computational Fluid Dynamics **13**, 251–257

631. Sagaut, P., Garnier, E., Tromeur, E., Larchevêque, L., Labourasse, E. (2004): Turbulent inflow conditions for large-eddy simulation of compressible wall-bounded flows. AIAA Journal **42**(3), 469–477
632. Sagaut, P., Grohens, R. (1999): Discrete filters for large-eddy simulation. Int. J. Numer. Methods Fluids **31**, 1195–1220
633. Sagaut, P., Labourasse, E., Quéméré, P., Terracol, M. (2000): Multiscale approaches for unsteady simulation of turbulent flows. Int. J. Nonlinear Sciences and Numerical Simulation **1**(4)
634. Sagaut, P., Lê, T.H. (1997): Some investigations on the sensitivity of large-eddy simulation. (Direct and Large Eddy Simulation II, Chollet, Voke and Kleiser eds.) Kluwer, 81–92
635. Sagaut, P., Montreuil, E., Labbé, O. (1999): Assessment of some self-adaptive SGS models for wall bounded flows. Aerospace Science & Technology **3**(6), 335–344
636. Sagaut, P., Troff, B. (1997): Subgrid-scale improvements for non-homogeneous flows. (Advances in DNS/LES, C. Liu, Z. Liu eds.) Greyden Press
637. Sagaut, P., Troff, B., Lê, T.H., Ta, P.L. (1996): Large eddy simulation of turbulent flow past a backward facing step with a new mixed scale SGS model. (Computation of three-dimensional complex flows, Notes on Numerical Fluid Mechanics 53, Deville, Gavrilakis and Rhyming eds.) Vieweg, 271–278
638. Saiki, E., Moeng, C.H., Sullivan, P. (2000): Large-eddy simulation of the stably stratified planetary boundary layer. Boundary-Layer Meteorol. **95**, 1–30
639. Salinas-Vasquez, M., Métais, O. (2002): Large-eddy simulation of the turbulent flow through a heated square duct. J. Fluid Mech. **453**, 201–238
640. Salvetti, M.V., Banerjee, S. (1994): A priori tests of a new dynamic subgrid-scale model for finite-difference large-eddy simulations. Phys. Fluids **7**(11), 2831–2847
641. Salvetti, M.V., Beux, F. (1998): The effect of the numerical scheme on the subgrid scale term in large-eddy simulation. Phys. Fluids **10** (11), 3020–3022
642. Salvetti, M.V., Damiani, R., Beux, F. (2001): Three-dimensional coarse large-eddy simulations of the flow above two-dimensional sinusoidal waves. Int. J. Numer. Meth. Fluids **35**, 617–642
643. Salvetti, M.V., Zang, Y., Street, R.L., Banerjee, S. (1997): Large-eddy simulation of free-surface decaying turbulence with dynamic subgrid-scale models. Phys. Fluids **9**(8), 2405–2419
644. Sandham, N., Yao, Y., Lawal, A. (2003): Large-eddy simulation sof transonic turbulent flow over a bump. Int. J. Heat Fluid Flow **24**, 584–595
645. Sarghini, F., De Felice, G., Santini, S. (2003): Neural networks based subgrid scale modeling in large-eddy simulations. Computers and Fluids **32**(1), 97–108
646. Sarghini, F., Piomelli, U., Balaras, E. (1999): Scale-similar models for large-eddy simulations. Phys. Fluids **11** (6), 1596–1607
647. Schilling, O., Zhou, Y. (2002): Analysis of spectral eddy viscosity and backscatter in incompressible, isotropic turbulence using statistical closure theory. Phys. Fluids **14**(3), 1244–1258
648. Schlichting, H., Gersten, K. (2000): Boundary layer theory, 8th revised and enlarged edition. Springer, Berlin
649. Schlüter, J.U., Pitsch, H., Moin, P. (2004): Large eddy simulation inflow conditions for coupling with Reynolds-averaged flow solvers. AIAA Journal **2**(3), 478–484
650. Schmidt, R., Kerstein, A., Wunsch, S., Nilsen, V. (2003): Near-wall LES closure based on one-dimensional turbulence modeling. J. Comput. Phys. **186**, 317–355

651. Schmidt, H., Schumann, U. (1989): Coherent structure of the convective boundary layer derived from large-eddy simulations. J. Fluid Mech. **200**, 511–562
652. Schmidt, S., Thiele, F. (2002): Comparison of numerical methods applied to the flow over wall-mounted cubes. Int. J. Heat Fluid Flow **23**, 330–339
653. Schumann, U. (1975): Subgrid scale model for finite difference simulations of turbulent flows in plane channels and annuli. J. Comput. Phys. **18**, 376–404
654. Schumann, U. (1995): Stochastic backscatter of turbulence energy and scalar variance by random subgrid-scale fluxes. Proc. R. Soc. Lond. A **451**, 293–318
655. Schumann, U. (1995): Boundary conditions at walls - The unsolved problem. Unpublished
656. Schumann, U., Dörnbrack, A., Dürbeck, T., Gerz, T. (1997): Large eddy simulation of turbulence in the free atmosphere and behind aircraft. Fluid Dyn. Res. **20**, 1–10
657. Schumann, U., Moeng, C.H. (1991): Plume fluxes in clear and cloudy convective boundary layers. J. Atmos. Sci. **48**(15), 1746–1757
658. Schumann, U., Moeng, C.H. (1991): Plume budgets in clear and cloudy convective boundary layers. J. Atmos. Sci. **48**(15), 1758–1770
659. Schwarz, K.W. (1990): Evidence for organized small-scale structure in fully developed turbulence. Phys. Rev. Letters **64**(4), 415–418
660. Scott Collis, S. (2001): Monitoring unresolved scales in multiscale turbulence modeling. Phys. Fluids **13**(6), 1800–1806
661. Scotti, A., Meneveau, C. (1997): Fractal model for coarse-grained nonlinear partial differential equations. Phys. Rev. Lett. **78**(5), 867–870
662. Scotti, A., Meneveau, C. (1999): A fractal model for large-eddy simulation of turbulent flow. Physica D **127**, 198–232
663. Scotti, A., Meneveau, C., Fatica, M. (1997): Dynamic Smagorinsky model on anisotropic grids. Phys. Fluids **9** (6), 1856–1858
664. Scotti, A., Meneveau, C., Lilly, D.K. (1993): Generalized Smagorinsky model for anisotropic grids. Phys. Fluids A **5**(9), 2306–2308
665. Sengupta, T.K., Nair, M.T. (1999): Upwind schemes and large-eddy simulation. Int. J. Numer. Meth. Fluids **31**, 879–889
666. Seror, C., Sagaut, P., Bailly, C., Juvé, D. (2000): Subgrid-scale contribution to noise production in decaying isotropic turbulence. AIAA Journal **38**(10), 1795–1803
667. Seror, C., Sagaut, P., Bailly, C., Juvé, D. (2001): On the radiated noise computed by large-eddy simulation. Phys. Fluids **13**(2), 476–487
668. Shah, K.B., Ferziger, J.H. (1995): A new non-eddy viscosity subgrid-scale model and its application to channel flow. Annual Research Briefs – Center for Turbulence Research, 73–91
669. Shao, L., Bertoglio, J.P., Cui, G.X., Zhou, H.B., Zhang, Z.S. (2003): Kolmogorov equation for large eddy simulation and its use for subgrid modeling. Proceedings of the 5th ERCOFTAC Workshop on Direct and Large Eddy Simulation, Munich, Germany
670. Shao, L., Sarkar, S., Pantano, C. (1999): On the relationship between the mean flow and subgrid stresses in large eddy simulation of turbulent shear flows. Phys. Fluids **11**(5), 1229–1248
671. Sheiki, M., Drozda, T., Givi, P., Pope, S. (2003): Velocity-scalar filtered density function for large-eddy simulation of turbulent flows. Phys. Fluids **15**(8), 2321–2337
672. Shen, L., Yue, D.K. (2001): Large-eddy simulation of free-surface turbulence. J. Fluid Mech. **440**, 75–116

673. Silveira Neto, A., Grand, D., Métais, O., Lesieur, M. (1993): A numerical investigation of the coherent vortices in turbulence behind a backward facing step. J. Fluid Mech. **256**, 1–25
674. Silvestrini, J. (1996): Simulation des grandes échelles des zones de mélange - Application à la propulsion solide des lanceurs spatiaux (in french). Thèse de Doctorat de l'INPG, Grenoble, France
675. Silvestrini, J.H., Lamballais, E., Lesieur, M. (1998): Spectral-dynamic model for LES of free and wall shear flows. Int. J. Heat and Fluid Flow **19**, 492–504
676. Smagorinsky, J. (1963): General circulation experiments with the primitive equations. I: The basic experiment. Month. Weath. Rev. **91**(3), 99–165
677. Smirnov, A., Shi, S., Celik, I. (2001): Random flow generation technique for large-eddy simulations and particle-dynamics modeling. J. Fluids Engng. **123**, 359–371
678. Sohankar, A., Davidson, L., Norberg, C. (2000): Large eddy simulation of flow past a square cylinder: comparison of different subgrid scale models. J. Fluids Engng. **122**, 39–47
679. Souliez, F.J., Long, L.N., Morris, P.J., Sharma, A. (2002): Landing gear aerodynamic noise prediction using unstructured grids. AIAA Paper 2002-0799
680. Spalart, P.R., Jou, W.H., Strelets, M., Allmaras, S.R. (1997): Comments on the feasability of LES for wings, and on a hybrid RANS/LES approach. (Advances in DNS/LES, C. Liu, Z. Liu eds.) Greyden Press, 137–147
681. Speziale, C.G. (1985): Galilean invariance of subgrid-scale stress models in the large-eddy simulation of turbulence. J. Fluid Mech. **156**, 55–62
682. Speziale, C.G. (1987): On nonlinear K-l and K-ϵ models of turbulence. J. Fluid Mech. **178**, 459–475
683. Speziale, C.G. (1991): Analytical methods for the development of Reynolds-stress closures in turbulence. Ann. Rev. Fluid Mech. **23**, 107–157
684. Speziale, C.G. (1998): A combined large-eddy simulation and time-dependent RANS capability for high-speed compressible flows. Journal of Scientific Computing **13**(3), 253–274
685. Speziale, C.G. (1998): Turbulence modeling for time-dependent RANS and VLES: a review. AIAA Journal **36**(2), 173–184
686. Spille-Kohoff, A., Kaltenbach, H.J. (2001): Generation of turbulent inflow data with a prescribed shear stress profile. (DNS/LES Progress and Challenges, Liu, Sakell and Beutner eds.), Greyden Press, 319–326
687. Sreedhar, M., Stern, F. (1998): Large eddy simulation of temporally developing juncture flows. Int. J. Numer. Meth. Fluids **28**, 47–72
688. Stanisic, M.M. (1985): The mathematical theory of turbulence. Springer-Verlag, Berlin
689. Staquet, C., Godeferd, F. (1998): Statistical modelling and direct numerical simulations of decaying stably stratified turbulence. Part 1. Flow energetics. J. Fluid Mech. **360**, 295–340
690. Stevens, B., Moeng, C.H., Sullivan, P. (1999): Large-eddy simulations of radiatively driven convection: sensitivities to the representation of small scales. J. Atmos. Sci. **56**, 3963–3984
691. Stolz, S., Adams, N.A. (1999): An approximate deconvolution procedure for large-eddy simulation. Phys. Fluids **11** (7), 1699–1701
692. Stolz, S., Adams, N.A., Kleiser, L. (1999): The approximate deconvolution model applied to LES of turbulent channel flow. (Direct and Large Eddy Simulation III, Voke, Sandham and Kleiser eds.) Kluwer, 163–174
693. Stolz, S., Adams, N., Kleiser, L. (2001): An approximate deconvolution model for large-eddy simulations with application to incompressible wall-bounded flows. Phys. Fluids **13**(4), 997–1015

694. Stolz, S., Adams, N., Kleiser, L. (2001): The approximate deconvolution model for large-eddy simulations of compressible flows and its application to shock-turbulent-boundary-layer interaction. Phys. Fluids **13**(10), 2985–3001
695. Strelets, M. (2001): Detached eddy simulation of massively separated flows. AIAA Paper 2001-0879
696. Su, C.W., Krueger, S., McMurty, P., Austin, P. (1998): Linear eddy modeling of droplet spectral evolution during entrainment and mixing in cumulus clouds. Atmospheric Research **47–48**, 41–58
697. Sullivan, P., Moeng, C.H., Stevens, B., Lenschow, D., Mayor, S. (1998): Structure of the entrainment zone capping the convective atmospheric boundary layer. J. Atmos. Sci. **55**, 3042–3064
698. Sullivan, P.P., McWilliams, J.C., Moeng, C.H. (1994): A subgrid-scale model for large-eddy simulation of planetary boundary-layer flows. Boundary-Layer Meteorol. **71**, 247–276
699. Sullivan, P., McWilliams, J.C., Moeng, C.H. (1996): A grid nesting method for large-eddy simulation of planetary boundary-layer flows. Boundary-Layer Meteorology **80**, 167–202
700. Tadmor, E. (1989): Convergence of spectral methods for nonlinear conservation laws. SIAM J. Numer. Anal. **26**(1), 30–44
701. Tafti, D. (1996): Comparison of some upwind-biased high-order formulations with a second-order central-difference scheme for time integration of the incompressible Navier–Stokes equations. Computers & Fluids **25**(7), 647–665
702. Tafti, D., Vanka, S.P. (1991): A numerical study of the effects of spanwise rotation on turbulent channel flow. Phys. Fluids A **3**, 642–656
703. Tao, B., Katz, J., Meneveau, C. (2000): Geometry and scale relationships in high Reynolds number turbulence determined from three-dimensional holographic velocimetry. Phys. Fluids **12**(5), 941–944
704. Tao, B., Katz, J., Meneveau, C. (2002): Statistical geometry of subgrid-scale stresses determined from holographic PIV measurements. J. Fluid Mech. **457**, 35–78
705. Temam, R. (1991): Approximation of attractors, large eddy simulations and multiscale methods. Proc. R. Soc. Lond. A **434**, 23–39
706. Temmerman, L., Leschziner, M., Melle, C., Froehlich, J. (2003): Investigation of wall-function approximations and subgrid-scale models in large-eddy simulation of separated flow in a channel with streamwise periodic constrictions. Int. J. Heat Fluid Flow **24**, 157–180
707. Templeton, J., Wang, M., Moin, P. (2002): Towards LES wall models using optimization techniques. Annual Research briefs - Center for Turbulence Research, 189–200
708. Tennekes, H., Lumley, J.L. (1972): A first course in turbulence. MIT Press
709. Terracol, M., Sagaut, P. (2003): A multilevel-based dynamic approach for subgrid-scale modeling in large-eddy simulation. Phys. Fluids **15**(12), 3671–3682a
710. Terracol, M., Sagaut, P., Basdevant, C. (2000): Une méthode multiniveau pour la simulation des grandes échelles des écoulements turbulents compressibles (in french). C. R. Acad. Sci. Paris, Série II.b **328**, 81–86
711. Terracol, M., Sagaut, P., Basdevant, C. (2001): A multilevel algorithm for large-eddy simulation of turbulent compressible flows. J. Comput. Phys. **167**, 439–474
712. Terracol, M., Sagaut, P., Basdevant, C. (2003): A time self-adaptive multilevel algorithm for large-eddy simulation with application to the compressible mixing layer. J. Comput. Phys. **184**, 339–365

713. Tessinici, F., Iaccarino, G., Fatica, M., Wang, M., Verzicco, R. (2002): Wall modeling for large-eddy simulation using an immersed boundary method. Annual Research briefs - Center for Turbulence Research, 181–188
714. Townsend, A. (1976): The structure of turbulent shear flows. Cambridge University Press, Cambridge
715. Tremblay, F., Manhart, M., Friedrich, R. (2001): LES of flow around a circular cylinder at high subcritical Reynolds number with Cartesian grids. (Direct and large-eddy simulation IV, Geurts, Friedrich and Métais eds.), Kluwer, 329–336
716. Troff, B., Lê, T.H., Loc, T.P. (1991): A numerical method for the three-dimensional unsteady incompressible Navier–Stokes equations. J. Comput. Appl. Math. **35**, 311–318
717. Tsubokura, M. (2001): Proper representation of the subgrid-scale eddy viscosity for the dynamic procedure in large eddy simulation using finite difference method. Phys. Fluids **13**(2), 500–504
718. Tsubokura, M., Kobayashi, T., Taniguchi, N., Jones, W.P. (2003): A numerical study on the structures on impinging jets excited at the inlet. Int. J. Heat Fluid Flow **24**, 500–511
719. Tucker, P. (2004): Novel MILES computations for jet flows and noise. Int. J. Heat Fluid Flow **25**(4), 625–635
720. Tucker, P., Davidson, L. (2004): Zonal k-l based large eddy simulations. Computers and Fluids **33**, 267–287
721. Tutar, M., Holdo, A.E. (2000): Large eddy simulation of a smooth circular cylinder oscillating normal to a uniform flow. J. Fluids Engng. **122**, 694–702
722. Tziperman, E., Yavneh, I., Ta'asan, S. (1993): Multilevel turbulence simulations. Europhys. Lett. **24**(4), 239–244
723. Ulitsky, M., Collins, L. (2000): On constructing relizable, conservative mixed scalar equations using the EDQNM theory. J. Fluid Mech. **412**, 303–329
724. Uzun, A., Blaisdell, G., Lyrintzis, A. (2003): Sensitivity to the Smagorinsky constant in turbulent jet simulations. AIAA Journal **41**(10), 2077–2079
725. van der Ven, H. (1995): A family of large eddy simulation (LES) filters with nonuniform filter widths. Phys. Fluids **7**(5), 1171–1172
726. Vandromme, D., Haminh, H. (1991): The compressible mixing layer. (Turbulence and coherent structures, O. Métais, M. Lesieur eds.) Kluwer Academic Press, 508–523
727. Vasilyev, O., Goldstein, D. (2004): Local spectrum of commutation error in large eddy simulations. Phys. Fluids **16**(2), 471–473
728. Vasilyev, O., Lund, T.S., Moin, P. (1998): A general class of commutative filters for LES in complex geometries. J. Comput. Phys. **146**, 82–104
729. Verstappen, R., Veldman, E. (1997): Direct numerical simulation of turbulence at lower costs. J. Engng. Math. **32**, 143–159
730. Verstappen, R., Veldman, E. (1998): Spectro-consistent discretization of Navier–Stokes: a challenge to RANS and LES. J. Engng. Math. **34**, 163–179
731. Verzicco, R., Fatica, M., Iaccarino, G., Moin, P., Khalighi, B. (2002): Large eddy simulation of a road vehicle with drag-reduction devices. AIAA Journal **40**(12), 2447–2455
732. Verzicco, R., Mohd-Yusof, J., Orlandi, P., Haworth, D. (2000): Large-eddy simulation in complex geometric configurations using boundary body forces. AIAA Journal **38**(3), 427–433
733. Vinokur, M. (1989): An analysis of finite-difference and finite-volume formulations of conservation laws. J. Comput. Phys **81**, 1–52
734. Visbal, M., Gaitonde, D. (2002): Large-eddy simulation on curvilinear grids using compact differencing and filtering schemes. J. Fluid Engng. **124**, 836–847

735. Voelkl, T., Pullin, D.I., Chan, D.C. (2000): A physical-space version of the stretched-vortex subgrid-stress model for large-eddy simulation. Phys. Fluids **12**(7), 1810–1825
736. Voke, P.R. (1990): Multiple Mesh simulation of turbulent flow. Technical report QMW EP-1082, Queen Mary and Westfield College, University of London, London, U.K.
737. Voke, P.R. (1996): Subgrid-scale modelling at low mesh Reynolds number. Theoret. Comput. Fluid Dynamics **8**, 131–143
738. Voke, P., Gao, S. (1998): Numerical study of heat transfer from an impinging jet. Int. J. Heat Mass Transfer **41**(4–5), 671–680
739. Voke, P.R., Gao, S., Leslie, D. (1995): Large-eddy simulations of plane impinging jets. Int. J. Numer. Meth. Fluids **38**, 489–507
740. Voke, P.R., Potamitis, S.G. (1994): Numerical simulation of a low-Reynolds-number turbulent wake behind a flate plate. Int. J. Numer. Meth. Fluids **19**, 377–393
741. Völker, S., Moser, R., Venugopal, P. (2002): Optimal large eddy simulation of turbulent channel flow based on direct numerical simulation statistical data. Phys. Fluids **14**(10), 3675–3691
742. von Kaenel, R., Adams, N.A., Kleiser, L., Vos, J.B. (2002): The approximate deconvolution model for large-eddy simulation of compressible flows with finite volume schemes. J. Fluid Engng. **124**, 829–835
743. von Kaenel, R., Adams, N.A., Kleiser, L., Vos, J.B. (2003): The approximate deconvolution model for large-eddy simulation of compresible flows with finite volume schemes. J. Fluids Engng. **125**, 375–381
744. von Kaenel, R., Adams, N.A., Kleiser, L., Vos, J.B. (2003): Effect of artificial dissipation on large-eddy simulation with deconvolution modeling. AIAA J. **41**(8), 1606–1609
745. Vreman, A. (2003): The filtering analog of the variational multiscale method in large-eddy simulation. Phys. Fluids **15**(8), L61–L64
746. Vreman, B., Geurts, B., Kuerten, H. (1994): Realizability conditions for the turbulent stress tensor in large-eddy simulation. J. Fluid Mech. **278**, 351–362
747. Vreman, B., Geurts, B., Kuerten, H. (1994): On the formulation of the dynamic mixed subgrid-scale model. Phys. Fluids **6**(12), 4057–4059
748. Waleffe, F. (1992): The nature of triad interactions in homogeneous turbulence. Phys. Fluids A **4**(2), 350–363
749. Waleffe, F. (1993): Inertial transfer in the helical decomposition. Phys. Fluids A **5**(3), 677–685
750. Waleffe, F. (2001): Exact coherent structures in channel flow. J. Fluid Mech. **435**, 93–102
751. Wang, M. (2000): Dynamic wall modeling for LES of complex turbulent flows. Proceedings of the Summer Program – Center for Turbulence Research, 241–250
752. Wang, L.P., Chen, S., Brasseur, J. (1999): Examination of hypotheses in the Kolmogorov refined turbulence theory through high-resolution simulations. Part 2. Passive scalar field. J. Fluid Mech. **400**, 163–197
753. Wang, L., Lu, X.Y. (2004): An investigation of turbulent oscillatory heat transfer in channel flows by large eddy simulation. Int. J. Heat Mass Transfer **47**, 2161–2172
754. Wang, M., Moin, P. (2000): Computation of trailing-edge flow and noise using large-eddy simulation. AIAA Journal **38**(12), 2201–2209
755. Wang, W.P., Pletcher, R. (1997): On the large eddy simulation of a turbulent channel flow with significant heat transfer. Phys. Fluids **8**(12), 3354–3366

756. Weinan, E., Shu, C.W. (1994): A numerical resolution study of high order essentially non-oscillatory schemes applied to incompressible flow. J. Comput. Phys. **110**, 39–46
757. Werner, H., Wengle, H. (1991): Large-eddy simulation of turbulent flow over and around a cube in a plate channel. (8th Symposium on Turbulent Shear Flows, Munich, Germany)
758. Wilson, P.G., Pauley, L.L. (1998): Two- and three-dimensional large-eddy simulations of a transitional separation bubble. Phys. Fluids **10**(11), 2932–2940
759. Winckelmans, G.S., Jeanmart, H. (2001): Assessment of some models for LES without/with explicit filtering. (Direct and large-eddy simulation IV, Geurts, Friedrich and Métais eds.), Kluwer, 55–66
760. Winckelmans, G.S., Jeanmart, H., Carati, D. (2002): On the comparison between Reynolds stresses from large-eddy simulation with those from direct numerical simulation. Phys. Fluids **14**(5), 1809–1811
761. Winckelmans, G.S., Jeanmart, H., Wray, A., Carati, D., Geurts, B.J. (2001): Tensor-diffusivity mixed model: balancing reconstruction and truncation. (Modern simulation strategies for turbulent flow, Geurts ed.), Edwards, 85–106
762. Winckelmans, G.S., Lund, T.S., Carati, D., Wray, A. (1996): A priori testing of subgrid-scale models for the velocity-pressure and vorticity-velocity formulations. Proceedings of the Summer Program – Center for Turbulence Research, Stanford, 309–329
763. Winckelmans, G., Wray, A.A., Vasilyev, O.V. (1998): Testing of a new mixed model for LES: the Leonard model supplemented by a dynamic Smagorinsky term. Proceedings of the Summer Program – Center for Turbulence Research, 367–388
764. Winckelmans, G.S., Wray, A.A., Vasilyev, O.V., Jeanmart, H. (2001): Explicit-filtering large-eddy simulation using the tensor-diffusivity model supplemented by a dynamic Smagorinsky term. Phys. Fluids **13**(5), 1385–1403
765. Wong, V.C. (1992): A proposed statistical-dynamic closure method for the linear or nonlinear subgrid-scale stresses. Phys. Fluids A **4**(5), 1080–1082
766. Wong, V., Lilly, D.K. (1994): A comparison of two subgrid closure methods for turbulent thermal convection. Phys. Fluids **6**(2), 1017–1023
767. Wu, X., Squires, K.D. (1997): Large eddy simulation of an equilibrium three-dimensional turbulent boundary layer. AIAA Journal **35**(1), 67–74
768. Wu, X., Squires, K.D. (1998): Prediction of the three-dimensional turbulent boundary layer over a swept bump. AIAA Journal **36**(4), 505–514
769. Wu, X., Squires, K.D. (1998): Prediction of the high-Reynolds-number flow over a two-dimensional bump. AIAA Journal **36**(5), 799–808
770. Wu, X., Squires, K.D. (1998): Numerical investigation of the turbulent boundary layer over a bump. J. Fluid Mech. **362**, 229–271
771. Wunsch, S., Kerstein, A. (2001): A model for layer formation in stably stratified turbulence. Phys. Fluids **13**(3), 702–712
772. Xiao, X., Edwards, J., Hassan, H., Baurle, R. (2003): Inflow boundary conditions for hybrid large eddy/Reynolds averaged Navier–Stokes simulations. AIAA Journal **41**(8), 1481–1489
773. Xu, C., Pasquetti, R. (2004): Stabilized spectral element computations of high Reynolds number incompressible flows. J. Comput. Phys. **196**(2), 680–704
774. Xu, H., Pollard, A. (2001): Large eddy simulation of turbulent flow in a square annular duct. Phys. Fluids **13**(11), 3321–3337
775. Yakhot, V., Orszag, S.A. (1986): Renormalization group analysis of turbulence. I. Basic Theory. J. Sci. Comput. **1**, 3–51

776. Yan, H., Knight, D. (2002): Large-eddy simulation of supersonic flat-plate boundary layers using the MILES technique. J. Fluid Engng. **124**, 868–875
777. Yang, Z. (2000): Large eddy simulation of fully developed turbulent flow in a rotating pipe. Int. J. Numer. Meth. Fluids **33**, 681–694
778. Yang, K.S. (2000): Large-eddy simulation of turbulent flows in periodically grooved channel. J. Wind Eng. Ind. Aerodyn. **84**, 47–64
779. Yang, K.S., Ferziger, J.H. (1993): Large-eddy simulation of turbulent obstacle flow using a dynamic subgrid-scale model. AIAA Journal **31**(8), 1406–1413
780. Yang, Z., Voke, P.R. (2000): Large-eddy simulation of separated leading-edge flow in general coordinates. Int. J. Numer. Mech. Engng. **49**, 681–696
781. Yao, F.W., Sandham, N.D. (2002): DNS of turbulent flow over a bump with shock/boundary-layer interactions. (Proceedings of the Fifth Engineering Turbulence Modelling and Measurments, Rodi and Fueyo eds.)
782. Yee, P.P., Domaradzki, J.A. (2001): The subgrid-scale estimation model for decaying isotropic turbulence. (Modern simulation strategies for turbulent flow, Geurts ed.), Edwards, 45–56
783. Yeung, P.K. (1994): Spectral transfer of self-similar passive scalar fields in isotropic turbulence. Phys. Fluids **6**(7), 2245–2247
784. Yeung, P.K., Brasseur, J.G. (1991): The response of isotropic turbulence to isotropic and anisotropic forcing at large scales. Phys. Fluids A **3**(5), 884–897
785. Yeung, P.K., Brasseur, J.G., Wang, Q. (1995): Dynamics of direct large-small scale couplings in coherently forced turbulence: concurrent physical- and Fourier-space views. J. Fluid Mech. **283**, 43–95
786. Yoshizawa, A. (1979): A statistical investigation upon the eddy viscosity in incompressible turbulence. J. Phys. Soc. Japan **47**(5), 1665–1669
787. Yoshizawa, A. (1991): Eddy-viscosity-type subgrid-scale model with a variable Smagorinsky coefficient and its relationship with the one-equation model in large eddy simulation. Phys. Fluids A **3**(8), 2007–2009
788. Yoshizawa, A. (1984): Statistical analysis of the deviation of the Reynolds stress from its eddy-viscosity representation. Phys. Fluids **27**(6), 1377–1387
789. Yoshizawa, A. (1988): Statistical modeling of passive scalar diffusion in turbulent shear flow. J. Fluid Mech. **195**, 544–555
790. Yoshizawa, A. (1989): Subgrid-scale modeling with a variable length scale Phys. Fluids A **1**(7), 1293–1295
791. Yoshizawa, A. (1991): A statistically-derived subgrid model for the large-eddy simulation of turbulence. Phys. Fluids A **3**(8), 2007–2009
792. Yoshizawa, A., Horiuti, K. (1985): A statistically-derived subgrid-scale kinetic energy model for the large-eddy simulation of turbulent flows. J. Phys. Soc. Japan **54**(8), 2834–2839
793. Yoshizawa, A., Kobayashi, K., Kobayashi, T., Taniguchi, N. (2000): A non-equilibrium fixed-parameter subgrid-scale model obeying the near-wall asymptotic constraint. Phys. Fluids **12**(9), 2338–2344
794. Yoshizawa, A., Tsubokura, M., Kobayashi, T., Taniguchi, N. (1996): Modeling of the dynamic subgrid-scale viscosity in large-eddy simulation. Phys. Fluids **8**(8), 2254–2256
795. Zahrai, S., Bark, F.H., Karlsson, R.I. (1995): On anisotropic subgrid modeling. Eur. J. Mech. B/Fluids **14**(4), 459–486
796. Zajaczkowski, F.J., Peltier, L.J. (2001): Energy-containing-range modeling of fully-developed channel flow using a hybrid RANS/LES technique. (DNS/LES Progress and Challenges, Liu, Sakell and Beutner eds.), Greyden Press, 823–828
797. Zandonade, P., Langford, J., Moser, R. (2004): Finite-volume optimal large-eddy simulation of isotropic turbulence. Phys. Fluids **16**(7), 2255–2271

798. Zang, T.A. (1991): Numerical simulation of the dynamics of turbulent boundary layers: Perspectives of a transition simulator. Philos. Trans. R. Soc. Lond. Ser. A **336**, 95–102
799. Zang, Y., Street, R.L., Koseff, J.R. (1993): A dynamic mixed subgrid-scale model and its application to turbulent recirulating flows. Phys. Fluids A **5**(12), 3186–3196
800. Zhang, W., Chen, Q. (2000): Large eddy simulation of indoor airflow with a filtered dynamic subgrid scale model. Int. J. Heat Mass Transfer **43**, 3219–3231
801. Zhao, H., Voke, P.R. (1996): A dynamic subgrid-scale model for low-Reynolds-number channel flow. Int. J. Numer. Meth. Fluids **23**, 19–27
802. Zhou, Y. (1990): Effect of helicity on renomalized eddy viscosity and subgrid scale closure for hydrodynamic turbulence. Phys. Rev. A **41**(10), 5683–5686
803. Zhou, Y. (1991): Eddy damping, backscatter, and subgrid stresses in subgrid modeling of turbulence. Phys. Rev. A **43**(12), 7049–7052
804. Zhou, Y. (1993): Degrees of locality of energy transfer in the inertial range. Phys. Fluids A **5**(5), 1092–1094
805. Zhou, Y. (1993): Interacting scales and energy transfer in isotropic turbulence. Phys. Fluids A **5**(10), 2511–2524
806. Zhou, Y., Brasseur, J.G., Juneja, A. (2001): A resolvable subfilter-scale model specific to large-eddy simulation of under-resolved turbulence. Phys. Fluids **13**(9), 2602–2610
807. Zhou, Y., Hossain, M., Vahala, G. (1989): A critical look at the use of filters in large eddy simulation. Phys. Lett. A. **139**(7), 330–332
808. Zhou, X., Luo, K., Williams, J. (2001): Study of density effects in turbulent buoyant jets using large-eddy simulation. Theoret. Comput. Fluid Dynamics **15**, 95–120
809. Zhou, Y., Vahala, G. (1990): Hydrodynamic turbulence and subgrid scale closure. Phys. Letters A **147**, 43–48
810. Zhou, Y., Vahala, G. (1992): Loca intercations in renormalization methods for Navier–Stokes turbulence. Phys. Rev. A **46**(2), 1136–1139
811. Zhou, Y., Vahala, G. (1993): Renormalization-group estimates of transport coefficients in the advection of a passive scalar by incompressible turbulence. Phys. Rev. E **48**(6), 4387–4398
812. Zhou, Y., Vahala, G., Hossain, M. (1988): Renormalization-group theory for the eddy viscosity in subgrid modeling. Phys. Rev. A **37**(7), 2590–2598
813. Zhou, Y., Vahala, G., Hossain, M. (1989): Renormalized eddy viscosity and Kolmogorov's constant in forced Navier–Stokes turbulence. Phys. Rev. A **40**(10), 5865–5874
814. Zikanov, O., Slinn, D., Dhanak, M. (2002): Turbulent convection driven by surface cooling in shallow water. J. Fluid Mech. **464**, 81–111

Index

Scientific Computation

A Computational Method in Plasma Physics
F. Bauer, O. Betancourt, P. Garabechan

Implementation of Finite Element Methods for Navier-Stokes Equations
F. Thomasset

Finite-Different Techniques for Vectorized Fluid Dynamics Calculations
Edited by D. Book

Unsteady Viscous Flows
D. P. Telionis

Computational Methods for Fluid Flow
R. Peyret, T. D. Taylor

Computational Methods in Bifurcation Theory and Dissipative Structures
M. Kubicek, M. Marek

Optimal Shape Design for Elliptic Systems
O. Pironneau

The Method of Differential Approximation
Yu. I. Shokin

Computational Galerkin Methods
C. A. J. Fletcher

Numerical Methods for Nonlinear Variational Problems
R. Glowinski

Numerical Methods in Fluid Dynamics
Second Edition M. Holt

Computer Studies of Phase Transitions and Critical Phenomena O. G. Mouritsen

Finite Element Methods in Linear Ideal Magnetohydrodynamics
R. Gruber, J. Rappaz

Numerical Simulation of Plasmas
Y. N. Dnestrovskii, D. P. Kostomarov

Computational Methods for Kinetic Models of Magnetically Confined Plasmas
J. Killeen, G. D. Kerbel, M. C. McCoy, A. A. Mirin

Spectral Methods in Fluid Dynamics
Second Edition C. Canuto, M. Y. Hussaini, A. Quarteroni, T. A. Zang

Computational Techniques for Fluid Dynamics 1 Fundamental and General Techniques Second Edition
C. A. J. Fletcher

Computational Techniques for Fluid Dynamics 2 Specific Techniques for Different Flow Categories Second Edition
C. A. J. Fletcher

Methods for the Localization of Singularities in Numerical Solutions of Gas Dynamics Problems
E. V. Vorozhtsov, N. N. Yanenko

Classical Orthogonal Polynomials of a Discrete Variable
A. F. Nikiforov, S. K. Suslov, V. B. Uvarov

Flux Coordinates and Magnetic Filed Structure: A Guide to a Fundamental Tool of Plasma Theory
W. D. D'haeseleer, W. N. G. Hitchon, J. D. Callen, J. L. Shohet

Monte Carlo Methods in Boundary Value Problems
K. K. Sabelfeld

The Least-Squares Finite Element Method
Theory and Applications in Computational Fluid Dynamics and Electromagnetics
Bo-nan Jiang

Computer Simulation of Dynamic Phenomena
M. L. Wilkins

Grid Generation Methods
V. D. Liseikin

Radiation in Enclosures
A. Mbiock, R. Weber

Large Eddy Simulation for Incompressible Flows An Introduction Second Edition
P. Sagaut

Higher-Order Numerical Methods for Transient Wave Equations
G. C. Cohen

Fundamentals of Computational Fluid Dynamics
H. Lomax, T. H. Pulliam, D. W. Zingg

The Hybrid Multiscale Simulation Technology An Introduction with Application to Astrophysical and Laboratory Plasmas A. S. Lipatov

Computational Aerodynamics and Fluid Dynamics An Introduction J.-J. Chattot

Scientific Computation

Nonclassical Thermoelastic Problems in Nonlinear Dynamics of Shells
Applications of the Bubnov–Galerkin and Finite Difference Numerical Methods
J. Awrejcewicz, V. A. Krys'ko

A Computational Differential Geometry Approach to Grid Generation V. D. Liseikin

Stochastic Numerics for Mathematical Physics
G. N. Milstein, M. V. Tretyakov

Conjugate Gradient Algorithms and Finite Element Methods
M. Křížek, P. Neittaanmäki, R. Glowinski, S. Korotov (Eds.)

Flux-Corrected Transport
D. Kuzmin, R. Löhner, S. Turek (Eds.)

Finite Element Methods and Their Applications Z. Chen

Mathematics of Large Eddy Simulation of Turbulent Flows
L. B. Berselli, T. Iliescu, W. J. Layton

Large Eddy Simulation for Incompressible Flows P. Sagaut

MIX
Papier aus verantwortungsvollen Quellen
Paper from responsible sources
FSC® C105338

If you have any concerns about our products,
you can contact us on
ProductSafety@springernature.com

In case Publisher is established outside the EU,
the EU authorized representative is:
Springer Nature Customer Service Center GmbH
Europaplatz 3, 69115 Heidelberg, Germany

Printed by Libri Plureos GmbH
in Hamburg, Germany